Springer Tracts in Civil Engineering

Springer Tracts in Civil Engineering (STCE) publishes the latest developments in Civil Engineering - quickly, informally and in top quality. The series scope includes monographs, professional books, graduate textbooks and edited volumes, as well as outstanding PhD theses. Its goal is to cover all the main branches of civil engineering, both theoretical and applied, including:

- Construction and Structural Mechanics
- Building Materials
- Concrete, Steel and Timber Structures
- Geotechnical Engineering
- Earthquake Engineering
- Coastal Engineering; Ocean and Offshore Engineering
- Hydraulics, Hydrology and Water Resources Engineering
- Environmental Engineering and Sustainability
- Structural Health and Monitoring
- Surveying and Geographical Information Systems
- Heating, Ventilation and Air Conditioning (HVAC)
- Transportation and Traffic
- Risk Analysis
- Safety and Security

Indexed by Scopus

To submit a proposal or request further information, please contact:
Pierpaolo Riva at Pierpaolo.Riva@springer.com (Europe and Americas) Wayne Hu at wayne.hu@springer.com (China)

More information about this series at http://www.springer.com/series/15088

Christian Hofstadler · Markus Kummer

Chances and Risks in Construction Management and Economics

A Systemic Approach to Dealing with Models and Uncertainties

Springer

Christian Hofstadler
Technische Universität Graz
Graz, Austria

Markus Kummer
Technische Universität Graz
Graz, Austria

ISSN 2366-259X ISSN 2366-2603 (electronic)
Springer Tracts in Civil Engineering
ISBN 978-3-030-64013-2 ISBN 978-3-030-64014-9 (eBook)
https://doi.org/10.1007/978-3-030-64014-9

This Springer imprint is published by the registered company Springer Nature Switzerland AG
The registered company address is: Gewerbestrasse 11, 6330 Cham, Switzerland

Preface

Prior to commencing construction work, it is imperative to identify the available resources in terms of funds and time in order to complete the project successfully. It is also essential to allow sufficient time to prepare and execute the project in a networked, integrated, and collaborative approach. At the project planning stage, the budget associated with specific use requirements should be calculated while at the same time determining the quantities and quality standards achievable within a predefined budget.

In early project phases, these will presumably be estimates, which are then replaced with sound calculations as the project progresses. Such calculation outputs usually lead to deterministic results that provide no indication of the risk taken or any existing chance potential.

Calculating construction costs and times is one of the most important and demanding tasks in construction management and economics. Valid data and information is constantly being sought for labor consumption rates, performance values, productivity levels, material consumption, quantities in stock (stock available and deployable on-site), number of transport cycles, and cost and time parameters that must be estimated and/or calculated *ex ante*. Ultimately, final cost and time parameters are determined on the basis of such considerations and calculations. *Ex post* and/or *inter actio* analyses are performed to check if actual values achieved in the construction phase are identical to the target values. In an ideal scenario, the productivity level would be higher and/or material consumption or equipment utilization lower than originally planned. However, the chances that calculated assumptions are exceeded in a positive sense are also associated with risks of non-compliance in a negative sense.

Accurate figures must be stated or submitted at the end of any analysis. These depend on the complexity of the building or structure and on the conditions prevailing at the actual work stages and rely on more or less uncertain input data. One possible solution to this issue is to consider ranges that can deliver final conclusions on the determined values. We focus on probabilistic calculation methods to systematically account for ranges of input parameters. Key outcomes of probabilistic calculations include histograms that are used to directly capture the chance/risk ratio relative to a specific (selected) value.

It is sometimes necessary to make decisions under very uncertain conditions because of frequently occurring inaccuracies of internal and external data, internal and external effects that are difficult to determine, and a lack of information in this context. Besides uncertainties, complex projects such as tunnels or power stations are often associated with partial insecurity and ignorance. These phenomena are very hard to quantify and assess in spite of applying a highly systematic approach. These effects are relevant to specialists who perform costing exercises and establish timelines in preparation of construction projects, as well as to process planners and people structuring, controlling, and managing the actual construction process.

During the construction phase, changes and disruptions to the process flow and logistics system may occur, most of which are difficult to predict. Such changes or disruptions are caused by nonlinearities in the process and by the complexity of the production factors, as well as by exogenous effects on the production system. Similar to a funnel tapering from the wide opening at its top to the neck at its bottom, such uncertainties will exhibit wider spreads in early project phases compared to the stage after the contract award, where an increasing number of specific project features will be added on the basis of the detailed design documents. As the degree of design detail increases, information on the building or structure will become increasingly accurate. Yet uncertainties remain with respect to the most appropriate mode of combining the production factors as well as to future conditions and circumstances prevailing on-site in the construction phase.

From the client's point of view, a short or very short construction time constitutes the optimal scenario because the client will be interested in being able to use or sell the building as soon as reasonably possible. This is why the client's optimal construction time will rarely be identical to that of the bidder/contractor because the latter will mainly be concerned with selecting and combining the production factors such that a "normal" productivity level is reached. Short or very short construction times favored by the client make it impossible for the contractor to achieve normal productivity so that the latter will be forced to take losses of productivity into account.

In all project phases, reliable data and empirical evidence are needed for the considerations, estimates, and calculations that become binding upon the parties when entering into the construction contract while also being key success drivers for the project. Particularly in this context, we should stress that data lacking information with respect to its generation will only be of minor value for subsequent use by third parties. The quality of the information associated with the data, which is also based on the systematic approach to capturing it, is a major criterion for achieving the defined targets as well as for the accurate and precise communication of facts. Quality criteria frequently applied to information include its structure, correctness, completeness, relevance, consistency, currency, allocability, clarity, reliability, validity, objectivity, and conciseness. The poorer the quality of the data and information, the greater the risk of misjudgment will be. High-quality data and information allow for a better evaluation and utilization of chances and a more accurate assessment and prevention of risks.

It is crucial to link the captured data and information to BIM models if we intend to pursue the goal of deriving plausible and reliable results for informing the decision-making process. Irrespective of the degree of sophistication of a BIM model, it will always require high-quality data (with appropriate temporal, spatial, and operational boundaries) with the associated information, including with respect to emergence and boundary conditions, in order to simulate construction times and costs. Only if these conditions are met will BIM models generate maximum benefit for all project phases and stakeholders, and thus for the entire life cycle of the building or structure. This also demonstrates the high added value expected to be generated by implementing and maintaining an integrated knowledge base within or across the organization.

This situational analysis highlights the urgent need for a systematic approach to dealing with chances and risks. We will be able to pave the road to success by implementing a predictive, smart chance and risk policy that defines the objectives and methods of the management of chances and risks. This book covers key aspects in this field in a transparent and plausible manner. The theoretical foundations will be complemented by worked examples that should enable and instigate independent thinking, application, and continuous improvement.

We demonstrate for both current and future projects how organizational units can effectively deal with chances and risks as part of their project portfolio management. This process requires continuous planning and prioritization as well as management and control across all phases and areas in order to ensure longer-term competitiveness. The main tasks and responsibilities comprise the assessment of potential future projects, the ongoing monitoring of projects currently being worked on, and information and knowledge management across projects.

We use worked examples to demonstrate how and in which project phases clients, bidders, and contractors can take uncertainties into account on a systematic basis. This discussion primarily aims to detect chances and risks as early as possible while improving the outputs of calculations as well as the decision-making process across the entire life cycle of the building if and when required. We explain the principle of the Monte Carlo simulation and its benefit for the key stakeholders involved in the planning and construction process on the basis of practical examples.

Any bidder that is awarded the contract in the course of the tendering process will become a contractor. In the next step, costing and pricing assumptions require scrutiny with respect to their specific implementation. This necessary in-depth analysis of the project reveals previously undetected risks while assumed potential chances can prove to be unrealizable with an increasing degree of detail in the evaluation. As a matter of course, it is possible to overestimate risks and underestimate chances when preparing the bid. In addition, partial combinations of previously outlined scenarios are possible. Forecasts turn into plans as the degree of detail increases and more specific information becomes available with respect to the type, number, and combinations of production factors. The related steps are always performed while considering the expected structural, production, and process conditions. After the macro-analyses dominating the bid preparation and submission process, the contractor will increasingly perform analyses at the micro-level after the contract award. Such analyses are indispensable especially for works or services with a tangible impact on costs and/or construction times.

The analysis will also include a wider perspective for the investigations pursued from the macro to the micro-level. Any macro-analysis of a project can consider the building in its entirety or divide it into the shell construction, finishing, building services, and façade sections. In contrast, micro-analyses aim to evaluate individual production units such as work sections in detail and deal with aspects of process optimization and logistics. The detection and utilization of chances is promoted and resilience against disruptions strengthened as a result of the in-depth management analyses performed particularly before but also during the construction phase.

This book also explains how trend analyses can be used to forecast trends with regard to construction progress and expended paid working hours. Such analyses rely on deterministic methods as well as on Monte Carlo simulations in order to fully utilize the added value of trend analyses and to increase forecast certainty and reliability. This approach makes it possible to identify required corrective actions at a very early stage and thus permits the planning and swift implementation of the most efficient and effective combination of production factors.

As an analytical step, the planning process fundamentally differs from forecasting in terms of its function. Planning serves an activity-based design function while forecasting constitutes an event-based flow of information. Agile forecasts primarily aim to create the ideal conditions for the planning process, which results in a chronological order. We can thus increase the potential of chances and reduce the risks if we apply these two methods of project management and construction management and economics correctly and accurately, as outlined above.

This book creates a particularly valuable, hands-on basis for identifying chances and risks in project management, construction management and economics in a plausible manner. Readers will obtain first-hand, specialist knowledge enabling them to draw the right conclusions in order to ensure project success. As a result, the process of preparing and making decisions will be shaped by a more systematic and target-driven approach while also improving decision certainty and transparency.

The aggregated knowledge creates the essential prerequisite required for successfully preparing the project and executing it efficiently and effectively.

In the future, digitization will have a growing influence on construction management. More generally, this process will shape all areas of society. Besides other areas of application, the construction industry should make increased use of the continuously improving set of digital tools in order to obtain more accurate data and information for the pricing and simulation of new projects with the aim of mitigating risks and utilizing chances more effectively. In current projects, collected real-time data and information can contribute to establishing more efficient management and control of construction processes. Ultimately, digitization can also contribute to improving health and safety at work by swiftly identifying potential hazards and risks using mixed-reality scenarios, for example by comparing the 3D design model of a scaffold with the actual installation on the construction site.

Any beneficial use of the elements and methods of digitization in construction management will not only enhance productivity and health and safety but also improve the operation and maintenance phases of buildings. One example are sensors and connected systems and processes that can play an important part in storing key building and process data, thus enhancing the operational productivity of buildings.

As a result of digitization, aspects related to the life cycle of the building will be considered more comprehensively. In this regard, it is advisable to implement systems of incentives (while expanding the value chain) in order to promote resource-saving design, construction, and operation.

Lean design largely relies upon these integral approaches and the resulting activities. This integral approach should be practiced not only at the planning and design stage so that all designers enter into a managed coordination process and adopt a collaborative planning mindset. In addition, construction itself should be perceived as an integral process in which all contractors follow this example and adopt the same managed approach to coordination, or collaborative construction, as a key success driver. The integrated combination of the discretionary production factors leads to an efficient and effective combination of their elementary counterparts. Lean construction is another method guided by these approaches and principles in order to support and complement agile process planning and process control, improvement, and management.

Since the client defines the total construction time or portions thereof, it also carries a major responsibility for the way in which resources are utilized. If the defined, exceedingly short construction time lies below a specific limit to be determined separately for each project, we would no longer be able to meet the requirements of resource-saving construction. It is thus advisable for the client to specify normal construction times in order to create the basis for enabling resource-saving construction in the first place.

Christian Hofstadler and Markus Kummer

Grafendorf and Velden a. W., Austria, April 2021

Acknowledgements

We would like to express our sincere gratitude to Steffen Walter for the outstanding support he provided by translating the manuscript from German into English and refining the English version. He made a valuable contribution to compiling this book thanks to his expertise and ability to always find the right word and phrase, as well as his extensive knowledge of the terminology used in the field.

The contents of this edition are largely based on the German version, "Chancen- und Risikomanagement in der Bauwirtschaft" by Christian Hofstadler and Markus Kummer, published by Springer-Verlag in 2017.

Table of Contents

Abbreviations, Acronyms and Notations

A

A	Area
A_{SF}	Standard floor area [m²]
$A_{SF,AVG}$	Average standard floor area [m²]
$A_{WA,MIN}$	Minimum work area [m²/WK]
$A_{WA,RED}$	Work area reduction [%]
$A_{WA,AV}$	Available work area [m²/AK]
ABGB	Allgemeines Bürgerliches Gesetzbuch ([Austrian] Civil Code)
ASCE	American Society of Civil Engineers

B

BGL	Baugeräteliste [List of Construction Equipment]
BGH	Bundesgerichtshof ([German] Federal Court of Justice)
BIM	Building Information Modeling
BKI	Baukosteninformationszentrum Deutscher Architektenkammern GmbH (Construction Cost Information Center of the German Chambers of Architects)
bSI	buildingSMART International
BVergG	Bundesvergabegesetz ([Austrian] Federal Procurement Act)
BoQ	Bill of quantities

C

°C	.... Degree celsius
c_A	.... Critical approximation [d]
cm	.... Centimeter
C	.... Crane(s)
$C_{V,TR}$	.... Standard volume of concrete delivered to the site per transport cycle [m³]
CC	.... Construction costs [€]
CC_{RED}	.... Crane capacity reduction [%]
CBRM	.... Change Based Risk Management
CIB	.... The International Council for Research and Innovation in Building and Construction
CIRS	.... Critical Incident Reporting System
CL	.... Client
Corr.	.... Correlation
Cov	.... Covariance
CPS	.... Cyber-physical system
CT	.... Contractor
CT	.... Construction time
CRMP	.... Chance and risk management process
C_{EQ}	.... Equipment costs [€/m³]
C_{MA}	.... Material costs [€/m³]
C_{RCO}	.... Cost of reinforced concrete operations [€]
C_V	.... (Reinforced) concrete volume [m³]

D

d	.... Working day(s)
det.	.... Deterministic
dx	.... Differential
D_{FO}	.... Duration of formwork operations [d]
D_{FI}	.... Duration of formwork operations (installation only) [d]
D_{FS}	.... Duration of formwork operations (stripping only) [d]
D_{CO}	.... Duration of concreting operations [d]
D_{RO}	.... Duration of reinforcing operations [d]
$D_{FW,U}$	.... Duration of formwork use [d]

D_{RCO} .... Duration of reinforced concrete operations [d]
D_{SP} .... Formwork stripping period [d]
D_{TOT} .... Total construction time [d]
D_{FP} .... Duration of final phase [d]
D_{MP} .... Duration of main construction phase [d]
D_{RP} .... Duration of ramp-up phase [d]
D_{SC} .... Duration of shell construction works [d]
DC .... Dismantling and demolition costs [€]
DIN .... Deutsches Institut für Normung (German Institute for Standardization)

E

et al. .... et alii (and others)
etc. .... et cetera
e.V. .... Eingetragener Verein (registered association)
$E_{i,j}$ .... Expert value j for the layout i
$E_{i,j,new}$ .... Scaled expert value j for the layout i
EU .... European Union
E(X) .... Expected value
EXP .... Exponent
Ed(s). .... Editor(s)
Eq. .... Equation

F

f. .... Following page
ff. .... Following pages
fr .... Formwork ratio [m^2/m^3]
fr_{bd} .... Formwork ratio for the entire building [m^2/m^3]
f_{WK} .... Proportional factor for workers [-]
f(x) .... Function of x
$f_{C,WK}$ .... Proportionality factor for cranes [WK/crane]
FA .... Formwork area [m^2]
FA_{BE} .... Break-even formwork area [m^2]
FA_{AD} .... Formwork area for assembly and disassembly [m^2]

FA_{TR} .... Formwork area delivered to the construction site per transport cycle [m²]

FA_{STR} .... Total formwork area of the structure [m²]

FIDIC .... Fédération Internationale des Ingénieurs Conseils (International Federation of Consulting Engineers)

FM .... Facility Management

FME(C)A .... Failure Mode and Effects (and Criticality) Analysis

Fig(s). .... Figure(s)

G

GCV .... Gross cubic volume [m³]

GFA .... Gross floor area [m²]

Ges.m.b.H. .... Gesellschaft mit beschränkter Haftung (Austrian company with limited liability; in Germany, abbreviated as GmbH)

GPS .... Global Positioning System

GC .... General Contractor

H

h .... Hour(s)

HACCP .... Hazard and Critical Control Point

HAZOP .... Hazard and Operability

HOAI .... Honorarordnung für Architekten und Ingenieure ([German] Official Scale of Fees for Services by Architects and Engineers)

HVAC .... Heating, Ventilation, and Air-Conditioning

HR .... Human resources

I

I .... Input

I_0 .... Investment amount or cost

i_i .... Imputed interest rate

Inc. .... Incorporated

incl. .... Including

ICB	.... IPMA Individual Competence Baseline
IFC	.... Industry Foundation Classes
IPMA	.... International Project Management Association
IQR	.... Interquartile range
ISO	.... International Organization for Standardization
IT	.... Information technology

K

k	.... Tuning constant (Huber's M-estimator)
K	.... Matrix
kg	.... Kilogram
km	.... Kilometer
KS	.... Kolmogorov-Smirnov test
KSA	.... Kolmogorov-Smirnov analysis

L

λ	.... Eigenvalue
L_i	.... Layout types
LB-HB	.... Leistungsbeschreibung-Hochbau (Specification of Works in Building Construction)
LCR	.... Labor consumption rate
$LCR_{ps,p,ws}$	.... Labor consumption rate [wh/UQ] (ps – process stage, p – process, ws – work section)
$LCR_{SC,AVG}$	.... Average labor consumption rate of shell construction works [wh/m³]
LCR_{FO}	.... Labor consumption rate of formwork operations [wh/m²]
LCR_{RO}	.... Labor consumption rate of reinforcing operations [wh/t]
LCR_{CO}	.... Labor consumption rate of concreting operations [wh/m³]
LCR_{RCO}	.... Labor consumption rate of reinforced concrete operations [wh/m³]
LCC	.... Life-cycle costs [€]
$\Delta LCR_{WA,INC}$	.... Labor consumption rate increase due to reduced available work area [%]
$\Delta LCR_{CC,INC}$	.... Labor consumption rate increase due to reduced crane capacity [%]

ΔLCR .... Labor consumption rate increase
ΔLCR_D .... Labor consumption rate increase due to non-normal construction time

M

μ .... Expected value
m .... Meter
m^2 .... Square meter
m^3 .... Cubic meter
min .... Minutes
mm .... Millimeter
mo .... Months
MAD .... Median of absolute deviations
MAX .... Maximum
MC .... Maintenance costs [€]
MIN .... Minimum
MSC .... Mathematics Subject Classification
MV .... Mean value
MW .... Mean wage [€/wh]

N

n .... Number of mentions per question [-]
n_{bs} .... Number of building sections [-]
n_f .... Number of floors [-]
n_{ws} .... Number of work sections per floor [-]
$n_{ws,s}$ .... Number of work sections to be worked on simultaneously [-]
NPV .... Net present value
No., NO .... Number
NO_C .... Number of cranes
$NO_{C,WK}$ .... Number of cranes relative to the number of workers
$NO_{C,OPT}$ .... Optimum number of cranes
$NO_{C,AV}$ .... Number of available cranes
$NO_{NEG,CORR}$ .... Number of negative correlations

O

O .... Output
OC .... Operating costs [€]
$OF_{FO,RO}$ .... Offset between formwork and reinforcing operations [d]
OEBGL .... Österreichische Baugeräteliste (Austrian List of Construction Equipment)
ÖGG .... Österreichische Gesellschaft für Geomechanik (Austrian Society for Geomechanics)
ON .... Österreichisches Normungsinstitut (Austrian Standards Institute)
OENORM .... Österreichische Norm (Austrian standard)
ONR .... ON-Regel (ON Rule)
OR .... Operations Research

P

p. .... Page(s)
p .... p value
p .... Percentage [%]
pc(s) .... Piece(s)
$p_{b,bs}$ .... Building sections to be worked on simultaneously [-]
$P_{L,NOR,MAX}$ .... Maximum normal labor productivity
$P_{L,NOR,AVG}$ .... Average normal labor productivity
PERT .... Program Evaluation and Review Technique
PF .... Production factor(s)
PC .... Price components
PB .... Pricing bases
PF_{MAX} .... Maximum utilization of production factors
PF_{AVG} .... Average utilization of production factors
PIL .... Patient information leaflet
PM .... Project management
PPH .... Project phase(s)
PPP .... Public Private Partnership
$PQ_{ps,p,ws}$ .... Production quantity (ps – process stage, p – process, ws – work section)
PV .... Performance (value)
PV_{FI} .... Daily formwork performance (installation only) [m^2/d]

PV_G .... Gross technical performance
PV_N .... Net technical performance
$PV_{SC,AVG}$ .... Average performance value of shell construction works [m^3/d]
P+S .... Planning and specification quality
PWH_{RCO} .... Paid working hours of reinforced concrete operations [wh]
$PWH_{ps,p,c}$ .... Paid working hours [wh] (ps – process stage, p – process, c – component)
$PO_{ps,p,c}$ .... Production output [UQ] (ps – process stage, p – process, c – component)
$PV_{RCO,AVG}$ .... Average performance value of reinforced concrete operations [m^3/d]
ΔPL .... Loss of productivity
ΔPL_D .... Loss of productivity due to non-normal construction time

Q

Q_p .... Quantile value
QG .... Quality Gate(s)

R

r .... Correlation coefficient
r_P .... Rate of placing
$r_{P,P}$ .... Permissible rate of placing
rr .... Reinforcement ratio [t/m^3] or [kg/m^3]
r_s .... Rank correlation
r_{xy} .... Product-moment correlation
R .... Risk
RF .... Reinforcement
$R_{Q,TR}$ .... Quantity of reinforcement delivered to the site per transport cycle [t]
RCO .... Reinforced concrete operations
REFA .... Verband für Arbeitsgestaltung, Betriebsorganisation und Unternehmensentwicklung (Association for Work Design, Business Organization and Corporate Development)
RVS .... Richtlinien und Vorschriften für das Straßenwesen ([Austrian] Guidelines and Regulations for the Planning, Construction and Maintenance of Roads)

RfP .... Request for Proposals

S

σ .... Standard deviation

Σ .... Sum

s .... Standard deviation

σ^2 .... Variance

s_x .... Standard deviation of x

s_{xy} .... Covariance

s_y .... Standard deviation of y

S_{FA} .... Standard floor area [m²]

SQ_F .... Stocking quantity (stock available and deployable on-site) of formwork [m^2]

SQ_{SF} .... Stocking quantity (stock available and deployable on-site) of slab formwork [m^2]

SQ_{WF} .... Stocking quantity (stock available and deployable on-site) of wall formwork [m^2]

SiGE .... Sicherheit und Gesundheitsschutz (Health and Safety)

Sim .... Simulation

Sim.No. .... Simulation number

Std.Dev. .... Standard deviation

StGB .... Strafgesetzbuch ([Austrian] Criminal Code)

StLB-Bau .... Standardleistungsbuch-Bau ([German] Standard Services Book for Construction)

SWOT .... Strengths, Weaknesses, Opportunities, and Threats

SC .... Soil class(es)

T

t .... Ton(s)

t .... Continuous index in period (e.g. years)

t_c .... Transport coefficient [1/UQ]

$t_{c,f}$ .... Transport coefficient of formwork [$1/m^2$]

$t_{c,r}$ .... Transport coefficient of reinforcement [1/t]

$t_{c,c}$ .... Transport coefficient of concrete [$1/m^3$]

t_I .... Idle time [h]

t_E .... Equipment time [h]
t_r .... Reference point in time
t_{AU} .... Additional utilization time [h]
t_{BE} .... Basic equipment utilization time [h]
T .... Reporting period, most recent period
TH .... Technische Hochschule (University of Applied Sciences)
TMP .... Temperature
TR .... Number of transport cycles
TU .... Technische Universität (University of Technology)

U

U .... Throughput
UQ .... Unit of quantity
UP .... Unit price(s)
UU .... Unit of use
UFS .... Usable floor space
UVS .... Unabhängiger Verwaltungssenat ([Austrian] Independent Administrative Senate)

V

Var(X) .... Variance
VDI .... Verein Deutscher Ingenieure (The Association of German Engineers)
VOB .... Vergabe- und Vertragsordnung für Bauleistungen (German Construction Contract Procedures)
VÖBV .... Verband Österreichische Biege- und Verlegetechnik (Austrian Association for Bending and Installation of Concrete Reinforcement)
vs .... versus
VwGH .... Verwaltungsgerichtshof ([Austrian] Supreme Administrative Court)
VSCT .... Very short construction time

W

WA_{AV} .... Available work area per worker [m²/WK]
WA_{MIN} .... Minimum work area required per worker [m²/WK]
WA_{RED} .... Work area reduction [%]
WA_{TOT} .... Total work area [m²]
WK .... Worker(s)
WK_{AVG} .... Average number of workers [wh/h]
WK_{MAX} .... Maximum number of workers [wh/h]
$WK_{RCO,AVG}$ .... Average number of workers for reinforced concrete operations [wh/h]
$WK_{RCO,MAX}$ .... Maximum number of workers for reinforced concrete operations [wh/h]
WP .... Work phase(s)
wh .... Paid working hours [wh]
ws .... Work section
WT .... Daily working time [h/d]
WT_{RCO} .... Daily working time of reinforced concrete operations [h/d]
WG .... Work group(s)

X

X_p .... Quantile value
$\bar{x}_i$ .... Mean value

Y

$\bar{y}_i$ .... Mean value

Z

ZAMG .... Zentralanstalt für Meteorologie und Geodynamik ([Austrian] Central Agency for Meteorology and Geodynamics)
ZT .... Ziviltechniker ([Austrian] chartered engineering consultant)

Authors

Assoc.Prof. Dipl.-Ing. Dr.techn. Christian Hofstadler

Chair of the
Institute of Construction Management and Economics
Graz University of Technology
Lessingstraße 25
8010 Graz, Austria
hofstadler@tugraz.at
www.bbw.tugraz.at
mail@christianhofstadler.at
www.christianhofstadler.at

Assoc. Prof. Christian Hofstadler is the Head of the Institute of Construction Management and Economics at Graz University of Technology in Austria. In his role as professor of construction management, he constantly strives to optimize the way in which knowledge is transferred on a sound and sustainable basis in university tuition and research to the benefit of the community of experts in both the private and the public sector. His endeavor has always been to generate and transfer networked knowledge and the associated know-why. His research findings are not only presented in tuition and research, they are also put to good use by construction practitioners in their daily work.

In addition, Prof. Hofstadler is putting great emphasis on engaging in a discourse and exchange with his colleagues of various academic disciplines. Besides his tuition activity at TU Graz, he delivers a large number of guest lectures as part of international study and research trips. He is a member of several expert committees and workgroups, in keeping with his mission of interlinking research with practice. In addition, he works as a court-appointed expert in the field of construction management and economics. This wide array of professional backgrounds provides the basis for him to co-author industry-relevant standards and guidelines. His academic activity encompasses the independent publication and (co)editorship of books as well as a large number of articles at the national and international level. His profound theoretical knowledge is complemented by his hands-on engineering expertise gained while working with a number of contractors in building and industrial construction, building conversion and redevelopment, and bridge construction. His research in the field of construction management and economics primarily covers the following areas worked on as part of joint research projects and own research initiatives:

- Agile construction management
- Improvement of production and logistics processes
- Management of chances and risks in organizations and for projects
- Life-cycle assessments and calculations
- Studies on productivity and construction process disruptions
- Systematic gathering of data, generation of information, and knowledge management for the construction projects of the future
- Consideration of uncertainties in diagnostic exercises and forecasting with the aid of Monte Carlo simulations
- Implementation of nonlinearities in the calculation of construction times and costs

- Resource-saving design-build-operate approaches
- Modeling, simulation, and processes
- Digitization of processes, buildings and structures, and production conditions
- Use of artificial intelligence in construction management and economics

Dipl.-Ing. Dr.techn. Markus Kummer

Deputy Chair of the
Institute of Construction Management and Economics
Graz University of Technology
Lessingstraße 25
8010 Graz, Austria
markus.kummer@tugraz.at
www.bbw.tugraz.at
info@markus-kummer.at
www.markus-kummer.at

Dr. Markus Kummer graduated from Graz University of Technology, obtaining a master's degree in industrial engineering/construction engineering sciences. His master's thesis was entitled "Application of the Monte Carlo Simulation for the Calculation of Building Costs and Building Time".

Since 2011, he has been working as an independent professional, preparing expert opinions in the field of construction management and economics. Since August 2013, he has also been in the role of university project assistant at Graz University of Technology, followed by a senior project scientist position at the Institute of Construction Management and Economics since August 2019. He has also been actively involved in academic teaching since 2016. He is the deputy head of the working group on the "Management of Chances and Risks, and Digitization" and is a member of the Faculty Board.

His doctoral thesis entitled "Aggregated Consideration of Losses of Productivity in Determining Construction Costs and Times – Deterministic and Probabilistic Analyses" dealt with nonlinearities and uncertainties in the calculation of construction costs and times. He obtained his doctorate in December 2015. His thesis received two awards, namely the gif (German Society of Property Researchers) Research Award and the DVP (German Association for Project Management in the Construction and Real Estate Industries) Sponsorship Award.

Dr. Markus Kummer is the author and co-author of several national and international articles in the construction engineering sciences field as well as co-host of scientific conferences and coeditor of compilations on selected topics in construction management and economics. He gained hands-on experience while working with construction contractors and engineering consulting firms prior to and during his degree course.

1 Introduction

Construction management heavily relies on the contextual selection of the production factors and their optimized combination. The way in which the production factors are combined, as well as the related points in time, has a significant effect on construction time and costs while at the same time impacting site logistics and the construction process as such. In construction economics, we use these management considerations and calculations for determining the costs of the selected types and combinations of production factors. The costs result from the consumption of production factors measured in monetary amounts. On this basis, operational and strategic analyses are carried out to translate the determined costs into prices.

All forecasts made and plans derived in this context are subject to temporal and budgetary constraints. The figures we obtain and submit after completion of any analysis need to be as accurate as possible for the investigated variables. Their accuracy and robustness will, however, depend on the complexity of the building or structure and on the conditions prevailing in the actual work phases or on a more or less comprehensive and accurate database. One possible solution to this issue is to consider ranges that can subsequently deliver valid conclusions on determined values. Probabilistic methods deliver valuable results that make it possible to systematically consider such input value ranges and correlations of key determinants. Key outcomes of probabilistic calculations include histograms that are used to capture or estimate the chance/risk ratio relative to a specific (selected) value in a reliable manner. Provided the underlying models and input data are sufficiently accurate, histograms establish a perfect basis for making informed and plausible decisions.

In each project phase, clients and contractors strive to determine possible combinations of production factors and identify their best combination in order to arrive at accurate results regarding construction time and costs. In so doing, the achievable degree of accuracy will always depend on how specific project and structural targets and objectives are, as well as on the quality of the available knowledge and various options for its contextual application to the current project. Only this step will add tangible value and knowledge aggregation if combined with near-real-time feedback loops and analyses.

The following section includes a situational analysis in order to discuss the combination of production factors and the complexity and uncertainties in systems. We also outline the structure of this book and identify its most relevant target audiences.

C. Hofstadler and M. Kummer, *Chances and Risks in Construction Management and Economics*, Springer Tracts in Civil Engineering,
https://doi.org/10.1007/978-3-030-64014-9_1

1.1 Production System and Processes

In construction management, production factors comprise elementary (labor, equipment and machinery, materials) and discretionary factors (project managers, site managers, technicians, etc.) performing a range of different activities, including planning, management, control, selection, organization, communication, and documentation, as well as knowledge work (see Fig. 1.1). In an ideal setting, near-real-time feedback loops are used to investigate the degree of productive interaction between the elementary production factors.

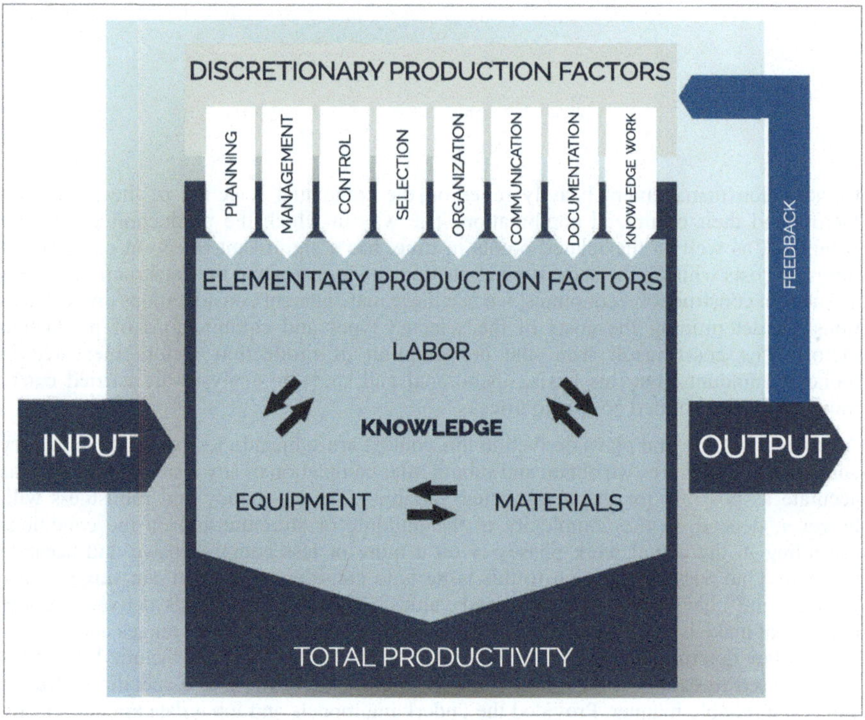

Fig. 1.1 Combination of production factors for achieving total productivity – Two-dimensional system[1)]

On the construction site, workers combine equipment and machinery with the aim of performing the specified construction operations at a "normal" productivity[2)] level. The time within which work is performed and the intensity of resource utilization have a major influence on the achievable productivity, and thus on the incurred costs. A deficit is generated if the contractor incurs higher costs than stated in the contract costing and

---

1) Cf. Hofstadler (2014), p. 15

2) The average normal productivity refers to the minimum human productivity that can be achieved on average in the long term by, and expected of, any competent worker after sufficient training and familiarization without harming their health if complying with the specified time for personal needs and recovery, if applicable. Note: Derived from the REFA normal performance definition. Cf. Hofstadler (2014), p. 67

pricing exercise. Depending on the party causing this difference and the provisions agreed upon in the construction contract, the contractor will be bound to bear this additional cost or may transfer the total cost or parts thereof to the client. In this context, the concept of costs refers to the consumption of goods and services expressed in monetary amounts.

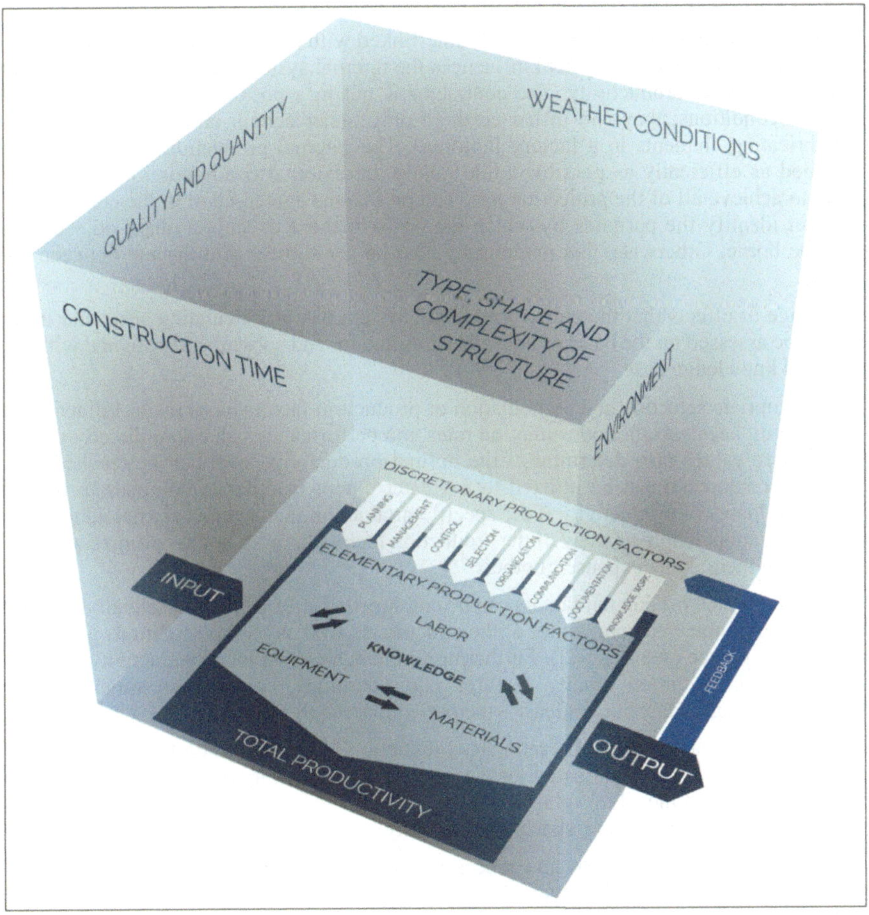

Fig. 1.2 Combination of production factors for achieving total productivity – Three-dimensional system[3)]

This two-dimensional system of total productivity needs to be complemented by an additional dimension because the production factors are influenced particularly by the conditions in the site surroundings and the environment, the type, shape, and complexity of the project, the qualities and quantities to be achieved, and – last but not least – the specified construction time (see Fig. 1.2).

[3)] Derived from Hofstadler (2014), p. 15

The internal production system refers to the isolated consideration of the type, number, and combination of elementary production factors. Such analyses focus on the workers, equipment, and materials needed to perform the specified works. It is absolutely necessary to coordinate time, work area, and work intensity so that the limits to losses of productivity are not exceeded. The discretionary production factors are crucial for ensuring the appropriate selection and functioning of these components.

The internal production system is closely interlinked with the external production system. In this setting, the external system influences the internal system and is a key determinant of the achievable productivity. The contractor is not in a position to freely choose the external conditions; they can influence them only under certain circumstances (e.g. by prefabricating elements in a factory building). The internal production system must be designed as efficiently as possible while taking the external conditions into account in order to achieve all of the project targets. The provisions agreed upon in the construction contract identify the party/ies by which the costs of these execution-phase adjustments shall be borne. Otherwise, this matter may also be the subject of a dispute between the parties.

The range of effects that the entire production system has on calculating costs and times should be assessed on the basis of data and information, and subsequently aligned with the enlarged knowledge base.

The appropriate selection and combination of production factors has a major influence on productivity because labor consumption rates and performance values are directly related to it. These values also determine if the normal productivity level can be reached. The available resources must be utilized to a greater extent in order to compensate for the generated losses of productivity if the specific production factor limits are not adhered to. We should adopt a holistic view of the production factors in order not to optimize only one or a few of them at the expense of others.[4)]

The way in which production factors can be combined cost-effectively essentially depends on the type, shape, and complexity of the structure as well as on the conditions governing work performance (see Fig. 1.2). Furthermore, the achievable total productivity is determined by the required quality features and quantities, construction time, the site environment, and – last but not least – the prevailing weather conditions.[5)]

Both the elementary and the discretionary production factors (i.e. their type, number, and combination) must be documented for the purpose of evaluating, assessing, and analyzing the achieved (or achievable) total productivity. In so doing, we should always establish relationships in terms of time, space, and intensity.

A holistic approach to systems and processes requires consideration of the effects of the individual subsystems when analyzing the overall system. The characteristics of the individual elements must be understood in the overall context.

The internal and external production conditions essentially determine the achievable productivity. Estimating this variable should take into account that any project is to be divided into ramp-up, main, and final phases. There will also be a final phase followed by a new ramp-up phase in the event of any disruptions, shutdowns, or winter breaks. Compared to the main construction phase, the productivity levels achieved in these ramp-up and final phases are usually lower.

---

4) Cf. Hofstadler (2014), p. 13

5) Cf. Hofstadler/Kummer (2015a), p. 52

We also need to make a distinction between different phases with respect to the types of work. Building construction projects are usually divided into the shell construction, finishing, services and utilities, and façade installation stages. We should also consider the internal sequence of these phases as well as the crucial compliance with the critical approximation.

1.1.1 Applying the Trapezoidal Model in Construction Management

On the one hand, the required resource utilization in terms of workers and equipment is a function of the available construction time and production performance. On the other hand, the shortest possible construction time results from the maximum of available and deployable resources with which the absolute upper performance limit can be reached. If this limit has already been reached with a particular combination of resources, and if these resources are increased further beyond this point, this will have a reversionary effect on the performance, potentially causing it to fall again. This phenomenon is due to the fact that losses of productivity occur that prevent any further increase in performance. This interplay between the efficient and effective number and combination of production factors and the principles governing productivity poses one of the biggest challenges in construction management and economics.

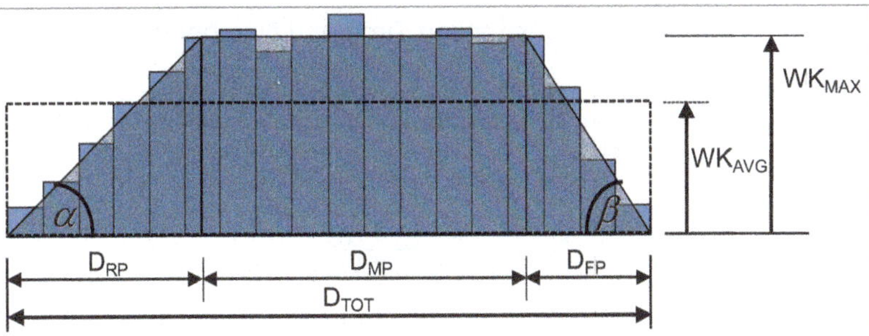

D_{RP} .... Ramp-up phase [d]

D_{FP} .... Final phase [d]

D_{MP} .... Main construction phase [d]

D_{TOT} .... Construction time [d]

WK_{MAX} .... Maximum number of workers [wh/h]

WK_{AVG} .... Average number of workers [wh/h]

Fig. 1.3 Application of the trapezoidal model indicating resource utilization – Development of production factors (in this case, workers) over the construction period (idealized model)[6]

[6] Hofstadler (2014), p. 42

Fig. 1.3 shows an idealized model of the trend in the number of workers required over the construction period (total construction time). A trapezoidal pattern was assumed for the sake of simplicity. Resource plans prepared for different building construction projects were evaluated in order to demonstrate the practicability of this simplified approach. In so doing, we found a good approximation to the trapezoidal model.

The total time to completion of a construction project comprises the period required for the ramp-up, main construction, and final phases. The maximum number of workers is required in the main construction phase.

1.1.1.1 Determining the Sum of Paid Working Hours

We apply Eq. (1-1) to calculate the sum of paid working hours PWH_{RCO} [wh] for reinforced concrete operations. To do so, we add up the total construction time (duration) D_{TOT} [d] and the duration of the main construction phase D_{MP} [d] and multiply the result with the maximum number of workers WK_{MAX} [wh/h] and the daily working time WT [h/d], followed by a division by 2.

$$PWH_{RCO} = \frac{1}{2} \cdot (D_{TOT} + D_{MP}) \cdot WK_{MAX} \cdot WT \tag{1-1}$$

1.1.1.2 Determining Construction Time and Number of Workers

At the preliminary planning stage, the gross cubic volume or reinforced concrete volume is used for determining the production output in the shell construction phase. If we apply the reinforced concrete volume, we obtain the required average number of workers WK_{AVG} [wh/h] from Eq. (1-2).

$$WK_{AVG} = \frac{C_V \cdot LCR_{RCO}}{D_{TOT} \cdot WT} \tag{1-2}$$

This equation includes a contractually defined construction period, whereas the quantity, total labor consumption rate, and working time are assumed or estimated. The numerator comprises the total concrete volume C_V [m³] and the total labor consumption rate of reinforced concrete operations LCR_{RCO} [wh/m³]. The denominator includes the duration of reinforced concrete operations D_{TOT} [d] and the average daily working time WT [h/d].

This approach should be applied accordingly if other units of quantity are used for representing production output, including the gross cubic volume or steel or masonry quantities. Any calculation of the shortest possible time should be preceded by determining the maximum possible average number of workers.

In building construction, the maximum number of available workers is defined by the number of available cranes as well as by the available workspace. The lower of these two values will determine the maximum total number.

When deviating from the optimal number of workers, the cost effectiveness of the work performed will decrease with the increase in the number of workers. Depending on the number of cranes, the maximum number of available workers WK_{MAX} (see Eq. (1-3)) follows from the product of the number of cranes NO_C [-] and the proportionality factor for cranes $f_{C,WK}$ [WK/crane]:

$$WK_{MAX} = NO_C \cdot f_{C,WK} \tag{1-3}$$

We obtain the maximum value from Eq. (1-4) if the number of workers is dependent on the work area.

$$WK_{MAX} = \frac{WA_{TOT}}{WA_{MIN}} \tag{1-4}$$

The numerator includes the total work area WA_{TOT} [m^2] that is available on average during the main construction phase; the denominator indicates the minimum work area required per worker WA_{MIN} [m^2/WK]. The lower of the values calculated applying Eq. (1-3) and Eq. (1-4) will determine the subsequent calculation steps.

If we apply the average standard floor area for calculating the existing or required work area, we obtain the number of work sections to be worked on simultaneously $n_{ws,s}$ [-] from Eq. (1-5).

$$n_{ws,s} = \frac{n_{ws} \cdot WK_{MAX} \cdot WA_{MIN}}{A_{SF}} \tag{1-5}$$

The numerator includes the product of the number of work sections per floor n_{ws} [-], the maximum number of workers WK_{MAX} [wh/h], and the minimum work area required per worker WA_{MIN} [m^2/WK]. The average standard floor area A_{SF} [m^2] is entered in the denominator.

According to the calculation, the maximum number of workers required follows from Eq. (1-6) in analogy to determining the area of an irregular trapezoid.

$$WK_{MAX} = \frac{D_{TOT} \cdot WK_{AVG}}{\left[D_{MP} + \frac{1}{2} \cdot (D_{RP} + D_{FP})\right]} \tag{1-6}$$

It is crucial to define a realistic maximum value in the main construction phase in order to arrive at equally realistic construction process and logistics planning assumptions. If we overestimate the maximum number of workers at the preliminary and/or detailed planning stages, this will inevitably lead to longer construction times if there is no option of adjusting timelines or capacities.

The duration of the ramp-up and final phases compared to the main phase will essentially depend on the building layout and design of its floors.

High-rise buildings require a longer ramp-up phase than buildings with only a few floors, such as factory buildings, hospitals, or warehouses, with an extensive footprint. In high-rise projects, it is hardly possible to perform work simultaneously in different work sections, except for constructing the core first or applying the top-down method. In contrast, lower buildings with a more extensive footprint generally permit simultaneous work right from the start, including, for instance, commencing work on two foundation slab sections at the same time.

The following formula applies to the ramp-up and final phases:

$$D_{RP} + D_{FP} = D_{TOT} - D_{MP} \tag{1-7}$$

The duration of the main construction phase is derived from Eq. (1-8) if the maximum duration as well as the average and maximum number of workers are known.

$$D_{MP} = \frac{2 \cdot D_{TOT} \cdot WK_{AVG}}{WK_{MAX}} - D_{TOT} \tag{1-8}$$

The main construction time calculated using Eq. (1-8) creates the basis for determining the duration of the ramp-up and final phases.

As a first approximation derived from analyzing the process of constructing several building shells, we can assume the final phase of reinforced concrete operations to last for about one third of the ramp-up phase.

Inserting Eq. (1-9) in Eq. (1-7)

$$D_{FP} = \frac{1}{3} \cdot D_{RP} \tag{1-9}$$

gives

$$D_{RP} = \frac{3}{4} \cdot (D_{TOT} - D_{MP}) \tag{1-10}$$

If the duration of the ramp-up and final phases calculated according to Eq. (1-9) and Eq. (1-10) is too short from a construction management point of view, the corresponding periods need to be prolonged at the expense of the main construction phase. A longer total construction time should thus be scheduled if and when required. The changed duration of the main phase is calculated by applying the values to Eq. (1-7). As a result, the value designating the required maximum number of workers will also increase.

Applying Eq. (1-7) to Eq. (1-6) delivers the following:

$$WK_{MAX} = \frac{D_{TOT} \cdot WK_{AVG}}{\frac{1}{2} \cdot (D_{TOT} + D_{MP})} \tag{1-11}$$

The calculation according to Eq. (1-11) is necessary if the total duration, main construction time, and average number of workers are known or have already been calculated, and if the maximum number of workers remains to be determined. The ratio of the average to the maximum number of workers usually lies between 0.65 and 0.85 and essentially depends on the duration of the ramp-up and final phases.[7)]

1.1.2 Efficiency and Effectiveness

Any effective or efficient construction process is characterized by optimally combining the individual production factors. This optimal combination is reached if the production factors can be utilized at a normal productivity level.

The notion of effectiveness describes the degree of target achievement, whereas the

7) Cf. Hofstadler (2007), p. 60

concept of efficiency refers to the degree of cost efficiency, or viability. In construction projects, the project targets or milestones that contractors are bound to achieve are defined in the construction contract. However, contractors are largely free in deciding how to achieve them. When selecting the production factors, contractors enjoy the greatest freedom in the way they combine them, including the construction process sequence, method, number of workers, and pieces of equipment. On-site activity is particularly efficient if the targets are achieved with the most economical type and combination of production factors.

Fig. 1.4 describes the relationship between efficiency and effectiveness by the utilization of production factors for achieving the project targets. This diagram includes four quadrants that highlight four different start-finish trajectories. The bottom left quadrant shows that many production factors were utilized. However, their combination was inappropriate, which made it impossible to achieve the target (i.e. to arrive at the finish). This is how targets can be missed in the production process despite high costs. Any path of this type is thus both inefficient and ineffective.

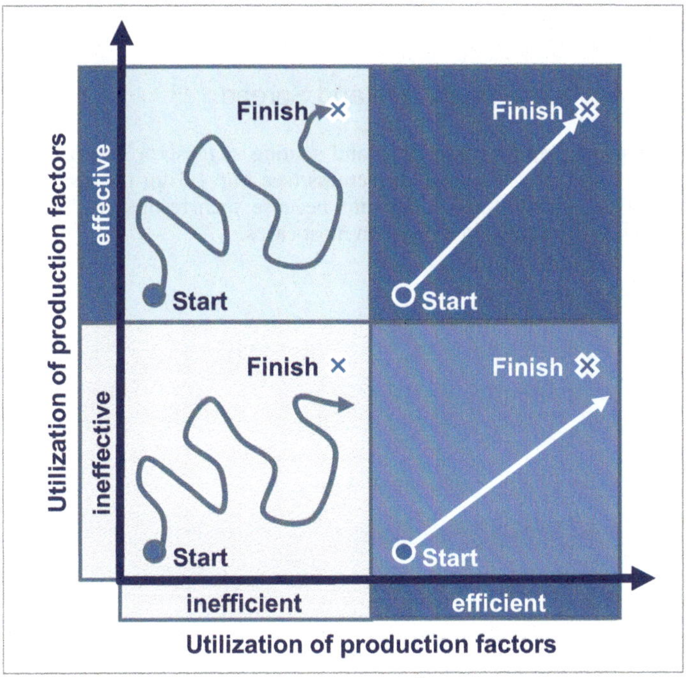

Fig. 1.4 Difference between effectiveness and efficiency[8)]

The upper left quadrant describes the relationship between successful target achievement (effective) and an uneconomical utilization of production factors (inefficient). In this scenario, the project was achieved at a very high (basically avoidable) cost. The other two cases describe the efficient utilization of production factors (i.e. their optimal combination at a normal productivity level), but with a different degree of target achievement. In the

[8)] Cf. Hofstadler/Kummer (2017a), p. 4

bottom right quadrant, the target was missed despite the efficient utilization of production factors, whereas, in the top right quadrant, their efficient utilization led to complete target achievement (effectiveness). Contractors strive to follow exactly this straight route from start to finish. They make every attempt to identify an efficient type and combination of production factors that leads to the effective achievement of the project targets.

It should be noted, though, that the client should/must not specify an exceedingly short construction time because, in this case, the production system would reach its limit of efficiency already before construction work commences. In addition, the system's resilience against process disruptions would be very low if the productivity limit were exceeded.

The interplay and interaction between the productivity factors is fraught with many uncertainties particularly in the early project phases because, at this stage, not much progress will have been made towards detailed planning and project management. We should assume for any deterministic calculations of values for future periods that the probability for their precise results to materialize is very low. Rather, these values will be either exceeded or underrun. In contrast, calculated ranges are usually consistent with the forecast reality, although they do not deliver precise results.

1.1.3 Accuracy, Distortion, and Spread

Stating deterministic results for pricing and costing suggests a very accurate forecast characterized by high precision and correctness (see Fig. 1.5 for the distinction between these terms). This situation poses problems because such results are likely to be very precise, yet incorrect and thus inaccurate in most cases.

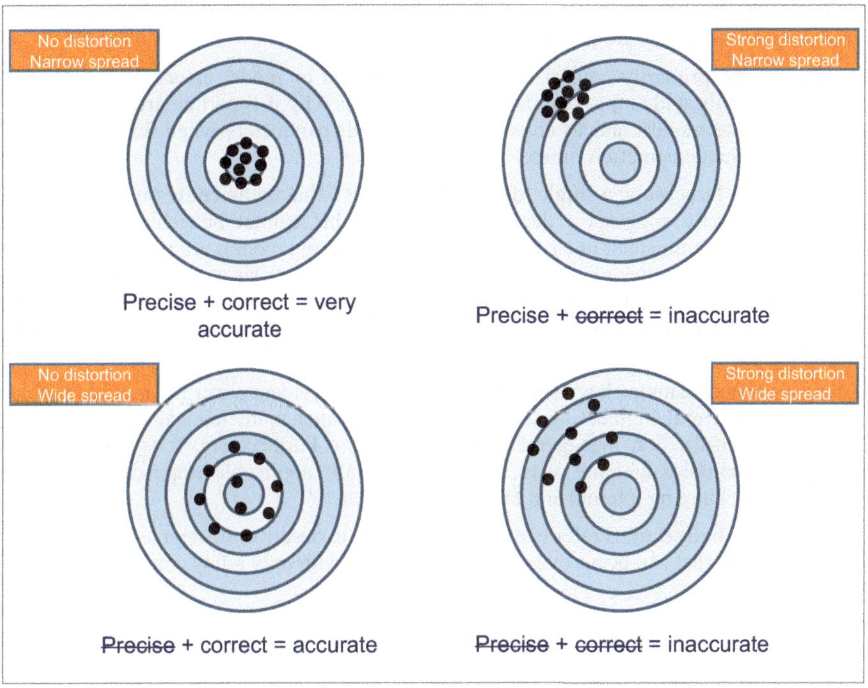

Fig. 1.5 Distinction between accurate, precise, and correct predictions[9]

Probabilistic calculations (i.e. calculations based on probability theory) deliberately integrate the existing uncertainties and visualize them as histograms when analyzing results. The aim is to obtain accurate results and to hit the "moving target" of costs and construction times as centrally as possible by freely choosing a corresponding chance/risk ratio (see Fig. 1.5). This target is considered to be moving because targets are not always clearly defined or may change over time, for instance owing to insufficient detailing in terms of planning and specification.

In the top-left part of Fig. 1.5, several tests (in the figurative sense, for instance for determining construction time) led to hitting the bull's eye of the target with a narrow spread and without distortion. This pattern of hits indicates very accurate results. "Very accurate" is synonymous with the "precise" and "correct" attributes. The top right target also shows hits within a narrow range, and thus a low spread. However, the problem lies in the major distortion of results because the aim was to hit the target in the center. Results show a lower spread, but they are subject to a strong distortion. These results can be considered to be inaccurate.

The hits shown on the bottom left target are all very close to the center, but they show a major spread. Although there is no distortion visible in this case, the results are also lacking precision. Results of this type can be considered to be accurate, as opposed to "very accurate", which necessitates correctness and precision. Although they lack

[9] See also DIN ISO 5725-1:1997, p. 15
Cf. Hofstadler/Kummer (2017a), p. 5

precision, they at least go in the right direction. On the bottom right target, the hits are located in the top left area, showing a wide spread and strong distortion. Such results are inaccurate because they are neither correct nor precise.

Distortions are primarily caused by unsuitable models or model errors. Spreads result from exceedingly wide input parameter ranges. This is why it is particularly important to create adequate models and to enter contextual input parameters when determining the project lead and construction times.

Calculating or forecasting a single (deterministic) cost value for an entire building or parts thereof incorrectly suggests a degree of accuracy that cannot be achieved owing to uncertain input parameters. In structural engineering, design models are used that consider spreads and uncertainties of material properties and loads as far as reasonably possible. This uncertainty is countered by applying the semiprobabilistic safety concept using partial safety factors. In costing and scheduling, these uncertainties are compensated by including appropriate buffers and contingencies.[10)]

Tried-and-tested material laws supported by empirical evidence can be applied particularly in the structural engineering sciences. In contrast, such mathematical laws are largely missing in construction management and economics. Furthermore, it is difficult, and exceedingly complex in some cases, to gather generally applicable empirical data owing to constantly changing internal and external influences.[11)]

The following two sections refer to examples in order to describe the degree of complexity and uncertainty that can occur even in simple systems with only a few components.

1.1.4 Complexity of Systems

A system is essentially composed of several structured elements with specific characteristics that are interrelated to one another (see the principle shown in Fig. 1.6). Networked systems functioning according to logical rules represent useful (in construction management, buildable) processes.

[10)] Cf. Hofstadler (2011a), p. 32

[11)] Cf. Kummer (2015a), p. 68

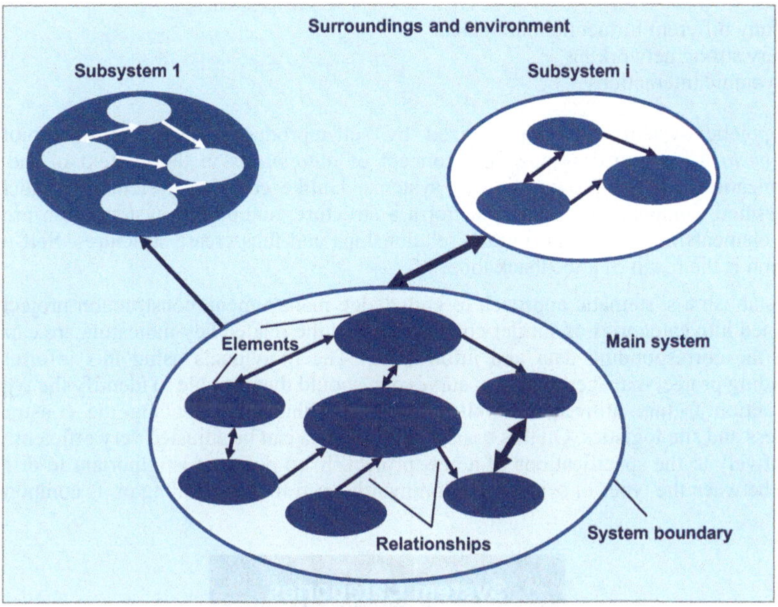

Fig. 1.6 Constituents of systems[12)]

Elements of a system are those components that are no longer separable within this set. The relationships between elements differ quantitatively (larger number of interactions) and qualitatively (higher yield of interactions) from their relationships to other entities (systems). This relationship differential constitutes a boundary that differentiates the system from its surroundings. Complex systems are self-organizing and self-referential. For this reason, any intervention in complex systems will be problematic and will often trigger surprising, unintended consequences.[13)]

Analyses and calculations need to be performed for any construction project. It is important to recognize if we deal with simple, complicated, or complex systems. Other types of systems expand themselves by self-reproduction.

Simple systems are characterized by:

- Few influential parameters
- Weak networking
- Low dynamics

Complicated systems are characterized by:

- Many different influential parameters
- Relatively strong networking
- Low dynamics

---

12) Cf. Hofstadler/Kummer (2017a), p. 6

13) Cf. http://wirtschaftslexikon.gabler.de/Archiv/3210/system-v12.html. Accessed December 21, 2020

Complex systems are associated with the following attributes:

- Many different influential parameters
- Very strong networking
- Dynamic interactions

Autopoietic systems are characterized by self-reproduction. Chilean neurobiologist *Humberto Maturana* developed the concept of *autopoiesis* in the context of the self-organization and self-preservation of systems. Unlike complex systems, in which the predefined elements are arranged to form a structure, autopoietic systems also produce new elements that enter into mutual relationships and thus create structures. Self-reproduction is the result of a feedback loop.[14)]

To establish a systematic approach to knowledge management, construction projects are grouped into categories of similar complexity, and the related key indicators are captured with the corresponding data and information. The individuals using this information, including project managers and cost surveyors, should then be able to identify the types of production factors utilized, as well as their combination, including the construction process and site logistics. On this basis, historical data can be adjusted very efficiently and effectively to the specifications of a new project. In so doing, it is important to differentiate between the types of systems predominantly used for the building or its components.

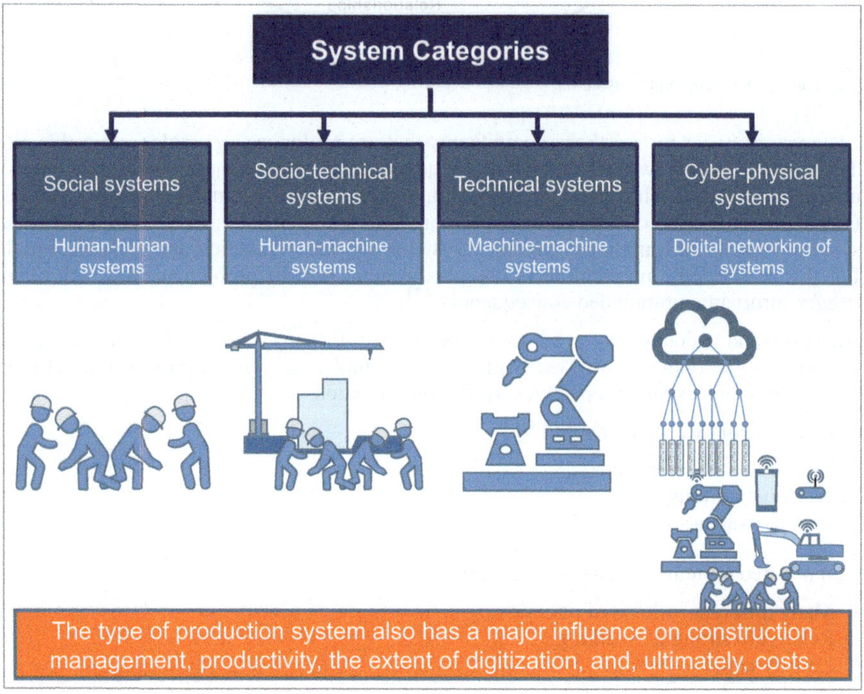

Fig. 1.7 Types of systems in construction management

14) Cf. Simon (2011), p. 32

Systems theory also provides the starting point in labor sciences and analyses related to construction management. The REFA methodology[15)] defines a system as an entity composed of elements whose relationships serve a certain purpose. Depending on the degree of detail, we can consider a single workplace, an entire company, a sector of the economy, or the national economy to be a system.

In business and economics, we start from the involved system elements (see Fig. 1.7) to distinguish between

- technical systems (machine systems)
- social systems (human-human systems)
- socio-technical systems (human-machine systems) and
- the cyber-physical (digital) networking of systems[16)]

In a cyber-physical system (CPS), IT and software components are interconnected with mechanical and electronic parts, which communicate through a data infrastructure such as the Internet. Cyber-physical systems are characterized by their high degree of complexity.

Construction projects are unique by nature and thus usually constitute a socio-technical system (at least when restricting our perspective to the construction site). Such systems are characterized by a high degree of complexity[17)] and project-specific uncertainties in system parameters. Quite often, this is caused by too many stakeholders and system components as well as temporal, country-specific, work-related, operational, regional, or seasonal factors. Conventional industrial operations are dominated by technical systems and, increasingly, cyber-physical systems. In the construction sector, the latter have already made inroads into some areas of prefabrication, structural engineering, and tunneling.

We describe a simple example with only a few components to highlight the complexity and uncertainties of a system. We assume that the work to be performed on a construction site involves creating a subgrade composed of several layers. Its overall design includes four different layers. The materials and pieces of equipment required for installing these layers are available. However, the workers are not in possession of any information clarifying the sequence in which the layers should be installed.

Applying the principles of combinatorics, the available four materials result in a total of 24 different options (permutations[18)]) for constructing the subgrade (see Fig. 1.8).

The number of possible combinations is derived from the factorial of the number of components n (see Eq. (1-12)). Accordingly, a four-layer subgrade would already provide 24 different permutations (= 1 * 2 * 3 * 4).

$$n! = 1 \cdot 2 \cdot 3 \cdot \ldots \cdot n = \prod_{k=1}^{n} k \qquad (1\text{-}12)$$

If only one of the permutations were permissible for installation, the probability for the workers to install the layers in the correct sequence would amount to only about 4.2%

15) The REFA methodology is based on systems theory. A system is defined as an entity composed of elements whose relationships serve a certain purpose. Depending on the degree of detail, we can consider a single workplace, an entire company, a sector of the economy, or the national economy to be a system. In business and economics, we start from the involved elements to define the following categories: technical systems (machine systems), social systems (human-human systems) and socio-technical systems (human-machine systems). Cf. Berg (1984), p. 51

16) Cf. ibid., p. 51

17) Complexity differs from complicatedness in that it includes a temporal component (dynamics).

18) Permutation can be defined as a rearrangement of things in a set where order matters.

(= 1 / 24 * 100%), provided that none of these options can be excluded on the basis of prior experience or additional information.

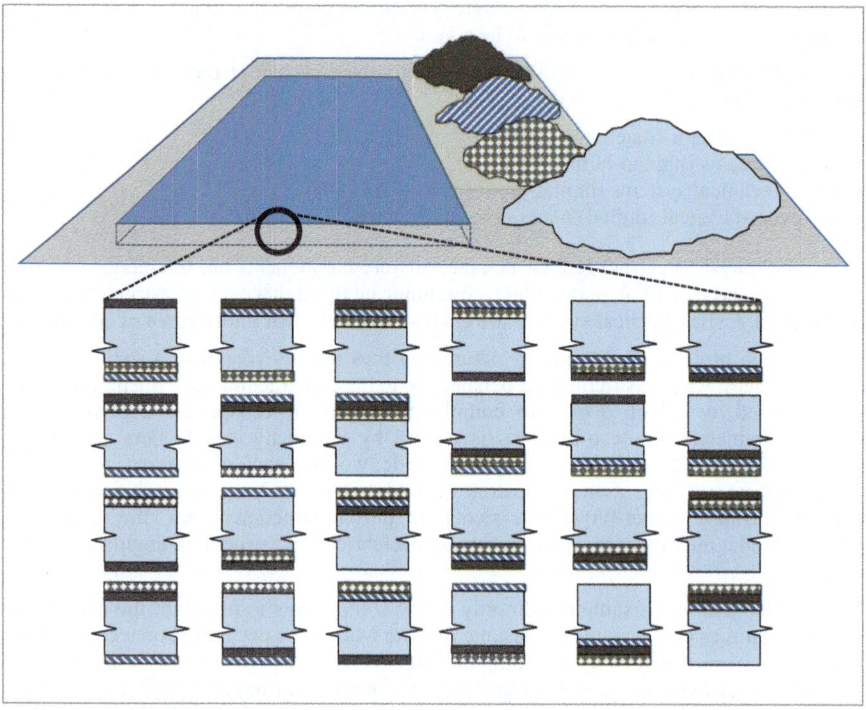

Fig. 1.8 Illustration – Complexity of systems – Constructing a four-layer subgrade[19)]

If the workers start installing the first layer while still waiting for the required information, there is a 25% probability for them to start with the correct layer (= 1 / 4 * 100%).

These percentages are objective probabilities that are easy to determine using the ratio of favorable to possible cases on the basis of the countable options. In practice, however, such an objective calculation of probabilities is possible only in extremely rare cases because any of the parameters can take a large number of different characteristics (uncountability). In most cases, we thus need to revert to subjective probabilities or qualified assessments.

This example illustrates that even simple systems (for instance those including only four components) can achieve a highly diverse set of combinations. Thus, this problem highlights that it is absolutely necessary, in construction projects, to deal with uncertainties on a systematic basis in order to cope with the existing complexity both from the client's and the contractor's point of view.

[19)] Cf. Hofstadler/Kummer (2017a), p. 8

1.1.5 Uncertainties in Systems

Besides complexity, uncertainties included in all systems and models are of particular importance for performing investigations and analyses.

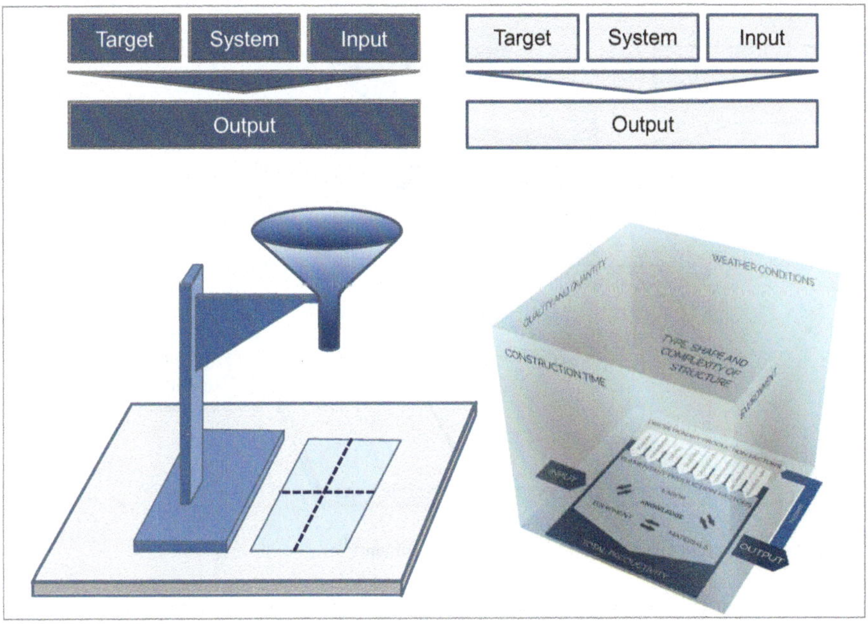

Fig. 1.9 Comparison – Funnel experiment versus production system[20)]

Representing the highly complicated or complex production system in construction management and economics, the uncertainties and their effects on results are illustrated by means of a funnel experiment. This meaningful comparison is possible because both systems are defined by a set target, an input, and an output that can be derived from it (see Fig. 1.9).

The funnel experiment uses a gallows setup on which a funnel is mounted. This assembly is put on a table on which a sheet of paper with a target is also placed. The funnel neck is oriented towards the center of the target paper. Spheres are thrown into the funnel in order to hit the center point (marked by a cross) on the paper.

The above system is fraught with uncertainties both with respect to the throw/input of the sphere and to the quality and stability of the gallows setup. It is also subject to additional effects from its immediate surroundings. Other uncertainties exist in relation to the orientation of the sheet of paper and the centering of the cross. The stability of the substructure and weather impact can have a direct effect on the target accuracy of the system (see Fig. 1.10).[21)]

[20)] Hofstadler/Kummer, derived from *Deming's* funnel experiment. http://www.denkeler-qm.de/Artikel/Trichter/trichter.html. Accessed December 21, 2020
Cf. Hofstadler/Kummer (2017a), p. 9

[21)] Cf. Hofstadler/Kummer (2019), p. 739ff.

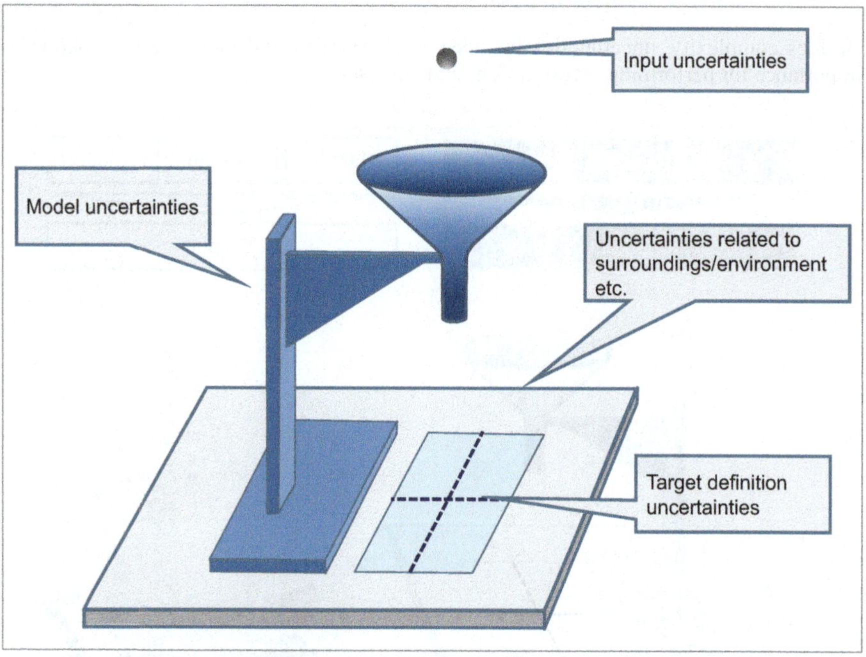

Fig. 1.10 Uncertainties of the system – Funnel experiment[22)]

Several attempts at throwing the sphere into the funnel will result in different hits on the target paper. These hits will occur within a certain area and are directly dependent upon the throw of the sphere into the funnel, the surface properties, the susceptibility of the gallows setup to vibration, the target definition, and external factors (such as wind or random shocks impacting the entire system). Marking every hit on the target paper results in a point cloud that spreads in either direction from the defined target.

Several options exist for reducing the uncertainties of the above system, including fixing the gallows setup on the tabletop and/or mounting a longer funnel neck in order to increase the length of the guided trajectory of the falling sphere (see Fig. 1.11, right).

[22)] Hofstadler/Kummer, derived from *Deming's* funnel experiment. http://www.denkeler-qm.de/Artikel/Trichter/trichter.html. Accessed December 21, 2020
Cf. Hofstadler/Kummer (2017a), p. 10

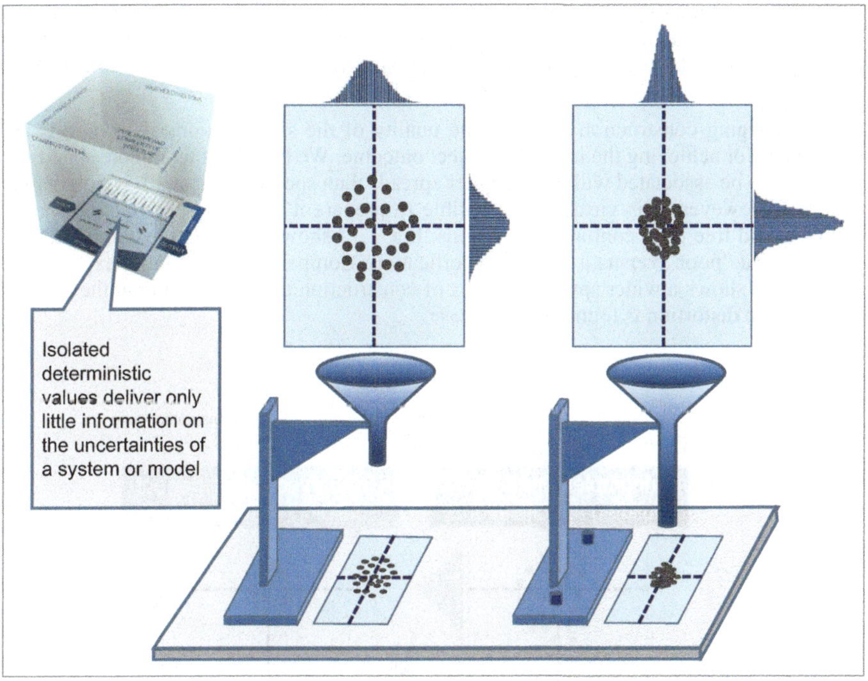

Fig. 1.11 Comparison of experimental results for systems with different uncertainties – Spreads but no distortion[23)]

Repeating this experiment using the improved system (with reduced uncertainties) involving throws of several spheres and plotting the hits will also result in a point cloud around the target cross. However, its spread will decrease compared to the first attempt, leading to a more accurate result.

Several experiments need to be carried out for obtaining a representative picture of possible results in order to compare the uncertainties associated with the two funnel systems shown in Fig. 1.11. If we performed just one experiment and accidentally hit the center of the system on the left-hand side of Fig. 1.11 while observing a deviation from the center in the system shown on the right, we might draw the incorrect conclusion that the system on the left is associated with less significant uncertainties than the system on the right.

Isolated deterministic results thus provide only little indication of the uncertainties of a system or model. Only by considering ranges or assessing possible spreads of results will we be able to arrive at sound conclusions regarding the corresponding system uncertainties. Fig. 1.11 shows the deviations of hits from the target center observed in the experiments in the form of histograms. We find that the system on the left is associated with greater uncertainties in either direction because the point cloud also exhibits a wider spread.

23) Cf. Hofstadler/Kummer (2017a), p. 11

Comparing the two systems illustrates that a single experiment or result will not permit any conclusion regarding uncertainties or potential chances and risks. Only if the ranges of the determining parameters are taken into account in a calculation model can we identify the probabilities of exceeding or underrunning specific values and thus make informed decisions.

When managing construction projects, the quality of the specifications is of significant importance for achieving the intended project outcome. We find that incomplete specifications tend to be associated with much wider spreads than specifications with a high degree of detail. However, it is virtually impossible to prepare a building specification that is complete and free from contradictory details. Fig. 1.12 shows a qualitative comparison of spreads for a "poor" versus a "good" specification. Compared to the funnel experiment, the left area shows a wider spread in terms of construction time and cost than the area on the right. No distortion is found in either case.

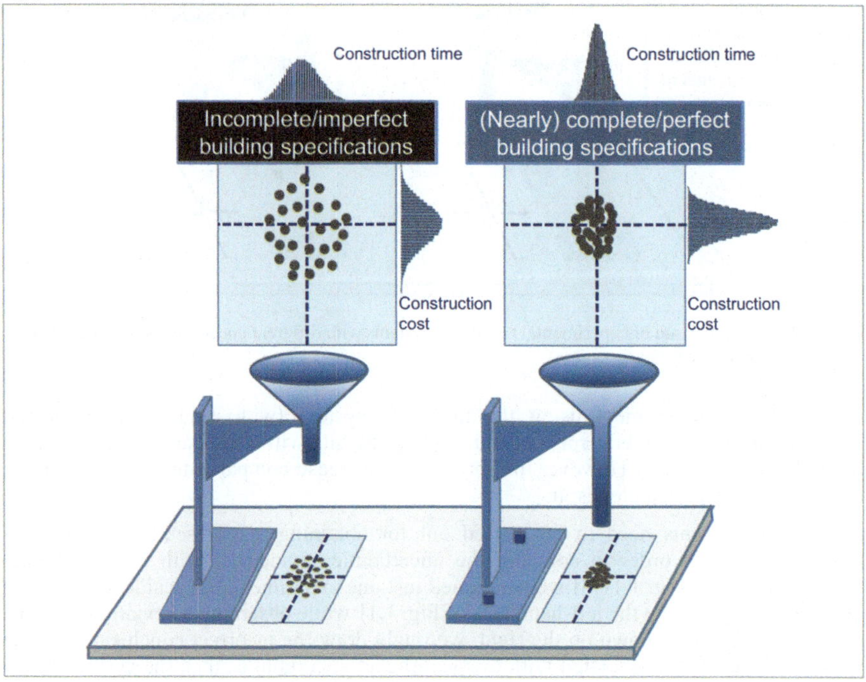

Fig. 1.12 Comparison of experimental results for systems with different uncertainties – Influence of the specification quality on the spread of construction time and cost – Spreads but no distortion

Any plan or forecast should aim to generate models for calculations designed to achieve narrower spreads while avoiding distortions. If the spread is very narrow in the bid phase, this usually indicates a very high quality standard in the project preparation phase that translated into a great degree of detail in the planning and specification documents. This is what we should aim for because such projects tend to be completed much more smoothly while also adhering to defined project targets.[24)]

[24)] Cf. Hofstadler/Kummer (2019), p. 739ff.

1.2 Objective

This book sets out to outline the fundamentals of managing chances and risks while making a difference between rough estimates and detailed analyses. The significance and the potential of managing chances and risks are indicated for the individual project phases. In so doing, we highlight he difference between deterministic and probabilistic calculation methods as well as the fundamentals of applying Monte Carlo simulations. Calculation accuracy essentially depends on the modeling approach, the uncertainties of input parameters, and the project phase under consideration.

The primary objective of this book is to outline and discuss a systematic approach to dealing with uncertainties and to illustrate the Monte Carlo simulation methodology in a clear, structured, and plausible manner. Particular emphasis will be put on explaining the interpretation of results, the factors influencing the shape of histograms, and the effects of correlations. We will also present the latest findings gathered from surveys conducted among construction experts. Readers of this book will be able to check the validity of collected data themselves on the basis of the transparent representation of key statistical parameters and the associated descriptive data analysis. Comparisons are made to discuss the reliability of the new approaches, as well as their deviations from those presented in the literature or implemented in construction practice.

Realistic calculation examples from both the client's and the contractor's perspective illustrate the modeling exercise and the interpretation of simulation results. The worked examples are based on specific issues encountered in construction practice. They demonstrate the wide range in which Monte Carlo simulations can be carried out in project management as well as in construction management and economics in different project phases.

1.3 Structure

This introduction is followed by the second chapter, which focuses on outlining the fundamentals of chances and risks as well as explaining key terminology and defining boundaries and limits.

This book defines these essential terms as follows: A risk (chance) refers to an expectation in the presence of an uncertainty that assumes a negative (positive) deviation from the target. Its impact and probability of occurrence are known either objectively or subjectively.

Particular emphasis is put on explaining the significance of expectations in conjunction with the chance/risk ratio. Any expectation of a positive or negative variance from the target essentially depends on the selected reference base. This means that any greater or smaller reference base for a labor consumption rate LCR [wh/UQ] will also result in comparatively greater or smaller probabilities of the actual labor consumption rate being lower than the defined reference base. Different designations and interpretations are used for statements and conclusions regarding future events depending on the degree of detail of the analysis and the considered timeline (short, medium, or long-term). By way of example, we include a list of different terms related to the future in conjunction with a non-exhaustive list of corresponding criteria.

The key terms in the field of statistics and probability theory that are most important for understanding relevant relationships and interpreting calculations and simulations are explained in detail while at the same time referring to practical examples. In addition, this book highlights the significance of key topics relevant to statistics, such as correlations, data fitting, and historical data.

The third chapter concentrates on explaining the basic significance of data, information, and knowledge. Construction contracts are contracts for work and give rise to a high degree of asymmetrical information depending on the type of specification and the contractual model. This asymmetrical information present in the relationship between the client and the contractor contributes to determining the magnitude of uncertainties and the dimensions of ranges. We also explain the fundamentals of adverse selection and the principal-agent theory and their effects on construction projects. We highlight the fact that the information differential between the creators and users of information is so large in the individual project phases that major deviations may occur in estimates and calculations if a non-systematic approach is chosen. In addition, it is demonstrated that smart data archiving and dealing proactively with this data can improve the accuracy of calculations in the individual project phases. The key quality criteria for information should be met in order to achieve the defined targets.

The individual processes of managing chances and risks are outlined in the fourth chapter. Identifying and seizing opportunities, or chances, is one of the most important objectives for organizations or companies. On the other hand, this is also true for detecting and preventing risks. We also demonstrate how early-warning systems contribute to improving processes and results for the entire company as well as for individual projects. A basis for the relevant consideration needs to be created for a quantitative analysis in order to clearly define the threshold from which a chance or a risk occurs. Analyses should also quantify the implications expected if a chance or risk materializes, such as exceeding or under-running costs or extending or shortening the construction period. The management of chances and risks either refers to the entire company (global view) or to individual projects (local view). The global analysis concentrates on the strategic focus. For this purpose, project portfolios are created in order to obtain a complete overview of key projects. The local analysis at the operational level creates meaningful portfolios of projects of the individual divisions and areas. A systematic approach to dealing with chances and risks needs to be promoted and demanded across the entire organization. The senior management team should lead by example in continuously managing chances and risks and applying the related methods and tools. As part of the process of managing chances and risks, we particularly refer to the application of feedback loops and their benefits.

The fifth chapter concentrates on the five project phases (PPH 1 to PPH 5) of construction projects. This temporal and task-based structure creates the basis for categorizing key content and worked examples that are discussed in Chapters 9 and 10 and supported by calculations and descriptions in order to add clarity for the reader. Project stakeholders from the client's and contractor's side are thus provided with relevant practical examples of applying Monte Carlo simulations for calculations and decision-making processes.

The sixth chapter outlines the fundamentals of the Monte Carlo method and Monte Carlo simulation while also highlighting their differences. Besides referring to the various subfields of mathematics to which the Monte Carlo simulation relates, we differentiate between scenarios and iterations. We include several examples of how to generate results in the form of histograms in order to prevent Monte Carlo simulations from becoming black boxes when applied by the user. The type and curve of the distribution functions selected for input parameters will affect the histograms representing the results. Yet simulation results are influenced not only by the selected distribution functions but also by the skewness of distributions, which is a key statistical variable. The following sections of this chapter concentrate on discussing this influence of the skewness of input parameters on the output values when applying the four basic arithmetic operations. We also outline the influence of correlations in several examples. The final section of this chapter illustrates the strengths and weaknesses, or chances and risks, for applying the Monte Carlo simulation method. Starting from this basic knowledge, the following chapters will be easier to understand so that they stimulate independent thinking and application.

The seventh chapter builds upon the fundamentals explained in Chapter 6 for outlining the practical application of Monte Carlo simulations, as well as providing guidance on how to interpret simulation results. Particular emphasis will be put on selecting distribution functions, data fitting (i.e. adjustment of distribution functions to an existing database), identifying correlations, and interpreting histograms. In addition, the application and benefits of other graphical representations of simulation results are outlined, including tornado, point, and spider diagrams. We also refer to determining realistic distribution functions for labor consumption rates and quantities on the basis of expert surveys. These realistic calculation inputs make a significant contribution to improving results. We highlight the significance and influence of correlations by explaining the example of calculating the total labor consumption rate of reinforced concrete operations.

The eighth chapter is dedicated to the essential influence of construction time on the achievement of project targets. The construction time defined by the client in the contract essentially determines if time is sufficient to achieve all defined or agreed project targets. This applies to all planning processes as well as to the project execution phase. Particularly demanding quality specifications, such as high-quality exposed concrete or waterproof structures, require sufficient time to be achieved owing to the related requirements in terms of manufacturing and technology. This chapter outlines the qualitative and quantitative relationship between construction time, labor consumption rate and productivity, and construction costs. Normal productivity will no longer be achievable if the production factors cannot be utilized efficiently, for instance owing to non-optimal team sizes or ratios between the number of workers and the number of pieces of equipment. Own approaches and methods known from pertinent literature are presented and critically reviewed to allow for quantitative assessments of losses of productivity. We illustrate the influence of construction time on costing and, consequently, on the actual construction process. Construction managers will be made aware of the necessity to closely monitor those construction-related factors that have an influence on construction time.

The ninth chapter describes and discusses worked examples from the client's perspective that can be used for various project phases and analytical objectives. We demonstrate not only deterministic approaches but also the systematic application of probabilistic methods. The worked examples are structured in a clear and transparent manner in order to add as much value as possible for users. Interim and final results are shown as histograms and interpreted comprehensively. A conclusion regarding the specific chance/risk ratio can be drawn for each of the selected deterministic values.

The tenth chapter concentrates on worked examples from the contractor's perspective. We discuss practical problems and solutions for the contract award (PPH 3) and project execution (PPH 4) phases. Monte Carlo simulations are carried out on the basis of the deterministic calculation models. Examples of selecting and combining the production factors should demonstrate the benefit of probabilistic calculations.

All worked examples should enable and encourage independent thinking and application so as to create a basis for systematically dealing with uncertainties even during current projects. The relevant chance/risk ratio should always be determined for all important result parameters and decisions. As shown in the examples, this ratio is determined by the selected base value.

This book concludes with lists of figures and tables, and a subject index.

1.4 Target Audiences

Project management starts as early as at the beginning of the project and sets the course for successfully achieving the defined project targets.

Starting from initial ideas for building use, project management proceeds to the actual design and structural analysis as well as the definition of quality standards. The related requirements become increasingly specific and give rise to detailed design documents and set deadlines. Accordingly, the uncertainties associated with costing and calculating the construction time are reduced as the design becomes more accurate and refined. The same applies to the ranges of selected input parameters although a certain degree of uncertainty is retained in each calculation step. This book explains how the application of Monte Carlo simulations makes it possible to systematically consider uncertainties in the individual project phases from the client's point of view.

Topics related to construction management and economics cover the entire range of the construction industry, including building construction, civil engineering, industrial construction, and infrastructural projects. They deal with the basic relationships and interactions with respect to selecting and utilizing elementary and discretionary production factors. In any subfield of construction, it is indispensable to systematically manage uncertainties and create an understanding of ranges and probabilities so that contractors can successfully plan and implement their projects. Applying the related methods and tools supports the assessment and analysis of uncertain system relationships and parameters and facilitates the process of preparing and making decisions.

This book thus primarily addresses clients and contractors, appointed experts, scientists, students, and other stakeholders interested in construction management and economics, as well as representatives of government agencies and authorities. However, the chapters on basics and fundamentals address a broader audience that goes beyond the parties involved in construction projects due to the universal applicability of mathematical and statistical methods.

1.5 Summary

Calculating construction costs and times accurately is one of the most important and demanding tasks in project management, construction management and economics. In this respect, the existing interdependencies between construction times and costs should be taken into account. To consider these interactions adequately, valid data and information is constantly being sought for labor consumption rates, performance rates, productivity, material consumption, stocks available and deployable on-site, number of transport cycles, and cost and time parameters in order to estimate or calculate them *ex ante*. The intended results include feasible (normal) construction times and affordable construction costs.

Exact figures must be provided or submitted at the end of any analysis. Their accuracy and robustness will, however, essentially depend on the complexity of the building or structure and on the conditions prevailing at the actual work stages or on a more or less accurate database. One possible solution to this issue is to consider ranges that can deliver valid conclusions on determined values. Probabilistic methods deliver valuable results that make it possible to systematically consider such input value ranges and correlations of key determinants. Key outcomes of probabilistic calculations include histograms that are used to capture or estimate the chance/risk ratio relative to a specific (selected) value in a reliable manner.

Recently, life-cycle costing exercises in the construction industry have attracted growing attention. Among other factors, this is due to the shift towards sustainability and increased quality requirements. This changing mindset is also associated with a growing interest in managing chances and risks because the associated methods permit more effective estimates of the impact of investments. At the same time, life-cycle costing provides an appropriate tool for interlinking changes in the planning phase with more comprehensive lean management methods. This approach is based on a holistic view that is to prevent "wastage" caused by subsequent modifications and additional costs by adding value through an optimized quality management system.

As part of these trends, the influence of digitization on construction management and related processes is becoming increasingly important. However, this progress necessitates the provision of even more accurate data and information for the pricing and simulation of future projects, whereby this database relies on the quality of the documentation of past projects. Chances can thus be seized more effectively and risks can be detected and assessed earlier on, which promotes the ultimate project success. In projects under construction, real-time data and information should contribute to establishing more efficient management and control of construction processes.

All of our activities in the field of prefabrication and in the on-site construction of buildings will inevitably require a comprehensive process understanding with respect to the characteristics and interactions of production factors. With this knowledge in place, a systemic management of chances and risks and the purposeful application of digitization methods will make a major contribution towards the resource-saving design, construction, and operation of buildings.

1.6 References

Berg, Gerhard (1984). REFA in der Baupraxis – Teil 1 Grundlagen. Ed.: REFA Verband für Arbeitsstudien und Betriebsorganisation e. V. Frankfurt am Main. ztv-Verlag. (ISBN 3-923597-12-6)

DIN ISO 5725-1 (November 1997). Genauigkeit (Richtigkeit und Präzision) von Meßverfahren und Meßergebnissen – Teil 1: Allgemeine Grundlagen und Begriffe. [Accuracy (trueness and precision) of measurement methods and results – Part 1: General principles and definitions]

Hofstadler, Christian (2007). Bauablaufplanung und Logistik im Baubetrieb. Berlin, Heidelberg. Springer-Verlag. (ISBN 978-3-540-34320-2)

Hofstadler, Christian (2011a). Berechnung der Bauzeit – Systematischer Umgang mit Projektunsicherheiten. In: Netzwerk Bau – Fachzeitschrift für Baumanagement und Bauwirtschaft – No. 15-011 – Wissen & Visionen – Chancen zur erfolgreichen Projektentwicklung erkennen und umsetzen – Fachsymposium 10 Jahre Netzwerk Bau – Tagungsband. Ed.: Stempkowski, Rainer. 32-41. Perchtoldsdorf. Stempkowski Baumanagement & Bauwirtschaft Consulting GmbH.

Hofstadler, Christian (2014). Produktivität im Baubetrieb – Bauablaufstörungen und Produktivitätsverluste. Berlin, Heidelberg. Springer-Verlag. (ISBN 978-3-642-41632-3)

Hofstadler, Christian; Kummer, Markus (2015a). Risiken und Chancen der Mischpreiskalkulation. In: Tagungsband – 13. Grazer Baubetriebs- und Bauwirtschaftssymposium – Preisermittlung und Vergabe in der Bauwirtschaft – Baubetriebliche, bauwirtschaftliche und rechtliche Aspekte. Eds.: Heck, Detlef; Mauerhofer, Gottfried; Hofstadler, Christian. 49-92. Graz. Verlag der Technischen Universität Graz. (ISBN 978-3-85125-385-9)

Hofstadler, Christian; Kummer, Markus (2017a). Chancen- und Risikomanagement in der Bauwirtschaft – Für Auftraggeber und Auftragnehmer in Projektmanagement, Baubetrieb und Bauwirtschaft. Berlin, Heidelberg. Springer-Verlag.
(ISBN 978-3-662-54318-4)

Hofstadler, Christian; Kummer, Markus (2019). Systematische Berücksichtigung von Unsicherheiten in der Preisermittlung – Einsatz der Monte-Carlo-Simulation. In: Aktuelle Entwicklungen in Baubetrieb, Bauwirtschaft und Bauvertragsrecht – 50 Jahre Institut für Baubetrieb und Bauwirtschaft der TU Graz. Ed.: Hofstadler, Christian. 739-768. Wiesbaden. Springer Vieweg (ISBN 978-3-658-27430-6)

http://www.denkeler-qm.de/Artikel/Trichter/trichter.html. Accessed December 21, 2020

Kummer, Markus (2015a). Aggregierte Berücksichtigung von Produktivitätsverlusten bei der Ermittlung von Baukosten und Bauzeiten – Deterministische und probabilistische Betrachtungen. Doctoral thesis. Graz. Institute of Construction Management and Economics, Graz University of Technology.

Simon, Fritz B. (2017). Einführung in Systemtheorie und Konstruktivismus. Heidelberg. Carl-Auer Verlag GmbH. (ISBN 978-3-89670-547-1)

Springer Gabler Verlag (Ed.), Gabler Wirtschaftslexikon, Keyword: System. Available online at http://wirtschaftslexikon.gabler.de/Archiv/3210/system-v12.html. Accessed December 21, 2020

2 Basics

This chapter provides definitions of, and detailed explanations on, key terms used in this book, with the primary aim of ensuring a sound and uniform understanding of the described relationships, methods, and use cases. In some of these, the definitions and interpretations given in the literature and in codes and standards differ from each other. This is why the meaning of the terms most frequently used in this book needs to be clarified beforehand, and the related concepts should be clearly delimited from one another by applying the developed definition criteria. Publications that dedicate only a small share of their content to explaining the pertinent terminology should generally be taken with the proverbial grain of salt since they tend to contribute to misleading or erroneous usage by generating ambiguity and vagueness, rather than promoting a coherent understanding within a defined field or discipline. We should make every attempt to prevent this from happening both in theory and practice.

2.1 Chance and Risk

The usage of the chance/risk concept is certainly controversial and requires in-depth consideration in order to arrive at precise conclusions that are relevant to actual practice. This applies particularly to the construction industry, which is considered to be a high-risk sector where the related terms have not yet been established and used clearly and unambiguously even though the management of chances and risks is still in the early stages of widespread implementation.

This chapter sets out to describe the usage of the concept of risk in general, non-technical language, in business and economics, and, more specifically, in construction management and economics. We consider this general introductory part necessary in order to arrive at a more refined understanding of the concept and thus to establish the required, theoretically founded delimitation.

The concepts of uncertainty or risk are used in many areas of everyday life, but also in a number of academic disciplines. However, we have not yet been completely successful in developing a clear definition or unambiguous usage of these terms. Given the wealth of different definitions and perspectives, it is necessary to clarify in advance what the key

C. Hofstadler and M. Kummer, *Chances and Risks in Construction Management and Economics*, Springer Tracts in Civil Engineering,
https://doi.org/10.1007/978-3-030-64014-9_2

concept of risk encompasses in this book, as well as related terms such as uncertainty, insecurity, hazard, or chance.

2.1.1 Etymology

It is not entirely clear where the concept of "risk" originates from. However, its etymology[1)] basically suggests two different explanations (see Fig. 2.1). On the one hand, it appears plausible to derive this term from the Italian verb *rischiare* (or *risicare* in Latin). This word refers to "the cliff to be circumnavigated", a "dangerous rock" or, more generally, a "daring venture". According to this interpretation, risk refers to something man-made, which implies a free choice of whether to circumnavigate a danger that is generally avoidable or not. The second possibility starts from the Greek term *riza* derived from the Arabic *rizq*. These terms refer to a predetermined superpower to which one is exposed, which is also known as fate. They comprise a mixture of the positive (chance) and the dangerous or risky, and are also closely related to the concept of fear. An essential distinguishing feature between the two etymological approaches is one's ability to influence[2)] them.

Historically, the concept of risk has also been associated with shipping and trade. Maritime trade insurers, in particular, had to deal with risks and draw up corresponding contracts. For this purpose, recourse was made to placing a bet as was customary already at the time of the Roman Empire. The random nature of events as well as betting on their occurrence or non-occurrence could also be transferred to maritime insurance.[3)]

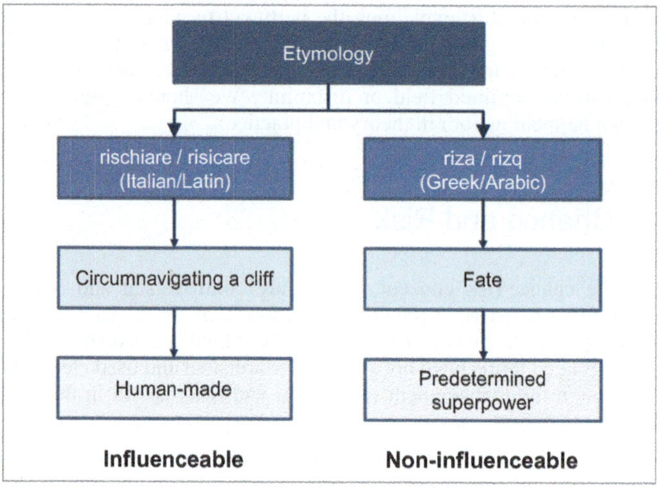

Fig. 2.1 Etymology of the concept of risk[4)]

[1)] Etymology is the study of the origin of words and the way in which their meanings have changed throughout history.

[2)] Cf. Jonen (2007), p. 4f.

[3)] Cf. Luhmann (1991), p. 18

[4)] Cf. Jonen (2007), p. 5

2.1.2 General Language

A glance at a reputable lexical reference, such as the Brockhaus in German-speaking countries or one of the Oxford or Cambridge dictionary editions, is often sufficient for learning more about the meaning of certain terms. In the Brockhaus, the concept of risk is defined as follows:

"[...] risks of losses, uncertainty and randomness factors associated with any economic activity. A distinction is made between natural risks (e.g. storm damage), technical risks (e.g. production defects), social risks (e.g. staff turnover), personal risks (e.g. illness), political risks (e.g. nationalization), and specific market risks (such as an economic downturn or industry crisis). [...]".[5)]

Gigerenzer notes that the concepts of risk and uncertainty are generally used synonymously in everyday language, but he also believes that thinking and logic are sufficient for making decisions in the presence of risks because the best possible decision can be calculated. Thus, it is assumed that the consequences and probabilities are known. However, these are unknown in uncertain situations, where rules of thumb and intuition are required for decision making. *Gigerenzer* goes on to say that games of chance such as slot machines or lotteries involve a risk because the probability of winning can be calculated. According to this view, the concept of uncertainty includes dealing with stocks, earthquakes, health, or the economy in general.[6)]

Risk expressed as an equation

Risk is often defined by the following, seemingly simple formula:

$$\text{Risk} = \text{Probability of occurrence} * \text{Consequence} \tag{2-1}$$

If we consider, for example, the production of ten exposed concrete walls for which one case of damage was documented, the probability of occurrence would be 10% (= 1 / 10 * 100%) based on historical data. The damage (i.e. the consequence) is estimated at €10,000. Applying Eq. (2-1), we thus obtain a risk of €1,000 (= €10,000 * 10% / 100%). This value could be included as a markup in the costing exercise. However, it should be noted that the real damage may well amount to the full €10,000. The calculated risk thus only reflects the average expected value. The actual damage will lie between 0 and €10,000 or even exceed this amount depending on the quality of the estimates.

Yoe proposes that this equation be treated as a conceptual model for exploring the concept of risk, rather than as a formula that actually makes risks calculable. What matters most is that either component of the above equation must be present for any risk to exist. Any consequence, no matter how severe, will not represent a risk if there is no associated probability of occurrence.[7)]

Yoe describes risk as a measure of the probability and consequence of uncertain future events. Thus, this opens up the possibility of an undesirable result. However, this "result" can take a variety of forms, such as a loss (for example due to fire, flood, illness, death, or financial setbacks) or any form of hazard[8)], but also a potential benefit that could not be realized, for instance because a new product failed to capture the anticipated market share or an investment delivered less than the expected return, or any other type of missed

5) Cf. F.A. Brockhaus GmbH (1999c), p. 464 (translation by Steffen Walter)

6) Cf. Gigerenzer (2013), p. 38

7) Cf. Yoe (2012), p. 1

opportunity. Any such potential adverse event can be attributed to the absence of information. This shortage of information is caused by a fundamental uncertainty inherent in future events.[9)]

Kaplan/Garrick also consider the characterization of risk according to Eq. (2-1) to be misleading. They justify this stance by stating that a scenario associated with a low probability of occurrence and high damage has the same risk as a scenario with a high probability of occurrence and comparatively low consequential damage (see Fig. 2.2). However, two such scenarios cannot at all be regarded as equally risky.[10)]

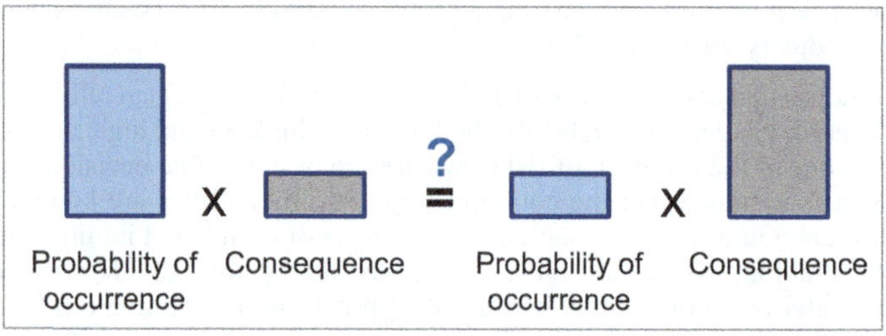

Fig. 2.2 Comparability of risks applying the risk formula[11)]

Risk vs. chance

In a number of languages and cultures, no clear distinction is made between the two terms. In the English-speaking part of the world, for example, existing chances and risks are often also referred to as upside risk and downside risk. The Chinese word for crisis, *weiji*, includes two characters which, when viewed individually, signify danger and opportunity. According to this view, any crisis or difficult situation will thus involve both a danger and an opportunity.

Risk vs. danger

A distinction between risk and danger is seen in the fact that, in the case of a risk, any potential damage can be attributed to a decision (choice), whereas, in the case of a danger, any potential damage will be caused by external (non-controllable) factors (see also the etymological information in Fig. 2.1). If, for example, we choose between two very similar alternatives (say, two airlines, and one of the planes crashes), the choice will not necessarily have been in favor of the riskier option. Instead, a danger has materialized. The alternatives must thus differ significantly in their extent or severity of damage.[12)]

Could we then turn any danger into a risk by making a decision? This question is not as easy to answer as it may seem because the answer will always depend on the point of view of the observer. What constitutes a risk for the decision maker taken on the basis of decisions (or non-decisions), will constitute a danger for those affected because they cannot counteract the potentially occurring damage by making an immediate decision.[13)]

---

8) Yoe describes the notion of hazard in a general sense as anything that could constitute a potential source of damage to a valued asset, including on the human, animal, economic, and social levels. This concept includes all biological, chemical, physical, and radiological causes as well as natural and/or anthropogenic events that could potentially trigger adverse effects on people, property, the economy, culture, social structures, or the environment. Cf. Yoe (2012), p. 2

9) Cf. ibid., p. 1

10) Cf. Kaplan/Garrick (1981), p. 13

11) Cf. Kummer (2015a), p. 19

12) Cf. Luhmann (1991), p. 30ff.

Those suffering from such damage did not create the danger threatening them by a decision, because it would be a risk in that case.[14)]

In either distinction, i.e. between risk and certainty and risk and hazard/danger, the concept of risk reflects a very complex set of circumstances, while its counterpart serves as a concept of reflection, or "mirror". Comparing risk/certainty is all about taking note of the measurement problem (operationalization), whereas the focus is on the free choice or decision in the case of the risk/danger comparison.[15)] For both distinctions, however, there is no such thing as risk-free behavior. In the former case, this means that there is no absolute certainty. In the latter case, this means that risks cannot be avoided. Strictly speaking, when heeding the advice of not overtaking in a narrow bend, we would run the risk of not advancing as fast as we would have if we dared making this move.[16)]

"One can only make risky decisions, or sit and wait. And the nature of risk means that waiting is also a risky decision."[17)]

The above definitions reflect a predominantly negative view of the concept of risk. On the cognitive level, there is thus obviously a link between risks and negative consequences, such as damage or loss, or a missed opportunity. But how is the concept of risk viewed and applied in business and economics? The following section will make an attempt at answering this question.

2.1.3 Usage in Business and Economics

When applying the concept of risk in an economic context, this is no longer just about the linguistic understanding of the word. Rather, the focus of the related definitions shifts to the measurability of its two components, i.e. probability of occurrence and consequence (see Eq. (2-1)).

Using a semantic box, *Jonen* examined a total of 329 definitions of risk taken from a wide variety of mainly German sources, including business books, journals, etc.[18)] He found that in recent years there has been a sharp increase in the attempts made at defining the concept of risk, which is indicative of its increasing significance.[19)] This study also proved an increase, from about 1940 onwards, in the subjective concept of probability incorporated in the definitions. Likewise, a moderate increase in the relative proportion of the mentions of the concept of chances was seen during this period. It should be noted, however, that about 87% of the definitions captured in the study came from the field of business administration and management.[20)]

In business and economics, *Knight's* definition is considered to be a certain type of dogma that is cited in a large number of publications on risk or risk management. *Jonen's* study also identified the definition provided by *Knight* as the one with the most references.[21)] As a result, conceptual innovations were reproached of having used these terms incorrectly.[22)] *Knight* considers uncertainty to be an umbrella term encompassing risk and

13) Cf. ibid., p. 36f.

14) Cf. ibid., p. 112

15) Cf. ibid., p. 32

16) Cf. ibid., p. 37

17) Cf. ibid., p. 81 (translation by Steffen Walter)

18) Cf. Jonen (2007), p. 39ff.

19) The examined definitions were taken from references published between 1903 and 2006.

20) Cf. Jonen (2007), p. 39ff.

21) Cf. Jonen (2007), p. 42ff.

insecurity. According to him, we refer to risk if quantitative measurements are possible, and to insecurity in all other cases (see Fig. 2.3).

Knight's concept excludes the possibility of specifying quantitative probabilities of occurrence for uncertain situations. This can be due to their unique appearance and/or too little empirical evidence.[23)]

Accordingly, these situations are referred to as insecurities if their possible effects are known but the related probabilities of occurrence can only be estimated.

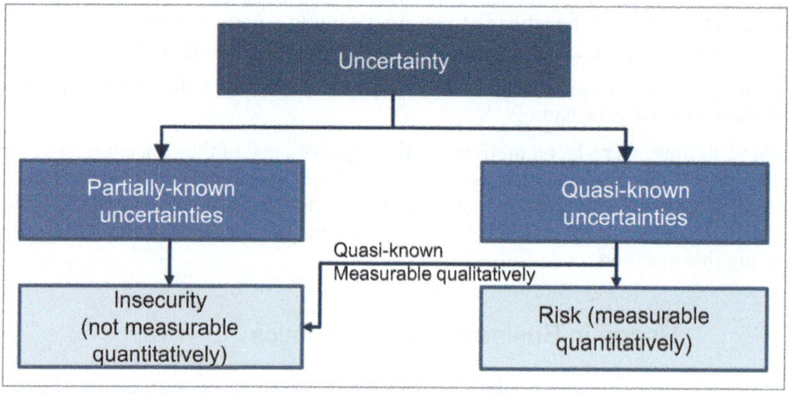

Fig. 2.3 Categorization of uncertainties according to *Knight*[24)]

Many attempts at arriving at a definition include the measurability or quantifiability of the probability of occurrence as a crucial component often regarded as a criterion for differentiating between risk and uncertainty.

The understanding of probability basically relies on two different approaches: First, the "objective" or "frequentist" view, which draws its understanding of probability from repetitive experiments, such as coin tossing. Second, the "subjective" understanding of probability, which reflects a level of knowledge, a degree of belief, or a level of confidence.[25)]

Knight goes on to define three types of probability situations:[26)]

- A priori probability (derived logically)
- Statistical probability (derived empirically)
- Estimated probability

Knight groups logical and empirical probabilities under the concept of risk. He refers to the estimated probability situation as true uncertainty.[27)]

To gain deeper insights into the scientific conceptualization of risks, we discuss below the national implementation of ISO 31000 (OENORM D 490x series, Risk Management for

22) Cf. Luhmann (1991), p. 9

23) Cf. Boeckelmann/Mildner (2011), p. 2

24) Cf. Wiggert (2009), p. 71

25) Cf. Kaplan/Garrick (1981), p. 17

26) Cf. Knight (1921), p. 224f.

27) Cf. Boeckelmann/Mildner (2011), p. 2

Organizations and Systems). The ISO 31000 international standard was to establish a risk management standard and a uniform language. OENORM D 4900:2021 describes the concept of risk as the effect of uncertainty on targets, activities, and requirements.

This notion includes the following aspects:

- "It is a combination of probability and effect.
- Effects can be positive (chance) or negative (threat or risk).
- System boundaries need to be defined in order to be able to classify the risk level (e.g. high, medium, low).
- Probabilities are used to estimate or determine the level of uncertainty or insecurity. Uncertainty or insecurity arises from contradictory or missing information, poor data availability, statistical deviations, knowledge gaps, blurred or ambiguous usage of terminology in the field, and/or insufficient communication regarding targets, activities, and requirements.
- Organizational targets comprise the entity's strategic development, including customer needs, innovation, and market share or position. Activities include operational activities, such as procurement, production, sales, and services. Requirements primarily refer to binding obligations of the organization, particularly with respect to workplace safety, property, and the environment.
- Risk is a consequence of events or developments."[28)]

Thus, probabilities can be both estimated (subjectively) and quantified (objectively), and risks refer to both sudden events and incremental developments. Uncertainties in which causal relationships and circumstances are known only partially, or just represent a degree of personal conviction or belief, are termed subjective probabilities. The term "likelihood" used in the English version of ISO 31000 covers both the objective and subjective understanding of probabilities.[29)]

The following section sets out to define the concept in the context of the specific discipline of construction management and economics.

2.1.4 Usage in Construction Economics

The question arises as to whether it is possible to apply *Knight's* narrow definition of risk, which calls for the objective and quantitative measurability of uncertainties, in construction management and economics in a meaningful manner. *Knight's* definition of risk and all other definitions derived from it are closely linked to statistical considerations, which require a correspondingly large amount of historical data to be plausible and reliable. In contrast, each construction project is unique, which is why there are no objective (measured) values for exactly the same project. This would mean that, in the course of costing and calculations of construction time, no reference should ever be made to risks, but only to uncertainties, because such a required data pool cannot be captured statistically. What is available, though, is the experience of the people involved, who are able to (and should) carry out analyses and evaluations based on similar projects.[30)]

---

28) Cf. OENORM D 4900:2021, p. 9f. (translation by Steffen Walter)

29) Cf. ONR 49000:2014, p. 7

30) Cf. Kummer (2015a), p. 26

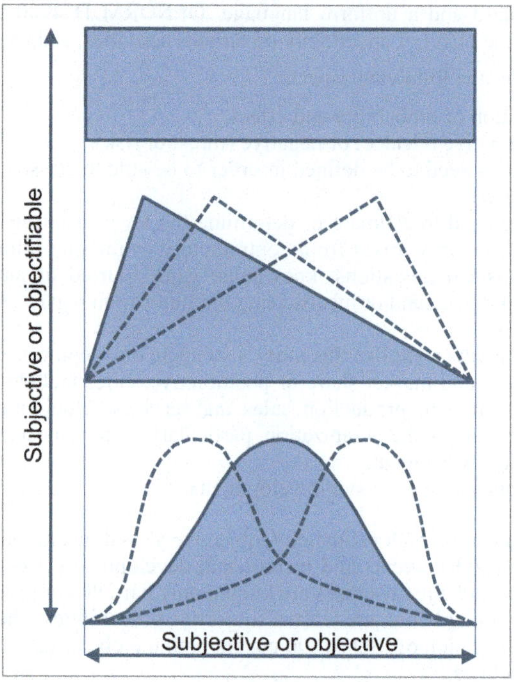

Fig. 2.4 Subjective or objective range – Subjective or objectifiable probability distribution[31)]

Ranges of uncertain parameters can either be determined objectively or applied on a purely subjective basis (see Fig. 2.4). In most cases, the probability for each value within the defined range is to be estimated subjectively by selecting distribution functions. Selecting distribution functions for input parameters becomes objectifiable if the curve of the corresponding distribution is determined on the basis of existing data, final costing exercises, results of expert surveys, or combinations of these methods. Software programs are used for adjusting such distributions to theoretical distribution functions, which is referred to as data fitting.

Section 7.1 discusses the selection of such distribution functions in detail.

Thus, risks should not just be limited to measurable factors because any type of risk management that only takes quantifiable risks into account would degenerate into a pure computational task. Yet it is the core purpose of risk management systems to also assess non-quantifiable risks because it is exactly this type of risk that has the greatest hazard potential due to the associated ignorance, or non-existing knowledge.[32)]

In his doctoral thesis, *Wiggert* examined German-language definitions of risk in the field of construction economics using the adapted semantic box method. He analyzed the evolution of the definitions of risk between 1971 (doctoral thesis by *Schubert*) and 2009 (doctoral thesis by *Riebeling*). Over time, this discussion got more thorough and intense in

31) Cf. Kummer (2015a), p. 27

32) Cf. Wiggert (2009), p. 112

construction economics (see Fig. 2.5). For this purpose, he examined a total of 69 pertinent literature references and 21 codes and standards with regard to the definitions of risk employed in these publications. Only one definition was allowed per author, with the exception of references published by groups of authors.[33)]

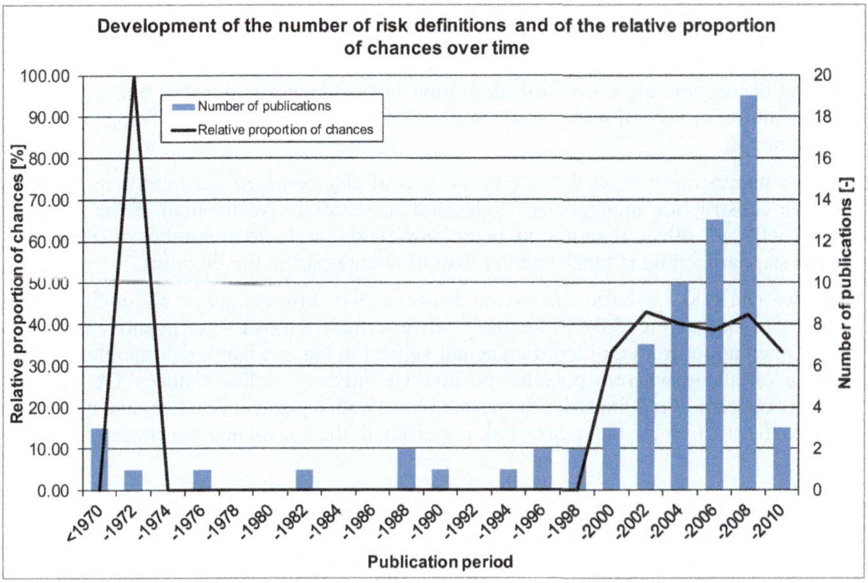

Fig. 2.5 Evolution of the number of risk definitions over time and of the relative proportion of chances[34)]

The explanation by *Link* (1999) has the most significant influence on the definitions of risk studied by *Wiggert* because five subsequent definitions can be traced back directly to it.[35)] Its wording is as follows:

"Risk is considered to be the possibility that the processes triggered by a decision do not necessarily lead to the desired outcome and that negative or positive deviations from the goal occur. Risk can be quantified by determining its scope and probability of occurrence."[36)]

Wiggert also explicitly examined the conceptual integration of the notion of opportunity, or chance, in the different definitions of risk. In his study, his focus was not only on the mere mention of the term, but also on assessing the overall concept employed in the definition of risk.[37)]

What is striking when considering the evolution of the definitions of risk in construction management and economics over time is that, although *Schubert* had incorporated the concept of risk profit as early as in 1971, it was only with *Link's* definition in 1999 that the

[33)] Cf. ibid., p. 102ff.

[34)] Cf. ibid., p. 104

[35)] Cf. Wiggert (2009), p. 105

[36)] Cf. Link (1999), p. 7 (translation by Steffen Walter)

[37)] Cf. Wiggert (2009), p. 107

notion of chance reemerged in the concept of risk used in construction economics. This marked the starting point for integrating the concept of chance into the definitions of risk. Nowadays, the proportion of definitions that incorporate the notion of chance fluctuates around 40%, regardless of the number of definitions considered (see Fig. 2.5[38]).

Based on definitions employed in construction economics, *Wiggert* compared the chance-driven view of risks with the approaches found in current standards relevant to construction management, as well as with definitions from general business administration and the financial sector. He found that approximately every third definition in construction economics, every fifth definition in the financial industry, and only every tenth definition in general business administration integrated the notion of chance into the concept of risk.

It is also interesting to note that the proportion of chance-based definitions in German-language construction management codes and standards is greater than in the literature (approx. 60% vs. 40%). It should be noted, however, that the total number of risk definitions in standardization is much smaller than its counterpart in the literature.

Thus, we can make a basic distinction between two different ways of looking at the concept of risk (see Fig. 2.6). Reference is always made to positive or negative variances of a planned target from an identified actual value. On the one hand, risk merely refers to negative variances, whereas potential positive variances are called chances. On the other hand, the concept of risk includes the possibility of both a positive (chance) and a negative variance (hazard). In the latter case, risk implies both the hazard and the chance.[39]

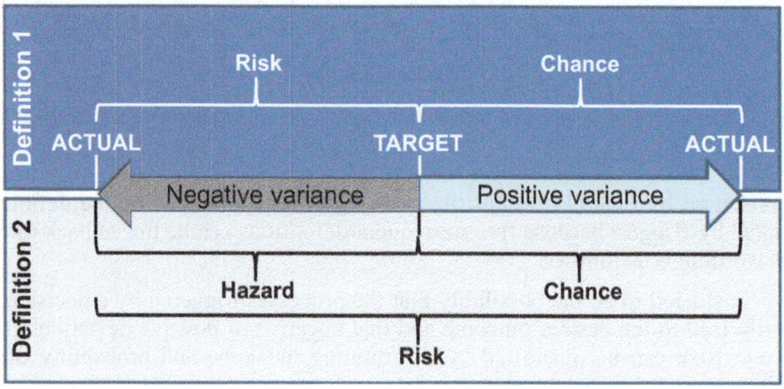

Fig. 2.6 Definition of risk

[38] The reduced number of risk definitions in the last class (i.e. 2009 to 2010) shown in Fig. 2.5 can be attributed to the fact that *Wiggert* submitted his doctoral thesis in September 2009 and that no newer definitions were included in the study.

[39] This is equivalent to the definition given in *Girmscheid/Motzko* (2013), p. 333 (translation by Steffen Walter): "In construction projects, the concept of risk refers to the possibility of deviating from specific project requirements in terms of costs, deadlines, and standards of workmanship, where potentially positive deviations are termed chance whereas potentially negative deviations are termed hazard."

2.1.5 Usage of the Concepts of Chance and Risk in this Book

The concepts of chance and risk are by no means clearly defined, as evidenced by the explanations and comments found in the literature. In the following part, we thus provide a clear conceptual delimitation and explanation of the meaning for the use of these terms, which we recommend to apply in subsequent theoretical papers owing to its high practical relevance.

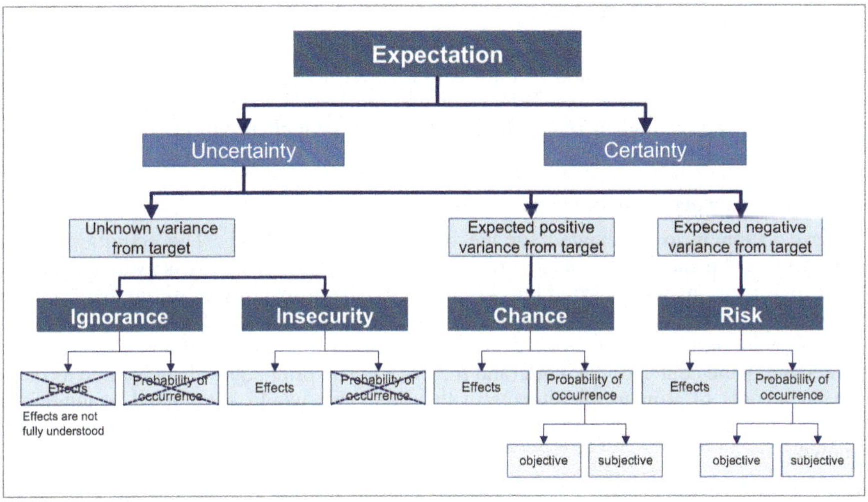

Fig. 2.7 Possible structure of expectations – Measurability of uncertainty

Fig. 2.7 differentiates the possible structure of expectations into uncertainty and certainty. There is generally no certainty because this would correspond to a 100% probability of occurrence. In the case of uncertainty, a distinction is made on the basis of the nature of the target deviation. If negative deviations from the target (i.e. damage or loss) are expected, we refer to this scenario as risk if both the consequences and the probabilities of occurrence are objectively known (in the presence of a sufficiently large database) or can be estimated subjectively on the basis of experience. This view contradicts *Knight's* definition, but it is plausible and conclusive given the current usage of the concept of risk in construction economics and the frequently observed scarcity of data. Conversely, we use the concept of chance if positive deviations from the target are expected and if the consequences and probabilities of occurrence are also known (objectively or subjectively). The notions of risk and chance are thus horizontally equivalent. Ignorance exists if both the consequences (or parts thereof) and the probabilities of occurrence (such as the possible positive or negative deviations from the target) are unknown. Uncertainty is the term used when the possible effects or consequences of an unknown deviation from the target are known, but the probability of their occurrence cannot be quantified.

Also worth mentioning in this context is the category of unknown unknowns, which refers to events whose occurrence and associated repercussions are unpredictable. Such "black swans"[40] will not be covered in this book.

---

[40] *Taleb* refers to extremely unlikely events as black swans. He refers to the fact that the complete population of swans was originally assumed to be white until black swans were discovered in Australia. See Taleb (2007), p. XVIIff.

Usage of the terms in everyday situations – Example

Let us assume you are flying to a conference in another country and did not have time to check the current weather forecast for the region before departure. There is thus no indication whatsoever of the probability of occurrence of a possible rain event, for example. Yet its consequences are well-known (you will get wet without an umbrella). In this case, you would act under the presumption of uncertainty. For example, many Europeans would be ignorant of the characteristics of a hurricane because its effects would not (or only partially) be known in Europe unless people have already gained relevant experience during such an event.

Provided you have looked at the weather forecast and the probability of rain was given as 70%, you will be in possession of information about both the probability of occurrence and the associated consequences. Without an umbrella, you will run the risk of getting wet (and this risk is quite high at 70%). On the other hand, your chance of getting to the conference with dry feet will be quite slim at 30%.

Usage of the terms in civil engineering – Example

An excavation is to be created for the basement of a building. The activities of the contractor will usually be associated with uncertainty if no soil survey is available and no prior investigations were carried out. For example, the probability of hitting hard rock is unknown, but the corresponding effects involving an increased effort required for loosening are quite familiar. If we were to come across the remains of an ancient Roman settlement in the course of the work (and if there were no prior signs of such a find as there are no records of Roman movements in this area), we would have acted in ignorance because neither the probability of occurrence nor the effects of such a find are fully known.

If, on the other hand, a soil survey is available in which the proportions of the soil classes are stated or if experience from adjacent construction projects can be used, the probability of occurrence (at least subjectively) and also the effects of a shift in the proportions between the soil classes are known. In this case, the soil survey and prior experience will serve as a starting point for evaluating the actual conditions and assessing the corresponding chances and risks depending on the expected deviation from the target.

Summary

There is no clear definition of the concept of risk, and its understanding and use vary across different disciplines, and even within the same discipline or field. The concept of risk and related terms were classified based on the overall field of expectation and represented schematically for the discipline of construction management because arriving at a global definition of the term appears to be unlikely in the near future. It becomes apparent that the concepts of chance and risk should be conceptually delimited due to their different connotations: Whereas, in practice, risk tends to be associated with a negative deviation from the target, chance is usually associated with a positive expectation. These two concepts are thus horizontally equivalent. It should also be noted that even with a subjective assessment of probabilities (for instance owing to poor data availability), using the concept of risk is considered to be legitimate. It is often necessary to resort to empirical evidence, particularly in construction economics but also in construction management. This experience is subsequently incorporated into calculations, construction time determinations, process decisions, etc., and thus becomes an integral part of the handling of construction projects from a construction management and economics as well as contractual point of view.

In this book, the two concepts of chance and risk are defined as follows (see also Fig. 2.7):

Risk refers to an expectation in the presence of an uncertainty that assumes a negative variance from the target. Its impact and probability of occurrence are known either objectively or subjectively.

Chance refers to an expectation in the presence of an uncertainty that assumes a positive variance from the target. Again, its impact and probability of occurrence are known either objectively or subjectively.

2.1.6 Selection of the Reference Base

Any expectation of a positive or negative variance from the target will essentially depend on the selected reference base (see Fig. 2.8). This means that any greater or smaller reference base for a labor consumption rate LCR [wh/UQ], for example, will also result in comparatively greater or smaller probabilities of the actual labor consumption rate being lower than the defined reference base.

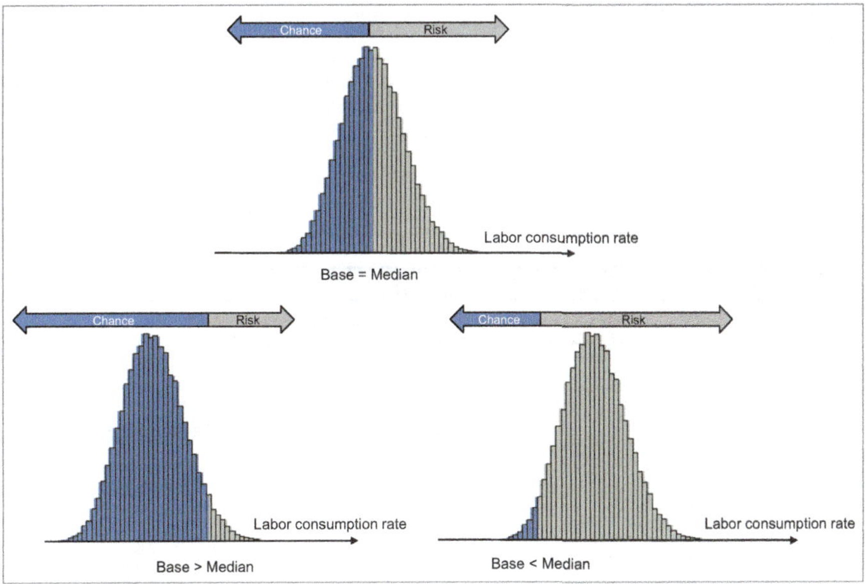

Fig. 2.8 Representation of chances and risks depending on the selected reference base[41)]

If, for instance, a histogram were derived for the labor consumption rate to be expected for reinforcing operations, the median value could be used as a baseline for the costing exercise (see Fig. 2.8, top). In this scenario, the probability of under- or overrunning the selected labor consumption rate is equally distributed (50% for either option). In other words, the possibility that a risk in terms of a higher labor consumption rate materializes is

[41)] Kummer/Hofstadler (2013), p. 180

just as high as the chance of generating a lower labor consumption rate during the construction phase. If, on the other hand, the baseline is assumed to be greater than the median (see Fig. 2.8, bottom left), the chance/risk ratio will shift towards a positive variance from the target. In this scenario, the chance of achieving a lower labor consumption rate is considerably greater than the risk of a higher labor consumption rate. Conversely, the risk will increase if a baseline labor consumption rate is selected that is lower than the median (see Fig. 2.8, bottom right). There is a correspondingly small chance for generating a labor consumption rate, at the execution stage, that is lower than the already small reference base. The chance/risk ratio depends on the readiness of the business to take risks as well as on strategic considerations. In practice, contractors usually operate closer to the risks than to the chances because they have to apply a comparatively low reference base due to the existing competitive environment.[42)]

It was *Schubert* who previously explored this topic of selecting an appropriate reference base and illustrated it by way of referring to the excavation performance.[43)]

Fig. 2.9 illustrates the extension of the understanding of risk depending on the respective reference base to other parameters. The left-hand side shows variables that ideally take a "large" numerical value, including performance and productivity. Conversely, the right-hand side includes those variables that are considered to be favorable for the course of a project in the case of "small" numerical values. These can include costs, prices, or labor consumption rates. The lower such values, the higher the risk taken by the bidder or contractor.

If only ideal values are applied, there is no more chance and the best possible result remains a situation in which no profit or loss is realized. This is in stark contrast to the interest of any company in generating revenue.

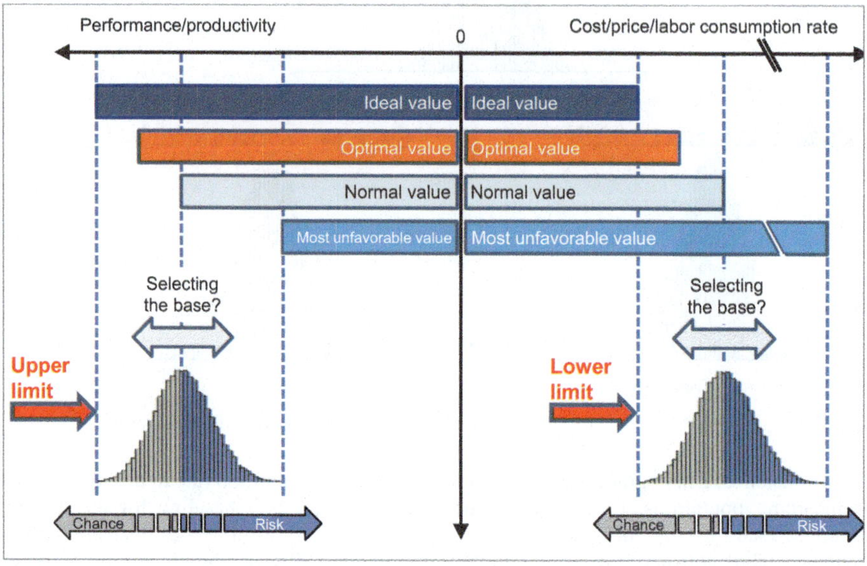

Fig. 2.9 Representation of chances and risks depending on the selected reference base[44)]

[42)] Cf. Kummer/Hofstadler (2013), p. 180

[43)] Cf. Schubert (1971), p. 10

It is also essential for the contractor to know its cost histogram in order to grant discounts or apply markups. The effect of a 10% discount, for example, on the chance/risk ratio would differ depending on the curve of the histogram. In histograms with a very narrow spread, an exceedingly steep discount can turn into a risk that is no longer acceptable because the chance of being able to execute the work at an even lower cost would rapidly diminish. Thus, for histograms with a small range, percentage discounts relative to the price have a greater influence on the chance/risk ratio than is the case for distributions with a large range. Without knowing the distribution, no conclusion can be drawn as to whether or not the risk is still acceptable with a discount or whether the new chance/risk ratio corresponds to the company's chance/risk policy.

Conversely, if a chance/risk ratio is specified, the maximum justifiable discount can be determined by means of a back-calculation. If, for example, the contractor wishes to deviate from the mean by a maximum of one standard deviation for their bid, the discount permissible up to this specific chance/risk ratio can be applied.

2.2 Assumptions Regarding the Future

A variety of designations and interpretations are used, or serve a different purpose, for assumptions regarding the future depending on the degree of detail of the analysis and the considered timeline (short, medium, or long-term). Fig. 2.10 includes a list of various terms related to the future together with an illustrative enumeration of corresponding criteria and examples.

For forward-looking statements, we first make a distinction between short- and medium-term and long-term horizons. Prophecies, utopias and visions are long-term assumptions whose contents are characterized by subjectivity and are intended to trigger different reactions among the recipients of these statements. The concept of prophecy is strongly reminiscent of ancient oracles that predicted a future event that cannot be influenced by activities in the present.

Trend assumptions, forecasts, and plans belong in the category of short- and medium-term assumptions regarding the future. They include past developments and/or decisions and generally contain less subjective elements than the aforementioned long-term assumptions. The predictive substance of forecasts relies on current and historical data. Forecasts are no plans, as aptly illustrated by the fact that assuming fair weather in the construction process planning phase will not prevent rainfall during construction. It is not possible to actively influence data or future actions on the basis of mere forecasts. However, forecasting is a tool for costing and work scheduling since it is obviously impossible to influence or modify the external project circumstances on this basis, yet these circumstances can be taken into account to an adequate degree. Cost surveying, for example, represents a mental preparation of future actions in which it is possible to arrive at greater cost certainty in line with project progress while the scope of action is narrowed at the same time.[45)]

[44)] Based on Schubert (1971), p. 11
See Čadež (1998), p. 55f., regarding the chance/risk ratio in relation to the target level.

[45)] Cf. Lechner (2010), p. 13

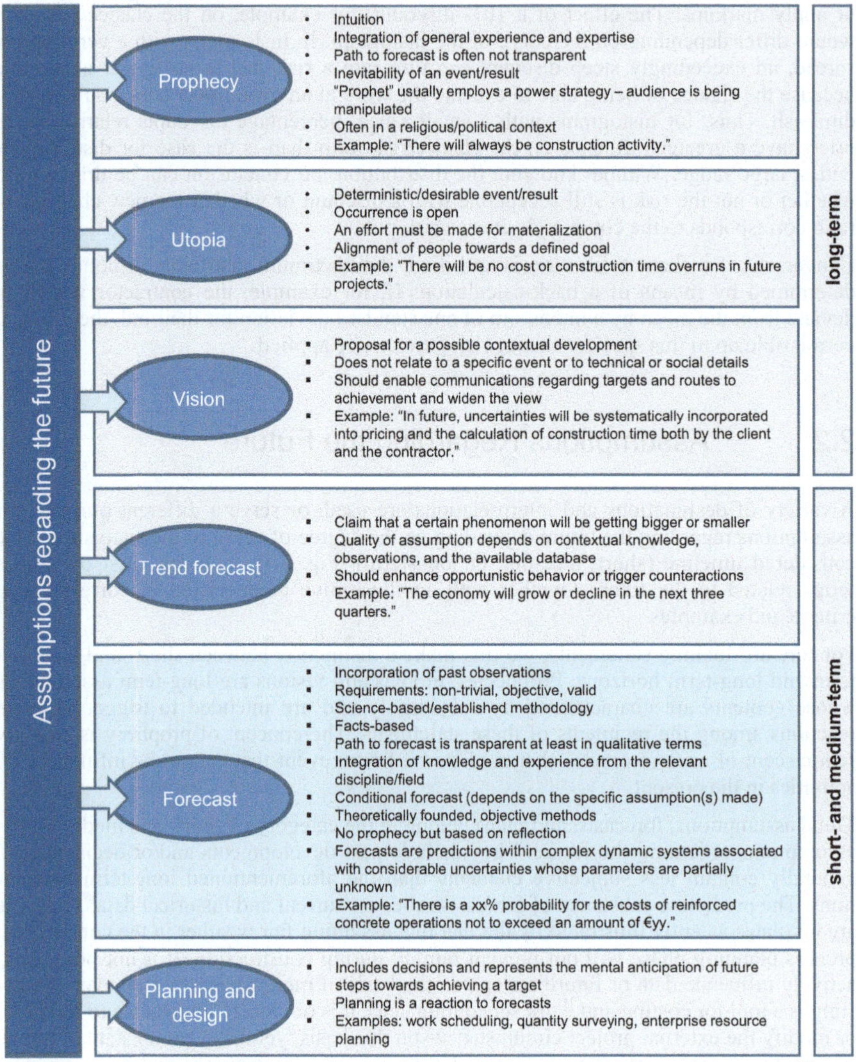

Fig. 2.10 Differentiation between various stages and types of assumptions regarding the future[46)]

However, it is possible to respond to the results or to an interpretation of a forecast by adjusting the planning and design exercise accordingly. Getting back to the rainfall example, the best response to forecast rainy patches would be to take an umbrella with you. Thus, by taking appropriate action, getting wet is no longer inevitable. Forecasts serve to identify possible risks and potential chances inherent to future events and processes as early as possible so as to be able to react to them.[47)]

[46)] Derived from Horx Zukunftsinstitut GmbH (2010) and http://wirtschaftslexikon.gabler.de/Archiv/4546/prognose-v11.html. Accessed December 19, 2020

Uncertainties in forecasting future events are inherent to the system and are also directly provided as information in certain areas. Weather forecasts, for example, use mathematical models for describing a highly complex and dynamic system (i.e. the weather) in order to arrive at assumptions regarding future events.

Even for temperature forecasts for a period of several days it has become common practice to display the ranges of possible temperature trends and to also represent them graphically in order to provide additional information.

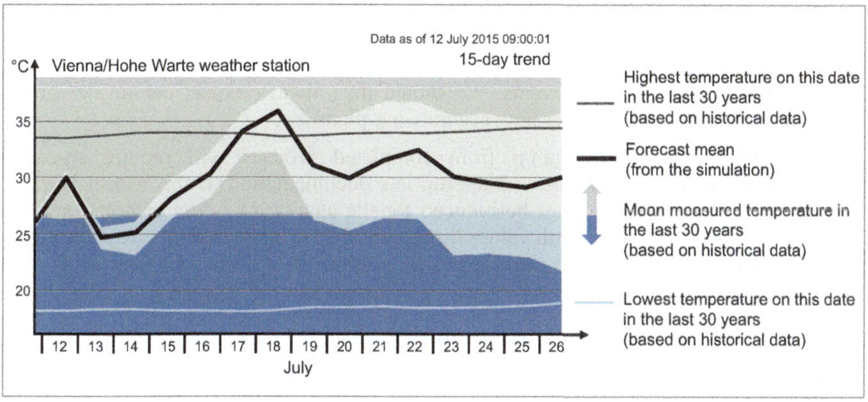

Fig. 2.11 15-day daytime temperature trend[48)]

Fig. 2.11 shows an example of the 15-day trend from 12 July 2015, 9:00 am. The abscissa lists the 15 days for which the forecast was made (i.e. 12 to 26 July) while the ordinate indicates the temperature in degrees Celsius. This forecast was made for the Vienna/Hohe Warte weather station. The dark line on the upper edge (between 30 and 35°C) and the bright line at the bottom edge (below 20°C) represent the highest and lowest temperature measured on the corresponding day of the last 30 years. The almost horizontal division of the diagram into two areas (approximately at 26.5°C) indicates the mean value of the measured temperatures from the past 30 years. Another striking feature is the forecast's increased scattering further into the future, as indicated by a so-called temperature funnel. Several model runs (iterations) were performed for generating the forecast. The mean value of these simulations is entered as a black line in Fig. 2.11. The upper and lower limits of the plotted uncertainty range are represented by the 5% and 95% quantiles derived from the simulations. This means, for example, that on 26 July the temperature will be between approximately 22 and 36°C with a probability of 90%. There is also a probability of more than 5% for the temperature on 26 July to exceed the maximum of the last 30 years. On 18 July, the mean of the simulation results is also clearly above the maximum of the last 30 years. In addition, the range of the values for this day is also smaller than for the following days due to the shorter forecast horizon of seven days.[49)]

Similarly to weather forecasts, the aim in the construction industry should be to derive assumptions regarding the future developments of a system while taking uncertain-ties into account. Again, the further into the future the events lie (in terms of the forecast

47) Cf. Kummer (2015a), p. 66

48) Cf. http://scoop.me/hashtags/orf. Accessed August 31, 2016

49) Cf. Wadsak (2016), response to a personal request submitted on March 31, 2016

horizon) or the earlier decisions have to be made about uncertain circumstances, the wider the range of possible results will be. This is the very nature of systems that are fraught with uncertainties. Performing deterministic calculations in this setting would mean to pretend a non-existent accuracy, which is usually impossible to achieve in early project phases and even during construction.

Cost and construction time forecasts or calculations and their underlying models are merely an approximation to reality, where the results can only be as good as the assumptions made and the basic data used. As such, the model does not have any intelligence, which is why it requires critical assessment and validation by specialists in each case.[50)]

In construction economics, forecasts and costing and/or pricing exercises aim to predict future events in terms of costs. They are thus characterized by uncertainties as well as unforeseen and unpredictable events. We should thus always expect certain deviations from such calculations, no matter how accurate our plans are.

Any intent to obtain information from completed projects will require so-called knowledge work in combination with appropriate documentation. If this aspect is disregarded, the uncertainties will not be reduced for the costing of future projects, and the forecast would be carried out with values that have "always been used this way". But how could we possibly arrive at an assumption for tomorrow without even knowing what happened yesterday?

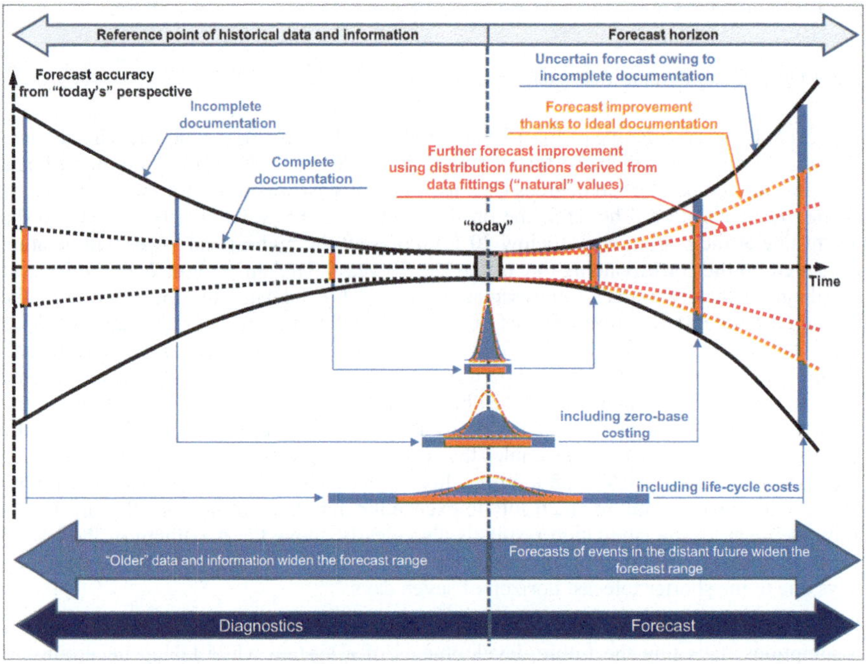

Fig. 2.12 Forecasts based on complete and incomplete documentation depending on the reference points in time of the data and information and the forecast horizons

[50)] Cf. Scherer (2010), p. 23

We can improve the accuracy of assumptions regarding the future on the basis of completed projects through systematic documentation and review. This approach makes it possible to narrow or, at least, estimate more accurately the ranges used for costing, but a certain degree of uncertainty will always remain.

In this respect, *Blecken* states that the assumption of a deterministic production process would oversimplify the production model and that the forecast quality of a deterministic model would be valid only under certain conditions.[51)]

The older the existing data and information and the more incomplete the documentation, the wider the forecast range will be. The same principle applies to exceedingly long forecast horizons (see Fig. 2.12).

2.3 Terminology in Statistics and Probability Theory

The following sections describe the key terms in the fields of statistics and probability theory that are most important for understanding relevant relationships and interpreting calculations and simulation results.

2.3.1 Stochastics

Stochastics is the mathematical theory of probability.[52)] Probability theory is a branch of mathematics that deals with the formal properties of probability measures, probability spaces, and probability distributions.[53)] It can also be referred to as the mathematics of randomness.[54)]

2.3.1.1 Probability Calculus

Probability calculus is a branch of stochastics concerned with the formalization and modeling of random processes. Probability can be defined according to the following approaches:[55)]

Probability according to *Kolmogorov*

Kolmogorov defined the following axioms:

1. Probability is not negative.
2. Probability is additive. If two events are mutually exclusive, the probability of either one or the other event occurring is equal to the sum of the individual probabilities.
3. Probability is normalized. This means that the sum of the probability of possible events is 1.

51) Cf. Blecken (1976), p. 208

52) Cf. Biermann/Grosser (1999), p. 174

53) Cf. ibid., p. 196

54) Cf. Henze (2012), p. 1

55) Cf. Biermann/Grosser (1999), p. 193ff.

Probability according to *Laplace*

The probability for an event to occur is the number of favorable cases divided by the number of all possible cases. This definition is only applicable to finite distributions because the probability would result in an undefined expression for infinite value ranges. The probability according to *Laplace* is a special case of *Kolmogorov's* axiomatic system. According to *Laplace*, randomness is the result of lack of knowledge or lack of information.[56)]

Statistical definition of probability

This definition refers to the relative frequency observed when a repetitive random experiment, such as the toss of a coin, is performed independently in stochastic terms. An event's probability is thus the limit of the relative frequency for an infinite repetition of the random experiment. The statistical definition of probability is another special case of *Kolmogorov's* axiomatic system.

2.3.1.2 Statistics

"Statistics is the science that develops rules and methods for collecting, describing, analyzing, and interpreting numerical data."[57)]

We can divide the field of statistics into the following subcategories:

- Descriptive statistics[58)]
- Inferential statistics[59)]
- Univariate statistics
- Hypothesis-generating (exploratory) statistics

The purpose of descriptive statistics is to describe and analyze the characteristics of a specific group. This must be done without drawing conclusions regarding other groups or the population as a whole. For illustrative purposes, results are often converted into graphs or diagrams.[60)]

In inductive statistics, an attempt is made to draw conclusions about the population based on information extracted from a sample.[61)]

Exploratory statistics has evolved to become a separate subfield covering "the transition between descriptive and inductive statistics".[62)] Its purpose is to identify structures, problems, and theoretically founded hypotheses. These results are then validated applying the methods of inductive statistics.

Another approach is to introduce a differentiation according to the number of characteristics considered. Univariate statistics only deal with one-dimensional random variables. Thus, there is only one variable per statistical unit, such as the body height of interviewed individuals.[63)] In contrast, bivariate methods use two variables and establish relationships

56) In this context, reference is made to the so-called *Laplacean demon*. According to *Laplace*, this demon was taken to be an intelligence knowing the state of motion of every single atom or molecule at any time, thus being capable of calculating the effects of interactions, and, consequently, the future course of events, in a deterministic approach.

57) Cf. Czado/Schmidt (2011), p. 1 (translation by Steffen Walter)

58) Also referred to as deductive statistics.

59) Also referred to as inductive statistics.

60) Cf. Biermann/Grosser (1999), p. 22

61) Cf. Biermann/Grosser (1999), p. 83

62) Cf. Duller (2019), p. 9 (translation by Steffen Walter)

between them. In the context of multivariate statistics, for example, it is examined whether different characteristics of the examined statistical units are related to each other (correlation) or whether they are independent of each other (such as the height and body weight of interviewed individuals).[64)]

2.3.1.3 Probabilistics/Probabilism

Under the influence of the natural sciences, the idea emerged in 17th century philosophy that everything that happens in the world, including human actions, is determined by unalterable natural laws.[65)]

In the field of philosophy, determinism (derived from Latin *determinare*, i.e. to delimit, to determine, to define) is "the doctrine of the unambiguous determinedness of all courses of events by causes, and of all later events by earlier events".[66)]

"*Newton's* laws are deterministic. If for an arbitrary system of bodies all state variables are known at the time t_1, then by means of the mechanical laws all state variables can be calculated for any earlier or later time t_2. The fact that state variables can never be known completely and with absolute accuracy has long been attributed to mere human imperfection. It was hoped that progress in measurement technology would enable the state variables to be determined with increasing accuracy. It was assumed that there was a convergence of the measurement results indicative of an increasingly accurate recording of the existing values [...]. From the assumption that all laws of nature are of a mechanical kind and that the world is a big 'world-machine' [...], the idea of a complete determinacy of the world arose. *Laplace* illustrated this with the idea of a superhuman intelligence[67)]: 'A spirit, which for a moment would know all the forces that animate nature, and the mutual position of all beings of which the world consists, would, if it were comprehensive enough to be able to subject all this data to mathematical analysis, have to understand in the same formula the motion of the greatest celestial bodies and of the lightest atoms, nothing would be uncertain for it, and future and past would be open to its eyes'."[68)]

Deterministic calculations assume that events in the future can be clearly determined. Uncertainties in the parameters or in the calculation model are not considered. This means that only deterministic numerical values are included in the calculations. If systems in nature were indeed deterministic, results would have to occur with 100% probability (i.e. with certainty). For complex systems with numerous dependencies, such as the erection of a structure under changing boundary conditions (including the weather) involving many stakeholders, we can assume that the occurrence of a deterministic forecast represents a special theoretical case. The realization that events cannot be predicted or calculated with absolute certainty led to probabilism, or mathematical probabilistics.

In the theory of knowledge and science, probabilism (derived from Latin *probabilis*, i.e. acceptable, probable, credible[69)]) describes "the view that there are no absolutely true but only probable propositions. – In quantum physics theories, probabilism expresses that certain events can only be predicted with a certain probability".[70)]

---

63) Cf. ibid., p. 185

64) Cf. ibid., p. 126f.

65) Cf. Historisches Wörterbuch der Philosophie – Band 2 (1972), p. 150

66) Cf. Brockhaus GmbH (1999b), p. 195 (translation by Steffen Walter)

67) Also referred to as the *Laplacean* demon.

68) Cf. Historisches Wörterbuch der Philosophie – Band 2 (1972), p. 155 (translation by Steffen Walter)

69) Cf. Stowasser et al. (1998), p. 404

"The adjective 'probabilistic' is used to describe [...] empirical theories that make use of probability theory concepts or statistical methods. The probabilistic nature of a scientific theory can be related to the degree of credibility of a subjective conviction with respect to a given set of data, but also to the objective properties of the objects or processes to be explained. Probabilistic lines of thought became widespread in the social and natural sciences in the 19th and early 20th centuries in various ways and led to fundamental doubts about *Laplacean* determinism [...]. This process is [...] referred to as 'probabilistic revolution'."[71)]

Voigt[72)] considers probabilistics to be a subfield of stochastics on the same level as statistics and combinatorics (see Fig. 2.13).

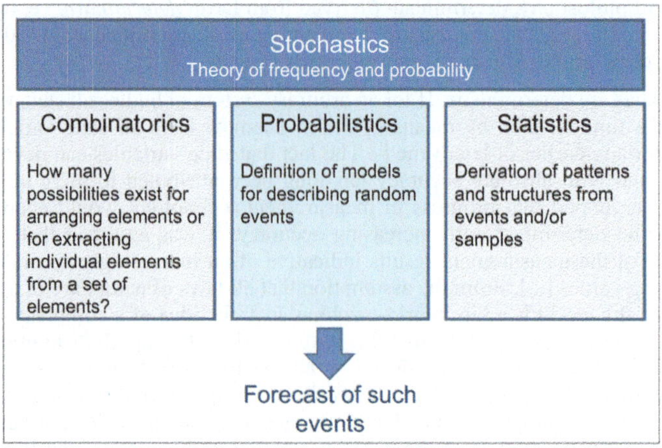

Fig. 2.13 Definitions of terms – Stochastics, combinatorics, probabilistics, and statistics[73)]

In probabilistic systems, it is thus impossible to predict a future event with absolute certainty. In contrast, we can only come up with the probability of a specific event, such as the cost of reinforced concrete operations.

From a terminological point of view, we recommend to describe random natural events or processes as stochastic and the mathematical analysis of such processes and their effects as probabilistic.[74)]

2.3.1.4 Combinatorics

Combinatorics is also a subfield of stochastics in which we examine, among other things, the number of different possible arrangements of the elements of a set or the number of possible new sets that can be formed using the elements of an initial set.[75)]

70) Cf. Brockhaus GmbH (1999c), p. 196 (translation by Steffen Walter)

71) Cf. Historisches Wörterbuch der Philosophie – Band 7 (1989), p. 1389 (translation by Steffen Walter)

72) Cf. Voigt (2008), Slide 3

73) Cf. ibid., Slide 3

74) Cf. http://www.businessdictionary.com/definition/probabilistic.html. Accessed November 12, 2016

75) Cf. Brockhaus GmbH (1998), p. 423

Section 6.2.5.1 includes a simple example of combinatorics. In this example, the Monte Carlo simulation method is applied for determining the range of possible dice scores.

In construction practice, combinatorial methods are applied, for instance, for determining the quantity proportions of different soil classes within an excavation, where the total amount of excavated material is constant but the composition of the proportions of individual soil classes can change (see also Section 6.2.5.2).

Similarly, the composition of the total amount of reinforcement for rebar in different diameters is to be understood as a combinatorial problem.

2.3.2 Random Variable

A random variable is a variable that takes a value depending on the occurrence of a random event. If a variable is functionally dependent on another random variable, it is also a random variable.[76)]

In a probabilistic calculation, random variables are all those parameters that are fraught with uncertainties and are included in the calculations or simulations as distributions rather than individual numerical (deterministic) values.

2.3.3 Distribution

A (probability) distribution indicates the probability with which a random variable such as the labor consumption rate of formwork operations can take certain characteristic values. Any distribution will be determined by its density[77)].[78)]

The density of a discrete random variable assigns to each value the corresponding probability with which it occurs. A discrete density can be represented in tabular form as long as the number of underlying values[79)] is not infinite. The sum of the probabilities must equal 1.

The density of a continuous random variable is a continuous function in each of its intervals, i.e. the graph of this function is "gapless" within each interval. Continuous densities cannot be captured in tables because the underlying set has continuum-many elements. The integral over the entire distribution (from $-\infty$ to $+\infty$) must result in 1 (see the third axiom, probability according to *Kolmogorov*, Section 2.3.1.1).[80)]

Fig. 2.14 illustrates the density functions of a discrete and a continuous function in qualitative terms.

76) Cf. Biermann/Grosser (1999), p. 210

77) Also referred to as density function or probability function.

78) Cf. Biermann/Grosser (1999), p. 190

79) Combining all elements of a set associated with a positive probability relative to a specific distribution results in the underlying set of said distribution.

80) Cf. Biermann/Grosser (1999), p. 40f.

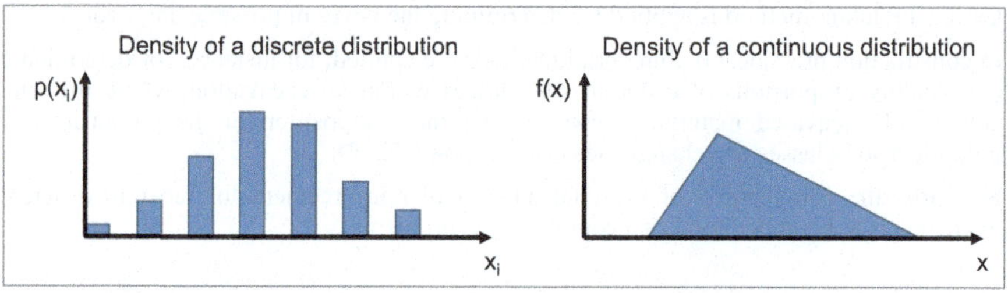

Fig. 2.14 Density function of a discrete (left) and continuous distribution (right) – Qualitative representation

Distributions can also be represented by sum curves by adding up (cumulating) the frequencies of the individual values (see Fig. 2.15).

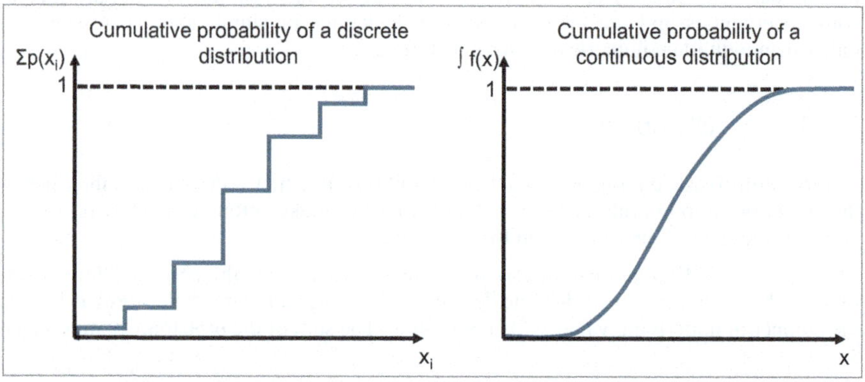

Fig. 2.15 Cumulative probability function of a discrete (left) and a continuous distribution (right) – Qualitative representation

Section 7.1 includes a more detailed discussion on selecting distribution functions for answering questions related to construction management and economics with the aid of probabilistic calculation methods.

2.3.3.1 Discrete Distributions

Discrete distributions can be usefully applied if the parameters can only take results within a subset of natural numbers when appropriately coded (for an example, see Fig. 2.16, right)[81)]. More specifically, the daily working time could range from 8 to 12 hours, and it is conceivable to subdivide this period into half-hour increments.

Likewise, the number of available cranes could also be modeled by a discrete distribution. For example, it would be plausible to assume the use of two, three, or four cranes on a single construction site.

81) Cf. Duller (2019), p. 211

In the @Risk software, a continuous distribution can easily be converted into an integer discrete distribution by entering the (ROUND) function provided in Excel. This function triggers rounding up or down of the intermediate values, and subsequent calculations are only performed using the rounded values.

The Bernoulli distribution is a special type of discrete distribution (see Fig. 2.16, left). This distribution takes only two values (either 0 or 1) and is therefore best suited for simulating the occurrence of certain events. A risk or a loss of productivity can either materialize (= 1) or not (= 0).[82)]

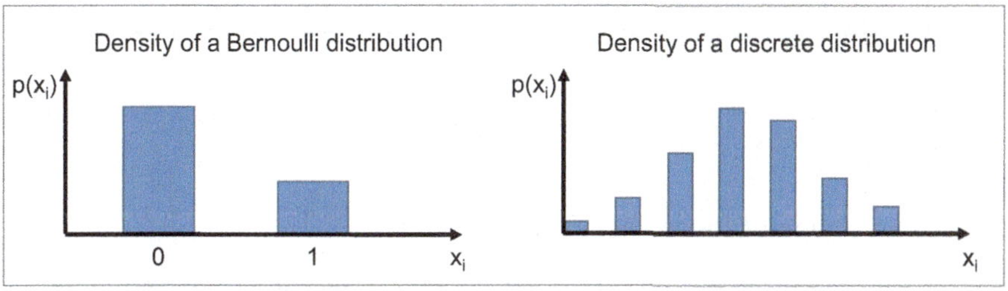

Fig. 2.16 Density function of a Bernoulli distribution (left) and a general discrete distribution (right) – Qualitative representation

2.3.3.2 Continuous Distributions

Continuous distributions are used if all possible numerical values can occur within a certain range (spread). "There are thus continuum-many infinite possible realizations."[83)] This assumption can reflect the real situation fairly accurately for cost values or consumption and performance rates, for example. Quite obviously, euro amounts with more than two decimal places cannot be represented. However, representing all possible euro amounts within a certain range by means of a discrete distribution would translate into an increased (and unnecessary) modeling effort, which is why we apply continuous distributions.

If the input parameter distribution curve is unknown, triangular distributions can be used as a theoretical approach (as shown, for example, in Fig. 2.14, right). At first glance, these distributions appear to be relatively artificial, and will not be encountered for processes or events in reality. One argument in favor of choosing triangular distributions, however, is their definition by only three values (minimum, most likely, and maximum value), which can be easily estimated by experts.

Skewed distributions are often used for labor consumption rates and cost parameters. On the one hand, this is due to the fact that human performance capacity tends to assume asymmetrical distributions. On the other hand, we can attribute this to the economic minimum principle, which is characterized by utilizing the smallest possible means or resources to achieve the defined target. In this context, we recognize a tendency towards applying right-skewed (positively skewed) distributions.[84)]

---

82) Possible practical example: In the management of chances and risks, it is possible to create a probabilistic extension to the "practitioner method" in which the probability of occurrence is multiplied by the consequence.

83) Cf. Duller (2019), p. 233 (translation by Steffen Walter)

84) Rohr (2005), p. 44, supports a similar view with respect to process durations.
Chau (1995a), p. 17, includes a similar view with regard to costs.

Other examples of (closed) continuous distributions include rectangular and PERT distributions (see Fig. 2.17).

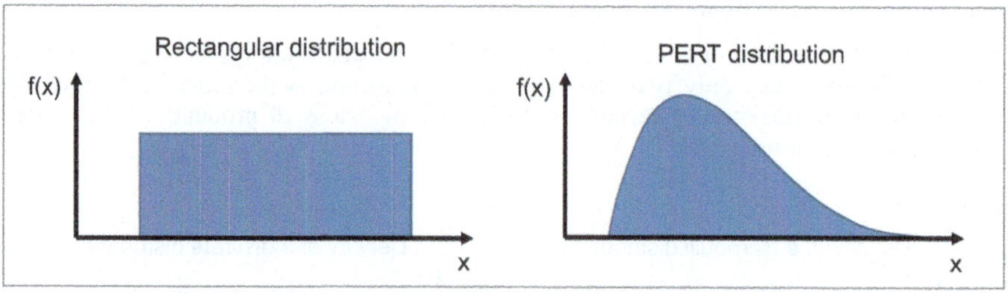

Fig. 2.17 Density function of a rectangular (left) and a PERT distribution (right) – Qualitative representation

2.3.3.3 Open/Closed Distributions

Closed distributions are those that have a defined, finite minimum and a defined, finite maximum. These include, for example, rectangular, triangular, and PERT distributions.

In the case of open distributions, we need to make a distinction between one-sided and two-sided open distributions. In two-sided open distributions, there is neither a finite minimum nor a finite maximum. In this case, both ends extend to $-\infty$ or $+\infty$.

One-sided open distributions include gamma, log-logistic, Weibull, and exponential distributions.

Examples of two-sided open distributions are normal, Laplace, and Student distributions.

Fig. 2.18 includes a qualitative representation of a one-sided open (left) and a two-sided open distribution (right).

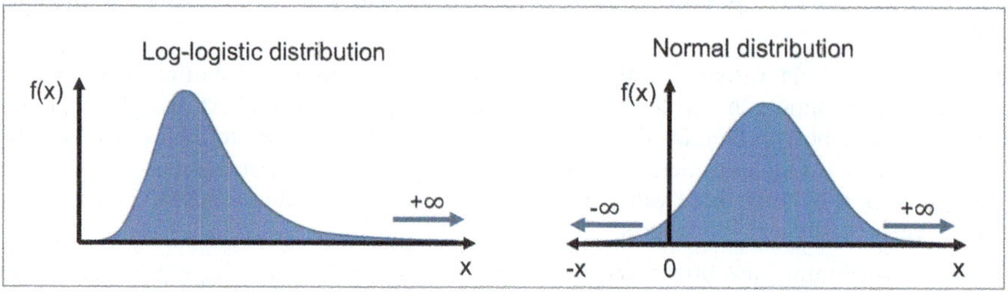

Fig. 2.18 Density function of a log-logistic distribution (left) and a normal distribution (right) – Qualitative representation

In two-sided open distributions, very high or very low values have a low probability of occurrence. For calculations in construction management and economics, using such distributions as input parameters is often inappropriate because we can usually specify a closed range for the input parameters. Likewise, entering negative values will not be permissible in most cases.

2.3.3.4 Unimodal/Multimodal Distributions

Non-constant distributions can be grouped into uni- and multimodal distributions. We refer to such distributions as unimodal if the density function has only a single maximum, such as in normal, log-logistic, PERT, and triangular distributions. If several values are associated with a higher probability (i.e. if there is more than one maximum), this type of distribution is called multimodal. Fig. 2.19 shows a qualitative representation of uni- and multimodal distributions.

Bimodal (two-peak) distributions are a specific subset of multimodal (multi-peak) distributions.

Multimodality is indicative of a stratified sample. We can obtain multimodal distributions, for example, in aggregating losses of productivity if, in the course of a Monte Carlo simulation, individual productivity losses either occur or do not occur in different iterations. This pattern will ultimately reach a limit state in which the number of occurring losses of productivity is not unambiguous.[85)]

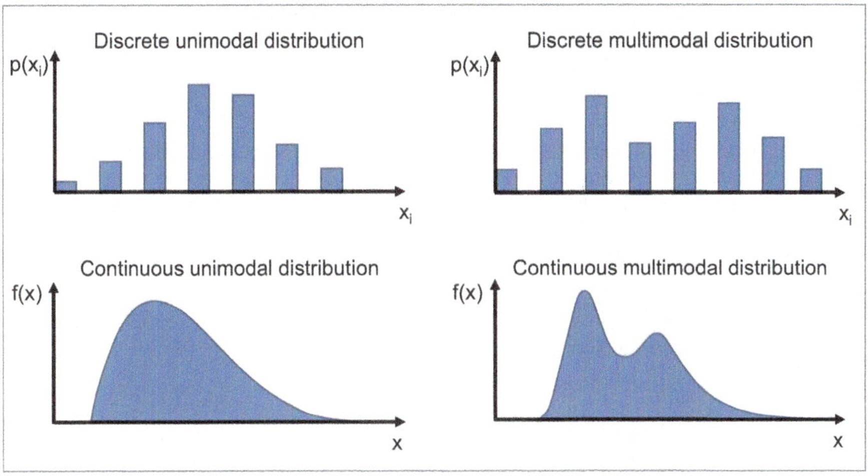

Fig. 2.19 Density function of discrete (top) and continuous (bottom) unimodal (left) and multimodal (right) distributions – Qualitative representation

2.3.3.5 Skewed/Symmetrical Distributions

Distribution curves can be further divided into symmetrical and skewed distributions. Symmetrical distributions have a vertical mirror axis (see, for instance, the normal distribution in Fig. 2.18). Skewed distributions exhibit an accumulation either near the minimum (right-skewed or positively skewed) or near the maximum (left-skewed or negatively skewed) (see Fig. 2.20).

---

85) Cf. Kummer (2015b), p. 167ff.

- For right-skewed (positively skewed) distributions, we can assume that modal value < median < mean
- For left-skewed (negatively skewed) distributions, we can assume that mean < median < modal value (see Fig. 2.20)

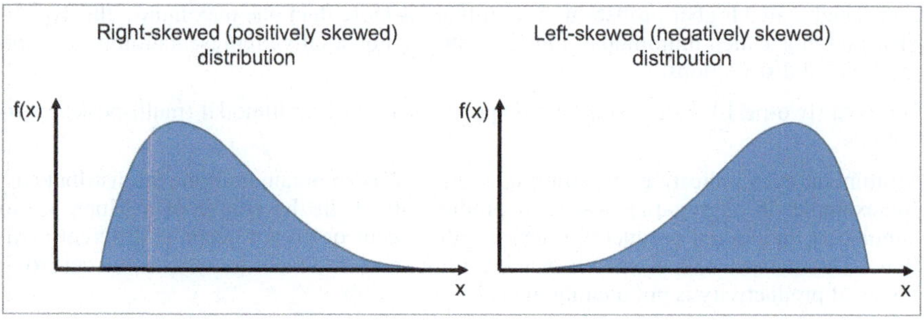

Fig. 2.20 Density function of a right-skewed (positively skewed) distribution (left) and a left-skewed (negatively skewed) distribution (right) – Qualitative representation

2.3.4 Histogram

Gathering a continuous characteristic usually requires discretization of this characteristic, which is achieved by forming classes. Representing the frequencies of classes (on the ordinate) by areas over the classes (abscissa) is referred to as a histogram (see Fig. 2.21). If the class width is non-equidistant, the height of the rectangles will not permit conclusions regarding the frequency of the individual values. In this case, frequency is expressed only by the area (class width times bar height).

The sum of the bar areas is normalized to 1 (see also the third axiom, probability according to *Kolmogorov*, Section 2.3.1.1).

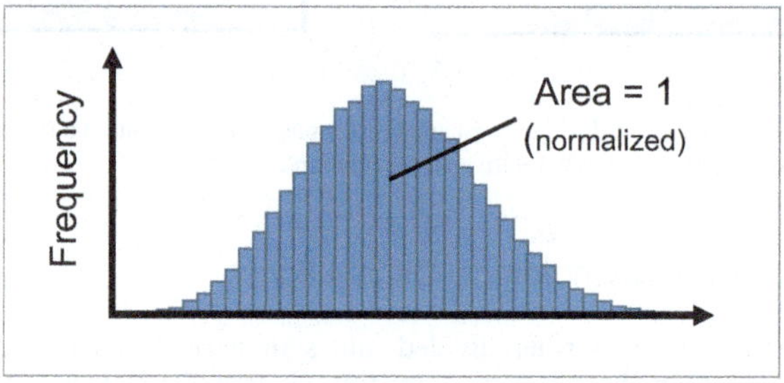

Fig. 2.21 Histogram – Qualitative representation

2.3.5 Location Parameters

In descriptive statistics, location parameters are used for describing the central position or the center of a distribution. These parameters condense the elements of the sample or population to a single number. The most significant of these values are the mean, mode, and median.[86)]

2.3.5.1 Mean Value

The arithmetic mean value is the most commonly applied value. It is repeatedly used in this book. This value is also referred to as average or mean and is only suitable for metric characteristics. It is not useful to calculate the arithmetic mean for ordinal and nominal characteristics.

Particularly in small samples, it is highly susceptible to possible outliers, which is why outliers may have to be eliminated depending on the use case and intended target. One way to take extremes (such as those stated by experts) into account without distorting the results is to apply the M-estimator method according to *Huber*. This approach assigns less weight to outliers and extremes included in the mean (see Section 2.3.5.5), which increases the robustness of the location parameter.

2.3.5.2 Modal Value

The modal value is also known as:
- Mode
- Most frequent value
- Anticipated value
- Expert value
- Most likely value
- Most probable value
- Planned value[87)]

"The [...] modal value [...] is the characteristic value with the highest (absolute or relative) frequency. For data grouped into classes, the modal class is the class with the highest population density."[88)]

The mode thus represents the maximum of the density function.

2.3.5.3 Median

The median, or central value, describes the value of a distribution for which 50% of the values are smaller and 50% of the values are greater than it. The median thus divides the distribution into two equally large areas. In symmetrical distributions, the median is identical to the arithmetic mean. Compared to the arithmetic mean, the median is more robust against outliers since they do not influence the height of the median as much as the arithmetic mean does.[89)]

---

86) Cf. http://de.statista.com/statistik/lexikon/definition/80/lageparameter/. Accessed November 14, 2016

87) Cf. Oepen (2012), p. 49

88) Cf. Sachs (2013), p. 35 (translation by Steffen Walter)

2.3.5.4 Expected Value

The expected value E(X) (or μ) can be interpreted as the number that the random variable takes on average. If the mean value is derived from the results of a test, it will converge against the expected value. The expected value represents a variable that can be expected if a large number of tests or experiments is conducted.

For an infinite number of repetitions of a random experiment, the arithmetic mean of the results is equivalent to the expected value of the distribution. In other words, the sample mean of a random variable approaches the expected value with increasing sample size (in accordance with the law of large numbers).

"The estimator function for the expected value is the arithmetic mean."[90)]

Summary of the most significant location parameters

The figure shown below includes a qualitative representation of the location parameters (i.e. mean, mode, and median) for skewed distributions. For right-skewed unimodal distributions (see Fig. 2.22, left), the mode assumes a smaller numerical value whereas the mean assumes a larger numerical value compared to the median on the abscissa. This relationship is inverse for left-skewed distributions (see Fig. 2.22, right).

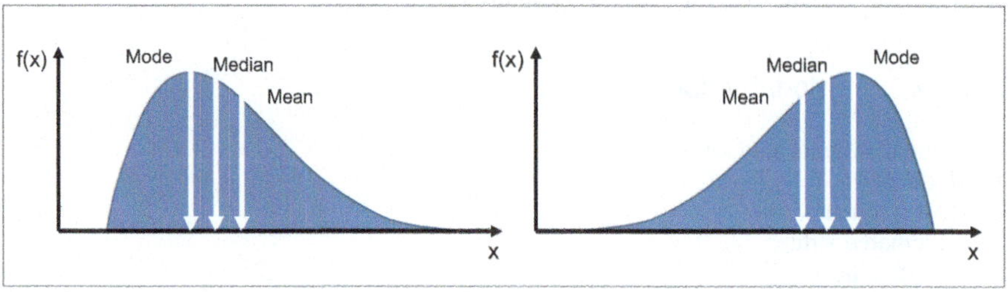

Fig. 2.22 Location parameters for skewed distributions – Qualitative representation

2.3.5.5 *Huber's* M-estimator

Huber's M-estimator is a robust location parameter less susceptible to distortion by outliers or extremes than, for instance, the arithmetic mean. When forming the M-estimator applying *Huber's* method, the collected values are weighted differently according to a function (see Fig. 2.23).[91)]

89) Cf. Bartsch (2004), p. 667

90) Cf. Suhl/Mellouli (2013), p. 283 (translation by Steffen Walter)

91) Cf. Hofstadler (2014), p. 105ff.

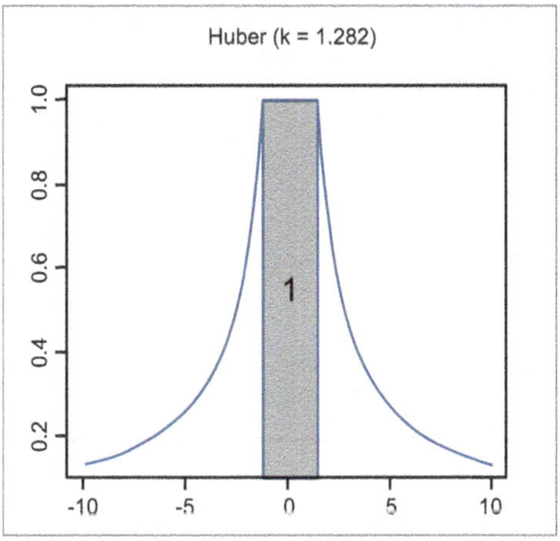

Fig. 2.23 *Huber's* M-estimator – Qualitative representation[92)]

Thus, outliers and extremes with low weight are included in the averaging step and do not need to be eliminated. A tuning constant k is applied to determine the range in which values are fully included in the calculation (i.e. with a weighting of 1). Table 2.1 provides an overview of commonly applied k values for *Huber's* M-estimator method. The specific k value chosen for the data analysis depends on the number, distance, and type of distribution of the outliers and extremes as well as on technical considerations.[93)]

Huber's M-estimators	**H19**	**H16**	**H12**	**H8**
Tuning constant k	1.960	1.645	1.282	0.842
Fully weighted range for averaging	~95%	~90%	~80%	~60%

Table 2.1 Tuning constant k for *Huber's* M-estimators ("*Huber* Proposal 2")[94)]

2.3.6 Variance

Variance is a measure of scattering or spread, describing the deviation of a random variable from its expected value. For discrete random variables, it is described as follows:

$$\mathrm{Var}(X) = \sum_{i=1}^{n} (x_i - \mu)^2 \cdot p_i \tag{2-2}$$

For continuous random variables, the variance is calculated according to Eq. (2-3).

$$\mathrm{Var}(X) = \int_{-\infty}^{\infty} (x - \mu)^2 \cdot f(x) \cdot dx \tag{2-3}$$

92) Cf. Hofstadler (2014), p. 106

93) Cf. ibid., p. 105ff.

94) Cf. ibid., p. 107

Fig. 2.24 provides a qualitative representation of the difference of the spread depending on a larger or smaller variance for two continuous symmetrical distributions with the same expected value. The flatter distribution exhibits a greater spread of possible characteristic values, and thus also a higher variance.

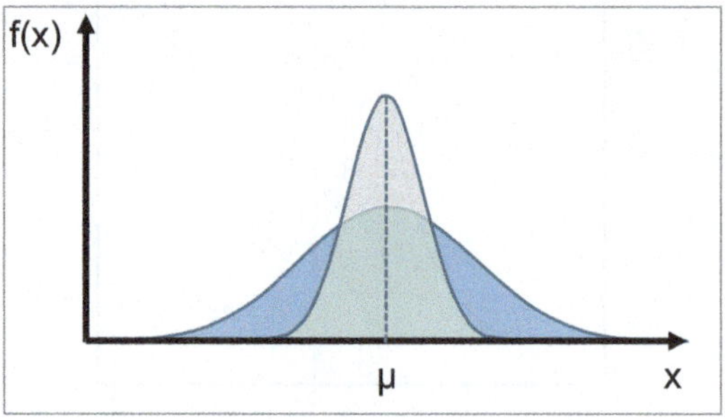

Fig. 2.24 Comparison of different spreads for an identical mean – Qualitative representation

2.3.7 Standard Deviation

If we calculate the square root of the variance, we obtain the standard deviation as an additional measure of scattering or spread.

$$\sigma = \sqrt{Var(X)} \tag{2-4}$$

Regardless of the numerical value of the mean and the standard deviation, a certain portion of the total area will always lie in symmetrical intervals ± a multiple of the standard deviation around the mean in the normal distribution (see Fig. 2.25):

- 68.27% of the values lie in the interval $\pm 1 * \sigma$ around the mean
- 95.45% of the values lie in the interval $\pm 2 * \sigma$ around the mean
- 99.73% of the values lie in the interval $\pm 3 * \sigma$ around the mean

For the normal distribution, the central 95% of the values are limited by the 1.96-fold standard deviation in the positive and negative direction from the mean. Accordingly, for the normal distribution, the central 99% of the values are limited by the 2.58-fold standard deviation in the positive and negative direction from the mean. It should be noted, however, that these assumptions only apply to normal distributions, but not to other distribution categories.

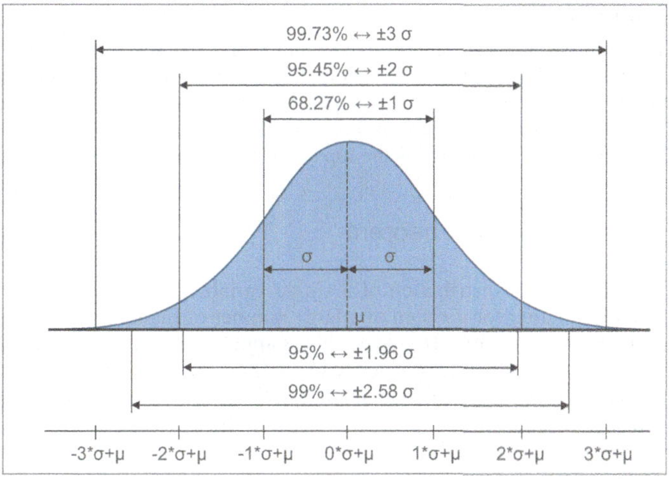

Fig. 2.25 Quantile values depending on the standard deviation – Normal distribution

2.3.8 Quantile Values and Ranges

A p quantile, also referred to as a fractile, represents that x value of a distribution below which the corresponding percentage p of values is located (see Fig. 2.26).[95)]

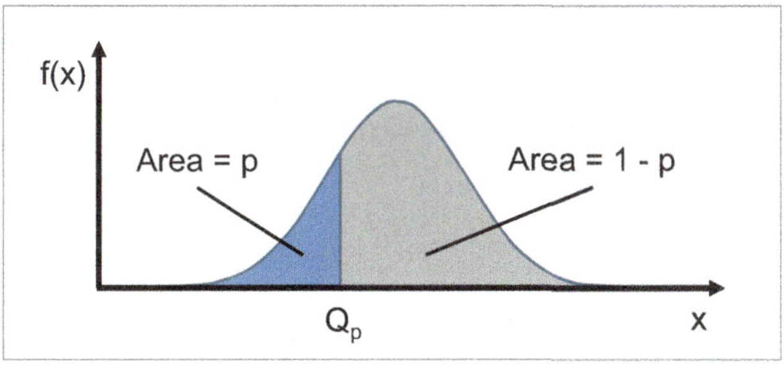

Fig. 2.26 Quantile value Q_p of a continuous distribution function

Quartiles (quarter values) are special quantile values in which the distribution function is divided into four equal ranges (i.e. 25% each). Thus, 25% of the values are located under the first quartile, 50% of the values are located under the second (also referred to as median), and so on.

Other special quantile values include quintiles (fifth values), deciles (tenths) and percentiles (hundredths).

The spread, or range, of a distribution results from the difference between the highest and the lowest value. We can thus specify a span only for two-sided closed distributions.

[95)] Besides Qp, the Xp symbol is also used.

A quantile range is defined individually and represents the difference between the quantiles Q_{1-p} and Q_p.

The quartile range is a special quantile range and represents the difference between the 75% quartile and the 25% quartile. It is a more stable indicator of the dispersion or spread of a distribution function than its span. The quartile range is also called interquartile range (IQR) and defines the area of the box in box plot diagrams. It separates a quarter of the values of the distribution both towards the top and towards the bottom.[96)]

2.3.9 Central Limit Theorem

Irrespective of the type of distribution of a single variable, we always obtain an approximate normal distribution for the sum of a large number of random variables in the limit case (i.e. n converges to infinity). However, this assumption is valid only if the individuall random variables are stochastically independent. For the product of several random variables, there is always an approximate logarithmic normal distribution for the limit case (where n converges to infinity). Again, this assumption is valid only if the individual random variables are stochastically independent.[97)]

No "genuinely" normal distribution (in the sense of a two-sided open distribution) can result from the addition of several random variables if the addends consist only of limited (closed) distributions. The same applies to the log-normal distribution[98)] as the product of several random variables. If the input parameters are defined by closed distributions, the sum and product of these will also have finite values, regardless of the number of variables.

2.3.10 Correlation and Correlation Coefficients

A correlation (interrelatedness) is a one-dimensional (standardized) measure of the linear relationship between two characteristics. Various types of bivariate[99)] correlations are used depending on the scaling of such characteristics[100)] (see Table 2.2).

---

96) Cf. Sachs (2013), p. 40

97) Cf. Lederer (1982), p. 58f.

98) Log-normal distribution = logarithmic normal distribution

99) Bivariate distributions are two-dimensional random variables for which both the characteristic x and the characteristic y can be represented by a distribution.

100) Types of scale include:
- nominal (variables without a natural order, such as places of birth, for example "Oxford/Birmingham/London")
- ordinal (variables for which a natural order exists where the difference between the individual variables cannot be quantified, such as high, medium, and low)
- interval (variables are represented by numerical values where ranking differences and intervals between values are calculable or measurable; there is no "true zero" value, such as in temperature scales)
- ratio (this type of scale can accommodate an absolute zero value while allowing for conclusions regarding proportions, such as in scales of physical parameters)
- dichotomous (variables can take only two different states, such as "true" or "false")

		Characteristic x	
		ordinal	metric
Characteristic y	ordinal	Rank correlation	Rank correlation
	metric	Rank correlation	Product-moment correlation

Table 2.2 Bivariate types of correlation[101)]

Other types of correlation include:
- Point-biserial correlation (dichotomous and interval-scaled characteristic)
- Φ coefficient (two dichotomous characteristics)
- Rank-biserial correlation (dichotomous and rank-scaled characteristic)

We can also make a distinction between inter-class and intra-class correlations:
- Inter-class correlations
  These describe relationships between variables of a different nature, such as:
  - Number of storks and number of births[102)]
  - Price and quality
  - Work expenditure and result
  - Revenue and economic cycle

- Intra-class correlations
  These describe relationships between variables of the same nature, such as:
  - From one work result to the next
  - From one daily turnover to the next
  - The development of share prices[103)]

Intra-class correlations are also referred to as inter-, intra-, or autocorrelations.[104)]

Costing and pricing is primarily concerned with inter-class correlations, such as:
- The relationship between formwork area and concrete volume
- The relationship between the formwork ratio and the labor consumption rate of formwork operations
- The relationship between the reinforcement ratio and the labor consumption rate of reinforcing operations

An existing correlation does not permit conclusions regarding the causality between two characteristic values. Rather, this assumption must be theoretically founded or derived.

---

101) Cf. Universität Siegen, Marktforschung Sommersemester 2011.
Available online at http://slideplayer.de/slide/644669/#. Accessed October 22, 2016

102) In this case, a collected database could give rise to a correlation, but this would not provide any evidence of a causal relationship.

103) Cf. Schulz (1980), p. 48

104) Cf. ibid.

The following causal relationships are possible:[105)]

- x causes y
- y causes x
- x and y are influenced by one or more other variables (false correlation or partial correlation)

2.3.10.1 Product-Moment Correlation

The product-moment correlation (also known as *Pearson* or *Bravais-Pearson* correlation) indicates the standardized relationship between two interval-scaled characteristics. This is the covariance of two characteristics s_{xy} divided by the product of the standard deviations (see Eq. (2-5)). This step standardizes the covariance and makes the correlation coefficients comparable.

$$r_{xy} = \frac{s_{xy}}{s_x \cdot s_y} = \frac{\sum_{i=1}^{n}(x_i - \bar{x}) \cdot (y_i - \bar{y})}{\sqrt{\sum_{i=1}^{n}(x_i - \bar{x})^2 \cdot \sum_{i=1}^{n}(y_i - \bar{y})^2}} \tag{2-5}$$

The correlation coefficient can only take values between -1 and +1 and reflects the power of the linear relationship between two variables.

A perfect negative correlation exists if the correlation coefficient equals -1. All measuring points lie on a straight line with a negative gradient.

A perfect positive correlation exists if the correlation coefficient equals +1. In this case, all measuring points lie on a straight line with a positive gradient.

No linear relationship exists if the correlation coefficient is zero. In other words, the variables are uncorrelated. However, this does not exclude potentially existing nonlinear relationships, such as a U-shaped correlation.

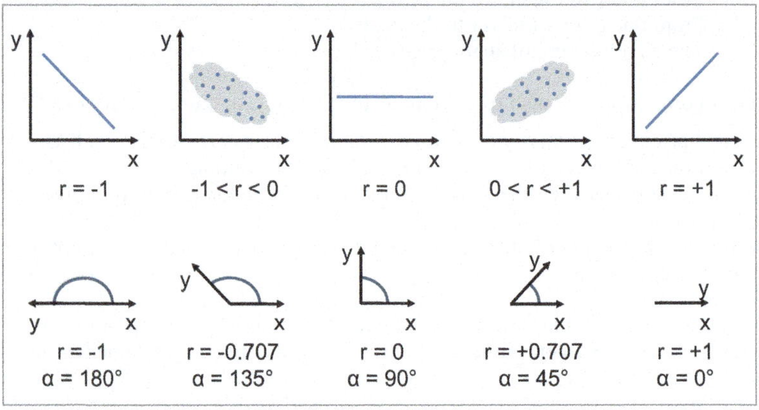

Fig. 2.27 Relationship between two variables x and y with different correlation coefficients – Top: Point clouds – Bottom: Vector diagrams[106)]

[105)] Cf. Universität Siegen, Marktforschung Sommersemester 2011.
Available online at http://slideplayer.de/slide/644669/#. Accessed October 20, 2016

Fig. 2.27 provides a graphical overview of the interpretation of correlation coefficients. The upper series of this diagram visualizes the relationships as point clouds while its lower counterpart shows vector pairs representing two variables (x and y). For the representation as a vector pair, the correlation coefficient r can be transformed into the cosine of α (see Eq. (2-6)).

$$r = \cos\alpha \tag{2-6}$$

The product-moment correlation is identical to the rank correlation, provided that interval-scaled characteristics are considered.[107)]

The @Risk software provides options for entering product-moment correlations as correlation matrices and thus considering them in the simulations.

2.3.10.2 Rank Correlation

The rank correlation is also referred to as the *Spearman* correlation coefficient and is applied in the presence of either two characteristics of an ordinal scale or one characteristic of an interval scale and one characteristic of an ordinal scale (see Table 2.2). Ranks are assigned to the characteristic values, and their difference d_i is calculated. The number of possible rank positions n corresponds to the number of existing pairs of measurement data. We calculate the rank correlation using Eq. (2-7).

$$r_s = 1 - \frac{6 \cdot \sum_{i=1}^{n} d_i^2}{n \cdot (n^2 - 1)} \tag{2-7}$$

2.3.11 Historical Data

The concept of historical data refers to a database that exists at a certain point in time, such as at the date of access or at the beginning of the year or current month. This database can be stored separately and used for defining uncertain parameters at a later point in time. Depending on the specific points of delimitation, historical data can be taken either from a current or from several completed projects. Among other things, such data is used for investigating the behavior or the expression of certain parameters or key indicators and for generating distribution functions for certain parameters (see also Section 2.3.12). In addition, the developments of key indicators or progress values (such as cost, deadline, content, or scope parameters) can be used for trend analyses. Such analyses should identify deviations from a planned target process, but they can also be used for making forecasts for a certain timeline or period.

The use of historical data essentially depends on an appropriate method of delimitation. Care should be taken to ensure that the database does not include any additional effects that may distort the values or assumptions made.

[106)] Cf. http://www.faes.de/Basis/Basis-Statistik/Basis-Statistik-Korrelation-Re/basis-statistik-korrelation-re.html. Accessed December 19, 2020
Cf. http://www.faes.de/Basis/Basis-Lexikon/Basis-Lexikon-Multivariate/Basis-Lexikon-Faktorenanalyse/basis-lexikon-faktorenanalyse.html. Accessed December 19, 2020
Cf. Schulz (1980), p. 78ff.

[107)] Cf. e.g. Krempl (2012), p. 56

2.3.12 Fitting

The transfer of such data records (e.g. from historical data) into distribution functions that can be described on a theoretical level is referred to as data fitting. The existing data is analyzed, and mathematical/statistical methods are applied for determining the theoretical distribution function that best reflects the existing data (i.e. with the smallest error).

A fit involves adjusting a function to any measurement errors or uncertainties related to the measuring points. The least-squares method is most commonly applied for data fitting purposes. A fitting exercise will always result in a family of curves within which the "true" functional correlation will lie with a certain probability.

Fitting aims to determine the parameters of a distribution function (displacement and curve parameters) in a numerical process such that the distribution reflects gathered raw data as accurately as possible.

The accuracy and applicability of the fitting results always depend on the quality of the basic data and the number of data points. The greater the number of data points, the greater the likelihood will be for the data fitting results to cover the population of values of a parameter.

The results of a data fitting exercise must always be checked for their plausibility. In most cases, this check will raise questions regarding the permissibility of negative value ranges and the formation of limits (i.e. open or closed).

2.4 Summary

Clear definitions of terms with a criteria-driven delimitation form the basis for a functioning (and thus functional) mode of understanding for a book and its readership, for the practical implementation of knowledge, and for the consistent communication within a discipline.

The concepts of chance and risk are central to this book, but there are two distinct, frequently applied definitions in the literature. For the purpose of this book, we define chances and risks as expectations in the presence of uncertainty. Besides uncertainties, any assumption regarding the future will be associated with partial insecurity and ignorance, whose characteristics differ from those of uncertainty.

Risk refers to an expectation in the presence of an uncertainty that assumes a negative variance from the target. Its impact and probability of occurrence are known either objectively or subjectively. In contrast, chance refers to an expectation in the presence of an uncertainty that assumes a positive variance from the target. Again, its impact and probability of occurrence are known either objectively or subjectively.

For all types of estimates and calculations of numerical values, the selected reference base is crucial for determining the chance/risk ratio taken. The considerations regarding the selection of the reference base are genuinely content-driven and thus of a strategic nature in construction management and economics. They will also strongly depend on the readiness of a business to take risks.

The characteristics of probabilities can be influenced by selecting specific distribution functions. Both discrete and continuous distributions are available for this purpose. Closed distributions should be used if the ranges include clear limits, whereas open distributions are recommended in all other cases. If there is a non-constant distribution function with only a single maximum, we refer to this distribution as unimodal while it is called multimodal in the absence of this condition. Distributions can take different shapes, namely symmetrical or skewed curves. Skewed distributions can be grouped into right- and left-skewed distributions. Besides other methods, historical data can be subjected to fitting and used for determining the shapes or curves of distribution functions.

Statistical parameters are available for further describing and interpreting distributions. These parameters were described in detail in this chapter. Dependencies between calculation parameters can be expressed by correlations. Integrating these relationships into the calculations is useful only if the causal nature of the interrelationships has been derived from reality in a theoretically sound manner.

2.5 References

Bartsch, Hans-Jochen (2004). Taschenbuch Mathematischer Formeln – 20th ed. Munich, Vienna. Fachbuchverlag Leipzig – Carl Hanser Verlag. (ISBN 3-446-22891-8)

Biermann, Bernd; Grosser, Karin (1999). Taschenlexikon Finanzmathematik/Statistik. Ed.: Eller, Roland. Stuttgart. Schäffer-Poeschel Verlag. (ISBN 3-7910-1462-5)

Blecken, Udo (1976). Die Produktions- und Kostentheorie im instationären Baubetrieb. In: Der Baubetriebsberater, Heft 4, 1/1976. 199-210. Wiesbaden, Berlin. Bau-Verlag.

Boeckelmann, Lukas; Mildner, Stormy-Annika (2011). Unsicherheit, Ungewissheit, Risiko – Die aktuelle wissenschaftliche Diskussion über die Bestimmung von Risiken. In: SWP-Zeitschriftenschau 2. Stiftung Wissenschaft und Politik – Deutsches Institut für Internationale Politik und Sicherheit. Berlin. Stiftung Wissenschaft und Politik, September 2011. (ISSN 1611-6380) Available online at http://www.swp-berlin.org/fileadmin/contents/products/zeitschriften-schau/2011zs02_bkm_mdn_ks.pdf. Accessed December 19, 2020.

BROCKHAUS GmbH (1998). Brockhaus in fünfzehn Bänden – Band 7 – Is-Kon. Leipzig, Mannheim. F.A. Brockhaus GmbH.

BROCKHAUS GmbH (1999b). Brockhaus in fünfzehn Bänden – Band 3 – Chl-Eir. Leipzig, Mannheim. F.A. Brockhaus GmbH.

BROCKHAUS GmbH (1999c). Brockhaus in fünfzehn Bänden – Band 11 – Pfe-Rog. Leipzig, Mannheim. F.A. Brockhaus GmbH.

BusinessDictionary.com. WebFinance, Inc. http://www.businessdictionary.com/definition/probabilistic.html. Accessed November 12, 2016.

Čadež, Ivan (1998). Risikowertanalyse als Entscheidungshilfe zur Wahl des optimalen Bauvertrags. Fortschritt-Berichte VDI, Reihe 4 – Bauingenieurwesen, Nr. 149. Düsseldorf. VDI Verlag. (ISBN 3-18-314904-4)

Chau, Kwong Wing (1995a): The validity of the triangular distribution assumption in Monte Carlo simulation of construction costs: empirical evidence from Hong Kong. In: Construction Management and Economics, Volume 13, Issue 1, 1995. 15-21. London. Taylor & Francis Group. (ISSN 0144-6193)

Czado, Claudia; Schmidt, Thorsten (2011). Mathematische Statistik – Statistik und ihre Anwendungen. Eds.: Dette, Holger; Härdle, Wolfgang. Berlin, Heidelberg. Springer Verlag (ISBN 978-3-642-17261-8)

Duller, Christine (2019). Einführung in die Statistik mit EXCEL und SPSS, 4th ed. Berlin. Springer-Gabler. (ISBN 978-3-642-37859-1)

Faes.de http://www.faes.de/Basis/Basis-Statistik/Basis-Statistik-Korrelation-Re/basis-statistik-korrelation-re.html. Accessed December 19, 2020

Faes.de http://www.faes.de/Basis/Basis-Lexikon/Basis-Lexikon-Multivariate/Basis-Lexikon-Faktorenanalyse/basis-lexikon-faktorenanalyse.html. Accessed December 19, 2020

Gigerenzer, Gerd (2013). Risiko – Wie man die richtigen Entscheidungen trifft – 3rd ed. Munich. C. Bertelsmann Verlag. (ISBN 978-3-570-10103-2)

Girmscheid, Gerhard; Motzko, Christoph (2013). Kalkulation, Preisbildung und Controlling in der Bauwirtschaft – Produktionsprozessorientierte Kostenberechnung und Kostensteuerung – 2nd ed. Berlin, Heidelberg. Springer Verlag (ISBN 978-3-642-36637-6)

Henze, Norbert (2012). Stochastik für Einsteiger – Eine Einführung in die faszinierende Welt des Zufalls – 9th ed. Wiesbaden. Vieweg+Teubner Verlag. (ISBN 978-3-8348-1845-4)

Historisches Wörterbuch der Philosophie – Band 2 (1972). D-F. Ed.: Ritter, Joachim. Basel, Stuttgart. Schwabe & Co Verlag. (ISBN 3-7965-0115-X)

Historisches Wörterbuch der Philosophie – Band 7 (1989). P-Q. Eds.: Ritter, Joachim; Gründer, Karlfried. Basel, Stuttgart. Schwabe & Co Verlag. (ISBN 3-7965-0115-X)

Hofstadler, Christian (2014). Produktivität im Baubetrieb – Bauablaufstörungen und Produktivitätsverluste. Berlin, Heidelberg. Springer Verlag (ISBN 978-3-642-41632-3)

Hofstadler, Christian; Kummer, Markus (2017a). Chancen- und Risikomanagement in der Bauwirtschaft – Für Auftraggeber und Auftragnehmer in Projektmanagement, Baubetrieb und Bauwirtschaft. Berlin, Heidelberg. Springer-Verlag. (ISBN 978-3-662-54318-4)

Horx Zukunftsinstitut GmbH (2010). Prognose – Prophezeiung – Vision – Semantische Unterschiede von Zukunfts-Aussagen. 2010. Available online at http://www.horx.com/Zukunftsforschung/Docs/01-G-09-Prognose-Vision-Prophzeiung.pdf. Accessed November 18, 2016.

http://scoop.me/hashtags/orf. Accessed August 31, 2016.

Jonen, Andreas (2007). Semantische Analyse des Risikobegriffs – Strukturierung der betriebswirtschaftlichen Risikodefinitionen und literaturempirische Auswertung – Beiträge zur Controlling-Forschung – No. 11 – 2nd ed. Ed.: Lingnau, Volker. Lehrstuhl für Unternehmensrechnung und Controlling der Technischen Universität Kaiserslautern, 08/2007. (ISSN 1612-3875)

Kaplan, Stanley; Garrick, John (1981). On The Quantitative Definition of Risk. In: Risk Analysis Vol. 1, No. 1. 11-27. Society of Risk Analysis.

Knight, Frank Hyneman (1921). Risk, Uncertainty and Profit. Boston, New York. Houghton Mifflin Company.

Krempl, Simone (2012). Korrelationen bei Aufwandswerten für Schal- und Bewehrungsarbeiten. Master's thesis. Graz. Institute of Construction Management and Economics, Graz University of Technology.

Kummer, Markus (2015a). Aggregierte Berücksichtigung von Produktivitätsverlusten bei der Ermittlung von Baukosten und Bauzeiten – Deterministische und probabilistische Betrachtungen. Doctoral thesis. Graz. Institute of Construction Management and Economics, Graz University of Technology.

Kummer, Markus (2015b). Aggregierte Berücksichtigung von Produktivitätsverlusten. In: Tagungsband – 26. BBB-Assistententreffen – Fachkongress der wissenschaftlichen Mitarbeiter der Bereiche Baubetrieb, Bauwirtschaft und Bauverfahrenstechnik. Ed.: Berner, Fritz. 167-183. Stuttgart. Institut für Baubetriebslehre der Universität Stuttgart. (ISBN 978-3-9814355-7-3)

Kummer, Markus; Hofstadler, Christian (2013). Einsatz der Monte-Carlo-Simulation zur Berechnung von Baukosten. In: bau aktuell, 4. Jahrgang, Heft Nr. 5, September 2013. 178-188. Vienna. Linde-Verlag Ges.m.b.H. (ISSN 2077-4737)

Lechner, Hans (2010). Kostenplanung, Normen, Regelwerke – Projektmanagement in der Bau- und Immobilienwirtschaft – Bauprojektmanagement LEVEL D – 1st ed. Eds.: Lechner, Hans; Heck, Detlef. Graz. Verlag der Technischen Universität Graz. (ISBN 978-3-85125-099-2)

Lederer, Siegfried (1982). Die Mittelbereitstellung für Bauleistungen unter Berücksichtigung der Unsicherheiten in der Planung. Doctoral thesis. Institut für Bauingenieurwesen IV Tunnelbau und Baubetriebslehre der Technischen Universität München.

Link, Doris (1999). Risikobewertung von Bauprozessen – Modell ROAD – Risk and Opportunity Analysis Device. Doctoral thesis. Vienna. Institut für Baubetrieb und Bauwirtschaft der Technischen Universität Wien.

Luhmann, Niklas (1991). Soziologie des Risikos. Berlin, New York. Walter de Gruyter Verlag. (ISBN 3-11-012939-6)

Oepen, Ralf-Peter et al. (2012). Risikoorientierte Bauprojekt-Kalkulation – Eine innovative Methode zur Risikobeherrschung und Eindämmung von Ausreißer-Projekten. Ed.: BRZ Deutschland GmbH. Wiesbaden. Vieweg+Teubner Verlag. (ISBN 978-3-8348-1892-8).

OENORM ISO 31000 (2018-09-01 edition) Risikomanagement – Leitlinien [Risk Management – Guidelines]

OENORM D 4900 (2021-01-01 edition) Risikomanagement für Organisationen und Systeme – Begriffe und Grundlagen

ONR 49000 (2014-01-01 edition) Risikomanagement für Organisationen und Systeme – Begriffe und Grundlagen – Umsetzung von ISO 31000 in die Praxis.

Sachs, Michael (2013). Wahrscheinlichkeitsrechnung und Statistik für Ingenieurstudenten an Fachhochschulen – 4th ed. Ed.: Engelmann, Bernd. Munich. Carl Hanser Verlag. (ISBN 978-3-446-43732-6)

Scherer, Raimar (2010). Mefisto – Partnerschaftliche Nutzung von Multi-Modellinformationen zur Steuerung, Simulation und Führung von Bauprojekten. In: MEFISTO: Management – Führung – Information – Simulation im Bauwesen – Tagungsband 1. Mefisto Kongress – October 21, 2010, Dresden – Veranstaltungen des Instituts für Bauinformatik – Heft 2. Eds.: Scherer, Raimar; Schapke, Sven-Eric. 9-36. Dresden. Institut für Bauinformatik der Technischen Universität Dresden. (ISBN 978-3-86780-187-4)

Schubert, Eberhard (1971). Die Erfaßbarkeit des Risikos der Bauunternehmung bei Angebot und Abwicklung einer Baumaßnahme – Heft 7. Ed.: Jebe, Hans. Lehrstuhl für Baubetrieb und Baubetriebswirtschaft der Technischen Universität Hannover. Düsseldorf. Werner Verlag. (ISBN 3-8041-3303-7)

Schulz, Josef (1980). Risikorechnung bei der Preiskalkulation – Ein Weg zur Ermittlung der maximalen Überschreitung von Kostenvoranschlägen. Wiesbaden, Berlin. Bauverlag GmbH. (ISBN 3-7625-1303-1)

Statista – Das Statistik-Portal:
http://de.statista.com/statistik/lexikon/definition/80/lage- parameter/.
Accessed November 14, 2016

Stowasser, Joseph M. et al. (1998). Stowasser – Lateinisch-deutsches Schulwörterbuch. Vienna, Munich. Verlag Hölder-Pichler-Tempsky und R. Oldenbourg Verlag GmbH. (Verlag Hölder-Pichler-Tempsky ISBN 3-209-01495-7; R. Oldenbourg Verlag ISBN 3-486-13405-1)

Springer Gabler Verlag (Ed.), Gabler Wirtschaftslexikon, Keyword: Prognose. Available online at http://wirtschaftslexikon.gabler.de/Archiv/4546/prognose-v11.html. Accessed December 19, 2020

Suhl, Leena; Mellouli, Taïeb (2013). Optimierungssysteme – Modelle, Verfahren, Software, Anwendungen – 3rd ed. Berlin, Heidelberg. Springer Verlag (ISBN 978-3-642-38936-8)

Taleb, Nassim Nicholas (2007). The Black Swan – The Impact of the Highly Improbable. New York. Random House. (ISBN 978-1-4000-6351-2)

Universität Siegen (2011). Lecture slides: Marktforschung – Sommersemester 2011. Available online at http://slideplayer.de/slide/644669/#. Accessed October 20, 2016

Voigt, Matthias (2008). Grundlagen der Probabilistik. Vortrag im Zuge des 1. Dresdner Probabilistik-Workshops der Professur für Turbomaschinen und Strahlantriebe am Institut für Strömungsmechanik. Technische Universität Dresden. October 9-10, 2008. Available online at
http://www.probabilistik.de/vortrag/Vortrag0102_-Voigt_TUD.pdf.
Accessed December 19, 2020.

Wiggert, Marcel (2009). Risikomanagement von Betreiber- und Konzessionsmodellen. Doctoral thesis. Graz. Institute of Construction Management and Economics, Graz University of Technology.

Yoe, Charles (2012). Primer on Risk Analysis – Decision Making under Uncertainty. Boca Raton, London, New York. CRC Press – Taylor & Francis Group. (ISBN 978-1-4398-5763-2)

3 Data, Information, Knowledge, and BIM

BIM[1)] models were developed in order to improve planning, forecasting, execution, and operational processes. In an ideal scenario, the holistic, digital, three-dimensional building model (an object-oriented 3D model with attributes) contains all current data and information relevant to all project stakeholders. Irrespective of the degree of sophistication of a BIM model, however, it will always require high-quality data (with appropriate temporal and operational boundaries) with the associated information, including the type and combination of production factors, boundary conditions, assumptions, and misjudgments, in order to simulate construction times and costs.

Data, information, and knowledge create the basic framework for fully achieving the targets defined by the client and the contractor. In addition, the knowledge generated from data and information serves to subject the individual elements and relationships within the entire production system to a continuous improvement process. Any identified potential for design optimization can thus be successfully integrated into future projects.

When managing chances and risks, it is crucial to know the chance/risk ratio associated with specific decisions made. For instance, a baseline value must always be chosen when calculating construction times and costs. The associated chance/risk ratio will depend on the selected baseline value. For this reason, and in order to make best use of chances while avoiding risks as far as possible, it is so important to acquire knowledge of the relationships in construction management and economics in conjunction with chance/risk considerations for ensuring project success and, ultimately, the success of the organizations involved. However, a practical challenge arises in this context when it comes to acquiring or accessing the corresponding knowledge bases.

This chapter will define the concept of knowledge used in this book as follows, adopting a cross-disciplinary approach: "Knowledge signifies the total body of knowledge and skills that individuals use for problem-solving. This definition comprises both theoretical knowledge and practical rules and instructions for everyday use. Knowledge relies on data and information, but unlike these, it is always linked to people. It is constructed by individuals and represents their expectations about cause-effect relationships in a specific context."[2)]

1) Building Information Modeling

C. Hofstadler and M. Kummer, *Chances and Risks in Construction Management and Economics*, Springer Tracts in Civil Engineering,
https://doi.org/10.1007/978-3-030-64014-9_3

It is not only knowledge but also the development of insights (i.e. a conclusion enabling an increase in knowledge) into a certain problem or a task that is crucial for successfully transforming knowledge into benefits or target-driven actions. Findings and conclusions add both contextual and interpretive value, unlike pure knowledge, which merely builds upon the latter.

A diagram with mere numerical values is of little or no relevance. Only the associated logical interpretation in the context of the mode of understanding will create access to relevant findings and thus enable knowledge aggregation.

According to *North*, the elementary basis for knowledge is formed by characters that can be arranged arbitrarily within a set of characters. These characters also form the first step of the knowledge stair according to *North* (see Fig. 3.2). A syntax is used to bring them into a meaningful context, which is how they are structured into data. Without a reference framework, however, the generated data will be arbitrary, isolated, and decontextualized.[3)] Their allocation to a specific topic or contextualization will result in oriented information. It is only through the meaningful combination of information and the integration of personal experience that knowledge considered to be true will emerge from this information.[4)] Combining new information with personal experience and existing knowledge allows for its comparison and evaluation. This step forms the basis on which to draw conclusions informing specific activities. Hence, generating knowledge is a continuous and iterative, rather than static, process. The concept of knowledge thus refers to a mixture of experience, values and beliefs, information, and expert skills and abilities. This blended scenario makes it possible to conceive a cognitive scheme for assessing and integrating new experiences and information.[5)] The acquired information can either be new to the information recipient and thus lead to new knowledge, or it can already be known, thereby verifying or falsifying existing knowledge.[6)] Skills and abilities (or know-how) emerge from the application of knowledge and its expansion by specific and practical references. If this process is linked to an intent to act, an action or activity will result from the skills and expertise. If, in turn, this action or activity takes place under the "right" conditions, this will lead to the emergence of competency, or competitiveness.[7)]

Knowledge of the relationships and interdependencies of systems and processes has a major influence on the management of chances and risks. It essentially determines the process of identifying and evaluating chances and risks. This process will inevitably have to be based on a credible presentation of the chance/risk ratio whenever a final decision is to be made for a value associated with a certain chance/risk ratio (such as the total project duration, construction time, project and construction costs). There are two possible approaches to solving this task: Either a desired chance/risk ratio will determine the selection of the corresponding value, or the knowledge of the chance/risk ratio associated with a (determined) value will govern the further strategic alignment of a project. As a result, the degree of individual or organizational risk appetite will also be of great significance for the actions taken in the context of chances and risks.

2) Cf. Romhardt (2001), p. 64 (translation by Steffen Walter)

3) Cf. North (2011), p. 36

4) Cf. Rathswohl (2014), p. 20

5) Cf. ibid., p. 20

6) Cf. Gust von Loh (2008), p. 119

7) Cf. Rathswohl (2014), p. 20

3.1 Types of Knowledge

In information science, the concept of knowledge is evaluated from a variety of perspectives. This setting results in differentiated approaches to categorizing types of knowledge (see Fig. 3.1). We can thus make a distinction between the epistemological dimension of knowledge, which deals with the articulability or documentability of knowledge, and its ontological dimension, which focuses on the number of experts, or knowledge bearers.[8)]

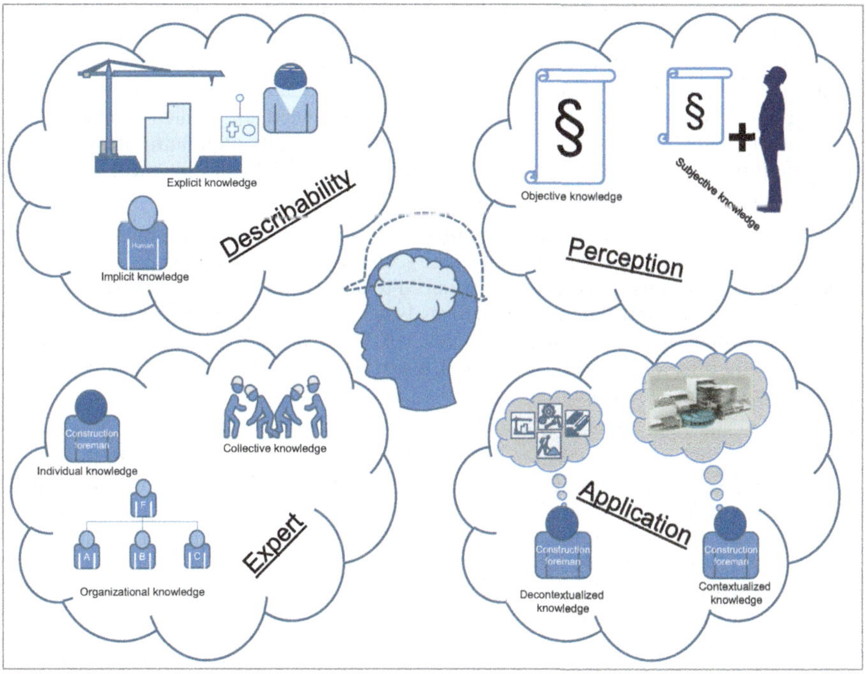

Fig. 3.1 Categorization according to types of knowledge[9)]

This book will clearly differentiate between the concepts of dispositional and orientational knowledge; these two types of knowledge are considered to be crucial for knowledge management. In the current age of digitization, the quantity of dispositional knowledge is rising to unprecedented levels. In the context of the Internet and rapidly advancing sensor technologies, there are constantly evolving and increasing options in the field of information and communications. However, this dispositional knowledge must be complemented by the dimension of orientation in order to be able to use it as guidance for action. *Mittelstraß* defines dispositional knowledge as a type of knowledge that "increases man's power of disposal over his world".[10)] This type of knowledge answers the question of what could be done, rather than what should be done. The latter approach requires orientational knowledge, which is a "regulative knowledge, a knowledge of goals and

8) Cf. ibid., p. 22

9) Based on Hofstadler/Ninaus (2020), Slide 9

10) Cf. Mittelstraß (1989), p. 19 (translation by Steffen Walter)

maxims".[11] Orientational knowledge is strongly action-driven and attempts to assess potential actions also in terms of ethics and values, an aspect that *Mittelstraß* criticizes as increasingly lacking in rational-technical cultures.[12] In the context of this book, orientational knowledge not only focuses on a primarily ethical "should-be" question. Rather, the orientation of knowledge also refers to the structuring, objectivization, filtering and, ultimately, refinement of the dispositional knowledge so as to ensure its meaningful use particularly in the field of construction management and economics. For this purpose, the complementary use of specific source knowledge ("know-where") appears to be a helpful means to orientate oneself correctly in a (digitized) world in which knowledge is becoming increasingly omnipresent and quickly available (but deceptive at the same time for exactly this reason). The ability to retrieve trustworthy (scientific) sources is a core skill – this refers to specific sources, locations, and access points to knowledge.[13] The inclusion of source knowledge makes it possible to contextualize, validate and systematize existing unstructured knowledge (dispositional knowledge), which supports the generation of valuable orientational knowledge. The availability of an intra-organizational or whole-project knowledge repository is hugely important in the practical context of construction management and economics.

State-of-the-art sensors provide a wide range of options for collecting innumerable data and information while enabling (in an ideal scenario) contextualization and sequential allocation. To make this scenario work, we need to answer the following questions: What do I have to analyze? What am I allowed to analyze? How should I go about analyzing this data and information? Without these answers, the data and information referred to above will be of only little value. In this context, it is extremely important to select and generate specific performance indicators.

Another relevant topic involves the evolution of knowledge into competency or skill. The mere existence of knowledge will not suffice for achieving defined targets. In the course of operational knowledge management (see also the knowledge stair according to *North* in Fig. 3.2), knowledge is transformed into competency or skill through creating a motivation for implementation and, subsequently, appropriate action. The know-how required for this purpose includes actionable knowledge, such as with regard to practices, methods, and strategies. This knowledge category is closely related to the know-why, which represents an explanatory, interpretational type of knowledge going beyond the mere awareness of the existence of certain facts or phenomena since cause-effect relationships can be identified. Elaborated forms of this type of knowledge include theories and models that create the basis for responsible actions.[14]

In addition to these differentiations of knowledge, the conceptualization of this term can also be viewed from an antagonist perspective, as it were. The scientific study of "non-knowledge" and its manifestations has received increasing attention in recent years and can help to concretize the concept of knowledge as such. Besides the intensified efforts to distinguish knowledge from non-knowledge, clarifying the terminology and establishing a typology of definitions of the related concepts of risk, uncertainty, and non-knowledge is considered to be equally crucial, especially for the interest in this subject area on the part of the social sciences and business administration disciplines.[15]

---

11) Cf. Mittelstraß (1989), p. 19 (translation by Steffen Walter)

12) Cf. Mittelstraß (1989), p. 19

13) Cf. Kiel/Rost (2002), p. 27

14) Cf. Kiel/Rost (2002), p. 27

15) Cf. Wehling (2001), p. 472

3.1.1 Implicit and Explicit Knowledge

Implicit knowledge involves any knowledge that can neither be formalized nor verbalized.[16] This concept refers to the expertise, or know-how, that an individual has acquired over time as a result of their experience and practice. The mentally stored implicit knowledge should be disclosed as far as reasonably possible because the related decisions made are often influenced in an undocumented (and unconscious) manner by the fears, hopes, predispositions, etc. contained therein.

Explicit knowledge is defined as knowledge that can be verbalized and documented, and thus be put into words relatively easily so as to communicate it clearly and unambiguously.[17] This type of knowledge can thus be easily transferred into manuals, work instructions, or daily construction reports.

3.1.2 Subjective and Objective Knowledge

Another way of differentiating the concept of knowledge results in the categories of subjective and objective knowledge. Subjective knowledge will always be bound to individuals. This type of knowledge is personal rather than being universally accessible. Objective knowledge is defined as the knowledge contained in databases, books, or other storage media. Objective knowledge is thus documented knowledge.[18] One of the quality criteria for objectivity refers to the fact that the documentation of knowledge and the production of results is as independent as possible of the individuals involved (cf. implementation, evaluation, and interpretation objectivity).[19]

According to *Nonaka/Takeuchi*, subjective knowledge is that experiential knowledge which is directly related to the performed activity and unconsciously supports the execution of that activity.[20] In contrast, objective knowledge refers to the theoretical knowledge stemming from the past.[21]

3.1.3 Individual, Collective, and Organizational Knowledge

While individual knowledge is bound to specific persons, collective knowledge emerges from combining the knowledge of individual experts within a community and thus creates new, independent knowledge.[22]

Organizational knowledge is the result of complementing individual or collective knowledge with corporate missions, objectives, and visions.[23]

[16] Cf. Rathswohl (2014), p. 22

[17] Cf. ibid., p. 23

[18] Cf. Stock (2007), p. 20

[19] Cf. Moosbrugger/Kelava (2012), p. 8

[20] Cf. Nonaka/Takeuchi (1997), p. 73

[21] Cf. Rathswohl (2014), p. 23

[22] Cf. ibid., p. 24

[23] Cf. Rathswohl (2014), p. 24

3.1.4 Contextualized and Decontextualized Knowledge

Contextualized knowledge is only of value in a specific situation or for a specific task or process.[24)] Such contextualized knowledge is difficult to apply beyond this specific situation. In contrast, decontextualized knowledge can be applied in various areas. This type of knowledge can be considered to be abstract, theoretical knowledge with no direct relation to the specific application.[25)]

3.2 General Considerations on Documentation and Knowledge Expansion

Documentation generally serves the purpose of making data and information easy to retrieve in order to record past or current facts and to pass them on or reproduce them.

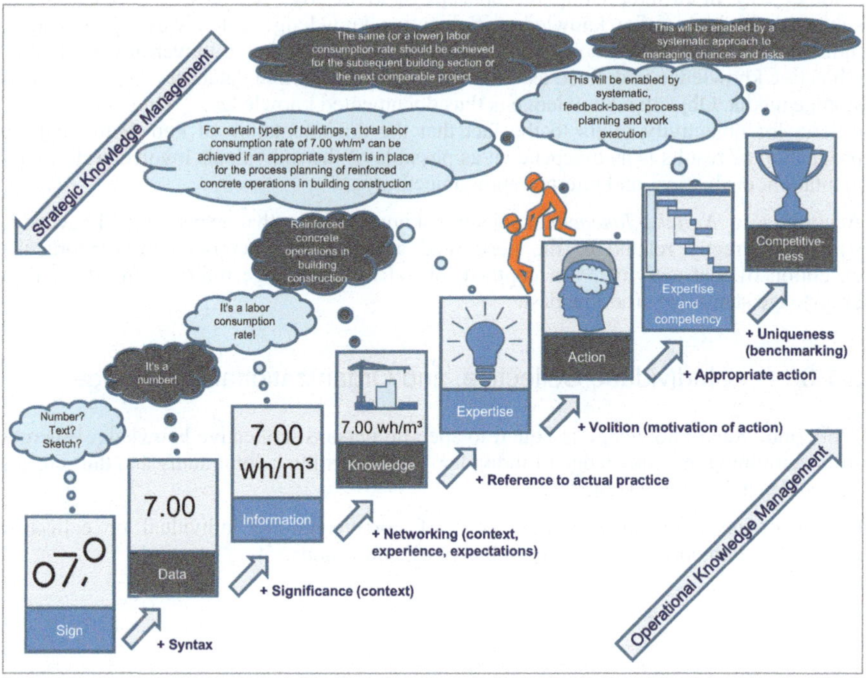

Fig. 3.2 Application of the knowledge stair according to *North* to construction management[26)]

Knowledge is transferred only if data and information are interlinked on a systematic basis and are put into the context of experience and expectations. Knowledge is the output of the

24) Cf. ibid., p. 25

25) Cf. ibid., p. 25

26) Cf. North (2011), p. 36; modified in Hofstadler/Kummer (2017a), p. 79

combination of data and information and relates to information in terms of purpose, perspective, context, and aspect. One and the same information can lead to different knowledge when viewed from another perspective or embedded in a different context. The quality of the documentation will depend on the knowledge level of the information generator and on how simple or complex the documentation process is. In addition, the quality of the documentation is influenced by the underlying rationale or motivation for documentation, which is also related to its potential use.

The following section describes an example of how to apply the knowledge stair according to *North* to the construction industry (see Fig. 3.2). Symbols and speech balloons explain the different levels that are discussed in more detail. The first step of the stair displays signs pointing to the numbers 7 and 0 or to the letter O. Yet these might as well be sketches.

Only the construction management syntax will establish a meaningful relationship. The signs are thus found to refer to the number 7.00, rather than to the number 0 or the letter O, or to sketches. At this stage, the initial step of the stair has been mastered, and we can exclude letters and sketches from the first interpretive attempt. However, considering just the number 7 is not yet meaningful. It could, for example, have a meaning in terms of the weather (such as referring to an average daytime temperature of 7°C), relate to the production factors (7 workers were used), or represent a period or duration (a specific process takes 7 working days).

The context of a system or subsystem is missing. Only by assigning a unit of measurement to the number does it become clear that the example is a relational number. Linking the number to the meaning leads to the conclusion that a consumption rate is meant. More specifically, this concerns the added "wh/m^3" part, which, as a unit, describes the number of paid working hours required per cubic meter. However, the context to which this cubic meter information refers is still missing since this could be the volume of a building, structural component, or construction material. Nor is it known, at this point, if the cubic meters refer to a dense or loosened volume. As such, the decontextualized information that the consumption rate amounts to 7 wh/m^3 is worthless for an uninvolved user because it cannot be linked to current or historical information. There is no possibility to allocate and categorize this piece of information.

In the next step, we need to identify the specific type of consumption rate to which the number refers. The consumption rate must again be related to an activity and to a type of building, to a structural component, and to specific works. Contextualization establishes a reference to the management of reinforced concrete operations to be performed for an entire building, which makes it possible to find out that the 7 wh/m^3 signifies the total labor consumption rate of reinforced concrete operations. This information provides a valuable enhancement of the previous knowledge level. Our networked thinking approach takes existing knowledge and experience into account, which is why it is becoming increasingly clear why a higher consumption rate is reached in comparison to a similar building.

The subsequently established reference to the specific application links details such as the type, form, and complexity of the building, as well as the relevant quantities and qualities, to the total consumption rate. This step makes it possible to assign the specific building to a building category. Users of this information will thus become increasingly aware of the circumstances and the production system with which the total consumption rate was reached. Know-how emerges from identifying the relationship between the previous project and the building currently under consideration. This networked approach will further improve the accuracy of assessing the estimated workload for the entire structure or components thereof.

Aiming to improve the combination of production factors for the current and the future project and thereby reducing the consumption rates gives rise to the stage of action or activity. This step constitutes a proactive attempt to transform knowledge into skills, leading to specific, measurable results. The resulting know-how can be applied in a targeted manner in order to execute the construction work more economically (stage of action).

The stage of competency is reached when the knowledge is applied intentionally. This is the case, for example, when realistic (accurate) consumption rates are determined for a project because the right conclusions have been drawn from analyzing the internal and external production systems. Systematic feedback processes are installed in the production system that provide near-real-time target-actual comparisons and enable countermeasures to be taken by acting correctly in the production process such that productivity is increased and costs are reduced. Taking the right action promotes the efficient selection and combination of production factors in order to meet the project targets.

Uniqueness is achieved when work is performed or services are provided that competitors are unable to offer. For example, systematically managing chances and risks and implementing chance/risk ratios in the costing exercise makes it possible to design the bid price in such a way that the contract is awarded while still ensuring the successful and commercially viable execution of the project. The core competencies[27)] of an organization are considered to be particularly advantageous for staying competitive.

3.3 Flow of Information

The targeted exchange of information is referred to as flow of information or information stream. In this process, the information generator conveys oral, written, or visual data to the information user.

In construction management and economics, the following situation appears to be conceivable (see Fig. 3.3): An employee from the costing department has been tasked to determine the pricing of reinforced concrete operations for a new building construction project. This person has little professional experience and turns to a colleague who worked in the construction management of a recently completed, similar project in order to find out about labor consumption rates and performance values, circumstances under which work was performed, problems encountered in the construction phase, etc.

The information that reaches its user will not be the same information that the information generator believes to have passed on. Part of the information will be lost due to omnipresent communication problems, such as those related to subjective perception, potential interface issues, or incomplete memory. This is why the employee in the costing department will get an incomplete picture of the utilization of the production factors.[28)]

27) "Core competencies are a combination of skills and technologies based on explicit and hidden knowledge and characterized by temporal stability and cross-product influence." North (2011), p. 38 (translation by Steffen Walter)

28) Cf. Hofstadler/Ninaus (2019). p. 887ff.

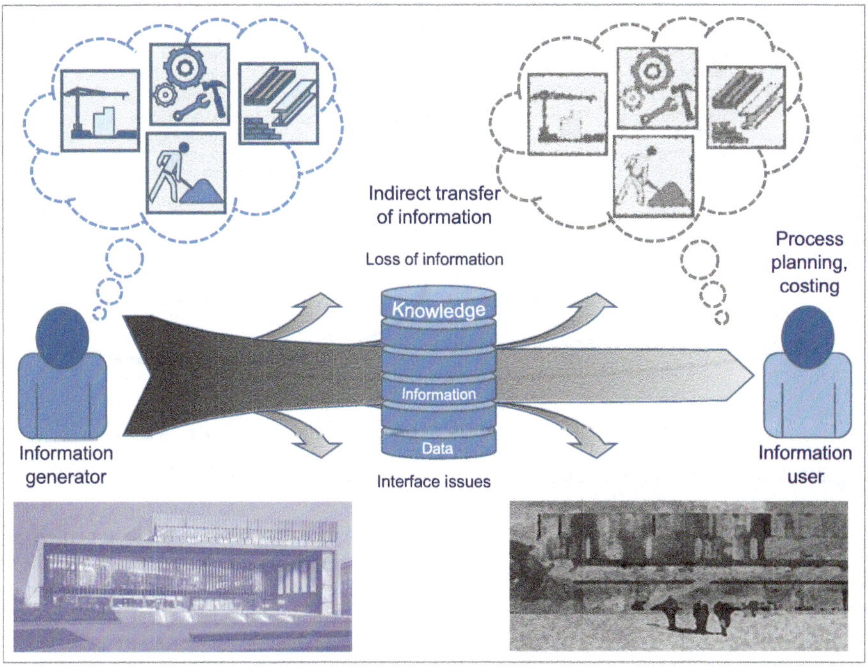

Fig. 3.3 Incomplete information flow[29)]

Documentation must be as systematic and structured as possible in order to ensure reliable pricing and calculations in the project preparation, bid pricing, process planning, and construction phases as well as in the course of claim management. The information content can be enhanced by photographs, videos, plans, process flow charts, etc.

The path of the information from the information generator to the information user must be designed so as to create an intuitively correct understanding of the collected data and information without requiring the information generator to provide explanations. Fig. 3.4 shows the qualitative flow of the documentation process. Creating and using smart knowledge repositories enables all project stakeholders to come up with more accurate construction cost and time calculations.

The circumstances under which consumption rates and performance values were reached should be recorded in order to document these values in a meaningful manner. Categorization according to building and site types, ground plan and elevations, degrees of difficulty, quality requirements, etc. can help to extract ranges or distribution functions for probabilistic calculations from the gathered data.

[29)] Cf. Hofstadler/Kummer (2016), p. 99

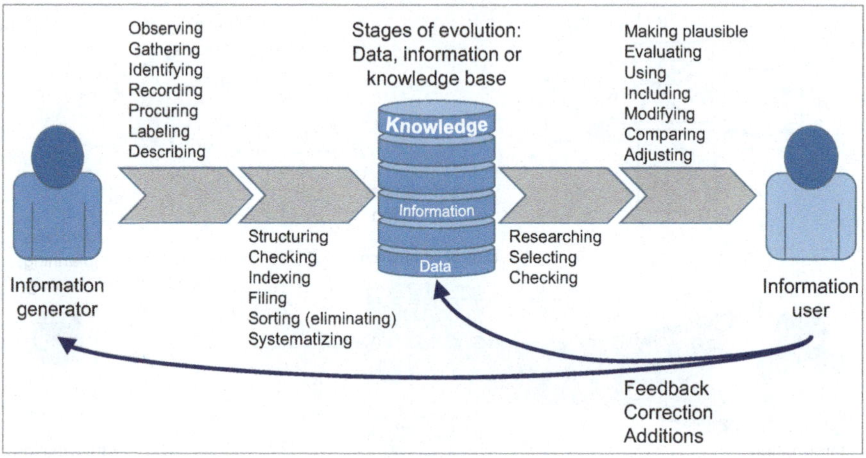

Fig. 3.4 Qualitative process of documentation and knowledge generation[30)]

Information relevant to construction management and economics is documented in various ways that require further differentiation. The following section describes these methods of documentation in more detail.

3.4 Information and Data Quality

The quality of information largely determines whether facts can be communicated precisely and correctly. This communication step is indispensable for achieving the defined targets. Quality criteria frequently applied to information include its correctness, completeness, relevance, consistency, currency, allocability, clarity, reliability, objectivity, and conciseness.

Wang/Strong apply the following categories and dimensions for determining the quality of information:

1. Access to information:
   system access, access security

2. Representation:
   interpretability, clarity, manipulability, integrity, and consistency

3. Information context:
   relevance, added value, currency, completeness, scope

4. Inherent value (inherent or intrinsic dimension):
   correctness, objectivity, credibility, reputation[31)]

30) Cf. Kummer (2015a), p. 260

31) Cf. Wang/Strong (1996), p. 5f.

People are constantly confronted with information in their private and professional lives. In private life, for example, this can include information contained in patient information leaflets explaining the administration of drugs; in the professional field, data such as labor consumption rates and performance values for calculating construction times and costs can be considered to be relevant information.

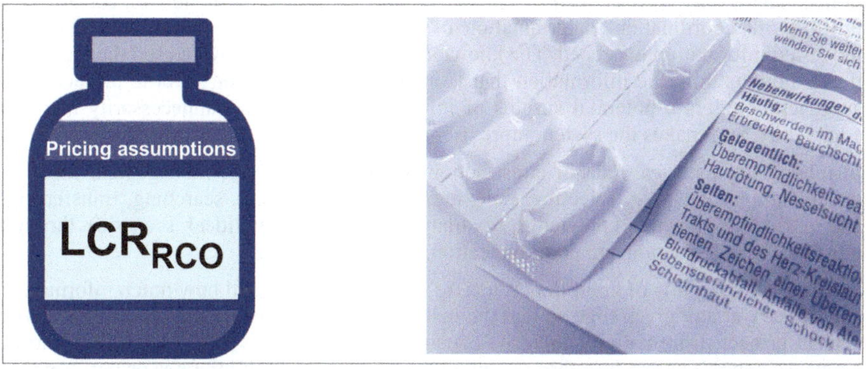

Fig. 3.5 Labor consumption rate, pharmaceutical product, and patient information leaflet[32)]

The patient information leaflet (PIL) is an important information medium for medicines (for an example, see Fig. 3.5). But what does the PIL wording have in common with construction management and economics? Like in PILs, the reliability and correctness of information is crucial for its use in the construction industry. This particularly applies if historical data is used as a basis for calculating construction costs and times, for instance. The quality of the information will essentially determine if its use leads to good or poor results.

It is problematic and risky to administer medicines for which no PIL is available. On the one hand, the drug may be inadequate for the disease to be treated; on the other hand, its dosage may be too high or too low. If the dosage is too low, the expected effect will not occur or will only appear much later. If the dosage is too high, the medical condition may worsen and unwanted adverse effects may occur. The same applies to PILs whose wording is unclear or misleading.

In construction management and economics, the same approach applies to the handling of historical data from past projects. Values will be useless and misleading in the absence of information (i.e. a "package insert text" for consumption rates, for instance; LCR_{RCO} = total labor consumption rate of reinforced concrete operations) on how and under what circumstances these values were determined. The way in which the production factors have been combined, as well as the related circumstances, need to be processed as comprehensible and plausible information. If the construction process is disrupted, it will be necessary to document and describe the symptoms as well as the "medical history" in order to arrive at a diagnosis. On this basis, effective and economically sensible countermeasures can be taken.

Consistently compiled, structured "PIL texts" are indispensable for ensuring reliable pricing and calculations in the project preparation, bid pricing, process planning, and construction phases as well as in the course of claim management. The information

[32)] Cf. Hofstadler/Kummer (2017a), p. 85

content can be enhanced by photographs, videos, plans, process flow charts etc. in order to increase information density. This is necessary in any case because it will prevent the disclosure of an unnecessarily large amount of information. In addition, great importance is attached to the central aspects of transparency and traceability of data, graphics, diagrams, etc. in the context of enhancing information density. Setting up filters enables systematic access to relevant information. The user of the information source indicates the project phase and type of analysis or calculation (pricing, calculation of construction times, work coordination, determination of losses of productivity, etc.) for which the requested information is needed. The knowledge management software used then automatically outputs only the information that is actually relevant to the specific project. This approach counters a potential surplus of information that would unnecessarily increase complexity, and makes the system more user-friendly.

In relation to historical data of construction projects, knowledge management includes documenting, analyzing, selecting, indexing, providing, reusing, searching, transferring, and retrieving relevant knowledge (available digitally in an ideal scenario) through processes of information and communication.

For users of data, it is of great importance to determine which and how much information is necessary at any given time so that the historical data can be applied meaningfully to a current project. Information should be considered useful if it is accompanied by known boundary conditions, such as those applicable to labor consumption rates or performance values. Only then will it be possible to determine the extent to which these values need to be adjusted (increased or reduced) to given facts and conditions. At any rate, it will be helpful in this context to interpret values in the form of ranges. An even better alternative is to apply probabilistic calculation methods, such as Monte Carlo simulations.

It is strongly discouraged to simply adopt data without first checking them for plausibility and applicability. In any case, black-box considerations are not recommended in this respect since only input-output relationships are measured. A black box is a system whose inner structure and workings are unknown or are not considered to be significant. The only matter of interest is the behavior of the black box ensuring a certain functionality via defined interfaces and guaranteeing a result.

In construction management, this would mean, for example, that only the number of production factors used during a certain period will be measured, as well as the output, quality, and quantity achieved by doing so. In contrast, it is unknown, or of no interest, which production factors were used, how they were combined, and what external and internal influences had an effect on the production system. As part of a target-actual comparison, it is easy to highlight deviations in labor consumption rates or performance values on the basis of the relevant figures. The related difficulty lies in the unambiguous, verifiable presentation of the cause-effect relationships as well as in clearly tracing the areas that have caused the deviations. In most cases, these efforts end up as a black box or are presented as such by the respective other party to the contract. The client should directly reject any additional cost claims drafted and justified in this way.

Any on-site observation of construction processes will be associated with underlying conditions that can easily be verbalized, such as the onset of rain at 09:00 a.m. and its end at 10:30 a.m. What is not possible, though, is to verbalize, let alone quantify, an observed sluggish start of reinforcing operations as part of the construction of a floor slab section, as well as the link to an ambiguous or even contradictory layout plan for the reinforcement. Although the resulting higher labor consumption rates are measurable, the true challenge lies in their clear and verifiable attribution to the root cause of the poor planning quality.

One of the problems to be solved when documenting facts is to apply selective perception[33] and, consequently, the transformation into specific, focused information. "The importance of the selection function of attention becomes apparent when we realize that at any given time a large number of auditory, visual, tactile, etc. stimuli act on our various sensory organs and set sensory reception processes in motion. However, we become aware of only a small portion of this huge amount of information. In other words, only a small part of this amount will determine our continuous interaction with the environment. This means that the relevant subset must always be selected from the total amount of incoming information as well as from the information stored in the memory in order to allow for efficient and trouble-free action."[34]

The key question is how the "information leaflet text" for useful costing approaches taken from completed construction projects must be worded such that historical data can be used efficiently and effectively for future projects. As the quality of the information increases, so does the quality in the application of the historical data, because, for example, the spread between the minimum and maximum values is narrowing.

Data quality criteria differ from those applicable to the quality of information and can be categorized as follows:

- "Correctness: The data must correspond to reality.
- Consistency: A data record may not contain any contradictions within itself or to other data records.
- Reliability: The origin of the data must be plausible and traceable.
- Completeness: A data record must contain all necessary attributes.
- Accuracy: The data must be available in the required accuracy (such as with regard to decimal places).
- Currency: All data records must correspond to the current state of the reflected reality.
- Freedom from redundancy: No duplicates may occur within the data records.
- Relevance: The information contained in the data records must meet the respective information needs.
- Uniformity: The information within a data record must be structured uniformly.
- Specificity: Each data record must be uniquely interpretable.
- Clarity: The terminology and structure of the data records must correspond to the terms and concepts applied in the relevant disciplines."[35]

3.4.1 Methods of Acquiring Information

Relevant information must be acquired and processed in order to be able to draw conclusions regarding future events based on past or present circumstances. Fig. 3.6 presents an overview of the relationships between different methods of acquiring information (referred to as information acquisition techniques by *Haberfellner et al.*).[36]

---

[33] Perception is the process, and the result, of acquiring and processing information from stimuli present in our environment and within the physical body of a living being.

[34] Cf. Hagendorf/Krummenacher/Müller/Schubert (2011), p. 179 (translation by Steffen Walter)

[35] http://www.computerwoche.de/a/gute-daten-schlechte-daten,1931857. Accessed December 19, 2020 (translation by Steffen Walter)

[36] Comment on the figure taken from *Haberfellner et al.*: Regarding the notion of "participatory observation", non-participatory observation would also be possible for the purpose of acquiring information. As for "intuitive forecasting", the methods listed in the figure are by no means purely "intuitive" in the sense of making decisions based on unconscious conclusions. Quite on the contrary, experts perform careful, rational considerations while relying on expert knowledge. For the sake of completeness, we nonetheless adopt the attribution chosen by *Haberfellner et al.*

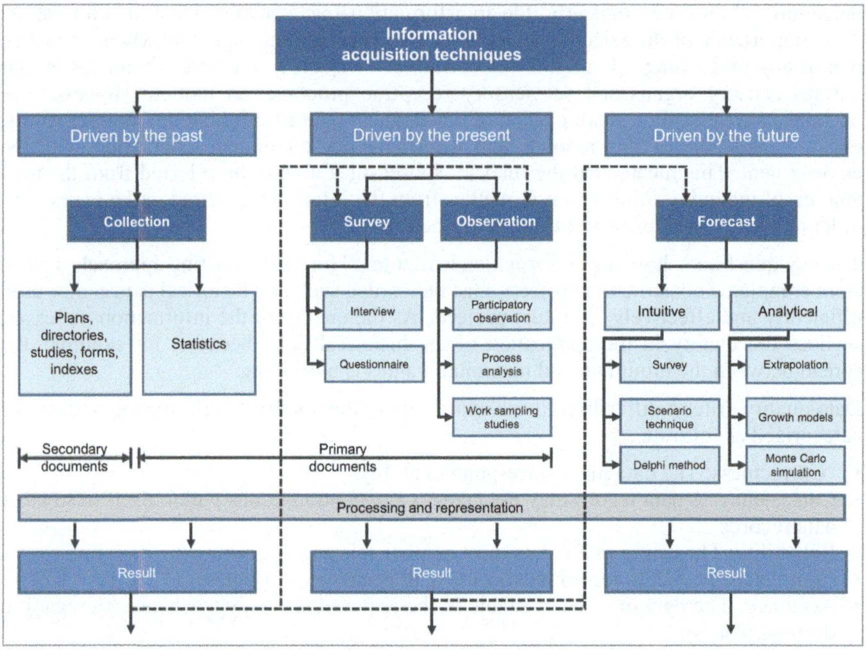

Fig. 3.6 Overview of methods of acquiring information according to *Haberfellner et al.*[37)]

Possible methods include collection, surveys, or observations, depending on the temporal aspect. Collected literature sources such as plans, directories, studies, forms, or indexes are so-called secondary documents. They are only indirectly related to the object under study or to the project. Primary data is made accessible purposefully through oral or written surveys, observations, or collections (statistics). Forecasting techniques are used to generate information on future developments and conditions. In this respect, a distinction can be made between intuitive and analytical methods.[38)]

The Monte Carlo method can be assigned to the model-based category since it incorporates information stemming from specific calculation parameters into the simulation through selecting distribution functions. In the absence of a systematically populated database, this method resorts to assessments provided by experts. These will define the input parameters on the basis of their experience and information gathered from personal observations in a non-systematic manner. The experts' knowledge with respect to ranges of certain parameters, including costs, labor consumption rates, performance values, and durations as well as their project- and situation-specific indications, can be converted into input parameter distribution functions in the course of the Monte Carlo simulation. If, on the other hand, a database is available (where particular emphasis should be placed on the appropriate delimitation of the data), data fitting can be applied for the purpose of identifying the theoretical distribution that best represents the corresponding data.

37) Cf. Haberfellner et al. (2002), p. 482

38) Cf. ibid., p. 481

3.4.2 Information Asymmetry

The following section sets out to discuss the general concept of information asymmetry. In a conversation about a subject between two people, an asymmetry of information will be present in most cases, i.e. one person is more informed than the other. This imbalance actually leads to the asymmetry. In an economic context, information asymmetry simply means that in a transaction or market activity with the involvement of two parties, one of them will be affected by an informational disadvantage. In other words, one of the two business partners is in possession of less information than the other, which is why it will inevitably be disadvantaged depending on the nature of the information.

Generally speaking, in economics, information or information flows are considered to be special goods. This becomes most apparent in private-sector research, where research is conducted within a defined discipline and access to the resulting information is initially restricted to the involved researchers and organizations. In relation to markets, we can thus conclude that information forms the basis of any economic activity.

In markets, information asymmetries will occur predominantly owing to the presence of private information in market activities. This type of information allows some market participants to gain an advantage over others. In such a case, the latter will act on the basis of incomplete information.

3.4.2.1 Reasons for Information Asymmetry

The fact that one of the parties or stakeholders has more information than its counterpart may be due to a variety of reasons.

Possible explanations for information asymmetries include:[39)]

- Hidden property or hidden characteristics: This term refers to a hidden property or characteristic of one of the stakeholders (or the goods or services it provides to the market) that cannot be observed or changed.
- Hidden action: This concept refers to a hidden, unobservable action taken by one of the stakeholders.
- Hidden information: Hidden information signifies a hidden, unobservable, constant condition or circumstance relating to the actions taken by the stakeholder.
- Hidden intention: This concept refers to a hidden, unobservable intention of one of the stakeholders.

3.4.2.2 Principal-Agent Theory

The principal-agent theory is the model most frequently used to describe the different distribution of information between the contracting parties.

In this case, the issue of information asymmetry lies in the division of ownership and control. Big corporations are often managed by an executive board, rather than by their owners themselves. In this scenario, the agents (managers) are thus responsible for representing the interests of the principal (owner).

When transferring this principle to the construction industry, we find that the client acts in the principal's role while the contractor takes on the agent role in the course of the contract

39) Cf. Arrow (1984), p. 3

award process. In this phase, the client is in possession of much more information about the construction project and has a huge advantage over the contractor in terms of available information and knowledge. After the contract award, the contractor starts performing on the contract by providing the construction operations owed under the contract.

Conversely, the client becomes increasingly familiar with the subsoil, the site surroundings, the building to be constructed, and the production system as a whole. As the process progresses, a reversal occurs with regard to the level of information on the structure and the circumstances governing work performance. To some degree, the contractor will thus assume the role of the principal while the client becomes the agent. It is also possible for the contractor to act in a principal role vis-à-vis its subcontractors. This situation can lead to a conflict of interest if the agent's objectives do not coincide with those of the principal.

According to the same principle, the representatives of the client also enjoy an information advantage over the bidders in the course of the tender procedure. The bidders will usually have only little time to review and analyze the tender documents in full and in detail. This is particularly true in view of the fact that only about one in ten bids will actually give rise to a contract. The costs of preparing the bids for about ten projects must therefore be allocated to a single project in order to recover these costs.

The client gets involved from the very first project idea and has already reached a level of information during the project preparation and planning phases that the bidders are unable to catch up with in the (usually short) bid phase. This relationship is reversed in the course of the construction process. The information advantage passes to the contractor because it is directly involved in the execution of the project and receives exclusive information from its subcontractors and suppliers.

3.4.2.3 Ex-ante Information Asymmetry

Ex-ante information asymmetry refers to an imbalanced distribution of information prior to entering into a contract. For a principal-agent relationship, we assume that the information advantage is on the side of the principal. Even if they are not able to change the information, it is at least known to them. In this setting, withholding relevant information (hidden information) may lead to the so-called adverse selection phenomenon (see Section 3.4.2.4)

3.4.2.4 Adverse Selection

Hidden information can lead to adverse (negative) selection[40]. The personal characteristics and traits of an individual constitute relevant information with an influence on a potential transaction, but, ultimately, they are difficult to identify for outsiders. For example, a seller will be in possession of more information regarding the characteristics of an item for sale than the buyer.

An article often cited on adverse selection is entitled "Market for Lemons" (referring to used cars) and was written by *Akerlof*.[41] The market for used cars includes good and bad vehicles. For the inquirer, however, it is not possible – or only at a considerable cost – to

[40] "Market failure phenomenon resulting from an information asymmetry between contracting parties prior to entering into the contract. High-quality suppliers withdraw from the market either partially or completely." http://wirtschaftslexikon.gabler.de/Definition/adverse-selection.html. Accessed December 19, 2020 (translation by Steffen Walter)

[41] Cf. Akerlof (1970), p. 488

determine what type of vehicle is involved. They will therefore assume an average vehicle quality. On the other hand, the seller of a car will be acutely aware of its quality. Thus, the seller has an information advantage over potential buyers. Since the potential buyers assume an average quality, they will only be prepared to pay an average price. The sellers of the good cars will subsequently not offer their cars at this undervalued price in order to ensure that they do not make a loss. In the end, this will leave only cars of a lower-than-average quality on the market.

Adverse selection in construction economics

The market for constructing buildings and structures includes both substandard and good or even outstanding contractors. For the client, however, it is not possible (or only at a considerable cost and enormous effort), particularly in the case of consumer transactions, to determine what quality of execution, reliability, and performance capacity to expect of the construction company. On the other hand, the representative of the construction company knows the quality of the company and its reliability and performance capacity very well. Thus, the contractor's representative has an information advantage over the potential client.

Since contractors do not currently enjoy a good reputation in the construction industry (see the media coverage on construction defects, cost overruns, wage and social dumping, construction scandals, unreliability, etc.), clients will only be prepared to pay an average or low price for construction operations. The representatives of construction companies acting with integrity, which employ their own highly skilled staff and use state-of-the-art equipment, will, in turn, no longer be prepared or willing to perform their high-quality work. They will no longer be able to cover their costs at the comparatively low price level referred to above. This means that only the construction companies offering lower quality (i.e. up to the average level) remain on the market because their operations will, quite simply, incur lower costs.

3.5 Systematic Knowledge Management

Documentation systems specially designed to indicate the individual influential factors, such as the type of production factors or the circumstances under which work is performed, play a crucial role in determining possible causes. What matters most in this context is to establish a cause-effect relationship. For successfully verifying construction process disruptions, it is necessary to document the individual building sections and works affected by such disruptions.

If the construction contract includes a bill of quantities, it must be shown for the relevant works which services in the bill of quantities are involved (for instance by indicating item numbers and plan references) and how work performance had originally been planned (referring to the time of the contract award). The representation of the target process shows the planned target production factors, particularly the standard of workmanship and employee skills and qualifications, as well as the target combination or "should-be" combination of the production factors, namely their numbers, temporal and spatial distributions, and sequencing relationships. The documentation must record those circumstances of work performance that have changed over time, as well as the observed effects on the actual production factors as such and on their actual combination, such as in the construction process or in logistics.[42)]

[42)] Cf. Hofstadler (2014), p. 62

Heck/Schubert cite timely documentation, in addition to the general organization of the construction process, as a major problem area in today's project management and execution. This is essentially due to the fact that the site managers have too little time to keep their work schedules updated in a timely manner. However, documenting the dynamic development of the actual construction processes is a key driver for successful verification for the purpose of enforcing contractor claims. Other relevant factors in this context include a consistent and, if necessary, detailed documentation of notices of obstruction, daily work reports, and scheduled delivery lists, as well as any other correspondence and recorded images and videos. The structure of the actual work schedule and the level of detail must be adapted to the agreed target schedule in order to create a consistent basis for comparison. If necessary, it will of course be helpful for the demonstration of evidence to strive for a higher density of the actual documentation. This should improve the probative value of the actual work schedule. At the same time, the very accurate representation of the evolution of the construction schedule over time provides a perfect controlling tool for both contractors and clients. However, this will inevitably require the timely adjustment of the construction schedules to the actual construction process. To ensure a systematic approach to verification, it is necessary to document the cause-effect relationships separately in the event of disruptions to work performance with an effect on the construction time.[43)]

Girmscheid/Motzko consider short intervals between site observations to be an essential requirement for targeted and systematic process control. Analyzing the relevant data, which is collected on a monthly basis according to generally accepted practice, is associated with the issue that no timely, and thus proactive, process control is possible because some of the considered processes will have been completed long before the analyzed data is made available. Data collection intervals thus need to be shortened, with the extent of the shortening to be determined specifically for each project and trade or discipline. Depending on the specific case, critical processes will at least require the documentation of work performance on a daily basis.[44)]

A functioning flow of information and associated systematic knowledge management is becoming increasingly important particularly because of the growing degree of specialization at the various departments of large construction companies. Insufficient information transfer often causes calculations, analyses, requests for information, and similar processes to be carried out more than once, which incurs additional, avoidable labor costs. This situation can be overcome by conceiving a project-related central knowledge management system, thus reducing duplicate work, errors, or incorrect interpretations of construction process disruptions owing to an insufficient, erroneous flow of information.

The process of passing on information on data and the circumstances of work performance is usually associated with information gaps. The information generator, who is directly involved in the observations and measurements, will not pass on their perceptions related to construction management and work circumstances "as is" to the information user or knowledge base, or will be unable to convert the collected impressions (or part of them) into relevant and usable information. These information losses occur particularly when transforming impressions of the observed construction process into usable information, such as descriptions, sketches, or process charts prepared for the information seeker who has not been involved in the creation of data and information. Such a loss will be particularly noticeable with respect to the scope, depth, precision, and clarity of the information. In this scenario, the observer will utilize a conscious and subconscious filtering process in

[43)] Cf. Heck/Schubert (2011), p. 189

[44)] Cf. Girmscheid/Motzko (2013), p. 132

order to convert only certain portions of the information into subjectively meaningful overall impressions.

In so doing, it is crucial whether the information generator describes the type and circumstances of the combination of production factors using its own, free wording or whether the organization provides information sheets containing boilerplate text (guided information[45]). Based on this guidance, the most appropriate descriptions can be selected by checking boxes, for example, which is also possible when using a smartphone or tablet. As a matter of course, specific choices not indicated in the forms can be covered by additional free text. It is obvious that with regard to information generation, those responsible will be more willing to provide complete documentation if the observer has to prepare only a small amount of text, and if standardized, tried-and-tested, ready-made systems and processes with low resource requirements can be used.

Regarding one's own contribution to generating information, a distinction is made between the following types (see Fig. 3.7):

- Independent generation of information
    - The information generator and supervisor have previously agreed upon the essential, priority, "nice to have", and non-targets
    - The information generator independently initiates the process of generating information
    - The information generator implements the process of generating information on the construction site
    - The supervisor provides support in clarifying interpretation issues and boundaries, also in advance if these are to be expected or if the information generator submits a related request

- Supported generation of information
    - The information generator and supervisor have previously agreed upon the essential, priority, "nice to have", and non-targets
    - The information generator independently initiates the process of generating information
    - The information generator implements the process of generating information on the construction site and contributes own ideas for improving the information system
    - The supervisor checks the suggestions for improvement and provides support in the case of problems of interpretation and delimitation, also in advance if these are to be expected

- Guided generation of information
    - The information generator and supervisor have previously agreed upon the essential, priority, "nice to have", and non-targets
    - The information generator carries out the generation of information using predefined information record sheets or a software. Previously compiled descriptions of certain production conditions enable the information generator to revert to predefined choices for specific questions
    - The possible choices allow for a systematic and comprehensive generation of information

[45] Guided information involves the use of an information record sheet or software in order to include boilerplate text referring to a selected range of issues, providing a number of choices for describing certain situations on the construction site, for example. The user is able to generate information quickly and easily by selecting a text field.

- The supervisor checks the suggestions for improvement and provides support in the case of problems of interpretation and delimitation, also in advance if these are to be expected

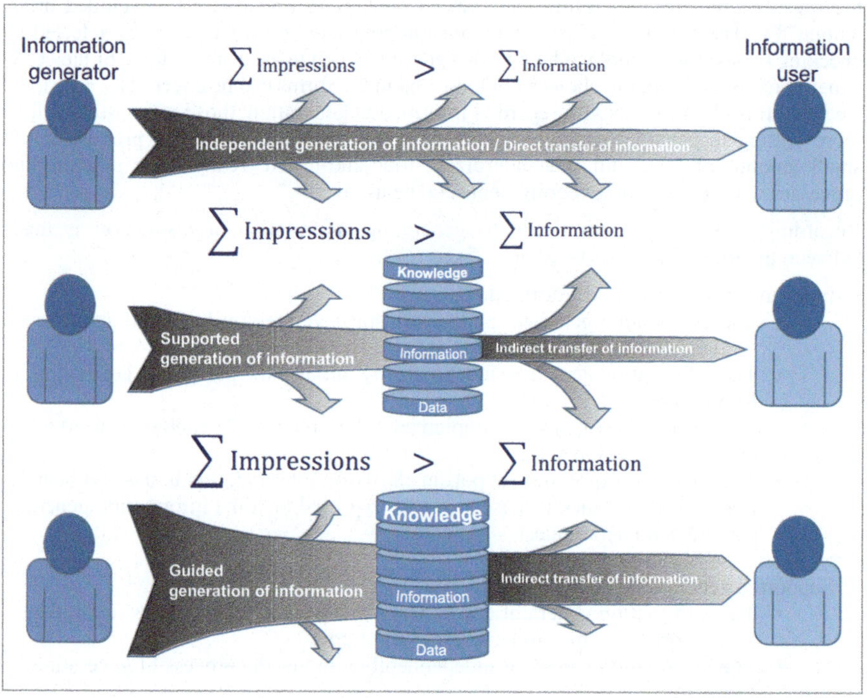

Fig. 3.7 Generation and transfer of information

In the course of the process of generating information, it should be noted that the information generator is usually no longer available to the subsequent information user for the purpose of interpreting data and information. However, irrespective of the quality of data collection and information generation, there will always be a residual information gap. This gap should be kept to a minimum.

The quality of the information will have a significant influence on achieving the defined targets and/or on the correct and precise communication of facts. Frequently applied quality criteria for information include its correctness, consistency, reliability, completeness, accuracy, currency, non-redundancy, relevance, uniformity, clarity, objectivity, and comprehensibility.

In recent years, more comprehensive research has been conducted into the simulation of construction processes. Such simulations allow for disruptions of various causes to be modeled and their effects displayed. These concepts are primarily geared towards optimizing the construction time and resource utilization in order to reduce costs without missing the quality targets. Essential factors for the quality and plausibility of such simulations, however, include input data reliability and a permanent target-actual comparison between the planning and execution processes. To allow for such a comparison, it is crucial to collect data on the use of people, equipment, and materials (i.e. the combination of production factors) on construction sites. One of the most important, ground-breaking research and development needs in the course of digitization deals with ways and means to carry out data acquisition and evaluation more efficiently through using innovative methods such as GPS acquisition, sensors, automatic image data analyses, etc. The fact that the cost-benefit ratio is improving constantly in comparison to conventional REFA analyses may give rise to a more widespread application in construction practice.

Analyses of the distribution of the hours that have to be spent on the REFA analysis show that the use of mobile devices (such as smartphones or tablets) as well as of the associated software that performs an automated analysis already during the data collection process can reduce the number of hours spent by more than 50%. In this respect, however, it should be noted that construction sites are usually exposed to weather conditions such as high or low temperatures, rain, wind, or snow, which makes data collection more difficult. In future, the ongoing process of digitization will increasingly provide opportunities to simplify such REFA analyses while rendering them even more systematic.

3.6 Data and Information for BIM

If the targets and objectives of BIM are fully translated into reality, a BIM building model will be generated that is binding upon all project stakeholders and will subsequently serve as a defined planning and change model for all project phases. This approach will result in a shared interface available to all project stakeholders for exchanging data and information. In this regard, it is important to define who needs to be informed of adjustments to be made so that subsequent, absolutely necessary change or revision activities are initiated for the relevant project stakeholders.

Change management certainly poses a major challenge both at present and in the future. In this process, a distinction should be made between changes to be implemented merely in response to requests submitted by the project stakeholders and changes that are mandatory owing to restrictions imposed by relevant authorities, laws and regulations, etc. Changes must be traceable at all times and must be verified and approved by a central entity. As a matter of course, this will require comprehensive knowledge of the planning and project relationships. A project emerges from a (project) idea and should be transferred very quickly into a 3D model so that the imagination of the project stakeholders becomes specific and thus more tangible even in early project phases. An integrated collaboration approach adopted by the project stakeholders to be involved will reduce idle time and planning errors. Difficulties and/or planning errors are easier to detect and eliminate on the basis of the 3D model particularly in the case of complicated and complex components, such as by identifying clashes of cable or service ducts and checking the model for consistency. A 3D model can also be used to further optimize buildings or components with regard to their use and operation phases as well as their commercially viable construction.

In all project phases, reliable data and information are needed for the considerations, estimates, and final calculations that become binding for the parties upon entering into the construction contract while also being key success drivers for the project. Particularly in this context, we should stress that data lacking contextual information will only be of minor value for subsequent use by third parties. The quality of the information associated with the data, which is also based on the systematic approach to capturing it, is a major criterion for achieving the defined targets or for the accurate and precise communication of facts. High-quality data and information allow for a better evaluation and utilization of chances and a more accurate risk assessment and prevention.

It is crucial to link the captured data and information to BIM models if we intend to pursue the goal of deriving plausible and reliable results for informing the decision-making process. Irrespective of the degree of sophistication of a BIM model, it will always require high-quality data (with appropriate temporal and contextual boundaries) with the associated information, including with respect to emergence and boundary conditions, in order to simulate construction times and costs. Only if these conditions are met will BIM models generate maximum benefit for all project phases and stakeholders, and thus for the entire life cycle of the building or structure.

This situational analysis and other contents of the book demonstrate the high demand for the systematic management of chances and risks and, subsequently, for a transparent and plausible integration into the entire BIM process.

3.6.1 Principles of BIM Models

The three-dimensional building model (i.e. an object-oriented 3D model with attributes) is the main component of the entire BIM process, which extends from the initial planning phase of a building to its use, reuse, and dismantling and demolition. The digital building model contains all current data and information for all project stakeholders.

For the 4D model, performance values and quantities are required so that process durations can be calculated (forward calculation). On the other hand, it is possible, from the bidder's/contractor's perspective, to use the total time or duration as a starting point for calculating the durations of "project segments" in order to determine the performance values to be achieved. The results of the BIM model provide information on the quantities, which ideally serve as the basis for generating KPIs in an automated processing step (for reinforced concrete structures, these can include, for instance, the formwork and reinforcement ratios). When assessing the complicatedness or complexity of structures, one factor of interest is how regularly or irregularly the structure develops over its ground plan and elevations.

The transition from the 4D to the 5D model is made by implementing costing details. The temporal relationship adds transparency with respect to the density of resources required for executing the project. Simulations with different specified construction times are carried out in order to highlight the related effects on the production factors. In so doing, it is important to take production-related and technological dependencies into account in order to exclude or reduce subsequent bottlenecks and overlaps in advance.

One of the core tasks of BIM is to fully capture data and information about the structural components of a building and to allow for an unrestricted, unambiguous, and consistent exchange with all parties involved over the entire life cycle.

3.6.2 Targets of BIM Models in the Project, Use, and Operation Processes

Effective construction and operation requires efficient tools and an optimized selection and combination of production factors (labor, equipment, and materials) in the project implementation and operation phases. BIM denotes a model-based, integrated work approach involving all project stakeholders (discretionary production factors) across all project phases and trades on a common basis and with defined interfaces. This method enables consistent planning, construction, and operation, provided the interfaces and the flow of data and information are defined and managed in a uniform manner.

In Chapter 5, the process associated with the construction project is divided into five project phases (PPHs) (see Fig. 5.1). The methodology to be developed in an incremental process and the corresponding BIM-related tools create a binding working basis for all project phases and tasks (see Fig. 3.8).

The project phases start with the release of a resource request and end with the project evaluation step. To unleash the full potential of BIM, a sixth project phase (called PPH 6 in Fig. 3.8) should be added in which the entire use and demolition/disposal phases can be taken into account.

To add value and to ensure the successful application of BIM models, particular emphasis should be placed on generating and exchanging realistic data and usable information without losses.

A 3D model with uniform interfaces for exchanging data and information (across all project phases) provides the basis for all project phases. Following analyses on both the macro and the micro-level, the other multidimensional models will gradually add data and information regarding quantities, labor consumption rates, performance and productivity values, cost estimates, information related to operation and use, chances/risks, and details on nonlinearities.

The IFC format developed by buildingSMART International (bSI) is used for the exchange of information in the construction industry independent of software manufacturers. Although the IFC[46)] file format generally allows for the exchange of data between programs supplied by different software manufacturers, a loss of data will still occur when importing IFC data from other programs.[47)]

In English-speaking countries, clearly defined levels (level of development, level of detail, or level of model definition) are used for describing model contents.[48)]

The following sections provide a basic description of the multidimensional data models[49)] with their properties and contents.

46) IFC: Industry Foundation Classes – an open standard in the construction industry used for digitally describing building models

47) Cf. Huhnt (2014), p. 13

48) Cf. Borrmann et al. (2015), p. 11

49) "Multidimensional data model – virtual building model providing more detailed information on the building and construction process on the basis of factual data." OENORM A 6241-2:2015, p. 5 (translation by Steffen Walter)

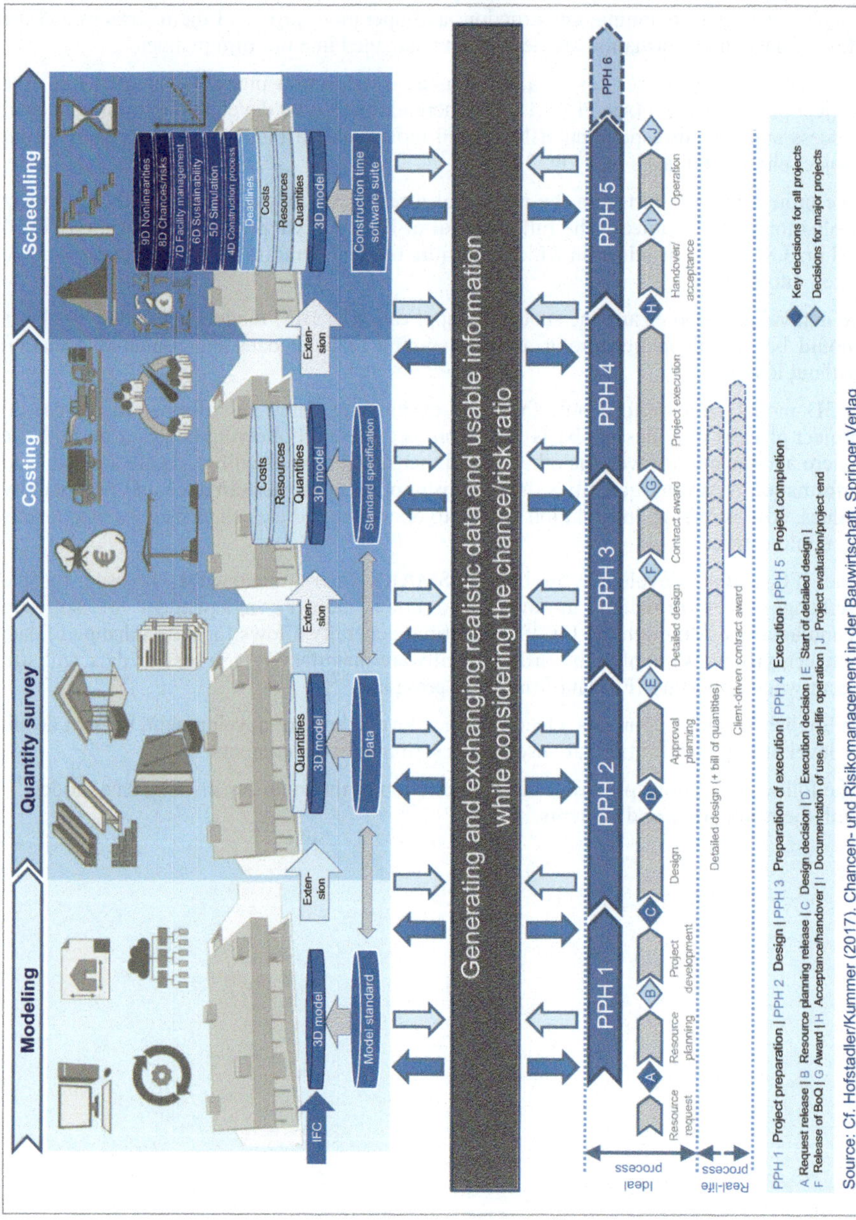

Fig. 3.8 BIM from the perspective of a construction contractor[50)] and exchange with the individual project phases – Present and future

---

50) Derived from Schreyer/Schütz/Hauber (2013), p. 88
Cf. Hofstadler/Kummer (2017a), p. 99

3.6.2.1 3D Model

The three-dimensional building model with its geometrical, physical, and functional properties creates the basis for coordinating individual trades and disciplines and for visualizing different design and construction phases. This is what the detailed design and construction phases will rely on.

The 3D project model is generated in response to the requirements defined by the client. This model comprises several submodels for the individual responsible parties (in Austria, in accordance with OENORM A 6241-1). Any of these submodels can again be subdivided into several models. The HVAC model, for example, can be split into heating, air conditioning, ventilation, sanitary installations, and electrical systems.

The submodels gradually evolve as a result of working on them in the individual project phases. The requirements described for the architectural submodel[51] apply analogously to all submodels.

Creating a comprehensive digital model of a building or structure relies on its geometrical characteristics but also on semantic[52] information required for enabling a systematic digital modeling of the structure.[53]

3.6.2.2 4D Model

The 4D model links the 3D model to construction schedules such that a logical relationship is established between process durations and elements or components. This logic should include all construction-related, technological, construction management, and building-specific interdependencies while pursuing the goal of representing a construction process that is realistic from the point of view of construction management and economics. The simulation based on this logic will integrate the dependencies of the processes on the performance values, labor consumption rates, productivity and consumption values assumed for the production factors.

3.6.2.3 5D Model

On the basis of the 3D model, all relevant process data generated in the planning, construction, and operating processes are stored in a dynamic database. The first module in this development process includes the determination of quantities and the related costing exercise. Many other applications have been developed in the meantime, leading up to the stage of digital fabrication (5D simulation).

Costs must be determined and allocated to actual works using (potentially) existing standardized work specifications (in Austria, in accordance with OENORM A 2063). Works or services not contained in the standardized specifications should be added as free-text items. Due to the continuous congruence of the project model with the building actually constructed (i.e. the as-built model), the consistency and transparency of the

[51] "The architectural model is a virtual building model that moves closer to reality over time. The graphical representation of all structural components known at the time of input as reflecting the actually intended construction (with their relevant properties, e.g. load-bearing/non-load-bearing, material, etc.) and structural component definitions must correspond to the element classification in their use." OENORM B 6241-2:2015, p. 6 (translation by Steffen Walter)

[52] "In the IT field, the concept of semantics refers to the meaning of data or information." Borrmann et al. (2015), p. 43 (translation by Steffen Walter)

[53] Cf. ibid., p. 43

processes and the evaluation options, it is possible to dispense with approximate quantity determination procedures as described in the standards governing contracts for work in the individual trades and to invoice the determined quantities according to actual numbers. In any case, this must be agreed upon separately in the contract.

3.6.2.4 6D Model

The sustainability assessment of buildings is carried out with regard to their environmental, social, and economic quality while taking into account their technical characteristics and functionalities. If we are to identify the contribution buildings make to ensuring sustainable development, sustainability itself is also put to the test. To this end, it is even more important to determine and assess life-cycle productivity compared to merely assessing the productivity in relation to the building or structure as such.

Productivity is a key indicator for assessing the effectiveness of individual work steps or of the entire production or economic process. It is expressed as the output/input ratio and characterized by complex relationships. The extent of life-cycle productivity in construction is determined by the effectiveness and efficiency in combining the elementary production factors. Elementary productivity comprises the productivity of labor, equipment, and materials. It is planned, organized, managed, controlled, etc. by the discretionary production factors in order to achieve the lowest life-cycle costs or the highest life-cycle productivity under the individual aspects of sustainability (i.e. ecological, economic, and socio-cultural and functional factors).

We should adopt a holistic, networked view of all components in order not to optimize only one or a few of them at the expense of others. It should be noted that the production factors and their utilization change over time. The project or operational targets usually remain unchanged and must be achieved with the aid of the different production factors. This poses a major challenge for the people involved in their decision-making processes and in dealing with life-cycle costs.

The individual parameters are calculated and compared with each other in the course of a comparison of different construction methods, for example, in order to determine life-cycle productivity. For an office building, for instance, the usable floor space measured in square meters can be used as the output, and the costs expressed in euros can be taken to represent the input. Life-cycle productivity is a key indicator that represents a relational number. It can thus be used as an essential assessment parameter besides life-cycle costs, which deliver absolute results.

$$\mathrm{LCC} = \mathrm{CC} + \mathrm{OC} + \mathrm{MC} + \mathrm{DC} \tag{3-1}$$

Life-cycle costs are calculated on the basis of the reduced basic formula in Eq. (3-1). These costs include the cost of a property (such as a production line or a building) and its components incurred over the entire life cycle during which they fulfill relevant use(r) requirements. Life-cycle costs (LCC) are composed of construction costs (CC), operating costs (OC), maintenance costs (MC), and dismantling and demolition costs (DC).

Special programs must be developed for the BIM model so as to allow for systematic and transparent life-cycle analyses and calculations. This would bring about benefits in the course of evaluating different structural and operational concepts as well as in the context of the bid comparison when identifying the best bidder. In addition, this approach can be used to step up efforts to design, build, and operate in a resource-saving manner.

3.6.2.5 7D Model

The 7D model creates the prerequisites for integrating essential aspects of Facility Management (FM). FM refers to the management and operation of land, buildings, equipment, and facilities. Operating structural and technical facilities entails very complex requirements in terms of planning, organization, coordination, management, and control. This is why it is ideal to integrate all related tasks and processes as well as data and information into a single BIM model.

Based on the tasks of the involved stakeholders and on the documents and data in the FM system, BIM-relevant processes such as task planning, creation and updating of the building documentation, and pricing and process planning are outlined, and corresponding BIM solutions are described in facility management, reaching from the creation of the model to its analysis. The basis for a uniform exchange of data and information is created and the tasks and specifications of the FM project stakeholders are integrated into the BIM process by defining standards and considering applicable codes in the BIM process. Current BIM trends such as the FM 3D process and data model, the BIM classification according to StLB-Bau[54)], and the National Library for BIM objects provide a very good foundation for the networking of stakeholders and processes over the entire life-cycle of buildings.[55)]

3.6.2.6 8D Model

The envisaged 8D model systematically considers chances and risks. This approach should make it possible to allocate chance/risk ratios to the resulting parameters. Essential or critical processes can thus be recorded and evaluated more effectively. If simulations show different construction times for a building, the individual corresponding chance/risk ratios can, and should, also be indicated.

During the construction phase, changes and disruptions to the process flow and logistics system may occur, most of which are difficult to predict. Such changes or disruptions are caused by the complexity of the production factors and by the associated nonlinearities, as well as by exogenous effects on the production system. Similar to a funnel tapering from the wide opening at its top to the neck at its bottom, such uncertainties will exhibit wider spreads in early project phases compared to the stage after the contract award, where an increasing number of specific project features will be added on the basis of the detailed design documents. The details pertaining to the building become increasingly specific as the planning process transitions from the macro to the micro-level. Yet uncertainties remain with respect to the most appropriate mode of combining the production factors as well as to future conditions and circumstances prevailing on-site in the construction phase. This is why there will always be a need for overcoming these challenges properly.

---

54) StLB-Bau: Standardleistungsbuch-Bau (Standard Services Book for Construction). "The StLB-Bau is prepared by the Joint Committee on Electronics in Construction [Gemeinsamer Ausschuss Elektronik im Bauwesen; GAEB], whose members represent public- and private-sector clients, architects, engineers, and the construction industry, in collaboration with the German Committee on the Award of Construction Contracts [Deutscher Vergabe- und Vertragsausschuss für Bauleistungen; DVA). It is published by Deutsches Institut für Normung e.V. (DIN; German Institute for Standardization). The StLB are structured according to work areas and contain standard descriptions, i.e. boilerplates for General Terms and Conditions for Work Specifications and for Supplementary Technical Regulations, as well as standard work descriptions, i.e. boilerplates for describing works or their sub-items. They contain information on the type of construction, component, building material, and dimensions for the production process and the quality of a work or service." Cf. Rösel/Busch (2008), p. 26 (translation by Steffen Walter)

55) Cf. Borrmann et al. (2015), p. 385

3.6.2.7 9D Model

The 9D model to be created should systematically account for the nonlinearities occurring in reality. For this purpose, it will be necessary, for example, to define threshold values for the production factors and their combination in the construction phase. These values describe operating conditions under which normal productivity can be expected. Losses of productivity will occur if the limits to the optimum range are underrun or exceeded. Penalty functions[56)] leading to a realistic reduction in productivity or to an increase in labor consumption rates should be defined in order to account for the resulting nonlinearities.

Binding start, interim, and end dates should also be linked to the BIM model so that the average quantities to be installed per unit of time can be identified at first glance.

Using the trapezoidal model, which establishes the practical relationship between time and resources, the relationships between linearities and nonlinearities in construction management can be mapped and considered in the BIM model. This step serves as a starting point for arriving at conclusions regarding the maximum resource requirements during the main construction phase (see Chapter 1 and Fig. 1.3). It is crucial that the entire construction process be aligned with these maximum resources.

After the ramp-up phase, the maximum number of production factors will have been reached that can be used with normal productivity on the construction site. The greatest potential of production factors is available during the main construction phase; it will then gradually fall back to zero, starting with the onset of the final phase. For this reason, the rectangle (see the rectangle with the length D_{TOT} and height WK_{MAX} in Fig. 3.9) can be used for maximum resource utilization only during the main construction phase. Under no circumstances should it be used for determining the total duration. This would require conversion of the maximum value of the resource potential to the average. This calculation step results in a rectangle with the length D_{TOT} and the height WK_{AVG}.

When calculating the construction time, the crucially important ratio f_{WK} (proportional factor for workers) between the average and maximum number of workers will determine the mean value, which, in turn, determines the construction time. Pure averaging relative to the maximum number of resources would lead to incorrect results because the curve replicating the trapezoidal model would not be considered. The performance would be too high and the calculated construction time would thus be much too short.

If the incorrect method for calculating the construction time is implemented in the BIM model, this will adversely affect the success of the project because the predicted total construction time can be adhered to only at a considerable additional cost. As a result, normal productivity can no longer be achieved, which means that the proportional relationship between performance and utilization of the potential is also lost.

---

56) Example of a penalty function: When calculating the construction time, penalty functions take constraints into account that reduce productivity and increase costs when deviating from the normal construction time (see Chapter 8, Section 8.6.2). This approach ensures that short/long construction times are realistically reflected in the calculations and comparisons.

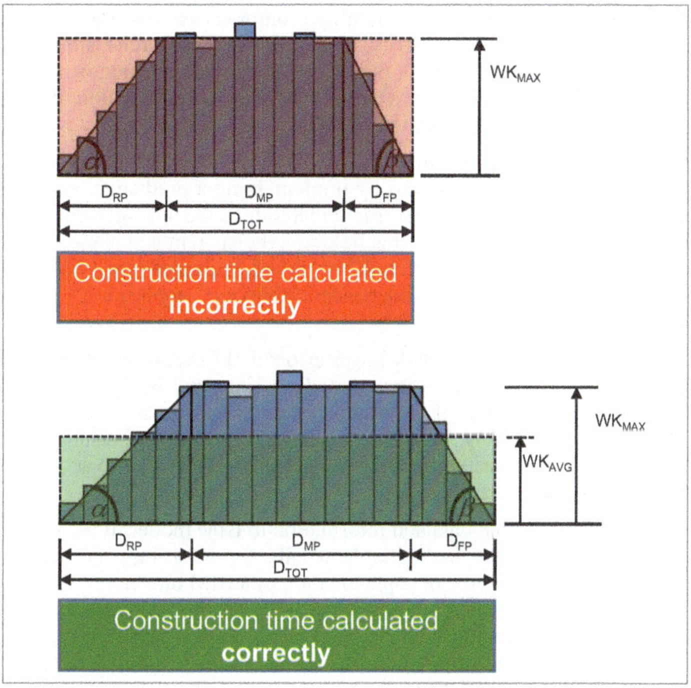

Fig. 3.9 Correct and incorrect application of the trapezoidal model

Any change in the construction time must automatically trigger a change in the number of resources in the BIM model. The resulting number of resources will then have an effect on the related costs.

Potential losses of productivity can be included in the calculations by taking nonlinear relationships into account. Based on this scenario, we obtain a realistic representation of the future possible combinations of production factors (such as in the course of calculating construction time and costs) and their cost effects.

3.7 Summary

Digitization will increasingly allow for generating "floods of data". This data will be of no value if it is not interrelated and is handled in an unstructured, uncontrolled, unreflected, and decontextualized manner. This holds true especially if essential information is lacking that would be required for ensuring the adequate collection, verification, traceability, comparability, and meaningful interpretation of the data. A prime example of this shortcoming are cases where data collection and use are separated due to system boundaries and where the data is to be subsequently used by third parties.

The quality of the information associated with the data, which is also based on the systematic approach to capturing it, is a major criterion for achieving the defined targets or for the accurate and precise communication of facts. Essential quality criteria for information include its correctness, consistency, reliability, completeness, accuracy, currency, non-redundancy, relevance, uniformity, clarity, objectivity, and comprehensibility. The poorer the quality of the data and information, the greater the risk of misjudgment will be. High-quality data and information allow for a better evaluation and utilization of chances and a more accurate assessment and prevention of risks.

There is an urgent need for a systematic approach to dealing with chances and risks. We will be able to pave the road to success by implementing a predictive, smart chance and risk policy that also defines the objectives and methods of the management of chances and risks. This book covers key aspects in this field in a transparent and plausible manner. The theoretical foundations are complemented by worked examples (see Chapters 9 and 10) designed to enable and stimulate independent thinking, application, and continuous improvement.

BIM aims to implement processes with fewer errors and to achieve higher-quality results so that construction projects can be completed on time and on budget. However, this endeavor will be successful only if high-quality data and information are made available to the BIM process interactively. To ensure high-quality simulation results, it is crucial to take interactions and nonlinearities between the production factors into account both in the applied assumptions and in the 3D model.

It is crucial to link the captured data and information to BIM models if we intend to pursue the goal of deriving plausible and reliable results for informing the decision-making process. Irrespective of the degree of sophistication of a BIM model, it will always require high-quality data (with appropriate temporal and operational boundaries) with the associated information, including with respect to emergence and boundary conditions, in order to simulate construction times and costs. Only if these conditions are met will BIM models generate maximum benefit for all project phases and stakeholders, and thus for the entire life cycle of the building or structure. This will set the course towards meeting the demands for resource-saving approaches to design, construction, and operation.

Systems and processes must be developed that allow for the simple and systematic generation, storage, and use of data in order to enhance the quality of systematic documentation and to create a knowledge management system interlinked with it. To this end, extensive research is being conducted at the Institute for Construction Management and Economics at Graz University of Technology in projects carried out jointly with the construction industry.

3.8 References

Akerlof, George (1970). The Market of "Lemons": Quality Uncertainty and the Market Mechanism. In: The Quarterly Journal of Economics. Vol. 84, No. 3 (August 1970). 488-500. MIT Press.

Arrow, Kenneth (1984). The Economics of Agency. Technical Report No. 451. A Report of the Center of Research on Organizational Efficiency. Institute for Mathematical Studies in the Social Sciences, Stanford University. Stanford, California.

Berg, Gerhard (1984). REFA in der Baupraxis – Teil 1 Grundlagen. Ed.: REFA Verband für Arbeitsstudien und Betriebsorganisation e. V. Frankfurt am Main. ztv-Verlag, 1984. (ISBN 3-923597-12-6)

Borrmann, André; König, Markus; Koch, Christian; Beetz, Jakob [Eds.] (2015). Building Information Modeling – Technologische Grundlagen und industrielle Praxis. Wiesbaden. Springer Vieweg. (ISBN 978-3-658-05605-6)

Girmscheid, Gerhard; Motzko, Christoph (2013). Kalkulation, Preisbildung und Controlling in der Bauwirtschaft – Produktionsprozessorientierte Kostenberechnung und Kostensteuerung – 2nd ed. Berlin, Heidelberg. Springer Verlag (ISBN 978-3-642-36637-6)

Gust von Loh, Sonja (2008). Wissensmanagement und Informationsbedarfsanalyse in kleinen und mittleren Unternehmen – Teil 1: Grundlagen des Wissensmanagements. In: Information Wissenschaft & Praxis, 59/(2008)2. 2008. 118-126. Available online at http://www.phil-fak.uni-duesseldorf.de/fileadmin/Redaktion/Institute/Informationswissenschaft/gust_von-loh/1204549266wissensman.pdf. Accessed December 19, 2020.

Haberfellner, Rainhard et al. (2002). Systems Engineering – Methodik und Praxis – 11th ed. Eds.: Daenzer, Walter; Huber, Fritz. Zurich. Verlag Industrielle Organisation. (ISBN 3-85743-998-X)

Hagendorf, Herbert; Krummenacher, Joseph; Müller, Hermann-Josef; Schubert, Torsten (2011). Wahrnehmung und Aufmerksamkeit. Berlin, Heidelberg: Springer Verlag. (ISBN-13: 978-3642127090)

Heck, Detlef; Schubert, Eberhard (2011). Der adäquat-kausale Nachweis von Bauablaufstörungen. In: Bauablaufstörungen – Baubetriebliche, bauwirtschaftliche und rechtliche Aspekte, 9. Grazer Baubetriebs- und Bauwirtschaftssymposium. Eds.: Heck, Detlef; Hofstadler, Christian; Lechner, Hans. Graz. Verlag der Technischen Universität Graz. (ISBN 978-3-85125-136-4)

Hofstadler, Christian (2014). Produktivität im Baubetrieb – Bauablaufstörungen und Produktivitätsverluste. Berlin, Heidelberg. Springer Verlag (ISBN 978-3-642-41632-3)

Hofstadler, Christian; Kummer, Markus (2016). Der Preis einer fast lückenlosen Dokumentation. In: Tagungsband – 14. Grazer Baubetriebs- und Bauwirtschaftssymposium – Belastbare Dokumentation in der Bauausführung – Baubetriebliche, bauwirtschaftliche und rechtliche Aspekte. Eds.: Heck, Detlef; Hofstadler, Christian; Kummer, Markus. 91-138. Graz. Verlag der Technischen Universität Graz. (ISBN 978-3-85125-442-6)

Hofstadler, Christian; Kummer, Markus (2017a). Chancen- und Risikomanagement in der Bauwirtschaft – Für Auftraggeber und Auftragnehmer in Projektmanagement, Baubetrieb und Bauwirtschaft. Berlin, Heidelberg. Springer-Verlag. (ISBN 978-3-662-54318-4)

Hofstadler, Christian; Ninaus, Cornelia (2019). Wissen als grundlegender Produktionsfaktor in der Bauwirtschaft. In: Aktuelle Entwicklungen in Baubetrieb, Bauwirtschaft und Bauvertragsrecht – 50 Jahre Institut für Baubetrieb und Bauwirtschaft der TU Graz. Ed.: Hofstadler, Christian. 887-908. Wiesbaden. Springer Vieweg (ISBN 978-3-658-27430-6)

Hofstadler, Christian; Ninaus, Cornelia (2020). Baubetrieb FS. Lecture slides. Graz. Institute of Construction Management and Economics, Graz University of Technology.

http://www.computerwoche.de/a/gute-daten-schlechte-daten,1931857. Accessed December 19, 2020

Hunt, Wolfgang (2014). Digitalisierung im Bauwesen: Thema für Baupraxis, Lehre und Forschung. In: Tagungsband – 1. Grazer BIM-Tagung – BIM – Werkzeug zur Optimierung der Planungs- und Bauprozesse. Eds.: Heck, Detlef; Fellendorf, Martin. 9-21. Graz. Verlag der Technischen Universität Graz. (ISBN 978-3-85125-369-6)

Kummer, Markus (2015a). Aggregierte Berücksichtigung von Produktivitätsverlusten bei der Ermittlung von Baukosten und Bauzeiten – Deterministische und probabilistische Betrachtungen. Doctoral thesis. Graz. Institute of Construction Management and Economics, Graz University of Technology.

Kiel, Ewald; Rost, Friedrich (2002). Einführung in die Wissensorganisation: Grundlegende Probleme und Begriffe. Würzburg. Ergon Verlag. (ISBN 3-89913-246-7)

Künstner, Gerhard (1984). REFA in der Baupraxis – Teil 2 Datenermittlung. Ed.: REFA Verband für Arbeitsstudien und Betriebsorganisation e. V. Frankfurt am Main. ztv-Verlag, 1984. (ISBN 3-923597-13-4)

Nonaka, Ikujiro; Takeuchi, Hirotaka (1997). Die Organisation des Wissens – Wie japanische Unternehmen eine brachliegende Ressource nutzbar machen. Frankfurt am Main. Campus Verlag GmbH. (ISBN 3-593-35643-0)

Mittelstraß, Jürgen (1989): Glanz und Elend in den Geisteswissenschaften, Oldenburg: Bibliotheks- und Informationssystem der Univ. Oldenburg.

Moosbrugger, Helfried; Kelava, Augustin [Eds.] (2012).Testtheorie und Fragebogenkonstruktion – 2nd ed. Berlin, Heidelberg. Springer Verlag (ISBN 978-3-642-20071-7)

North, Klaus (2011). Wissensorientierte Unternehmensführung – Wertschöpfung durch Wissen – 5th ed. Wiesbaden. Gabler Verlag | Springer Fachmedien. (ISBN 978-3-8349-2538-1)

OENORM A 2063 (2009-06-03 edition) Austausch von Leistungsbeschreibungs-, Ausschreibungs-, Angebots-, Auftrags- und Abrechnungsdaten in elektronischer Form.

OENORM A 6241-1 (2015-07-01 edition) Digitale Bauwerksdokumentation – Teil 1: CAD-Datenstrukturen und Building Information Modeling (BIM) – Level 2.

OENORM A 6241-2 (2015-07-01 edition) Digitale Bauwerksdokumentation – Teil 2: Building Information Modeling (BIM) – Level 3-iBIM.

Probst, Gilbert; Raub, Steffen; Romhardt, Kai (2012). Wissen managen – Wie Unternehmen ihre wertvollste Ressource optimal nutzen – 7th ed. Wiesbaden. Springer Gabler. (ISBN 978-3-8349-4562-4)

Rathswohl, Stefan (2014). Entwicklung eines Modells zur Implementierung eines Wissensmanagement-Systems in kleinen und mittleren Bauunternehmen. Schriftenreihe Bauwirtschaft – I Forschung 28. Ed.: Institut für Bauwirtschaft der Universität Kassel. Doctoral thesis. Kassel. Kassel University Press GmbH. (ISBN 978-3-86219-824-5)

Romhardt, Kai (2001). Organisation aus der Wissensperspektive – Möglichkeiten und Grenzen der Intervention. Doctoral thesis. Université de Genève. Available online at http://www.romhardt.com/tl_files/romhardt/Artikel%20und%20Ressourcen/DissKRomhardt.pdf. Accessed December 19, 2020.

Rösel, Wolfgang; Busch, Antonius (2008). AVA-Handbuch – Ausschreibung – Vergabe – Abrechnung – 6th ed. Wiesbaden. Springer Gabler. (ISBN 978-3-8348-0348-1)

Schreyer, Marcus; Schütz, Maximilian; Hauber, Thomas (2013). BIM in der Bauausführung – mit virtuellen Modellen in der Realisierung Prozesse optimieren und Risiken managen. In: BIM – Building Information Modeling – Ernst & Sohn Special – November 2013 – A61029. Berlin. Ernst & Sohn – Verlag für Architektur und technische Wissenschaften GmbH & Co. KG.

Springer Gabler Verlag (Ed.), Gabler Wirtschaftslexikon, Keyword: Adverse selection. Available online at http://wirtschaftslexikon.gabler.de/Definition/adverse-selection.html. Accessed December 19, 2020.

Stock, Wolfgang (2007). Information Retrieval – Informationen suchen und finden. Munich. Oldenbourg Wissenschaftsverlag GmbH. (ISBN 978-3-486-58172-0)

Wang, Richard; Strong, Diane (1996). Beyond Accuracy: What Data Quality Means to Data Consumers. In: Journal of Management Information Systems, Vol. 12, No. 4 (Spring, 1996). 5-33. M.E. Sharpe, Inc. Available online at http://mitiq.mit.edu/Documents/Publications/TDQMpub/14_Beyond_Accuracy.pdf. Accessed December 19, 2020.

Wehling, Peter (2001). Jenseits des Wissens? Wissenschaftliches Nichtwissen aus soziologischer Perspektive. In: Zeitschrift für Soziologie, Jg. 30, Heft 6, 465-484

4 Management of Chances and Risks

Systematically implementing chance and risk management establishes the basis for conducting a comparative analysis of chances and risks (historical, project-related, etc.) within the organization and for optimally utilizing the existing potential on the basis of this knowledge aggregation. Its primary aim is to identify potential chances and risks early on in order to derive appropriate measures to increase/utilize them (chances) or to reduce/avoid/mitigate them (risks). We recommend applying the chance and risk management process described in detail in this chapter in order to ensure that the discovery and handling of chances and risks is not random and unsystematic. The size, strategic alignment, and philosophy of an organization influence whether and to what extent the management of chances and risks is utilized in terms of its intensity and consistency.

In the construction industry, the greatest chances and risks involve the development of the overall volume of construction contracts and of the construction market, as well as the correlating construction prices. On the one hand, the client will make the decision if and when to build only after a thorough analysis of these factors. On the other hand, bidders rely on the current chance/risk ratio for determining if participating in a competition for a construction contract can at all be considered to be commercially viable. The current order backlog determines the extent to which the costs derived from zero-base costing will be underrun or exceeded in the course of preparing the bid. Any major decision will inevitably require profound knowledge of the respective chance/risk ratio as a prerequisite for ensuring the long-term commercial success of the company.

The management of chances and risks is primarily geared towards future developments. The chances and risks for contract wins, construction, and financing are examined from a company perspective. On the project level, the focus is on how the chances and risks involved in selecting and utilizing the production factors can be assessed for future projects, and what productivity and price levels need to be applied in order to achieve the defined targets. In this context, it is essential to determine and evaluate the specific ratio between chances and risks prior to making important decisions that will have an impact on either the entire organization or on the specific project. Adopting a holistic and networked view, any business should always be aware of the risk level tolerable in the period under consideration. The project portfolio should be structured so as to ensure a balanced ratio between chances and risks, at least in the longer term. The benefit from chances will

C. Hofstadler and M. Kummer, *Chances and Risks in Construction Management and Economics*, Springer Tracts in Civil Engineering,
https://doi.org/10.1007/978-3-030-64014-9_4

always be limited, but the associated risks can result in deficits leading up to the insolvency of the company. The optimized application of the chance and risk management system is a dynamic process that should be adapted to current developments as quickly as possible.

In recent years, the importance of a systematically implemented chance/risk management system has increased significantly. Depending on the type of organization (corporate governance), applying such a management tool can also be defined as a mandatory requirement. However, the management of chances and risks should not only be pursued out of obligation. Rather, it should be in the best interest of the stakeholders involved in the construction process because it is not only an essential factor for the continued existence of a company, but it also makes a valuable contribution to its success and growth. Evaluating chances and risks in a structured and systematic manner according to the control loop principle plays a major role in achieving defined corporate goals. Besides the strategic, operational, and financial targets of the business, health and safety as well as environmental requirements must also be met. Even more importantly, aspects of sustainability should be incorporated in the target definition, such as a resource-saving design-build-operate approach. In construction projects, sustainability aspects must already be integrated in the early stages of the project because this is when far-reaching decisions need to be made regarding the structural design of the building as well as its fit-out and finish and mechanical services.

Recognizing and quickly seizing opportunities and identifying and counteracting risks is of crucial importance for any organization or company. For this purpose, installing an early-warning system covering the entire organization as well as individual projects will be beneficial. A basis for the relevant consideration needs to be defined for a quantitative analysis in order to clearly define the threshold from which a chance or a risk will occur. Analyses should also quantify the implications, or consequences, expected if a chance or risk materializes, such as exceeding or underrunning costs or extending or shortening the construction period.

The management of chances and risks either refers to the entire company (global view) or to individual projects (local view). The global analysis is concerned with the strategic focus since it assesses the project portfolio. In contrast, the local view is centered on analyzing the operational level of the business. A systematic approach to dealing with chances and risks should be promoted and demanded across the entire organization. In addition, senior corporate management should lead by example by ensuring the seamless application of the chance and risk management system.

4.1 Management of Chances and Risks in Accordance with ISO 31000

The application of chance and risk management extends to private, public, and charitable organizations as well as to systems (i.e. technical systems, processes, and projects). As explained in Section 2.1, the concept of risk is by no means clearly defined. There are generally two different definitions of the term. On the one hand, it is associated with a purely negative target deviation whereas a positive target deviation is referred to as a chance. On the other hand, risk is considered to be an umbrella term encompassing both hazard and chance. This second definition includes positive as well as negative consequences.

These terminological discrepancies are also reflected in the conceptual characteristics of the systems used for managing chances and risks: Chances and risks are either treated equally, or, in other systems, the concept of risk is used as an umbrella term to refer to both positive and negative deviations from the defined targets. On a practical level, such systems predominantly focus on the hazard or danger components while treating chances only marginally, if at all. This book adopts a view that treats chances and risks equally, thus intending to add value through the proactive handling of chances. In many companies, the management of chances is often part of strategic management, corporate and product development, marketing, and the continuous improvement process. The management of chances usually follows a less systematic pattern than the risk management system as such. However, in order to make the best possible use of potential chances, we recommend that the management of chances also be carried out in a systematic and targeted manner and that the interaction of corporate goals, chances, and risks be taken into account.

The following part outlines the principles of (chance)/risk management in accordance with the ISO 31000 standard.

The focus of this standard is on the management of risks. Chances are only mentioned in passing. As part of the description of the risk control system, the standard stipulates that taking and/or increasing a risk serves as a means to seize a chance. In the context of identifying risks, it is important to also highlight those risks that result from missed chances.

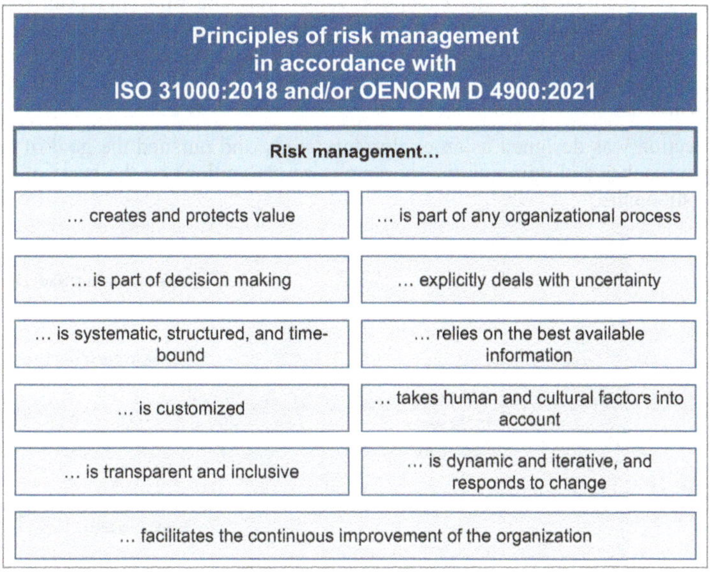

Fig. 4.1 Principles of risk management[1)]

The ISO 31000 standard says that the primary task of risk management is to identify and manage risks so as to achieve the defined project targets. Fig. 4.1 illustrates the principles of risk management.[2)]

[1)] Based on OENORM ISO 31000:2018 and OENORM D 4900:2021 (national implementation of ISO 31000 in Austria).

ISO 31000 was developed for organizations in general and thus does not refer to a specific field or industry. It can be applied to the entire life cycle of an organization, but its practical implementation needs to be adjusted to the respective requirements and characteristics (in relation to goals, structure, processes, functions, products, projects, etc.). Although this standard includes general guidelines, it does not intend to standardize risk management.[3] It also contains a definition of the activity-related aspects of the risk management process as well as a graphical representation of its iterative structure.

4.2 Situational Analysis with Respect to Chance and Risk Behavior

Below we present some of the results of a survey on the chance and risk behavior in construction companies. We also refer to the chance and risk behavior relative to competitive intensity as well as to the quality of the planning and specification process.

4.2.1 Chance and Risk Behavior in Construction Companies

Specific studies on the current approaches to chance and risk management in construction companies have so far been the exception rather than the rule, and even more so in German-speaking countries. The currency of this topic and the limited availability of empirical data gave rise to an expert survey entitled "Risk Management in Construction Companies" that was conducted at Graz University of Technology in 2013 (authored by *Alber/Hofstadler*).

Data collection was designed as an exploratory study and pursued the goal of collecting current approaches and opinions (i.e. best practice perception) on the basis of a standardized questionnaire.

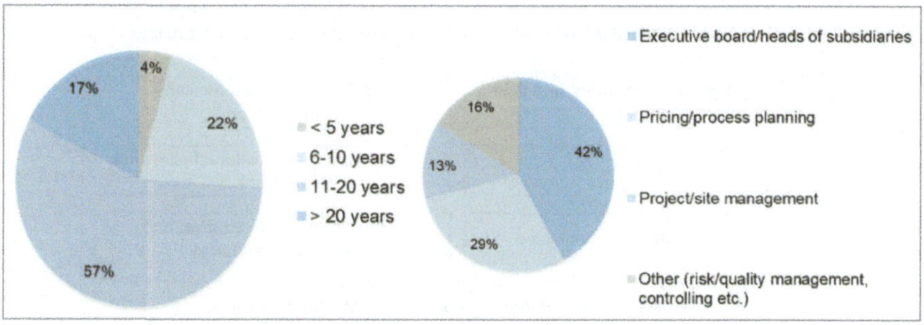

Fig. 4.2 Professional experience and responsibilities of experts

The statistically analyzed primary data includes information from 31 personally contacted respondents working in different positions at medium-sized and large construction contractors. Fig. 4.2 shows the sample of the interviewed experts, grouped according to

2) Cf. Wiggert (2009), p. 118

3) Cf. OENORM ISO 31000:2010, p. 6

years of professional experience as well as tasks and responsibilities in their current roles. The survey's exploratory nature was due to the relative scarcity of empirical research into this topic in the construction industry as well as to the sampling based on different predefined criteria (more specifically, professional experience, expertise in the management of chances and risks, and industry affiliation). The composition of the sample was also based on the necessary knowledge and expertise in the respective fields.

The experts were asked to adopt the contractor's perspective and to indicate, in this role, how they assess the potential for given external and internal risk fields. They were also given the option of evaluating the relationship between the project phase and the error proneness.

4.2.1.1 Relationship between Risk Potential and Risk Fields

27 of the 31 interviewed experts assessed the risk potential of external and internal risk fields. The related scale included the option of selecting values ranging from -1 (very low) to 1 (very high), where 0 stood for risk-neutral.

The respondents identified the price battle and industry competition as the main overarching risk field and rated it high on average. They also identified a high risk potential for the costing, pricing, and bid preparation as well as for contractual arrangements and liability. Besides these aspects related to construction economics, technical and construction management risk fields, such as construction process planning and logistics, also fall into the high-risk category. According to the experts, IT is the only low-risk field for the construction industry (with a negative average value). Some of the respondents also mentioned country risk as an additional category, which is particularly relevant for groups of companies doing business on an international scale. In lesser-known markets, these country risks are difficult to assess and cover almost all of the listed risk fields.

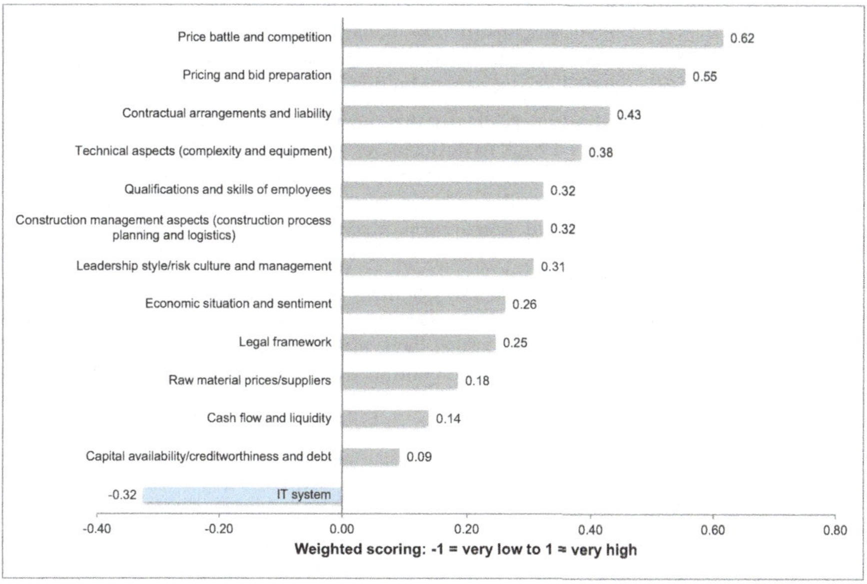

Fig. 4.3 Risk potential of risk fields from the contractor's perspective[4)]

This question aimed to establish a hierarchical ranking of the most important risk fields with regard to the experts' risk assessments (see Fig. 4.3). In practice, however, it is difficult to make a generally valid distinction because dependencies may exist between the areas mentioned. Thus, it appears plausible that the project- and phase-independent price battle and competition risk field is associated with the highest risk potential, and will have repercussions on all other fields. In terms of phases, the experts identified the highest risks prior to the construction process in the bid preparation and contract management fields. The more extensive and complex a risk field is, the higher the risk potential assigned by the respondents will be, which renders the sequence of risk areas plausible, some of which can only be differentiated to a limited extent.[5)] Costing and pricing, for instance, builds upon the forecast of technical and construction management as well as contractual aspects. The risk potential resulting from the qualifications and skills of employees is also categorized as high. A possible relationship stems from the illustrative fact that the appropriate qualification and training of the workforce is a substantial cost item while at the same time forming an elementary basis of functioning business processes. A cross-industry study conducted in 2009 found that a shortage of skilled workers in the construction industry was one of the main risks for 76% of the companies surveyed.[6)] Compared to the financial services sector and the manufacturing industry, IT systems for construction companies are associated with relatively low risks. The risk fields that the interviewees assessed and grouped into a realistic ranking indicate the areas on which the focus should or could be set within the risk management system implemented by construction companies.

4.2.1.2 Phase Dependency of Risk Materialization

Survey participants were also requested to produce a phase-related assessment of when the highest number of errors is committed in the course of the project. Fig. 4.4 shows the phase dependency of risk materialization from the respondents' perspective. The 31 experts were asked the specific question in which project phase the frequency of errors subsequently causing problems in construction management was the highest. Each of the experts had the opportunity to select three answers from the specified range of options. Of the 31 respondents, 30 added their assessment on this question.

The analysis of the survey shows very clearly that most of the problems (or cases in which risks materialized) can be traced back to deficiencies in early project phases. Experts saw the greatest error potential in the design phase followed by the specification and bid phase, which are usually attributable to the client's scope. A similarly high potential was indicated for the pricing and bid preparation phase, which is obviously part of the contractor's responsibilities.

Errors in the contractual negotiation and contract award stages were identified as a frequent cause of subsequent difficulties. Besides bid preparation, the quality of the contract management that many companies have implemented to a highly professional standard also plays a crucial role. After the contract award decision, the risk potential for

[4)] Cf. Alber (2013), p. 131

[5)] The exception to this rule is the comprehensive and complex risk field of economic situation and sentiment, which tends to be located in the lower third of the resulting hierarchy. Again, this appears to apply independently of specific projects. However, a relatively stable situation has evolved over the past few decades, particularly in the western industrialized countries, which might be the reason for this low risk classification. As can be concluded from the COVID-19 crisis, however, some of these developments have a massive impact on the economy, and should thus not be neglected. In this case, we can thus assume very severe consequences of the risk, but the risk frequency is rather on the low side.

[6)] Cf. Mesaros/Vanselow/Weinkopf (2009), p. 17

the contractor increases again in the process planning phase because important decisions need to be made under great time pressure with regard to selecting the construction method(s) as well as for construction process planning, logistics, and site mobilization and equipment.

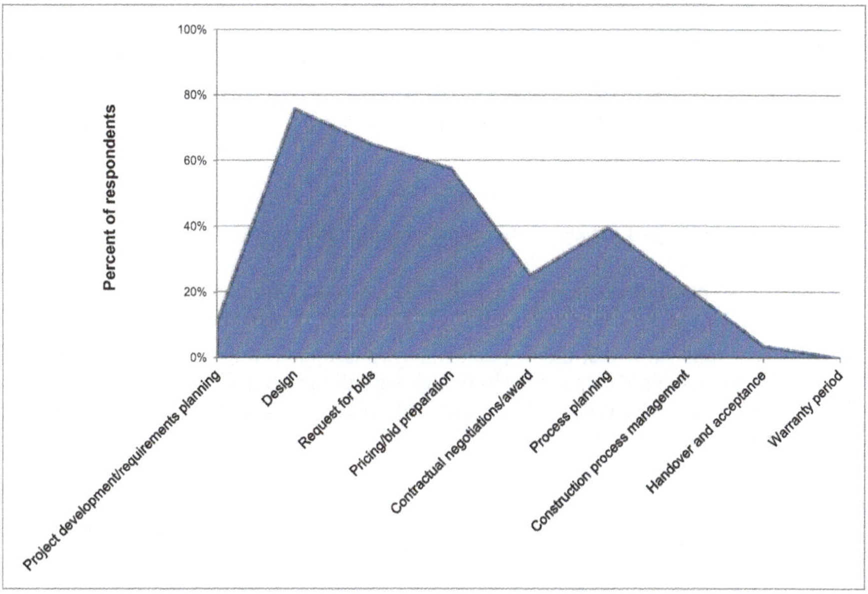

Fig. 4.4 Project phases as sources of error and risk[7)]

In addition to the predefined response categories, the interviewees also mentioned the field of human resource development as a major source of problems and risks. This again provides ample proof of the high significance of human resource risks within the hierarchical assessment of the risk fields.

4.2.1.3 Risk Affinity

In the survey, we were also interested in whether or not the company's readiness to take risks would differ from that of their employees. The following response options were available for assessing one's own as well as the company's risk affinity:

- Risk-averse
- Rather risk-averse
- Neutral
- Rather risk-taking
- Risk-taking

Fig. 4.5 shows the result of the analysis of 24 expert responses. Overall, 11 respondents (equivalent to 46%) revealed no difference between their own risk affinity and that of the

7) Cf. Alber/Hofstadler (2014), p. 132f.

company. Interestingly, more than half of them (54%) did indicate such a difference. In seven companies (approx. 29%), the risk affinity was higher at the corporate level compared to that of its employees. One quarter of the experts (6 answers) stated a higher risk affinity compared to that of their companies.

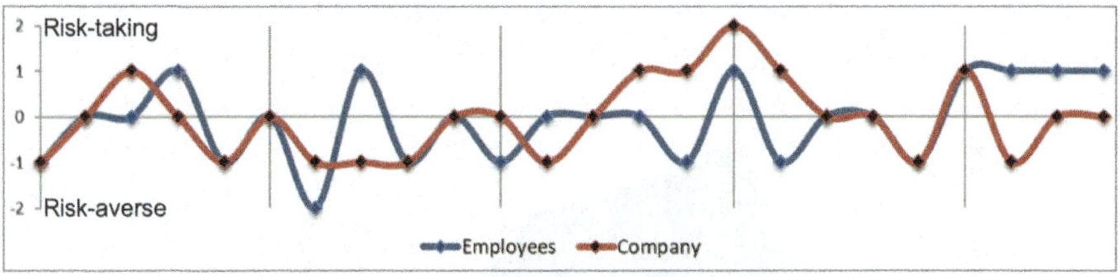

Fig. 4.5 Risk affinity – Difference between employees and companies

In contrast, two respondents (approximately 8%) indicated a more pronounced risk-taking attitude compared to the risk-averse policy adopted by their companies.

Most of the results show that the respondents' own attitude towards risks is perceived as deviating from the company's risk affinity and way of managing risks. This conclusion is also an indication of the fact that there are often no uniform guidelines for risk management (i.e. the risk policy of the company), which can be assumed to be the cause of the different risk behavior.

4.2.2 Chance and Risk Behavior and Competitive Intensity

Bid prices are calculated using a zero-base costing[8)] exercise to allow for a sound assessment of the chances and risks with respect to bid pricing and final contract pricing. The subsequent calculation of prices relies on the assumption that all production factors are utilized normally. In these cases, losses of productivity need not be considered *ex ante* because normal productivity can be achieved in the construction process. This deterministically calculated value represents the most likely value stated by the expert, for instance by the bidder's cost surveyor. For a normal construction time, this price is combined with strategic considerations, if required, and leads directly to the final bid price, which must be an exact figure when the contract is awarded. It is necessary to compute a histogram of the price distribution of the zero-base costing exercise in order to derive the corresponding chance/risk ratio. For this purpose, in the case of a unit-price contract, distribution functions should be assigned to the derived prices wherever uncertainties are suspected for calculating the unit prices so that the histogram of the price distribution can be determined, for instance by means of a Monte Carlo simulation.

For the continuous distribution in Fig. 4.6, the chance/risk ratio for calculated prices is shown on the horizontal axis, whereas the related probability density is plotted on the vertical axis. Fig. 4.6 illustrates the prevailing competitive intensity by a (randomly selected) dashed line with a gradient of about 20° (Scenario 1). Horizontal shifting of the straight line accounts for the existing competitive intensity (termed "Competitive intensity 1"). The fiercer the competition (many bidders, but only a few requests for proposals), the more the straight line has to be shifted toward the left, with the result that it is increasingly

8) Zero-base costing: "Determination of prices that cover all the costs, excluding speculative elements, of construction operations to be performed at a later stage." Oberndorfer/Jodl et al. (2001), p. 36 (translation by Steffen Walter)

moving into the area of commercially unviable prices[9)] (i.e. below the minimum acceptable price level). At the same time, the risk potential increases, depicted by the arrow pointing to the right that runs parallel to the abscissa. There is thus a decreasing probability that a chance or opportunity can be grasped. It will then be highly improbable that the costs on which the agreed prices are based will be adhered to.

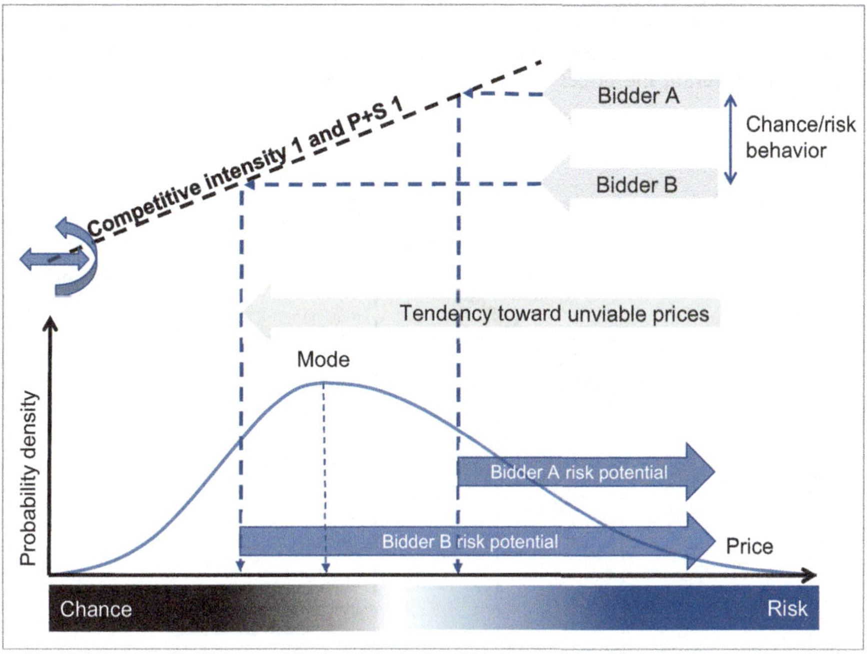

Fig. 4.6 Chance and risk potentials depending on the competitive situation (mainly influenced by competitive intensity and quality of planning/specification) – Scenario 1

The slope of the straight line can be equated with a forecast with respect to the essential boundary conditions to costing/pricing associated with the specific construction project (planning and specification quality = P+S 1). The greater the accuracy of the calculated prices (as a result of fewer uncertainties in quantities and costing assumptions), the steeper the slope of this straight line will be, corresponding to a comparatively tight spread between bid prices. This effect will also materialize if two bidders are farther apart from each other in terms of their chance/risk behavior (the vertical distance between Bidders A and B that represents the chance/risk behavior remains constant). In this diagram, Bidder B is clearly more willing to take risks than Bidder A. This can be explained, for example, by the fact that Bidder B urgently needs to win a contract in order to achieve the planned annual turnover, or by its generally prevailing risk-taking attitude. Another reason could be that Bidder B wants to prevent Bidder A from entering this market segment by all means.

9) The tendency toward unviable prices becomes more pronounced as the bidder moves, in the course of its bid pricing, from full cost coverage to partial cost coverage compared to zero-base costing.

Fig. 4.7 shows another scenario derived from Fig. 4.6 that involves both competitive intensity and the quality of the planning and specification process. This diagram includes an unchanged chance/risk behavior of the two bidders. The straight line was shifted further to the right due to a more favorable competitive situation for the bidders (labeled "Competitive intensity 2 and P+S 2"). This setting represents a lower competitive intensity, which is why bid prices should tend to be higher. Furthermore, the slope of the straight line was increased to consider a higher quality of planning and specification (planning and specification quality = P+S 2) compared to the previous example in Fig. 4.6. This shift reduces the probability of planning and specification errors and thus diminishes the potential for speculation. Overall, the prices quoted by Bidder A and Bidder B are located closer to each other.

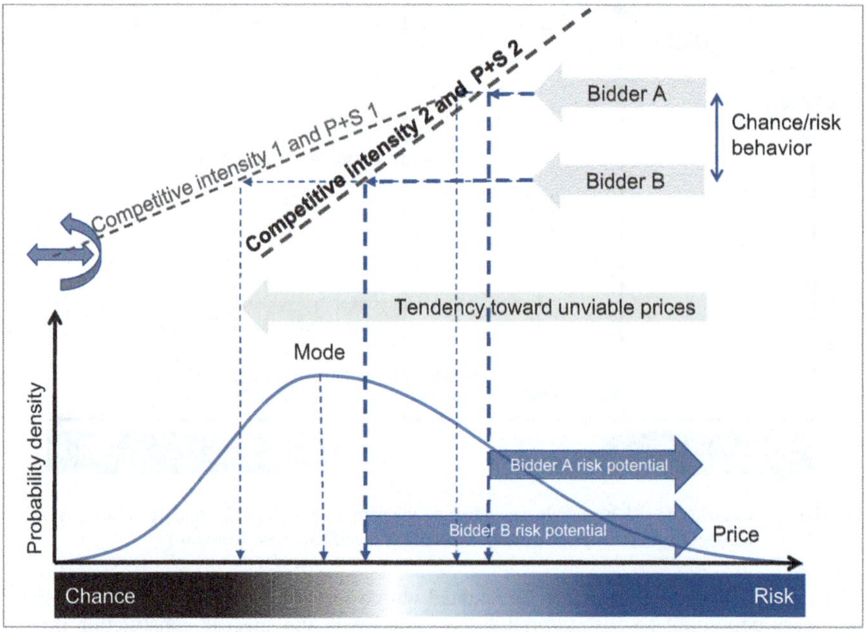

Fig. 4.7 Chance and risk potentials depending on the competitive situation (mainly influenced by competitive intensity and quality of planning/specification) – Scenario 2

Bidders are thus taking lower risks and are more likely to follow the assumption that they will adhere to the calculated construction costs during the construction phase. The subsequent contractor thus enjoys a greater chance potential. For the client, this scenario brings the advantage of a high probability of the building or structure to be completed within the scheduled timeframe while complying with the agreed targets in terms of quality and budget.

Scenario 3 highlights the role that the construction time plays in relation to the chance/risk ratio (see Fig. 4.8). For this purpose, another straight line is incorporated in the diagram that is shifted to the right, running parallel to the previously introduced straight line labeled "Competitive intensity 2 and P+S 2". This new line is labeled "Planning and specification quality = P+S 2 and VSCT". All other relationships remain unchanged compared to Scenarios 1 and 2.

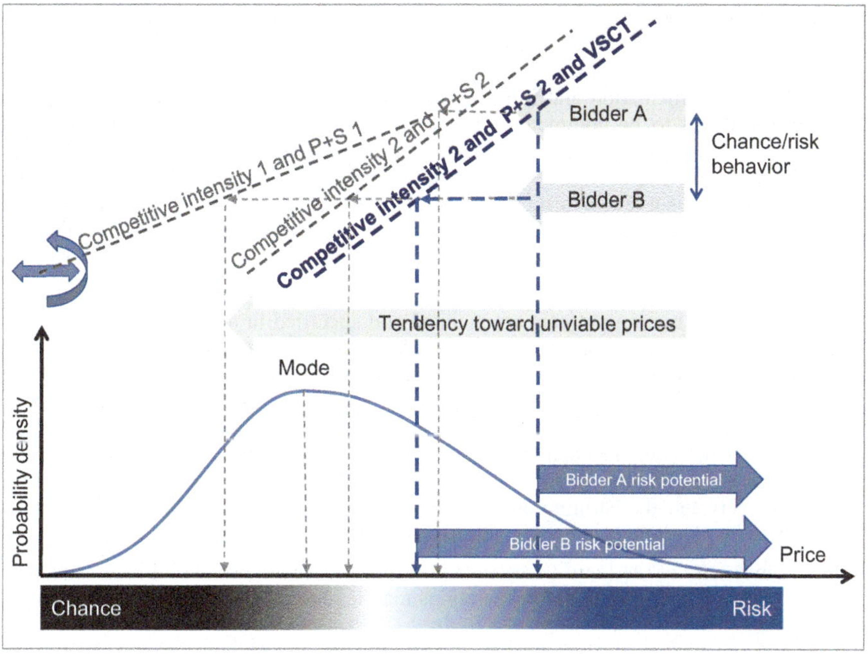

Fig. 4.8 Chance and risk potentials depending on the competitive situation (mainly influenced by competitive intensity and quality of planning/specification as well as construction time) – Scenario 3

VSCT stands for a very short construction time specified for the project. In this case, the "normal" construction time[10)] is underrun, and losses of productivity occur in the construction phase because more production factors have to be utilized than would be economically justifiable from the perspective of the contractor. More specifically, exceedingly short construction times disrupt the appropriate interaction of the elementary production factors. This effect is amplified in line with the increase in resources utilized beyond the normal level of productivity. This effect is also referred to as economies of scale and causes even more pronounced losses of productivity.

If all bidders are aware of the relationship between underrunning the normal construction time and the resulting losses of productivity, this will lead to (estimated) additional costs of the relevant works or services. Consequently, the prices quoted by both bidders will increase in order to reduce the correspondingly higher risk the individual bidder is exposed to. If one of the two bidders were unaware of these dependencies, there would be no shift of the straight line to the right. This bidder/contractor would thus be exposed to an increased risk owing to the specified very short construction time if the contract were awarded to it.

There are limits to changing the slope of the straight line. Slopes of 0 and 90° are of a purely theoretical nature since they cannot occur in actual construction practice. A vertical straight line would correspond to a fully planned, seamlessly specified, contradiction-free

[10)] Chapter 8 extensively deals with the concept of normal construction time.

construction project. In addition, the parties to the contract would have to be in a position to access complete and accurate information[11)], which is virtually impossible in the case of contracts for work in particular. There is thus only a theoretical possibility for a vertical straight line because the bidders would quote different bid prices for this scenario as well, although the documentation and information provided for costing and pricing purposes can be considered to be complete and free from contradictions. This phenomenon is mainly due to the unique nature of construction projects as well as to subjective differences in the assessment of costs in relation to production factors (internal production system). Yet cost surveyors also arrive at different assessments with respect to the complexity and circumstances under which work is to be performed (external production system). This phenomenon results from a different handling of information as well as from the varying chance/risk affinity.

Cases where construction projects are planned and specified in an entirely inappropriate manner do not (or hardly) occur in practice. There are binding standards and guidelines defining a minimum quality standard to be adhered to in the planning and specification phase. This means that it is also impossible to arrive at a horizontal straight line. However, an exceedingly poor quality of planning and specification will result in only a small slope of the straight line, and thus lead to a wider bid price spread. This phenomenon is mainly caused by an exceedingly large room for interpretation that may potentially result in major differences between the assumptions applied by the bidders. The entire system is thus characterized by a pronounced information asymmetry.

Even if the results of the (zero-base) costing exercises performed by all bidders are very similar, strategic or speculative considerations may lead to huge differences in the quoted bid prices.

Although Figs. 4.6 to 4.8 include only an idealized scenario of the key mechanisms between competition, planning and specification accuracy, prices, construction time, and chances/risks, they provide a very good overview of the economic relationships and constraints prevailing in construction management.

4.3 Chance and Risk Policy

Any organization will derive its chance and risk policy from its vision, values, and mission. This policy represents the standards, attitudes, and decision-making policies of an organization in dealing with chances and risks, even if not explicitly documented. Fig. 4.9 shows that this internal organizational alignment creates the basis for implementing the chance and risk management system.

[11)] One of the key assumptions of neoclassical economic models is that of complete information, which means that the parties involved are aware of all states and conditions in their environment and are in a position to observe the activities of the parties to their contracts. Information is available free of charge, contracts are complete, and contractual performance can be observed and fully enforced at no cost. Despite this unrealistic assumption, the use of neoclassical models can lead to solid results in some areas.

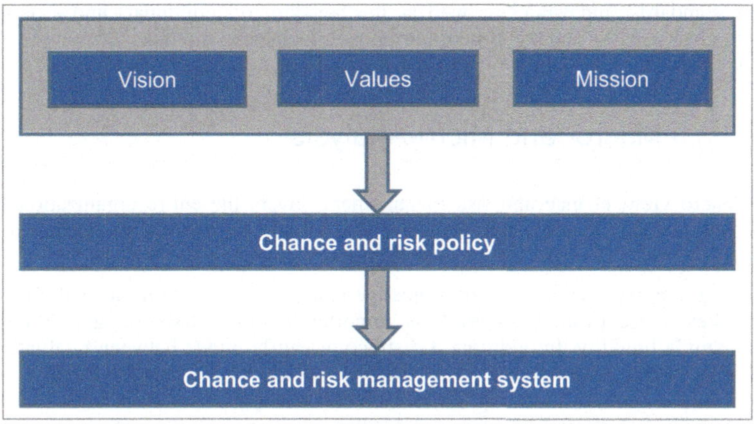

Fig. 4.9 Significance of the chance and risk policy[12)]

In an ideal scenario, the top management of the organization will ensure that its chance and risk policy

- is appropriate for the purpose of the organization
- encompasses, as a minimum, those chances/risks that are significant for ensuring the organization's continued existence or target achievement
- considers commercial, technical, ecological, and social aspects within the organization's environment
- indicates success drivers or threats to potential success
- takes legal and regulatory requirements into account
- identifies the chances/risks that the organization itself is willing to take
- ensures the mandatory implementation of the chance and risk management system across the entire sphere of responsibility
- is continuously reviewed and, if and when required, revised to ensure its effectiveness and continued appropriateness[13)]

The risk policy (risk affinity) will be determined prior to identifying and assessing the risks. At this stage, fundamental strategic assumptions and knock-out criteria (or "go" and "no-go" criteria) are elaborated either for the entire company or for a specific project. This approach makes it possible to cancel a project even before its start if it fails to meet the requirements and thus represents an intolerable risk for the company. Risk policy is usually an executive management responsibility, but it should be set forth in writing for all parties involved in order to raise awareness and to establish a risk culture within the company.

The notion of chance and risk policy generally refers to the guidelines issued by senior management on the principles, implementation, integration, and communication of chance and risk management. In this context, the chance and risk policy will be driven by the strategies and targets of the organization. In companies predominantly doing project business, these strategies and targets will have to be defined at the corporate level, and project managers should interpret and specify them for the respective project on the basis of the company-wide chance and risk policy.

[12)] Cf. OENORM S 2410:2010, p. 10

[13)] Cf. OENORM S 2410:2010, p. 11

The chance and risk culture (corporate culture) describes attitudes (basic convictions and beliefs), standards, and values as well as the behavior of individuals and groups with regard to chances/risks and the system for managing chances and risks.[14)]

4.4 Macro- and Micro-Analysis

In the macro view, chance and risk management covers the entire organization; in the micro view, it refers to individual projects. The organization should implement a systematic project portfolio management system between these two levels.

In so doing, the overarching primary objective is to secure long-term profitability, or the continued existence, of the business. Most importantly, this includes the goal of retaining the company's liquidity. In addition, a distinction can be made between qualitative and quantitative target requirements. The former include, for example, quality, market position, and communication; the latter include measurable variables such as returns, costs, a low ratio of warranty claims, and deadlines.

Successful implementation of chance and risk management systems in the construction sector is the key to achieving the following targets and objectives:

- Proactive rather than reactive corporate governance, exerting an active influence on and consistently identifying chances and risks
- Awareness raising within the organization in conjunction with clearly assigning tasks and responsibilities
- Compliance with external requirements, laws and regulations, and requirements arising from codes and standards
- Building trust by clearly indicating the specific risk exposure
- Traceable sharing of risks (such as in a consortium of bidders/contractors) and influence on rating processes
- Creating bases for decision making and the risk-based selection of projects with a suitable risk/return profile
- Improving operational performance capacity and planning reliability by reducing the volatility of profit and cash flow

What these goals have in common is the ultimately intended "output" in the form of a specific, long-term commercial benefit arising from the system of managing chances and risks.[15)]

---

14) Cf. Spang/Gerhard (2016), p. 425f.

15) Cf. Alber (2013), p. 16f.

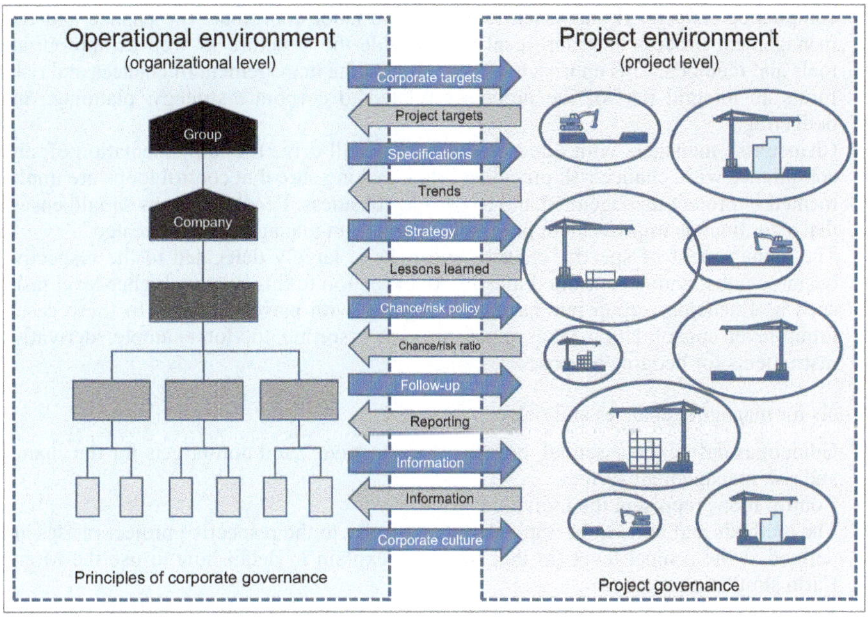

Fig. 4.10 Chance and risk management on different levels – Interaction between operational and project environment

Fig. 4.10 includes an example to illustrate the interaction between the operational and the project environment. In the construction industry, operating results are achieved on the basis of different projects (in terms of total turnover, private vs. public clients, etc.). According to the Pareto principle[16)], about 80% of the turnover is generated from about 20% of the projects. It is precisely these projects that the process of managing chances and risks should focus on.

The chance and risk management process comprises the following essential elements:

- Integrated process for capturing, assessing, monitoring, and documenting chances and risks, and to provide reflection with feedback
- Distributed management of specific chances and risks
- Aggregation to the segment, group, and project levels

Organization and responsibilities:

- The management of chances and risks is an executive-level responsibility. Company management will define the chance and risk management process for the corporate organizational units (finance, controlling etc.), the divisions (building construction, railroad construction, civil engineering, etc.), and the individual projects.

16) The discovery of this rule dates back to Italian economist *Vilfredo Pareto* (1848-1923), who found, at the beginning of the 20th century, that 80% of the national wealth of Italy was attributable to only 20% of its citizenry. The Pareto rule (also known as 80/20 rule) is an idealized rule not applicable to each and every case. This principle was generalized and developed further to say that about 80% of a result are attributable to 20% of the cost or effort. In construction projects, this can mean, for instance, that 20% of the items of unit-price contracts account for 80% of the bid price.

- Executive management is supported by the organizational units as well as by compliance officers, if and when required. The latter coordinate the chance and risk management process at group level and provide the structure as well as appropriate tools and methods. This approach will ensure that the management of chances and risks forms an integral part of the processes related to corporate strategy, planning, and budgeting.
- Chance/risk managers with direct responsibility will drive the implementation of, and compliance with, chance/risk processes while making sure that control loops are implemented in project management and day-to-day business. Feedback loops should ensure that a continuous improvement process (towards lean management) is created.
- The management of specific chances and risks is largely delegated to the respective business units, and is controlled locally. An exception to this rule are higher-level tasks such as financing, group purchasing, or dealing with currency risks. In these cases, group-level consolidation takes place prior to resorting to, for example, derivative instruments for hedging purposes.

Tools for managing chances and risks:

- Guidelines define the essential, priority, "nice to have", and non-targets for the chance and risk management system.
- Control loops represent the individual stages.
- The methods and tools to be applied systematically to the respective project phases are defined at the project level (in this book, we explain in detail how to use the Monte Carlo simulation method).
- A central knowledge repository allows for the sharing of processes, methods, and tools as well as data and information.
- A catalog that lists categories of chances and risks makes it possible to identify all relevant chances and risks as completely as possible.
- Standardized data collection and reporting tools are suitable for identifying and analyzing chances and risks. Aggregating chances, risks, and sensitivities at business unit and group level using the Monte Carlo simulation method helps identify company-wide effects and trends.
- Operational chances and risks (covering a period of up to one year) are communicated to the executive level in a monthly management report prepared by corporate controlling. In addition, the corporate controlling and finance group functions provide information on the aggregate chance/risk exposure of the group, divisions, units, etc. on a semi-annual basis. Moreover, there is a direct reporting duty for newly arising individual risks whose impact on earnings exceeds a defined amount, as well as for reputational risks.
- As part of strategy development, the strategic planning and controlling units perform strategic chance/risk analyses for a ten-year period. These analyses are reviewed annually as part of the strategic controlling process, and revised if necessary.
- A company-wide compliance program is designed to ensure that legal requirements and internal company guidelines are adhered to. A global code of conduct for all employees ensures that these standards are not only adhered to but also practiced in day-to-day business. What is important in this context is that executives lead by example and visibly demonstrate that the applicable rules and processes are complied with.

4.4.1 Macro-Analysis – Organizational Level

Businesses should implement a system for managing chances and risks in their own best interest. Yet they often overlook that chances and risks are of a different nature and thus have different characteristics (Table 4.1 shows some of the differences between chances and risks), and it is not always appropriate to apply a "one-size-fits-all" methodology to deal with the specific situation or context.

No.	Criterion	Characteristics of risks	Characteristics of chances
0	A	B	C
1	Probability of occurrence	One of the unpleasant characteristics of risks is that they are recurrent. This is why it makes sense to assess the probability of occurrence of risks.	Chances are often difficult-to-grasp, one-off phenomena. They are usually associated with a decreasing time value, and can be realized only within a given time window. We talk about letting a chance go by. In this context, the following saying appears to be fitting: "Whoever comes late will be punished by life."
2	Perception	Risks can materialize regardless of whether we identify or perceive them, or not.	Chances can be taken only if we recognize them.
3	Need for action	We are not free with respect to our exposure to risks. In many cases, the respective risk can only be reduced. A residual risk will continue to exist, and its occurrence cannot be prevented with absolute certainty.	We are free in deciding whether we want to try to take a chance or not. The management of chances is strategic in nature. It aims to capture existing options for action, and thus the degree of freedom in our decision making. It supports the question of what we want to achieve in the future. The opportunity/ability to take a chance is not always present. Taking or realizing chances is thus associated with risks.
4	Handling	Handling of risks: • A risk can be prevented completely. • The probability of occurrence of a risk can be reduced. • The severity expected if the risk materializes can be reduced. • Risks can be accepted.	Handling of chances: • A chance can be realized to its full extent. • The probability of realizing a chance can be increased. • A chance can be realized partially. • Chances can be ignored.
5	Interaction of chances/risks	Risks inherently hold chances.	Chances are associated with risks. Chances cannot be realized without consciously taking risks.

Table 4.1 Differences between chances and risks[17)]

Risks have the unpleasant characteristic of being recurrent, whereas chances are usually of a one-off nature, which is why they have a "time value" only within a limited window of opportunity (see, for example, the real options theory). The following example appears to be appropriate to illustrate this: After its launch in the 1990s, the early adoption of the Internet technology opened up a great added-value opportunity for some companies.

[17)] Cf. Secricon GmbH (2011), http://www.secricon.com/pdf/Chancenmanagement.pdf. Accessed December 21, 2020

Nowadays, for many companies, this technology is just an essential requirement for doing business.[18)]

The management of chances and risks considers all organizational units to form a single coherent system. All organizational units are considered to be interdependent, which is why they must not be viewed in isolation.

Top management ensures that a system for managing chances and risks is implemented, documented, and kept operational. Executive board members also continuously improve the effectiveness of the chance and risk management system.

This system can be divided into responsibilities and elements, as shown in Fig. 4.11.[19)]

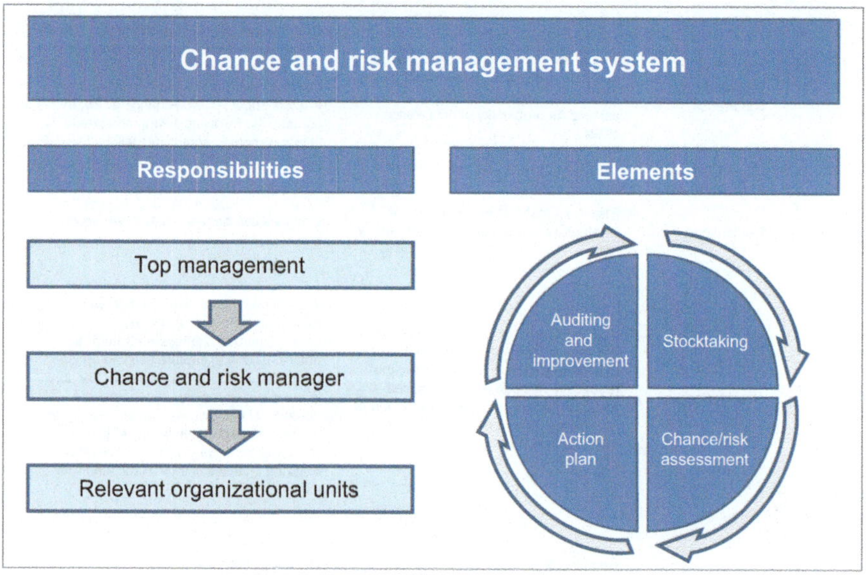

Fig. 4.11 Chance and risk management system[20)]

The chance and risk management system provides guidance to those responsible in organizations in order to meet the wide variety of economic, legal, and normative requirements.

The chance and risk management system essentially aims to[21)]

- ensure the continued existence of the organization
- provide protection against the loss of, or damage to, information, assets, or corporate reputation
- establish a risk culture or risk awareness across the business
- apply a structured approach in order to identify chances and risks with an influence on corporate targets and objectives
- reduce the probability of insolvency
- create strategic flexibility by enhancing the adaptability of the organization

[18)] Cf. Secricon GmbH (2011), http://www.secricon.com/pdf/Chancenmanagement.pdf. Accessed December 21, 2020

[19)] Cf. OENORM S 2410:2010, p. 11

[20)] Cf. OENORM S 2410:2010, p. 12

[21)] Cf. OENORM S 2410:2010, p. 12

- open up additional potential for success
- comply with legal and normative requirements
- increase efficiency by optimizing processes within the organization
- improve strategic and operational planning
- establish a chance- and risk-based management and control of the organization
- adjust the business model and market activity to future developments
- ensure continued, value-driven development and increase the market value of the organization

The macro-analysis of the organization sets the course for implementing a systematic management of chances and risks at the micro-level. This is where, from the contractor's point of view, the relevant chances and risks should be identified and dealt with for the individual projects on the basis of a predefined process encompassing the phases of bid preparation, process planning, and construction as well as additional cost claims and project handover and completion.

4.4.2 Micro-Analysis – Project Level

The local view puts the focus on construction projects. In the first step, the chances and risks taken in the assessed project are analyzed. This is when the important underlying decision needs to be made whether to carry on with the project at all or to stop it because, for example, a technical risk has been identified as a knock-out criterion and assessed as such.

Relevant targets, the timeframe, the parties involved in the analysis, and the methods[22] to be applied should be defined in advance of examining a construction project with regard to its chances and risks. An ABC analysis should initially be conducted to determine the very important, important, and less important areas so that the priorities can be set correctly. For Project Phases 3 and 4, the cycle is applied in a targeted manner on the basis of the production cube (see Chapter 1, Fig. 1.2). It should be noted that this is an iterative process structured as a control loop. When applied to a construction project, this is also quite logical because numerous decisions have to be made in the course of such a project, but different risks will materialize in the shell construction phase than, for example, during the finishing or use of the building.

Newly arising situations on the construction site or changed circumstances under which work is performed need to be analyzed, optimized, monitored, and tracked continuously with regard to the associated chances and risks.

4.5 Chance and Risk Management Process

Fig. 4.12 shows the chance and risk management process for working on projects. The overall system and its subsystems, including the identification of chances and risks, should be regarded as control loops and, depending on their status, form the input or output for subsequent processing steps or control loops.

[22] The concept of method should not be equated with the notion of a tool. The applied method determines the range of tools as well as the way in which they are used. Tools are means for achieving a set of targets, and their combination results from the selected method.

At the outset, a chance and risk management process will be triggered by an order/query. Within the chance/risk analysis block, the main objectives are to identify and select the relevant chances and risks and to assess their degree of relevance. Depending on the project phase, more or less detailed documents will be available for this purpose.

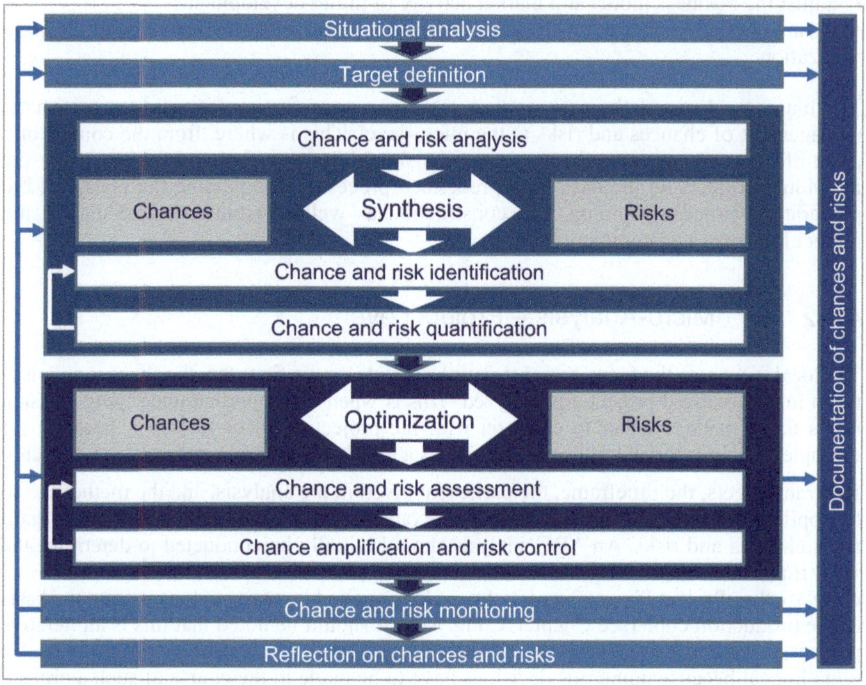

Fig. 4.12 Chance and risk management process

A process describes the dynamic interconnection of systems. Processes should be designed such that the people who work on tasks based on these processes are happy to perform them and are also motivated to continuously improve the processes. The overriding objective of the above chance and risk management process is to identify the relevant chances and risks promptly and to deal with them efficiently.

For any target-driven, systematic chance and risk management process, control loops are recommended, which we explain and describe below. A control loop essentially comprises a cycle and is composed of the reference element, the control element, the controlled system, and the measuring element. Control loops operate on the basis of positive and negative feedbacks. The targets are defined by the reference element, while the control element refines them for the controlled system. Running through the cycle several times increases the probability that no significant chances and risks are overlooked or assessed incorrectly.

Depending on the size of the organization or project, the individual elements of the control loop can be formed by teams or individuals. In small organizations and projects, the entire control loop is reduced to a single person.

4.5.1 Analysis and Synthesis of Chances and Risks

In the first step, the analysis of chances and risks is concerned with identifying and collecting the complete range of possible chances and risks. Possible chances and risks are recorded for the various production factors and stakeholders, as well as for the project surroundings and the environment. In addition, quantity, quality and, in particular, complexity should be taken into consideration. Whether and how many chances and risks will be identified largely depends on the experience of those involved and on the tools and methods used. Not all collected data or information will be relevant for further investigations. At the synthesis stage, the chances and risks identified in the analysis are filtered on the basis of the target definition and subjected to thorough processing in the subsequent process steps. As part of the assessment, it is very helpful to separately record the chances and risks that the bidder/contractor can influence directly, indirectly, or not at all.

Subsequently, the optimization process will focus on the significant chances and risks (aligned with project targets). To this end, the chances and risks are assessed first, and then suitable measures are sought to increase chances and to avoid risks.

4.5.1.1 Identification of Chances and Risks

The section below outlines and describes the objectives of identifying chances and risks, as well as the control loop for their identification.

4.5.1.1.1 Objective of Chance and Risk Identification

Identifying chances and risks is an important step of the operational chance and risk management process, with the aim of recording all relevant project uncertainties comprehensively and systematically that have been identified in the course of the analysis and filtered out at the synthesis stage. The process of identifying chances and risks should also be extended to the corporate level in order to capture chances and risks relevant to the company as a whole, besides project-specific uncertainties. Furthermore, identification is a critical step because unidentified chances and risks will not be included in subsequent analyses, and thus either remain unutilized or cannot be avoided.

When identifying chances and risks, a distinction can be made between non-systematic (intuitive) and systematic methods.

Another way of categorizing identification methods is to divide them into collection methods (such as checklists, morphological boxes) and search methods (such as an analytical questionnaire or creativity techniques including brainstorming and brainwriting). Selecting the most appropriate methodology will essentially depend on the specific risk profiles.

It is also conceivable to use combinations of the available methods for identifying chances and risks as accurately as possible.

The identification of chances and risks essentially aims to
- capture current and future chances and risks to the highest possible degree of completeness
- identify them as early as possible during the process (allocation)
- ensure profitability and efficiency

In this setting, profitability is contradictory to the first two requirements. This is why it is often considered useful to apply the Pareto principle, for instance in an ABC analysis, prior to analyzing chances and risks in more detail. Table 4.2 lists a number of methods that can be used for the entire chance and risk management process in the field of construction. Processing takes place either intuitively or systematically.

Additional methods available for the chance and risk management process include:

- Pondering[23)]
- Structured analysis of project documents
- Employee surveys
- Brainwriting
- 6-3-5 method[24)]
- Checklists
- Morphological box
- Indicator analysis
- Functional analyses
- Expert survey

The perception of chances and risks varies greatly depending on the specific person, position, and qualifications and skills. Where some recognize a risk, others do not see any risk at all, or even a chance. However, a chance can never be realized without having taken a risk before. The temporal aspect also plays an important role since the analysis will be less profound if only little time is available. Such situations are often associated with a tendency towards risk-taking.

The step of identifying (or detecting) chances and risks aims to analyze all chances and risks of a project captured in the synthesis, and to select those considered to be relevant. Only the latter should be assessed and ultimately taken into account in the decision-making process.

The individual methods can also be combined if and when required.

23) Pondering is the simplest intuitive and unstructured method to identify risks. It only requires a person, a pen, a sheet of paper, and the project documents.

24) Creative brainwriting method. The name of this method stems from the number of participants (six would be most appropriate), where each of them produces three initial ideas. Each of these ideas is passed on five times in order to develop ideas building upon or derived from them (six participants, three ideas each, passing them on five times).

No.	Category	Method	Risk management process: Identification	Assessment: Consequences	Assessment: Probability	Degree of risk	Coping	Possible application in the construction industry
0	A	B	C	D	E	F	G	H
1	Creativity techniques	Brainstorming	+++	+	+		+	- Project development/planning - Variant analysis
2	Creativity techniques	Delphi method	(++)	++	++		++	- Evaluation of new methods, techniques, and processes
3	Creativity techniques	World Café	+++	+	+			- Project development/planning - Variant analysis
4	Scenario analyses	Citizen Conference	+++	++	+		+	- Project development/planning - Variant analysis
5	Scenario analyses	Damage analysis or cause-effect diagram	++	+	+		++(+)	- Analysis of execution, design and specification errors
6	Scenario analyses	London Protocol	+++	+			+++	- Health and safety/accident prevention
7	Scenario analyses	Error tree and process analysis		++	+++	+	+	- Logistics - Scheduling
8	Scenario analyses	Scenario analysis	+++	+++	++	++	++(+)	- Dealing with design, execution and investment risks
9	Indicator analyses	CIRS (Critical Incident Reporting System)	+++		+		+(+)	- Target-actual comparisons
10	Indicator analyses	CBRM (Change Based Risk Management)	+++	+		(++)		- Dealing with changed market prices (e.g. steel)
11	Indicator analyses	Horizon scanning	+++	+		++		- Early strategic identification of economic, social, technological, political, and ecological changes
12	Functional analyses	FME(C)A (Failure Mode and Effects (and Criticality) Analysis)	+++	++	++	+	++(+)	- Interface management - Functional and spatial planning - Execution of complex construction methods
13	Functional analyses	Risk assessment	++	+++	++	++	++(+)	- Health and safety
14	Functional analyses	HAZOP (Hazard and Operability Study)	+++	+++	++	+	++(+)	- Scheduling (assessment of the critical path)
15	Functional analyses	HACCP (Hazard and Critical Control Point)	++	++			++(+)	- Concrete and asphalt production
16	Statistical analyses	Standard deviation		++	+++	++		- Costing, pricing
17	Statistical analyses	Confidence interval		++	+++	++		- Instrument of risk policy - Ratio between chance and risk
18	Statistical analyses	Monte Carlo simulation	+	++	+++	++		- Applicable to any calculations involving uncertain factors
19	Statistical analyses	Qualitative analyses		++	+++	++		- Assessment of chances and risks using qualitative methods
20		Key	+ suitable	++ well suited	+++ very well suited			
			(...) assessment differs in the ONR 49002-2:2010 and 2014 versions and in OENORM D 4900:2021					

Table 4.2 Overview of methods in the chance and risk management process[25)]

Besides the methods, the temporal aspect of risk identification must also be considered. Identification can take place either before or after the occurrence of deviations. Accordingly, available methods are often divided into *ex ante* (progressive) and *ex post* (retroactive) procedures. The former include the methods of early identification and response, early detection, and early warning. The latter category refers to, for instance, the final costing exercise.

The approaches pursued in the identification process vary in their scope:

- Early warning: early localization of risks
- Early detection: early localization of chances and risks
- Early identification and response: early localization of chances and risks and initiation of measures in response

On the operational level, processes and methods can be implemented either in a top-down or in a bottom-up approach. One top-down example would be that chance or risk areas or categories are defined by senior company management. In an alternative context, the assessment of chances and risks is also referred to as top-down if the focus is on known consequences of identified chances and risks (effect-related). In contrast, the bottom-up approach relies on a chance/risk analysis at the employee or organizational unit level (below the senior management level). The idea behind this method is to consolidate, for example, individual construction site risks so as to arrive at a holistic view of a total project or company risk. This cause-related approach will usually be associated with much higher chances and/or risks than the effect-related method. In addition, it is based on complex procedures such as scenario analyses, which justifies its higher cost or required utilization of resources. This process results in chance/risk totals and key metrics made available to senior management. It is particularly suitable for construction projects because it will be possible to identify and assess the prevailing chances and risks in a project-specific manner only if starting from the expertise available in the respective subfield or discipline.

In general, chances and risks should be identified systematically, but the related process should not be overformalized. An experience-based approach ensures that creativity and expertise will play their part in localizing chances and risks in advance of or in addition to systematic approaches.[26)]

4.5.1.1.2 Chance and Risk Identification as a Control Loop

The control loop shown in Fig. 4.13 is recommended for a target-driven, systematic identification process. The cycle can be run several times if required, thus increasing the probability that no significant chances and risks are overlooked.

The networked view of a system involving elements and relationships results in a control loop that maps an iterative solution, decision making, or retrieval process. A control loop is basically a closed cycle of effects, which means that the effect of a system element influences this system element itself as well as the elements with which it interacts.

Control loops are suitable for working on tasks or problems of a quantitative and qualitative nature. A target-actual comparison can be carried out in order to determine the

[25)] Cf. ONR 49002-2:2010, p. 5
Cf. ONR 49002-2:2014, p. 5
CF. OENORM D 4900:2021, p. 6

[26)] Cf. Alber (2013), p. 33ff.

degree of target achievement for quantitative variables. If the control loop yields qualitative solutions, the target-actual comparison will no longer be simple, which is why it is complemented by subjective assessments of the measuring element or of the reference and control elements.

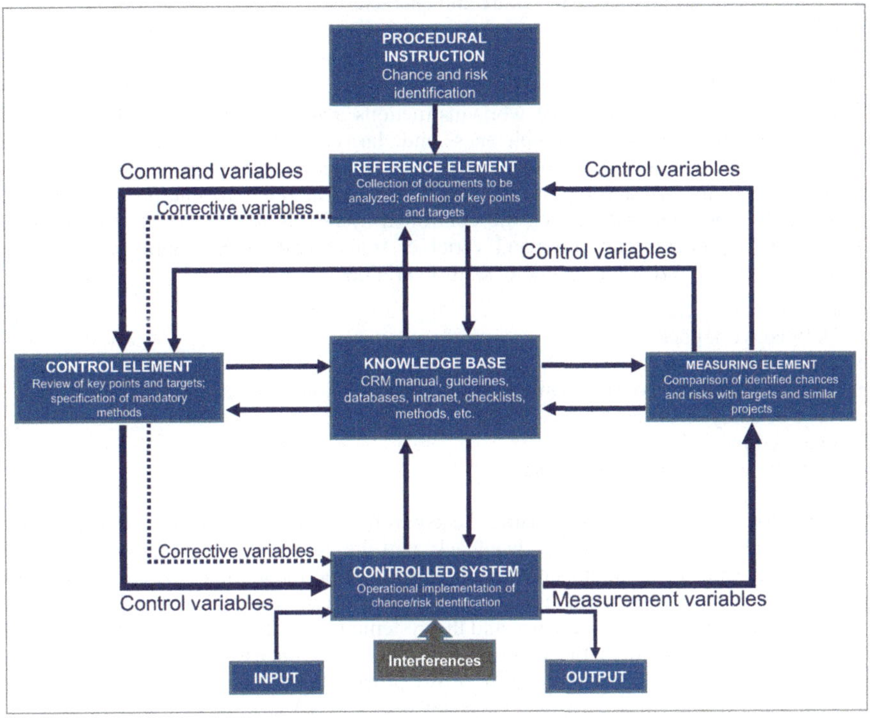

Fig. 4.13 Control loop for the identification of chances and risks

Key benefits and characteristics of a control loop include:

- A better understanding of complex real processes
- Consideration of dynamic behavior
- Detection of complex relationships and interactions
- Mental processing of basic system characteristics as a prerequisite for ensuring that actions only result in intended changes
- Avoidance of monocausal cause-and-effect thinking
- Detection and visualization of feedback effects and dependencies
- Interactive access to a knowledge base

The following section describes essential elements of the control loop in more detail.

Reference element

The reference element provides the work instructions for the control loop and defines the targets and processing priorities. It also provides the relevant documents for the analysis. Preselection of the documents by the reference element is very helpful because this step will ensure that only essential information will be passed on and idle time can be reduced.

To support the process, the reference element accesses the knowledge repository to search for best practices that have proven their worth in other projects. In so doing, both information on best practices and lists of errors can prove to be helpful and supportive. Such lists should be structured so as to make it abundantly clear which errors have occurred at which points and how they can be avoided in the future. The reference element should also adapt the work instructions to the nature and characteristics of the project or project stakeholders.

Control element

The control element receives the work instructions and documents from the reference element and checks them for completeness and clarity. Feedback[27)] will be given to the reference element if ambiguities occur or if the documentation appears to be incomplete. Depending on the size and type of the project as well as the available processing time, methods are specified with which the controlled system is to run the chance and risk identification process. The command variables transmitted to the control element are processed and converted into editable control variables.

Control variables can include:
- Defined work packages
- Defined scope of analysis
- Time to completion of the chance and risk identification process
- Type of analysis of results
- Involved individuals
- Applied identification method(s)

These control variables should enable the controlled system to perform the processing steps efficiently and effectively. Feedback will be given to the control element if ambiguities occur or if the documentation appears to be incomplete. Taken together, all control variables and documents form the input for the controlled system in which chances and risks are queried. The experience and the systematic and structured way of working of the employees involved in the controlled system plays an essential role for ensuring processing quality. Not only the quality of the processing will depend on this factor but also the number and scope of the identified chances and risks.

Controlled system

The controlled system uses the input and information from the knowledge repository or from own experience for identifying chances and risks. Limited time and resources are available for processing. This is why it is important to use the right tools when launching the process of chance/risk identification. A chance/risk checklist made available as a software, for instance, can be a useful tool that provides various blocks with a range of chance/risk criteria for selection purposes. This checklist can be used for selecting the appropriate criteria and sub-criteria for the project to be worked on. There is thus already a large structured selection of chances/risks that can be expanded on a project-specific basis applying the hermeneutic loop principle. With each project, this checklist stored in the knowledge repository will grow further as extended criteria or new chances and risks will be implemented when new projects are added. This existing range of options will minimize the probability of forgetting or overlooking chances and risks. Previously

[27)] Feedback is a mechanism in signal-enhancing or information processing systems in which a certain portion of the output variable is fed back to the system input directly or in modified form. Negative feedback leads to a self-regulated stabilization of the system within predefined limits. The system would explode or come to a complete standstill if there were only positive feedbacks.

prepared checklists can be used, expanded and adjusted for new projects. This will allow for synergies between similar projects to be utilized while minimizing the processing effort and optimizing the workflow. The larger the "pool" of possible chances and risks, the lower the proportion of unidentified chances and risks will be. The output at the end of the controlled system can be a chance/risk checklist adjusted to the current project, serving as the basis for the chance and risk assessment control loop. As a matter of course, it is also possible (and often necessary) to return to the control loop of chance and risk identification in order to identify previously neglected or overlooked chances/risks on the basis of the broadened understanding.

Interferences with the controlled system can be both external and internal. If the underlying conditions of the entire project or parts thereof change, this will also have an impact on the chances and risks to be identified. Internal interferences can occur due to the loss of experienced employees, inappropriate priorities, unsuitable identification methods, or inadequate forms of graphical representation. Interferences influence the controlled system and thus the process of chance and risk identification and must be managed either by the controlled system itself or by the reference or control element.

Measuring element

The measuring element checks whether the specifications resulting from the command and control variables have been adhered to. Since these variables are difficult to measure, process compliance is monitored and the type and number of identified chances and risks are presented. Using the knowledge repository increases the probability of identifying all relevant chances and risks. If data and information from comparable projects are available, it is advisable to make comparisons in order to better assess the material chances and risks. For this purpose, the measuring element ideally accesses a knowledge base from which the specific data and information can be retrieved. The knowledge repository is not only a passive element available for the entire control loop. Rather, the active control loop also supplies new data and information to it. The knowledge repository includes suitable methods for identifying chances/risks that have already proven their worth in other projects. Ideally, it will also provide descriptions of how to proceed efficiently and which errors to avoid.

Depending on the internal rules (limits to self-organization), the measuring element will transfer the check variables to the reference or control element. The check variables serve to determine whether the defined targets have been achieved or whether there is a need for corrective actions. Adequate corrective variables are determined in the latter case.

In the context of the control loop described above, these could include:

- Specifying (a) different identification method(s)
- Specifying a different graphical representation of chances and risks
- Extending/limiting the scope of analysis
- Setting a different focus for the analyses
- Changes in the individuals involved in the controlled system (e.g. consultation of experts)

The controlled system uses the corrective variables to initiate a new processing cycle. The measuring element will again analyze its result and pass on the resulting check variables. The control loop is executed until the expected or intended quantity and quality of identified chances and risks has been reached and/or the time for processing has expired. These analyses are carried out under enormous time pressure and have a great impact on the commercial success of the project. Any overlooked relevant chances or risks may lead to material adverse effects.

4.5.1.2 Quantification of Chances and Risks

The following section describes the key objectives of the chance and risk quantification process.

After their identification, chances and risks must be quantified and assessed. In so doing, the following criteria should be applied:

- Significance for the project
- Controllability (direct, indirect, non-controllable)
- Change over time
- Consequence
- Probability of occurrence
- Corrective measures in the case of risks
- Amplification measures in the case of chances
- Dependency between individual chances or risks
- Influence on the chance/risk ratio

4.5.2 Optimization of Chances and Risks

No clear distinction is often made between the notions of analyzing and assessing chances and risks. The analysis of chances and risks aims to create an understanding of chances and risks while determining efficient strategies and methods for their further treatment. Identifying and analyzing chances and risks in a cause-related approach provides the input for the assessment of chances and risks that should inform the decision-making process. This includes, for example, the preferential treatment of chances and risks that contribute to achieving the two basic objectives of informing decision makers promptly about the possible scope of chances and risks and unleashing the potential for increasing efficiency in chance and risk management.

A methodological separation of these process steps is often only implied in the description of their practical application contained in the literature. Instead, various methods (i.e. qualitative, semi-quantitative, and quantitative approaches) are referred to and presented in the context of explaining the pertinent terminology.

The phase of optimizing chances and risks aims to actively promote the occurrence of chances while reducing the deleterious effects of risks. The members of the project team decide how to deal with the identified and analyzed chances and risks. This process includes the following steps:

- Checking the chances and risks for existing options for action
- Decision in favor of one of these options
- Implementing this option, i.e. taking the required actions

4.5.2.1 Assessment of Chances and Risks

Fig. 4.14 shows a systematic classification of various approaches to assessing chances and risks. It should be noted that methods such as Monte Carlo simulations or sensitivity analyses are not necessarily dependent on expert surveys based on the Delphi method, as the graph suggests.

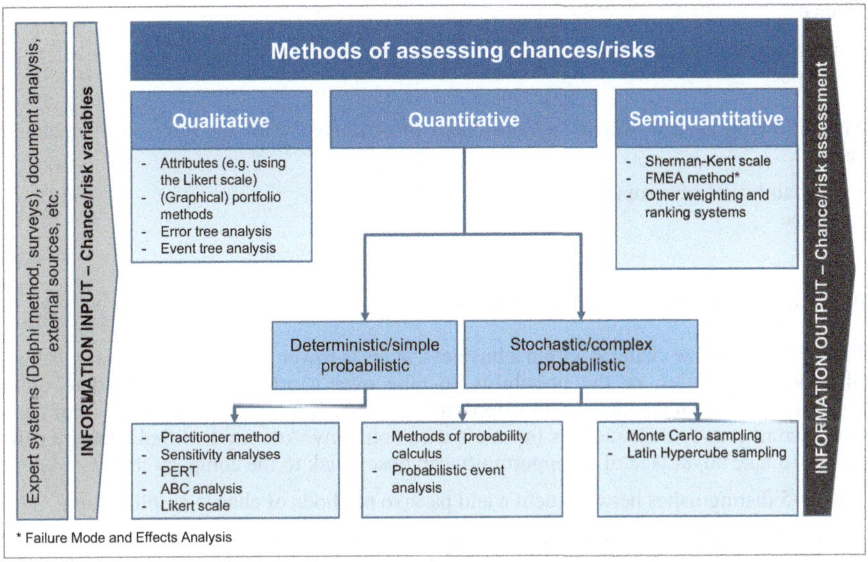

Fig. 4.14 Methods of assessing chances and risks (non-exhaustive list)[28)]

Any assessment of chances and risks should start with the areas that appear to be the most significant. Special attention must be paid to the criteria used for the assessment.

The identified chances and risks should be assessed according to the following criteria:

- Significance for the project
- Controllability (direct, indirect, non-controllable)
- Change over time
- Impact/consequence
- Probability of occurrence
- Corrective measures in the case of risks
- Amplification measures in the case of chances
- Dependency between individual chances or risks
- Influence on the chance/risk ratio

4.5.2.2 Amplification of Chances and Risk Control

The amplification of chances comprises the selection and utilization of opportunities to aggregate one or more chances. Risk control, on the other hand, comprises the selection and implementation of one or more options to counteract the risks.

In principle, a distinction can be made between active (cause-related) and passive (effect-related/corrective) measures.

[28)] Cf. Alber (2013), p. 88

The following aspects should be considered predominantly for the purpose of amplifying chances:

- Probability of occurrence
- Scope

The following aspects should be considered predominantly for the purpose of reducing risks:

- Probability of occurrence
- Scope

4.5.2.2.1 Amplification of Chances

Failing to recognize chances within a business, such as lower costs of market entry, launch of new technologies, or the possibility to take over a competitor, can lead to other companies taking advantage of these opportunities, which from an economic point of view can have negative consequences for one's own company. Any unidentified chance or the failure to take advantage of an opportunity can pose a risk to the company itself.

Fig. 4.15 distinguishes between active and passive methods of chance amplification.

If the opportunity arises to utilize a chance, it is advisable to promote the related area in order to increase the chance potential as much as possible.

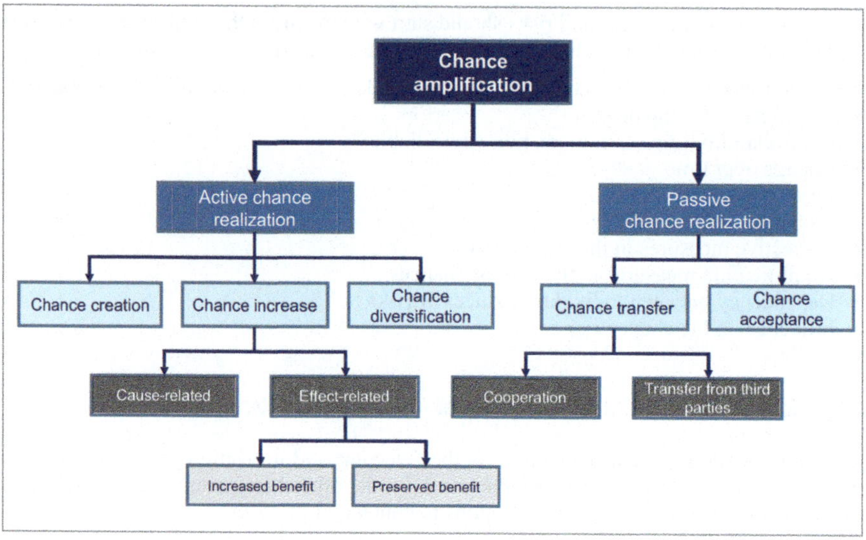

Fig. 4.15 Methods of chance amplification

The following section discusses the key elements of chance amplification taken from Fig. 4.15 in more detail.

Chance creation

A chance is created, for instance, if a different type of structural component design is proposed to the client, making it possible to utilize the production factors more economically. This type of value engineering[29)] will also create opportunities for the client. Yet contractors are equally capable of generating additional chances through submitting alternative bids. The new design options and/or construction methods referred to above allow for the type of production factors and their combination to be rendered more efficient.

Chance increase

In the areas in which the potential for chance increase appears to be the greatest, measures should be implemented in order to utilize or further expand these chances.

Chance diversification

Diversifying chances involves expanding the company's offering to new products or markets in order to amplify existing chances. Diversification can be internal or external. Internal diversification refers to the in-house development of new construction methods and equipment. External strategies involve entering into cooperation agreements, joint ventures, consortia, etc.

Chance transfer

Chances are transferred to third parties either completely or as part of a cooperation agreement. In this setting, such third parties are in a better position to utilize the chance potential.

Chance acceptance

Chance acceptance means to utilize occurring chances passively. This also means that no steps are taken to optimize the utilization of chances.

4.5.2.2.2 Risk Control

Appropriate control measures are chosen primarily depending on the controllability of the risk. This selection is derived from the adopted risk strategy or policy of the company. Objective decision making is possible on the basis of decision tree methods, decision tables, or sensitivity, utility value, or SWOT analyses.

Any of the risk control alternatives will incur costs either with certainty (e.g. insurance) or with a certain probability (e.g. the damage resulting from risk materialization). Thus, the preferred risk control measure should be selected on the basis of the specific situation. Besides costs, the effects on other parameters such as deadlines and quality should also be considered. Fig. 4.16 shows one possible approach to risk control.

An important aspect of risk control is contingency planning, which is also applied if unidentified risks occur after implementing the options for action referred to above. In addition, it is a frequently neglected practice to reassess risks in combination with possible control measures, such as insurance, since, for example, this can have a significant influence on project profitability (involving the notions of gross and net risk positions of a project).

[29)] Value engineering is designed to improve the combination of production factors in order to achieve targets more efficiently in the context of technical, organizational, and process optimization in the construction of buildings/structures or structural components.

For the sake of brevity, we do not include a general list and description of risk control methods at this point. Following the selection and implementation of suitable risk control measures, their effectiveness and efficiency must be verified continuously as part of risk monitoring and risk controlling. In the worst case, inappropriate or ineffective risk control measures can even constitute risks themselves (secondary risks).

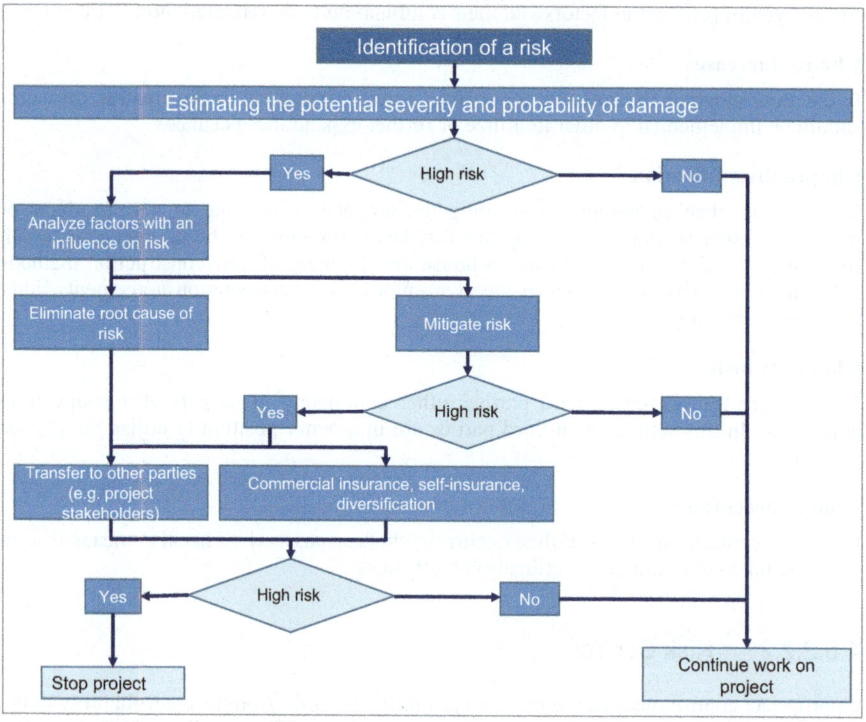

Fig. 4.16 Risk control process[30)]

Once the potential risks for a project are known, the aim is to avoid the full consequences (in the negative sense) if they occur. Various options exist for dealing with or controlling risks (see Fig. 4.17).

[30)] Cf. Alber (2014), p. 33

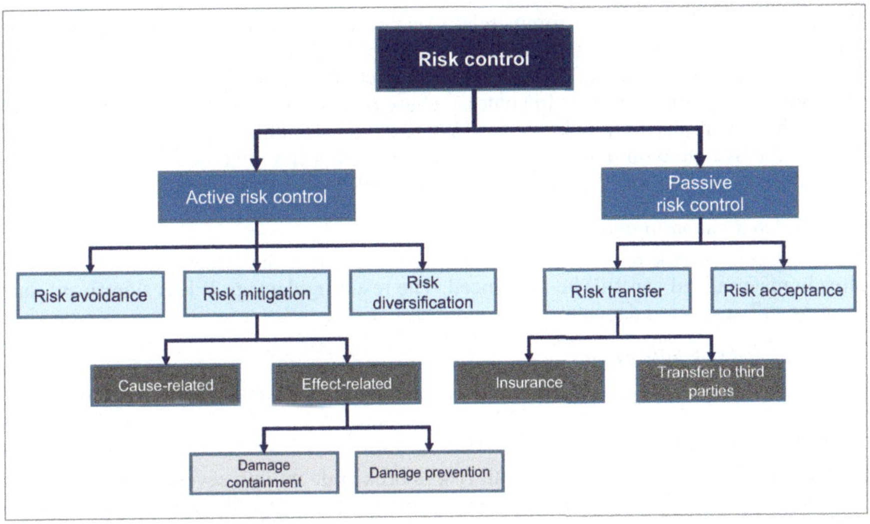

Fig. 4.17 Risk control methods

For risk control purposes, a risk strategy must be defined on the basis of which a decision will be made regarding the methods of active or passive risk control to be applied in each case. The control measures must be based on the goals and objectives of the company or on the project targets and must be selected with a view to cost and benefit.

A risk strategy that is perfectly suited for risk control contributes to securing the continued existence of the company and increasing its enterprise value. The way of dealing with the identified risks is derived from this strategy, i.e. whether they should be accepted, transferred, diversified, reduced, or avoided. The following section discusses the key elements of risk control taken from Fig. 4.17 in more detail.

Risk avoidance

In the course of the bidding process, the analysis of the building specifications considers the works that can generally be performed by the company itself. For these considerations, it is important to determine if your own service level is sufficient to perform the work to the specified standard of workmanship. Subsequent considerations aim to identify the chance/risk ratio to be applied to the different work items in order to have at least a minor chance to emerge as the best bidder in the course of the contract award procedure. Items associated with very high risks will be avoided as far as possible by substituting them with other items.

Risk avoidance occurs, for example, if a company decides not to carry out planned activities, such as investing in new construction equipment, or to refrain from submitting a bid because the bid price required for a successful contract award would have to be 20% lower than the result of the zero-base costing exercise, for example. This example of risk avoidance describes the complete abandonment of a risky (potential) contract. This approach should be chosen only if the existing poor chance/risk ratio can no longer be improved. This type of risk control reduces the probability of occurrence or damage of a specific risk to zero. Risk avoidance thus means to avoid the damage or loss or the occurrence of the risk completely.

Risk mitigation

If a risk cannot be avoided, commercially viable risk mitigation measures should be implemented in order to reduce the risk to an acceptable level. The related actions can be cause- or effect-based. Any effect-related risk reduction will involve either a limitation or a prevention of damage or loss. In contrast, cause-related risk mitigation measures aim to reduce the probability for a risk to occur. What is crucial in this respect is to improve the information available on uncertain parameters. Selecting the reference base (see Section 2.1.6) also has a significant influence on the probability of risk occurrence.

Effect-related risk mitigation

For effect-related risk reduction measures, we assume that the effects of a risk that has already materialized can still be influenced. The resulting damage is less significant than the originally expected damage.

Cause-related risk mitigation

Cause-related risk mitigation measures aim to reduce the probability for risks to occur.

Risk diversification

Risk diversification occurs if an overall risk is broken down into several individual risks that are not positively correlated with each other (as far as possible), thus resulting in a wide spread. This is the case, for example, with loan portfolios that ideally have a high level of detail and specificity while posing low concentration risks (the portfolio theory is based on the principle of not sourcing all materials from the same supplier). Within a group of companies, mutually independent risks can be diversified by region, property, or people:

- Regional diversification is achieved by manufacturing the same product in different locations (simultaneous production)
- Property-related diversification is achieved, for example, by creating several similar production facilities (redundancy)
- People-related diversification involves, for instance, several people in the construction management team assuming site management functions

Risk diversification serves to regulate risks. However, rather than minimizing the probability of occurrence of individual risks, it has an effect on the extent of the damage or loss. Since a simultaneous occurrence of all risks in their entirety is very unlikely, the danger of dependencies should be avoided by making sure that several suppliers are available and selecting business partners on a comparative basis.

Risk transfer

Risks that can neither be avoided nor reduced further must be borne by one of the project stakeholders. Risk sharing or risk transfer (either to an insurance company or to third parties, such as subcontractors) can be stipulated in the contract. It should be noted that costs will be incurred in either case, either as insurance or as a risk premium. As a rule, risk transfer is tolerable only for risks with an "insignificant" or "minor" extent of damage or loss combined with an "unimaginable" or "unlikely" probability of occurrence.

At the end of the risk control measures, any risks that could not be avoided, mitigated, or transferred must be borne by the company itself as residual risks in addition to the risks not considered or not identified. In so doing, the company may well decide to also shoulder transferable risks because this could be the less costly option under certain conditions (e.g. by seizing an opportunity).

Risk transfer involves transferring existing risks to other economic entities that can control these risks at least as efficiently as the entity they originate from. Risk transfer reduces the consequences of the occurrence of a risk, but not its probability of occurrence. This primarily involves insurers that provide cover for damage to property (such as fire or business interruption insurance), credit insurance (for suppliers) or export credit insurance (for exporters). The entire insurance business is considered to be a risk transfer between the insurer and the policyholder against payment of premiums (risk transfer concept). In this setting, a viable cost-benefit ratio is of the essence. Another type of risk transfer is risk sharing through joint ventures, consortia, or reinsurance. Outsourcing (such as in IT or accounting) also constitutes a risk transfer.

Risk acceptance (risk allowance)

Risk acceptance or allowance consists of accounting measures with relevance to the risk-bearing company. The principle of prudence requires that all risks and losses be adequately accounted for in the financial statements. Thus, provisions, write-downs, and write-offs must be made in accordance with applicable accounting legislation and standards. Balance sheet risk provisioning also includes the creation of an adequate equity base to cover any losses that may occur and to create cash reserves to avoid insolvency. In risk provisioning, the risk is borne by the company itself. This option falls under the category of passive risk control exclusively related to the consequences of risk occurrence, which need to be cushioned by the financial position of the company.

4.5.3 Monitoring of Chances and Risks

The aim of this exercise is to continuously track, in the subsequent project phases, the chance and risk situation defined in the course of the risk identification, analysis, and control processes. In this case, the focus is on monitoring the effectiveness of problem-solving measures by means of a target-actual comparison and appropriate documentation. For example, risk controlling can be integrated into the existing project controlling by adding another control loop.

In current research in construction economics, the risk controlling process step plays only a minor role. Among other factors, this might be due to the blurred transition between "conventional" project controlling and its logical extension by specific aspects of risk monitoring.

The chance/risk management process implemented in the organization must be continuously reviewed to ensure that it is always up-to-date and target-driven. For all relevant projects, the chances and risks captured and dealt with on the target side must be documented, as well as the degree of utilization of chances and avoidance or reduction of risks on the actual side. This comparison creates the basis on which to reconsider applied tools and methods and to initiate corrective actions for new projects if necessary. Risk reporting includes ongoing reporting to the relevant stakeholders on the chances and risks identified and assessed, as well as on the measures implemented in order to take advantage of chances and to control risks.

Risk reporting must ensure that the findings of risk control are documented as early as possible and in a comprehensible, meaningful form. It encompasses the presentation and assessment of the risk situation and, if necessary, proposals for action to improve the risk position, thus ensuring appropriate communication within the company and allowing for informed decision making with regard to required risk control measures.

Risk reporting also covers the accountability (proof of dutiful conduct), safety and security (measures to prevent errors), and auditability functions (providing the basis for the audit conducted by the supervisory board, internal audit department, etc.).

4.5.4 Documentation of Chances and Risks

A seamless documentation system reflects the processing steps and results of the entire chance and risk management process and provides a solid basis for the systematic expansion of the internal knowledge repository. Furthermore, the meaningful documentation of completed projects can lead to a significant competitive advantage for the contractor in the execution of follow-up projects. Documentation generally serves the purpose of making data and information easy to retrieve in order to record past or current facts and disseminate or reproduce them. Knowledge[31)] is transferred only if data and information are interlinked on a systematic basis and are put into the context of experience and expectations. The output of the combination of data and information generates knowledge that, compared to information, is purpose-related, perspective-based, contextual, and dependent on specific aspects. One and the same information can lead to different knowledge when viewed from another perspective or embedded in a different context. Aspects of lean management should also be considered in this regard.

4.5.5 Reflection on Chances and Risks

Reflecting on chances and risks involves the results of the assessment of chances and risks, the chances and risks that have actually occurred, and the measures taken in response in order to arrive at a holistic view. To this end, the assessments carried out before and after the measures are compared with the actual values. The project team should discuss the specified process and its application in a debriefing session while the project is still in progress as well as post-completion. The debriefing session provides the opportunity to reflect on what was adequate in the chance and risk management process and what can be improved. Problems are highlighted and discussed, and appropriate solutions are sought. The outcomes of the debriefing exercise contribute to the continuous improvement of the process and of the applied methods and tools.

4.6 Summary

This chapter concentrated on promoting a chance- and risk-based mindset, which is a valuable basis for efficient and effective action.

The systematic management of chances and risks is the prerequisite for using the corresponding tools and methods in the chance and risk management process to efficiently optimize the utilization of chances and to control risks. This type of support makes it possible to effectively achieve the higher-level corporate objectives as well as the more specific project targets.

31) "Knowledge signifies the total body of knowledge and skills that individuals use for problem-solving. This definition comprises both theoretical knowledge and practical rules and instructions for everyday use. Knowledge relies on data and information, but unlike these, it is always linked to people. It is created by individuals and represents their expectations about cause-effect relationships." Probst/Raub/Romhardt (2012), p. 23 (translation by Steffen Walter)

Key results of an expert survey on the topic of chance and risk behavior in construction companies were incorporated in this chapter. In this context, explanations were given on how the risk potentials of the risk fields covered in the survey were assessed from the contractor's perspective. We also indicated the (selected) project phases that are associated with the highest potential for errors and risks.

In addition, we used a distribution and the chance/risk ratio taken by bidders to demonstrate qualitatively to what extent the chance and risk behavior is linked to competitive intensity and the quality of planning and/or specification. The chance and risk behavior of an organization or of the involved stakeholders is essentially influenced by the overall chance and risk policy and the implemented system for managing chances and risks.

The management of chances and risks pertains to the entire organization on the macro-scale whereas it relates to specific projects on the micro-scale. In order to provide a timely overview of key projects, a systematic project portfolio management should be implemented between the two levels.

In so doing, the overarching primary objective is to secure long-term profitability, or the continued existence, of the business. Most importantly, this includes the goal of preserving the company's liquidity. A distinction is made between qualitative and quantitative targets. The former include, for example, quality, market position, and communications; the latter include measurable variables such as returns, costs, a low ratio of warranty claims, and deadlines.

This chapter outlined and described the concept of a feedback loop to exemplify the successful implementation of a target-driven, systematic process of managing chances and risks. A control loop essentially comprises a cycle and is composed of the reference element, the control element, the controlled system, and the measuring element. The operation of control loops is triggered by positive and negative feedbacks. The targets are defined by the reference element, while the control element refines them for the controlled system. Running through the cycle several times increases the probability that no significant chances and risks are overlooked or assessed incorrectly. At the core of the control loop is the knowledge repository, which provides valuable data and information for the various processing phases. Any systematic knowledge management will require the knowledge repository to be fed with newly acquired data and information from project work.

The management of chances and risks is an indispensable, state-of-the-art management discipline for clients and contractors alike. Provided an effective link is created between the final process steps implemented by the employees on the project and company levels, the conscious management of chances and risks will add value for all stakeholders.

4.7 References

Alber, Alexander (2013). Risikomanagement in Bauunternehmen – Eine Analyse von Theorie und Praxis. Master's thesis. Graz. Institute of Construction Management and Economics, Graz University of Technology.

Alber, Alexander; Hofstadler, Christian (2014). Risikomanagement in Bauunternehmen – Eine Analyse in Theorie und Praxis. In: bau aktuell, 5. Jahrgang, Heft Nr. 2, March 2014. 42-53. Vienna. Linde-Verlag Ges.m.b.H. (ISSN 2077-4737)

Hofstadler, Christian; Kummer, Markus (2017a). Chancen- und Risikomanagement in der Bauwirtschaft – Für Auftraggeber und Auftragnehmer in Projektmanagement, Baubetrieb und Bauwirtschaft. Berlin, Heidelberg. Springer-Verlag. (ISBN 978-3-662-54318-4)

Mesaros, Leila; Vanselow, Achim; Weinkopf, Claudia (2009). Fachkräftemangel in KMU – Ausmaß, Ursachen und Gegenstrategien – Eine Literaturanalyse. Institut Arbeit und Qualifikation, Universität Duisburg-Essen. Duisburg. Available online at http://www.fes.de/wiso/pdf/mittelstand/2009/040609/expertise.pdf. Accessed November 22, 2016.

Oberndorfer, Wolfgang; Jodl, Hans Georg et al. (2001). Handwörterbuch der Bauwirtschaft – ON V 208 – 2nd ed. Vienna. ON Österreichisches Normungsinstitut. (ISBN 3-85402-072-4)

OENORM ISO 31000 (2018-09-01 edition) Risikomanagement – Leitlinien [Risk Management – Guidelines]

OENORM D 4900 (2021-01-01 edition) Risikomanagement für Organisationen und Systeme – Begriffe und Grundlagen

OENORM S 2410 (2010-01-01 edition) Chancen- und Risikomanagement – Analyse und Maßnahmen zur Sicherung der Ziele von Organisationen.

ONR 49000 (2014-01-01 edition) Risikomanagement für Organisationen und Systeme – Begriffe und Grundlagen – Umsetzung von ISO 31000 in die Praxis.

ONR 49002-2 (2010-01-01 edition) Risikomanagement für Organisationen und Systeme – Teil 2: Leitfaden für die Methoden der Risikobeurteilung – Umsetzung der ISO 31000 in die Praxis.

Probst, Gilbert; Raub, Steffen; Romhardt, Kai (2012). Wissen managen – Wie Unternehmen ihre wertvollste Ressource optimal nutzen – 7th ed. Wiesbaden. Springer Gabler. (ISBN 978-3-8349-4562-4)

Secricon GmbH – Risk and Security Consulting (2011). Available online at http://www.secricon.com/pdf/Chancenmanagement.pdf. Accessed December 21, 2020.

Spang, Konrad; Gerhard, Michael (2016). Risikomanagement. In: Projektmanagement von Verkehrsinfrastrukturprojekten. Ed.: Spang, Konrad. 419-453. Berlin, Heidelberg. Springer Verlag (ISBN 978-3-662-46457-1)

Wiggert, Marcel (2009). Risikomanagement von Betreiber- und Konzessionsmodellen. Doctoral thesis. Graz. Institute of Construction Management and Economics, Graz University of Technology.

5 Applying the Management of Chances and Risks to Project Phases

A building or structure is always the product of striving to satisfy a particular need. It is, or should be, linked to the ambition of ensuring its sustainable (re)use and resource-saving dismantling and demolition at the end of its use phase. The construction of such a building is commissioned by either a public or a private client. Any public entity or other client of this category, including sectoral clients such as utilities, is subject to the laws and regulations governing the tendering and contract award procedures. Private clients can also be consumers whose activities are governed by consumer protection laws.

Any building or structure will involve different project phases (PPHs) and can be constructed in many different ways. Such a project will start from an idea that, again, is to be developed across various project phases in order to define a specific object of the contract and to ultimately realize it in a usable property. Depending on the type of client, a number of restrictions exist in the individual project phases with respect to compliance with defined processes, laws, and standards.

Clients are involved in all phases of the construction project and make decisions with regard to qualities and quantities, construction time (or scheduled deadlines), and the earmarked budget. By contrast, contractors[1)] usually join the project only at the end of PPH 3 when it comes to preparing and revising the bid. They will exit the project in their role as active stakeholders upon handover/acceptance (with the exception of warranty periods), i.e. at the end of PPH 4. Contractors' own projects are an exception to this rule.

In the individual project phases, clients and contractors are confronted with uncertainties while having to make important, far-reaching decisions under these circumstances. The underlying factors for decision making are based on uncertain parameters and figures (where the extent of such uncertainties depends on the available database and the circumstances under which the specific project is being constructed), or on rough estimates. We thus recommend to clients and contractors to apply probabilistic calculation methods in a targeted manner in order to systematically deal with such uncertainties and the ranges resulting for key metrics and values. This approach will increase the level of decision confidence. The following sections outline tools and methods and their specific application with the aim of enhancing decision confidence.

[1)] In this context, the term "contractors" refers to construction contractors.

C. Hofstadler and M. Kummer, *Chances and Risks in Construction Management and Economics*, Springer Tracts in Civil Engineering,
https://doi.org/10.1007/978-3-030-64014-9_5

In early project phases, the spreads of value-related quantity, productivity, and cost ranges will be comparatively wide since key parameters will be associated with greater uncertainties because the project has not yet been sufficiently detailed and refined. The opportunities for the client to influence the project will decrease in line with project progress and the associated specification.

For bidders, and subsequent contractors, the degree of planning and design completion and the quality of the specification documents are key factors with an influence on the submission of their bids and resource planning processes. If, for instance, a large number of estimates need to be included in the costing and pricing exercise (as is the case, for example, for combined price items), any assumptions based on these estimates will inevitably be more uncertain.

This chapter discusses the individual phases of a construction project and demonstrates where clients and contractors can apply probabilistic calculations on a systematic basis. In so doing, we limit our discussion to those options for application in relation to project phases for which we present worked examples in the following chapters. These possible applications are described in more detail in conjunction with the individual examples.

5.1 Phases of a Construction Project

For the purpose of (construction) project management, the process associated with the construction project is divided into five project phases (PPHs) (see Fig. 5.1). These phases start with the release of a resource request and end with the project evaluation step.

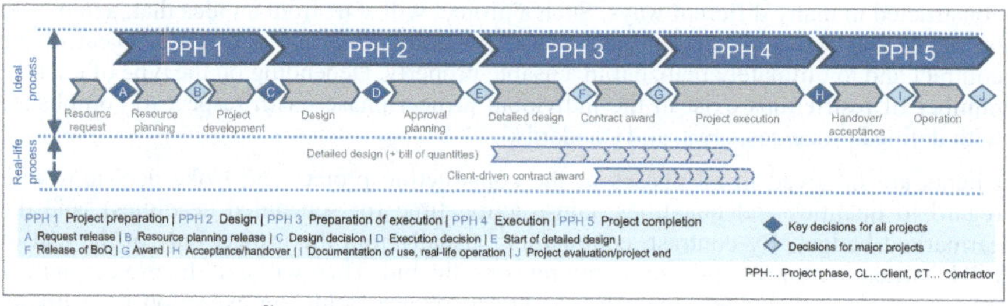

Fig. 5.1 Project phases[2)]

The five project phases, namely *project preparation*, *planning and design*, *preparation of execution*, *project execution*, and *project completion*, can be structured into additional sublevels. So-called Quality Gates (QGs) mark the transitions from one project phase to the next. These quality gates ensure that work on the following project phase begins only after the key decisions for the current project phase have been made.

Prior to releasing the specifications for a new project, a basic project idea must obviously have been developed, followed by the related request release (QG A). The project preparation phase (PPH 1) is subdivided further into the phases of resource planning (QG B) and project development. PPH 1 is finalized by the design decision (QG C), followed by

2) Cf. Lechner (2014). http://www.pmtools.eu/download/seminar/Zeitstrukturmodell_PPH_LPH-A_TW_TA.pdf. Accessed December 21, 2020
Cf. Hofstadler/Kummer (2017a), p. 158

initiating the planning and design phase (preliminary design and design; PPH 2). The implementation decision (QG D) is followed by the application for building permission that also forms part of PPH 2. Its completion constitutes the last but one step of PPH 2, which is finalized by initiating the building permission process.

When the building permission has been granted, work will begin on preparing the detailed design or execution drawings (QG E), which also marks the starting point of PPH 3 (preparation of execution). Upon release of the execution drawings, the associated bills of quantities will be prepared in order to initiate the request for proposals process for the work (QG F) and select the most appropriate bidder (contract award decision, QG G).

In an ideal process, the individual project phases will occur one after the other. According to this assumption, both the detailed design phase and the award of all works to the contractor would be finalized before the actual construction phase starts. In reality, however, the detailed design phase is usually still underway when construction starts; it then continues simultaneously with the construction process. In addition, the client will usually not have completed the award of works in all trades and disciplines at this stage. This will primarily depend on the type of contract award (GC model[3]) vs. award of different works to individual contractors).

All project phases constitute start-end relationships. The transition to the next phase begins when the relevant quality gate has been met. Overlapping the individual phases would obviously save time. Whether this is practicable and useful will depend on the type of the project, its volume and, even more importantly, its complexity.

The total time to project completion can be reduced by overlapping PPH 3 with PPH 4, which is beneficial to the client from an overall perspective because they will be able to start using the property at an earlier point in time. However, this approach will also give rise to certain issues because of the high time pressure usually associated with the design phase progressing simultaneously with the construction process. On the one hand, the fact that the design phase has not yet been completed makes it possible to respond more flexibly to changing project requirements. On the other hand, design changes will usually result in additional costs and/or non-compliance with set deadlines, which could have been avoided if the design were complete at the start of construction.

In addition, there will always be the risk of plans and drawings not being available on time for the respective work step on-site or of a poorer quality standard of such plans and drawings owing to scarce time resources.

However, if no further design changes can be implemented during the construction process, this will also prevent any flexible response to new needs or findings. Any need for change will primarily be the result of new priorities defined in terms of the client's requirements. For multi-year projects in particular, the intensity of design activities during the construction process will increase, as will the proneness to errors or disruptions, for instance in the event of delays of scheduled deliveries. The same applies to buildings where decisions regarding building services or medical equipment can be made only when a relatively long period has passed after commencement of construction. Prime examples include airport terminal and hospital buildings. Also, rapid technological progress and quickly changing rules and regulations will enable decision making only at a very late point in time, which can result in massive modifications to the structural framework, the routing of ducts, and building services. This will inevitably affect construction time and costs to a significant extent.

3) GC stands for "General Contractor".

As explained above, PPH 4 design activities during the construction process are associated with high time pressure, which is why they are also more prone to errors than those prior to the contract award (PPH 3). Likewise, coordinating the planning and design of different trades and disciplines is more complex, and thus associated with greater error proneness, during the construction phase. In this context, the use of BIM[4] provides a number of advantages because it permits more effective coordination of the interfaces between the technical designers involved in the project. A big effort is currently being made in order to further advance the widespread use of BIM so as to unleash its full potential.

The bidder, or subsequent contractor, if selected, will join the project by studying the tendering documents and specifications and deciding in favor of preparing a bid. Compared to the contractor, the client is in possession of much more project-related information in this phase. It is thus imperative for the client to specify the work owed under the contract in a clear, complete, and neutral manner in order to inform the costing and pricing, bid preparation, and process planning exercises to be completed by the contractor. Bidders must be able to rely on the details provided in the specification documents and to perform their costing and pricing exercises on the basis of their experience and expertise while dealing with uncertainties in a systematic manner.

After the contract award decision and commencement of construction in PPH 4, the amount of knowledge and information available for the contractor will continuously increase during the remaining phases of the project. This is why the contractor will "overtake" the client in terms of the information they possess. In the construction phase, the contractor will continuously optimize its production system and make every attempt to perform the specified works to the agreed quality standard and at the lowest possible cost.

Work completion is followed by the handover (from the contractor's perspective) or acceptance (by the client) (QG H). Reaching this quality gate is also referred to as technical project completion. This milestone is followed by PPH 5 (project completion), which includes the handover and acceptance procedure during which any remaining invoices are settled and defects remedied. When these works have been completed (also referred to as commercial project completion), the next step is to perform the documentation of use followed by the trial run or real-life operation phase (QG I). This project stage is the latest point at which contractors will complete and conclude the project unless they also act as the operators of the property, such as in PPP projects[5]. Acceptance triggers the commencement of warranty periods during which the contractor may again get actively involved in the project if defects need to be remedied.

---

4) Building Information Modeling, see Section 3.6

5) Public-private partnership projects

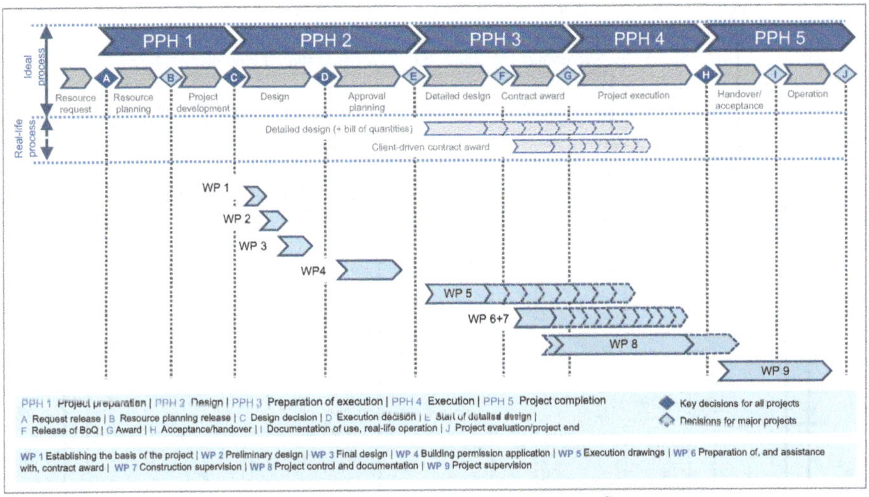

Fig. 5.2 Comparison – Project phases and work phases for designers[6)]

The stage of project completion (end of PPH 5) will inevitably have to include a project evaluation exercise. This evaluation gathers and documents key data and parameters so that empirical values can be systematically fed into new projects or into the client's internal rules and guidelines (QG J).

If several projects are being completed at the same time, these will usually have different volumes, will have progressed to different project stages, and will be associated with deviating chance/risk ratios. Selected individual projects can be analyzed and evaluated holistically adopting a tailored project portfolio management approach, which applies to both the client and the contractor.

Besides being structured into project phases (PPHs), projects can also be divided into work phases (WPs) for the individual design disciplines. Thus, a total of ten work phases (0 to 9) is allocated to Project Phases 1 to 5 outlined above (see Fig. 5.2). Work Phase 0 should be considered to be merely a notional phase.

6) Cf. Lechner (2014). http://www.pmtools.eu/download/seminar/Zeitstrukturmodell_PPH_LPH-A_TW_TA.pdf. Accessed December 21, 2020
Cf. Hofstadler/Kummer (2017a), p. 160

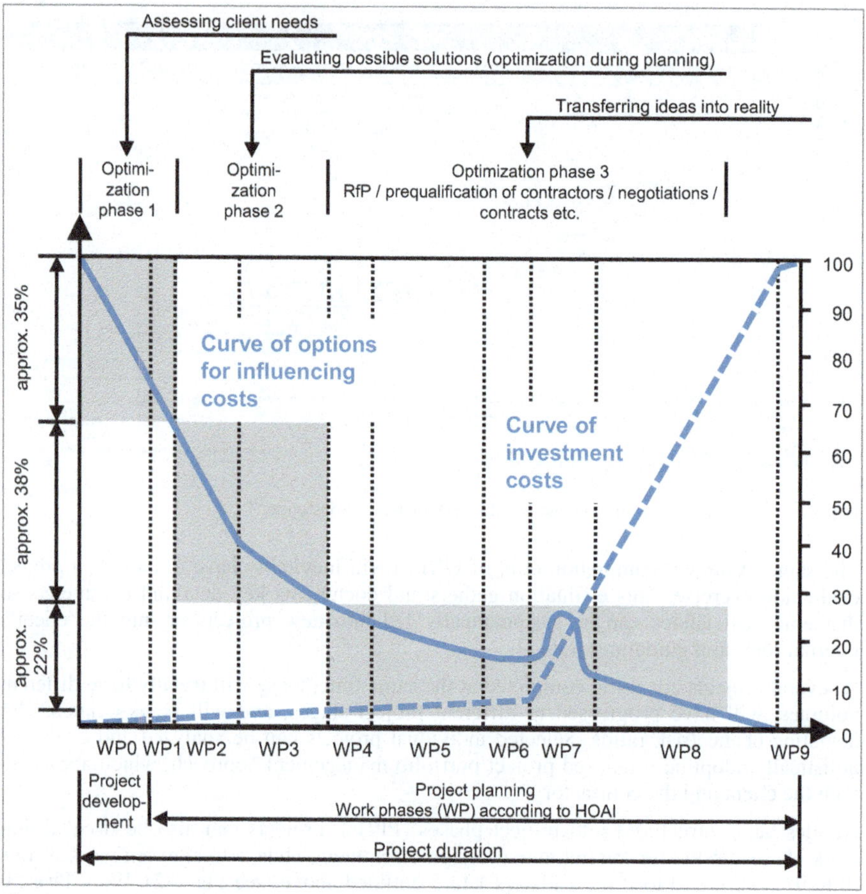

Fig. 5.3 Controllability of costs over the project duration[7)]

Project costs can be controlled most effectively in the initial work phases (see Fig. 5.3); this effectiveness of cost control then decreases over time. Compared to the project preparation phase (WP 0), cost controllability will already be reduced significantly in the phase of preparing the contract award (WP 6) as well as during the construction phase (WP 7). However, at this stage, the information on construction costs will, on the basis of a detailed specification, for the first time reach a degree of accuracy within a commonly applied tolerance range (see the cost funnel with variation tolerances in Fig. 5.5). In addition, Fig. 5.3 illustrates that the execution phase (WP 7) is again associated with an increase in cost controllability, for instance because of supplementary works, claiming optional and/or contingency items, or contract assignments.[8)]

---

7) Cf. Volkmann (2003), p. 115
Cf. Hofstadler/Kummer (2017a), p. 161

8) Cf. Lechner (2010), p. 11

The bidders, or subsequent contractors, if selected, will join the project only at the bid preparation stage or when being awarded the contract. They are bound to adapt their project targets, i.e. combining various production factors in a commercially viable manner, within the range that remains controllable at this point in time. This is why contractors will be particularly interested in performing their work at the lowest possible expenditure level (minimum principle) and thus cost. Conversely, it is also in the interest of the client to keep construction costs as low as possible without compromising the specified quality standard during subsequent use in order to prevent budget overruns.

Considering and evaluating various options with respect to their effects on cost, benefit, and construction time should already be aimed for in early project phases. At any rate, they will contribute to reducing disruptions and errors in the detailed design documents (for instance in the form of value engineering[9)]). The greater the number of scenarios investigated systematically at the early planning stages, the more effectively plans and design documents can be adjusted to essential and/or non-targets. Clear illustrations of cause-effect relationships enable the client to identify and correct diverging or lacking concepts of the envisaged reality at an early stage. An early, integrated approach to planning and design[10)] strives to reduce the number of fundamental design changes becoming necessary as the design approaches the scheduled commencement date of the construction project.

At any rate, an effective integrated planning methodology implemented across the entire organization will improve the constructability of the building. This process is geared towards facilitating work performance, ensuring quality compliance, enhancing both safety at work and productivity, optimizing construction times (by specifying normal construction times) for all stakeholders, eliminating or reducing the need for remedying defects, and mitigating environmental impact. As a result, aspects of sustainability in particular will be considered more comprehensively. In this area, it is particularly advisable to implement systems of incentives (by extending the value chain) in order to promote resource-saving design, construction, and operation.

5.2 Using Tools and Methods of Chance and Risk Management in Project Phases

Fig. 5.4 shows the five project phases for clients and bidders/contractors as well as tools and methods of probabilistic chance/risk management discussed in this book. This provides added guidance and permits targeted access to worked examples and related information.

Both for the client and the bidder, or subsequent contractor, dealing with uncertainties on a systematic basis constitutes an essential part of their chance/risk management during the project, with the spread of related ranges narrowing in line with project progress. This is due to the fact that the level of available information and knowledge in relation to project-specific data will increase in the course of the project. Whereas empirical values and

9) Mental method to improve and further develop products, tasks, workflows, etc. with the aim of optimizing their value or benefit at the lowest possible level of resource utilization.

10) Integral planning combines the creative process with the planning processes of all project stakeholders proactively, simultaneously, and across disciplines. This interaction takes place between clients/users and architects, structural engineers, and technical designers responsible for building services and utilities. To add maximum value, it is preferable to involve experts in the field of project execution in a consulting role in order to define and achieve the project targets.

metrics need to be used in early project phases, project-specific historical data[11] can be accessed in later phases.

Clients and contractors can apply probabilistic calculation methods or tools and methods of chance/risk management in the individual project phases to support their analyses, in their pricing exercises as part of their decision-making process, or for supervision and control purposes. Parameter-based analyses will be carried out up to and including WP 4. Binding bids will have been submitted from WP 6.

In addition, clients and contractors have access to different degrees of knowledge and information when they join the project in the individual phases. Clients will have to systematically account for a greater degree of uncertainty of data and information particularly in the project preparation and planning and design phases, as well as in the run-up to the commencement of construction, but their decisions must be sound and viable nonetheless. One of the key advantages, however, from the client's perspective is that they are still in a position to respond relatively flexibly in early project phases and to manage and control further project progress, for instance with regard to choosing the most appropriate construction site, defining requirements for use and reuse, selecting quality standards as well as the specific award procedure and contract design, and determining construction times and important project milestones.

Only at a later point in time (usually in PPH 3) will contractors join the project. They are then able to respond to the more or less complete client-driven planning and design process and its outcomes, for example in respect of the structural design and specific dimensions. At this stage, contractors will have to rely on the specifications and requirements contained in the tender documents in order to arrive at conclusions regarding the future type and combination of their production factors. In addition, they are confronted with the challenge of having to compete with other bidders amid a fierce price battle. When preparing the bid, it is essential for the contractor to identify potential chances and risks with respect to its own zero-base costing exercise in order to have sufficient room to maneuver on the pricing side. In the bid preparation phase, contractors analyze key metrics of the construction project in order to reach conclusions regarding resource utilization (distribution and intensity) as well as the degree of complexity of the project.

[11] The concept of historical data refers to a database that exists at a certain point in time, such as at the date of access, at the beginning of the year or of the current month (see also Section 2.3.11).

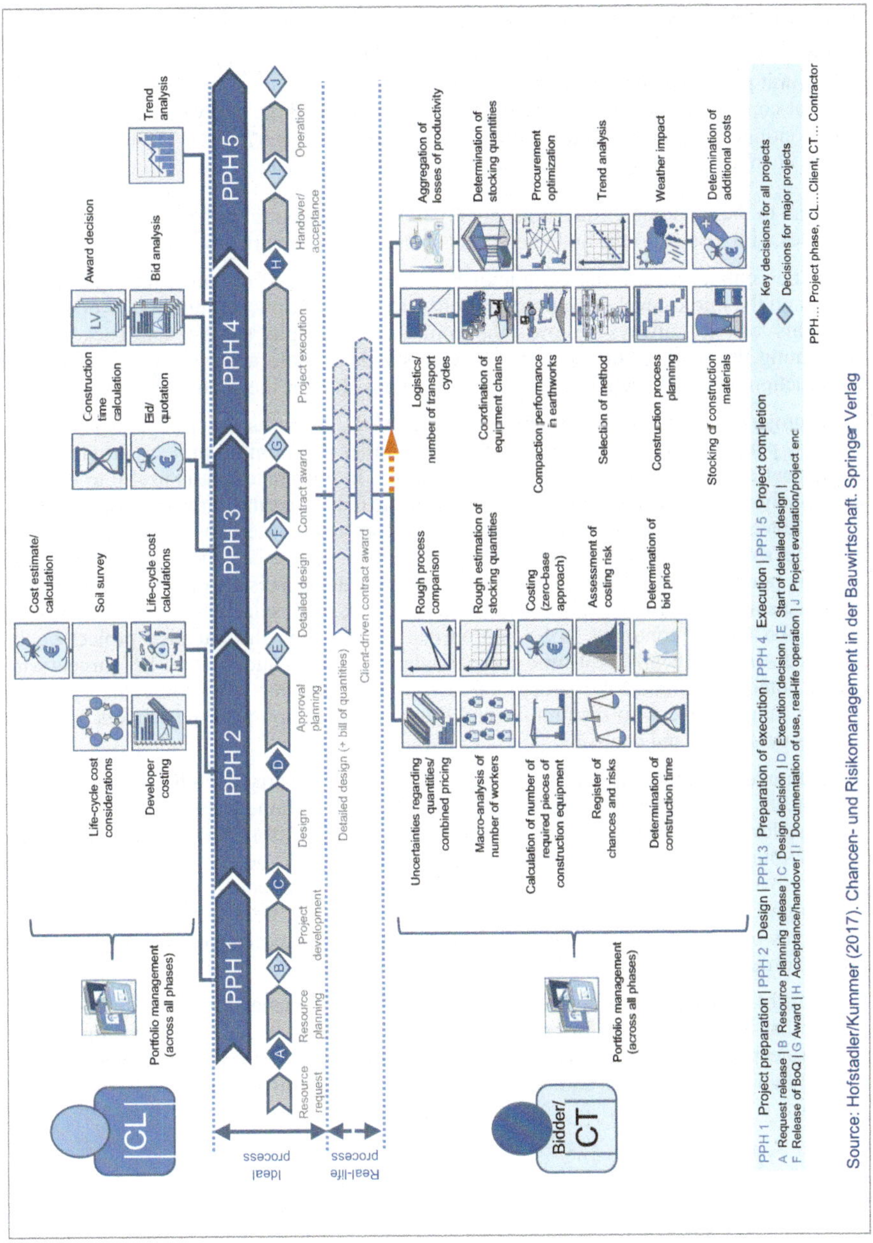

Fig. 5.4 Use of tools and methods of chance/risk management for clients and contractors – Allocation to project phases[12)]

---

12) Cf. Hofstadler/Kummer (2017a), p. 164

After the bid submission phase, it is the client's responsibility to check the received bids for plausibility and appropriateness on the basis of the following questions:

- Would it still be possible, in the detailed design phase, to respond to particularly high or low unit prices stated in the bids?
- What consequences will bid prices have for project funding, and what financial buffer or contingencies does the client include in the project budget?
- Can changes in volumes lead to a shift to a competing bidder?

Answers to these and other questions are crucial for the client in order to avoid unwanted surprises with respect to the budgeted costs and specified construction time.

After the contract award, the contractor begins to adjust and optimize their production system on the micro-level, including the construction workflow, definition of work cycles etc. This activity always follows the minimum principle against the backdrop of performing the work to the specified quality standard while adhering to the scheduled construction time and keeping the cost as low as possible for the contractor.

In the project execution or construction phase (PPH 4), clients as well as contractors continuously perform target-actual comparisons in order to monitor project progress and implement corrective actions if and when required. In this context, it is useful to conduct trend analyses because they rely on project-specific data to permit forecasts or projections of the developments to be expected over the course of a certain period. By way of example, such analyses can cover indicative qualities, performance values, costs, or paid working hours. Another factor to be considered in the analysis is the quality or standard of workmanship in order to conclude, in a timely manner, if the selection and combination of production factors continues to be appropriate for reaching the intended project goals. Near-real-time or, ideally, real-time comparisons of essential and "should-be" targets with actual outcomes provide the benefit of identifying deviations early on. This approach enables rapid feedback to disruptions and thus creates an optimal environment for agile process evaluation and control.

From an overall perspective, it is crucial for contractors to carry out a holistic analysis of all projects currently under construction (with different volumes, chance/risk ratios, and progress levels) in order to generate a balanced project portfolio in the medium to long term. Within this portfolio, project chances and risks should be combined in an optimal fashion so as to ensure long-term commercial viability. It is of course possible to include a few "high-risk" or "landmark" projects in the portfolio in order to pursue the strategic objectives of the company[13)]. Overall, however, the share of chances should exceed the risks in order to operate successfully in the longer term.

The portfolio management approach is also an effective means for major public-sector clients that initiate requests for bids for infrastructural projects, where, for example, new road projects are divided into several lots or subprojects, because it ensures that the total project budget will be met and that the project can be split into different construction sections.

---

13) Such as entering new markets

5.3 Costs of Buildings/Structures – Cost Funnel – Contingencies

At the beginning of the project, designers are unable to predict technical details or "very accurate" costs[14] within the system of work phases. Technical design is a gradual process responding to the increasingly specific requirements defined by the client across work phases 0 to 9 and up to the completed project. In the first few work phases, emerging hypotheses or assumptions are used and verified or falsified in the subsequent planning steps. This methodology ensures that the project is defined increasingly clearly and to a greater degree of detail – a process that can be likened to the hermeneutic loop[15]. As the project progresses, the controllability of the process and the available room to maneuver decrease for clients and designers.

Applying this methodology and initiating this gradual process creates a situation in which initial cost projections will inevitably deviate from the actual costs recorded at the end of the project. Any identical figures would be the result of a random event. This is due to the fact that the cost projection prepared by the designers will be replaced with the (basically unpredictable) wide range of prices quoted by the bidders (a spot market[16] involving highly significant assessment factors with respect to the circumstances under which the bidders should perform their work).

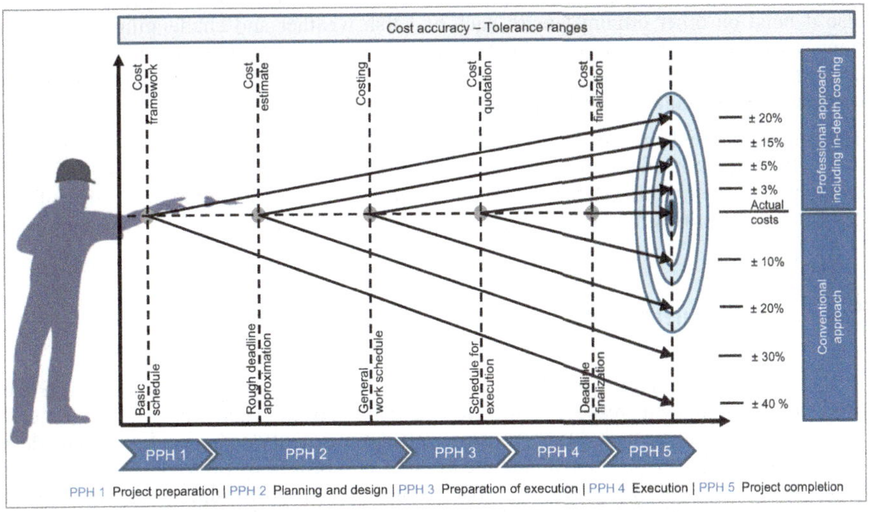

Fig. 5.5 Cost tolerances for professional, in-depth costing and work scheduling versus the conventional approach[17][18]

[14] "Very accurate" cost projections are those that lead to precise and correct results.

[15] Starting from the level of knowledge known at the time t = 0, insights are continuously expanded by knowledge aggregation (through the study of the project documents, codes and standards, as well as of pertinent literature, and by holding and organizing meetings and processes for generating new ideas) as well as critical feedback loops.

[16] Spot pricing market, also referred to as cash market or effective market. A marketplace in which cash transactions are effected, such as the purchase or sale of goods within a short period, as opposed to forward or future transactions. This can be either a cash exchange or an over-the-counter market. Cf. Springer Gabler (2014), p. 304

[17] Based on Lechner (2007), Slide 44

[18] Cf. Hofstadler/Kummer (2017a), p. 187

In Austria, for example, this dilemma is countered by the provisions stipulated in OENORM B 1801-1. This standard includes cost group 9 in which contingencies should be included to account for the volatility of indices or design changes. At the various stages of planning, the A, B, and C final cost items are clarified so that a rule can be defined for a possible projection. The cost funnel cited in the literature appears to be a useful tool for this purpose.

The diagram shown in Fig. 5.5 clearly illustrates that we should account for wider ranges in costing in early project phases. A systematic chance/risk management approach is necessary to obtain reliable information on the specific chance/risk ratio associated with the selection of calculated values. This method makes it possible to deal with uncertainties very effectively and efficiently even at the early stages of the project.

Cost projections can be likened to shooting at a target where the probability of hitting the bull's eye increases – with the aim to predict the final construction costs – as we get closer to the target, i.e. the project completion stage. Fig. 5.5 shows the tolerance ranges for costing that should be achieved in the professional approach, as opposed to those where engineers and/or designers would be held liable under German law in the event of any overrun (conventional approach[19)]).

The same applies to construction time forecasts. Such projections are based on performance values that are derived, for labor-intensive activities, from consumption rates, the number of workers, and the number of working hours per day. In the case of equipment-intensive work, the actual performance will usually depend on the available chains of equipment, particularly on pieces of equipment prone to bottlenecks. Performance values also depend on other circumstances, such as harsh weather and challenging soil conditions, as well as on the skill set and motivation of workers or equipment operators. In this context, all input parameters are characterized by the fact that they are fraught with major uncertainties particularly in early project phases.

Thus, the situational analysis carried out with respect to cost and construction time projections clearly demonstrates that uncertainties in the individual project phases should be taken into account on a systematic basis in order to prevent gaps in the perception of the respective chance/risk ratio and the associated negative implications.

5.4 Summary

In the individual project phases, it is crucial to analyze, synthesize, and evaluate the production system with respect to the envisaged objectives. This applies to all project-related considerations, including estimates, calculations, and decisions. All stakeholders should thoroughly study and evaluate the systems and processes relevant to construction in order to achieve a high quality of forecasts and, thus, project implementation. Appropriate time and funds as well as the integration of contextual models are necessary to achieve a sufficiently high quality in construction project management, planning, and specification.

An optimally implemented construction planning process has a direct influence on the quality of the specification, and thus determines the quality standard achievable in planning and specification activities. This quality standard, in turn, creates the basis for bid pricing and is considered to be an essential factor determining the costing and pricing risks on the bidder's side.

[19)] Based on Lechner (2007), Slide 44

It is thus of considerable relevance to the client to thoroughly deal with the production systems selected by the future contractors. The more effectively the client considers specific project workflows and potential restrictions to implementation in its plans and projections, the greater the probability of achieving the envisaged project targets will be.

At any rate, an effective integrated planning and coordination methodology implemented across the entire organization will enhance the constructability of the building as well as the temporary measures implemented with the aim of supporting the construction process. This process of optimizing the constructability of the building is geared towards facilitating work performance, ensuring compliance with the specified standard of workmanship, enhancing safety at work and productivity, optimizing construction times (by specifying normal construction times) for all stakeholders, eliminating or reducing the need for remedying defects, and mitigating environmental impact. As a result, aspects related to the life cycle of the building will be considered more comprehensively. In this regard, it is advisable to implement systems of incentives (extending the value chain) in order to promote a life cycle-driven, resource-saving design, construction, and operation. Lean design largely relies upon these holistic approaches and the resulting activities. This integrated approach should be practiced not only at the planning and design stage so that all designers enter into a managed coordination process and adopt a collaborative planning mindset. In addition, construction itself should be considered to be an integrated process in which all contractors follow this example and adopt the same managed approach to coordination, or collaborative construction, as a key success driver. Combining elementary production factors effectively and efficiently will be successful particularly if accompanied by an integral combination of discretionary production factors. Lean construction is a method based on these approaches and principles.

A building or structure can be constructed in many different ways. This process starts from a project idea to which more and more detail is added over the course of the project phases; it then becomes the object of the contract at the end of PPH 3 (QG G).

In the individual project phases, clients and contractors have to rely on assumptions for costing and pricing, such as performance or cost values, that are associated with more or less wide ranges of uncertainties. Like a funnel tapering from the wide opening at its top to the neck at its bottom, such uncertainties will exhibit wider spreads in early project phases compared to the contract award stage, where an increasing number of specific project features will be added on the basis of the detailed design documents. As the degree of design detail increases, the information the client provides on the building or structure will also become more comprehensive and accurate. Yet uncertainties remain with respect to the most appropriate mode of combining the production factors as well as to future conditions and circumstances prevailing on-site in the construction phase.

In early project phases, the ranges of quantities as well as productivity and cost values are comparatively wide because the greater potential influence on the overall project must be taken into account at this stage. In later project phases, the construction project becomes increasingly specific whereas the influence of the client on project execution decreases.

Decisions must be made on the basis of uncertain parameters or values, or even rough estimates. Applying the Monte Carlo simulation method permits a systematic approach to dealing with uncertainties and spreads of values. We recommend the targeted application of probabilistic calculation methods both on the client's and the contractor's side in order to increase the level of decision confidence.

This book describes methods and tools applied in combination with Monte Carlo simulations in the first four phases of construction projects.

5.5 References

Hofstadler, Christian; Kummer, Markus (2017a). Chancen- und Risikomanagement in der Bauwirtschaft – Für Auftraggeber und Auftragnehmer in Projektmanagement, Baubetrieb und Bauwirtschaft. Berlin, Heidelberg. Springer-Verlag. (ISBN 978-3-662-54318-4)

Lechner, Hans (2007). Vergleichende Beurteilung von (Bau-) Kostenaussagen in frühen Planungsphasen. Presentation. Graz. Institute of Construction Management and Economics, Graz University of Technology. April 17, 2007.

Lechner, Hans (2014). http://www.pmtools.eu/download/seminar/Zeitstrukturmodell_PP-H_LPH-A_TW_TA.pdf. Accessed December 21, 2020.

OENORM B 1801-1 (2015-12-01 edition). Bauprojekt- und Objektmanagement – Teil 1: Objekterrichtung.

Springer Gabler (2014). Kompakt-Lexikon Wirtschaft – 12th ed. Wiesbaden. Springer Gabler. (ISBN 978-3-658-05790-9)

Volkmann, Walter (2003). Projektabwicklung für Architekten und Ingenieure – Handbuch für die planerische und baupraktische Umsetzung – Methoden, Arbeitshilfen, Vordrucke – 2nd ed. Essen. Verlag für Wirtschaft und Verwaltung Hubert Wingen GmbH + Co. (ISBN 3-8028-0513-5)

6 The Fundamentals of Monte Carlo Simulations

Potential users may not necessarily have (in-depth) knowledge of the Monte Carlo simulation method because it heavily relies on probability theory. This chapter describes its basic principles and includes worked examples that provide further explanation. These hands-on examples provide a fundamental understanding of the method to create the basis for the following chapters. At the same time, they should stimulate independent thinking and application of the method.

6.1 Origins and Evolution of the Method

The Monte Carlo method uses random numbers to enable calculations with the aim of delivering approximate solutions. It is based on the law of large numbers and on the principles of probability theory. It is suitable for determining approximate solutions to mathematical problems or calculations for which it is difficult or impossible to apply analytical methods. The related computation process relies on repetitive steps termed iterations, which are performed using random numbers.

The concept referred to as the Monte Carlo method denotes not an algorithm but a number of numerical methods that use random variables to derive approximate solutions or to simulate a wide variety of processes.[1)]

Random events were used for solving numerical problems as early as in the 18th century. One pertinent example is how an approximation of the number π was obtained by randomly dropping needles onto a sheet of lined paper.[2)]

However, the Monte Carlo method was not widely applied until and during the Second World War. Further research on the Monte Carlo method was conducted in the Manhattan Project in the course of conceiving the first atomic bomb.[3)]

1) Cf. Theis/Kernbichler (2002), p. 2

2) This method of estimating π is known as *Buffon's* needle problem. It is calculated using the angle the needles form with the parallel lines printed on the sheet of paper.

3) In this case, random processes regarding the diffusion of neutrons in fissile material were simulated. Cf. Hengartner/Theodorescu (1978), p. 12f.

C. Hofstadler and M. Kummer, *Chances and Risks in Construction Management and Economics*, Springer Tracts in Civil Engineering,
https://doi.org/10.1007/978-3-030-64014-9_6

John von Neumann and *Stanislaw Ulam* are reported to have developed the original method, though its name was borrowed from the famous casino in Monte Carlo where the results of roulette games had been published on a regular basis. Roulette involves a simple mechanical means for generating (discrete) random numbers.

Metropolis/Ulam[4)] wrote the first publication on this topic entitled "The Monte Carlo Method". It was published in 1949, which is also considered to be the "year of birth" of this method.[5)]

Random numbers generated by (mechanical) experiments involving dice or roulette games, for example, are referred to as discrete random numbers because any such attempt is associated with only a limited number of possible outcomes. Dice, for instance, have six different faces showing values from one to six. French roulette only involves integer values from 0 to 36.

Continuous random numbers are created by random generators, which apply an algorithm for delivering numbers in the interval between zero and one. This is why this method did not become more widespread until computers became available for generating such numbers.[6)]

More detailed information on the generation of random numbers and on sampling methods is provided in Sections 6.3 and 6.3.1.

6.2 Distinction between Method and Simulation

It is not always possible to make a clear distinction between the Monte Carlo method and Monte Carlo simulations. Nonetheless, this section sets out to explain the main differences between the Monte Carlo method, simulations in general, and Monte Carlo simulations.

6.2.1 Monte Carlo Method

The term Monte Carlo method refers to the use of iterations as a method for ascertaining the long-term outcomes of an event. In this process, stochastic techniques are employed to solve a deterministic problem.[7)]

One example of applying the Monte Carlo method, where iterations are used to arrive at an approximate solution to a problem, is the determination of the area of any closed figure. For this purpose, this figure is drawn into a geometric square with a side length of 1. Generating two random numbers between one and zero (on the horizontal and vertical axis) makes it possible to determine random coordinate points in the square area, which can then be entered in the diagram (see Fig. 6.1). Such points are either "hits", such as the $(x_2|y_2)$ coordinate in Fig. 6.1, or "misses", such as the $(x_1|y_1)$ coordinate in Fig. 6.1, depending on the coordinate point being located either inside or outside the figure under consideration.

4) Metropolis/Ulam (1949)

5) Cf. Hengartner/Theodorescu (1978), p. 17

6) Cf. ibid., p. 12f.

7) Cf. Sawilowsky (2003), p. 218f.

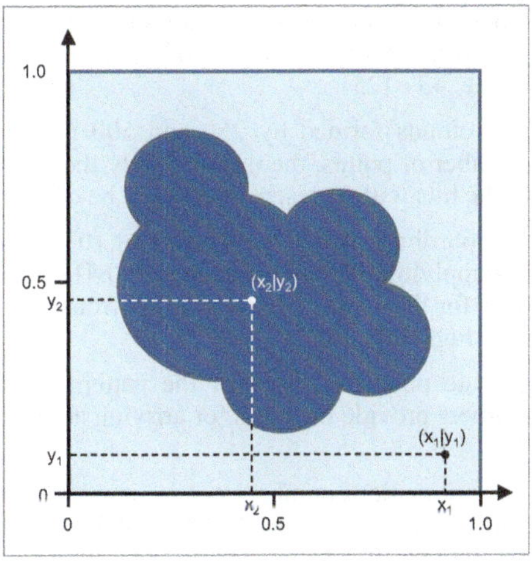

Fig. 6.1 Example – Area calculation using the Monte Carlo method

This step needs to be repeated several times to obtain the approximate area of the figure. We derive this area from the ratio of the number of hits to the total number of generated coordinate points (see Eq. (6-1)).

$$A = \frac{NO_{hits}}{NO_{hits} + NO_{misses}} \tag{6-1}$$

The larger the number of randomly generated coordinates and the more uniform their distribution across the entire square, the more accurate the resulting area of the figure will be.

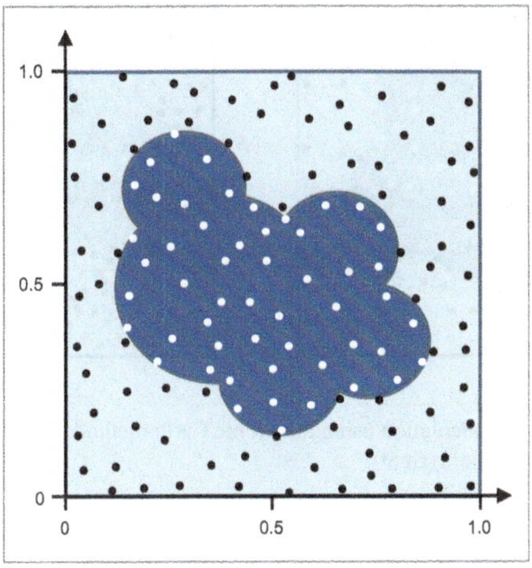

Fig. 6.2 Example – Area calculation using the Monte Carlo method – 125 coordinate points

If only 125 points are generated, for instance, these will form the point cloud shown in Fig. 6.2. This example includes 43 hits located within the relevant area. Correspondingly, there are 82 misses located outside the area (125 - 43 = 82). We use Eq. (6-1) and get the area of the figure: 0.344 (i.e. 43 / 125).

Fig. 6.3 shows the point clouds formed by 250 and 500 random coordinates. Consequently, the larger the number of points, the more densely they will cover the given area, and the more accurately the hits will represent the area to be determined.

In the example with 250 coordinate points, we can count 103 hits (see Fig. 6.3, left). We then calculate the area by applying Eq. (6-1), resulting in 0.412 (i.e. 103 / 125). This value is an approximate solution for the actual area that differs from the result achieved by only 125 coordinate points (see Fig. 6.2).

Accordingly, 500 coordinate points will deliver the pattern shown in Fig. 6.3 (right). Randomly generated numbers provide the basis for arriving at 197 hits. The resulting area is 0.394 (i.e. 197 / 500).

This example illustrates how random numbers can be employed to determine an approximate solution to a problem. This outcome is essentially dependent on the principle of iteration. The greater the number of repetitive steps or iterations (in this case, the greater the number of generated coordinate points), the more accurate the approximated solution will be because the Monte Carlo method is based on the law of large numbers.[8)]

However, it should be noted that the mere use of random numbers and iterations for resolving this example does not make it a true Monte Carlo simulation. This is because a mathematical computation model does not reflect reality or a real system, as would be necessary for a genuine simulation.

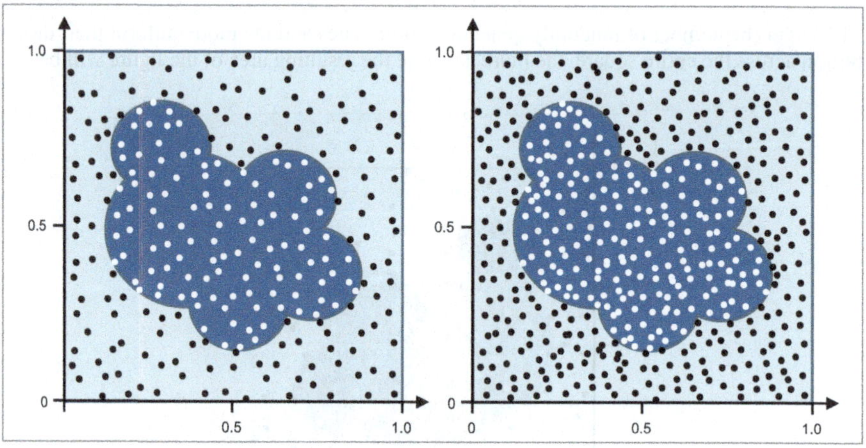

Fig. 6.3 Example – Area calculation using the Monte Carlo method – 250 coordinate points (left) – 500 coordinate points (right)

[8)] According to the law of large numbers, the relative incidence of a random result usually stabilizes around the theoretical probability value if the underlying random experiment or simulation is repeatedly performed under identical conditions (iterations).

6.2.2 Simulation

In this book, the concept of simulation refers to any numerical replication of real and/or planned situations employing a computer-based calculation model. Simulations are useful whenever real-life tests are prohibitively expensive, would potentially destroy the object to be tested, or put human lives at risk. One classic example are flight simulators. Simulations are also applied if real problems can be described by a mathematical model while the application of analytical methods is either impossible or would require excessive computing power.[9)]

In addition, simulations with systematic parameter variations serve to derive findings that are transferable to real-life circumstances or situations.[10)] Here, simulations are used to replicate and structure the space of possibilities or chances.[11)]

Fig. 6.4 illustrates how experimenting options and simulation models can be categorized and classified. Of the sub-items, only those are explained in detail (and marked by solid-line arrows) that are of relevance to this book and its topic.

Starting from the system[12)] under consideration, we can group experimenting options into real-system experiments and experiments based on system models. For this purpose, either physical or mathematical models can be used. This book uses mathematical models in cases where analytical solutions to the examples outlined in the following chapters would require a disproportionately high effort and incur excessive costs. In the case at hand, Monte Carlo simulations are applied.

In a first step, simulation models can be divided into static and dynamic models, with the former completely neglecting the temporal aspect. Dynamic models replicate processes or workflows and can be further subdivided into continuous and discrete models. Discrete simulations are used in production and logistics to plan, assess, improve, and control systems and processes.[13)]

Continuous models can be described by physical or biological laws; they are represented by differential equations.[14)]

Discrete simulation models can be broken down into the event-driven and time-dependent subcategories. In time-dependent simulations, the simulation period is gradually extended, and model components are updated on the basis of status changes recorded in the preceding period. Typical applications include environmental forecasting models, macro-economic simulations, and meteorological models. Event-driven simulations involve status changes that are triggered by events. Such simulations can be subdivided into transaction-based and process-based models.[15)]

---

9) Cf. Suhl/Mellouli (2013), p. 13f.

10) Cf. Wenzel et al. (2008), p. 1

11) Cf. Gottschalk-Mazouz (2011), p. 503

12) "System" generally denotes a totality of elements that are interrelated and interact with one another.

13) Cf. Wenzel et al. (2008), p. 1

14) Cf. Raffel (2001), p. 4

15) Cf. Mattern (1996), p. 4ff.

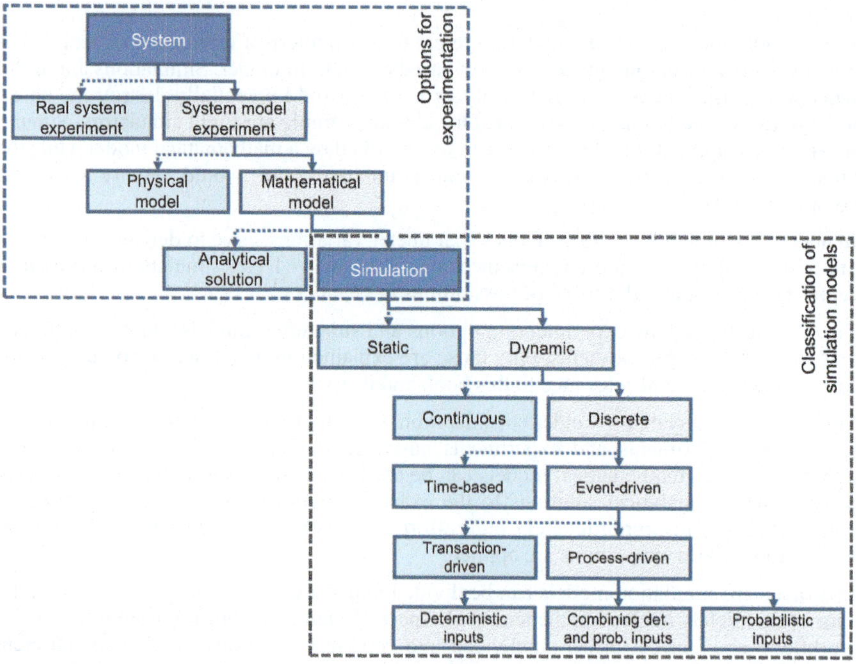

Fig. 6.4 Categorization of experimenting options and classification of simulation models[16)]

Transaction-based simulations are particularly well-suited for queuing systems, where a system of networked stations enables the transfer of work units (transactions) from one station to the next.[17)]

Process-based models merge activities related to an object into a process. Processes can deactivate themselves and activate other processes, such as the occurrence of losses of productivity.[18)]

A final distinction is made between different types of inputs. Once established, the mathematical model will allow either deterministic or probabilistic inputs, or a combination of both. Probabilistic considerations provide the advantage of being able to incorporate uncertainties directly into the calculation process by selecting appropriate distribution functions.

Simulations make it possible to conduct cost-efficient experiments, sensitivity analyses, assessments of available options, and optimization steps (operations research) using a mathematical model. In so doing, it is crucial to ensure that the outcomes of the simulation are indeed transferable to real systems.[19)]

---

16) Cf. Böhnlein (2004), p. 1ff.
Cf. Böhnlein (2010), p. 95ff.
Cf. Raffel (2001), p. 4

17) Cf. Mattern (1996), p. 4f.

18) Cf. Raffel (2001), p. 4

19) Cf. Suhl/Mellouli (2013), p. 269

By way of example, a mathematical model is to be developed to simulate a coin toss. For this purpose, a random generator creates a number between zero and one. If this number is smaller than or equal to 0.50, this means that "heads" (the front of the coin) is the outcome of the simulated toss. If, on the other hand, this random value is greater than 0.50, the simulation result is "tails" (the reverse of the coin).

This example of a simulation uses a random number generator. However, it is not a Monte Carlo simulation because it considers just a single coin toss, rather than involving repetitive (iterative) steps. To be able to draw conclusions from probability theory on the basis of the law of large numbers, several simulation runs (iterations) would have to be performed and the totality of their outcomes would have to be analyzed and assessed, for example by means of histograms.[20)]

6.2.3 Monte Carlo Simulation

Combining a mathematical computation model (simulation) whose input parameters are defined by random numbers (in the form of distribution functions) with the repetitive calculation of random results (Monte Carlo method) using a suitable software is referred to as a Monte Carlo simulation. This numerical method determines the outcome of each simulation step (also called iteration) as a random variable applying a predefined computation model. Monte Carlo simulations perform a number of iterations defined on a case-by-case basis.

The aim of determining the distribution of the sum of several independent random variables, for instance, is achieved by convolving the individual distributions. If several addends are processed, convolution quickly results in complex integrals that need to be solved. Convolution will rapidly reach its limits if additional computations are performed using a range of variables where individual parameters are interdependent (correlations). Monte Carlo simulations are a very efficient means to arrive at approximate solutions to problems such as these, e.g. in probabilistic cost calculations.

Monte Carlo simulations rely on probability theory[21)], statistics, and probabilistics; they are also used in operations research. These applications pertain to subfields of stochastics, which, in turn, is a subfield of mathematics.[22)]

Fig. 6.5 shows the relevant subfields of mathematics outlined in Section 2.3, as well as the subfields into which Monte Carlo simulations, such as for determining construction costs and times, can be classified.

---

20) Cf. Sawilowsky (2003), p. 218f.

21) Probability calculus also encompasses the field of decision theory.

22) Cf. Kummer (2015a), p. 114
The 2010 Mathematics Subject Classification (MSC2010) provides another option for classification into the subdisciplines of mathematics. It includes 62 main categories and more than 6,130 sub-categories. The MSC2010 aims to classify the literature pertaining to mathematics so as to enable interested readers to quickly locate new information and findings in a specific field. Cf. http://www.ams.org/mathscinet/msc/pdfs/classifications2010.pdf. Accessed December 21, 2020.

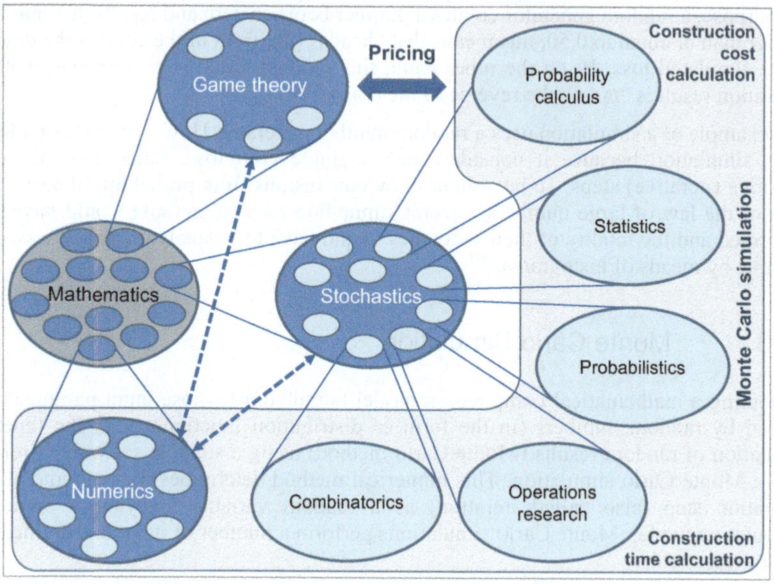

Fig. 6.5 Monte Carlo simulations within subfields of mathematics[23)]

As illustrated in this figure, Monte Carlo simulations belong both in the numerics and stochastics fields because they use random numbers. In other words, the Monte Carlo approach constitutes a numerical method within stochastics.

However, whereas this book refers to Monte Carlo simulations as a means for managing chances and risks, additional strategic considerations would be necessary for pricing, to name but one example. The decision-making process can be optimized primarily by combining results of Monte Carlo simulations with considerations based on game theory. Reference is made specifically to *Werkl's* doctoral thesis[24)] with respect to a more thorough analysis of the bases for decision-making processes as well as the risk/reward behavior of economic stakeholders.

6.2.4 Scenario vs. Iteration

The concepts of iteration and scenario are often used interchangeably, yet they describe different phenomena in conjunction with calculations. Scenarios relate to the (different) contextual conditions of a calculation, whereas iterations describe computation processes performed either in an incremental pattern or repetitively.

23) Cf. Kummer (2015a), p. 113
24) Cf. Werkl (2013)

6.2.4.1 Iteration

In the case of iterations, a distinction should be made between incremental and repetitive iterations. Fig. 6.6 shows a schematic comparison of these two categories.

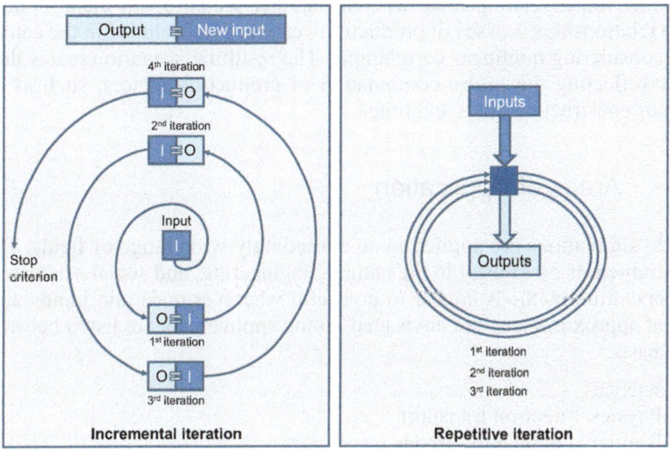

Fig. 6.6 Schematic of incremental and repetitive iterations[25)]

Incremental iterations are applied for aggregating productivity losses or as part of optimization processes, to name but two examples. Here, the results of any given iteration build upon the results of previous iterative steps. The output (O) of an iterative step becomes the new input (I) of the next computation n+1 (see Fig. 6.6, left). Thus, this method allows a stepwise approximation of the output to be determined. Computations are continued until a stop criterion is met, such as a time limit, a predefined number of iterations, or no significant change in the output despite an increasing number of iterations.[26)]

Repetitive iterations are performed in simulations according to the Monte Carlo method. In this case, the random input numbers change independently of the result of the preceding iteration. Individual results of repetitive iterations can be aggregated and shown in a histogram.[27)]

6.2.4.2 Scenario

A scenario constitutes the framework in which the process or workflow is embedded. It refers to the totality of circumstances and conditions under which an event is unfolding.[28)]

When performing computations in relation to future events, for example, we can distinguish between a scenario with certain expectations (i.e. a deterministic computation) and a scenario with uncertain expectations (such as a probabilistic computation).

[25)] Cf. Kummer (2015a), p. 137

[26)] Cf. ibid.

[27)] Cf. ibid.

[28)] Cf. Duden (2007), p. 1654

A scenario thus defines the underlying conditions in which a computation is performed. Other possible scenarios associated with Monte Carlo simulations include computations with independent variables or computations that define dependencies between parameters by means of correlations.

When calculating construction time, for instance, one of the selected scenarios may exclusively consider linear relationships, whereas another scenario can incorporate nonlinearities in the relationships. Losses of productivity can thus be included in the calculations as a result of considering nonlinear correlations. The resulting scenario creates the basis for realistically reflecting the future combination of production factors, such as during the calculation of construction costs and times.

6.2.5 Areas of Application

Monte Carlo simulations are applied in an exceedingly wide range of fields. This mathematical instrument is often used in the natural, engineering, and social sciences[29)]. Probabilistic computation methods appear to be useful wherever uncertain inputs are assumed or numerical approximations are envisaged. Some applications are listed below on a non-exhaustive basis:

- Natural sciences
    - Physics – neutron transport
    - Biology – population trends
    - Medicine – drug development

- Engineering sciences
    - Structural engineering – determination of partial safety factors, material behavior
    - Industrial engineering – determination of costs and times
    - Construction management – determination of resources, logistics, analyses for selecting processes and methods, site mobilization, scoping
    - Energy sector – estimating crude oil reserves and resources, nuclear fission and fusion processes
    - Logistics-related analyses
    - Water supply and sanitation, transportation – modeling and analysis of networks
    - Queueing models
    - Life-cycle cost analyses
    - Optimization processes

- Social sciences
    - Business and finance – structuring of securities, investment calculus
    - Macro-economics – economic forecasts
    - Costing, pricing
    - Demographics

[29)] This also includes economics as a subfield of social sciences. Construction management and economics are application-oriented scientific fields that can be grouped into both the engineering and the social sciences.

6.2.5.1 Worked Example – Sum of Die Values

To demonstrate the working principle of Monte Carlo simulations, the probability of the sum of the values of two dice is determined both analytically and by means of a Monte Carlo simulation.

For this purpose, we assume two dice, each having six faces, and equal probability for each of the faces to lie face-down after the throw. We further assume that the two dice have identical characteristics in terms of their surface, weight, and shape. Our computation thus gives a probability of one sixth, or 16.67%, for each of the faces.

The sum of the values of two dice can lie within a range from two to twelve. Each of the values within this range can result from different pairs of values:

- $1 + 1 = 2$
- $1 + 2 = 2 + 1 = 3$
- $1 + 3 = 2 + 2 = 3 + 1 = 4$
- $1 + 4 = 2 + 3 = 3 + 2 = 4 + 1 = 5$
- $1 + 5 = 2 + 4 = 3 + 3 = 4 + 2 = 5 + 1 = 6$
- $1 + 6 = 2 + 5 = 3 + 4 = 4 + 3 = 5 + 2 = 6 + 1 = 7$
- $2 + 6 = 3 + 5 = 4 + 4 = 5 + 3 = 6 + 2 = 8$
- $3 + 6 = 4 + 5 = 5 + 4 = 6 + 3 = 9$
- $4 + 6 = 5 + 5 = 6 + 4 = 10$
- $5 + 6 = 6 + 5 = 11$
- $6 + 6 = 12$

This simple example demonstrates the possibility of arriving at a large number of combinations of possible results. Although this system involves only two discrete random variables, listing the complete set of possible combinations already requires considerable time and effort.

Overall, the above example includes 36 potential combinations, which define the so-called event space. There is a 100% probability of one of the combinations occurring during random rolls of the dice.

Eq. (6-2) can also be applied to determine the number of possible combinations. In this equation, n represents the number of faces of a die whereas k stands for the number of dice.

$$\text{Number of possible combinations} = n^k \tag{6-2}$$

Since only one possible combination exists for either of the sums two and twelve, their probability of occurrence equals 1/36, or 2.78%.

The same pattern can also be used to determine the probabilities for the other possible sums of values. There is a probability of 6/36, or 16.67%, of getting a sum of seven from two dice. This is the highest probability of all possible results.

As an alternative to the analytical approach, applying a Monte Carlo simulation for determining probabilities is also shown. Any simulation starts from a modeling exercise in which a mathematical computation model is applied to capture a real or planned condition or system behavior in an abstract form. This is why we first need to simulate the possible outcomes of rolling a die in a model or by means of a distribution function.

The equal probability of 16.67% determined for any of the die values means that only discrete values are possible. Thus, a discrete rectangular distribution (see Fig. 6.7) appears to be appropriate for this modeling step.

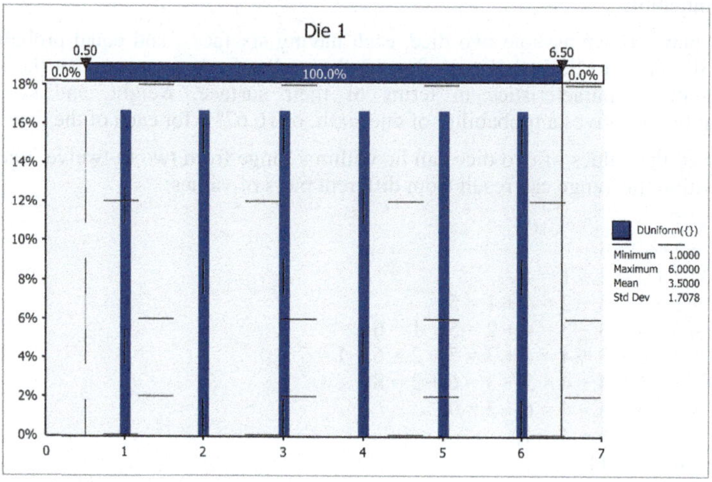

Fig. 6.7 Probability distribution – Possible sums of die values

The diagram shows the die values (possible results) on the horizontal axis and their associated probabilities of occurrence on the vertical axis. The six bars illustrate that the probability of occurrence (16.67%) is identical for each of the values.

In the next step, the computation model defines two independent random variables by applying the distribution shown in Fig. 6.7. The sum of the two random variables is defined as the output of the simulation. After several repetitions (iterations) of the die experiment and gathering of results, this output can be represented in a histogram that reflects approximate probabilities of occurrence for the individual possible results (see Fig. 6.8).

For discrete distributions as shown in Fig. 6.8, the probability of occurrence of each outcome can be read directly on the vertical axis. For continuous distributions, the probability of over- or underrunning a certain deterministic value can be stated. If an output occurring within a certain range of values is of interest, we can define this range by applying two deterministic values and derive the corresponding probability.

Fig. 6.8 illustrates how slide controls can be used to define a range from 1.00 to 13.00 on the upper edge of the image. There is a 100% probability for the sum of values from two dice to lie within this range.[30)]

In the course of the simulation, a value between one and six is randomly selected for each of the two defined random variables (cf. Fig. 6.7), and their sum is calculated. Individual results are aggregated in a histogram, as shown in Fig. 6.8. The larger the number of iterations, the closer the result will get to the outcome of the analytical calculation. In the example referred to above, 50,000 iterations were performed using the Latin Hypercube sampling method.

30) In the @Risk software, slide controls can be moved randomly, directly indicating the probability for the three defined areas.

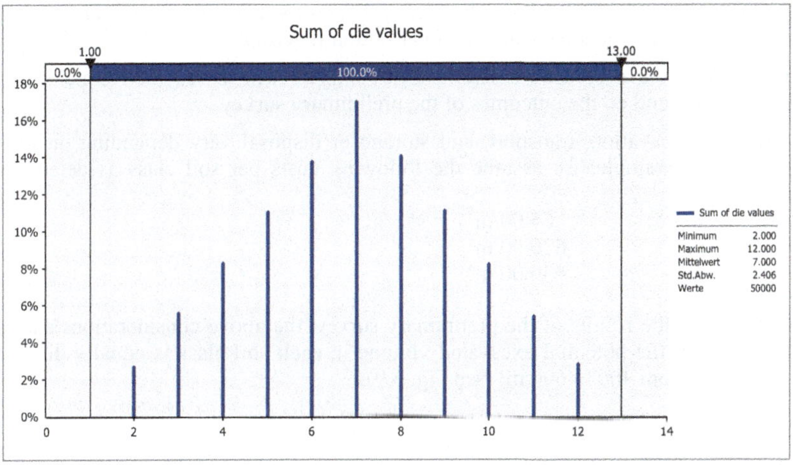

Fig. 6.8 Probability distribution – Sum of values of two dice[31)]

Fig. 6.8 shows the resulting histogram representing the sum of values obtained in the die experiments after 50,000 iterations. The fact that the output is a discrete distribution permits the conclusion that a sum of seven is associated with the highest probability of occurrence.

In the simulation, additional dice can be considered without much additional effort because only one other independent distribution needs to be integrated in the summation. For the analytical solution, however, the new event space comprising the number of possible combinations and the associated probabilities would have to be determined in the initial step.

Compared to analytical methods, the benefits of numerical approaches become particularly apparent in highly complex computation models and the correspondingly large number of required calculation steps.

6.2.5.2 Worked Example – Excavations

The advantages of probabilistic calculations are best highlighted by another hands-on example dealing with the volume excavated from a construction site. We assume 1,500 m^3 of solid excavated material showing no signs of loosening or loss of density. Overall, three soil classes are expected to occur in the excavation area. Preliminary tests permit the conclusion that one-third of the total excavated volume can be allocated to each of the soil classes.

However, this expert opinion is fraught with uncertainties because it is impossible to perform a complete soil survey to eliminate all uncertainties.

---

31) Even though 50,000 iterations using the Latin Hypercube sampling method were performed for deriving the histogram, comparing the results for the values of six and eight reveals a slight difference in the bar height that is attributable to the numerical calculation. Analytical calculation of the probabilities results in exactly the same probabilities for die values of six and eight (5/36, or 13.89%).

We thus assume a spread of ±20% for the volume allocated to each of the soil classes. Furthermore, the probability should be identical for all values that lie within this defined range. We therefore select a rectangular distribution function.

Figures result from subjective assessments by experts who based their conclusions on their experience and on the outcomes of the preliminary survey.

The costs of excavation, transport, and storage or disposal vary depending on the soil class. For this example, we assume the following costs per soil class as deterministic values:

- Soil classes 3-5: €5.00/m³
- Soil class 6: €20.00/m³
- Soil class 7: €30.00/m³

Combined with the results of the preliminary survey, the above considerations suggest a model in which the potential excavated volume in each soil class is equally distributed across a range from 400 to 600 m³ (see Fig. 6.9).

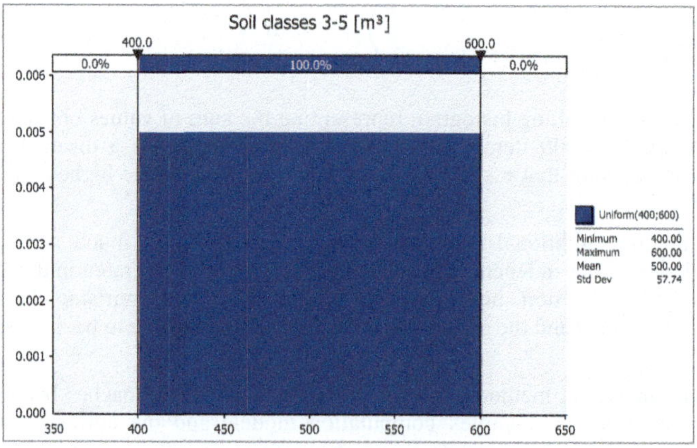

Fig. 6.9 Distribution function for excavated volumes in soil classes 3-5

Unlike in the die example above (see Section 6.2.5.1), each of the values in the given range (i.e. 400 to 600 m³) is equally probable, which applies to any volume between the minimum and the maximum. Identical distribution functions are assumed for the remaining two soil classes.[32)]

In each calculation step (iteration) of the Monte Carlo simulation, a random value is chosen from the ranges determined for the three soil classes, and multiplied by the respective unit price. The resulting sum represents a distribution for the price to be expected for the total excavated volume.

It should be noted, however, that deviations from the initially assumed excavated volume of 1,500 m³ will occur during the simulation, which are due to the random distribution of the excavated material across the three soil classes. On either end of the distribution range, extremes may occur, i.e. a minimum of just 1,200 m³ (= 3 * 400 m³) and a maximum of

32) If, for example, an increased probability can be allocated to a specific quantity value, a unimodal distribution (e.g. a triangular or PERT distribution) could also be applied when selecting the distribution functions for input parameters.

1,800 m³ (= 3 * 600 m³). The modeling exercise thus includes normalization of volumes in each iterative step to prevent unwanted and/or unrealistic total volumes from occurring in the iterations. Accordingly, randomly chosen values are normalized to the total excavated (fixed) volume of 1,500 m³ and subsequently used to establish the price of excavated material.

Fig. 6.10 shows the histogram representing the price of the total excavated volume. This illustration reveals that a significant spread of the total price occurs that is exclusively attributable to the shifts possible in the volume ratios within the soil classes. There is a 90% probability for the price to lie between €25,770 and €29,233. Correspondingly, there is a probability of 5% for either over- or underrunning these two prices (see slide controls on the top edge of the histogram shown in Fig. 6.10).[33)]

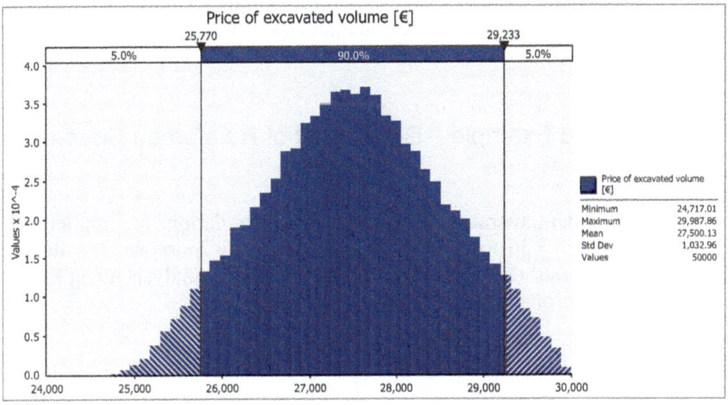

Fig. 6.10 Histogram – Price of total excavated volume

In the above example, the number of possible combinations increases not only as a result of the three considered soil classes but also due to the continuous distributions applied to the subsets. Within the defined ranges, any of the values (with any number of places behind the decimal point) can be used for establishing possible combinations. Any analy-

33) The minimum excavation price can also be determined analytically based on theoretical considerations developed for this simple example. If the amount of excavated material is the largest in the least expensive soil class and the smallest in the most expensive soil class, this analysis will result in a minimum price of €25,000 (= 600 m³ * €5.00/m³ + 500 m³ * €20/m³ + 400 m³ * €30/m³). However, the histogram also includes prices below €25,000 (the minimum is €24,717.01). This outcome is due to the quantity proportions being normalized to 1,500 m³. In rare cases, normalization will lead to an overrun of the ranges originally defined for the soil classes. In the case at hand, this phenomenon would occur if a value close to 600 m³ were selected for soil classes 3-5 and a value close to 400 m³ were selected for each of the soil classes 6 and 7 in a single iteration (resulting in a total of approx. 1,400 m³). Normalization converts the randomly selected values in such a way that we obtain a total excavated volume of 1,500 m³. According to the applied system, this volume is composed of the following: Soil classes 3-5: 642.86 m³ (= 600 m³/1,400 m³ * 1,500 m³) | Soil classes 6 and 7: 428.57 m³ (= 400 m³/1,400 m³ * 1,500 m³). If we multiply these volumes by the deterministic prices stated above, we obtain a total price of €24,642.86 (= 642.86 m³ * €5.00/m³ + 428,57 m³ * €20/m³ + 428.57 m³ * €30/m³), which is thus below the theoretical minimum of €25,000. Values below €25,000 are thus attributable to a pragmatic modeling approach that is tolerated with a view to the inherently uncertain parameters. In the case at hand, only 70 out of 50,000 iterations involve a combination of volumes resulting in a price below €25,000. This is equivalent to a percentage of 0.14% (= 70/50,000 * 100%).
The above-described effect can be countered already at the modeling stage by means of a step-by-step iteration as part of the normalization process, or by introducing fixed limits for the individual volume proportions. Alternatively, a filter can be applied to the simulation result in order to exclude values above and below the theoretical extremes. The larger the number of individual components (in this case, soil classes) considered in the model for calculating a total quantity, the more improbable such a special case will become.

tical calculation of this example that involves only few variables and simple distribution functions will already require in-depth mathematical knowledge.

Performing a Monte Carlo simulation with a sufficiently large number of iterations (in this case, 50,000) will provide a very good approximation to the results.

Construction management and economics focus on defining input parameters, deciding if and when correlations should be considered, modeling, and interpreting results, rather than attempting to solve complex equations, integrals, and problems related to probability theory.

In this context, Monte Carlo simulations provide a tool that includes numerical computations based on random numbers in order to come as close as possible to an accurate analytical solution to a mathematical problem. A large number of iterations and the selection of "smart" sampling methods, such as Latin Hypercube sampling, make it possible to derive sufficiently accurate solutions in construction management and economics while requiring only little time and effort.

6.2.5.3 Worked Example – Direct Cost of Reinforced Concrete Operations

This worked example demonstrates how Monte Carlo simulations are applied as part of a computation exercise in construction economics. For this purpose, the direct cost of reinforced concrete operations is determined in a preliminary analysis using Eq. (6-3) with both deterministic and probabilistic values.

$$C_{RCO} = (MW \cdot LCR_{RCO} + C_{EQ} + C_{MA}) \cdot C_V \qquad (6\text{-}3)$$

We obtain the direct cost of reinforced concrete operations C_{RCO} [€] by multiplying the mean wage MW [€/wh] by the total labor consumption rate of reinforced concrete operations LCR_{RCO} [wh/m³] and adding equipment costs C_{EQ} [€/m³] and material costs C_{MA} [€/m³]. The sum is then multiplied by the reinforced concrete volume C_V [m³].

If purely deterministic values (i.e. specific figures) are applied to the individual computation parameters (see Column D in Table 6.1), this computation will again result in a specific figure (Cell D6). Even though this single value has been calculated correctly, it fails to consider the uncertainties associated with these computation parameters. If this result is to be used for forecasting future costs, this will most likely lead to this value being missed, even if only by a slight margin. Deterministic calculations do not form an appropriate basis for drawing conclusions with respect to uncertainties, ranges, and probabilities.

No.	Calculation parameters		Unit	MOST LIKELY
0	A	B	C	D
1	Mean wage	MW	€/wh	34.00
2	Total labor consumption rate	LCR_{RCO}	wh/m³	5.90
3	Cost of equipment	C_{EQ}	€/m³	65.00
4	Cost of materials	C_{MA}	€/m³	180.00
5	Concrete volume	C_V	m³	38,000.00
6	**Direct cost – reinforced concrete operations**	C_{RCO}	**€**	**16,932,800.00**

Table 6.1 Deterministic computation of the direct cost of reinforced concrete operations

If we use the same deterministic approach according to Eq. (6-3), but replace the deterministic inputs for the computation parameters with distributions, we can perform a Monte Carlo simulation with several thousand individual results yielding a histogram that can be used to derive conclusions with regard to probabilities.

In Table 6.2, a triangular distribution was assumed for each of the inputs (Rows 1 to 5). This distribution is defined by minimum, most likely, and maximum values (Columns D to F). Column G shows the distributions as such. The figures in Columns G1 to G5 represent the mean values of the defined distributions. When applying these means in Eq. (6-3), we get the direct cost value entered in Cell G6.[34)]

However, performing a Monte Carlo simulation will not just result in a single value but, in the case at hand, in 50,000 distinct results (i.e. 50,000 repetitive iterations). Cell G6 shows the histogram resulting from these iterations[35)] that reflects the range of possible scenarios. For the sake of clarity, the histogram is shown in a separate figure (see Fig. 6.11, bottom), rather than being included in Table 6.2.

No.	Calculation parameters		Unit	MIN	MOST LIKELY	MAX	Distribution
0	A	B	C	D	E	F	G
1	Mean wage	MW	€/wh	32.00	34.00	36.00	34.00
2	Total labor consumption rate	LCR_{RCO}	wh/m³	5.30	5.90	6.80	6.00
3	Cost of equipment	C_{EQ}	€/m³	61.00	65.00	69.00	65.00
4	Cost of materials	C_{MA}	€/m³	165.00	180.00	192.00	179.00
5	Concrete volume	C_V	m³	36,600.00	38,000.00	39,200.00	37,933.33
6	**Direct cost – reinforced concrete operations**	C_{RCO}	€				**16,994,133.33**

Table 6.2 Probabilistic computation of the direct cost of reinforced concrete operations

Fig. 6.11 illustrates the working principle of the simulation. In each of the iterations, random values from within the range of the applied triangular distributions are selected (represented by white arrows), and the previously established Eq. (6-3) is used to calculate a result for the direct cost of reinforced concrete operations. Summarizing these distinct results in a histogram allows a range to be established within which, for instance, the central 90% of all calculated values are located (range between the 5% and the 95% quantile).

The white arrow shown in the direct cost histogram represents the value calculated according to the deterministic method, i.e. €16,932,800.00. Only a Monte Carlo simulation will identify the probability of exceeding or underrunning this single value, and/or where it is located within the determined range.

34) Minor deviations from a "manual" calculation may occur owing to the fact that the values entered in the table were calculated in MS Excel (i.e. with more decimal places than actually shown).

35) Specific feature of the @Risk software used in the simulations.

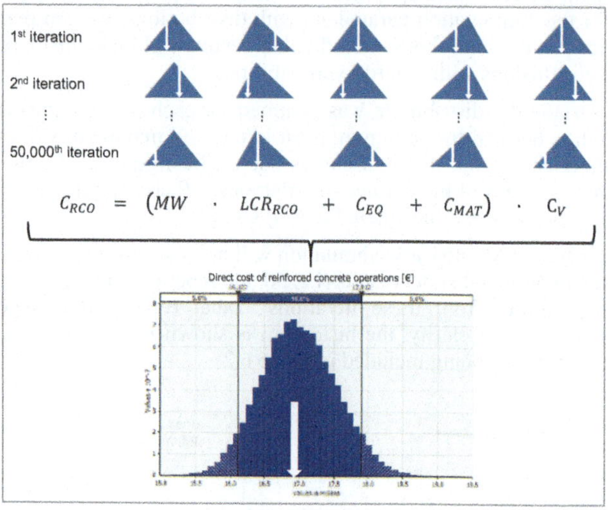

Fig. 6.11 Working principle of Monte Carlo simulations

In the above example, we assumed input parameters to be independent of one another. It appears plausible to use correlations in order to even better reflect and consider relationships between parameters (see Section 6.5).

The type of results derived from of Monte Carlo simulations might initially be unusual for engineers because they describe the uncertainties, or blurred boundaries, of a computation, rather than providing a clear-cut picture. Such uncertainties can be taken into account in subsequent considerations and decision-making processes by representing them as histograms. A defined certainty level makes it possible to quantify the risk taken or the chance expected of a decision. Decisions can thus be based on objective criteria.[36)]

The following section will deal with generating random numbers. In so doing, we distinguish between true and pseudo-random numbers. In addition, different sampling methods are highlighted and explained, including the Monte Carlo and Latin Hypercube sampling techniques, with the aim of creating a better understanding of the processes and workflows that are usually run automatically in the background by an appropriate software package.

6.3 Generation of Random Numbers

Random numbers are a crucial component of Monte Carlo simulations, but they are also essential to various subfields of statistics as well as cryptography. Random numbers create an artificial sample determined by means of a random generator. In today's applications, these generated numbers are not true random numbers resulting from random experiments, such as roulette games, but instead, so-called pseudo-random numbers generated on the basis of a given algorithm.[37)]

---

36) Cf. Lederer (1982), p. 45

37) Cf. Hengartner/Theodorescu (1978), p. 12f.

State-of-the-art software suites commonly include the option of selecting one out of several algorithms (random generators[38]). Here, a distinction is generally made between non-deterministically and deterministically generated random numbers.

Under identical starting conditions, deterministic random generators will always deliver the same sequence of random numbers. In contrast, non-deterministic random generators always create different random numbers, even under identical conditions. Software programs that always apply a deterministic approach must involve an external (e.g. physical) process for implementing a non-deterministic random generator.

If the number of iterations is sufficiently large and combined with high-quality random numbers (see below for relevant quality criteria), applying a deterministic random generator may even be associated with certain advantages. Deterministic random generators make it possible to repeat simulations in order to replicate the same results. This is why it is not always necessary to store all of the iteration results. Rather, they can be resimulated even after closing and subsequently restarting the software.

A random number generator is considered to be non-deterministic if it delivers different values even under identical starting conditions. Since implemented software procedures always apply a deterministic approach, an external (e.g. physical) process must be involved for creating a non-deterministic random number generator.

Software suites often rely on non-deterministic random generators that use "quasi-random" events as inputs, such as the system time until a user action occurs. A previously non-deterministic random generator becomes deterministic if the seed for generating random numbers is determined as a fixed value.[39]

In practice, random numbers are thus generated by a computer on the basis of an algorithm. These numbers are not "true", but so-called pseudo-random numbers because they result from applying deterministic methods. Random numbers are primarily characterized by their independence and uniform distribution. For pseudo-random numbers, the random generator and the underlying algorithm must meet these and other conditions. Other requirements to be met by pseudo-random numbers include:[40]

- Efficiency: Random numbers should be generated quickly without requiring much storage space.
- Reproducibility: Defining a seed, i.e. a starting value for generating random numbers, makes it possible to restore the random numbers for resimulation (deterministic random numbers).
- Long cycles: True random number sequences exist if such sequences cannot be described by a law. This is why random numbers generated by algorithms (pseudo-random numbers) are not truly random, and are characterized by cycles that should be as long as possible to avoid the repeated generation/use of the same (pseudo) random numbers.

[38] Version 6 of @Risk includes the following eight random generators: RAN3I; MersenneTwister; MRG32k3a; MWC; KISS; LFIB4; SWB; KISS_SWB. MersenneTwister is used as the default for the simulations. This algorithm has a long period (return cycle of ~ 4,3 * 10^6001) and a very low correlation between consecutive values occurring in the output sequence. MersenneTwister is available online as a freeware program: http://www.math.sci.hiroshima-u.ac.jp/~m-mat/MT/emt.html. Accessed December 21, 2020.

[39] If no seed is defined, results of individual simulation runs will always slightly deviate from each other despite unchanged inputs and modeling because random values are applied. Selecting a larger number of iterations can reduce the variances between simulations because the stability of results increases with the number of iterations.

[40] Cf. Suhl/Mellouli (2013), p. 280f.

Simulation experiments require generating random variables with characteristics that correspond to the selected input distributions. Although random generators are suitable for generating (pseudo) random numbers, special sampling methods are necessary for selecting random values within defined distributions that can subsequently be used for computation steps as part of the simulation.

6.3.1 Sampling Methods

The @Risk software used in this example provides two different sampling methods for selecting random values for a (repetitive) iteration from the predefined distributions of input parameters:

- Monte Carlo sampling method
- Latin Hypercube sampling method

Either of these methods starts from the cumulative probability distribution (sum curve) of the inputs with selecting random numbers from the given range. These randomly "scanned" values are then used for performing the intended calculations.

6.3.1.1 Monte Carlo Sampling Method

Fig. 6.12 illustrates the working principle of the Monte Carlo sampling method. First, for each iteration, a random value (pseudo-random number) between 0 and 1 is generated by means of a random generator (see Section 6.3). These random values are then localized on the vertical axis of the cumulative probability distribution, and a horizontal line is drawn to get an intersection with the sum curve. Extension of a vertical line from this intersecting point to the horizontal axis delivers a random value within the defined range. This value can be used as an input for the iteration. This step is repeated until the number of iterations defined for the simulation has been reached. The flaw of this method becomes apparent if only a few iterations are completed. In this case, clustering of scanned values may occur on the horizontal axis (see Fig. 6.12). Consequently, any extremes located in the boundary zones are hardly (or not at all) considered in the simulation. This applies particularly to values located in areas characterized by a flat sum curve (representing a low probability of occurrence), which are captured less frequently in the case of only a few iterations. Generally speaking, this means that the Monte Carlo sampling method requires more time (in other words, a larger number of iterations) in order to reflect the defined probability functions with their random values.

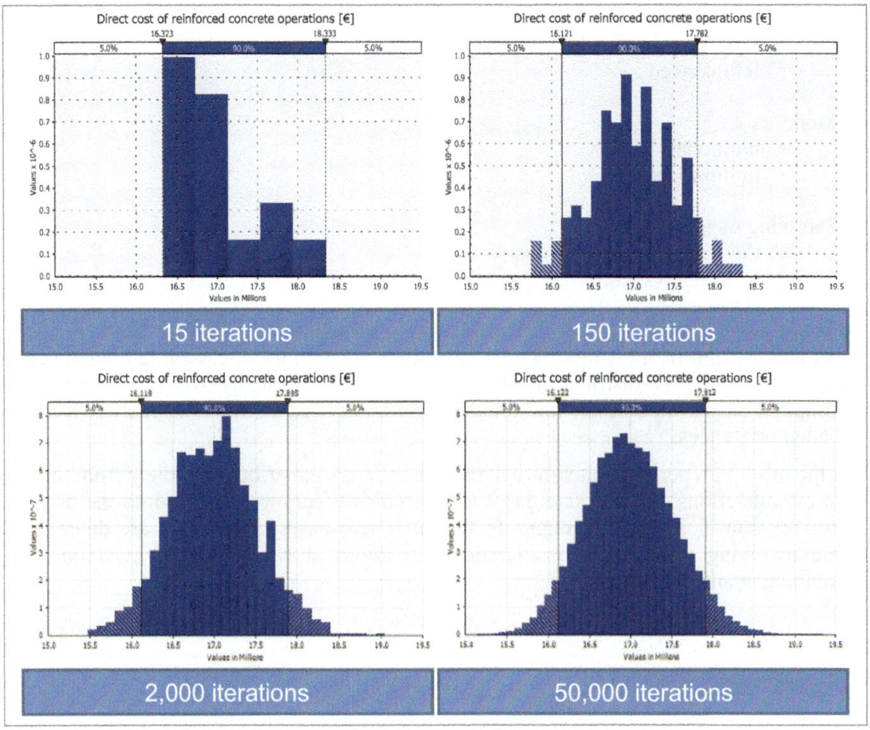

Fig. 6.14 Influence of the number of iterations on the representation of the results of a Monte Carlo simulation (Latin Hypercube sampling)

6.4 Distribution Functions

To be able to perform Monte Carlo simulations, the modeling step determining calculation rules and methods must be followed by the allocation of distribution functions to input parameters associated with uncertainties.

Table 6.3 to 6.5 below include a comparison of selected approaches to the distribution functions of various parameters taken from the literature.

The following differentiation criteria are applied:

- Curve
    - discrete
    - continuous

- Type of boundaries
    - closed
    - one-sided open
    - two-sided open

- Skewness
    - symmetrical
    - right-skewed
    - left-skewed

- Mode
    - unimodal
    - multimodal

- Sampling method
    - theoretical considerations
    - historical vs. collected data
    - surveys

What is most significant is the frequent indication of continuous, closed, unimodal distributions. Right-skewed distributions are also mentioned repeatedly. Discrete distributions are applied only in rare cases; multimodal and left-skewed distributions are not listed in the cited references.[45)]

For the most part, pertinent literature derives the shape of distributions solely from theoretical considerations. Only in rare cases is an empirical survey conducted on the basis of historical data or survey outcomes. In the latter two cases, parameters are determined while involving theoretical considerations mostly in a complementary fashion, thus combining available methods.[46)]

45) Cf. Kummer (2015a), p. 166f.

46) Cf. ibid., p. 167

No.	Author	Parameters	Curve		Limits			Skewness			Mode		Sampling method			Specific distribution
			discrete	continuous	closed	one-sided open	two-sided open	symmetrical	right-skewed	left-skewed	unimodal	multimodal	theoretical considerations	historical vs collected data	survey	
0	A	B	C	D	E	F	G	H	I	J	K	L	M	N	O	P
1	Hruschka (1968)	Labor consumption rate – production of fresh concrete		X					X		X			X		n/a
2		Labor consumption rate – reinforcing operations		X					X		X			X		n/a
3	Schubert (1971)	Risk of weather-induced work disruptions		X		X			X		X		X	X		Gamma distribution
4		Risk of lower-than-expected output		X		X			X		X		X			Gamma distribution
5		Quantified damage		X		X			X		X				X	Exponential distribution
6			X			X			X		X				X	Poisson distribution
7		Risk arising from changed scope of work		X		X			X		X		X			Gamma distribution
8		Price risk		X		X			X		X		X			Gamma distribution
9		Risk arising from construction process delays and disruptions		X		X			X		X		X			Exponential distribution
10			X			X			X		X		X			Poisson distribution
11	Golenko (1972)	Process durations		X	X						X		X	X		Beta distribution
12				X	X								X			Rectangular distribution
13				X		X			X		X		X			Log normal distribution
14				X	X						X		X			Triangular distribution
15	Dinort (1977)	Distribution of precipitation		X		X			X		X			X		Log normal distribution
16		Number of frost and ice days	X		X				X		X			X		Binomial distribution (p = 0.1)
17			X		X				X		X			X		Poisson distribution
18				X		X			X		X			X		Chi-square distribution
19		Disruption times		X		X			X		X		X	X		Gamma distribution
20				X		X					X		X	X		Gamma/Erlang distribution
21				X		X			X		X		X	X		Chi-square distribution
22	Schulz (1980)	Costing assumptions		X	Must be considered						X			X		Normal distribution
23	Lederer (1982)	Process durations		X		X			X		X		X			n/a
24	Strehl (1983)	Labor consumption rates – formwork operations (columns, walls, floors)		X		X			X		X			X		Gamma distribution
25		Labor consumption rates – concreting operations (columns, walls, floors)		X		X			X		X			X		Gamma distribution
26	Liberda (1985)	Labor consumption rates		X		X			X		X			X		Gamma distribution
27		Mean gross wage		X			X	X			X		X			Normal distribution
28	Iwan (1992)	Process durations		X		X			X		X		X			Exponential distribution
29				X	X								X			Rectangular distribution
30				X			X	X			X		X			Normal distribution
31				X		X			X		X		X			Erlang distribution
32		Disruption times		X		X			X		X		X	X		Exponential distribution
33		Intermediate disruption times		X		X			X		X		X	X		Erlang distribution
34	Chau (1995a)	Construction costs		X		X			X		X		X		X	n/a
35				X	X				X		X		X		X	Log triangular distribution
36	Raftery (1999)	Distributions applied in risk assessments		X	X								X			Rectangular distribution
37				X	X						X		X			Triangular distribution
38				X	X								X			Trapezoidal distribution
39				X	X								X			Incremental rectangular distribution
40			X		X								X			Discrete distribution

Table 6.3 Analysis of selected approaches to the distribution functions of various parameters in pertinent literature – Part 1[47)]

47) Cf. Kummer (2015a), p. 167

No.	Author	Parameters	Curve		Limits			Skewness			Mode		Sampling method			Specific distribution
			discrete	continuous	closed	one-sided open	two-sided open	symmetrical	right-skewed	left-skewed	unimodal	multimodal	theoretical considerations	historical vs collected data	survey	
0	A	B	C	D	E	F	G	H	I	J	K	L	M	N	O	P
41	Link (1999)	Life cycle of workpieces		X		X			X		X		X			Exponential distribution
42		Inflation rate		X			X	X			X		X			Normal distribution
43		Price trend of fuels		X			X	X			X		X			Normal distribution
44		Fuel consumption		X	X						X		X			Triangular distribution
45		Construction materials, units sold per week		X	X						X		X			Triangular distribution
46		Defective bricks/blocks in production	X		X			X			X		X			Binomial distribution
47		Telephone calls per minute / number of typing errors on a single page	X			X			X		X		X			Poisson distribution
48		Number of dry wells (oil exploration)	X			X			X		X		X			Geometrical distribution
49		Number of defective products in one pack	X		X						X		X			Hypergeometric distribution
50	Werner (2002)	Time consumption rates and costs		X	X		X				X		X	X		PERT distribution
51				X			X	X			X		X	X		Normal distribution
52				X	X						X		X			Rectangular distribution
53	Raaber (2003)	Estimated quantities and costs		X	X						X		X			Rectangular distribution
54				X		Must be considered					X		X			Normal distribution
55				X	X						X		X			Triangular distribution
56				X	X						X		X			Parabolic distribution (not otherwise specified)
57	Rohr (2005)	Process durations		X	X				X		X		X	X		Beta distribution
58				X		X			X		X		X	X		Three-parameter gamma distribution
59				X		X			X		X		X	X		Three-parameter inverse Gaussian distribution
60				X		X			X		X		X	X		Three-parameter log normal distribution
61				X		X			X		X		X	X		Three-parameter Weibull distribution
62	Schach (2005)	Costing		X	X								X	X	X	Rectangular distribution
63				X	X				X		X		X	X	X	Triangular distribution
64				X	X				X		X		X	X	X	Beta distribution
65				X	X				X		X		X	X	X	PERT distribution
66	Feik (2006)	Singular risks		X	X						X		X			Triangular distribution
67	Girmscheid/ Motzko (2007)	Quantified damage		X	X				X		X		X			PERT distribution
68	Girmscheid/ Busch (2008b)	Probability of occurrence of an event – rough model		X	X						X		X			Beta distribution
69		Process durations and expert opinions		X	X						X		X			PERT distribution
70		Finance theory – profit and cash flow		X			X	X			X		X			Normal distribution
71		Expert opinions		X	X						X		X			Triangular distribution
72		Life cycle of machines with constant probability of failure		X		X			X		X		X			Exponential distribution
73		Rough modeling of risks		X	X								X			Rectangular distribution
74		Probability of breakdown	X		X						X		X			Binomial distribution
75		Risk occurrence	X		X								X			Discrete distribution
76		Probability of breakdown	X			X			X		X		X			Poisson distribution
77	Demmler (2009)	Tunneling – gross advance rate		X	X								X	X		Rectangular distribution
78		Disruption times		X	X						X		X	X		Triangular distribution

Table 6.4 Analysis of selected approaches to the distribution functions of various parameters in pertinent literature – Part 2[48)]

[48)] Cf. Kummer (2015a), p. 168

No.	Author	Parameters	Curve		Limits			Skewness			Mode		Sampling method			Specific distribution
			discrete	continuous	closed	one-sided open	two-sided open	symmetrical	right-skewed	left-skewed	unimodal	multimodal	theoretical considerations	historical vs collected data	survey	
0	A	B	C	D	E	F	G	H	I	J	K	L	M	N	O	P
79	Hauwermeiren/ Vose (2009)	Cost of risks		X		X			X		X		X			Exponential distribution
80				X		X			X		X		X			Log gamma distribution
81				X		X			X		X		X			Log logistic distribution
82				X		X			X		X		X			Log normal distribution
83				X		X			X		X		X			Pareto distribution
84		Expert opinions	X		X								X			Bernoulli distribution
85			X		X								X			Discrete distribution
86				X	X						X		X			PERT distribution
87				X	X						X		X			Triangular distribution
88				X	X								X			Rectangular distribution
89	Gleißner (2011)	Probability of occurrence	X		X								X			Binomial distribution
90		Number of incidents	X			X			X		X		X			Poisson distribution
91		Quantified damage		X			X	X			X		X			Normal distribution
92		Income and product prices		X		X			X		X		X			Log normal distribution
93		Quantitative risk assessment by user		X	X						X		X			Triangular distribution
94		Life cycle of pieces of equipment		X		X			X		X		X			Exponential distribution
95		Financial market returns		X		X			X		X		X	X		Pareto distribution
96		Modeling of variables with little available information		X	X								X			Rectangular distribution
97	Sander (2012)	Expert opinions		X	X						X		X			Triangular distribution
98				X	X						X		X			PERT distribution
99		Quantity and cost uncertainties		X	X				X		X		X			Triangular distribution
100				X	X				X		X		X			PERT distribution
101				X	X								X			Rectangular distribution
102	Mayer (2013)	Basic unit prices		X			X	X			X		X			Normal distribution
103		Quantity values		X			X	X			X		X	X		Normal distribution
104	Hofstadler (2014b)	Labor consumption rates – formwork operations		X	X				X		X		X		X	Log logistic-like distribution
105	Schach/ Flemming (2014)	Risk modeling		X	X								X			Rectangular distribution
106				X	X						X		X			Triangular distribution
107				X	X								X			Trapezoidal distribution
108			X		X								X			Discrete distribution

Table 6.5 Analysis of selected approaches to the distribution functions of various parameters in pertinent literature – Part 3[49)]

The above analysis of pertinent literature allows to derive a tendency towards using continuous, closed, right-skewed, unimodal distributions for parameters in construction management and economics as well as in the management of chances and risks. In the literature, these indications are usually justified by theoretical considerations partially supported by historical data and analyses taken from empirical studies.[50)]

Section 7.1 explains the principle of how to systematically select distribution functions for modeling purposes, complemented by a description of how to determine distribution functions on the basis of data records. However, we first use an example to demonstrate the way in which the shape of input parameter distribution functions may influence results at all.

49) Cf. Kummer (2015a), p. 169

50) Cf. ibid.

6.4.1 Influence of the Shape of Distribution Functions

This section outlines the influence of different input distributions on the output.[51)] The following input parameter distribution functions are compared on the basis of calculating the total labor consumption rate of reinforced concrete operations:

- Triangular distribution
- PERT distribution
- Rectangular distribution
- Normal distribution
- Log-triangular distribution according to *Chau*[52)]
- Log-logistic-like distribution according to *Hofstadler*[53)]

The total labor consumption rate of reinforced concrete operations LCR_{RCO} [wh/m³] results from the labor consumption rates of formwork operations LCR_{FO} [wh/m²], reinforcing operations LCR_{RO} [wh/t], and concreting operations LCR_{CO} [wh/m³] as well as from the formwork ratio fr [m²/m³] and the reinforcement ratio rr [kg/m³] according to Eq. (6-4).

$$LCR_{RCO} = LCR_{FO} \cdot fr + LCR_{RO} \cdot rr + LCR_{CO} \tag{6-4}$$

This example starts from performing a deterministic calculation of the total labor consumption rate using the figures stated in Table 6.6. We apply Eq. (6-4) to obtain a deterministic total labor consumption rate of 6.10 wh/m³.

No.	Calculation parameter		Unit	MOST LIKELY
0	A	B	C	D
1	Formwork ratio	fr	m²/m³	3.00
2	Reinforcement ratio	rr	kg/m³	115.00
3	Labor consumption rate of formwork operations	LCR_{FO}	wh/m²	1.20
4	Labor consumption rate of reinforcing operations	LCR_{RO}	wh/t	13.00
5	Labor consumption rate of concreting operations	LCR_{CO}	wh/m³	1.00
6	**Total labor consumption rate**	**LCR_{RCO}**	**wh/m³**	**6.10**

Table 6.6 Example: Deterministic calculation of the total labor consumption rate of reinforced concrete operations

These deterministic inputs are subsequently replaced with different distribution functions. For each simulation, the same distribution functions are applied to all variables in order to identify the influence of different input distribution functions on the results.

For triangular distributions, the deterministic values indicated in Table 6.6 are assumed to be the most likely values, and additional minimum and maximum values are used for establishing the corresponding range.

Table 6.7 contains the minimum, most likely, and maximum values for the five input parameters in Columns D, E, and F.

51) See also a similar study conducted by Hofstadler (2011).

52) Cf. Chau (1995b), p. 378

53) Cf. Hofstadler (2014b), p. 194ff.

No.	Calculation parameter		Unit	MIN	MOST LIKELY	MAX
0	A	B	C	D	E	F
1	Formwork ratio	fr	m²/m³	2.70	3.00	3.30
2	Reinforcement ratio	rr	kg/m³	100.00	115.00	140.00
3	Labor consumption rate of formwork operations	LCR_{FO}	wh/m²	1.00	1.20	1.60
4	Labor consumption rate of reinforcing operations	LCR_{RO}	wh/t	12.00	13.00	16.00
5	Labor consumption rate of concreting operations	LCR_{CO}	wh/m³	0.90	1.00	1.30

Table 6.7 Example: Calculation of the total labor consumption rate of reinforced concrete operations – Triangular distributions

PERT distributions require the same input values as used for the triangular distributions. Thus, the same minimum, most likely, and maximum values are applied.

Rectangular distributions are defined by the respective minimum and maximum values indicated. No most likely values are required in this case because the probability is identical for each value within the defined range.

Normal distributions are defined on the basis of the mean and standard deviations. For the purpose of this comparison, the mean values and standard deviations of the applied triangular distributions (see Table 6.7) are determined and used as inputs for the normal distributions.

The minimum, most likely, and maximum values of the triangular distributions are also used for the log-triangular distribution according to *Chau*[54)], and the natural logarithm is derived from each of these values. The triangular distribution formed by these log values is then exponentiated to arrive at the distribution defined by *Chau*. The following input is made in the @Risk software: =EXP(RiskTriang(LN(D);LN(E);LN(F))). The D, E, and F parameters refer to the input parameter columns in Table 6.7.

The log-logistic-like distribution according to *Hofstadler*[55)], which had originally been determined for labor consumption rates of formwork operations only, is also applied to the remaining input parameters of Eq. (6-4). Both a minimum and a most likely value are necessary for defining this type of distribution. These two values are, in turn, taken from the initially defined triangular distributions.

For comparing the outputs, the same types of distribution functions (triangular, rectangular, PERT, etc.) are applied to labor consumption rates as well as the formwork and reinforcement ratios. We also assume input parameters to be independent of one another. We refer to Section 6.5 as regards information on the influence of correlations and their adequate consideration.

Fig. 6.15 includes a comparison of the results obtained for the total labor consumption rate of reinforced concrete operations applying different input distributions. This simulation was performed with the Latin Hypercube sampling method, including 50,000 iterations. Results are shown as lines (splines) rather than conventional histograms in order to provide an optimized visualization of the comparison.

54) Cf. Chau (1995b), p. 378

55) Based on Hofstadler (2014), p. 194ff. For the purpose of this comparison, *Hofstadler* modified his method such that only a minimum and a most likely value are required as input parameters for defining the log logistic-like distribution. In this case, these two values must not be identical since otherwise the distribution would not be defined. Section 7.2 includes a more detailed description of this method.

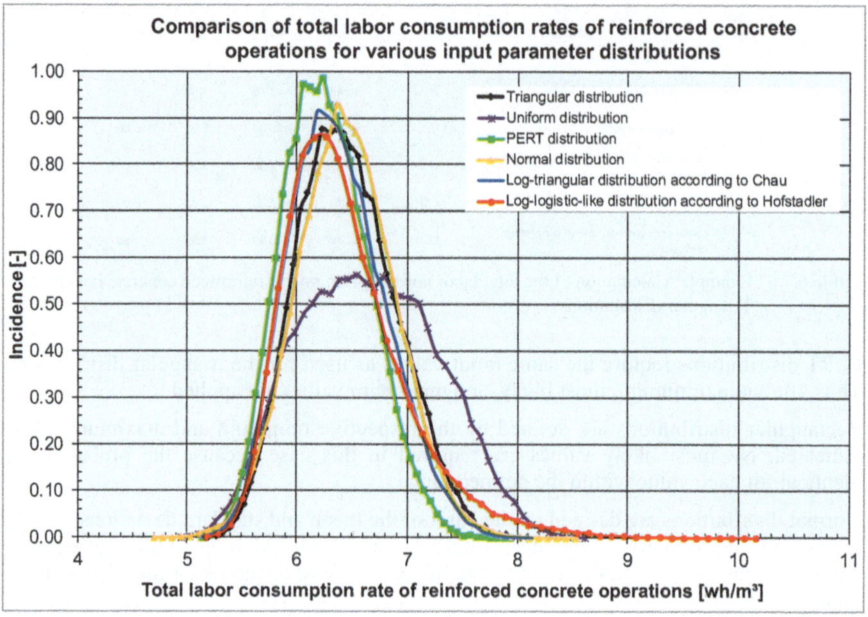

Fig. 6.15 Comparison of the total labor consumption rate of reinforced concrete operations for different input parameter distributions

We find that the outputs are influenced by the different types of distributions; some of the results differ greatly from one another. The differences in the resulting spreads are the most striking.

These variances are particularly large for results where log-logistic-like distributions according to *Hofstadler* were used as inputs. However, it should be noted that labor consumption rates in excess of 9 wh/m³ are very unlikely in this example, and can thus be virtually excluded on the basis of theoretical considerations alone.

Output variances are also attributable to the fact that only few mathematical operations are required for calculating the total labor consumption rate. If a simulation involves a large number of computation steps, the outputs of additions and multiplications will approximate a normal and a logarithmic normal distribution, respectively (central limit theorem). For this to occur, no correlation must exist between the variables.

It is worth noting that the shape of the distribution of the total labor consumption rate, where rectangular distributions were used as inputs, differs particularly strongly from the results obtained for the other distributions. This is due to the fact that all input parameter values are associated with an identical probability of occurrence, which means that the total labor consumption rate derived from the simulation also tends to exhibit a uniform distribution.

This comparison demonstrates that the types of distributions applied to input parameters have a major influence on simulation outcomes.

6.4.2 Influence of the Skewness of Distribution Functions

One of the key factors influencing the results of simulations is the skewness of certain distributions. This section explains how marked this influence is when applying the four basic arithmetic operations.

For each of these operations, two triangular distributions with different skewnesses are combined with each other and graphically compared in the form of histograms (results of Monte Carlo simulations with 50,000 iterations – Latin Hypercube sampling).[56] In the diagrams, the skewness of outputs is also stated as a figure[57] to ensure a more appropriate comparison of visually similar distributions.

Distribution 1 is characterized by a range from 2 to 10. Within this range, the most likely value is varied so that a right-skewed (most likely value = 4), a symmetrical (most likely value = 6), and a left-skewed distribution (most likely value = 8) is created.

Distribution 2 is characterized by a range from 4 to 20 (i.e. double the spread of Distribution 1). Within this range, the most likely value is again varied so that a right-skewed (most likely value = 8), a symmetrical (most likely value = 12), and a left-skewed distribution (most likely value = 16) is created.

6.4.2.1 Addition

The central limit theorem applies to the addition of uncorrelated random variables with a finite and positive variance. This theorem states that the sum of these variables shows an approximately normal distribution. If, however, only few parameters are added up, the skewness of the random variable does have an influence on the sum's skewness. This is where the ratio of an addend to the sum constitutes a crucial factor that determines the expected skewness of the result. Fig. 6.16 shows the input distributions (Distributions 1 and 2) with different skewnesses (right-skewed, symmetrical, left-skewed) and the related sums as histograms. We find that the sum's skewness is more similar to that of Distribution 2, which exhibits a wider range and larger values overall. If, for instance, Distribution 2 were left-skewed, the sum would also tend to exhibit a similar skewness (i.e. positive skewness values between +0.27 and +0.34). If we add up distributions with different skewness characteristics, the results will show a tendency towards a symmetrical distribution. If both inputs are symmetrical, the histogram of the sum will also tend to be symmetrical (skewness = +0.0034).

[56] Cf. example in Kummer (2015a), p. 146ff.

[57] Skewness is a statistical indicator representing the type and significance of a distribution's asymmetry. If the skewness is smaller than zero, we refer to the distribution as left-skewed or positively skewed. If the skewness is greater than zero, the distribution is termed right-skewed or negatively skewed. A skewness of zero indicates a symmetrical distribution. @Risk automatically calculates skewness for defined outputs.

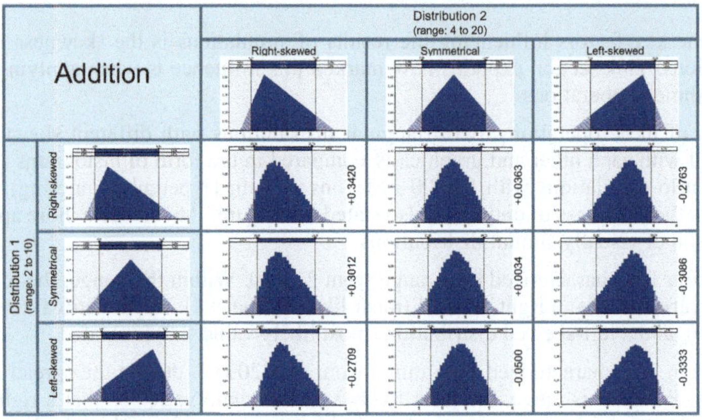

Fig. 6.16 Addition of two distribution functions with different skewness characteristics[58)]

6.4.2.2 Subtraction

In this operation, we subtract the distribution with the smaller total values within the defined range (Distribution 1) from the other distribution (Distribution 2). The resulting curve is similar to the trend described for addition. In subtraction, the minuend influences the skewness of the difference. Unlike in addition, different input skewnesses will increase the skewness of the output (see Fig. 6.17). Likewise, symmetrical inputs to subtraction will cause the result to trend towards a symmetrical distribution (skewness = -0.0093).

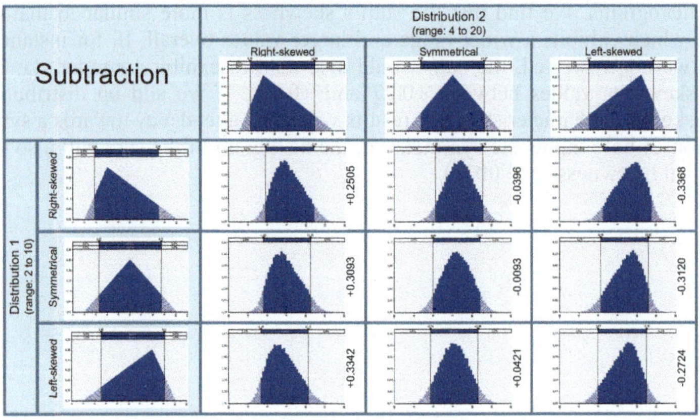

Fig. 6.17 Subtraction of two distribution functions with different skewness characteristics[59)]

58) Cf. Kummer (2015a), p. 147

59) Cf. ibid., p. 148

6.4.2.3 Multiplication

Multiplication of independent random variables with a finite and positive variance will result in an approximately logarithmic normal distribution. Fig. 6.18 shows that this is already true for multiplying just two factors. All histograms of the products are right-skewed, even though their skewness ranges approximately from +0.99 to +0.18.

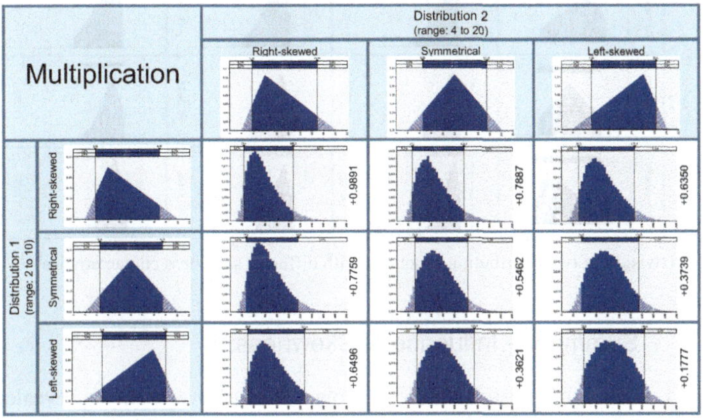

Fig. 6.18 Multiplication of two distribution functions with different skewness characteristics[60)]

If we multiply two left-skewed distributions, this operation will result in a less pronounced skewness compared to multiplying two right-skewed distributions. The product will also be right-skewed if two symmetrical distributions are multiplied.

6.4.2.4 Division

In this operation, we divide the distribution with the greater total values within the defined range (Distribution 2) by the other distribution (Distribution 1). The resulting curve is similar to the trend described for multiplication. Irrespective of the skewness of inputs, the quotient will always be a right-skewed distribution whose skewness varies between approximately +0.94 and +1.70 (see Fig. 6.19). Like in multiplication, the quotient will also exhibit a pronounced right-skewed curve when performing division for symmetrical distributions (skewness = +1.3448).

60) Cf. Kummer (2015a), p. 148

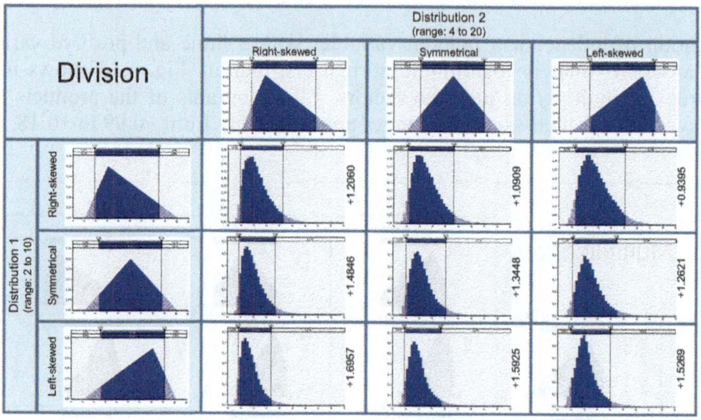

Fig. 6.19 Division of two distribution functions with different skewness characteristics[61)]

6.4.2.5 Summary – Influence of Skewness

Combining two triangular distributions, as shown in the above example, revealed a major influence of input parameter skewness on the skewness of outputs. The smaller the number of computing operations, the greater this influence will be.

The above graphical comparison of the addition, subtraction, multiplication, and division of two random variables with different skewnesses can be used for checking the plausibility of simulation models. If the sequence of mathematical operations and the input parameter curve are known, the above diagrams permit conclusions regarding the curve of the desired output.

6.5 Correlations

In Section 2.3.10, we provided an introduction to correlations and correlation coefficients.

Correlations are usually identified indirectly through analyzing existing data series taken from measurements, surveys, experiments, or other sources to check them for linear correlations (see Fig. 6.20, right).

Correlation coefficients can also be captured directly by conducting surveys among experts. Section 7.3 presents the results of such a survey for the total labor consumption rate of reinforced concrete operations.

61) Cf. Kummer (2015a), p. 149

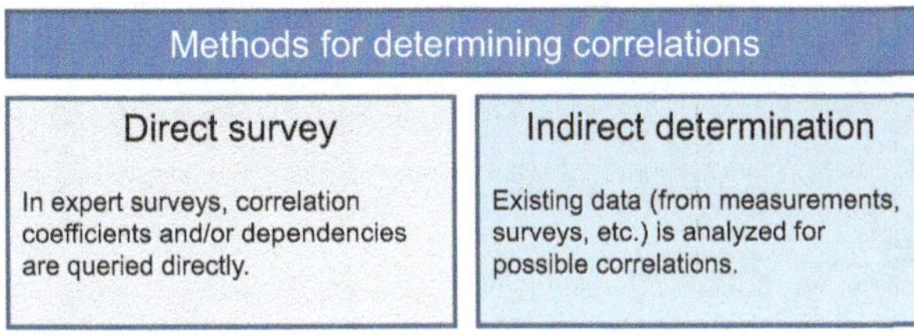

Fig. 6.20 Methods for identifying correlations[62]

6.5.1 Correlations in the Literature on Construction Management and Economics

Tables 6.8 and 6.9 compare selected statements on correlations taken from the literature on construction management and economics, to the extent to which we were able to identify details on the type and magnitude of possible correlation coefficients. The following differentiation criteria are applied:[63]

- Type of correlation
  - positive
  - negative
  - independent

- Quantification of correlation coefficients (as a single value or as a range)
- Sampling method
  - theoretical considerations
  - historical vs. collected data
  - surveys

- Conclusions regarding the significance of correlations

Authors frequently refer to positive correlations although they are rarely applied in simulations, which is justified by the argument of the high computing capacity necessary for modeling and assessing the values correlation coefficients can take. Some authors neglect negative correlations altogether.

[62] Cf. Kummer (2015a), p. 134

[63] Cf. ibid., p. 198f.

No.	Author	Area of application	Parameters with assumed correlations		Correlations: positive	Correlations: negative	Correlations: independent	Correlations: Statement of specific correlation coefficients	Sampling method: theoretical considerations	Sampling method: historical vs collected data	Sampling method: survey	Significance of correlations
0	A	B	C	D	E	F	G	H	I	J	K	L
1	Hruschka (1968)	Construction process planning and costing	Labor consumption rate of concreting operations	Formwork ratio	X					X		
2			Total labor consumption rate of reinforced concrete operations	Formwork ratio	X					X		
3			Paid working hours per m³ of gross volume	Construction material weight per m³ of gross volume	X					X		
4	Schulz (1980)	Costing	Economic transactions		X				X			Highly significant from a theoretical standpoint
5	Lederer (1982)	Critical path analysis	Volume	Process durations	X				X			Highly significant
6			Volume	Process outputs	X				X			
7			Construction progress	Production system	X				X			
8	Liberda (1985)	Costing	Mean gross wage	Labor consumption		X		-0.95	X			Correlations can be excluded by decisions
9			Labor consumption	Material consumption					X			
10			Material cost	Material consumption					X			
11			Labor consumption rates		X				X			
12	Touran/Wiser (1992)	Costing	Specific correlation matrix for cost components – 12-by-12 matrix		X	X		between -0.31 and +0.75		X		Significant for predicting cost variance
13			Specific correlation matrix for cost components – 6-by-6 matrix		X			between +0.28 and +0.75		X		
14	Chau (1995a)	Costing	Specific correlation matrix for cost components – 19-by-19 matrix		X	X		between -0.81 and +0.95		X		Independent input parameters distort simulation results
15			Specific correlation matrix for cost components (electrical installations) – 10-by-10 matrix		X	X		between -0.79 and +0.89		X		
16	Link (1999)	Costing and risk management	Inflation	Energy cost	X				X			Considering correlations allows for a significant increase in forecast accuracy
17			Product price	Sales figures		X			X			
18			Occurrence of adverse weather	Shift of completion date	X				X			
19			Required extent of documentation	Risk of disputes over invoicing and handover		X			X			
20	Yang (2005)	Costing	Specific correlation matrix for cost components – 8-by-8 matrix		X		X	between +0.10 and +0.64	X	X		Influence on both spread and mean of output
21	Girmscheid/ Busch (2008b)	Corporate risk management Internal influential factors	Risks within the project arising from management and support processes, e.g. financing, IT and HR		X			possible: 0.00 to +1.00 probable: +0.50 to +1.00	X			Correlations generally influence the spread around the expected value and affect the amount of risk markups in the risk aggregation process Singular risks and projects are considered to be independent of one another for practical reasons
22			Risks arising from project-specific construction methods and techniques as well as processed materials and construction products		X			probable: 0.00 to +1.00	X			
23		Corporate risk management External influential factors	Project-specific risks arising from the unique nature of projects, e.g. in relation to the client and other external stakeholders (e.g. architects, technical designers, subcontractors), geological and local conditions etc.		X			probable: 0.00 to +1.00	X			
24			Demand risks due to the geographical distribution of projects		X	X		possible: -1.00 to +1.00 probable: -0.50 to +1.00	X			
25	Sander (2012)	Risk management and costing	Detailed assessment of risks		X			+1.00	X			Using correlations should improve the result without overwhelming the user

Table 6.8 Comparison of selected approaches to correlations in pertinent literature – Part 1[64)]

[64)] Cf. Kummer (2015a), p. 200

No.	Author	Area of application	Parameters with assumed correlations		Correlations				Sampling method			Significance of correlations
					positive	negative	independent	Statement of specific correlation coefficients	theoretical considerations	historical vs collected data	survey	
0	A	B	C	D	E	F	G	H	I	J	K	L
26			Labor consumption rate of formwork operations – simple layout	Labor consumption rate of formwork operations – layout with medium complexity	X			+0.36			X	
27			Labor consumption rate of formwork operations – simple layout	Labor consumption rate of formwork operations – layout with high complexity	X			+0.63			X	
28			Labor consumption rate of formwork operations – layout with medium complexity	Labor consumption rate of formwork operations – layout with high complexity	X			+0.44			X	
29	Krempl (2012)	Costing – formwork and reinforcing operations for floor slabs	Labor consumption rate of formwork operations – floor slab on rectangular columns	Labor consumption rate of formwork operations – floor slab on round columns	X			+0.90			X	
30			Labor consumption rate – reinforcing operations	Reinforcement ratio	No detectable generally positive correlation						X	
31			Labor consumption rate of formwork operations	Labor consumption rate of reinforcing operations – steel rebar	X			+0.89			X	
32			Labor consumption rate of formwork operations	Labor consumption rate of reinforcing operations – steel mesh	X			+0.86			X	
33	Mayer (2013)	Costing	Basic unit price	Markup factor		X	X	Interrelatedness can be excluded; negative correlation appears possible	X			Negative correlations are disregarded completely as a result of the narrowing spread
34			Markup factor	Quantity			X		X			
35			Basic unit price	Quantity			X		X			
36	Schach/ Flemming (2014)	Risk modeling	Concrete volume	Reinforcement ratio	X				X			Modeling dependencies is necessary to reflect reality as accurately as possible

Table 6.9 Comparison of selected approaches to correlations in pertinent literature – Part 2[65)]

Over time, incorporating correlations has become much simpler by applying simulation techniques as opposed to analytical computations (convolution). Some authors stress the influence of correlations on output values. However, specific values describing the strength of the correlation are stated only in rare cases. Instead, most conclusions are restricted to stating the trend of the correlation (i.e. positive, negative, independent).

To a certain extent, correlation matrices are derived from historical data from completed projects or on the basis of expert surveys. If correlations are to be incorporated, this will inevitably also require theoretical considerations in order to verify detected dependencies and identify cause-effect relationships.

Generally speaking, one should be aware of the influence of correlations when conducting Monte Carlo simulations and interpreting simulation results. Outputs to be assessed with respect to possible spreads or maximum and/or minimum values require an in-depth analysis of correlation coefficients. Ranges for different correlation coefficients can also be studied for the purpose of performing sensitivity analyses, for instance in iterative steps from -1 to +1.

[65)] Cf. Kummer (2015a), p. 200

In this context, simulation results derived from independent variables can be compared with results based on a particularly large number of negative and/or positive correlations.

The study of pertinent literature reveals a tendency towards positive correlations in the management of chances and risks as well as in costing. However, further analyses are necessary to arrive at generalizable conclusions. Correlations need not be considered for relevant parameters if dependencies can be captured directly in the computation model.

6.5.2 Influence of Correlations

Probabilistic calculations usually start from the assumption of independent input parameters. In the literature, this approach is justified by the difficulties involved in identifying correlations subjectively. In addition, the data required for the mathematical determination of correlations is often missing. Moreover, additional effort is needed to integrate correlations into simulation models.

This section includes simple worked examples for highlighting the influence correlations can have on results for different basic arithmetic operations.[66)]

Chau used triangular distributions to investigate the effect on the sum of two variables.[67)]

In addition, we will also calculate the difference, the product, and the quotient while varying correlation coefficients, and evaluate the effects of applied correlations. The two triangular distributions have a range from 2 to 10. The most likely value is assumed to be 6, which results in symmetrical distributions.

For the subsequent analyses, the coefficient of the correlation between the two variables is varied between -1 and +1, and 25 separate simulations with 50,000 iterations each are carried out applying the Latin Hypercube sampling method.[68)] The value of the seed is defined as 1 to be able to replicate the simulation results if and when required. We deliberately apply symmetrical distributions because the influence of correlations on results is the only parameter to be verified. Section 6.4.2 already outlined the effects of skewness.

6.5.2.1 Addition

We calculate the sum of two identical distribution functions (i.e. triangular distributions defined by the values 2 | 6 | 10) with the aid of a Monte Carlo simulation. The related diagram (see Fig. 6.21) includes the following statistical indicators to show the change in output in relation to the correlation coefficient:

- 5% quantile
- 95% quantile
- Simple standard deviation from mean (bidirectional)
- Mean

The dashed (horizontal) line in Fig. 6.21 (at 12.00) represents the sum expected for a deterministic calculation applying the values expected for the two variables (= 6 + 6). The outermost curves delimit the area between the 5% and 95% quantiles that results from different correlation coefficients. The curves labeled "Mean + std.dev." and "Mean - std.dev." cover the

66) Cf. example in Kummer (2015a), p. 178ff.

67) Cf. Chau (1995b), p. 371

68) In @Risk, symmetrical matrices are used for entering correlation coefficients. For example, a 3-by-3 matrix is entered to reflect dependencies between three parameters.

simple standard deviation from the mean. The straight black line represents the mean of the sum of the two assumed random variables; it is independent of the correlations. The symmetrical input distributions result in the two straight lines for the most likely and the mean values (taken from the simulations) being identical at 12.00.

At a correlation coefficient near -1, the range of results is very narrow. It becomes wider as the coefficient value moves closer to +1.

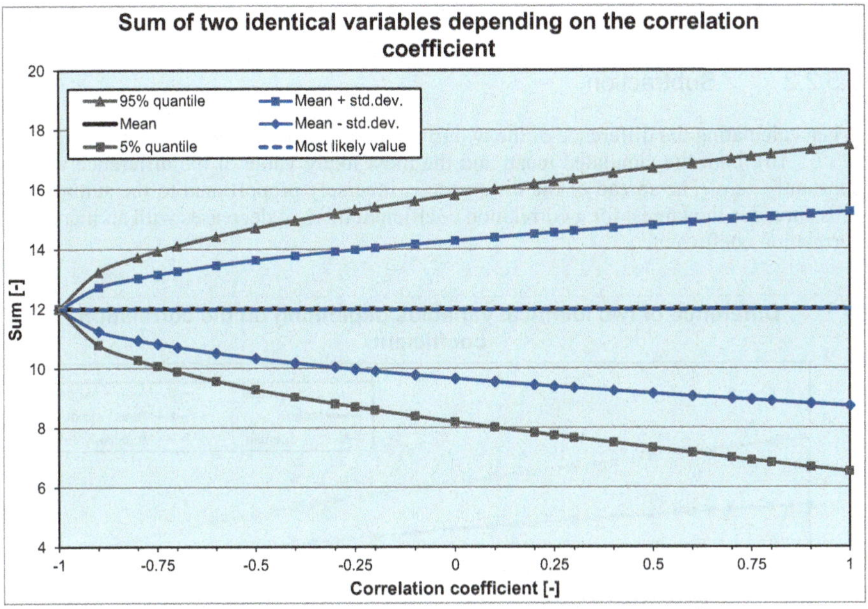

Fig. 6.21 Sum of two identical variables depending on the correlation coefficient[69)]

We calculate the variance of the sum of two variables x and y as follows:

$$\mathrm{Var}(x+y) = \mathrm{Var}(x) + \mathrm{Var}(y) + 2 \cdot \mathrm{Cov}(x, y) \tag{6-5}$$

Var(x + y) .... Variance of the sum of x and y

Var(x) .... Variance of x

Var(y) .... Variance of y

Cov(x,y) .... Covariance of x and y

The covariance is positive in the case of a positive correlation between the two variables. Correspondingly, it is negative for a negative correlation. Var(x + y) is the largest (smallest) if a positive (negative) covariance prevails. The covariance is zero for two independent random variables. Consequently, the sum's variance is equivalent to the sum of the two individual variances of x and y (see Eq. (6-6)).

69) Cf. Kummer (2015a), p. 179

$$\mathrm{Var}(x + y) = \mathrm{Var}(x) + \mathrm{Var}(y) \tag{6-6}$$

For the example above, it is important to note that the two addends were modeled with symmetrical distributions. If skewed distributions are applied, the expected result of a deterministic calculation may deviate from the mean of the simulations, which would depend on the location of the most likely values. Depending on the correlation coefficients, this will then also change the probability of exceeding or underrunning the most likely value.

6.5.2.2 Subtraction

When calculating the difference of the two triangular distributions assumed to be identical (2 | 6 | 10), both the simulated mean and the most likely value of the difference will be constantly zero. The spread of the difference is inversely proportional to the summation. This spread is the largest for a correlation coefficient of -1; it decreases with an increasing correlation coefficient.

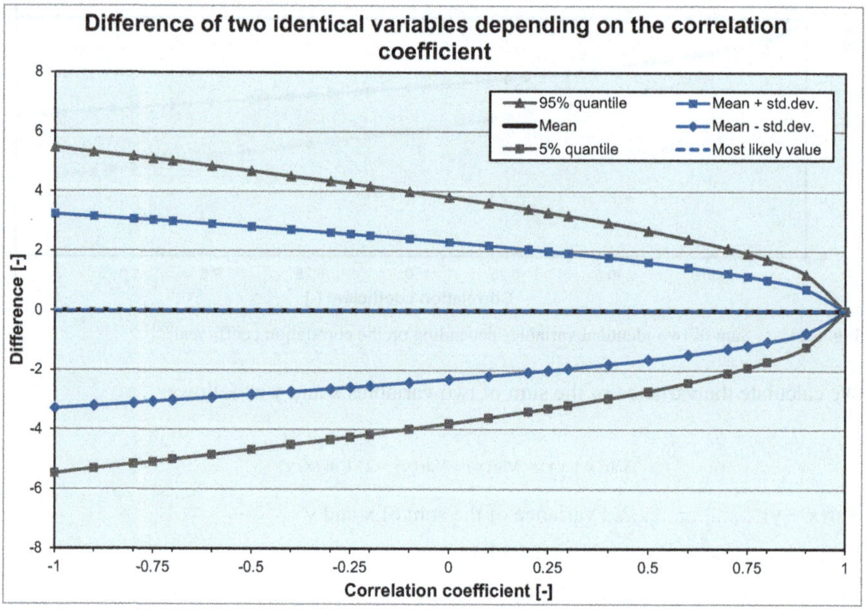

Fig. 6.22 Difference of two identical variables depending on the correlation coefficient[70)]

This is due to the fact that, for a correlation coefficient of +1, values from similar locations within the identical ranges of input distributions are selected for calculating the difference, which reduces the difference spread as the correlation coefficient increases. The extreme case would involve choosing exactly the same random values from the two distributions because the difference of these two variables would always be zero. In the case of a negative correlation, values from opposite areas of the distributions are selected to

70) Cf. Kummer (2015a), p. 180

calculate the difference. If we take, for instance, a high random value from the first distribution, this will lead to a high probability of choosing a correspondingly low value from the second distribution at a correlation near -1, resulting in a comparatively large difference. Conversely, we obtain a highly negative difference if a high value from the second distribution is subtracted from a low value from the first distribution. Fig. 6.22 illustrates this relationship as a spread whose variance depends on the correlation coefficients.

6.5.2.3 Multiplication

Calculating the product of the two triangular distributions (2 | 6 | 10) results in the curve for the considered statistical parameters shown in Fig. 6.23 for different correlation coefficients between the two variables.

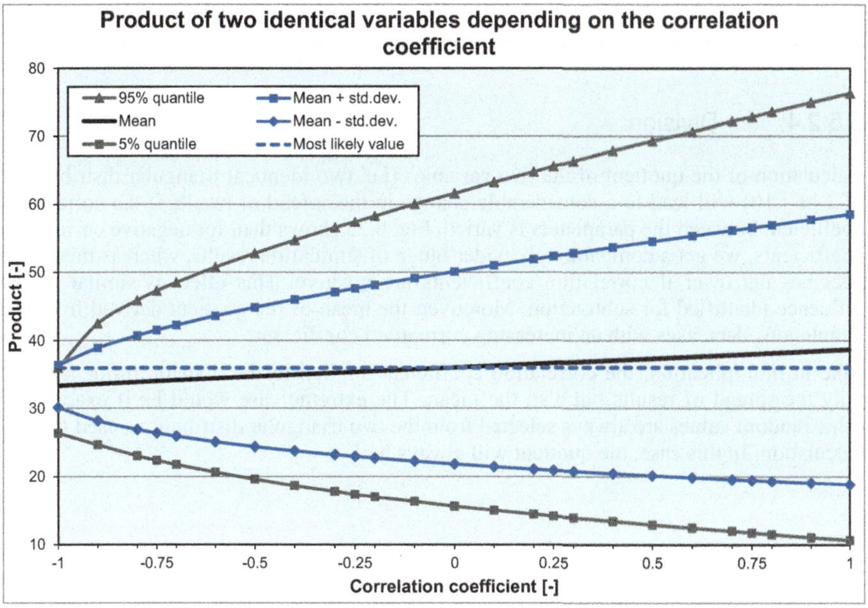

Fig. 6.23 Product of two identical variables depending on the correlation coefficient[71)]

The range of results becomes continuously wider in line with the increasing correlation coefficient, similar to the effect identified for the addition. We note that, for a correlation coefficient of -1, the 95% quantile lies below the mean, increased by a standard deviation – see the curve labeled "Mean + std.dev.". This result is attributable to the pronounced left skewness of the histogram that forms the basis for the data points shown in the diagram (see Fig. 6.24).

In addition, Fig. 6.23 shows that the mean derived from the simulations also increases with the rising correlation coefficient. In this case, correlations influence not only the spread of results but also the mean.

71) Cf. Kummer (2015a), p. 181

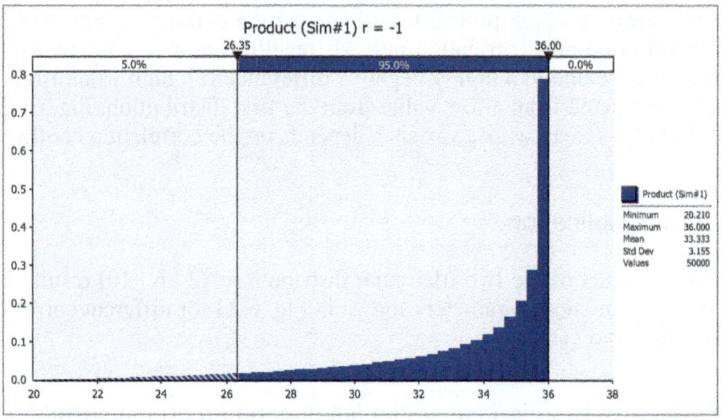

Fig. 6.24 Product of two triangular distributions (2 | 6 | 10) with a correlation coefficient of r = -1[72)]

6.5.2.4 Division

Calculation of the quotient of the two variables (i.e. two identical triangular distributions of 2 | 6 | 10) will lead to a considerable change in the spread of results if the correlation coefficient between the parameters is varied. Fig. 6.25 shows that, for negative correlation coefficients, we get a comparatively wider range of simulation results, whereas this range becomes narrower if correlation coefficients are positive. This effect is similar to the influence identified for subtraction. Moreover, the mean of the quotient derived from the simulations decreases with an increasing correlation coefficient.

Like in multiplication, the correlation coefficient applied to the division influences not only the spread of results but also the mean. The extreme case would be if exactly the same random values are always selected from the two triangular distributions used for this calculation. In this case, the quotient will always be 1.

72) Cf. Kummer (2015a), p. 182

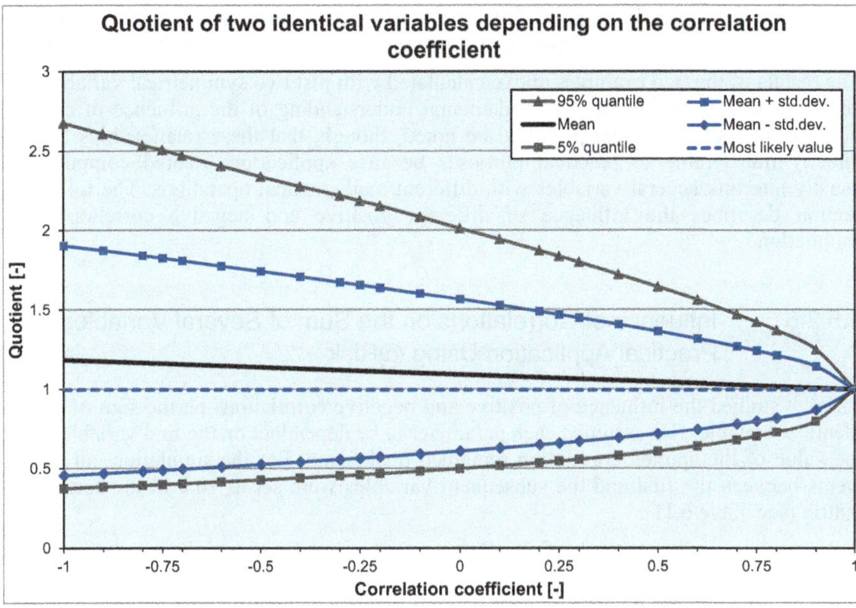

Fig. 6.25 Quotient of two identical variables depending on the correlation coefficient[73)]

6.5.2.5 Summary – Influence of Correlations on Two Variables

In the addition and multiplication of two random variables, the range of results will increase with the correlation coefficient and reach its maximum at r = +1. Conversely, in subtraction and division, the range of results will decrease as the correlation coefficient increases, and reach its minimum at r = +1. In addition and subtraction, the mean derived from the simulations will remain constant even if the correlation coefficient is varied. In multiplication and division, the mean of the outputs will also change depending on the correlation coefficients. Table 6.10 includes an overview of the effects of correlations on the results of the four basic arithmetic operations for two symmetrical variables.

Arithmetic operation		**Addition**		**Multiplication**		**Subtraction**		**Division**	
Correlation		negative	positive	negative	positive	negative	positive	negative	positive
Spread of results gets	wider		X		X	X		X	
	narrower	X		X			X		X
Influence of correlations on mean	Yes			X				X	
	No	X				X			

Table 6.10 Influence of correlations on spread and mean shown for the four basic arithmetic operations[74)]

[73)] Cf. Kummer (2015a), p. 182

[74)] Cf. Kummer (2015a), p. 183

Calculations in construction management and economics often require addition and multiplication operations, which, in combination with positive correlations, tend to widen the spread of results compared to operations performed for uncorrelated input variables.

The results of the two examples above calculated with just two symmetrical variables are very illustrative for providing a fundamental understanding of the influence of correlations on simulation results. It should be noted, though, that these relationships are not directly transferable to practical purposes because application-oriented computations usually interlink several variables with different mathematical operations. The following section describes the influence of different positive and negative correlations on summation.

6.5.2.6 Influence of Correlations on the Sum of Several Variables – Practical Application Using @Risk

Chau[75] studied the influence of positive and negative correlations on the sum of eleven identical variables. He assumed each parameter to be dependent on the first variable while the value of the applied correlation remained unchanged. For the simulation, all coefficients between the first and the subsequent variables were set to +0.9 in the correlation matrix (see Table 6.11).

@Risk only permits positive definite and positive semidefinite correlation matrices. The input matrix is checked for consistency by calculating its eigenvalues. The @Risk software performs such a check prior to any simulation. A positive definite matrix has only positive eigenvalues, whereas a positive semidefinite matrix has eigenvalues greater than or equal to zero. Consequently, any matrix with at least one negative eigenvalue is not accepted for the subsequent computation steps, and must thus be redefined or adjusted.[76]

The specific value used in the worked example (see Table 6.11) results in a nonconsistent correlation matrix. The remaining fields of the symmetrical matrix must be filled, and the previously entered correlation coefficients need to be adjusted if required.

		1	2	3	4	5	6	7	8	9	10	11
	$NO_{NEG,CORR} = 0$	A	B	C	D	E	F	G	H	I	J	K
1	A	1.00										
2	B	0.90	1.00									
3	C	0.90	0.86	1.00								
4	D	0.90	0.79	0.72	1.00							
5	E	0.90	0.65	0.86	0.79	1.00						
6	F	0.90	0.81	0.80	0.83	0.81	1.00					
7	G	0.90	0.78	0.73	0.71	0.79	0.83	1.00				
8	H	0.90	0.81	0.79	0.82	0.80	0.75	0.82	1.00			
9	I	0.90	0.80	0.78	0.81	0.79	0.77	0.81	0.78	1.00		
10	J	0.90	0.81	0.79	0.82	0.80	0.76	0.81	0.77	0.78	1.00	
11	K	0.90	0.81	0.78	0.82	0.80	0.75	0.82	0.77	0.78	0.77	1.00

Table 6.11 Definition of a correlation matrix for the correlations between the first and subsequent variables set to +0.9[77]

[75] Cf. Chau (1995b), p. 372

[76] Cf. http://kb.palisade.com/index.php?pg=kb.page&id=74. Accessed December 21, 2020

[77] Cf. Kummer (2015a), p. 184

The @Risk software, which was also used by *Chau* in his studies,[78] automatically checks entered correlation matrices for consistency. If users enter a matrix as shown in Table 6.11 in @Risk, they will automatically be queried if this matrix should be adjusted. If "Yes" is chosen as an answer, @Risk continues the computation process with the automatically adjusted matrix shown in Table 6.12.[79] This matrix can be termed "next best" because it meets no particular requirements.[80]

	$NO_{NEG,CORR} = 0$	1 A	2 B	3 C	4 D	5 E	6 F	7 G	8 H	9 I	10 J	11 K
1	A	1.00										
2	B	0.90	1.00									
3	C	0.90	0.86	1.00								
4	D	0.90	0.79	0.72	1.00							
5	E	0.90	0.65	0.86	0.79	1.00						
6	F	0.90	0.81	0.80	0.83	0.81	1.00					
7	G	0.90	0.78	0.73	0.71	0.79	0.83	1.00				
8	H	0.90	0.81	0.79	0.82	0.80	0.75	0.82	1.00			
9	I	0.90	0.80	0.78	0.81	0.79	0.77	0.81	0.78	1.00		
10	J	0.90	0.81	0.79	0.82	0.80	0.76	0.81	0.77	0.78	1.00	
11	K	0.90	0.81	0.78	0.82	0.80	0.75	0.82	0.77	0.78	0.77	1.00

Table 6.12 Correlation matrix automatically adjusted by @Risk – Relative to the matrix in Table 6.11[81]

We use an example to show how automatic adjustment of a nonconsistent matrix works. We assume a 3-by-3 matrix (K) taken from Eq. (6-7), yielding the following eigenvalues for this matrix:

$$K = \begin{matrix} & A & B & C \\ A & 1 & 0.9 & 0.8 \\ B & 0.9 & 1 & 0 \\ C & 0.8 & 0 & 1 \end{matrix} \tag{6-7}$$

The eigenvalues for this matrix are as follows:

$\lambda_1 = -0.204159$

$\lambda_2 = 1$

$\lambda_3 = 2.204159$

Since one of the eigenvalues is negative, this matrix is nonconsistent and needs to be adjusted. For this purpose, the smallest eigenvalue (λ_1) is multiplied by the identity matrix and subtracted from the original correlation matrix (K):

$$K' = \begin{bmatrix} 1 & 0.9 & 0.8 \\ 0.9 & 1 & 0 \\ 0.8 & 0 & 1 \end{bmatrix} - \begin{bmatrix} -0.204159 & 0 & 0 \\ 0 & -0.204159 & 0 \\ 0 & 0 & -0.204159 \end{bmatrix}$$

[78] *Chau* used @Risk Version 1.55.

[79] If either "No" or "Cancel" is selected, no simulation can be performed, and @Risk will abort the process.

[80] The notion of special requirements refers to locking inputs using an adjustment matrix. This step prevents manually entered coefficients from being changed during automatic adjustment of an inconsistent matrix.

[81] Cf. Kummer (2015a), p. 185

$$K' = \begin{bmatrix} 1.204159 & 0.9 & 0.8 \\ 0.9 & 1.204159 & 0 \\ 0.8 & 0 & 1.204159 \end{bmatrix}$$

In the next step, we divide the new matrix K′ by the value $(1 - \lambda_1)$. The resulting new correlation matrix K″ is already identical to the adjusted matrix suitable for continuing the simulation in @Risk.

$$K'' = \begin{bmatrix} 1.204159 & 0.9 & 0.8 \\ 0.9 & 1.204159 & 0 \\ 0.8 & 0 & 1.204159 \end{bmatrix} / (1 + 0.204159)$$

$$K'' = \begin{bmatrix} 1 & 0.747409 & 0.664364 \\ 0.747409 & 1 & 0 \\ 0.664364 & 0 & 1 \end{bmatrix}$$

The new eigenvalues for this matrix are as follows:

$\lambda_1 = 0$

$\lambda_2 = 1$

$\lambda_3 = 2$

Such automatic adjustment was also performed for the 11-by-11 correlation matrix used for calculating the sum of eleven variables. Table 6.12 shows the new correlation matrix.

This matrix, or the correlation values in its first column, are quite far removed from the originally used correlations, which had been strongly positive. This yields a distorted result for the sum of the eleven variables.

To prevent this from happening while still obtaining a consistent, valid correlation matrix, we can apply an adjustment-factor matrix to "fix" the user-entered correlations in the course of automatic adjustment. Adjustment factors are dimensionless and can have values between 0 and 100. The higher this factor, the closer the automatic adjustment result will remain to the entered correlation. Thus, the automatically adjusted correlation matrix will correspond to the relationships intended when entering individual correlation coefficients. The adjustment-factor matrix for the correlation matrix in Table 6.11 looks as follows:

$NO_{NEG.CORR} = 0$		1 A	2 B	3 C	4 D	5 E	6 F	7 G	8 H	9 I	10 J	11 K
1	A	1.00										
2	B	0.90	1.00									
3	C	0.90	0.86	1.00								
4	D	0.90	0.79	0.72	1.00							
5	E	0.90	0.65	0.86	0.79	1.00						
6	F	0.90	0.81	0.80	0.83	0.81	1.00					
7	G	0.90	0.78	0.73	0.71	0.79	0.83	1.00				
8	H	0.90	0.81	0.79	0.82	0.80	0.75	0.82	1.00			
9	I	0.90	0.80	0.78	0.81	0.79	0.77	0.81	0.78	1.00		
10	J	0.90	0.81	0.79	0.82	0.80	0.76	0.81	0.77	0.78	1.00	
11	K	0.90	0.81	0.78	0.82	0.80	0.75	0.82	0.77	0.78	0.77	1.00

Table 6.13 Adjustment-factor matrix – Relative to the matrix in Table 6.11[82)]

The new, automatically adjusted correlation matrix taking into account the adjustment-factor matrix is equivalent to that in Table 6.14. We recognize that the correlation coefficients originally entered in the first column were retained even during automatic adjustment.

$NO_{NEG.CORR} = 0$		1 A	2 B	3 C	4 D	5 E	6 F	7 G	8 H	9 I	10 J	11 K
1	A	1.00										
2	B	0.90	1.00									
3	C	0.90	0.86	1.00								
4	D	0.90	0.79	0.72	1.00							
5	E	0.90	0.65	0.86	0.79	1.00						
6	F	0.90	0.81	0.80	0.83	0.81	1.00					
7	G	0.90	0.78	0.73	0.71	0.79	0.83	1.00				
8	H	0.90	0.81	0.79	0.82	0.80	0.75	0.82	1.00			
9	I	0.90	0.80	0.78	0.81	0.79	0.77	0.81	0.78	1.00		
10	J	0.90	0.81	0.79	0.82	0.80	0.76	0.81	0.77	0.78	1.00	
11	K	0.90	0.81	0.78	0.82	0.80	0.75	0.82	0.77	0.78	0.77	1.00

Table 6.14 Adjusted correlation matrix, taking into account the adjustment-factor matrix in Table 6.13 – Relative to the correlation matrix in Table 6.11[83)]

Combined with the eleven variables, this new correlation matrix can be used for a Monte Carlo simulation (Latin Hypercube sampling) in order to calculate a sum distribution. 50,000 iterations were completed in the case at hand. Symmetrical triangular distributions (minimum: 2; most likely value: 6; maximum: 10) are applied for summation.

In the subsequent studies, the positive correlations in the first column of Table 6.11 were gradually replaced with negative correlations ($r = -0.9$) until all of the correlations were negative. The adjustment-factor matrix (see Table 6.13) was applied to automatically generate consistent correlation matrices. No changes were made to the eleven identical distribution functions of input parameters (triangular distributions: 2 | 6 | 10) in the summation step. The calculated sums are represented in a diagram on the basis of the

82) Cf. Kummer (2015a), p. 186

83) Cf. ibid., p. 187

different correlations. Besides the mean and most likely values of the deterministic computation, the following additional statistical indicators are shown for each simulation (see Fig. 6.26):

- 5% quantile
- 95% quantile
- Simple standard deviation from mean (bidirectional)

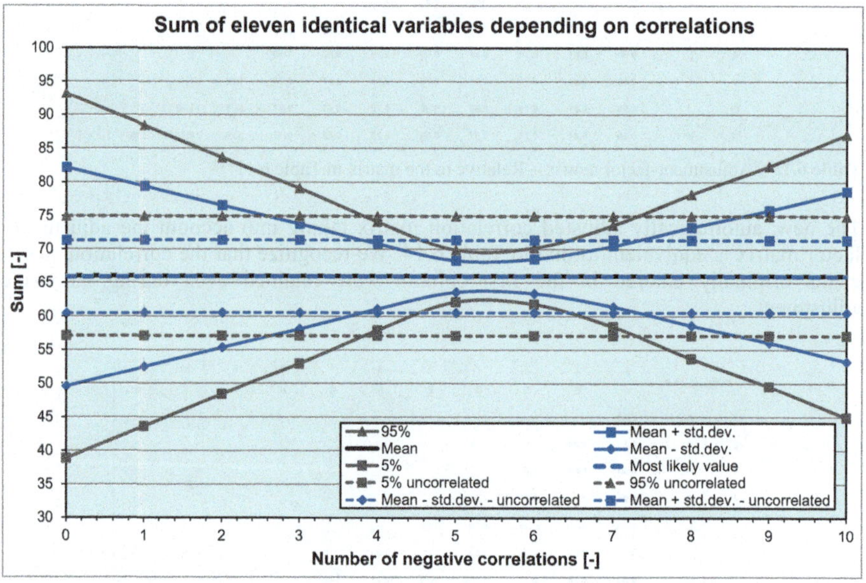

Fig. 6.26 Sum of eleven identical variables depending on the correlations, compared to the uncorrelated result[84)]

For comparison, the same statistical indicators are also shown for an uncorrelated computation (i.e. parameters are independent from one another) as a dashed line.

The narrowest range of the sum of eleven identical variables is recorded for a balanced ratio between positive and negative correlations. Only the spread of results is tightened multiple times by selecting the corresponding correlations, yet the range of input parameters remains unchanged. Compared to uncorrelated computations, wider or narrower ranges for the results follow, depending on the correlation coefficient selected.

For the example above, it is important to note that all addends were modeled with symmetrical distributions. If skewed distributions were applied, the expected result of a deterministic calculation would deviate from the mean of the simulations.

84) Cf. Kummer (2015a), p. 188

6.6 Comparison – Deterministic and Probabilistic Calculations

This section compares the problems of deterministic calculations with the results of a Monte Carlo simulation and describes the strengths, weaknesses, opportunities, and threats associated with this type of simulation (SWOT analysis).

6.6.1 Critical Review of Deterministic Calculations

In practice, deterministic calculations are the rule rather than the exception when it comes to determining construction time and costs. Probabilistic methods are applied only in rare cases. Nor are they widely known among experts.[85)]

One of the problems arising with deterministic calculations is that their input parameters are usually fraught with major uncertainties. They must thus be determined by means of qualified estimates. At best, deterministic calculations systematically account for uncertainties only by introducing percentage markups or safety margins. This approach makes it impossible to quantify chance/risk ratios.

Furthermore, although deterministic calculation results appear to be very accurate at first glance, their validity is limited. A deterministic result (i.e. a single number – see calculation mode 1 in Fig. 6.27) is not an indicator of probability. A deterministic approach also fails to utilize the knowledge about possible input parameter ranges and their relationships. Nor are other calculation modes, such as enhanced deterministic, PERT[86)], or simplified probabilistic methods, suitable for deriving conclusions with respect to probabilities of occurrence. In any case, each of the possible scenarios can be reflected by just a single figure, or, at best, by minimum, most likely, and maximum values. A comparison of five calculation modes showed that only the results of a probabilistic computation method (Monte Carlo simulation) made it possible to assess the quality of results obtained with other methods (see Fig. 6.27).[87)]

---

85) Cf. Alber (2013), p. 165ff.

86) PERT stands for Program Evaluation and Review Technique.

87) Cf. Kummer/Hofstadler (2013), p. 183ff.

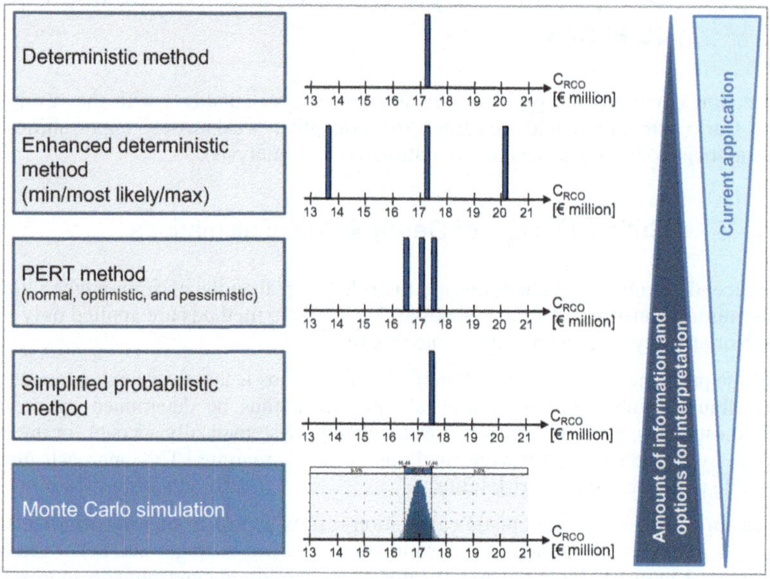

Fig. 6.27 Comparison of five calculation modes – Determination of the direct costs of reinforced concrete operations

This problem has been discussed in the literature for many years, and some criticism of deterministic computation methods with regard to their suitability for determining construction times and costs has been expressed.

Schach considers the deterministic calculation of investment costs to be problematic. In a deterministic approach, cost overruns can only be investigated by means of simplified methods involving, for instance, sensitivity or extreme analyses. Monte Carlo simulations routinely deliver such sensitivity or extreme analyses. *Schach* strongly recommends that expected costs should in future be determined with the aid of Monte Carlo simulations.[88)]

According to *Schneider/Spiegl/Sander*, demanding an exact figure directly contradicts the requirement of an exceedingly accurate forecast. The frequently applied deterministic methods lead to results of extremely limited validity.[89)]

Schulz mentions that historical data can be used to derive incidence distributions, which permit conclusions as to the probability distributions associated with similar processes. He analyzed 712 bids from submissions in the construction industry and calculated average bid amounts and percentage deviations for each of the submissions. Results were plotted in a histogram, with a spread from 55% to 165% documented between the highest and the lowest bid, where 100% represented the mean of the bids per submission. In other words, the highest bid quoted about three times the amount of the lowest. *Schulz* further states that a larger number of analyzed bids would presumably have expanded the spread even more. He emphasizes that absolute spread limits are not useful.[90)]

---

88) Cf. Schach (2005), p. 124

89) Cf. Sander/Spiegl/Schneider (2010), p. 606

90) Cf. Schulz (1980), p. 29ff.

Schulz outlines the following benefits of knowing the spreads of cost values:[91)]

- Cost surveyors would be able to compare their own costing spreads with those of their colleagues or competitors.
- Construction contractors can become aware of potential losses if they are awarded a contract at a comparatively low price.
- Construction contractors would have to question their performance capacity and productivity if they always end up far above the lowest bid in spite of aggressive pricing.
- Conversely, clients could ask themselves by how much bid prices would decline if they increased the number of requested bids.

In this context, *Lederer* concurs that a specific figure (which is almost certainly inaccurate) is not exactly helpful for clients either when it comes to financial planning. We would thus require specific reliability indicators since isolated figures alone are insufficient for deriving accurate forecasts.[92)]

Criticisms expressed against Monte Carlo simulations are that they are more resource-hungry and require mathematical knowledge and expertise to ensure appropriate inputs. At any rate, a related understanding is necessary for modeling, defining inputs, and interpreting simulation results.

6.6.2 SWOT Analysis for Monte Carlo Simulations

In the individual project phases, Monte Carlo simulations are a useful tool for clients and contractors alike for informing the decision-making process, carrying out sensitivity analyses, and implementing optimization steps.

A SWOT analysis[93)] studies the benefits and disadvantages associated with applying Monte Carlo simulations.

In this context, *Wenzel et al* state the following with respect to "simulation worthiness": Any task will be worthy of being simulated if the solution to a problem can only be identified by means of a simulation. This is the case, for instance, if dynamic processes or stochastic effects play a non-negligible role within the system under study. Simulation worthiness can also be assumed if a simulation model makes it much easier to arrive at a solution, even if this were also possible on the basis of other mathematical methods. If specific requirements are defined for communicating and visualizing results, developing a simulation model is also justifiable even if the underlying problem could be solved more easily or quickly by applying other methods.[94)]

SWOT analyses are a strategic management tool that includes four categories for identifying and recording the strengths, weaknesses, opportunities, and risks of a given method, process, or measure.

The following section describes the approach to performing such a SWOT analysis for the use of Monte Carlo simulations in construction management and economics. The individual items listed below form a non-exhaustive list of associated benefits and disadvantages.

---

91) Cf. ibid., p. 34

92) Cf. Lederer (1982), p. 11f.

93) Acronym of Strengths, Weaknesses, Opportunities, and Threats

94) Cf. Wenzel et. al. (2008), p. 15

6.6.2.1 Strengths

We can identify the following strengths of Monte Carlo simulations compared to deterministic calculations:

- Ranges can be taken into account on a systematic basis.
- Considerations with respect to uncertain parameters are directly incorporated in the calculations; they can be subsequently reproduced from the simulation model.
- Distribution functions are used to counter the effect of the non-systematic selection of deterministic input parameters.
- Both deterministic and probabilistic inputs can be applied within a single simulation.
- Monte Carlo simulations automatically reflect best- and worst-case scenarios.
- A single simulation run is sufficient for analyzing a large number of possible scenarios.
- Monte Carlo simulations create a broader base for expert knowledge.
- Sensitivity analyses are part of the simulation because uncertain input values are randomly considered within the defined ranges in the individual iterative steps.
- The need for analytical solutions based on convolution is eliminated because random numbers are used for the simulation. Compared to such analytical methods, probabilistic calculations are relatively simple to perform by means of simulations.
- Uncertainties associated with specific parameters can easily be incorporated in the calculations.
- Computation models remain largely unchanged compared to the deterministic approach.
- Results can be graphically represented in histograms.
- Software suites for performing Monte Carlo simulations are commercially available, which is why no in-depth programming knowledge and expertise is needed for their use.
- Most commercially available software suites come with a comprehensive manual, technical support, online tutorials, discussion boards, and other sources of useful information.

6.6.2.2 Weaknesses

We can identify the following weaknesses of Monte Carlo simulations compared to deterministic calculations:

- Users need a basic mathematical understanding both for the entry of inputs and the interpretation of results.
- A commercial software solution is needed in the absence of a proprietary program. Such solutions are usually add-ins to MS Excel.
- Input is more sophisticated than in the case of deterministic calculations.
- Subjective elements of the calculations are retained in the Monte Carlo simulation, but they are systematically integrated into the calculations by distribution functions.

6.6.2.3 Opportunities

Performing Monte Carlo simulations is associated with the following opportunities:

- Simulation results can be interpreted with the aid of sensitivity analyses such as tornado, point, and spider diagrams.
- Simulation models can potentially make a major contribution to implementing probabilistic optimization (operations research).
- Dependencies between inputs can be incorporated in the calculations by means of correlations.

- Probability theory information and statistical indicators enhance the validity of calculation results.
- Decisions can be made on the basis of a selected chance/risk ratio.
- Input parameter ranges can be derived directly from practice (data collections and/or expert surveys).
- Together with theoretical considerations and characteristics, combinations of historical data, in-house documentation and expert knowledge can be translated into distribution functions and thus be incorporated into the computation process.
- Any systematic data collection involves the option of linking this data to existing databases to fit distribution functions to individual input parameters in order to ensure more accurate, constantly updated equations.

6.6.2.4 Threats

The following threats exist in the context of applying Monte Carlo simulations:

- The quality of the outputs will always depend on the quality of the inputs and of the computation model, which is very much in line with the GIGO (Garbage In, Garbage Out) principle.
- Simulation results appear as a product of a "black box" in the absence of any indication and description of the input parameters used.
- If no adjustment-factor matrix is defined when considering correlations, automatic adjustment of a nonconsistent correlation matrix may result in a matrix incompatible with the relationships intended at the input stage.

6.7 Summary

Performing probabilistic calculations poses a challenge to students, researchers, and practitioners. In the presence of a basic mathematical understanding and after deterministic modeling, Monte Carlo simulations make such calculations easier. Simulations deliver numerical solutions that enable users to invest more time and effort in modeling and selecting appropriate distribution functions as well as correlations.

Potential users may not necessarily possess (in-depth) knowledge of the Monte Carlo simulation method because it heavily relies on probability theory. This chapter described its basic principles and included worked examples for further explanation. It thus provided basic knowledge to create an understanding of the following chapters while encouraging independent thinking and application of the method.

This chapter described the origin and evolution of the Monte Carlo method while also making the distinction between method and simulation. In addition, the problem of generating random numbers was discussed in detail. Worked examples were used to visualize and interpret the effects of different distribution functions and correlations on simulation results.

The following chapter concentrates on the systematic selection of distribution functions as an essential prerequisite for modeling. It also explains how data records can be fitted for generating distribution functions for input parameters. Furthermore, it also stresses the significance of correlations and outlines ways of identifying them by means of expert surveys. In its last part, the next chapter deals with the interpretation of simulation results, which is a crucial component of the entire process.

6.8 References

Alber, Alexander (2013). Risikomanagement in Bauunternehmen – Eine Analyse von Theorie und Praxis. Master's thesis. Graz. Institute of Construction Management and Economics, Graz University of Technology.

Böhnlein, Claus-Burkard (2004). Simulation in der Betriebswirtschaft. In: Experiences from the Future – New Methods and Applications in Simulation for Production and Logistics – Tagungsband 11. ASIM-Fachtagung Simulation in Produktion und Logistik, Berlin, October 4-5, 2004. Eds.: Mertins, Kai; Rabe, Markus. 1-22. Stuttgart. Fraunhofer IRB Verlag. (ISBN 3-8167-6640-4)

Böhnlein, Claus-Burkard (2010). Simulationsunterstützte Spezifikation und Analyse von Geschäftsmodellen und Geschäftsprozessen. Available online at http://www.asim-fachtagung-spl.de/asim2010/papers/E9_Boehnlein.pdf. Accessed December 21, 2020.

Chau, Kwong Wing (1995a): The validity of the triangular distribution assumption in Monte Carlo simulation of construction costs: empirical evidence from Hong Kong. In: Construction Management and Economics, Volume 13, Issue 1, 1995. 15-21. London. Taylor & Francis Group. (ISSN 0144-6193)

Chau, Kwong Wing (1995b): Monte Carlo simulation of construction cost using subjective data. In: Construction Management and Economics, Volume 13, Issue 5, 1995. 369-383. London. Taylor & Francis Group. (ISSN 0144-6193)

Demmler, Markus (2009). Risikomanagement im internationalen Tunnelbau unter Anwendung der Vertragsform FIDIC Red Book. Doctoral thesis. Darmstadt. Fachbereich Bauingenieurwesen und Geodäsie der Technischen Universität Darmstadt. (ISBN 978-3-00-027533-3)

Dinort, Günter (1977). Berücksichtigung stochastischer Einflüsse in der Kostenrechnung einer Bauunternehmung. Doctoral thesis. Munich. Lehrstuhl für Tunnelbau und Baubetriebslehre der Technischen Universität München.

Duden (2007). Deutsches Universalwörterbuch. Mannheim. Bibliographisches Institut & F. A. Brockhaus AG. (ISBN-10: 3-411-05506-5)

Feik, Roland (2006). Elektronisch gestütztes Risikomanagement im Bauwesen – Ein Konzept eines elektronischen Chancen- und Gefahrenmanagementsystems für Auftraggeber. Doctoral thesis. Innsbruck. Fakultät für Bauingenieurwissenschaften der Leopold-Franzens-Universität Innsbruck.

Girmscheid, Gerhard; Motzko, Christoph (2007). Kalkulation und Preisbildung in Bauunternehmen – Grundlagen, Methodik und Organisation. Berlin, Heidelberg. Springer-Verlag. (ISBN 978-3-540-36694-2)

Girmscheid, Gerhard; Busch, Thorsten A. (2008b). Unternehmensrisikomanagement in der Bauwirtschaft. Berlin. Bauwerk Verlag GmbH. (ISBN 978-3-89932-184-5)

Gleißner, Werner (2011). Grundlagen des Risikomanagements im Unternehmen – Controlling, Unternehmensstrategie und wertorientiertes Management – 2nd ed. Munich. Verlag Franz Vahlen. (ISBN 978-3-8006-3767-6)

Golenko, D. I. (1972). Statistische Methoden der Netzplantechnik. Stuttgart, Leipzig. B. G. Teubner Verlagsgesellschaft. (ISBN 3-519-02013-0)

Gottschalk-Mazouz, Niels (2011). Risiko. In: Handbuch Ethik. Eds.: Düwell, Marcus; Hübenthal, Christoph; Werner, Micha H. Stuttgart, Weimar. Metzler-Verlag. (ISBN 978-3-476-02388-9)

Hauwermeiren, Michael Van; Vose, David (2009). A Compendium on Distributions. eBook. Available online at http://www.vosesoftware.com/content/ebook.pdf. Accessed October 21, 2016.

Hengartner, Walter; Theodorescu, Radu (1978). Einführung in die Monte-Carlo-Methode. Munich, Vienna. Carl Hanser Verlag. (ISBN 3-446-12331-8)

Hofstadler, Christian (2011). Application of the Monte-Carlo Method to Determine the Costs for Construction Projects – Influence of Ranges and Correlations on Probability Distribution. Presentation – 2011 Palisade Risk Conference – Risk Analysis, Applications & Training. Amsterdam.
Available online at http://www.palisade.com/downloads/UserConf/EU11/2011-Palisade@RISK-Hofstadler.pdf. Accessed December 21, 2020.

Hofstadler, Christian (2014). Produktivität im Baubetrieb – Bauablaufstörungen und Produktivitätsverluste. Berlin, Heidelberg. Springer-Verlag. (ISBN 978-3-642-41632-3)

Hofstadler, Christian (2014b). Monte-Carlo-Simulation im Baubetrieb – Verteilungsfunktion für Aufwandswerte bei Schalarbeiten – Herleitung der Verteilungsfunktion aus den Ergebnissen einer Expertenbefragung. In: Bauingenieur, Band 89, May 2014. 189-201. Düsseldorf. Springer-VDI-Verlag GmbH & Co. KG. (ISSN 0005-6650)

Hofstadler, Christian; Kummer, Markus (2017a). Chancen- und Risikomanagement in der Bauwirtschaft – Für Auftraggeber und Auftragnehmer in Projektmanagement, Baubetrieb und Bauwirtschaft. Berlin, Heidelberg. Springer-Verlag. (ISBN 978-3-662-54318-4)

Hruschka, Johann (1968). Die Anwendung des Normprodukts in der Baubetriebsplanung, Erläutert am Beispiel der Rohbauarbeiten im Hochbau. Doctoral thesis. Munich. Fakultät für Bauwesen der Technischen Hochschule München.

http://www.math.sci.hiroshima-u.ac.jp/~m-mat/MT/emt.html. Accessed December 21, 2020.

Iwan, Gerhard (1992). Beitrag zur Anwendung der Simulation und Reihenfolgeoptimierung bei der Analyse von Bauproduktionsprozessen. Fortschrittsberichte VDI – Reihe 4: Bauingenieurwesen, Nr. 113. Düsseldorf. VDI-Verlag. (ISBN 3-18-141304-6)

Krempl, Simone (2012). Korrelationen bei Aufwandswerten für Schal- und Bewehrungsarbeiten. Master's thesis. Graz. Institute of Construction Management and Economics, Graz University of Technology.

Kummer, Markus (2015a). Aggregierte Berücksichtigung von Produktivitätsverlusten bei der Ermittlung von Baukosten und Bauzeiten – Deterministische und probabilistische Betrachtungen. Doctoral thesis. Graz. Institute of Construction Management and Economics, Graz University of Technology.

Kummer, Markus; Hofstadler, Christian (2013). Einsatz der Monte-Carlo-Simulation zur Berechnung von Baukosten. In: bau aktuell, 4. Jahrgang, Heft Nr. 5, September 2013. 178-188. Vienna. Linde-Verlag Ges.m.b.H. (ISSN 2077-4737)

Lederer, Siegfried (1982). Die Mittelbereitstellung für Bauleistungen unter Berücksichtigung der Unsicherheiten in der Planung. Doctoral thesis. Institut für Bauingenieurwesen IV Tunnelbau und Baubetriebslehre der Technischen Universität München.

Liberda, Walter (1985). Ermittlung des Kalkulationswagnisses im Bauwesen durch stochastische Berechnung der Herstellkosten. Doctoral thesis. Graz. Fakultät für Bauingenieurwesen der Technischen Universität Graz, Erzherzog-Johann-Universität.

Link, Doris (1999). Risikobewertung von Bauprozessen – Modell ROAD – Risk and Opportunity Analysis Device. Doctoral thesis. Vienna. Institut für Baubetrieb und Bauwirtschaft der Technischen Universität Wien.

Mattern, Friedmann (1996). Modellbildung und Simulation. Fachbereich Informatik der Technischen Hochschule Darmstadt, 1996. Available online at http://www.vs.inf.ethz.ch/publ/papers/Mod_Sim.pdf. Accessed December 21, 2020.

Mayer, Franz Xaver (2013). Kostensicherheit zum Zeitpunkt der Realisierungsentscheidung – Entwicklung eines Kosten-Prognose-Modells für Bauwerkskosten im Hochbau. Doctoral thesis. Munich. Lehrstuhl für Bauprozessmanagement und Immobilienentwicklung der Technischen Universität München.

Metropolis, Nicholas; Ulam, Stanislaw (1949). The Monte Carlo Method. In: Journal of the American Statistical Association, Vol. 44, No. 247, September 1949. 335-341. Oxford. Taylor & Francis, Ltd. (ISSN 01621459)

MSC2010: http://www.ams.org/mathscinet/msc/pdfs/classifications2010.pdf. Accessed December 21, 2020.

Naumann, René (2007). Kosten-Risiko-Analyse für Verkehrsinfrastrukturprojekte. Aus Forschung und Praxis – Band 8 – Schriftenreihe des Instituts für Baubetriebswesen der Technischen Universität Dresden. Ed.: Schach, Rainer. Dresden. Expert Verlag. (ISBN-13: 978-3-8169-2701-3)

Palisade Corporation (2013). Benutzerhandbuch – @Risk – Risikoanalysen- und Simulations-Add-In für Microsoft Excel – March 6, 2013.

Palisade Corporation: 5.3. How @Risk Tests a Correlation Matrix for Validity. Palisade Knowledge Base. http://kb.palisade.com/index.php?pg=kb.page&id=74. Accessed December 21, 2020.

Raaber, Norbert (2003). Beitrag zu Ausschreibung und Vergabe von Bauleistungen. Schriftenreihe des Instituts für Baubetrieb und Bauwirtschaft der Technischen Universität Graz – Heft 15. Ed.: Stadler, Gert. Graz. Verlag der Technischen Universität Graz. (ISBN 3-901351-76-0)

Raffel, Wolf-Ulrich (2001). Simulationstheorie. Institut für Informatik der Freien Universität Berlin. Available online at http://www.wuraffel.de/wissenschaft/publikationen/Simulationstheorie.pdf. Accessed November 1, 2016.

Raftery, John (1999). Risk Analysis – In: Project Management. London, New York. E & FN Spon, an Imprint of Routledge. (ISBN 0-419-18420-1)

Rohr, Stefan (2005). Optimierung der Bauproduktion durch Simulation von Prozessen. Doctoral thesis. Munich. Institut für Geologie, Geotechnik und Baubetrieb der Technischen Universität München.

Sander, Philip (2012). Probabilistische Risiko-Analyse für Bauprojekte – Entwicklung eines branchenorientierten softwaregestützten Risiko-Analyse-Systems. Doctoral thesis. Innsbruck. Fakultät für Bauingenieurwissenschaften der Leopold-Franzens-Universität Innsbruck.

Sander, Philip; Spiegl, Markus; Schneider, Eckart (2010). Die exakte Zahl – Gedanken zum Umgang mit Unschärfen. In: Die Wirtschaftliche Seite des Bauens – Festschrift zum 60. Geburtstag von Rainer Wanninger. Ed.: Wanninger, Rainer. Braunschweig. Institut für Bauwirtschaft und Baubetrieb der Technischen Universität Braunschweig.

Sawilowsky, Shlomo S. (2003). You Think You've Got Trivials? In: Journal of Modern Applied Statistical Methods, Vol. 2, No. 1, May 2003. 218-225. JMASM Inc. (ISSN 1538-9472) Available online at http://digitalcommons.wayne.edu/coe_tbf/21. Accessed December 21, 2020.

Schach, Rainer (2005). Stochastische Investitionskostenberechnung für Verkehrsinfrastrukturprojekte. In: Tagungsband zur 5. Dresdner Fachtagung Transrapid am September 29, 2005. 107-125. Ed.: Schach, Rainer. Dresden. Verlag der Technischen Universität Dresden. (ISBN-13: 978-3860054901)

Schach, Rainer; Flemming, Christian (2014). Grundlagen der Kostensimulation unter Einbeziehung von Risiken. In: Informationssysteme im Bauwesen 2 – Anwendungen. Eds.: Scherer, Raimer J.; Schapke, Sven-Eric. 253-267. Berlin, Heidelberg. Springer-Verlag. (ISBN 978-3-662-44760-4)

Schubert, Eberhard (1971). Die Erfaßbarkeit des Risikos der Bauunternehmung bei Angebot und Abwicklung einer Baumaßnahme – Heft 7. Ed.: Jebe, Hans. Lehrstuhl für Baubetrieb und Baubetriebswirtschaft der Technischen Universität Hannover. Düsseldorf. Werner-Verlag. (ISBN 3-8041-3303-7)

Schulz, Josef (1980). Risikorechnung bei der Preiskalkulation – Ein Weg zur Ermittlung der maximalen Überschreitung von Kostenvoranschlägen. Wiesbaden, Berlin. Bauverlag GmbH. (ISBN 3-7625-1303-1)

Strehl, Gustav (1983). Ein Modell für die Ermittlung eines Angebotspreises von Bauleistungen bei stochastischer Kostenermittlung. Doctoral thesis. Vienna. Technische Universität Wien.

Suhl, Leena; Mellouli, Taïeb (2013). Optimierungssysteme – Modelle, Verfahren, Software, Anwendungen – 3rd ed. Berlin, Heidelberg. Springer-Verlag. (ISBN 978-3-642-38936-8)

Theis, Christian; Kernbichler, Winfried (2002). Grundlagen der Monte Carlo Methoden. Graz University of Technology, Institut für Theoretische Physik – Computational Physics. Available online at https://itp.tugraz.at/MML/MonteCarlo/MCIntro.pdf. Accessed December 21, 2020.

Touran, Ali; Wiser, Edward P. (1992). Monte Carlo Technique with Correlated Random Variables. In: Journal of Construction Engineering and Management, Vol. 118, No. 2, June 1992. 258-272. American Society of Civil Engineers (ASCE). (ISSN 0733-9364)

Wenzel, Sigrid et al. (2008). Qualitätskriterien für die Simulation in Produktion und Logistik – Planung und Durchführung von Simulationsstudien. Berlin, Heidelberg. Springer-Verlag. (ISBN 978-3-540-35276-1)

Werkl, Michael (2013). Risiko- und Nutzenverhalten in der Bauwirtschaft – Eine entscheidungstheoretische Betrachtung im institutionenökonomischen Kontext. Doctoral thesis. Graz. Institute of Construction Management and Economics, Graz University of Technology.

Werner, Andreas (2002). Datenbankgestützte Risikoanalyse von Bauprojekten – Eine Methode zur rechnergestützten Monte-Carlo-Simulation des Bauablaufes für die Risikoanalyse im Bauunternehmen. Doctoral thesis. Fakultät für Ingenieurwissenschaften der Universität Rostock.

Yang, I.-T. (2005). Simulation-based estimation for correlated cost elements. In: International Journal of Project Management, Volume 23, Issue 4, May 2005. 275-282. Elsevier Ltd. and IPMA. (ISSN 0263-7863)

7 Application of Monte Carlo Simulations

This chapter builds upon the fundamentals explained in Chapter 6 for outlining the practical application of Monte Carlo simulations, as well as providing guidance on how to interpret simulation results. Particular emphasis will be put on selecting distribution functions, data fitting (i.e. adjustment of distribution functions to an existing database), identifying correlations, and interpreting histograms and other graphical representations of simulation results.

7.1 Selection of Distribution Functions

Any probabilistic calculation requires the selection of distribution functions for uncertain input parameters. When trying to select the "right" distribution, users are constantly confronted with the issue of which of the many available theoretical (mathematically described) distributions[1)] would best reflect the characteristics of a parameter. After selecting the type of distribution, the corresponding curve parameters must be defined in order to be able to perform a Monte Carlo simulation.

Three methods (and their combinations) can generally be used to select distribution functions (see Fig. 7.1):

- Theoretical considerations based on known or assumed characteristics
- Gathering empirical data
- Expert surveys

Selecting distribution functions for uncountable parameters fundamentally differs from the procedure to be applied to countable parameters, for which probabilities can be calculated. The probability of occurrence of a die value can be accurately determined owing to the limited number of possibilities and due to the fact that any of the values is equally likely to occur. Any completion of a building or structure will be associated with certain

1) Including triangular, rectangular, normal, PERT, log-logistic, binomial, and Bernoulli distributions.
It is also possible to freely define distributions for input parameters. Such distributions can consist of a combination of different continuous functions, or be applied as freely defined discrete distributions.

C. Hofstadler and M. Kummer, *Chances and Risks in Construction Management and Economics*, Springer Tracts in Civil Engineering,
https://doi.org/10.1007/978-3-030-64014-9_7

upper and lower limits for performance values and labor consumption rates. Yet, for each of the values within the resulting range, the probability of occurrence will not be identical. In addition, selecting distribution functions for describing future events always involves a subjective component. This purely subjective approach can be partially compensated for by relying on historical data or expert opinions, which will, however, at best result in intersubjectivity[2)] rather than objectivity in the true sense of the word.[3)]

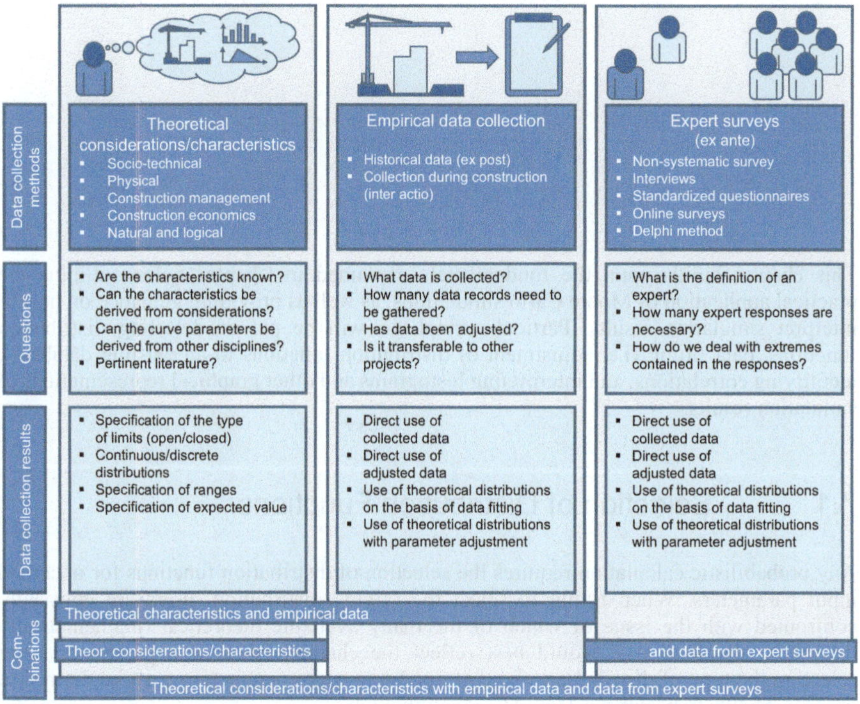

Fig. 7.1 Methods for determining distribution functions

If no, or only an insufficient, database exists, employees experienced in the field are usually able to state or estimate values for individual parameters within a practicable range.

The following sections include a more comprehensive description of data collection methods and of the approach to generating distribution functions.

2) Intersubjectivity forms an intermediate level between subjectivity and objectivity. It describes the consistency of certain perceptions, lines of thought, and experiences across different individuals, irrespective of their specific differences.

3) Cf. Kummer (2015a), p. 170

7.1.1 Theoretical Considerations Based on Known or Assumed Characteristics

The process of selecting a distribution function can start from basic theoretical considerations and include known or assumed characteristics derived from the socio-technical, physical, construction management and economics, natural or logical properties of the input parameter. For instance, it is not useful to assume negative prices or an infinite number of cranes. In most cases, certain parameters can be stated within ranges that are realistic from a construction management point of view. It is thus usually unhelpful to choose one-sided or two-sided open distributions for input variables in construction management and economics[4)] because these would also permit values that are infinite both on the negative and on the positive side.[5)]

Theoretical considerations and the definition of theoretical characteristics aim to exclude implausible or impossible results right from the start by selecting appropriate distribution functions. This approach should establish the best possible basis for decision making when interpreting the results of a Monte Carlo simulation while fully utilizing the available computing capacity with a view to applying optimization methods (in the sense of operations research) because possible solutions are permissible only within certain realistic ranges.

In addition, we should ascertain whether the circumstances and underlying conditions can best be reflected by discrete or continuous distributions, depending on the specific issue and input parameter. If the input is defined by a simple binary relationship (yes/no or true/false), a discrete (Bernoulli) distribution should be used. If only certain values are possible for a parameter, this again warrants the use of a discrete distribution (such as the specific number of cranes in a certain construction phase). A continuous distribution should be selected whenever a parameter can be a random decimal (such as for labor consumption rates, or for costs and prices applying a simplified method). Continuous distributions can be preferred over discrete distributions for calculating average values, such as the average number of workers, because the actual number of workers can only be expressed as an integer whereas averages can also be decimals.

The distribution will show a unimodal or multimodal curve[6)] if a higher probability of occurrence is assigned to one or more values of a parameter.

Further considerations with respect to the possible curves and characteristics of parameter-dependent distribution functions can also be derived from other disciplines or from pertinent literature.

The process of selecting distribution functions is also influenced by subjective assessments as well as the risk attitude or risk preference (i.e. risk neutrality, risk affinity, or risk aversion) of those individuals or groups of people who perform the simulations or modeling exercises. The degree of risk affinity can thus impact the selection of the skewness of distribution functions.[7)]

It is assumed that risk-neutral individuals would preferably select symmetrical distribution functions with equal probabilities of exceeding or underrunning the most likely value (provided such distributions are unimodal).

---

4) In the context of selecting the distribution functions, it is assumed that the building or structure is actually constructed and completed.

5) Cf. Kummer (2015a), p. 171

6) Pertinent literature does not include examples of multimodal input parameters – see also Tabels 6.3 to 6.5.

7) Cf. Hofstadler/Kummer (2017b), CPM-26

Conversely, there is reason to believe that individuals with high risk affinity would tend to apply right-skewed distributions (in which high values are relatively unlikely to occur) to parameters such as labor consumption rates or costs. In this case, a rather optimistic most likely value in the lower portion of the range will be selected (see Fig. 7.2, left).[8)]

Any risk-averse selection of distribution functions will tend to assign more weight (or a comparatively high probability of occurrence) to unfavorable values within a defined range. A rather pessimistic most likely value in the upper portion of the range will be selected for labor consumption rates, costs, etc. This tends to result in a left-skewed distribution (see Fig. 7.2, right).[9)]

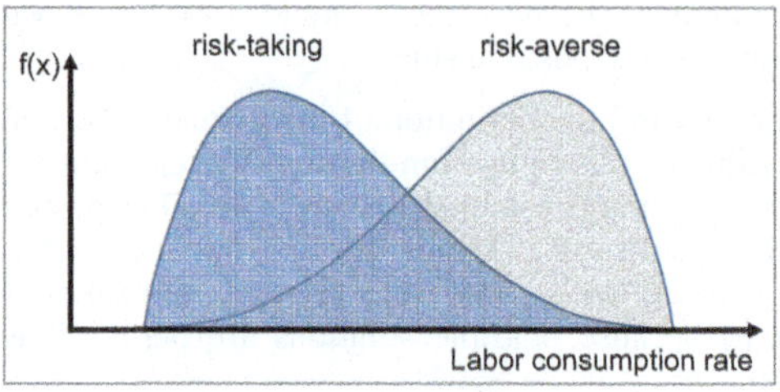

Fig. 7.2 Influence of risk attitude on the skewness of distribution functions

Fig. 7.3 shows possible criteria and alternatives according to which the curves of distributions can be defined in a gradual process. This figure also includes qualitative graphical representations of the individual distribution curves. We first need to make a distinction between deterministic and probabilistic inputs, depending on whether they are associated with uncertainties or not. For a deterministic input, a single number rather than a distribution function will be entered. Monte Carlo simulations provide the option of combining deterministic and probabilistic input parameters in a single simulation. Values known with certainty need not be "artificially" linked to distributions.

In the case of an uncertain parameter, we first need to decide if a discrete or a continuous distribution is to be selected for the probabilistic input. Discrete distributions will be used whenever the parameter can be reflected by distinct values. The distribution will be continuous if any of the characteristics may occur within a certain value range. In addition, limits can be grouped into different types (two-sided open, one-sided open, and two-sided closed). In the last step, characteristics are selected within the defined range. Two-sided closed distributions may also be associated with a constant curve within the specified range. All other distributions are likely to show a unimodal or multimodal curve. The Bernoulli distribution is a special case in the discrete distribution category since it permits only two states (zero and one), which is why it is suitable for yes/no or true/false inputs, such as the probability of occurrence of risks.

In addition, a distinction needs to be made between symmetrical and asymmetrical (skewed) distributions for the two-sided open and two-sided closed (non-constant) categories (not shown in Fig. 7.3).

[8)] Risk-taking individuals tend to prefer left-skewed distributions for output and similar values.

[9)] Risk-averse individuals tend to prefer right-skewed distributions for output and similar values.

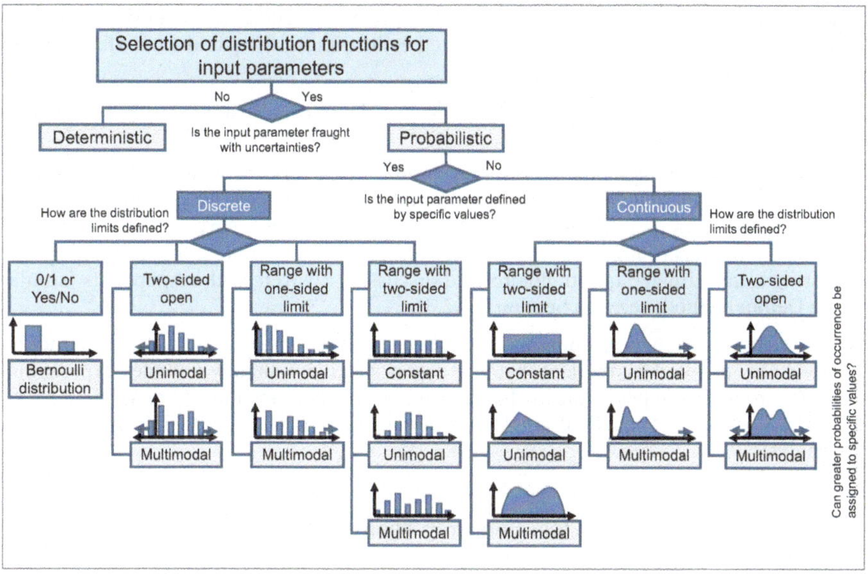

Fig. 7.3 Qualitative representation of possible characteristics of distribution functions – Decision-making structure[10)]

In this regard, *Raaber* states that triangular, parabolic or, in some cases, rectangular distributions should be used for construction-related analyses that are almost always associated with clearly visible limits. This is because of the generally observed phenomenon of significant clustering around the most likely value, with the related range mostly being asymmetrical.[11)]

7.1.2 Gathering Empirical Data

The second basis for selecting an appropriate distribution function is gathering data directly from practice. In this case, an additional distinction is made between historical data (*ex post* – after the construction phase, for instance information taken from final costing or daily work reports) and data collected during the construction phase (*inter actio* – such as from direct site observations).

When continuously collecting data during the construction phase, appropriate boundaries or limits must be defined in terms of space, time, and content, and data quality must be ensured. Prior to any analysis, collected data must also be assessed as to whether it can be applied directly or if there is any need for fitting or adjustment. In this respect, we refer particularly to the issue of collecting data and generating information (see Chapter 3). Only if data has been put into, or linked to, the appropriate context can information be generated from it that makes this data usable for additional analyses and calculations.

10) Cf. Kummer (2015a), p. 172

11) Cf. Raaber (2003), p. 167ff.

Another key criterion that determines data quality is the number of data points, or number of observations. The smaller the sample and the greater the standard deviation of gathered data, the greater the inaccuracy of the derived estimation of the parameters of the underlying population will be.

AbouRizk/Halpin state that it is not possible to categorize distributions on the basis of fewer than 20 observations[12)].[13)]

It should be noted, though, that 20 responses from expert surveys provide a higher quality of information than 20 data points taken from observations, which, for instance, have been extracted from various projects using related logs and records. This principle holds true particularly if special emphasis is placed on selecting appropriate experts, drafting and preparing a standardized questionnaire, and discussing answers with respondents (similar to the Delphi method; see also Section 7.1.2.3).

This preference for expert opinions over specific values measured in the construction phase is also justified by the fact that any expert statement already includes experiential evidence from several past projects. In addition, respondents are able to link objective data to (known) complex contextual conditions in construction practice, which adds value to the analysis.

7.1.2.1 Distributions Derived from Historical Data (*Ex Post*)

Dinort already identified the potential and the usefulness of historical data for determining distribution functions. He stated:

"Any operational analysis must aim to determine the probability distribution of the occurrence of the different possible values."[14)]

If data from past projects is available, inference statistics can use significance tests in order to permit conclusions regarding theoretical distributions of the statistical population. In this setting, data from past projects forms the sample from which generic conclusions are derived. The existing database must also be assessed for its actual suitability for determining distribution functions. In addition, the prevailing contextual conditions for gathering historical data must be verified with respect to their comparability with the current project, thus justifying the application of the identified distribution.

It should be noted that any data gathered from practice will only represent a sample extracted from the population. Also, the random nature of the collected sample needs to be ascertained in order to determine if it actually permits conclusions regarding the parent population (in this case, completed construction projects).

Strictly speaking, the methodology of inference statistics may only be applied to random samples. Possible sampling methods include:[15)]

- Random sample (probability sample)
- Judgment sample (e.g. quota sample)
- Convenience sample (e.g. according to the snowball principle[16)])

12) The notion of observations is taken to mean "time measurements for specific construction-related activities".

13) Cf. AbouRizk/Halpin (1992), p. 538

14) Cf. Dinort (1977), p. 16 (translation by Steffen Walter)

15) Cf. Diekmann (2007), p. 378

16) The mode of dissemination of a message or similar communication where each person passes the received information on to several other individuals.

Even if the formal requirements for random samples cannot be met (compliance would be virtually impossible due to the type and nature of the existing database related to construction projects), the gathered data can at least be utilized for determining a distribution curve, which can then be used for further considerations. However, any uncritical use of the derived distributions should be viewed with skepticism because historical data relates to projects and circumstances that are generally not directly comparable with a current object under study, such as a project under construction. In such a case, utilizing historical data in a direct and uncritical way may give rise to incorrect results and interpretations. If, on the other hand, we appropriately adjust the curve parameters of the derived distributions to the prevailing circumstances, this will improve results significantly compared to selecting distribution functions in a purely subjective manner based on theoretical considerations and characteristics.

Data fitting is applied in order to determine the theoretical distribution that captures an existing database, such as historical data records, with the least possible error.

Prior to fitting, historical data must be checked for integrity, plausibility, and usability. If the conditions and circumstances in which this data was gathered are unknown, it should be subjected to fitting only after a more thorough evaluation.

Direct application of final costing values, for instance for preparing a bid, is possible only if certain conditions are met because the construction process may have been associated with changes in scope, work disruptions, productivity losses, etc. Prior to being used, the historical data must be adjusted for such effects and allocated to the individual work items (less any supplementary quotations and additional work). While this data processing step requires additional time and effort, it must be completed since calculation results would otherwise be likely to be distorted.

Song/AbouRizk postulate that historical data must be accurate, consistent, and comprehensive for the user to derive a benefit from it in combination with a calculation model. The authors associate the following issues with historical data:[17)]

- Many businesses have not implemented a formal process for capturing project progress and costs within their organization. Simply speaking, historical data is often unavailable.
- Data collection methods vary across individual projects. Consequently, there is no uniform and consistent data pool.
- Data collection in paper-based systems requires time-consuming manual data extraction and is thus associated with high resource intensity.
- Many companies collect data only on a global basis. For productivity measurements and forecasts, however, more detailed data would have to be gathered and analyzed.

Hutzelmeyer also stresses the risk of using inappropriate comparative data:[18)]

When transmitting cost data from current or completed projects to a new project, care must be taken to ensure comparability in the following areas:

- Technical similarities of the structure
- Equivalent quality
- Situation on the regional construction market

Uncritically applying mean or suggested values may lead to utterly inappropriate results.

[17)] Cf. Song/AbouRizk (2008), p. 788

[18)] Cf. Hutzelmeyer (1984), p. 70

BIM models[19] will improve the database only if data and information is gathered at a central point in the organization. Most importantly, the appropriate allocation of collected data to specific work items is necessary. BIM models will not add value to 4D or 5D analyses unless the requirements for such a consistent data collection and information system have been met.

7.1.2.2 Data Collection in the Construction Phase (*Inter Actio*)

Collecting data, such as performance values and/or labor consumption rates, in the construction phase can provide a sound basis for forecasting completion dates and expected itemized costs, as well as for performing target-actual comparisons and trend analyses.

This type of data collection makes it possible to document any changes in the scope of work or unexpected/changed circumstances dictating work performance, and to then compare actual figures with pricing assumptions. To justify additional cost claims by the contractor or to allow the client to defend itself against such claims, the gathered data can be compared with initial costing assumptions, and the associated chances and risks can be assessed with the aid of probabilistic calculations.

Data should be gathered on a systematic basis, and the amount of collected data should be sufficient for the intended purpose. However, only if data and its significance (i.e. context) are embedded in an intelligent knowledge management system will it be of practical relevance, which will expand existing experience and expertise. Data on its own is useless without the corresponding context. Only its significance will add information that can be interlinked in order to generate knowledge (see Fig. 3.2, knowledge stair according to *North*).

Even at the construction stage, the documentation of collected data may be influenced by internal and external contextual conditions, including familiarization effects, disruptions to the construction process, and adverse weather conditions. If, for instance, labor consumption rates are measured in summer for deriving a forecast for work to be performed in the winter season, the related data must be delimited and adjusted as appropriate.[20]

Studies conducted at Graz University of Technology showed that the REFA system provides an effective, plausible, and relatively simple method for collecting data in the construction phase. Snapshots were taken in several instances to capture individual activities of a team using a counting method. This system allows distinguishing between primary and secondary activities while also capturing additional work and interruptions due to the construction process, disruptions or obstructions, recovery phases/breaks taken by workers, and personal reasons. It thus provides a means of appropriate delimitation of activities. However, creating and maintaining such records in a systematic way is very time-consuming and labor-intensive.

Labor consumption rates for the actual construction phase can be derived directly from practice from documenting team sizes, hours worked, and work performed (such as the amount of installed reinforcement).

Again, only in exceptional cases can the theoretical distribution determined by data fitting be directly applied to subsequent calculations even when gathering data in the

[19] BIM = Building Information Modeling

[20] Cf. Kummer (2015a), p. 175

construction phase. It should be noted, though, that applying the parameter-specific distribution function, modified by specifically defined curve parameters (such as minimum and most likely values) as well as by theoretical considerations, will definitely improve the final result.

Xie/AbouRizk/Fernando describe an example of how to integrate collected construction data into a simulation model.[21)] In this specific case, advance rates of a tunnel-boring machine were recorded, and data fitting was carried out in order to apply the identified distribution functions for performing additional calculations and deriving forecasts. The larger the amount of data, the more stable and reliable data fitting will be.

7.1.2.3 Expert Surveys (*Ex Ante*)

Expert surveys can generally be conducted in writing and/or as oral interviews. The latter can be face-to-face or telephone interviews.

In empirical social research, a distinction is often made between qualitative and quantitative approaches. The former method uses open-ended surveys, whereas the latter is highly standardized, including the use of questionnaires.

It is useful to conduct written surveys (e.g. by completing a standardized questionnaire) in the presence of researchers if only a few respondents are likely to take part and if expert knowledge is queried. If this is not possible, it is suggested to discuss responses either in a face-to-face setting or in a telephone conversation after completion of the questionnaire. This approach is associated with the benefit that misunderstandings or lack of clarity can be easily overcome, which increases the validity of the survey (i.e. the quality of the measurement). This also gives respondents the opportunity to revise and correct their answers (similar to the Delphi method) or to respond to previously unanswered questions in the oral setting.

Due to the limited amount of available resources, the willingness of experts to respond, and for pragmatic reasons, only a selected sample of all experts can be interviewed within a single field. This is why sampling is performed so as to be able to derive generic conclusions.

However, before contacting suitable individuals, we first need to clarify what the concept of a subject matter expert entails. Such experts are individuals who possess specific knowledge in a clearly delineated field and act as a source of such knowledge in the survey.[22)]

According to this definition, experts possess a certain level of knowledge as well as skills and abilities in a specific field that clearly surpass the average level. Expert knowledge usually comprises exceedingly large amounts of information, including simplifications, lesser-known facts, rules of thumb, and heuristics that enable efficient problem solving (in the relevant field).[23)]

The decision-making criteria for selecting experts rely on prior theoretical considerations and certain fundamental questions, including:[24)]

---

21) Cf. Xie/AbouRizk/Fernando (2011), p. 3457f.

22) Cf. Gläser/Laudel (2010), p. 117ff.

23) Cf. http://wirtschaftslexikon.gabler.de/Archiv/55229/expertenwissen-v9.html. Accessed December 21, 2020

24) Cf. Gorden (1969), p. 115ff.

- Who is in possession of the relevant information?
- Who is most capable of providing accurate information?
- Who is most willing to provide information?
- Which of these experts are available?

The professional experience that experts have gained in a distinct field should be considered another key criterion for assessing and evaluating expert responses.

A standardized questionnaire is an appropriate tool for gathering quantitative data. Data fitting can be applied to quantitative expert responses in order to generate theoretical distribution functions, such as for quantity deviations or labor consumption rates (see Section 7.2). Only in exceptional cases will such functions correspond to the complete range of theoretical considerations and characteristics of the relevant calculation parameter (see Section 7.1.1). The results of data fitting must thus be combined with theoretical considerations and characteristics for defining the curves of distributions.

7.1.3 Combination of Data Collection Methods

Each of the three data collection methods referred to above is associated with certain benefits and disadvantages, which is why these methods should be combined.

Like others, *Yang* states that a combination of historical data and subjective assessments (ideally based on expert knowledge) will be necessary in order to identify distribution functions for different cost components. Historical data will not be available for all parameters, or the evaluation of limits or boundaries (ranges) for a specific project may differ from the conclusions drawn from historical data.[25)]

The following combinations appear to be useful:

- Theoretical considerations and characteristics combined with empirical data
- Theoretical considerations and characteristics combined with data gathered from expert surveys
- Theoretical considerations and characteristics combined with empirical data and data from expert surveys

Merely combining data from construction practice with the outcomes of expert surveys does not appear to be ideal because the theoretical considerations and characteristics arising from the nature of the parameters would not inform the process of selecting the distribution functions. It is precisely such theoretical considerations that allow the exclusion or modification of illogical or nonsensical distribution curves, such as with regard to distribution boundaries.

In many data fittings, collected values are best represented by open distributions, such as log-logistic or normal distributions. In reality, however, these values are located within a certain (closed) range that results from actual construction activity. Very high (positive infinite) values can often be limited whereas negative (negative infinite) values can mostly be excluded.

When combining empirical data collected *ex post* or *inter actio* with data from expert surveys, it is crucial to ensure appropriate delimitation and adjustment. Can different sets of data be integrated into a data pool, or do they have to be processed separately? If data cannot be merged directly, it is possible to perform two fitting operations and to compare

25) Cf. Yang (2005), p. 276

the distribution functions to each other on the basis of ratios (percentages) or solely based on the curves of distributions. On this basis, the most appropriate distribution function can be selected while also incorporating theoretical considerations and relevant characteristics.

Results should improve significantly if, compared to selecting the input parameter distribution functions on a purely subjective basis, the relevant distribution curve is applied while ensuring appropriate adjustment of the parameters that determine locations and ranges to the prevailing circumstances, as well as combining these parameters with theoretical considerations and characteristics. Monte Carlo simulations thus permit a chance/risk-based approach to considering parameters and issues related to construction management and economics.

7.2 Data Fitting

The issue of which distribution functions to apply to individual calculation parameters essentially determines the quality of the output of a Monte Carlo simulation. There is no "one-size-fits-all" answer to this question. In the literature, considerations regarding the "nature" and possible characteristics of specific parameters are discussed in conjunction with utilizing historical data and outcomes of expert surveys.

The concept of transferring such data records (e.g. from *ex ante*, *ex post*, or *inter actio* surveys or observations) into distribution functions that can be described on a theoretical level is referred to as data fitting. In this process, existing data is first analyzed graphically in histograms and then assessed using mathematical/statistical methods in order to identify the theoretical distribution function that best reflects existing data (i.e. with the smallest deviation). The shape or curve of distributions of uncertain calculation parameters is of greater interest than absolute values because distributions need to be adjusted to project-specific circumstances and conditions.

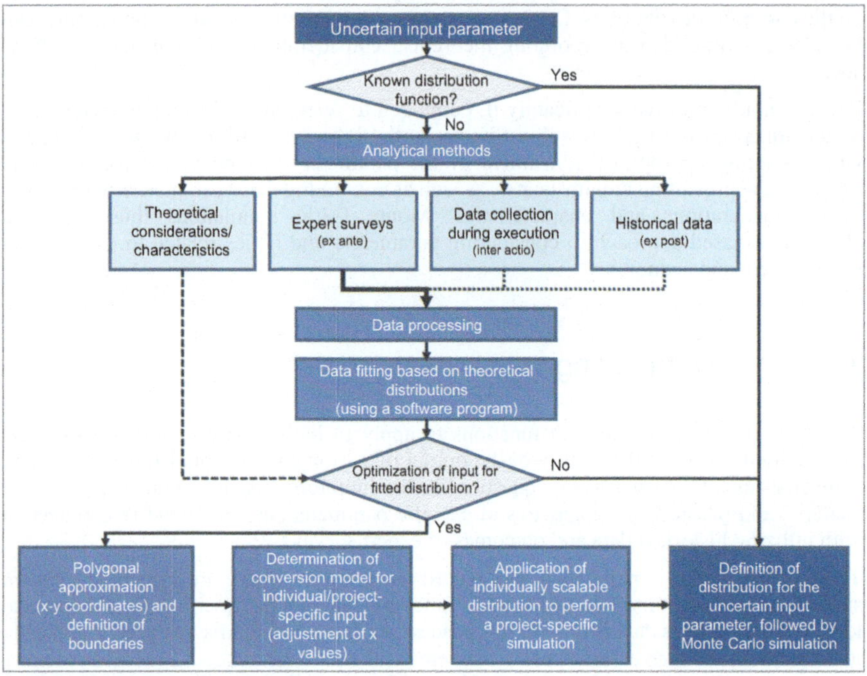

Fig. 7.4 Flowchart – Data fitting

Fig. 7.4 illustrates the process of determining distribution functions with the aid of data fitting for an uncertain input parameter with an unknown distribution curve.

Expert surveys are recommended as the method of choice for selecting the database because the *inter actio* and *ex post* gathering of data may be associated with delimitation and allocation issues.

Data processing is followed by actual data fitting, which is described in more detail in Section 7.2.1.2 by means of a worked example. As part of this fitting step, the adjusted data is compared with theoretical (mathematically described) distributions and checked for consistency. This theoretical distribution can then either be applied directly to the uncertain input parameter, or the input is optimized for the fitted distribution. This optimization step offers the advantage that the boundaries of the distribution can be modified on the basis of theoretical considerations and characteristics, which significantly simplifies the input (of minimum and most likely values, for example) for the user (see Section 7.2.1.3). In addition, such optimization makes it possible to define a conversion scheme to be used for scaling and adjusting the identified distribution function to project-specific circumstances.

In this respect, a distinction must be made between data fitting and a mere regression analysis. Regression and fitting differ in the nature of the underlying problem and the type of data to be analyzed.

Regression investigates possible correlations between data points with an assumed internal (e.g. linear) relationship. Data points are taken to be constant and fixed (measurement errors are usually disregarded or considered to have only a negligible influence in the regression analysis). We apply a continuous function, such as a straight-line equation reflecting a linear correlation, to capture existing data as effectively as possible (for instance by using the least-squares method). Regression analyses are also useful for analyzing trends as part of target-actual comparisons (see Section 10.4.10).

In contrast, fitting involves adjusting a function to any measurement errors or uncertainties related to measuring points. In this case, the resulting parameters of the adjusted function will be associated with uncertainties in much the same way as the input data. The least-squares method is most commonly applied for data fitting purposes. A fitting exercise will always result in a family of curves within which the "true" functional correlation will lie with a certain probability.

Fitting aims to determine the parameters of a distribution function (displacement and curve parameters) in a numerical process so that the distribution reflects the gathered raw data as accurately as possible.

7.2.1 Labor Consumption Rates

Hofstadler conducted an expert survey at the Institute of Construction Management and Economics at Graz University of Technology with the aim of deriving a distribution function of the labor consumption rates of formwork operations. For this purpose, 19 experts were asked to complete a standardized questionnaire. They were to provide information on different parameters, including consumption rates of formwork operations for concrete floor slabs. Experts were able to discuss any questions directly with the researchers. The experts were provided with information on the dimensions and characteristics of the structural components, which included vertical structural members (columns) and horizontal structural members (floor slabs). They were then asked to quantitatively assess labor consumption rates for seven different floor layouts separately for the two work steps of formwork installation and stripping.

This survey aimed to derive theoretical distribution functions from the expert statements by means of data fitting. On this basis, practical recommendations for applying the derived distribution functions can be given.

Layout classes were derived from the results of the expert survey in order to present the findings for the individual floor layouts in a more generic form. The seven different layouts were categorized according to their complexity and subsequently grouped into three classes (see Fig. 7.5). This step does not necessarily increase the precision of the derived distribution functions but renders them more resistant to fluctuations and ranges resulting from the expert survey.

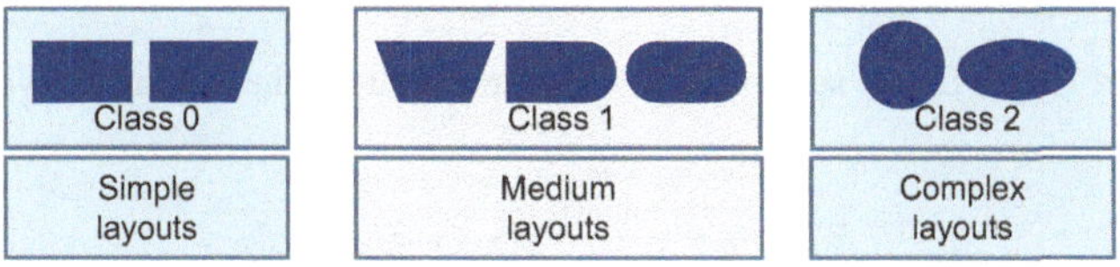

Fig. 7.5 Categorization into layout classes[26)]

26) Cf. Hofstadler (2014b), p. 193

Each of the data records gathered from the survey incorporates labor consumption rate estimates provided by various experts for a certain layout. Fitting a distribution function on the basis of the criteria and considerations described in the previous sections also makes it possible to use this data indirectly.

7.2.1.1 Data Processing

Classification is achieved through data aggregation. The data records derived from the expert survey are biased due to their different levels of complexity. This bias needs to be corrected in the data aggregation process.

Overall, the following three classes C_c (c = 0, 1, 2) are defined:

- Simple C_0
- Medium C_1
- Complex C_2

The above classes are derived from the data records of the layout types as follows:

- C_0 is composed of layout types L_1 and L_2
- C_1 is composed of layout types L_3, L_4, and L_5
- C_2 is composed of layout types L_6 and L_7

Each of the layout types i is represented by 19 expert values j (E_{ij}).

Survey data records for the individual classes are aggregated according to the following sequence:

1. In the first step, the average minimum is calculated from the expert values of the related layout types of Class C_c (c = 0, 1, 2).

$$\min C_c = \frac{\sum_{i \in C} \min(L_i)}{\text{number of layout types included in Class}} \tag{7-1}$$

2. We calculate the spread of expert values within the individual layout types of the respective class (i ϵ C).

$$s_{L_i} = \max(L_i) - \min(L_i) \tag{7-2}$$

3. We calculate the average spread in Class C_c.

$$s_c = \frac{\sum_{i \in C} s_{L_i}}{\text{number of layout types included in Class}} \tag{7-3}$$

4. The expert values E_{ij} are scaled to the j-th respondent for the i-th layout type within the k-th class.

$$E_{ij,new} = \frac{E_{ij} - \min(L_i)}{s_{L_i}} \cdot s_c + \min C_c \tag{7-4}$$

Completing this sequence results in an average spread assigned to the classified expert values $E_{ij,new}$.

The following example calculated for the medium class C_1 illustrates this approach:

Data records of layout types L_3, L_4, and L_5 are applied to Class C_1 (floor slabs are supported on columns with a circular cross-section). These data records are listed in Table 7.1 below:

L3	E1	E2	E3	E4	E5	E6	E7	E8	E9	E10	E11	E12	E13	E14	E15	E16	E17	E18	E19
Labor consumption rate [wh/m²]	0.55	0.65	0.58	0.80	0.70	0.65	0.60	0.65	0.60	0.50	0.55	0.55	0.55	0.60	0.70	0.60	1.00	1.10	0.45
L4	E1	E2	E3	E4	E5	E6	E7	E8	E9	E10	E11	E12	E13	E14	E15	E16	E17	E18	E19
Labor consumption rate [wh/m²]	0.65	0.70	0.66	0.75	0.70	0.75	0.70	0.70	0.60	0.60	0.62	0.60	0.60	0.65	0.70	0.65	1.00	1.10	0.55
L5	E1	E2	E3	E4	E5	E6	E7	E8	E9	E10	E11	E12	E13	E14	E15	E16	E17	E18	E19
Labor consumption rate [wh/m²]	0.65	0.70	0.66	0.90	0.80	0.85	0.80	0.80	0.65	0.70	0.67	0.70	0.70	0.70	0.75	0.75	1.15	1.30	0.55

Table 7.1 Data records from the expert survey for layout types L_3, L_4, and L_5[27)]

Expert value minima $\min(L_i)$ are taken from each layout type, added up, and divided by the number of layout types included in Class C_1 (in this case, three).

$$\min C_1 = \frac{\min(L_3) + \min(L_4) + \min(L_5)}{3} \tag{7-5}$$

$$\min C_1 = \frac{0.45 + 0.55 + 0.55}{3}$$

$$\min C_1 = 0.517 \, \frac{\text{wh}}{\text{m}^2}$$

In the next step, the spreads of expert values are calculated for each layout type:

$$s_{L_3} = \max(L_3) - \min(L_3) \tag{7-6}$$

$$s_{L_3} = 1.10 - 0.45$$

$$s_{L_3} = 0.65 \, \frac{\text{wh}}{\text{m}^2}$$

$$s_{L_4} = \max(L_4) - \min(L_4) \tag{7-7}$$

$$s_{L_4} = 1.10 - 0.55$$

$$s_{L_4} = 0.55 \, \frac{\text{wh}}{\text{m}^2}$$

$$s_{L_5} = \max(L_5) - \min(L_5) \tag{7-8}$$

$$s_{L_5} = 1.30 - 0.55$$

$$s_{L_5} = 0.75 \, \frac{\text{wh}}{\text{m}^2}$$

27) Cf. Hofstadler (2014b), p. 193

The third step involves calculating the average spread in Class C_1, as follows:

$$s_1 = \frac{s_{L_3} + s_{L_4} + s_{L_5}}{3} \tag{7-9}$$

$$s_1 = \frac{0.65 + 0.55 + 0.75}{3}$$

$$s_1 = 0.650 \frac{\text{wh}}{\text{m}^2}$$

Finally, all expert values E_{ij} of Class C_1 are scaled in order to obtain a new data record for subsequent fitting. The following equations illustrate the scaling of a single value:

$$E_{ij,new} = \frac{E_{ij} - \min(L_i)}{s_{L_i}} \cdot s_c + \min C_c \tag{7-10}$$

$$E_{3,1,new} = \frac{E_{3,1} - \min(L_3)}{s_{L_3}} \cdot s_1 + \min C_1$$

$$E_{3,1,new} = \frac{0.55 - 0.45}{0.65} \cdot 0.650 + 0.517$$

$$E_{3,1,new} = 0.62 \frac{\text{wh}}{\text{m}^2}$$

Table 7.2 shows the complete, scaled data record for Class C_1.

L3	E1	E2	E3	E4	E5	E6	E7	E8	E9	E10	E11	E12	E13	E14	E15	E16	E17	E18	E19
Labor consumption rate [wh/m²]	0.62	0.72	0.65	0.87	0.77	0.72	0.67	0.72	0.67	0.57	0.62	0.62	0.62	0.67	0.77	0.67	1.07	1.17	0.52
L4	E1	E2	E3	E4	E5	E6	E7	E8	E9	E10	E11	E12	E13	E14	E15	E16	E17	E18	E19
Labor consumption rate [wh/m²]	0.63	0.69	0.65	0.75	0.69	0.75	0.69	0.69	0.58	0.58	0.60	0.58	0.58	0.64	0.69	0.63	1.05	1.17	0.52
L5	E1	E2	E3	E4	E5	E6	E7	E8	E9	E10	E11	E12	E13	E14	E15	E16	E17	E18	E19
Labor consumption rate [wh/m²]	0.60	0.65	0.61	0.82	0.73	0.78	0.73	0.73	0.60	0.65	0.62	0.65	0.65	0.65	0.69	0.69	1.04	1.17	0.52

Table 7.2 Scaled data records for layouts L_3, L_4, and L_5[28)]

These scaled records can then be applied for subsequent fitting using the @Risk software.

7.2.1.2 Software-Based Data Fitting

When the data has been adjusted, checked for plausibility, and prepared for further processing, the @Risk software is used for determining the most appropriate distribution function.

This software uses the Levenberg-Marquardt method, which is based on the following simplified algorithm:

1. First, the start values for the parameters of the function to be adjusted are selected.
2. A chi-squared test is performed to determine the error margin by which the function deviates from the raw data.
3. The function's parameters are modified by the factor λ.

28) Cf. Hofstadler (2014b), p. 194

4. The error margin is determined again, repeating the second step.
5. If the parameters modified in the third step produce a smaller error margin, these parameters are accepted, and the factor λ is reduced.
6. In contrast, the factor λ is increased if the error got larger after the third step.
7. We then apply step three once again.

The algorithm ends when the error changes by only a very small margin.

Obviously, there will still be a residual error after performing the above processing steps. This error can be quantified by different statistical tests, including the chi-squared, Kolmogorov-Smirnov, and Anderson-Darling methods.

All potential distribution functions provided in @Risk were considered for performing the statistical tests. The distribution with the smallest error margin will best reflect the existing database by a theoretical distribution.

For the purpose of analyzing expert surveys, the error is determined by the Kolmogorov-Smirnov test because this test is best suited to analyzing only a small number of data records.

The Kolmogorov-Smirnov adjustment test (KS or KSA test; named after *Andrei Nikolaevich Kolmogorov* and *Nikolai Vasilyevich Smirnov*) is a statistical test for checking the congruence of two probability distributions. This test can involve a comparison of the distributions of two samples (two-sided KS test), but also testing if and to what extent a sample replicates a previously assumed probability distribution (one-sided KS test). In the latter case in particular, the KS test is also suitable for small samples ($n < 50$), as opposed to the chi-squared test.

In the fitting step, both the data records for the individual layouts and the scaled data records of the three classes (simple, medium, and complex) were analyzed in order to determine the distribution functions for the labor consumption rates of formwork operations performed for different floor layouts. In all cases, the log-logistic distribution proved to be adequate.

Log-logistic distributions are generally described by Eq. (7-11). For $x > 0$, the probability density of this type of distribution is defined as:

$$f(x) = \frac{\left(\frac{\beta}{\alpha}\right) \cdot \left(\frac{x}{\alpha}\right)^{\beta - 1}}{\left[1 + \left(\frac{x}{\alpha}\right)^{\beta}\right]^2} \tag{7-11}$$

In this equation, $\alpha > 0$ and $\beta > 0$ are parameters of the distribution.

The distribution is determined by a displacement parameter and a curve parameter. Displacement indicates the location of the data mean (α), which is identical to the expected value for many distributions. The curve parameter represents the breadth of the distribution (β).

For convergence testing, the *Kolmogorov-Smirnov* criterion was applied as a statistical test because this method is also suitable for comparatively small samples. This approach involves comparing the expert survey results with different distribution functions and subjecting them to the test. The distribution with the smallest deviation is indicative of the highest convergence. In the analyzed cases, this holds true for the log-logistic distribution.

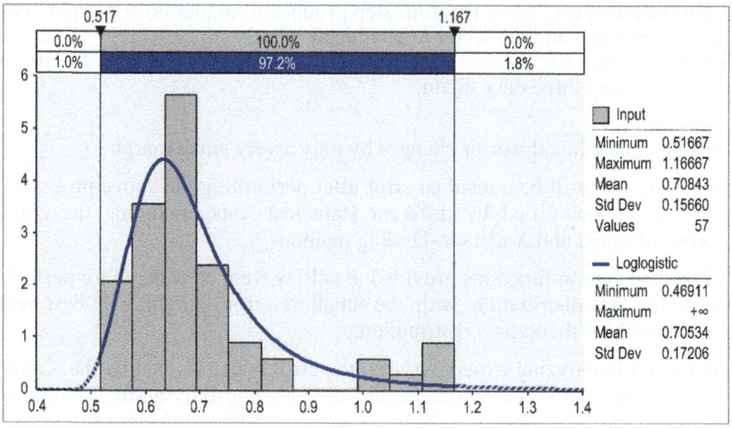

Fig. 7.6 Fitting – Class 1: Formwork operations for floor slabs supported on columns with a circular cross-section

Fig. 7.6 shows the distribution function determined for Formwork Class 1. The steepness of the log-logistic functions results from the fact that the majority of experts estimated labor consumption rates in a narrower spread within the lower value range. However, top-end outliers cannot be entirely excluded, yet their probability of occurrence is much smaller than that documented for lower consumption rates.

If we were to apply a triangular distribution instead of a log-logistic distribution to the case at hand, a greater probability of occurrence would be assigned to the values located near the maximum. This would, however, contradict the results of the expert survey (see Fig. 7.7). In Fig. 7.7, the minimum values of the log-logistic and the triangular distribution are set to be identical for comparison purposes. The mean of the log-logistic distribution is equivalent to the most likely value of the triangular distribution.

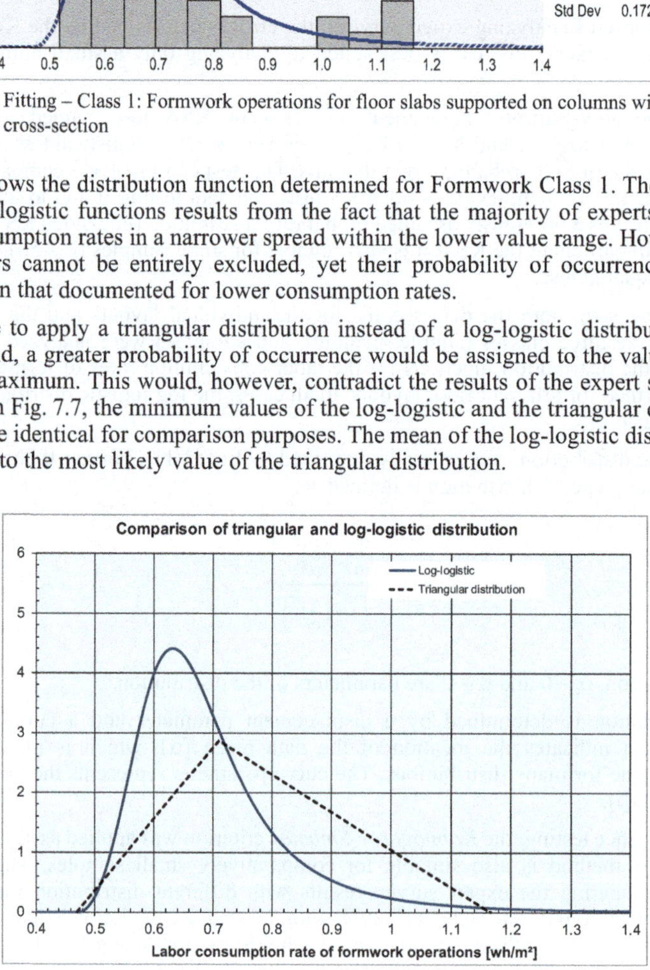

Fig. 7.7 Comparison of a triangular distribution with the log-logistic distribution generated by fitting

The two parameters for the curve and displacement of the distribution (α, β) must be entered in order to define a log-logistic distribution as an input to a Monte Carlo simulation. However, this approach does not correspond to the previously applied method, where minimum, most likely, and maximum values were stated for the triangular distributions. Since a and b are very hard to estimate for practitioners, additional considerations are necessary in order to retain the log-logistic curve of the distribution while also providing the option of entering minimum, most likely, and maximum input values.

7.2.1.3 Optimization of Inputs for the Fitted Distribution

Theoretical considerations are incorporated for modifying the log-logistic distribution generated in the fitting step.

We assume that, in general, a continuous distribution reflects the labor consumption rate of formwork operations more effectively and accurately than a discrete distribution. The following additional assumptions apply:

- The building or structure is actually constructed.
- The building or structure is actually completed.

We can thus exclude distributions with no upper limit. Likewise, distributions without a lower limit can be excluded because a minimum labor consumption rate exists that cannot be underrun.[29] Nor are negative values permissible. Experienced cost surveyors can assign a higher probability of occurrence to a certain value, as opposed to others (also referred to as expert value or most likely value). The distribution curve will thus be unimodal.

Purely theoretical considerations permit the conclusion that a continuous, two-sided closed, unimodal distribution can realistically reflect the labor consumption rates of formwork operations (see Fig. 7.3). The log-logistic distribution meets these requirements, with the exception of the two-sided closed criterion, which is why it needs to be modified.

The curve of the log-logistic distribution is substituted by a similar, polygonally approximated distribution using connected straight lines in order to be able to apply the log-logistic distribution determined by data fitting to simulations for different labor consumption rates and calculation scenarios. On the one hand, this approximated distribution does not tend towards infinity; on the other hand, defining this distribution will not require curve and location parameters (α and β), as is the case for the initial log-logistic distribution. Instead, a user-defined minimum value and a most likely value are applied. In @Risk, the so-called general distribution is used for approximating the distribution function.

Fig. 7.8 shows the polygonally approximated log-logistic distribution derived from the fitting for Class 1.

[29] If a reverse calculation of item prices and of the mean wage price results in labor consumption rates below the absolute minimum possible in terms of construction management and work scheduling, this is an indication of illegal practices, such as excessively low wages or social dumping. Cf. Hofstadler et al. (2016).

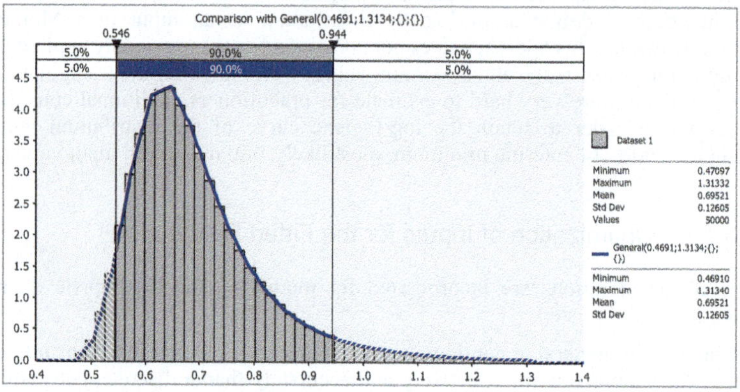

Fig. 7.8 Polygonally approximated distribution

Coordinate points define x and y values for the approximated distribution. Any choice of individual limits for input values will thus necessitate conversion of the x values to the new input in order to retain the curve of the distribution function defined in the fitting step. @Risk automatically normalizes the area under the curve to 1.

No.	Coordinates of data points of the polygonally approximated distribution	
	X	Y
0	A	B
1	0.4691	0.0000
2	0.4691	0.0000
3	0.5043	0.0810
4	0.5395	0.3480
5	0.5746	0.7100
6	0.6098	0.9670
7	0.6450	1.0000
8	0.6802	0.8620
9	0.7154	0.6670
10	0.7505	0.4900
11	0.7857	0.3530
12	0.8209	0.2530
13	0.8561	0.1830
14	0.8913	0.1340
15	0.9264	0.1000
16	0.9616	0.0750
17	0.9968	0.0570
18	1.0320	0.0450
19	1.0671	0.0350
20	1.1023	0.0280
21	1.1375	0.0220
22	1.1727	0.0180
23	1.2079	0.0150
24	1.2430	0.0120
25	1.2782	0.0100
26	1.3134	0.0090
27	1.3134	0.0000

Table 7.3 Coordinates of the polygonally approximated log-logistic distribution (for Class C_1) [30)]

For the purpose of this conversion, the minimum value and the most likely value (i.e. the value with the highest probability or with the greatest associated y value) are first derived from the polygonally approximated distribution (see Table 7.3):

Minimum x value of the polygonally approximated distribution: 0.4691

Most likely x value of the polygonally approximated distribution: 0.6450 (see Row 7 in Table 7.3 – associated y value = 1.0000)

The interval between the two x values is termed r and is the result of the difference between the most likely and the minimum x values, i.e. 0.1759 (= 0.6450 - 0.4691).

When wanting to adjust the log-logistic-like distribution to a specific construction project, the cost surveyor only needs to state a minimum and a most likely value to determine new x values for the polygonal distribution, as follows:

By way of example, we assume that a user wishes to apply a minimum labor consumption rate of 1.100 wh/m² and a most likely labor consumption rate of 1.200 wh/m² to his costing exercise. These two values result in a new r value (r_{new}) of 0.100 wh/m² (= 1.200 wh/m² - 1.100 wh/m²). This value is required for the subsequent conversion steps.

In @Risk, any distribution can be defined as a "general distribution" by the frequency and characteristics of specific values. The minimum x value of the new distribution (see Table 7.4, Column C, Row 1) corresponds to the user's input (= 1.100 wh/m²) and the associated y value (see Table 7.4, Column B, Row 1) taken from the initial approximation (= 0.000). The remaining x values must be converted using the following equation:

$$x_{new,i} = Input_{minimum} + (x_i - x_{min,polygon}) \cdot \frac{r_{new}}{r} \tag{7-12}$$

We perform this conversion for the third x value (Row 3 in Table 7.4):

$$x_{new,3} = 1.1000 + (0.5043 - 0.4691) \cdot \frac{0.1000}{0.1759} = 1.12001$$

The most likely value in Row 7 of Table 7.4 can be used as a control value for the newly determined x values. Both for the original and new x values, the seventh data point constitutes the most likely value (with the highest probability or the greatest associated y value). This is where the most likely value entered by the user (= 1.2000 wh/m²) should be found (cf. Table 7.4).

---

30) These values are taken from the free-form distribution (non-normalized polygonal approximation) automatically generated in @Risk. The y values of the distribution shown in Fig. 7.8 have been adjusted accordingly (the area under the distribution curve is 1). The values listed in Table 7.3 can nonetheless be used for calculation purposes because any distribution entered in @Risk will be automatically normalized to an area of 1 for the subsequent analyses and graphical representations.

No.	Coordinates of data points of the polygonally approximated distribution		
	X	Y	X_{new}
0	A	B	C
1	0.4691	0.0000	1.1000
2	0.4691	0.0000	1.1000
3	0.5043	0.0810	1.1200
4	0.5395	0.3480	1.1400
5	0.5746	0.7100	1.1600
6	0.6098	0.9670	1.1800
7	0.6450	1.0000	1.2000
8	0.6802	0.8620	1.2200
9	0.7154	0.6670	1.2400
10	0.7505	0.4900	1.2600
11	0.7857	0.3530	1.2800
12	0.8209	0.2530	1.3000
13	0.8561	0.1830	1.3200
14	0.8913	0.1340	1.3400
15	0.9264	0.1000	1.3600
16	0.9616	0.0750	1.3800
17	0.9968	0.0570	1.4000
18	1.0320	0.0450	1.4200
19	1.0671	0.0350	1.4400
20	1.1023	0.0280	1.4600
21	1.1375	0.0220	1.4800
22	1.1727	0.0180	1.5000
23	1.2079	0.0150	1.5200
24	1.2430	0.0120	1.5400
25	1.2782	0.0100	1.5600
26	1.3134	0.0090	1.5800
27	1.3134	0.0000	1.5800

Table 7.4 New coordinates of the polygonally approximated log-logistic distribution (for Class C_1)

Fig. 7.9 shows the new log-logistic-like distribution with user-specific input data for the minimum and most likely values. This distribution can be used for probabilistic calculations, such as for construction cost and time.

Users thus need to enter just two realistic, project-specific input values. By comparison, for defining an asymmetrical triangular distribution, a minimum, most likely, and maximum value need to be entered.

The key shows that the minimum and the mode (the latter is equivalent to the most likely value) correspond exactly to the input data chosen in the worked example.

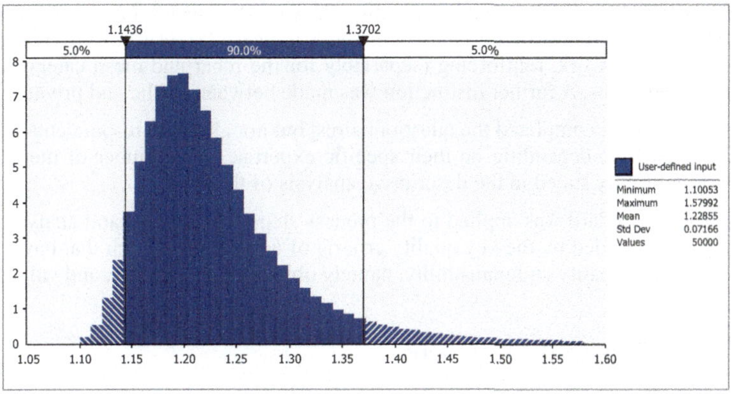

Fig. 7.9 Log-logistic-like distribution with user-defined input data

This approach can be applied in very much the same way to any right-skewed unimodal distribution function. Its advantage is that the distribution curve can be derived from data fitting applied to realistic values from expert surveys, data collected in the construction phase, or historical data. However, theoretical considerations and characteristics are incorporated in the subsequent work steps. In addition, it is more convenient, in practical terms, to define the applied distribution function by means of just two "tangible" values (i.e. minimum and most likely), rather than having to deal with complex curve and location parameters that can be estimated only on the basis of in-depth mathematical knowledge and skills, or need to be obtained by conversion. The curve of the distribution derived from data fitting can also be applied to other projects. Users will then adjust it to specific project parameters while also relying on their understanding and expertise in construction management and economics in order to influence calculation results. Selecting input parameters (i.e. minimum and most likely values) constitutes another means for determining fundamental distribution characteristics. Only the curve will be derived from several expert assessments or historical data taken from previously completed projects, thus supporting practitioners when performing analyses with the aid of Monte Carlo simulations.

The next section will describe how information from expert surveys was used for deriving distributions for quantity deviations in reinforced concrete and masonry operations in building construction.

7.2.2 Quantity Deviations

An expert survey conducted by *Kummer/Hofstadler* at Graz University of Technology in 2015/16 collected responses with respect to the magnitude of expected deviations from quantities stated in the specifications of formwork and masonry operations for public and private clients.

The experts were not queried for absolute quantities, but instead, for percentage deviations from the specified quantity, which was defined as 100%. The respondents determined the lower limit of quantity deviations (quantity reduction) by indicating a percentage value smaller than or equal to 100% and the upper limit (quantity increase) by indicating a percentage value greater than or equal to 100%. For instance, stating a lower limit of 80% would correspond to a 20% quantity reduction. Conversely, specifying an upper limit of, say, 115% would be equivalent to a 15% quantity increase. The survey covered quantity deviations for formwork, reinforcing (separately for the rebar and mesh categories), and concreting operations. A further distinction was made between public and private clients.

A total of 29 experts completed the questionnaires, but not all of the respondents answered all of the questions, depending on their specific expertise. The number of mentions per question is explicitly stated in the descriptive analysis of the survey.

A high quality standard was applied to the process steps of gathering and analyzing data. The survey was guided by the key quality criteria of empirical research that have a major influence on data quality and plausibility, namely objectivity, reliability, and validity.

7.2.2.1 Formwork Operations

In the expert survey, 23 responses were received with respect to quantity deviations in formwork operations (see Table 7.5). The figures stated for public and private clients differed only to a minor extent, which permits the conclusion of a similar overall degree of accuracy of the specifications for formwork operations. For formwork operations, the mean quantity underrun drops to about 93% (see Table 7.5, Row 2, Cells B and D, and Row 14, Cells B and D). Owing to the bimodal curve, the mode of quantity reductions amounts to 90% and 95%, respectively, in relation to responses regarding private clients (each with six responses). The greatest specified quantity reduction falls to 80% in the case of public and to 85% in the case of private clients. Thus, quantity reductions in formwork operations tend to be greater in projects specified by public clients. The experts justified this finding by the fact that public-client specifiers tend to include an additional buffer when specifying the quantities of items contained in structural specifications. If the installed quantities are lower than originally specified, the difference between invoiced prices can be set off against any additional cost claims submitted by the contractor for other items. This makes it easier to stand one's ground in discussions with the client because greater quantities stated in the bid usually tend to result in a bid price increase, which creates a certain leeway compared to invoicing that can be utilized for negotiations between representatives of the client and the contractors without having to modify the project funding model.

Mean quantity overruns are expected to range from 7% to 8% (see Table 7.5, Row 2, Cells C and E, as well as Row 14, Cells C and E). A slightly larger quantity increase relative to mean values and robust M-estimators should be assumed for public clients. This tendency is also confirmed when considering the maximum values stated for quantity increases. Experts stated quantity increases of up to 25% and 40%, respectively, for private and public clients. However, the mode equals 100% in both categories. This means that most respondents did not assume any potential for quantity overruns.

We also find that negative deviations are associated with a left-skewed distribution of responses, whereas positive deviations result in a right-skewed distribution of responses (see Table 7.5, Row 13). Relative to the complete range of answers, we can thus assume a "cluster" (i.e. a comparatively large number) of responses near the 100% mark.

No.	Parameter	Formwork operations – public clients		Formwork operations – private clients	
		Negative deviation	Positive deviation	Negative deviation	Positive deviation
0	A	B	C	D	E
1	n	23.00	23.00	23.00	23.00
2	Mean	92.87	108.17	93.17	107.61
3	Standard error of mean	1.03	1.93	0.86	1.54
4	Standard deviation	4.96	9.26	4.11	7.37
5	Coefficient of variation	5.34%	8.56%	4.41%	6.85%
6	Median	93.00	105.00	95.00	107.00
7	Mean absolute deviation	3.00	5.00	3.00	6.00
8	Robust coefficient of variation	3.23%	4.76%	3.16%	5.61%
9	Mode	90.00	100.00	90.00 or 95.00	100.00
10	Minimum	80.00	100.00	85.00	100.00
11	Maximum	100.00	140.00	100.00	125.00
12	Spread	20.00	40.00	15.00	25.00
13	Skewness	-0.65	1.95	-0.16	0.76
14	M-estimator (H12 (80%))	93.18	107.01	93.25	106.91
15	Standard deviation (H12 (80%))	4.92	7.52	4.68	7.32
16	Robust coefficient of variation	5.28%	7.03%	5.01%	6.85%
17	Normal distribution (Shapiro-Wilk, $p < 0.05$)	Yes	No	Yes	No

Table 7.5 Descriptive representation of data pertaining to quantity deviations in formwork operations[31)]

Data fitting was performed on the basis of the expert responses given for public and private clients using the @Risk software (see Fig. 7.10) in order to provide realistic recommendations for possible distribution functions of the quantities for formwork operations. In both cases (i.e. public and private clients), log-logistic distributions reflect the expert statements very accurately, as confirmed by a KS test.[32)]

[31)] Cf. Kummer (2018), p. 184

[32)] For private-sector clients, the log-logistic distribution provides the best adjustment (KS test). For public clients, only an "extreme value distribution" provides a slightly better adjustment of the existing database. However, this type of distribution is two-sided open, which is not plausible for quantity deviations, particularly for underruns of more than 100%.

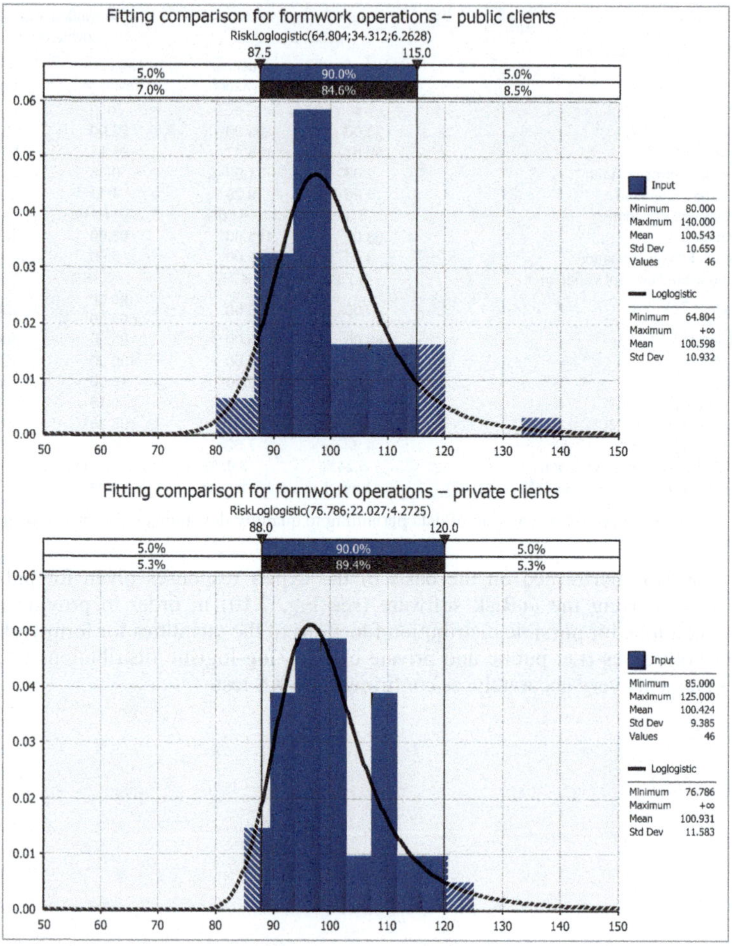

Fig. 7.10 Data fitting – Quantity deviations in formwork operations – Top: Public clients – Bottom: Private clients

7.2.2.2 Reinforcing Operations – Steel Rebar

Like in the survey for formwork operations, 23 responses were received in the expert survey on quantity deviations in reinforcing operations using steel rebar (see Table 7.6). The figures stated for public and private clients differed only to a minor extent, which permits the conclusion of a similar overall degree of accuracy of the specifications for reinforcing operations using steel rebar. For reinforcing operations using rebar, the mean quantity underrun drops to about 92% (see Table 7.6, Row 2, Cells B and D, and Row 14, Cells B and D). The mode of quantity reductions amounts to 90% and 95%, respectively, in relation to responses regarding private and public clients (each with six responses). Yet there is also a difference in the modes recorded for quantity increases: 110% for public clients and 100% for private clients. The greatest specified quantity reduction falls to 60% in the case of private and to 70% in the case of public-sector clients. Thus, steel rebar quantity reductions tend to be greater in projects specified by private clients. The experts justified this finding by stating that most private clients base their specifications on a preliminary structural verification that is subsequently optimized at the project execution stage.

Respondents stated maximum quantity increases to 140% and 135% for public and private clients, respectively. The potential for exceeding a specified quantity is thus greater for public clients. However, the mean quantity increase ranges between 7% and 8% for both client categories.

In addition, the outcomes revealed that negative deviations were associated with a left-skewed distribution of responses, whereas positive deviations resulted in a right-skewed distribution of responses (see Table 7.6, Row 13). Relative to the complete range of answers, we can thus again assume a "cluster" (i.e. a comparatively large number) of responses near the 100% mark.

No.	Parameter	Reinforcing operations – steel rebar – public clients		Reinforcing operations – steel rebar – private clients	
		Negative deviation	Positive deviation	Negative deviation	Positive deviation
0	A	B	C	D	E
1	n	23.00	23.00	23.00	23.00
2	Mean	91.57	108.30	91.30	108.70
3	Standard error of mean	1.57	1.81	1.77	1.90
4	Standard deviation	7.53	8.66	8.48	9.10
5	Coefficient of variation	8.22%	7.99%	9.28%	8.37%
6	Median	95.00	110.00	95.00	110.00
7	Mean absolute deviation	5.00	5.00	5.00	6.00
8	Robust coefficient of variation	5.26%	4.55%	5.26%	5.45%
9	Mode	95.00	110.00	90.00	100.00
10	Minimum	70.00	100.00	60.00	100.00
11	Maximum	100.00	140.00	100.00	135.00
12	Spread	30.00	40.00	40.00	35.00
13	Skewness	-1.25	2.24	-2.45	1.26
14	M-estimator (H12 (80%))	92.51	107.26	92.65	107.54
15	Standard deviation (H12 (80%))	6.69	6.73	5.78	7.99
16	Robust coefficient of variation	7.24%	6.27%	6.24%	7.43%
17	Normal distribution (Shapiro-Wilk, $p < 0.05$)	No	No	No	No

Table 7.6 Descriptive representation of data pertaining to quantity deviations in reinforcing operations – Steel rebar[33)]

Data fitting was performed on the basis of the expert responses given for public and private clients using the @Risk software (see Fig. 7.11) in order to provide realistic recommendations for possible distribution functions of the quantities for reinforcing operations using rebar. For public clients, a Weibull distribution was found to be the best approximation, followed by a log-logistic distribution (shown at the top in Fig. 7.11). Data fitting applying the KS test delivered a logistic distribution as the closest result for quantity deviations recorded for private clients.[34)]

[33)] Cf. Kummer (2019), CPM-10

[34)] Similarly to the log-logistic distribution, the Weibull distribution includes a defined minimum, but has no upper limit.

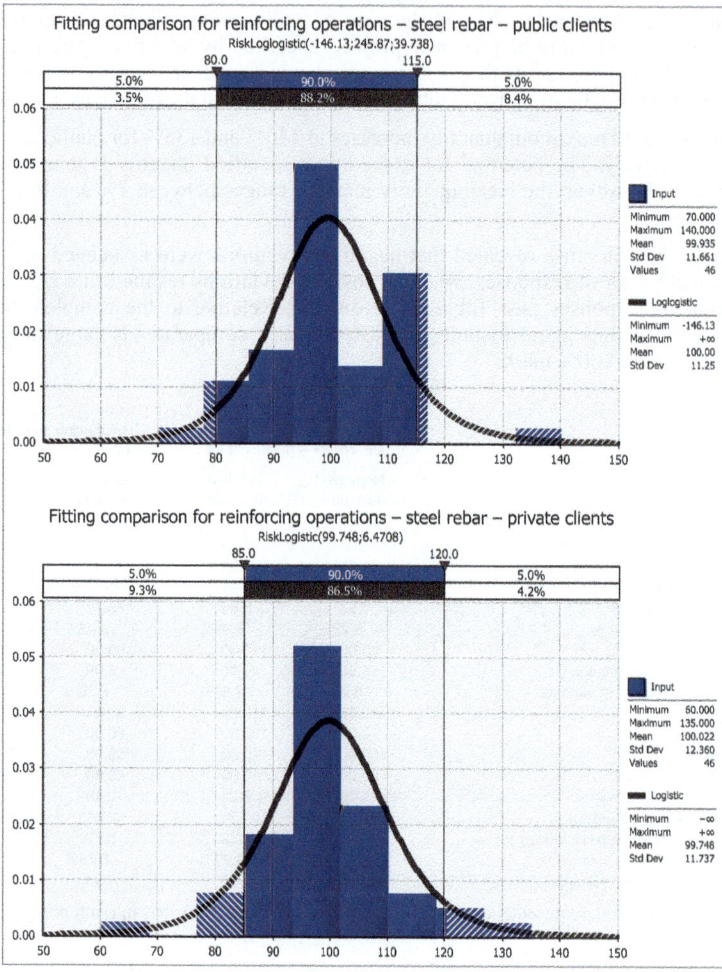

Fig. 7.11 Data fitting – Quantity deviations in reinforcing operations using steel rebar – Top: Public clients – Bottom: Private clients

7.2.2.3 Reinforcing Operations – Steel Mesh

In the expert survey, 21 responses were received with respect to quantity deviations in steel mesh reinforcing operations (see Table 7.7). Again, the figures stated for public and private clients differed only to a minor extent, which permits the conclusion of a similar overall degree of accuracy of the specifications for steel mesh reinforcing operations. In general, however, experts stated a larger range of quantity deviations for public clients. For reinforcing operations using steel mesh, the mean quantity underrun drops to about 90% to 91% (see Table 7.7, Row 2, Cells B and D, and Row 14, Cells B and D). The mode of quantity reductions amounted to about 95% in responses given for both private and public clients. However, the modes recorded for quantity increases revealed a difference, i.e. 110% for public clients and 100% for private clients. The greatest specified quantity reduction declines to 70% in the case of public and to 60% in the case of private clients. Thus, quantity reductions in steel mesh reinforcing operations tend to be greater in projects specified by private clients. The experts justified this finding by stating that most private clients base their specifications on a preliminary structural verification that is subsequently optimized at the project execution stage.

Respondents stated maximum quantity increases to 140% and 120% for public and private clients, respectively. The potential for exceeding a specified quantity is thus greater for public clients. However, the mean quantity increase ranges from 6% to 8% for public clients and from 6% to 7% for private clients.[35)]

We also find that negative deviations were associated with a left-skewed distribution of responses, whereas positive deviations resulted in a right-skewed distribution of responses (see Table 7.7, Row 13). Relative to the complete range of answers, we can thus again assume a "cluster" (i.e. a comparatively large number) of responses near the 100% mark.

No.	Parameter	Reinforcing operations – steel mesh – public clients		Reinforcing operations – steel mesh – private clients	
		Negative deviation	Positive deviation	Negative deviation	Positive deviation
0	A	B	C	D	E
1	n	21.00	21.00	21.00	21.00
2	Mean	90.57	107.76	89.24	106.76
3	Standard error of mean	1.60	2.00	1.87	1.59
4	Standard deviation	7.33	9.15	8.56	7.28
5	Coefficient of variation	8.09%	8.49%	9.59%	6.82%
6	Median	92.00	108.00	90.00	105.00
7	Mean absolute deviation	3.00	3.00	5.00	5.00
8	Robust coefficient of variation	3.26%	2.78%	5.56%	4.76%
9	Mode	95.00	110.00	95.00	100.00
10	Minimum	70.00	100.00	60.00	100.00
11	Maximum	100.00	140.00	100.00	120.00
12	Spread	30.00	40.00	40.00	20.00
13	Skewness	-1.24	2.32	-2.12	0.70
14	M-estimator (H12 (80%))	91.49	106.26	90.5	106.17
15	Standard deviation (H12 (80%))	6.41	6.19	6.05	7.55
16	Robust coefficient of variation	7.01%	5.82%	6.68%	7.11%
17	Normal distribution (Shapiro-Wilk, $p < 0.05$)	No	No	No	No

Table 7.7 Descriptive representation of data pertaining to quantity deviations in reinforcing operations – Steel mesh

Data fitting was performed on the basis of the expert responses given for public and private clients using the @Risk software (see Fig. 7.12) in order to provide realistic recommendations for possible distribution functions of the quantities for reinforcing operations using steel mesh. For public clients, a log-logistic distribution was found to be the best approximation (see Fig. 7.12, top). Data fitting applying the KS test delivered a logistic distribution as the closest result for quantity deviations recorded for private clients (see Fig. 7.12, bottom).

35) See M-estimators and means in Table 7.7.

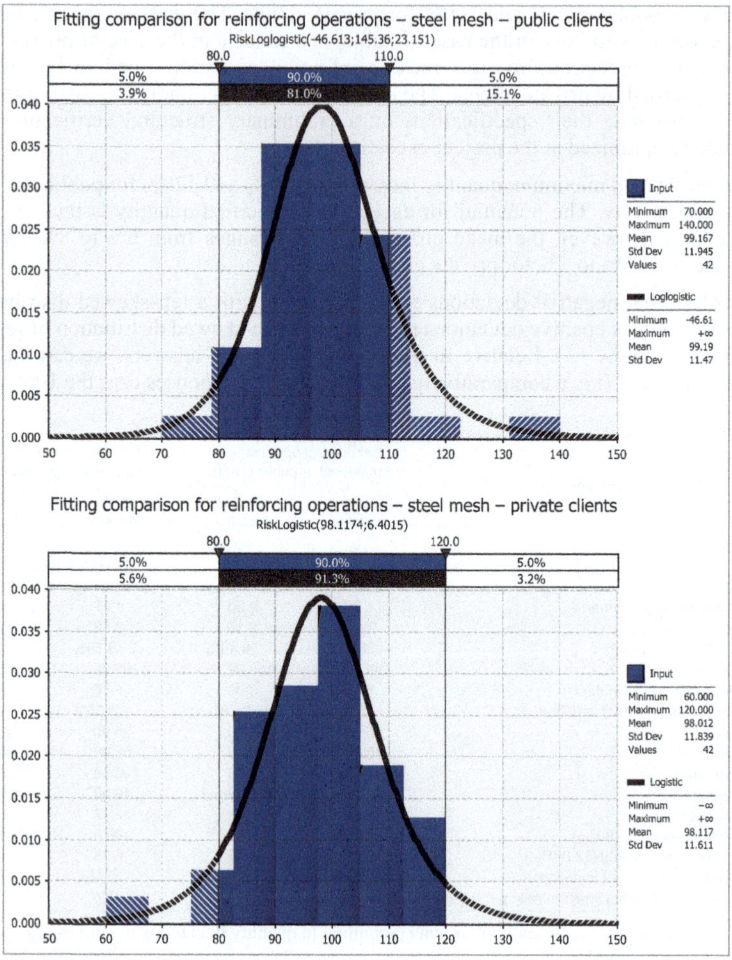

Fig. 7.12 Data fitting – Quantity deviations in reinforcing operations using steel mesh – Top: Public clients – Bottom: Private clients

7.2.2.4 Concreting Operations

In the expert survey, 23 responses were received with respect to quantity deviations in concreting operations (see Table 7.8). The figures stated for public and private clients differed only to a minor extent, which permits the conclusion of a similar overall degree of accuracy of the specifications for concreting operations. For concreting operations, the mean quantity underrun drops to about 93% (see Table 7.8, Row 2, Cells B and D, and Row 14, Cells B and D). Owing to the bimodal curve, the mode of quantity reductions amounts to 90% and 95%, respectively, in relation to responses regarding private clients (each with five responses), whereas 95% were documented for public clients. A maximum quantity reduction to 80% was recorded for both client categories.

Mean quantity overruns are expected to range from 6% to 8% (see Table 7.8, Row 2, Cells C and E, as well as Row 14, Cells C and E). A maximum quantity increase to 125% was stated for private clients, whereas the same value equaled 130% for public clients. For private clients, a bimodal curve was recorded for quantity increases, i.e. 100% and 110% (each with seven mentions). The mode amounted to 100% in the public client category. This means that most respondents did not assume any potential for quantity overruns.

We also find that negative deviations are associated with a left-skewed distribution of responses, whereas positive deviations result in a right-skewed distribution of responses (see Table 7.8, Row 13). Relative to the complete range of answers, we can thus assume a "cluster" (i.e. a comparatively large number) of responses near the 100% mark.

No.	Parameter	Concreting operations – public clients		Concreting operations – private clients	
		Negative deviation	Positive deviation	Negative deviation	Positive deviation
0	A	B	C	D	E
1	n	23.00	23.00	23.00	23.00
2	Mean	92.61	106.35	93.00	107.52
3	Standard error of mean	1.25	1.39	0.93	1.61
4	Standard deviation	6.00	6.65	4.45	7.72
5	Coefficient of variation	6.48%	6.25%	4.79%	7.18%
6	Median	95.00	105.00	95.00	107.00
7	Mean absolute deviation	3.00	5.00	3.00	6.00
8	Robust coefficient of variation	3.16%	4.76%	3.16%	5.61%
9	Mode	95.00	100.00	95.00	110.00
10	Minimum	80.00	100.00	80.00	100.00
11	Maximum	100.00	130.00	100.00	125.00
12	Spread	20.00	30.00	20.00	25.00
13	Skewness	-1.02	2.01	-1.17	0.84
14	M-estimator (H12 (80%))	93.52	105.59	93.42	106.62
15	Standard deviation (H12 (80%))	5.12	5.42	3.70	7.38
16	Robust coefficient of variation	5.48%	5.13%	3.96%	6.93%
17	Normal distribution (Shapiro-Wilk, $p < 0.05$)	No	No	No	No

Table 7.8 Descriptive representation of data pertaining to quantity deviations in concreting operations

Data fitting was performed on the basis of the expert responses given for public and private clients using the @Risk software (see Fig. 7.13) in order to provide realistic recommendations for possible distribution functions of the quantities for concreting operations. In both cases (i.e. public and private clients), log-logistic distributions reflect the expert statements very accurately, as confirmed by a KS test.[36)]

[36)] For public clients, a Weibull distribution ensures a slightly better adjustment of the existing database compared to the indicated log-logistic distribution (KS test). For private clients, only an "extreme value distribution" provides a slightly better adjustment of the existing database. However, this type of distribution is two-sided open, which is not plausible for quantity deviations, particularly for underruns of more than 100%.

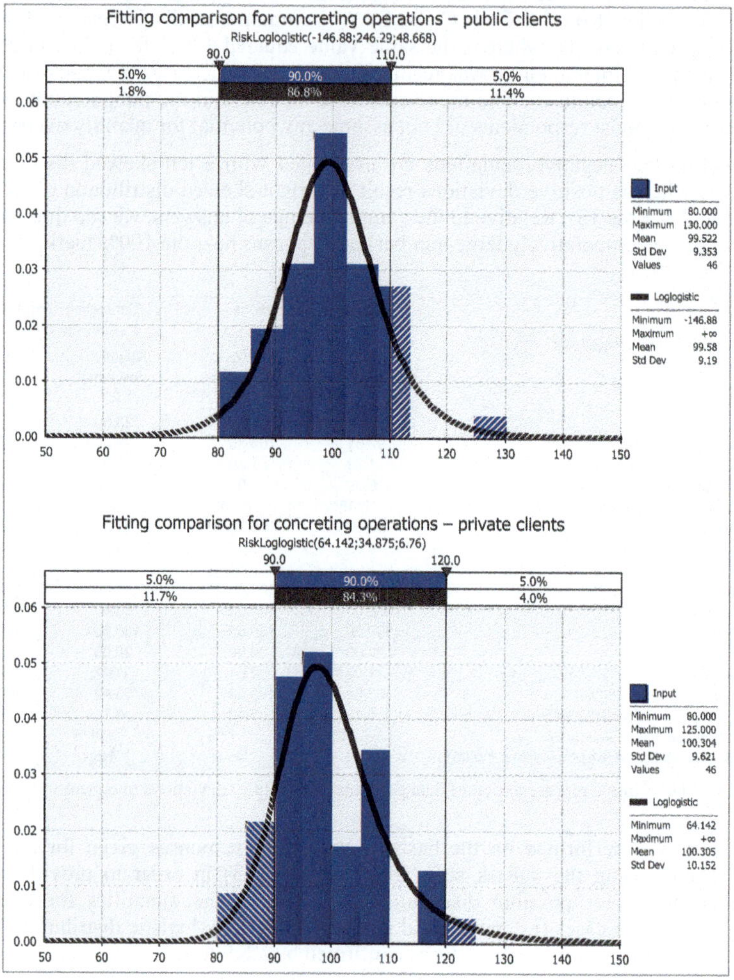

Fig. 7.13 Data fitting – Quantity deviations in concreting operations – Top: Public clients – Bottom: Private clients

7.2.2.5 Masonry Operations

In the expert survey, 23 responses were received with respect to quantity deviations in masonry operations (see Table 7.9). The figures stated for public and private clients differed only to a minor extent, which again permits the conclusion of a similar overall degree of accuracy of the specifications for masonry operations. In this case, the mean quantity underrun falls to about 94% to 95% (see Table 7.9, Row 2, Cells B and D, and Row 14, Cells B and D). The mode of quantity reductions amounted to about 95% in responses given for both private and public clients. A maximum quantity reduction to 80% was recorded for both client categories.

Mean quantity overruns are expected to range from 5% to 8% (see Table 7.9, Row 2, Cells C and E, as well as Row 14, Cells C and E). A maximum quantity increase to 125% was recorded for both client categories. The mode of quantity increases equaled 100% for both public and private clients. This means that most respondents did not assume any potential for quantity overruns.

Overall, we find that negative deviations were associated with a left-skewed distribution of responses, whereas positive deviations resulted in a right-skewed distribution of responses (see Table 7.9, Row 13). Relative to the complete range of answers, we can thus assume a "cluster" (i.e. a comparatively large number) of responses near the 100% mark.

No.	Parameter	Masonry operations – public clients		Masonry operations – private clients	
		Negative deviation	Positive deviation	Negative deviation	Positive deviation
0	A	B	C	D	E
1	n	23.00	23.00	23.00	23.00
2	Mean	93.74	106.04	94.22	107.57
3	Standard error of mean	1.23	1.39	1.18	1.65
4	Standard deviation	5.90	6.66	5.66	7.89
5	Coefficient of variation	6.30%	6.28%	6.00%	7.34%
6	Median	95.00	105.00	95.00	107.00
7	Mean absolute deviation	5.00	5.00	4.00	7.00
8	Robust coefficient of variation	5.26%	4.76%	4.21%	6.54%
9	Mode	95.00	100.00	95.00	100.00
10	Minimum	80.00	100.00	80.00	100.00
11	Maximum	100.00	125.00	100.00	125.00
12	Spread	20.00	25.00	20.00	25.00
13	Skewness	-1.08	1.33	-1.34	0.94
14	M-estimator (H12 (80%))	94.56	105.17	94.99	106.61
15	Standard deviation (H12 (80%))	5.16	5.71	4.80	7.31
16	Robust coefficient of variation	5.46%	5.43%	5.05%	6.86%
17	Normal distribution (Shapiro-Wilk, $p < 0.05$)	No	No	No	No

Table 7.9 Descriptive representation of data pertaining to quantity deviations in masonry operations

Data fitting was performed on the basis of the expert responses given for public and private clients using the @Risk software (see Fig. 7.14) in order to provide realistic recommendations for possible distribution functions of the quantities for masonry operations. In both cases (i.e. public and private clients), log-logistic distributions reflect the expert statements very accurately, as confirmed by a KS test.[37)]

[37)] For both private and public clients, only an "extreme value distribution" provides a slightly better adjustment of the existing database (KS test). However, this type of distribution is two-sided open, which is not plausible for quantity deviations, particularly for underruns of more than 100%.

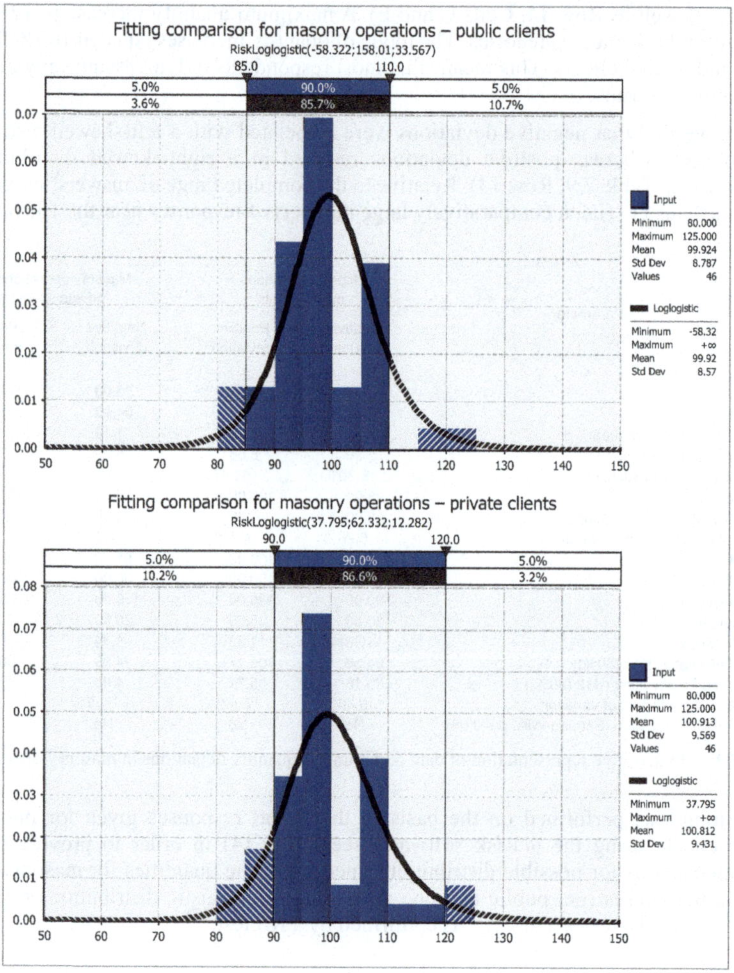

Fig. 7.14 Data fitting – Quantity deviations in masonry operations – Top: Public clients – Bottom: Private clients

7.2.2.6 Comparison of Quantity Deviations

A direct comparison of the ranges of quantity deviations derived from the expert responses shows that they expected a wider range for public clients, especially for quantity increases in formwork operations. Fig. 7.15 shows the spreads for all mentions (bright bar) and the ranges between the derived M-estimators for quantity increases and reductions (dark bar) for formwork, reinforcing, concreting, and masonry operations, as well as for public and private clients.

For masonry and concreting operations, similar results were recorded in terms of quantity deviations irrespective of the client category. Public clients tend to be associated with slightly narrower M-estimator spreads for these two work categories.

Irrespective of whether rebar or mesh is used, the range of quantity reductions in reinforcing operations was generally found to be wider for private clients. The largest spread of all requested quantity deviations (i.e. 75%; ranging from 60% to 135%) was identified for reinforcing operations using rebar in the case of private-client projects. For both rebar and mesh, the identified range of M-estimators for reinforcement quantities was lower for public than for private clients.

For formwork operations, the experts generally considered quantity increases to be more likely than quantity reductions. However, the range of all responses received for public clients revealed a much greater spread of quantity deviations (60%) than for private clients (40%). The calculated range of M-estimators was almost identical for public and private clients.

The experts expected the smallest probability for quantity reductions for formwork, whereas they indicated the smallest likelihood of quantity overruns for mesh reinforcement (in both cases in relation to private clients). On the other hand, they identified the greatest uncertainties for quantity reductions of mesh and rebar in private-client projects, as well as for quantity increases of mesh and rebar as well as formwork, in public-client projects.

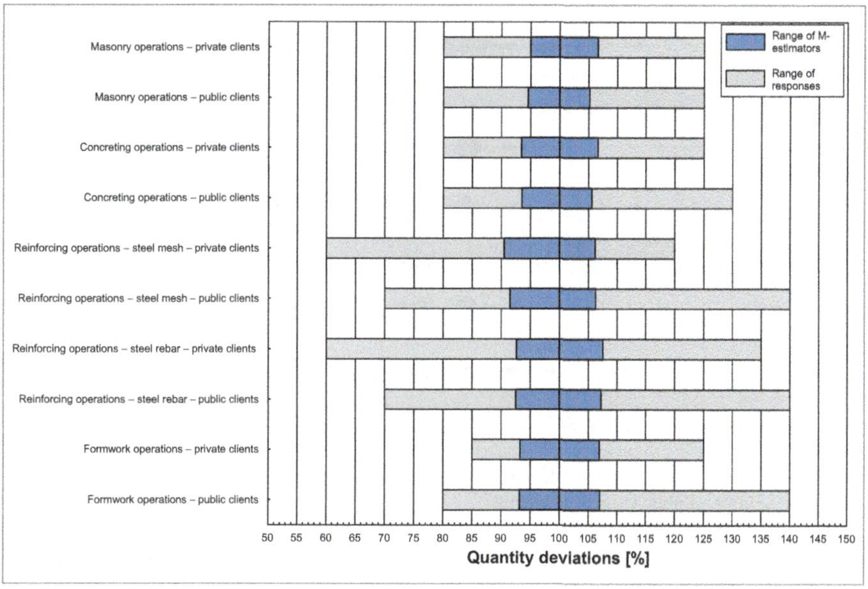

Fig. 7.15 Quantity deviations – Spreads of expert responses and ranges of M-estimators

Fig. 7.16 includes not only the range, or spread, of mentions, but also the violin plots of expert responses. Violin plots are an extension to the known box plot representation. They combine the information of a density function (or of a smoothed histogram) with that of a box plot in a single diagram. Their advantage is that the graphical representation includes not only box plot information but also distribution characteristics. In the diagram, the density curve is plotted symmetrically above and below the (vertical) box plot.[38] The two halves of the violin plot differ only in the direction in which density is plotted. They are thus mirrored along the axis on which the relevant parameter is shown. This symmetrical representation facilitates interpretation and comparison of several distributions plotted adjacent to each other.[39]

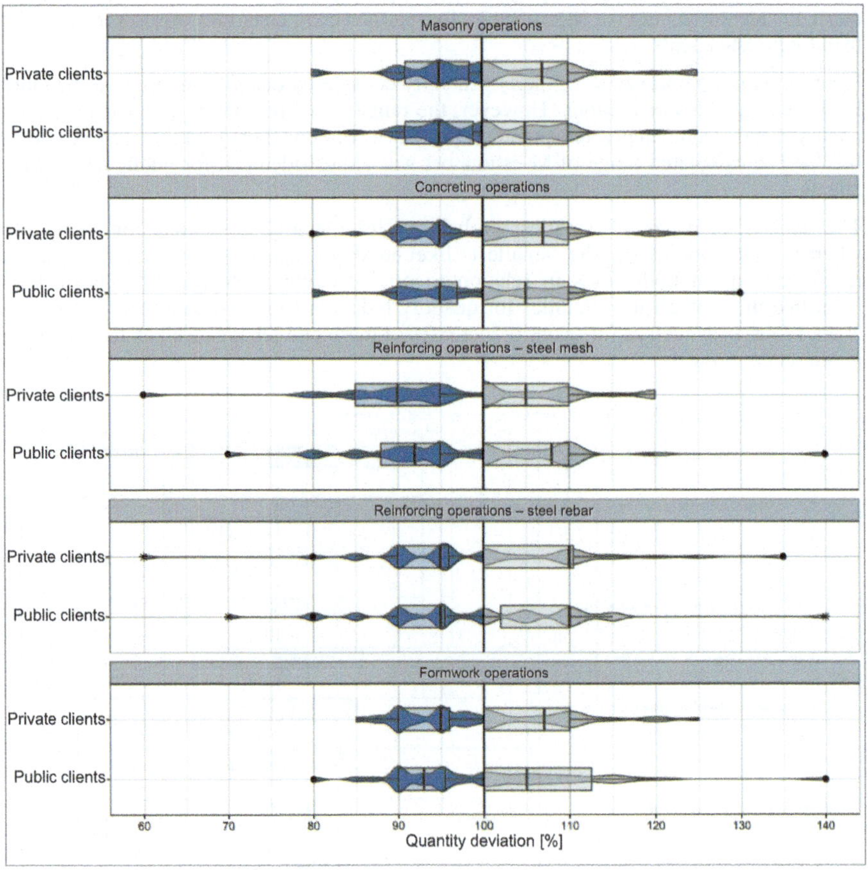

Fig. 7.16 Violin plots of quantity deviations – Formwork, reinforcing, concreting, and masonry operations – Public vs. private clients

Another benefit of violin plots is that the curve of the distribution or the presence of any accumulations or clusters within certain value ranges are easier to visualize. This type of information can get lost in conventional box plots. The greater the frequency of tapered zones in the violin plot, the greater the number of mentions within certain value ranges. For quantity increases in masonry operations in private-client projects, for example, the

38) By way of analogy, box and violin plots can also be represented in the vertical direction.

39) Cf. Hintze/Nelson (1998), p. 181f.

expert responses were found to be clustered around 100% and 110%. Between these two values, the narrow violin plot is rather flat, which permits the conclusion of an approximately equal distribution of responses within this area. If there are no responses in a certain value range, the area of the violin plot will be reduced to zero in this range, and is represented as a line. See, for example, the quantity reduction to about 85% recorded for masonry operations for the private client category. Outliers are shown as circles/dots in Fig. 7.16, whereas extremes are marked by stars. The number of responses cannot be derived directly from the violin plot because the number of underlying responses is not stated either. This is why descriptive data analyses were also included in the previous sections. Fig. 7.16 includes not only violin plots but also conventional box plots. The thicker line within the box represents the median.

Incorporating the data taken from the expert survey makes it possible to optimize, and increase the user friendliness of, the input of parameters of the fitted distribution functions for the quantity deviations (see Section 7.2.1.3).

Besides the selection of the distribution functions, however, correlations (dependencies) between input parameters also influence the determined outputs of a probabilistic calculation.

7.3 Identifying Correlations

Correlations are usually determined indirectly by analyzing existing data series gathered from measurements, surveys, experiments, etc. examined with respect to their linear relationship (see Fig. 6.20, page 189).

Another option is to directly determine correlation coefficients based on expert surveys. In 2015/16, *Kummer/Hofstadler* conducted a survey at Graz University of Technology in which experts provided estimates of the coefficients corresponding to the correlations between certain parameters. These estimates were given in the context of determining the total labor consumption rate of reinforced concrete operations. Analyses were carried out using violin plots and descriptive statistics. In addition, mean values and M-estimators were generated that delivered sound recommendations for considering correlations when determining the total labor consumption rate of reinforced concrete operations.

7.3.1 Total Labor Consumption Rate of Reinforced Concrete Operations

The total labor consumption rate LCR_{RCO} [wh/m³] expresses the labor intensity of the reinforced concrete operations for a building or structure, or for individual structural components. This metric describes the amount of labor required per cubic meter of concrete in the reinforced concrete operations. The total labor consumption rate is the average value determined for all of the reinforced concrete operations within a single building or structure, or for groups of components. The total labor consumption rate of reinforced concrete operations must consider all of the activities carried out within the system, such as formwork, reinforcing, and concreting operations.[40)]

[40)] Cf. Hofstadler (2008), p. 302

For the global consideration of the labor intensity required for a specific building or structure during the preliminary planning phase, specific labor consumption rates for formwork LCR_{FO} [wh/m²], reinforcing LCR_{RO} [wh/t], and concreting LCR_{CO} [wh/m³] operations are added up in accordance with Eq. (7-13) to obtain the total labor consumption rate.

$$LCR_{RCO} = LCR_{FO} \cdot fr + LCR_{RO} \cdot rr + LCR_{CO} \tag{7-13}$$

In the first term, the mean labor consumption rate of formwork operations [wh/m²] is multiplied with the formwork ratio fr [m²/m³]. Then, multiplying the mean labor consumption rate of reinforcing operations [wh/t] with the reinforcement ratio rr [kg/m³] will result in the second term. The final step will yield the mean labor consumption rate of concreting operations [wh/m³].

Eq. (7-13) includes a total of five calculation parameters that the resulting total labor consumption rate will depend on. Ten different dependencies, or correlations, are possible between these five input variables (see Fig. 7.17).

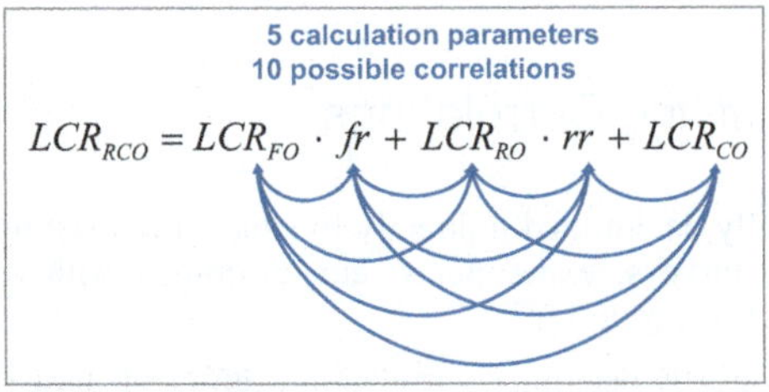

Fig. 7.17 Possible correlations between the parameters used for calculating the total labor consumption rate of reinforced concrete operations

In the survey, experts were asked to estimate the dependencies between the individual parameters for calculating the total labor consumption rate. Experts were able to select correlations between -1.0 and +1.0 in increments of 0.1. A correlation of 0.0 describes complete independence between two calculation parameters, which means that any change in one of these input parameters will have no effect on the other.

In addition, the positive and negative correlations were described in detail in order to provide the experts with a better understanding.

In this context, positive correlations mean that any increase in one parameter will also lead to an increase in the other. Any change in a parameter will lead to a change in the correlated parameter in the same direction. For instance, any increase in the formwork ratio will also cause a rise in the labor consumption rate of formwork operations.

Accordingly, negative correlations mean that any increase in one parameter will lead to a decrease in the other (and vice versa). Any change in a calculation parameter will thus lead to an inverse change in the second parameter. If, for example, the reinforcement ratio increases, the labor consumption rate of reinforcing operations will decrease.

The magnitude of the relevant quantitative correlation can be described as follows:

- 0.9 to 1.0 very strong
- 0.7 to 0.8 strong
- 0.5 to 0.6 medium
- 0.3 to 0.4 weak
- 0.1 to 0.2 very weak
- 0.0 independent
- -0.1 to -0.2 very weak
- -0.3 to -0.4 weak
- -0.5 to -0.6 medium
- -0.7 to -0.8 strong
- -0.9 to -1.0 very strong

The next section presents and describes the results of the expert survey in diagrams and tables using descriptive statistics.

7.3.2 Expert Survey Results

Correlations are generally identified by performing statistical analyses of available data. However, there is usually no such database available in construction management and economics, although dependencies between individual calculation parameters do exist and can be partially reproduced on the basis of their characteristics relevant to construction management and economics. The objective of the expert survey was to evaluate if and to what extent experts were capable of stating and reasonably justifying correlation values. For this purpose, each of the experts was interviewed to discuss the questionnaire in order to clarify questions and to resolve any misunderstandings.

Most of the experts indicated positive correlations for the dependencies covered in the questionnaire. Correlations between individual parameters were explained or justified as follows:[41)]

Formwork ratio fr [m²/m³] and reinforcement ratio rr [kg/m³]

Experts indicated a positive correlation between formwork and reinforcement ratios. The formwork ratio describes the relationship between the formwork surface area and the volume of concrete, thus permitting conclusions as to the amount of detail of a building or component. The majority of experts stated that any change in the formwork ratio would cause a similar change in the reinforcement ratio. Thus, any increase in the former parameter is expected to lead to a corresponding increase in the latter. Conversely, experts indicated that any decrease in the formwork ratio would also result in a decrease in the reinforcement ratio. Respondents thus generally associated highly detailed components with a trend towards greater amounts of reinforcement per cubic meter of concrete, and vice versa.

Formwork ratio fr [m²/m³] and labor consumption rate of formwork operations LCR_{FO} [wh/m²]

Owing to their high formwork ratio, more detailed buildings, work sections, or structural components tend to be associated with a higher consumption rate of formwork operations (as part of reinforced concrete operations) compared to buildings or structures with a

41) Cf. Kummer (2015a), p. 246ff.

lower formwork ratio. In this case, most experts indicated a mid-range positive correlation. A high formwork ratio usually results in an increased proportion of mating surfaces and formwork pieces fabricated in a conventional process, such as fillers and blockouts, which increases the labor intensity of this work step.

Formwork ratio fr [m²/m³] and labor consumption rate of reinforcing operations LCR_{RO} [wh/t]

In addition, a high formwork ratio generally leads to an increase in the amount of labor required for reinforcing because, in this case, we would expect a larger number of reinforcement locations, a higher proportion of rebar with complex bend patterns, a lower mesh ratio, a large number of stirrups, and a more sophisticated reinforcement design within more confined spaces.

Formwork ratio fr [m²/m³] and labor consumption rate of concreting operations LCR_{CO} [wh/m³]

Experts also indicated a positive correlation between the formwork ratio and the labor consumption rate of concreting operations. Bulky elements with a low formwork ratio are usually easier to produce than more detailed components. Among other factors, this is due to integrated blockouts or openings that render concrete pouring and compaction more difficult. Likewise, the rate at which the concrete rises in the form will determine pouring speed and thus the labor consumption rate of concreting operations, particularly for slender vertical elements.

Reinforcement ratio rr [kg/m³] and labor consumption rate of formwork operations LCR_{FO} [wh/m²]

Any change in the reinforcement ratio is expected to cause a corresponding change in the labor consumption rate of formwork operations. Thus, any increase in the reinforcement ratio will tend to result in an increased consumption rate for formwork operations (and vice versa). This assumption is supported by the fact that vertical elements can cause difficulties when attempting to close the form. In addition, infills for rebar connections are generally more labor-intensive, and rework may be required to comply with concrete cover specifications.

Reinforcement ratio rr [kg/m³] and labor consumption rate of reinforcing operations LCR_{RO} [wh/t]

The majority of experts responded very clearly when queried for the correlation between the reinforcement ratio and the labor consumption rate of reinforcing operations. In this case, they assumed a positive correlation and justified this response with the fact that a higher reinforcement ratio would tend to require a larger number of reinforcement items, more complex rebar designs, greater rebar density, and a larger amount of labor for correcting installation errors.

However, this question also yielded two outliers and one extreme value for which independence or a negative correlation between the two parameters were assumed. To justify this assumption, experts stated that the reinforcement ratio would also increase simply because greater rebar diameters are selected. In general, greater rebar diameters are associated with a lower consumption rate (wh/t) than smaller rebar diameters, provided an identical rebar design is chosen.

This is why a negative correlation also appears plausible as long as the change in rebar diameters is the only variable considered. In most cases, however, any change in the reinforcement ratio will also result in modifications of rebar and stirrup designs, which will then also cause the labor consumption rate of reinforcing operations to move in the same direction (i.e. positive correlations).

Reinforcement ratio rr [kg/m³] and labor consumption rate of concreting operations LCR_{CO} [wh/m³]

According to the experts, any decrease in the reinforcement ratio will generally also reduce the amount of labor required for concrete pouring because a sufficient number of openings is available for pouring, the concrete can be compacted more easily, and the risk of internal vibrators getting stuck is reduced. In addition, the concrete recipe in terms of maximum aggregate size, consistency, etc. is not as critical for comparatively low reinforcement ratios as for elements with a relatively high reinforcement ratio.

Labor consumption rate of formwork operations LCR_{FO} [wh/m²] and labor consumption rate of reinforcing operations LCR_{RO} [wh/t]

Experts assumed a slightly positive correlation between the labor consumption rate of formwork operations and the corresponding value determined for reinforcing operations. If the consumption rate increases for formwork operations, this will likely also result in an increase in the consumption rate of reinforcing operations. Experts justified their responses by stating that they had also assumed positive correlations

- between the formwork ratio and the reinforcement ratio
- between the formwork ratio and the consumption rates of formwork and reinforcing operations, and
- between the reinforcement ratio and the consumption rates of formwork and reinforcing operations.

Consequently, a positive value should also be expected for the relationship between the consumption rates of formwork and reinforcing operations.

Labor consumption rate of formwork operations LCR_{FO} [wh/m²] and labor consumption rate of concreting operations LCR_{CO} [wh/m³]

Experts followed roughly the same line of thought for their assumptions in respect of the labor consumption rates of formwork and concreting operations. According to them, any increase in the formwork ratio is expected to be associated with increased consumption rates for formwork operations and concrete pouring. The same reasoning applies if the reinforcement ratio increases. Conversely, experts also expected a positive correlation between the consumption rates of formwork operations and concreting.

Labor consumption rate of reinforcing operations LCR_{RO} [wh/t] and labor consumption rate of concreting operations LCR_{CO} [wh/m³]

Since the majority of experts assumed a positive correlation between the reinforcement ratio and the consumption rates of reinforcing and concreting, they also expected a positive correlation between the two consumption rates as such.

Tables 7.10 and 7.11 illustrate the descriptive statistics of the correlations identified between the individual calculation parameters. However, not all of the respondents answered the complete set of questions in the 29 submitted questionnaires (for the number of mentions, see Row 1 in Tables 7.10 and 7.11).

No.	Parameter	$fr - rr$	$fr - LCR_{FO}$	$fr - LCR_{RO}$	$fr - LCR_{CO}$	$rr - LCR_{FO}$
0	A	B	C	D	E	F
1	n	25.00	26.00	25.00	25.00	25.00
2	Mean	0.52	0.49	0.48	0.51	0.36
3	Standard error of mean	0.05	0.08	0.06	0.06	0.06
4	Standard deviation	0.26	0.41	0.31	0.29	0.32
5	Coefficient of variation	49.96%	83.58%	64.77%	56.74%	88.57%
6	Median	0.50	0.60	0.50	0.50	0.30
7	Mean absolute deviation	0.20	0.20	0.20	0.20	0.30
8	Robust coefficient of variation	40.00%	33.33%	40.00%	40.00%	100.00%
9	Mode	0.50	0.70	0.50	0.60	0.00
10	Minimum	0.00	-0.60	-0.20	0.00	0.00
11	Maximum	1.00	1.00	1.00	1.00	1.00
12	Spread	1.00	1.60	1.20	1.00	1.00
13	Skewness	-0.55	-1.06	-0.57	-0.24	0.48
14	M-estimator (H12 (80%))	0.54	0.54	0.50	0.52	0.34
15	Standard deviation (H12 (80%))	0.23	0.38	0.32	0.32	0.35
16	Robust coefficient of variation	41.88%	71.32%	65.00%	60.62%	101.68%
17	Normal distribution (Shapiro-Wilk, $p < 0.05$)	Yes	No	Yes	Yes	No

Table 7.10 Descriptive representation of correlations indicated between parameters for the purpose of calculating the total labor consumption rate of reinforced concrete operations – Part 1

No.	Parameter	$rr - LCR_{RO}$	$rr - LCR_{CO}$	$LCR_{FO} - LCR_{RO}$	$LCR_{FO} - LCR_{CO}$	$LCR_{RO} - LCR_{CO}$
0	A	B	C	D	E	F
1	n	27.00	26.00	24.00	25.00	25.00
2	Mean	0.62	0.59	0.39	0.32	0.52
3	Standard error of mean	0.08	0.04	0.07	0.06	0.05
4	Standard deviation	0.42	0.21	0.34	0.28	0.24
5	Coefficient of variation	67.27%	35.65%	87.36%	87.88%	45.27%
6	Median	0.70	0.60	0.40	0.40	0.60
7	Mean absolute deviation	0.20	0.15	0.30	0.20	0.10
8	Robust coefficient of variation	28.57%	25.00%	75.00%	50.00%	16.67%
9	Mode	0.80	0.60	0.00	0.00	0.50
10	Minimum	-0.80	0.20	0.00	0.00	0.00
11	Maximum	1.00	0.90	1.00	1.00	1.00
12	Spread	1.80	0.70	1.00	1.00	1.00
13	Skewness	-2.04	-0.27	0.15	0.43	-0.57
14	M-estimator (H12 (80%))	0.70	0.60	0.39	0.31	0.53
15	Standard deviation (H12 (80%))	0.26	0.25	0.41	0.32	0.22
16	Robust coefficient of variation	37.68%	42.00%	105.10%	101.50%	41.36%
17	Normal distribution (Shapiro-Wilk, $p < 0.05$)	No	Yes	No	No	Yes

Table 7.11 Descriptive representation of correlations indicated between parameters for the purpose of calculating the total labor consumption rate of reinforced concrete operations – Part 2

The standard errors of the means are relatively low so that we can assume an estimate that is roughly consistent with expectations. The coefficient of variation ranges from about 36% to 89%. The smallest robust coefficients of variation[42)] (based on the applied M-estimators) are obtained for the following correlations:

- Reinforcement ratio rr and labor consumption rate of reinforcing operations LCR_{RO} (Table 7.11 – Column B)
- Labor consumption rate of reinforcing operations LCR_{RO} and labor consumption rate of concreting operations LCR_{CO} (Table 7.11 – Column F)
- Formwork ratio fr and reinforcement ratio rr (Table 7.10 – Column B)
- Reinforcement ratio rr and labor consumption rate of concreting operations LCR_{CO} (Table 7.11 – Column C)

Most responses reveal a left-skewed distribution. These results mean that responses are clustered near the higher correlation values.

42) In this case, expert estimates tended to be the most consistent. This is where responses show the narrowest spread.

Fig. 7.18 shows the violin plots of expert responses with respect to the correlations between the calculation parameters used to determine the total labor consumption rate of reinforced concrete operations. The diagram reveals only isolated negative correlation values whereas the majority of responses refers to positive correlations. Some experts also stated that they assumed no correlations at all for some parameters (r = 0).

Fig. 7.18 indicates bimodal distributions for the fr – LCR_{FO}, LCR_{FO} – LCR_{RO}, and LCR_{FO} – LCR_{CO} correlations. At r = 0, responses are also clustered near the positive correlation values. An in-depth analysis did not reveal any relationship between professional experience and the correlations mentioned in the responses to this question.

Isolated extremes and outliers are visible, which distort the mean values of correlations accordingly. This is why applying *Huber's* M-estimators is recommended.

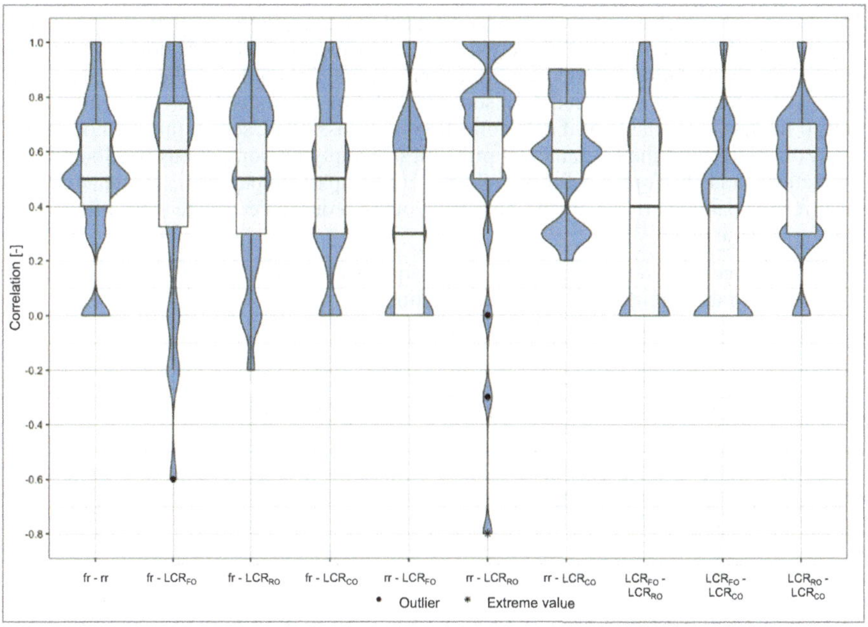

Fig. 7.18 Correlations between the parameters of the total labor consumption rate of reinforced concrete operations[43)]

The M-estimators applied to the correlations listed in Tables 7.10 and 7.11 lead to the correlation matrix for the parameters for calculating the total labor consumption rate of reinforced concrete operations shown in Table 7.12. This key result of the expert survey makes it possible to systematically consider the real correlations between the input parameters of the total labor consumption rate of reinforced concrete operations in the simulations.

---

43) Cf. Kummer/Hofstadler (2017), CPM-23

Correlations	LCR_{FO}	fr	LCR_{RO}	rr	LCR_{CO}
LCR_{FO}	1.00				
fr	0.54	1.00			
LCR_{RO}	0.39	0.50	1.00		
rr	0.34	0.54	0.70	1.00	
LCR_{CO}	0.31	0.52	0.53	0.60	1.00

Table 7.12 Correlation matrix with the M-estimators used in the expert survey

As part of the analysis of expert responses, the full correlation matrices stated by the experts were also checked for consistency. Out of a total of 24 complete sets (stating all 10 correlation coefficients), 14 matrices were found to be consistent and thus suitable for simulations without requiring adjustment.

This finding shows that experts are indeed capable of estimating correlation coefficients. These estimates can then be converted into a consistent matrix. Although the above derivation of correlation coefficients from expert surveys should not take preference over mathematical calculations based on a solid database, it can indeed deliver useful indications if no data is available or if data collection is impossible. Users of this method should always bear in mind the fundamental principles of applying correlations to Monte Carlo simulations, and check results for plausibility. For comparison purposes, a simulation with uncorrelated parameters can always be carried out in order to determine their influence on the output parameters.[44)]

This type of comparison is illustrated by a worked example calculating the total labor consumption rate of reinforced concrete operations.

7.3.3 Effects of Correlations on the Total Labor Consumption Rate of Reinforced Concrete Operations

A worked example applying the M-estimators derived for the correlations between the input parameters for calculating the total labor consumption rate of reinforced concrete operations LCR_{RCO} [wh/m³] demonstrates the influence that these correlations – as compared to uncorrelated parameters – have on the output of a Monte Carlo simulation (Latin Hypercube sampling with 50,000 iterative steps). To simplify the process, triangular distributions are allocated to the individual input parameters. Table 7.13 lists the minimum, most likely, and maximum values entered for each of the five input parameters.

No.	Parameter		Unit	MIN	MOST LIKELY	MAX
0	A	B	C	D	E	F
1	Labor consumption rate of formwork operations	LCR_{FO}	wh/m²	1.00	1.20	1.60
2	Formwork ratio	fr	m²/m³	2.70	3.00	3.30
3	Labor consumption rate of reinforcing operations	LCR_{RO}	wh/t	12.00	13.00	16.00
4	Reinforcement ratio	rr	kg/m³	100.00	115.00	140.00
5	Labor consumption rate of concreting operations	LCR_{CO}	wh/m³	0.90	1.00	1.30

Table 7.13 Input data for calculating the total labor consumption rate of reinforced concrete operations

44) Cf. Kummer (2015a), p. 250

In order to establish a deterministic computation model for the simulation, we apply the equation used for calculating the total labor consumption rate (see Eq. (7-13)). Calculations are performed first with uncorrelated parameters ($r = 0$) and then with the correlation coefficients that were derived as M-estimators as part of the expert survey. Table 7.12 shows the correlation matrix used in this example.

Fig. 7.19 shows that the spread of the total labor consumption rate becomes wider when incorporating the correlations. This is due to the fact that only positive correlations are applied and that the deterministic calculation model (see Eq. (7-13)) includes only addition and multiplication operations. Combined with positive correlations, these two operations result in an increased range of simulation results for identical input parameters (see also Section 6.5.2, page 192ff.).

Compared to a simulation with correlated input parameters, an uncorrelated analysis would assign insufficient weighting to high and low labor consumption rate values and thus distort any subsequent calculations. Overall, correlated input parameters for a simulation reveal a wider range of possible results.

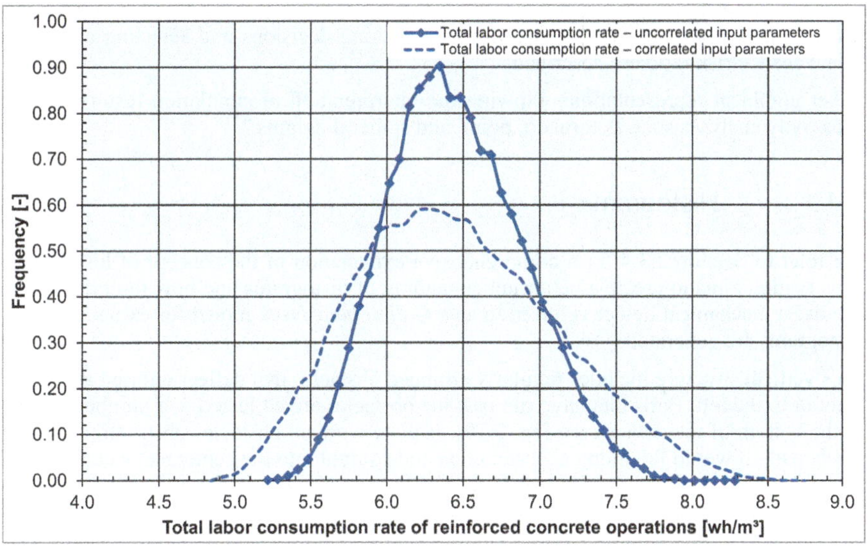

Fig. 7.19 Comparison of the total labor consumption rate of reinforced concrete operations using uncorrelated and correlated input parameters

Neglecting correlations and using uncorrelated input parameters leads to a narrower range of results, and thus also to an underestimation of existing uncertainties.

If no correlations between input parameters can be defined or estimated, sensitivity analyses help capture their effects on simulation results, particularly on their ranges, and support their consideration in subsequent analyses, calculations, and decisions.

7.4 Interpretation of Simulation Results

At first glance, the results of probabilistic calculations may appear unfamiliar especially to technicians or engineers because a distribution of possible results, or histogram, is determined, rather than a single figure. This type of representation requires in-depth considerations on how to interpret results. In this regard, the risk affinity of the user or decision maker is also crucial.

It is essential to interpret results correctly in order to add real value to probabilistic calculations. Simulation results are generally shown in the form of histograms that must be interpreted accordingly in order to obtain the information required for further considerations and decisions.

This section sets out to create a better understanding of histograms and uses the example of a random game using a mechanical device with balls (*Galton board*) to demonstrate how distributions are formed and what conclusions can be drawn with respect to probabilities of occurrence. This example is then applied to the results of numerical simulations.

Particular emphasis is placed on quantile values as well as over- and underrun probabilities. In addition, realistic considerations for making decisions and selecting an appropriate chance/risk ratio are described.

Other graphical representations allowing the interpretation of simulation results include sensitivity analyses such as tornado, point, and spider diagrams.[45)]

7.4.1 Histograms

We refer to Section 2.3.4 for a comprehensive explanation of the concept of histograms. This section aims to create a better understanding of histograms and how to derive them. We use a mechanical device referred to as a *Galton board* as a model for explaining how histograms are generated.

This vertical structure includes regularly arranged obstacles that deflect falling balls to the right or to the left. Balls that have run past the obstacles are collected in a number of slots at the bottom of the device (see Fig. 7.20). If there were no obstacles obstructing a ball's direct path, it would fall along a vertical line and straight into the center slot of the bottom structure. This scenario would result in a deterministic state allowing direct, reliable conclusions as to the outcome of the experiment.

The obstacles incorporated in the device represent disruptive influences and random events affecting the result, i.e. the slot into which the ball actually falls. These disruptions and random events can either accumulate along the trajectory of a ball, thus leading to a significant overall deviation of the result from an experiment conducted without obstacles, or cancel each other out.

If we repeat the experiment with several balls, we will find different filling heights of the individual slots at the bottom. These heights result from different paths of the balls variously deflected by the obstacles. If the number of repetitions is sufficiently high, this pattern will deliver a discrete distribution whose shape resembles a continuous normal distribution (see the dashed line in Fig. 7.20). A finite number of deflections and slots will result in a binomial distribution.

[45)] Cf. Palisade Corporation (2013), p. 52ff.

Fig. 7.20 Qualitative representation of a *Galton board*

If we assume, for the device shown in the figure, that there is an equal probability for the ball to be deflected to the left or to the right at each obstacle, this will result in a pattern allowing the calculation of the probability for each slot into which the ball may fall (see Fig. 7.21).

At the first obstacle, this probability equals 50% (or 1/2) for each ball to be deflected to the left or to the right. The ball can be deflected again in the second row of obstacles. Since there are two routes/possibilities leading to the gap between the obstacles in the second row, the probability of the ball continuing on its path from the center of the second row of obstacles is also 50% (or 2/4). The calculated probability of a ball to pass the second row of obstacles on the far right or far left side equals 25% (or 1/4) for either of these cases.

These steps can be continued until either the bottom row of obstacles or the slots have been reached. This exercise results in a probability of approximately 27.34% (70/256) for the center slot and in a probability of just 0.39% for each of the two outer slots.

For the *Galton board* shown in the figure, it is thus highly improbable for a ball to be deflected only to the left or to the right in several consecutive steps while on its downward path so that it ends up in one of the outer slots of the bottom structure. On the other hand, the probability of the ball ending up in one of the three central slots after having passed the obstacles is relatively high, i.e. approx. 71.09% (= 21.88% + 27.34% + 21.88%).

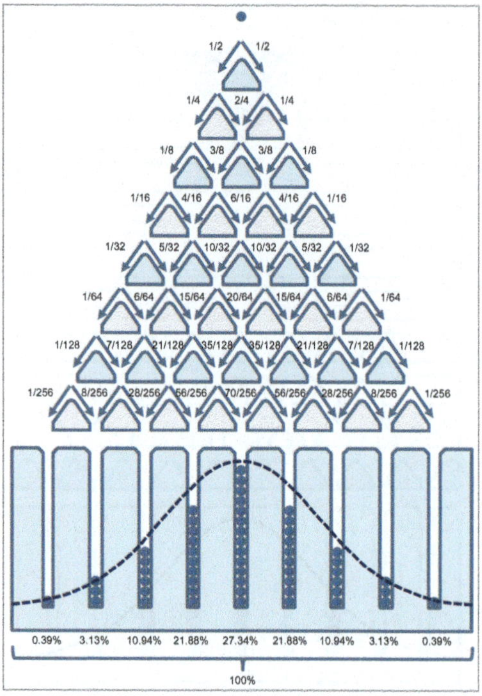

Fig. 7.21 Probabilities for the route of a ball past the obstacles[46)]

7.4.1.1 Over- and Underrun Probability

Fig. 7.22 illustrates the conversion of the outcomes of several *Galton board* experiments into a histogram, where classes are equally wide and do not reveal any gaps in-between. The diagram also includes the normal distribution curve as a dashed line superimposed on the histogram.

Comparing the results of a Monte Carlo simulation with those of a *Galton board* appears to be particularly appropriate because, in either case, random events and input parameter uncertainties will ultimately yield a similar picture after several thousand iterations, i.e. a histogram covering the range of possible calculation or experimental results.

The heights of the bars in the histogram, or the related areas, can be interpreted in analogy to the filling heights of the *Galton board* slots. This approach makes it possible to easily indicate the probability of over- or underrunning specific values, as well as the probability of values lying within a certain range (see Fig. 7.22).

However, there is also a big difference between *Galton board* experiments and Monte Carlo simulations (such as those performed for construction costs), namely that applying (continuous) distributions as input parameters will result in not just two (left/right), but a

[46)] The percentages for the probabilities in each discipline were rounded to two decimal places. All decimal places must be included when calculating the total of 100%.

large number of possible outcomes. In addition, the probability of occurrence of any of these outcomes need not be equal on either side (left: 50%; right: 50%). Instead, it is also possible to apply discrete or continuous as well as uni- or multimodal distributions with different skewness parameters and limits.

Thus the results of a Monte Carlo simulation include numerous "obstacles" leading to different individual results for each iteration. Their multiple repetition and collection will result in a histogram that can be interpreted with regard to the probabilities of occurrence.

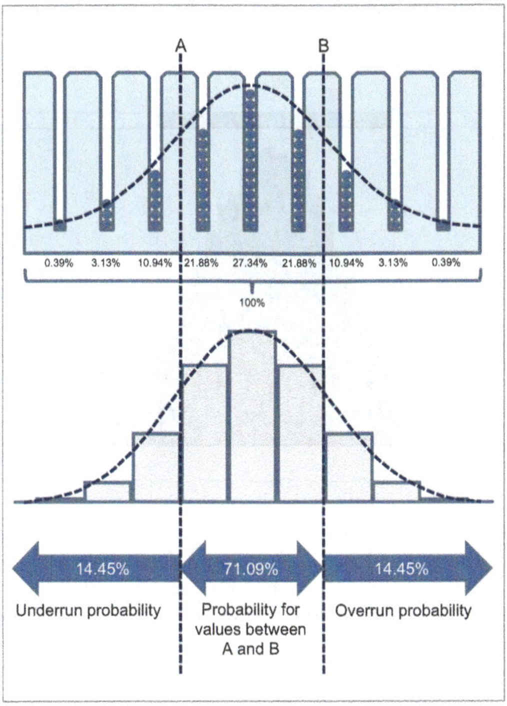

Fig. 7.22 Schematic conversion of the results of several *Galton board* experiments into a histogram[47)]

The @Risk software includes the option of defining both interim and final results as outputs and of representing them in the form of histograms (see Fig. 7.23) or sum curves (see Fig. 7.24).[48)]

Possible characteristics of specific outputs are shown on the abscissa whereas their associated frequencies, or incidence values, are plotted on the ordinate. In this regard, it should be noted that the area under the histogram has been normalized to 1. Including several histograms with different ranges in a single diagram may thus result in different graphical representations seemingly distorted in the vertical direction.

47) The percentages for the probabilities in each discipline were rounded to two decimal places. All decimal places must be included when calculating the total of 100%.

48) The histogram shown in Fig. 7.23 is related to the histogram in Fig. 7.22 only qualitatively, and not in terms of its substance.

At its top, the diagram includes so-called slide controls, which can be used to adjust the display to intended x values or specific probabilities. These sliders permit conclusions regarding the probability of over- or underrunning specific values as well as the probability of values lying within defined boundaries.

It is, however, also possible to submit reverse queries. For instance, the 5% or 95% quantile can be specified so that the probability of over- or underrunning the value equals 5%. This feature permits direct conclusions as to the corresponding deterministic values, which define the boundaries between the lowest and highest 5% and the central 90% of the area of the histogram (see the sliders at the top of the diagram shown in Fig. 7.23).

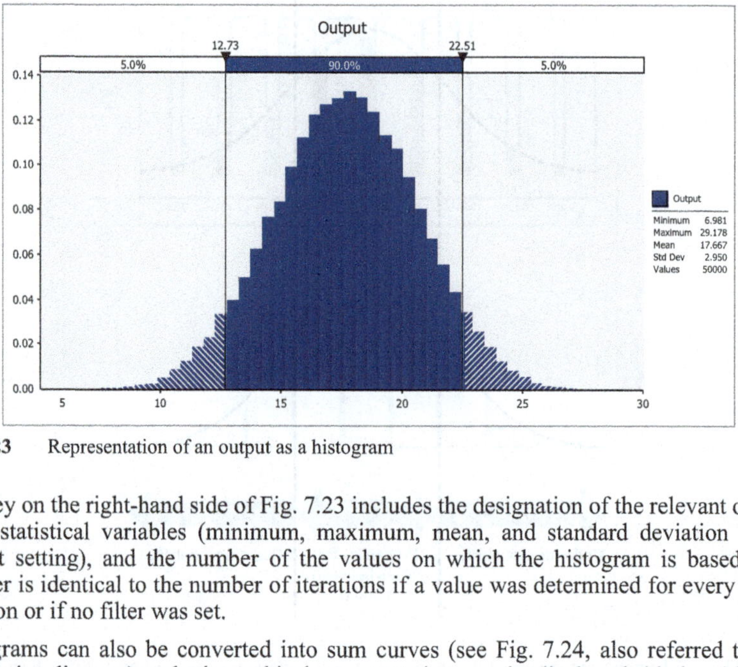

Fig. 7.23 Representation of an output as a histogram

The key on the right-hand side of Fig. 7.23 includes the designation of the relevant output, some statistical variables (minimum, maximum, mean, and standard deviation in the default setting), and the number of the values on which the histogram is based. This number is identical to the number of iterations if a value was determined for every single iteration or if no filter was set.

Histograms can also be converted into sum curves (see Fig. 7.24, also referred to as a cumulative diagram), or both graphical representations can be displayed side-by-side. The slope of the sum curve indicates the frequency or incidence in the relevant value ranges. The steeper the slope of the sum curve, the greater the probability will be for values to lie in this range.

The sum curves of discrete distributions exhibit a step-like pattern where the horizontal portions of the curve indicate a probability of 0% for values to lie in the relevant range (see, for example, Fig. 2.15, page 50).

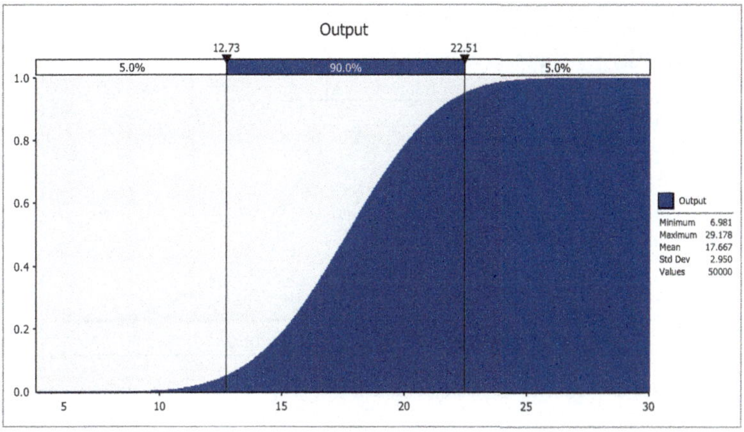

Fig. 7.24 Representation of an output as a sum curve

Within the range of a continuous distribution or continuous histogram, a single value will always be associated with a probability of 0% because a value of, say, 24.02 will always be accompanied by 24.023 or 24.02300000000001, for example. Any probability can be described only by the area of a portion of a probability distribution, i.e. the integral within a defined area. What we can state, though, is the over- or underrun probability for a single value because this involves determining the area of the distribution/histogram to the left or to the right of the defined value.

7.4.1.2 Selection of a Chance/Risk Ratio

Decision making ultimately requires specific values to be stated or selected within the defined ranges on the basis of a chance/risk ratio. One of these options would be the median, which is associated with a 50% probability of either exceeding or underrunning the value. In this case, there would be an equilibrium of the chances and risks taken.

Yet, construction contractors are often forced to take higher risks owing to an increasingly competitive environment in the construction industry. One realistic approach would be to consider the (upward or downward, depending on the specific situation) standard deviation σ from the mean μ.

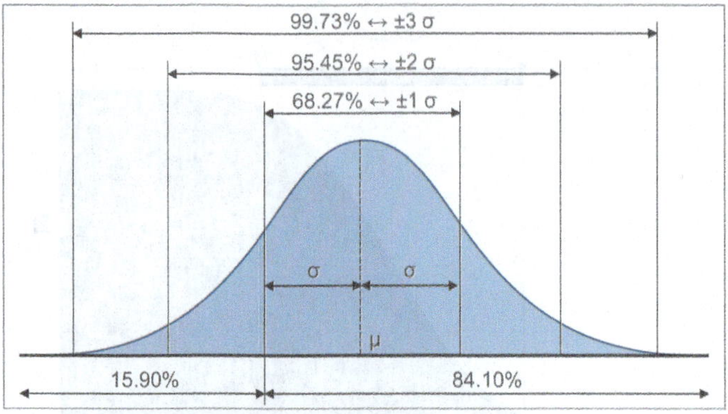

Fig. 7.25 Normal distribution – Standard deviation[49)]

By way of example, we can apply a normal distribution in order to estimate the related chance/risk ratio. In this distribution, about 68.27% of all values lie within the simple standard deviation from the mean, whereas approximately 95.45% of all values lie within the double standard deviation from the mean. Subtracting the simple standard deviation from the mean results in an underrun probability of approximately 15.90% and an overrun probability of about 84.10% for this value (see Fig. 7.25). This ratio will reverse if we add the simple standard deviation to the mean.[50)]

In addition, in the normal distribution, each of the inflection points of the density function is located one standard deviation away from the mean.

Defining or selecting a chance/risk ratio is a practice-driven approach that does not depend on any probabilistic target. The selection of the right chance/risk ratio will ultimately depend on the market price level or market situation, the willingness to take risks, and other strategic considerations. Care should be taken to consider the chance/risk ratio both with respect to winning the bid and to working profitably after the contract award (see Fig. 7.26).[51)]

49) Hofstadler/Kummer (2015a), p. 87

50) Cf. ibid., p. 86ff.

51) Cf. Kummer (2017), CPM-24
Cf. Hofstadler/Kummer (2018), p. 222ff.

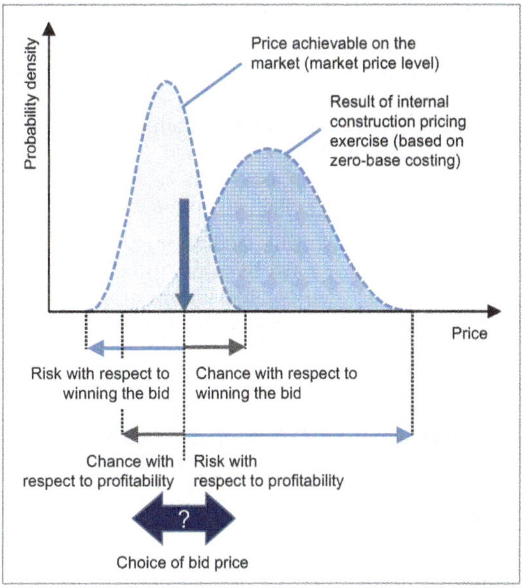

Fig. 7.26 Selection of the bid price with respect to profitability and probability of contract award[52)]

Owing to the nature of probabilistic calculations, we cannot accurately predict the outcomes of a construction project, for instance in terms of construction costs and times. However, it is possible to draw conclusions regarding the expected chance/risk ratio.

When performing bid-related calculations, for instance, each bidder will arrive at their own estimates of the ranges of input parameters (such as cost, labor consumption rate, and performance values). For one and the same project, this scenario will result in different cost or price histograms with varying spreads, skewnesses, mean values, etc. In other words, the estimated risk or chance is different for each bidder. These differences are due to different assessments, construction processes and methods, levels of knowledge and expertise, etc. At identical prices, bidders will thus take different risks and enjoy different chances in order to adhere to, or underrun, the bid price in the construction phase.

When bids are opened, each bidder will only see the prices quoted by the other bidders, without any information on how they arrived at the underlying values. However, evaluating bid prices based on a histogram and the corresponding probabilistic information will permit conclusions with respect to speculative elements or possible technological advantages. Any bidder that applies probabilistic calculation methods will thus be able to determine if they would have been able to cover their full cost based on their assumptions while executing the contract at the price of a cheaper competitor, and to identify the magnitude of the risk they would have taken in doing so.

52) Kummer (2015a), p. 75

7.4.1.3 Superposition of Several Histograms

Merely comparing mean values is not sufficient when having to decide in favor of one of several options. Even if the mean of Option A is smaller than that of Option B, this will not deliver any conclusion with respect to the significance of such differences because no information on the associated standard error was included in these considerations. The difference of the output values must be determined (either as a distribution or as a histogram) in order to compare two options with each other. This step will give an estimate of the "true" difference between the two options.[53)]

If several histograms with different spreads and strongly deviating nominal values are represented in a single diagram, care must be taken to ensure that the area under each histogram is normalized to 1. This is why such diagrams may result in graphical representations of these histograms that seem distorted in the vertical direction.

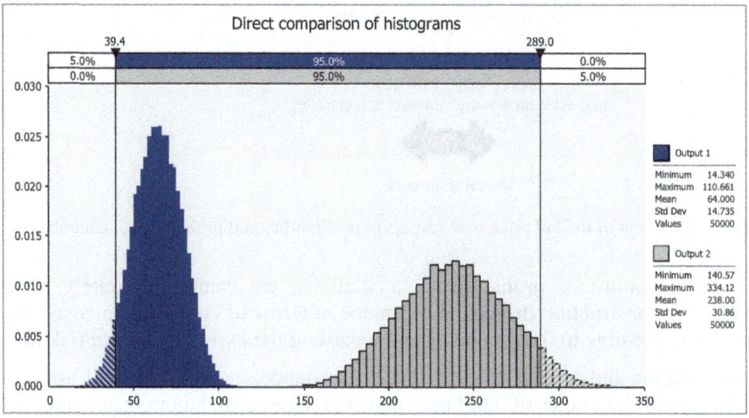

Fig. 7.27 Representation of several histograms within a single diagram

Fig. 7.27 includes two histograms with different ranges in relation to both nominal values and to the minimum and maximum. At first glance, we might assume a greater probability for the values in the histogram on the left compared to the histogram on the right-hand side. However, this is merely a visual illusion because the area under each of the histograms has been normalized to 1.[54)]

The slide controls shown at the top of Fig. 7.27 should be used for assessing the over- and underrun probability for specific deterministic values. A separate slider can be displayed for each of the histograms incorporated in the diagram.

7.4.1.4 Decision Making in the Case of Several Options

Distributions or histograms (especially those whose curve resembles a normal distribution) can be accurately described by the expected value m and the standard deviation σ,

[53)] Cf. Suhl/Mellouli (2013), p. 285

[54)] To check this finding, we perform an approximate calculation:
We obtain the approximate area of the left histogram in Fig. 7.27 as follows: (110 - 14) * 0.025 * 0.40 ≈ 1.0.
We obtain the approximate area of the right histogram in Fig. 7.27 as follows: (334 - 140) * 0.013 * 0.40 ≈ 1.0.

or the variance σ^2. If a decision is to be made in favor of one of several options, and if a distribution or histogram has been derived for each of these options, the decision maker can act in a risk-neutral, risk-averse, or risk-taking manner.

Risk-neutral individuals will completely disregard the spread of results. Their decision making will be based solely on expected values. In the case of two design options with different cost implications, risk-neutral decision makers will thus only compare expected values. Fig. 7.28 shows two such cost distributions from the client's perspective. Compared to Option 2, Option 1 is associated with a smaller expected value. A risk-neutral decision would thus favor Option 1.

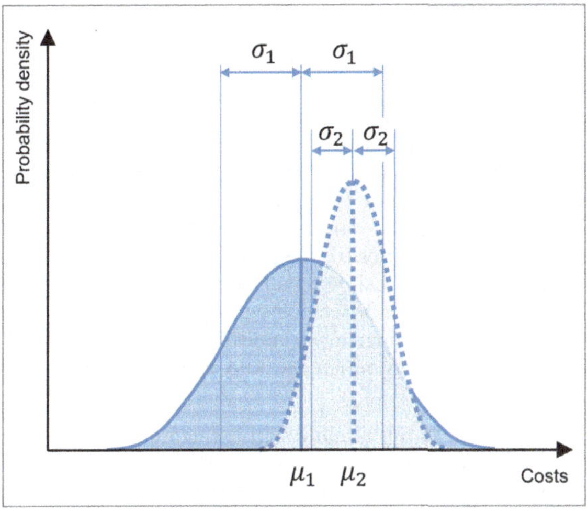

Fig. 7.28 Decision between options applying the μ-σ rule

In the case of risk aversion, a decision maker, from a certain level of approximation of the expected values associated with Option 2, would favor Option 2 because the narrower spread of possible results would be seen as a positive factor in the decision-making process, which is why this option is preferred to an option associated with a comparatively large spread.

Conversely, decision makers with a risk-taking attitude will consider the greater spread of Option 1 to be a positive factor because this will also be associated with the chance of the costs being significantly below the expected value. However, this setting will also pose an equal risk for expected costs to be exceeded, provided the distribution is symmetrical.

For a decision maker applying the μ-σ rule, a system of indifference curves can be used to express the dependency of preferences on just two parameters. An indifference curve is the geometrical location or graphical representation of all μ-σ combinations that the decision maker considers to be equivalent, i.e. resulting in the same benefit. The arrows in the diagrams shown in Figs. 7.29 and 7.30 signify the directions in which indifference curves with greater preferences occur.[55)]

[55)] Cf. http://wirtschaftslexikon.gabler.de/Archiv/830359611/erwartungswert-varianz-prinzip-v2.html. Accessed December 21, 2020

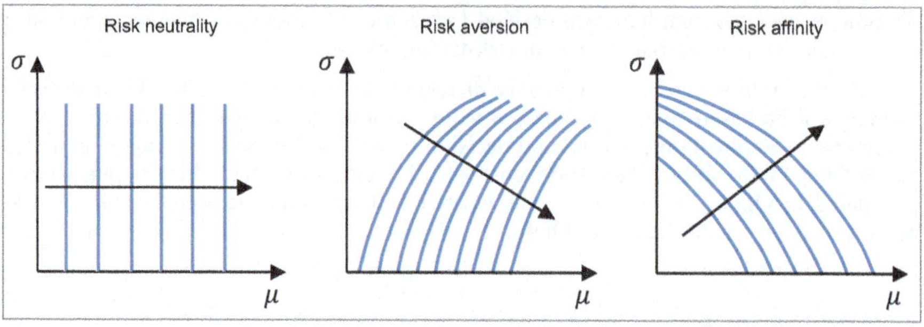

Fig. 7.29 Decision between options applying the μ-σ rule, e.g. in relation to output and productivity values[56)]

Here, the determined variables should be examined as to whether they are, ideally, "large" or "small" numbers. For instance, comparatively large numbers should be preferred for performance and productivity values (see the direction of preference indicated by the arrows in Fig. 7.29). Conversely, small expected values are desirable for costs, consumption rates, and similar metrics (dashed arrows mark the direction of greater preference in Fig. 7.30). In this context, it should be noted that not only the directions of arrows change but also the bends of the indifference curves. Whereas Fig. 7.29 describes risk-averse behavior with degressively increasing indifference curves, Fig. 7.30 illustrates risk aversion with progressively decreasing curves. Risk affinity is associated with the reverse pattern.

Generally speaking, any risk-taking attitude will associate a greater benefit with options exhibiting a larger spread or range (i.e. a higher σ). Preference arrows thus point upward for risk affinity, as shown in Fig. 7.29 (right) and Fig. 7.30 (center). In the case of risk-averse behavior, the benefit will increase if the spread becomes more narrow. Preference arrows thus point downward for risk aversion, as shown in Fig. 7.29 (center) and Fig. 7.30 (right).

In reality, implementing the μ-σ rule is associated with the problem that the utility functions according to which one of several options is selected are largely unknown.[57)]

[56)] Cf. ibid.

[57)] We refer to the doctoral thesis by Werkl (2013), which includes an empirical survey of utility functions for clients and contractors.

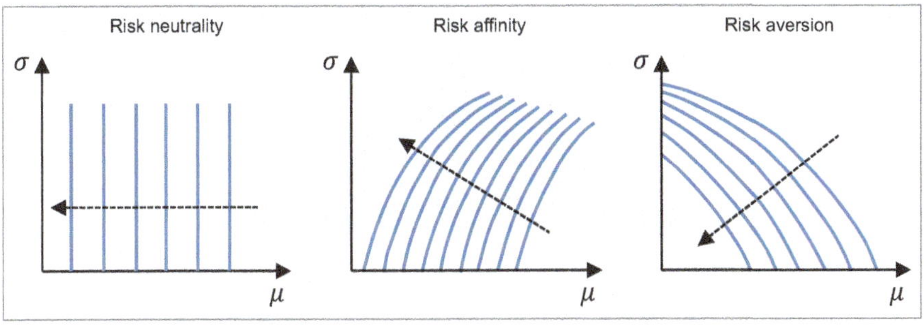

Fig. 7.30 Decision between options applying the μ-σ rule, e.g. in relation to costs and consumption rates[58)]

7.4.2 Tornado Diagrams

If the calculation of an output value is informed by several input parameters, it can be of interest to find out which input parameters have the most significant effect on the result. Deterministic verification by varying the inputs requires a fair amount of computing power. In contrast, any probabilistic calculation can easily be complemented by a sensitivity analysis using tornado diagrams as part of a Monte Carlo simulation. Tornado diagrams show the ranking of input distributions that have an influence on outputs. The @Risk software provides the option of obtaining a tornado diagram for any defined output.

The sensitivity analysis ("Tornado – Change in output mean") in @Risk represents the absolute change of the output relative to an input value in the ascending and descending directions toward the mean of the output.

In the first step, the random values of the individual iterations completed for a specific input are sorted in ascending order to determine the corresponding deviations from the output base value. If, for instance, 50,000 iterations were completed, the resulting 50,000 values will subsequently be sorted by size and grouped into 10 bins[59)] of equal size (i.e. the default setting used in @Risk). This means that 10% of the lowest values and 10% of the highest values of an input parameter will be allocated to the first and to the last bin, respectively. All other iteration values (in this case, 80%) will be uniformly distributed across the remaining (eight) bins. For a symmetrical triangular distribution, this means that the bins in the upper and lower value ranges have a greater class width than the bins in the central portion of the distribution (see Fig. 7.31).

58) Cf. http://wirtschaftslexikon.gabler.de/Archiv/8303596l1/erwartungswert-varianz-prinzip-v2.html. Accessed December 21, 2020

59) When data is grouped into classes, such classes are also referred to as bins. Plotting bins next to each other as rectangles results in a histogram. The width of bins can generally be constant or variable. This book consistently uses constant class widths for histograms. However, a different method is applied for calculating the change in the output of the means for sensitivity analyses using tornado diagrams. In this case, the default setting groups the inputs into ten bins of an identical area (i.e. different class widths).

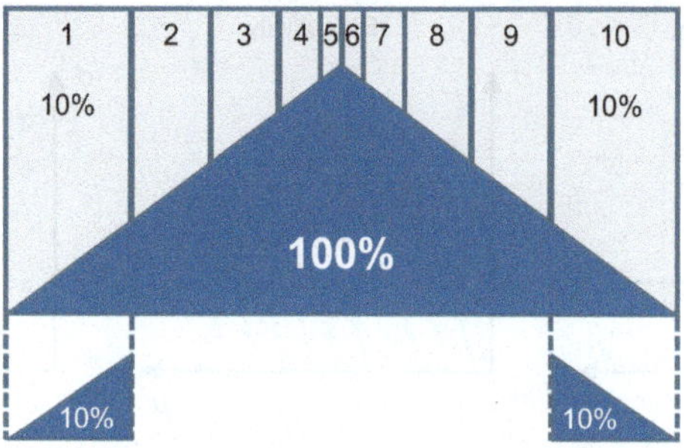

Fig. 7.31 Qualitative categorization of input parameters for the purpose of calculating tornado diagrams

Next, from each of these bins, a mean is derived from the random numbers. The resulting ten mean values are used for calculating an output. The results of these operations will then also be used for deriving spider diagrams (see Section 7.4.4). In the set of equations, the remaining input values are applied as means of their defined distributions, and kept constant. This calculation step results in ten outputs, one of which represents the minimum and another the maximum. In the tornado diagram, these two outputs will be entered as the left and the right bar, starting from the base value. However, this representation does not permit conclusions as to whether the mean of the first, second or, say, last bin has resulted in the minimum or maximum output values. This would require in-depth analyses of the effects of specific inputs on outputs (one possible means to do so would be scatter plots).

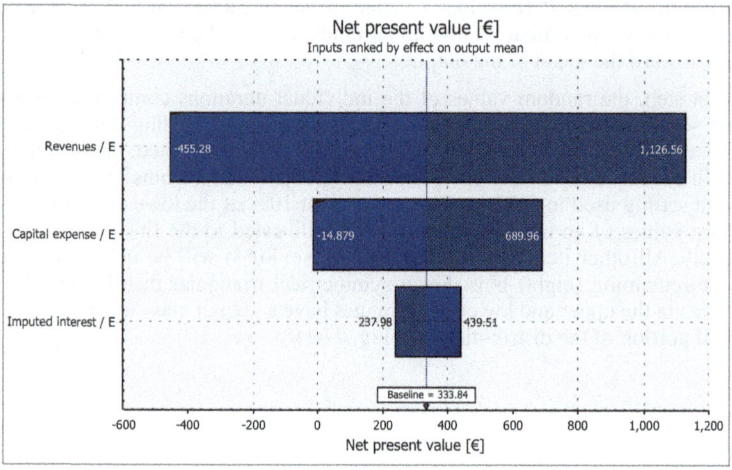

Fig. 7.32 Example – Tornado diagram ("Tornado – Change in output mean") – Net present value

Fig. 7.32 shows a tornado diagram for the net present value resulting from the cash inflows and outflows generated in the course of an investment project. The range of minimum and maximum net present values is shown as bars on the abscissa. According to the described approach, these ranges result from varying the relevant uncertain input parameter plotted on the ordinate. This simple example applies distributions to just three input parameters (thus making them uncertain) for the purpose of determining the net present value.

Revenues are demonstrated to have the greatest influence on the net present value. In addition, they can also lead to a large absolute spread of results. In the worked example, the sensitivity of the net present value to the uncertainties assumed for the imputed interest rate is relatively low, however.

Thus, the diagram shown in Fig. 7.32 does not directly reveal if increasing an input parameter would have a beneficial or adverse effect on the output. Other tornado diagrams will deliver the related information. @Risk distinguishes between three additional types of tornado diagrams:[60)]

Tornado diagram for regression sensitivity ("Regression coefficients")

In regression coefficient tornado diagrams, the length of the bar shown for each input distribution is based on the coefficient value calculated for the relevant input. The numbers plotted on the ordinate of the tornado diagram are the coefficient values (see Fig. 7.33).[61)]

This type of diagram shows the change in the standard deviation of the output if the input is increased by one standard deviation. For instance, a regression coefficient of 0 means that there is no significant input-output relationship. A regression value of +1 or -1 will result if an input change by one standard deviation is also directly associated with the output change by one standard deviation (+1 designates a change in the same direction, -1 stands for a change in the opposite direction).

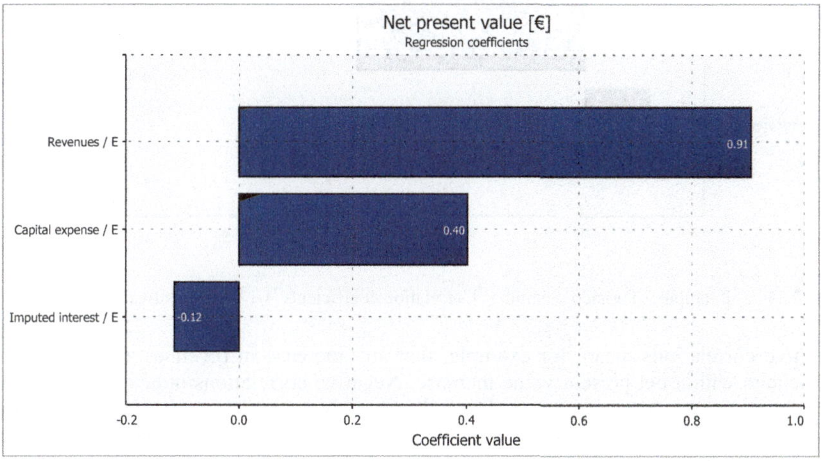

Fig. 7.33 Example – Tornado diagram ("Regression coefficients") – Net present value

To be able to perform a non-normalized interpretation of the results of this diagram, the standard deviations of the relevant input and of the output must be known. For the revenues, this means that the regression coefficient (= 0.91) is first multiplied with the

60) A tornado diagram or chart can include up to 16 bars. Cf. Palisade Corporation (2013), p. 278.

61) Cf. ibid., p. 277

standard deviation of the net present value (= 504.44) and then divided by the standard deviation of the revenues (= 163.3).[62] This calculation results in a non-normalized coefficient of 2.81 (= 0.91 * 504.44 / 163.30). In other words, any increase in revenues by one unit will cause the output to increase by 2.81 units.[63]

In this diagram, it seems noteworthy that an increase in capital expenses also increases the net present value. This is due to the fact that capital expenses are fed into the net present value (NPV) calculation as negative values (output), whereas revenues are expressed as positive values. A formal increase in the parameter must thus be equated with a reduction in capital expenditure. The visualization shown in this as well as in the following tornado diagrams is thus plausible in terms of its substance.

Tornado diagram for correlation sensitivity ("Correlation coefficients")

In correlation coefficient tornado diagrams, the length of the bar shown for each input distribution is based on the coefficient value calculated for the relevant input. The values plotted on the abscissa of the tornado diagram represent the determined rank correlations (cf. Section 6.5) between inputs and outputs (see Fig. 7.34).

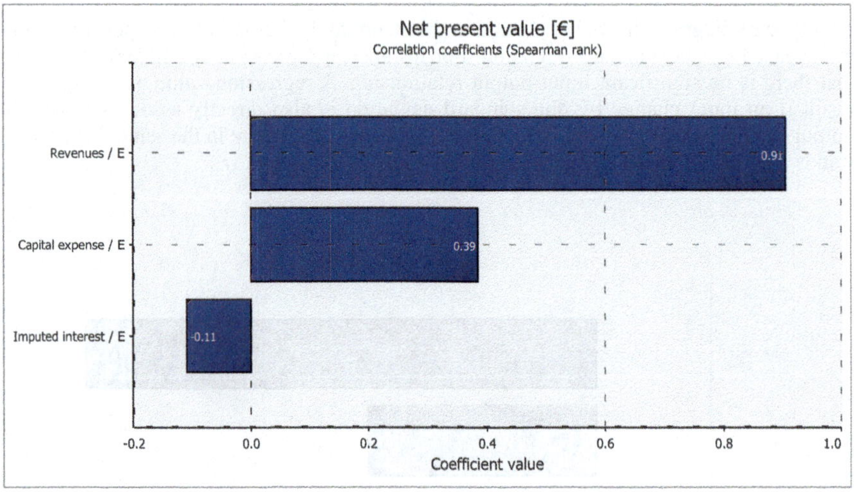

Fig. 7.34 Example – Tornado diagram ("Correlation coefficients") – Net present value

Positive correlations mean, for example, that any increase in revenues also tends to be associated with a net present value increase. Negative correlations indicate, for instance, that an increased imputed interest rate will probably lower the net present value. Such diagrams need to be interpreted with a view to the uncertainties associated with the individual calculation parameters.

62) In @Risk, the standard deviations for inputs and outputs can be taken directly from the key to the relevant distributions.

63) Cf. http://kb.palisade.com/index.php?pg=kb.page&id=138. Accessed December 21, 2020

Tornado diagram for regression sensitivity showing input values rather than coefficients at the bars' top ends ("Regression – mapped values")

In this diagram type, the lengths of the bars for the individual input distributions represent the magnitude of the change in the output that has resulted from a +1 change in the standard deviation at the input stage. In other words, the values shown on the tornado diagram bars are output values for an input standard deviation of +1. Accordingly, if the input plotted on the ordinate increases by one standard deviation, the output will change by the abscissa value (in the positive or in the negative direction) that also corresponds to the length of the relevant bar (see Fig. 7.35).[64]

We find that an increase in imputed interest will adversely affect the net present value. Conversely, increasing revenues by one standard deviation will also increase the NPV.

The bar representing the investment, which was incorporated into the calculation as a negative value because of it being an expense, needs to be interpreted such that any increase in the value of the investment is substantially equivalent to a decrease in the capital expense. In other words, the lower the capital expense (in monetary units), the greater the increase in the net present value will be.

Any interpretation of tornado diagrams must consider whether parameters were integrated into the calculations with a plus or a minus prefix.

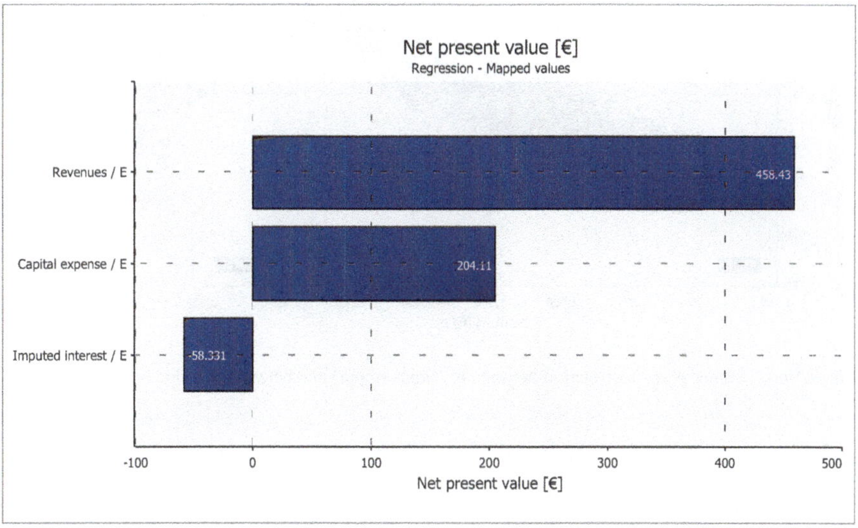

Fig. 7.35 Example – Tornado diagram ("Regression – mapped values") – Net present value

7.4.3 Scatter Plots

Scatter plots can be used to represent the relationship between a simulated output and the sample values taken from an input distribution. Such diagrams are x-y diagrams in which the identified input value is compared to the calculated output value for each iteration of

64) Cf. Palisade Corporation (2013), p. 278

the simulation. An ellipsis identifies the range into which x-y values will fall for a defined confidence level. Scatter plots can also be standardized to make it easier to compare values from several inputs in a given scatter plot.[65)]

Fig. 7.36 illustrates a comparison of the "revenues" (input) and "net present value" (output) parameters. Each of the points included in the diagram corresponds to one out of a total of 50,000 completed iterations. In this example, the ellipsis covers 99% of the points. The orientation of the point cloud indicates a positive correlation between the two parameters, as was expected from the interpretation of the tornado diagrams. The key to the diagram also includes the rank correlation ["Corr. (Rank)"] (see Fig. 7.34).[66)]

The delimiters shown in the diagram represent one deterministic input and one deterministic output value. The percentages in the quadrants indicate the over- or underrun probability for these values. Each of the delimiters marked by arrows can be shifted manually, similarly to the slide controls included in @Risk histograms.

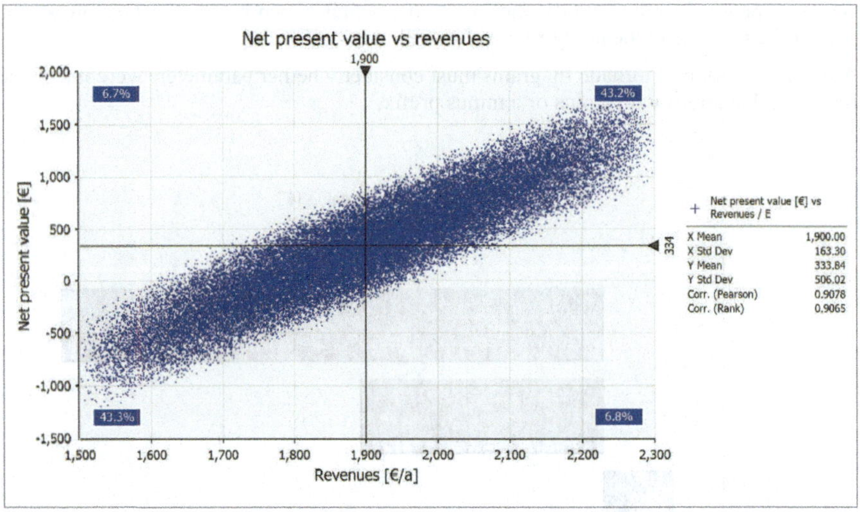

Fig. 7.36 Scatter plot – Correlation between revenues (input) and net present value (output)

7.4.4 Spider Diagrams

The results of sensitivity analyses can also be shown in so-called spider diagrams. As shown in the qualitative analysis in Fig. 7.31 of Section 7.4.2, the distributions of the uncertain input parameters are first divided into bins (ten bins in the default setting) with an identical probability of occurrence. The mean of each bin is then entered in the relevant equations, with all other input parameters required for the calculation being kept constant. This step will result in a number of outputs that corresponds to the number of bins. These outputs are plotted in a diagram as a function of the percentile for the respective input and connected to each other by a line.

65) Cf. Palisade Corporation (2013), p. 84

66) For representing scatter plots, the @Risk software automatically calculates *Spearman's* rank correlation coefficient and *Pearson's* product-moment correlation coefficient.

A spider graph indicates how the statistical output changes with each change in the identified input value. The steeper the slope of the line, the greater the effect of the input on the output will be. Compared to tornado diagrams, spider diagrams provide added information value because they deliver details on the output value change rate in relation to input changes.[67)]

Fig. 7.37 includes the net present value as an output plotted on the ordinate. The different slopes of the curves included in the diagram indicate the input with the most pronounced effect on the output, as well as the input value range in which this influence is particularly large. Curves forming a straight line mean that the input variation (from one bin mean to the next) has a linear influence on the output. S-shaped curves indicate either a nonlinear behavior of the computation model in relation to input change, or strongly varying intervals between the bin means (e.g. owing to an exceedingly flat input distribution curve). Fig. 7.37 shows, for the revenue curve, that a much steeper slope of output results should be expected for the first and last bin. This effect is based on the relatively flat curve of the distribution at its upper and lower edges (see, for example, the symmetrical triangular distribution in Fig. 7.31). A horizontal graph in Fig. 7.37 would mean that the relevant parameter has no influence on the output within the relevant value range.

In addition, we find that a reduction in the net present value should be expected if the imputed interest rate increases. The capital expenses have been integrated as negative values into the computation model. This is why increasing values (representing a cost reduction) should also be indicative of a net present value increase.

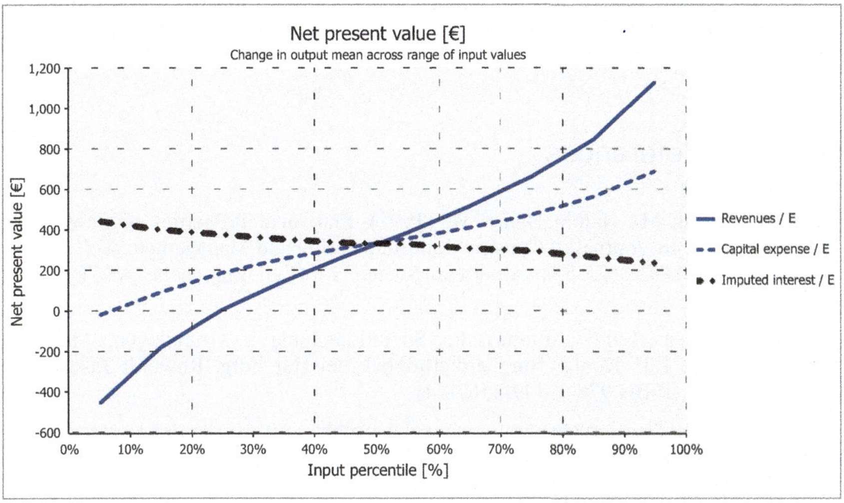

Fig. 7.37 Spider diagram – Correlation between the change in inputs and the change in output (net present value)

67) Cf. Palisade Corporation (2013), p. 279

7.5 Summary

Monte Carlo simulations can be applied correctly and purposefully only if "realistic" distribution functions are selected for uncertain input parameters. The various methods for selecting distribution functions constituted one of the main aspects of this chapter.

This chapter also concentrated on incorporating theoretical considerations based on known or assumed characteristics, and on analyzing data from site or expert surveys. The availability of a contextual database will allow data fitting for the purpose of direct conversion into a distribution function. Using such a fitted distribution as an input for a Monte Carlo simulation will make it possible to incorporate theoretical considerations, such as with respect to identifying the distribution boundaries, and to optimize inputs by applying, as far as reasonably possible, a combination of minimum, most likely, and maximum values rather than mere curve parameters.

This chapter explicitly listed distribution functions for labor consumption rates determined for formwork operations and for quantity deviations for reinforced concrete operations, which were based on the expert surveys conducted by the authors. Expert surveys also delivered correlation coefficients for the input parameters for calculating the total labor consumption rate of reinforced concrete operations.

For probabilistic calculations to add as much value as possible, selecting realistic distribution functions and, if applicable, correlation coefficients is crucial. The final sections of this chapter primarily dealt with the appropriate interpretation of simulation results. In this context, the significance of quantile values and over- and underrun probabilities was thoroughly discussed. For a realistic approach to decision making, it is crucial to select an appropriate number that will subsequently quantify the related chance/risk ratio.

7.6 References

AbouRizk Simaan M., Halpin Daniel W. (1992). Statistical Properties of Construction Duration Data. In: Journal of Construction Engineering and Management, Vol. 118, No. 3, September 1992. 525-544. American Society of Civil Engineers (ASCE). (ISSN 0733-9364)

Diekmann, Andreas (2007). Empirische Sozialforschung – Grundlagen, Methoden, Anwendungen. Ed.: König, Burghard. Reinbek bei Hamburg. Rowohlt Taschenbuch Verlag GmbH. (ISBN 978-3-499-55678-4)

Dinort, Günter (1977). Berücksichtigung stochastischer Einflüsse in der Kostenrechnung einer Bauunternehmung. Doctoral thesis. Munich. Lehrstuhl für Tunnelbau und Baubetriebslehre der Technischen Universität München.

Gläser, Jochen; Laudel, Grit (2010). Experteninterviews und qualitative Inhaltsanalyse als Instrumente rekonstruierender Untersuchungen – Lehrbuch – 4th ed. Wiesbaden. VS Verlag für Sozialwissenschaften, Springer Fachmedien GmbH. (ISBN 978-3-531-17238-5)

Gorden, Raymond L. (1969). Interviewing – Strategy, Techniques, and Tactics. Homewood, Illinois. The Dorsey Press.

Hintze, Jerry L.; Nelson, Ray D. (1998). Violin Plots: A Box Plot-Density Trace Synergism. In: The American Statistician, Vol. 52, No. 2. 181-184. An Official Journal of the American Statistical Association. Taylor & Francis. (ISSN 0003-1305; Online 1537-2731)

Hofstadler, Christian (2008). Schalarbeiten – Technologische Grundlagen, Sichtbeton, Systemauswahl, Ablaufplanung, Logistik und Kalkulation. Berlin, Heidelberg. Springer-Verlag. (ISBN 978-3-540-85178-3)

Hofstadler, Christian (2014b). Monte-Carlo-Simulation im Baubetrieb – Verteilungsfunktion für Aufwandswerte bei Schalarbeiten – Herleitung der Verteilungsfunktion aus den Ergebnissen einer Expertenbefragung. In: Bauingenieur, Band 89, May 2014. 189-201. Düsseldorf. Springer-VDI-Verlag GmbH & Co. KG. (ISSN 0005-6650)

Hofstadler, Christian; Kummer, Markus (2015a). Risiken und Chancen der Mischpreiskalkulation. In: Tagungsband – 13. Grazer Baubetriebs- und Bauwirtschaftssymposium – Preisermittlung und Vergabe in der Bauwirtschaft – Baubetriebliche, bauwirtschaftliche und rechtliche Aspekte. Eds.: Heck, Detlef; Mauerhofer, Gottfried; Hofstadler, Christian. 49-92. Graz, Verlag der Technischen Universität Graz. (ISBN 978-3-85125-385-9)

Hofstadler, Christian et al. (2016). Einfluss von Lohn- und Sozialdumping auf den Wettbewerb in der Bauwirtschaft – Forschungsprojekt. Graz, Institute of Construction Management and Economics, Graz University of Technology.

Hofstadler, Christian; Kummer, Markus (2017a). Chancen- und Risikomanagement in der Bauwirtschaft – Für Auftraggeber und Auftragnehmer in Projektmanagement, Baubetrieb und Bauwirtschaft. Berlin, Heidelberg. Springer-Verlag. (ISBN 978-3-662-54318-4)

Hofstadler, Christian; Kummer, Markus (2017b). Calculation of the labor consumption rate for shuttering works whilst considering uncertainties – Influence of distribution functions on probabilistic calculations. In: Resilient Structures and Sustainable Construction – ISEC-9. Eds.: Pellicer, Eugenio; Adam, José M.; Yepes, Víctor Yepes; Singh, Amarjit; Yazdani, Siamak; CPM-26. Fargo. ISEC Press. (ISSN 2644-108X; ISBN 978-0-9960437-4-8)

Hofstadler, Christian; Kummer, Markus (2018). The Dilemma of Pricing Against the Backdrop of the Chance/Risk Ratio. In: The Tenth International Conference on Construction in the 21st Century (CITC-10). Eds.: Ahmed, Syed M.; Shah, Attaullah; Azhar, Salman; Smith, Norma A.; Campbell, Shaunna C.; Mahaffy, Kelly; Saul, Amelia. 222-231. Greenville, North Carolina, USA. (ISBN 978-1-7324416-1-3)

Hutzelmeyer, Hannes (1984). Die Kostensteuerung des Bauauftraggebers. In: Praxis-Kompendium Baubetrieb – Leitfaden, Arbeitsunterlagen und Nachschlagewerk für Praktiker und Studenten – Band 2. Eds.: Schub, A.; Meyran, G. 37-79. Wiesbaden, Berlin. Bauverlag GmbH. (ISBN 3-7625-2163-8)

Kummer, Markus (2015a). Aggregierte Berücksichtigung von Produktivitätsverlusten bei der Ermittlung von Baukosten und Bauzeiten – Deterministische und probabilistische Betrachtungen. Doctoral thesis. Graz. Institute of Construction Management and Economics, Graz University of Technology.

Kummer, Markus (2017). Calculation of the labor consumption rate for shuttering works whilst considering uncertainties – Significance of the choice of the chance/risk ratio. In: Resilient Structures and Sustainable Construction – ISEC-9. Eds.: Pellicer, Eugenio; Adam, José M.; Yepes, Víctor Yepes; Singh, Amarjit; Yazdani, Siamak; CPM-24. Fargo. ISEC Press. (ISSN 2644-108X; ISBN 978-0-9960437-4-8)

Kummer, Markus; Hofstadler, Christian (2017). Probabilistic calculation of total labor consumption rate of reinforced concrete works – with and without correlations. In: Resilient Structures and Sustainable Construction – ISEC-9. Eds.: Pellicer, Eugenio; Adam, José M.; Yepes, Víctor Yepes; Singh, Amarjit; Yazdani, Siamak; CPM-23. Fargo. ISEC Press. (ISSN 2644-108X; ISBN 978-0-9960437-4-8)

Kummer, Markus (2018). Quantity Uncertainties in Shuttering Works – Comparison of Public versus Private Clients. In: The Tenth International Conference on Construction in the 21st Century (CITC-10). Eds.: Ahmed, Syed M.; Shah, Attaullah; Azhar, Salman; Smith, Norma A.; Campbell, Shaunna C.; Mahaffy, Kelly; Saul, Amelia. 180-188. Greenville, North Carolina, USA. (ISBN 978-1-7324416-1-3)

Kummer, Markus (2019). Quantity Uncertainties in Reinforcement Works – Comparison of Public versus Private Clients. In: Interdependence between Structural Engineering and Construction Management – ISEC-10. Eds.: Ozevin, Didem; Ataei, Hossein; Modares, Mehdi; Gürgün, Asli Pelin; Yazdani, Siamak; Singh, Amarjit. CPM-17. Fargo. ISEC Press. (ISSN 2644-108X; ISBN 978-0-9960437-6-2)

Palisade Corporation (2013). Benutzerhandbuch – @Risk – Risikoanalysen- und Simulations-Add-In für Microsoft Excel – Version 6 of March 2013.

Palisade Corporation: 7.7. Interpreting Regression Coefficients in Tornado Graphs. Palisade Knowledge Base. http://kb.palisade.com/index.php?pg=kb.page&id=138. Accessed December 21, 2020.

Raaber, Norbert (2003). Beitrag zu Ausschreibung und Vergabe von Bauleistungen. Schriftenreihe des Instituts für Baubetrieb und Bauwirtschaft der Technischen Universität Graz – Heft 15. Ed.: Stadler, Gert. Graz. Verlag der Technischen Universität Graz. (ISBN 3-901351-76-0)

Song, Lingguang; AbouRizk, Simaan M. (2008). Measuring and Modeling Labor Productivity Using Historical Data. In: Journal of Construction Engineering and Management, Vol. 134, No. 10, October 2008. 786-794. American Society of Civil Engineers (ASCE). (ISSN 0733-9364)

Springer Gabler Verlag (Ed.), Gabler Wirtschaftslexikon, Keyword: Expertenwissen. Available online at http://wirtschaftslexikon.gabler.de/Archiv/55229/expertenwissen-v9.html. Accessed December 21, 2020.

Springer Gabler Verlag (Ed.), Gabler Wirtschaftslexikon, Keyword: Erwartungswert-Varianz-Prinzip. Available online at http://wirtschaftslexikon.gabler.de/Archiv/830359611/erwartungswert-varianz-prinzip-v2.html. Accessed December 21, 2020.

Suhl, Leena; Mellouli, Taïeb (2013). Optimierungssysteme – Modelle, Verfahren, Software, Anwendungen – 3rd ed. Berlin, Heidelberg. Springer-Verlag. (ISBN 978-3-642-38936-8)

Werkl, Michael (2013). Risiko- und Nutzenverhalten in der Bauwirtschaft – Eine entscheidungstheoretische Betrachtung im institutionenökonomischen Kontext. Doctoral thesis. Graz. Institute of Construction Management and Economics, Graz University of Technology.

Xie, Hua; AbouRizk, Simaan; Fernando, Siri (2011). Integrating Realtime Project Progress Input into a Construction Simulation Model. In: Proceedings of the 2011 Winter Simulation Conference – December 11-14, 2011. Eds.: Jain, S. et al. 3448-3459. Phoenix, AZ. IEEE. (ISBN 978-1-4577-2107-6)

Yang, I.-T. (2005). Simulation-based estimation for correlated cost elements. In International Journal of Project Management, Volume 23, Issue 4, May 2005. 275-282. Elsevier Ltd. and IPMA. (ISSN 0263-7863)

8 Influence of Project Lead Time and Construction Time on Project Targets

Clients are the main stakeholders and decision makers when planning and implementing the project and building targets. They define milestones and related schedules, which is why they have a major influence on the quality and standard of workmanship in the individual project phases, with repercussions on the overall quality of the building or structure. The quality of the project documents depends on a sufficient (normal) project lead time and essentially determines the frequency and magnitude of disruptions and additional costs. For the client, this setting results in a major "apex" that determines the controllability of the project. This point in time is determined by the contract award. The quality standard achieved up to this point in terms of planning, tendering, and contract design will be crucial in enabling project success in the ensuing construction and operation phases. The client will subsequently have to bear any additional costs and/or deal with poorer quality resulting from prior deficiencies or shortcomings. When awarding the construction works, the commencement date and time to project completion, or construction time, will be agreed upon as the key binding contractual components.

All project stakeholders need to consider the specified construction time as a major factor influencing the achievement of defined and agreed project targets. Any exceedingly high quality standard specified in the design brief, for instance with respect to exposed concrete, will require a period of time commensurate with these requirements.

Construction time has a major influence not only on the cost-efficient execution of construction works but also on the overall project outcome that the client aims to achieve. Clients would do themselves a disservice if they specified short or extremely short construction periods for their projects. The same applies to time made available for project preparation, particularly for the design and bid phases. Any poor project design, tendering, and execution due to time pressure will adversely affect workmanship but also the use and reuse of the building or structure. Cost, time, and quality targets are at risk owing to the resulting disruptions to the construction process.

Productivity is adversely affected not only by unduly short but also excessively long construction times. If the construction time is short or very short, this will have major repercussions on the efficient and effective utilization of the production factors. This applies to both elementary (workers, materials, equipment and machinery) and discretionary (project manager, site manager, technicians, etc.) production factors. In many

C. Hofstadler and M. Kummer, *Chances and Risks in Construction Management and Economics*, Springer Tracts in Civil Engineering,
https://doi.org/10.1007/978-3-030-64014-9_8

cases, the client will not realize this effect or consider it only to an insufficient extent, or disregard it completely.

This chapter outlines the basic relationship between construction time, labor consumption rates, and productivity as well as construction cost. Own approaches and methods known from pertinent literature are presented and critically reviewed in order to be able to quantitatively assess losses of productivity. In addition, it sets out to determine the influence of construction time on costing and, consequently, on the construction process. Construction managers will be made aware of the necessity to closely monitor those construction-related factors that have an influence on construction time.

Another factor that deserves increasing attention is the gradual ("notional"[1)]) shortening of construction time. More often than not, the scope of work to be performed will increase while the construction time remains unchanged, thus adversely affecting the circumstances under which the contractor is bound to perform their work. Contractors will thus be forced to accelerate their workflows in small increments to be able to adhere to the specified construction time in spite of process disruptions. In some cases, this will incur significant additional costs that will usually have to be borne by the contractor if the deviation was not previously notified to the client in line with the periods provided for in the contract.

This book essentially aims to create a better understanding of factors promoting and inhibiting productivity while raising awareness of possible construction process disruptions. On the one hand, this will further enhance the efficient and effective utilization of the production factors; on the other hand, it will become easier to identify, and to subsequently assess, parameters and values exceeding the thresholds to losses of productivity.

This approach will strengthen the position of contractors in their interactions with clients; justified additional cost claims can be identified more quickly and enforced more easily.

Even prior to initiating the bid phase, clients are in a position to assess the effects of the specified construction time and, if required, compare the expected additional costs of work execution with the benefit of an earlier project completion date. Any extra costs incurred during the use and reuse phases of the building owing to an exceedingly short specified construction time must be considered separately in such a comparison.

8.1 System of Targets for Construction Projects

Just considering the well-known "magic triangle" of cost, quality, and time is clearly insufficient for achieving the defined project and operational targets. Achieving project success on all levels essentially requires expansion of this triangle to a hexagon so as to take additional relevant factors into account for the purpose of planning and forecasting. This holistic view also integrates the criteria of process quality, resilience to disruptions, and quantity into the system of targets.

In an integrated approach, quality, time, cost, quantity, resilience to disruptions, and process quality will be considered to be the key decision-making or design criteria that need to be analyzed more thoroughly in order to be able to assess a construction project in

1) There is no actual shortening of the construction time in the schedule. How-ever, resource utilization needs to be increased owing to the higher consumption rate and/or additional work to be able to adhere to the agreed construction time. The construction time would have to be extended in the case of an unchanged resource utilization. To prevent this from happening, the notionally extended construction time is shortened to the originally agreed duration. This step results in a notional shortening of the construction time, which is equivalent to accelerating the individual works.

sufficient detail. In this setting, the quality of the project is the determining success factor because it will have repercussions on the realizable market price of the property. In addition, the quality of the project is the key driver of additional costs because a higher quality standard will usually imply longer maintenance cycles (for instance thanks to the robustness and ease of maintenance of surfaces) and thus optimize building use. Relevant decision-making examples include sheet steel versus copper roofings or water-based versus mineral-based paints. When analyzing the temporal aspect, it is crucial to consider that ever-shorter product life cycles will also incur higher costs. In this context, it should be noted that the potential costs are limited by the credit lines provided by the banks. In the construction industry, there is an increasing demand for adopting a holistic project management view, which requires expanding the conventional magic triangle of costs, deadlines, and quality beyond the project target in order to define a new, more comprehensive set of operational targets. These essentially comprise life-cycle costs, sustainable quality standards, and long-term targets related to the temporal aspect, including warranty, service agreements, and other contracts. We thus find that the currently defined set of project targets have proven to be perfectly appropriate in terms of quality and deadlines because costs can essentially be considered to be a parameter predefined by the client. In addition, a life-cycle approach should be chosen in order to prevent the above, rigidly focused target-setting process and to move towards a sourcing strategy linked to long-term targets, as shown in Fig. 8.1. It is also necessary to assess the parameters of process quality and (non)susceptibility to disruptions in the construction project and operational phases.

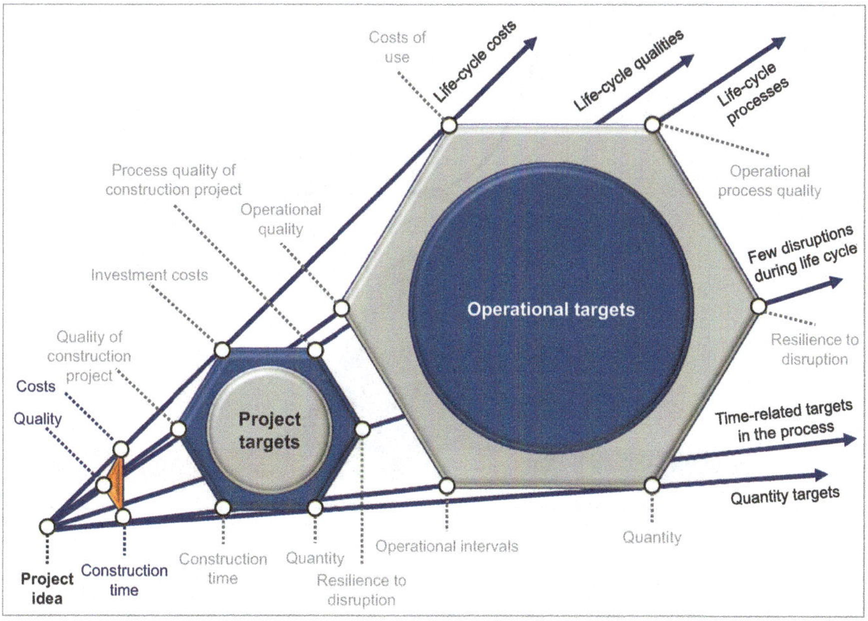

Fig. 8.1 System of targets for the life cycle of buildings[2)]

[2)] Cf. Hofstadler (2014b), p. 145

The selected chance/risk ratio is essential for achieving the defined targets. If, for instance, a very short project lead time were chosen combined with a very tight planning-ahead budget, there would be an extremely high chance of very low project preparation costs incurred. However, this approach would be associated with a high risk of underachieving the defined project targets, or not achieving them at all, thus failing to realize the initially expected benefit.

8.2 Project Lead Time and Project Targets

Fig. 8.2 includes a qualitative representation of the relationship between the estimated project lead time and the achievable project, building, operational, and use quality. This diagram shows the degree of target achievement over time. A normal project lead time should be aimed for depending on the type of the project. The range of the "normal" project preparation time describes the lead time range suitable for preparing and/or processing the documents on which to base the planning and decision-making process to a sufficiently high quality standard. The 100% indicated on the ordinate mean that the defined targets can be achieved at minimum costs, disruptions, disputes, etc.

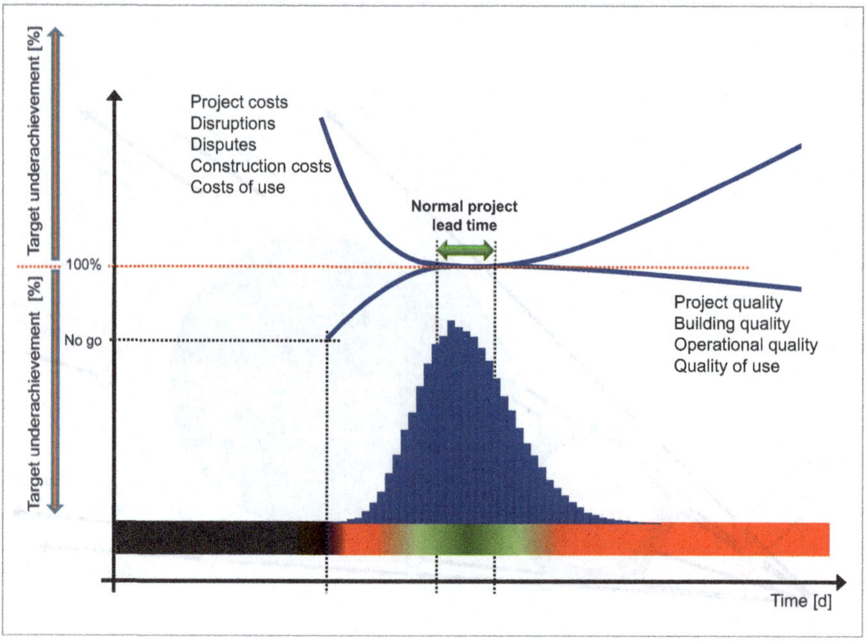

Fig. 8.2 Relationship between project lead time and project targets

Costs will increase if the project lead time is too short or too long. There will be a growing number of disruptions and disputes especially if the lead time is too short, or is shortened on repeated occasions. This scenario leads to higher construction and project costs, and thus increased costs of use. While the client may have initially drafted a sound construction contract package, this contract may be unsuitable for subsequently solving each and every problem because the associated project management, planning and bidding, and process organization fails to meet the requirements. The crux of the matter is that it is always people who plan, organize, and coordinate and who will require an adequate time budget for achieving a normal productivity level. For it is the existing "project pressure", in combination with the defined exceedingly tight construction period and budgeted hours, that often makes it impossible to incorporate important coordination steps and to provide guidance for interaction between the individual disciplines.

It is very difficult to quantify the additional costs incurred by an exceedingly short project lead time *ex ante*. Associated with this is the big problem of having to "discount" these costs to the reference point in time when applying the net present value method. Nor is it possible, in this setting, to compare the total cost of normal and short project lead times.

There will also be a cost increase if the project lead time is too long because of required additional planning and organizational activities that will not lead to a further improvement of the final project outcome, including considering new rules and guidelines, changes in the client structure, or amendments to the financing agreements. Quite on the contrary, an exceedingly long lead time may result in a lower quality standard, for instance by overstretching the planning assumptions.

The production system is the focus of our attention during the project execution phase. The achievable productivity level is essentially determined by the efficiency and effectiveness of the production system.

8.3 Productivity and Production System

Productivity is a key performance indicator in construction management and economics. It constitutes a crucial parameter for planning, costing, and implementing construction projects. It is used for evaluating the effectiveness of individual work steps or of the entire production or economic process and is expressed by the output/input ratio.

8.3.1 Production System

Besides the elementary production factors (i.e. labor, equipment, and materials), total productivity is also influenced indirectly by the discretionary production factors, including activities related to planning, management, control, selection, organization, communication, documentation, and knowledge work (see the illustration of the production system in Chapter 1, Fig. 1.2).[3)]

In this context, it is crucial to ensure the appropriate selection and combination of the production factors. These factors have a direct influence on the consumption and performance rates, and thus on the productivity level. These values also determine if the "normal" productivity[4)] level can be reached. The available resources must be utilized to a greater extent in order to compensate the generated losses of productivity if the specific

---

3) Cf. Hofstadler (2014), p. 13

4) The average "normal" productivity refers to the minimum human productivity that can be achieved on average in the long term by, and expected of, any competent worker after sufficient training and familiarization without harming their health if complying with the specified time for personal needs and recovery, if applicable. Note: Derived from the REFA definition of normal performance. Cf. Hofstadler (2014), p. 67

production factor limits are not adhered to. We should adopt a holistic view of the production factors in order not to optimize only one or a few of them at the expense of others.[5)]

The way in which production factors can be combined cost-effectively essentially depends on the type, shape, and complexity of the building or structure as well as on the conditions governing work performance. In addition, the achievable total productivity is determined by the required quality features and quantities, construction time, the site environment, and – last but not least – by prevailing weather conditions.[6)]

For any building or structure, the effects on the production system should be assessed in a more or less detailed analysis depending on the project phase. Fig. 8.3 shows major aspects with an influence on productivity.

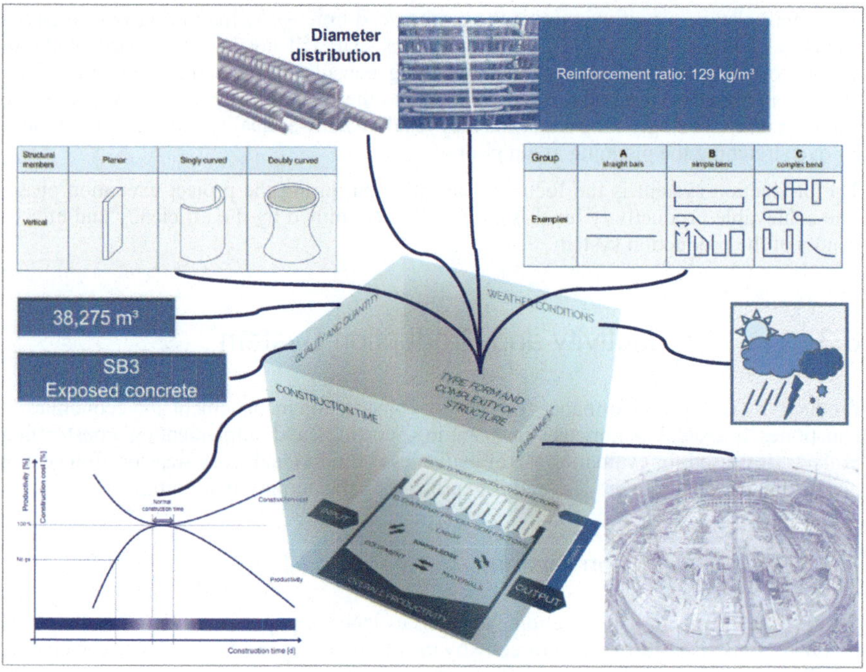

Fig. 8.3 Assessment of the production system with respect to external effects

In relation to the elements of the external production system, the following questions need to be answered when preparing the pricing of the bid (non-exhaustive list):

- Construction time: Has a "normal" construction time been specified or not? If not, how significantly does the specified time exceed or underrun the normal duration?
- Project environment: How should the project environment within the site mobilization area and the adjacent storage and handling areas be assessed?
- Weather conditions: Are critical operations, such as pouring exposed concrete, scheduled for unfavorable seasons?

5) Cf. Hofstadler (2014), p. 13ff.

6) Cf. Hofstadler/Kummer (2015a), p. 52f.

- Quality: Which requirements exist for concrete surfaces, for example Class SB3 specifying exposed concrete?
- Quantities: Which quantities need to be installed (for instance insertion of a total reinforcement quantity of 5,000 tons)?
- Building type, form, and complexity: What does the building layout and design of its floors look like? How regular/standard or irregular/non-standard are the structural components and/or building sections? How high is the reinforcement ratio, and which bend shape distribution should be expected?

Both the elementary and the discretionary production factors (i.e. their type, number, and combination) must be documented for the purpose of evaluating, assessing, and analyzing the achieved (or achievable) total productivity. In so doing, we should always establish relationships in terms of time, space, and intensity.

8.3.1.1 Internal Production System

The internal production conditions refer to the isolated consideration of the type, number, and combination of elementary production factors. Such analyses focus on the workers, equipment, and materials needed to perform the specified works. It is absolutely necessary to coordinate the associated time, work area, and work intensity so that the thresholds to losses of productivity are not exceeded. The discretionary production factors are crucial for ensuring the appropriate selection and functioning of these components.

The internal production system is closely interlinked with the external production system.

8.3.1.2 External Production System

The external production conditions influence the internal system and constitute a key determinant of the achievable productivity. The contractor is not in a position to freely choose these conditions; they can influence them only under certain circumstances (e.g. by prefabricating elements in a factory building). The internal production system should be adapted in the context of these defined conditions. The provisions agreed upon in the construction contract identify the party/ies by which the costs of these execution-phase adjustments shall be borne. Otherwise, this matter may also be the subject of a dispute between the parties.

The external production system comprises the type, form, and complexity of the building, weather conditions, specified qualities and quantities, and the project environment, as well as the specified construction time. The latter should be regarded as the determining factor for selecting the type, number, and combination of production factors.

8.3.1.2.1 Construction Time

The productivity level is strongly influenced by the construction time. The balance between the production factors will be disturbed in the case of a deviation from the "normal" construction time, particularly if the construction time is too short. Likewise, this balance will also be disturbed if the construction time is too long, albeit to a lesser extent. The following sections describe and discuss the influence of construction time in more detail.

8.3.1.2.2 Project Environment

The project environment constitutes the context within which a project is being planned and implemented. It has a strong influence on the project and also includes interactions.

In addition, the ICB - IPMA[7] (Competence Baseline) describes that the project environment performs a determining and assessing function, which means that it is both the origin and the assessing entity of the project. The ICB mentions a number of factors determining the interactions between the project and its environment. More specifically, these include physical, ecological, social, psychological, cultural, political, economic, financial, legal, contractual, organizational, technological, and aesthetic factors.

A project environment analysis is performed in order to define the specific project environment. This analysis forms part of the specifications for the bidder/contractor.

8.3.1.2.3 Weather

Weather conditions include the general, average, or prevailing nature of the weather in a specific place over a period of several days or weeks. This approach particularly considers tangible elements, such as precipitation, temperature, wind, and humidity.

Weather characterizes the state of the atmosphere in a certain location and at a defined point in time. Key determining factors include the meteorological components of radiation, atmospheric pressure, air temperature, and wind, as well as the derived elements of cloudiness, precipitation, and visibility. Weather is a snapshot of a process (i.e. changes over time) that primarily takes place in the troposphere. It may change several times a day.

Any detailed analysis of weather effects needs to rely on weather data gathered over a certain period.

8.3.1.2.4 Quality and Quantity

As a determinant of a construction-related production system, quality refers to the execution of the project in technical and aesthetic terms. There is substantial evidence that any increase in the work speed caused by a specified short or very short construction time will also lead to an increase in the error rate. This phenomenon is due to the higher work speed, but also to the fact that more workers will be required. Thus, there will no longer be an optimal alignment of the workforce in both quantitative and, even more importantly, qualitative terms.

Quantity influences the extent to which familiarization effects materialize as well as the average performance required for adhering to the contractually agreed construction schedule and project milestones. Sourcing terms will also essentially be influenced by the required quantities.

8.3.1.2.5 Building Type, Form, and Complexity

When considering the type of the building or structure, we need to distinguish between, for instance, multi-story buildings, engineering structures, and industrial estates. Specific

[7] International Project Management Association

effects result from the type of the structural components, their dimensions, and the design of the structural framework. In buildings constructed with reinforced concrete, the formwork ratio and the reinforcement ratio have a major influence on the achievable productivity level.

In multi-story buildings, the degree of repetition of individual construction sections, the spans of the structural framework, and the axial grid of the building are of significant importance. In addition, the heights of the structural components or of the entire building require particular attention in this context. As a rule of thumb, we can assume the amount of labor required for the construction work to grow in line with the increasingly delicate architectural arrangement and confined site situation, for instance owing to a comparatively high formwork ratio reaching over 5 m^2/m^3 in building construction.

All project stakeholders will have to demonstrate in-depth, networked knowledge in the fields of structural design, formwork and concrete technology, and construction management particularly if sophisticated exposed concrete requirements are specified for individual structural components or for the entire building. In this setting, it is also crucial to ensure a high degree of flexibility in arranging construction joints. The greater the freedom of choice contractors enjoy in defining the size and number of work sections, the more efficient and effective the construction process will be. Monolithic construction requires more time and effort than a division into several sections because concealing joints within the formwork systems is very equipment- and labor-intensive, and thus expensive.

For the contractor, section-by-section completion will be more viable than a monolithic approach. The former is the preferred method because it requires less time and money. In terms of time, this is due to the fact that the individual sections are smaller, which renders the construction process more efficient. In addition, the familiarization effects brought about by a large number of work sections have a positive influence on time, quality, and cost. Monolithic construction is associated with limitations in terms of architectural and structural design as well as site management and construction economics. Designers should be aware of these limitations already when preparing their drawings in order not to give the wrong impression of what is feasible and commercially viable.

The aspect of quality is also governed by the dimension of monolithic construction. The larger the concreting sections in the construction process, the lower the accuracy with which the combination of production factors can be observed in the formwork. Monitoring the pouring of concrete into the formwork and its spreading and compaction is crucial for ensuring the quality of the finished product. Quality control will be particularly difficult in work sections with heights in excess of 3 m and inclined structural components, which is why work can be supervised only to an insufficient extent in these areas. This shortcoming results in a less uniform and thorough concrete distribution and compaction, with an immediate adverse effect on the end result.

It is necessary, for assessing the degree of difficulty of the production system, to thoroughly consider as many key characteristics of the building as possible in order to be able to estimate the related effects as realistically as possible as early as at the preliminary pricing stage. The project information gathered in this early phase will become binding and contractually relevant in the event of a subsequent contract award. Any later assessment of the pricing risk taken by the contractor will thus have to include a verification of the factors that the contractor had to take into account on the basis of the bid specifications, information provided by the client, plans and drawings, etc. Particular attention should be paid to the plans and drawings for the purpose of assessing the effects of the type, form, and complexity of the building and/or of its structural components.

Only the plans will reveal whether the structural components are arranged regularly or irregularly. Contractors will also be able to find out if component surfaces are planar, curved, or doubly curved, for instance. The drawings will also provide indications of any inclined support areas as well as clearances. In order to ultimately mitigate the pricing risk, the building specifications and plans need to be checked for conformity as part of the pricing exercise irrespective of the sequence of the contractual elements that form part of the tender. Plans and drawings should also be checked for potential inconsistencies between the specifications and the planning documents as part of the bidder's precontractual duty to warn and inform.

8.3.2 Labor Consumption Rates

The labor consumption rate is the single most important indicator for preparing forecasts in construction management and economics as well as assessing labor-intensive activities. This metric establishes a direct link to the requirements that can be derived from the production system.

Labor consumption rates are used as key parameters for estimating and calculating costs as well as for determining construction times. The uncertainties and ranges of input parameters can be reduced as the knowledge of building data and information and of their effects on the potential behavior of the production system increases.

Any in-depth bid assessment will have to thoroughly consider labor consumption rates, particularly for labor-intensive types of work. The wage component is derived from the product of the mean wage price and the consumption rate for all unit price contracts. More specifically, disclosure of the costing form used for calculating the mean wage price makes it possible to divide the wage component by the mean wage price to obtain the labor consumption rate applied by the bidder to the specific work item. If these calculated labor consumption rates result in a very small number of paid working hours per unit of quantity [wh/UQ], this is potentially indicative of excessively low wages and/or social dumping. However, this may also be due to the inappropriate inclusion of pricing components in a different item. There are process-related and technological minimum thresholds for labor-intensive activities that cannot be underrun (serving as a measure for "substandard" labor consumption rates). In a project-specific analysis, labor consumption rates will generally be much higher than these lower limits owing to effects stemming from the categories of weather, environment, quality and quantity, project environment, and type, form, and complexity of the building.

8.3.2.1 Calculation

Relative to the labor productivity, the labor consumption rate $LCR_{ps,p,c}$ [wh/UQ] is a key indicator for pricing and scheduling labor-intensive types of work. In a construction management and economics context, labor consumption rates are used for calculating wage costs as well as performance values and construction times. We obtain the consumption rate by dividing the sum of paid working hours $PWH_{ps,p,c}$ [wh] by the production output $PO_{ps,p,c}$ [UQ] (see Eq. (8-1)).[8)]

[8)] Cf. Hofstadler (2014), p. 46f.
Cf. Hofstadler/Franzl (2011), p. 129
ps – process stage, p – process, c – component

$$LCR_{ps,\,p,\,c} = \frac{PWH_{ps,\,p,\,c}}{PO_{ps,\,p,\,c}} \qquad (8\text{-}1)$$

The magnitude of labor consumption rates influences labor productivity and thus essentially the performance generated in the individual process steps and sections. Labor productivity is influenced by the normal performance that the workers can achieve for the types of work under study. When relating the concept of normal performance to labor consumption rates, we can refer to a "normal consumption rate" in this context.

In the bid phase, labor consumption rates create the main basis for calculating cost and time. At the process planning stage, labor consumption rates form an integral part of calculating the duration of the individual process steps, and thus of the entire production process and associated resource utilization. Data taken from the construction phase is recorded systematically, for instance for final costing and site controlling purposes. After capturing the total hours and specific production performance values for a specific type of work, we use Eq. (8-1) to calculate the consumption rate of an activity (or process).[9)]

The time required for a specific task depends on a large number of influential factors, including variances in the performance of individual workers, type and condition of utilized equipment, materials used, workplace conditions, and organizational quality. The intensity of these influences may change over time, and they may also interact with each other. This means that the time required for a specific type of work will vary from project to project, even under identical working conditions.[10)]

From the contractor's point of view, influences arising from the internal and external production systems should always be considered when calculating labor consumption rates in the bid phase. The final value should always be established with a view to assessing the chance/risk ratio. If an exceedingly low value is chosen, the baseline and thus the potential chance of achieving or even underrunning this value at the subsequent execution stage will also be very low.

9) Cf. Hofstadler/Franzl (2011), p. 129

10) Cf. Spranz (1975), p. 5

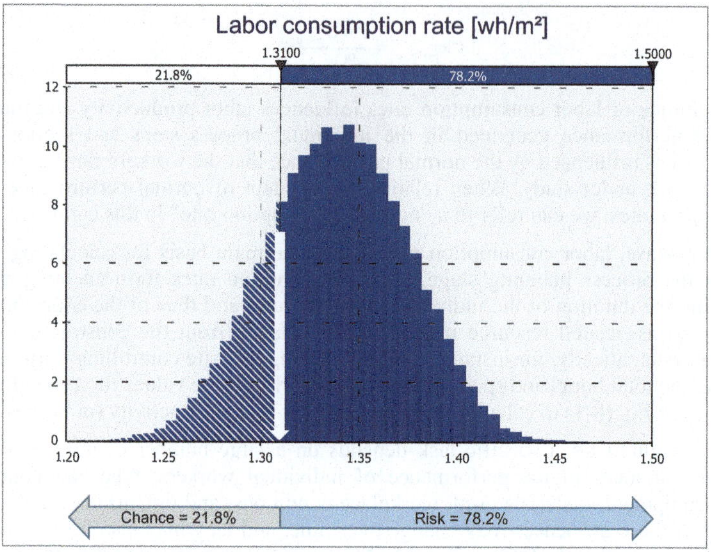

Fig. 8.4 Histogram for the mean total labor consumption rate of formwork operations – Selected consumption rate with corresponding chance/risk ratio

To illustrate this relationship, we refer to the histogram that the bidder derived in its (zero-base) costing exercise for the consumption rate of formwork operations in a building construction project (see Fig. 8.4). If, for example, a consumption rate of 1.31 wh/m² is chosen, there is a chance of only approximately 22% for the bidder to reach or underrun this value during the execution phase. In about 78% of cases, the applied assumptions (not explained in this section) result in a higher mean labor consumption rate of formwork operations. The bidder (or subsequent contractor) would thus run an increased risk if they adopted this approach, but would still be able to prove the plausibility of the bid from a construction management point of view.

To get a contract in the first place, bidders inevitably prepare a tight cost schedule and thus select labor consumption rates associated with low chance/risk ratios. In the project execution phase, such exceedingly tight cost schedules require maximum efficiency when combining the production factors.

8.3.2.2 Differentiation of Labor Consumption Rates

When calculating and using labor consumption rates, these must be differentiated according to the item considered, i.e. wage costs versus construction time, and the scope of work needs to be delimited accordingly.[11)]

Hoffmann/Motzko/Corsten point out that a distinction should be made between the consumption rate related to the required time (indicative (piecework) consumption rate) and the indicative rate used for the costing exercise.[12)]

[11)] Cf. Motzko/Boska (2013), p. 18

[12)] Cf. Hoffmann/Motzko/Corsten (2012), p. 23

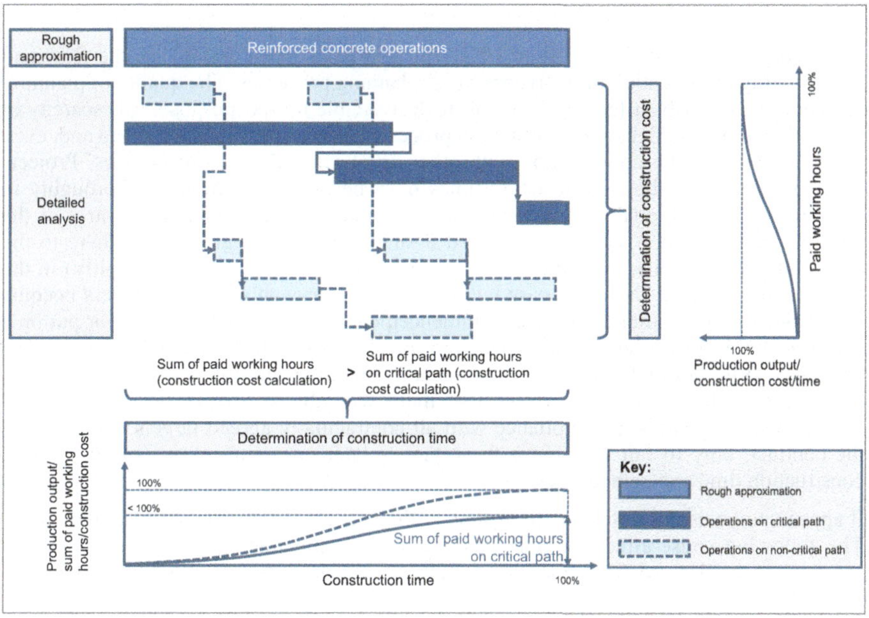

Fig. 8.5 Labor consumption rates – Differentiation according to degree of detail[13)]

Accordingly, we basically need to distinguish between labor consumption rates used for calculating the construction cost and labor consumption rates that are essential for determining the construction time (see Fig. 8.5). In the costing exercise, the total paid working hours expended for all activities of a contractually agreed work must be captured, irrespective of their relationships or time offsets. When determining the construction time or preparing the project schedule, the focus needs to be on the time expended for the individual activities along the critical path. In this context, offsets between such activities must be taken into account. For the purpose of work scheduling, it is essential to consider the critical path that would impact the final completion date in the event of any delay. Off-peak hours[14)] worked that are unrelated to the critical path only play a marginal role for work scheduling. This is why the resulting consumption rate for scheduling (less times along the non-critical path) will be lower than the rate determined for activities along the critical path.[15)]

[13)] Cf. Kummer (2015a), p. 60

[14)] "Off-peak hours: hours necessary for preparatory and final works or for work steps repeated several times (for instance cleaning the formwork before its reinstallation). Such hours are a portion of total productive hours. When calculating labor consumption rates (e.g. in wh/m²) or applying them, care must be taken to verify if off-peak hours have actually been considered as a portion of the total labor consumption rate." Oberndorfer/Jodl et al. (2001), p. 128 (translation by Steffen Walter)

[15)] Cf. Kummer (2015a), p. 59

8.4 Influence of Construction Time on Planning Processes

If the construction time is too short, this will also have repercussions on the flow and susceptibility to disruption of the (supporting) planning processes. The quality of planning and coordination will suffer if less time is made available for this purpose. Any scarcity of time will massively affect the coordination process within a specific work group and, even more importantly, the coordination with other involved trades and disciplines. Projects with specified very short construction times must be prepared extremely thoroughly in order to ensure high-quality structuring and subsequent management and control of the planning and execution interfaces. Even if a single designer is unable to adhere to the agreed project schedule, disruptions will begin to pile up and aggregate (amplify) in the construction process. The worst case will occur if one or even several designers become absent due to illness after the project commencement date. This will obstruct, or put on a complete hold, all designers at or near the interface as well as the dependent trades. Clients need to thoroughly analyze the pros and cons of specifying a short or very short construction time. Experience shows that there will almost always be designers and contractors that guarantee compliance with all contractually agreed targets upon signing the contract, only to fail to achieve these targets later on owing to the specified short construction time and agreed prices.

It appears to be useful to take a closer look at the past for the purpose of planning ahead. The following cause-effect relationships and recommendations for the desired project success can be derived from completed and ongoing projects and should be taken into account.

The shorter the construction time,

- the greater the project target underachievement will be (particularly in the case of poor project preparation)
- the poorer the planning and/or tender quality will be (resulting, for instance, in losses of productivity not compensated by additional cost claims)
- the earlier designers need to coordinate their works and services, and the better the organization and management of interfaces needs to be
- the larger the number of meetings to be held per unit of time
- the shorter the meeting and thus decision-making intervals
- the shorter the time left for decision making will be, and more decisions need to be made within a shorter period
- the higher the degree of project planning completeness needs to be at the beginning of the bid phase
- the greater the degree of detail of project management required even prior to commencement of construction
- the more thoroughly the soil and the building (in the case of conversions of or extensions of existing buildings) will have to be surveyed
- the more susceptible the project will be to disruptions and additional costs as well as quality deficiencies
- the shorter the response and processing times will be in the event of planning and design errors
- the lower the resilience to construction process disruptions
- the larger the number of processes or work steps will be that have to be run simultaneously
- the shorter the time left for corrective actions will be, and the less effective these actions will be
- the higher the invoiced total project costs will be

8.5 Significance of Construction Time

Construction time influences the project in many different ways. This parameter is relevant not only in terms of construction management and economics but also from a legal point of view. Construction time influences both the total quality and the individual quality features as well as the number of disruptions to the construction process, and thus additional cost claims (see Fig. 8.6).

Generally speaking, sufficiently long periods must be scheduled for preparing bids and construction activities and for actual work performance on the job site. Additional factors to be considered when determining normal construction time include seasonal effects and specific site conditions as well as other influential factors and predictable difficulties.

The combination of a very short construction time, high building quality, and building under unfavorable seasonal conditions is particularly problematic. This combination is the perfect recipe for a project disaster.

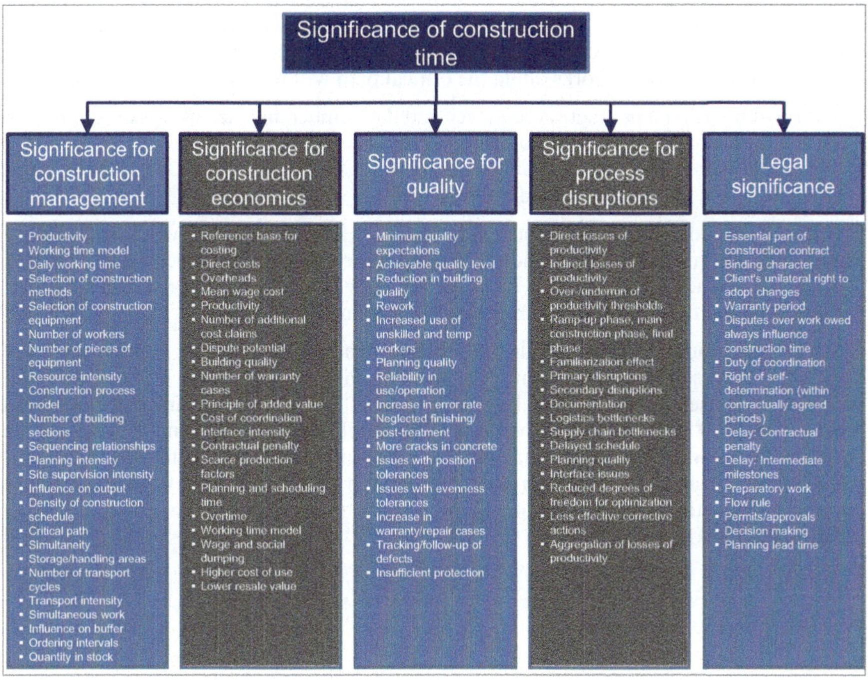

Fig. 8.6 Significance of construction time for planning and construction processes (non-exhaustive list)

8.5.1 Significance for Construction Management

The following section discusses the influence of the construction time on productivity, the construction process, logistics, and site mobilization.

8.5.1.1 Productivity

If the construction time becomes increasingly shorter, the productivity thresholds are either exceeded or underrun, resulting in losses of productivity. Losses also occur during longer construction periods, albeit to a lesser extent because more time can be invested in work, or the proportion of non-productive and distribution times increases.

8.5.1.2 Construction Process

During a normal construction period, the number of simultaneous processes can be limited to the required minimum. Activities taking place simultaneously are planned where it makes sense in terms of production and technology. The shorter the specified construction time, the more sensitively works along the critical path will respond to disruptions.

The three-dimensional production sequence and the number and size of work sections also have a significant influence on labor productivity. Productivity benefits will be generated if the production factors or parts thereof are implemented several times at the same working level (in the sense of the same floor level).

Productivity can only reach a normal level if the workflow is designed such that downtime is reduced to a minimum. In addition to planning the construction process and logistics, this includes an adequate design of the work sections in terms of their size and sequence, for instance, and the appropriate combination of production factors.

If the construction time for a building is reduced steadily, the resource requirements will inevitably increase, as will the number of workers to be deployed per unit of time. This will also increase the need for a larger number of work sections and more generous working space. If there is too little working space available, the work cannot be carried out in the usual workflow, and mutual obstructions will occur. For extensive horizontal components, an average minimum work area per worker is necessary for achieving a “normal” productivity level. For flat vertical components, a minimum working length per worker is required for this purpose.

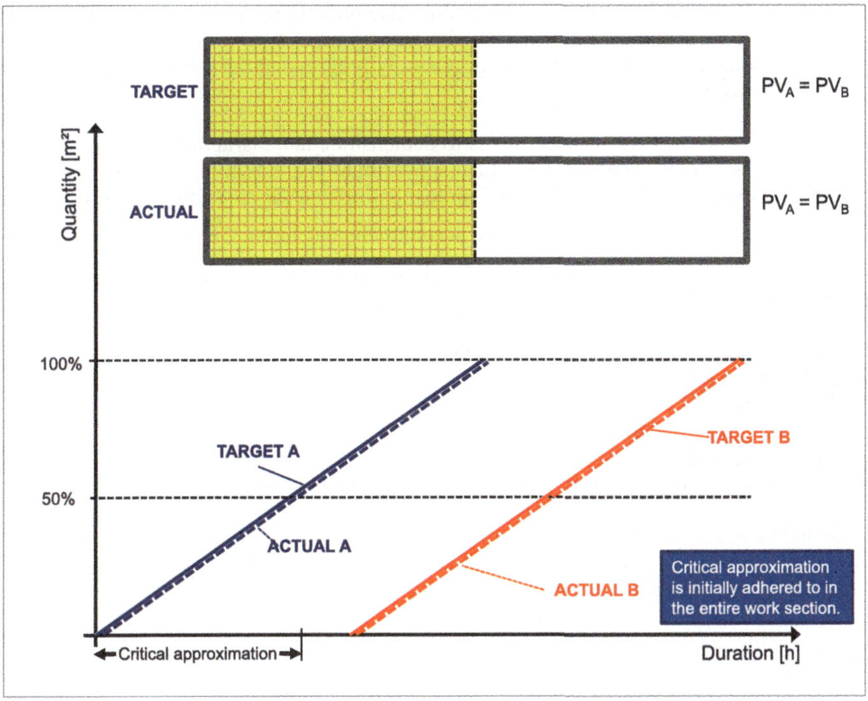

Fig. 8.7 The floor is constructed in two work sections: Example of complying with the critical approximation between two processes

Between operations, care must be taken to ensure that the critical approximation is adhered to (rather than underrun) because, otherwise, the production processes will be mutually obstructed. Fig. 8.7 shows a floor slab that is constructed in two work sections. The area is plotted on the ordinate, and the duration in hours [h] is shown on the abscissa. Work starts with the slab formwork installation process. A sufficiently large formwork area must have been installed before the subsequent reinforcing process can start, in line with the logic of an economical site operation. This minimum gap between the start of the two processes is essentially determined by the number of workers, the minimum work area per worker, and the respective performance values. The working speed and thus the gradient of the two straight lines depends on the performance values. This is where both straight lines ($PV_A = PV_B$) have the same gradient and thus indicate the same working speed. It is obvious that the target performance is identical to the actual performance. This is the ideal scenario that is so difficult to achieve. It is not optimal that no distance/time lag to the critical approximation is considered in order to integrate a buffer for possible disruptions. In any case, planning a production sequence without buffers is not advisable because disruptions can result from the internal and/or external production conditions.

For instance, disruptions due to mutual obstruction occur if

- the installation of the reinforcement is progressing faster than the installation of the formwork, thus underrunning the critical approximation

- the installation of the formwork is progressing more slowly while the reinforcing operations are completed according to plan, thus also underrunning the critical approximation
- the installation of the reinforcement is generally progressing faster than the installation of the formwork

Irrespective of the type of activity, this is to demonstrate that disruptions will occur if the second process already starts at the critical approximation and is completed more quickly. Likewise, attention must be paid to the implications arising for the subsequent process/es.

The following aspects must be taken into account when planning the workspace:

- Safe working should be possible in all phases (formwork installation and stripping, cleaning, revitalization, rebuilding, relocation, etc.)
- Occupational health aspects
- Critical approximation to the precursor/s
- Critical approximation by successors
- Thresholds from which losses of productivity will occur
- Technological aspects
- Production-related aspects
- Dimensions of the equipment and materials to be manipulated by the workers

It should be noted that any increase in the number of workers will generally also result in a growing number of work sections and thus of construction joints, which will, in turn, increase the required amount of labor.

8.5.1.3 Logistics

The intensity of transport will increase as construction time becomes shorter. This results in more transport cycles being necessary per working day because the number of transport processes is spread over fewer days. In addition, material bottlenecks during short construction periods (for instance if more workers are affected by waiting times) have a much more negative effect on the entire construction process than in the case of a normal construction time.

The following cases always resulting in increased transport intensity may occur if the construction time becomes shorter:

- The total number of transport processes remains largely constant.
- The total number of transport processes becomes larger, for example because the work sections become smaller due to the shorter construction time, leading to an increase in the number of work sections.

8.5.1.4 Site Mobilization

The shorter construction period necessitates larger storage and handling areas. If the number of workers also increases, additional site containers, sanitary facilities, parking spaces, etc. must be provided on the construction site. Bottlenecks in logistics and site operations may occur and adversely affect total productivity, particularly where space is confined, as is often the case in inner-city construction projects. Short construction times may also make it necessary to install more pieces of equipment in the site mobilization area. For example, any increase in the number of workers will also require a larger number of cranes if the available crane capacity is to be retained. Alternatively, other construction methods or types of lifting equipment should be used in order to compensate for a possible crane capacity bottleneck if construction time is too short.

8.5.2 Significance for Construction Economics

The type of work, the circumstances in which the work is performed, the quantity and quality as well as the construction time are factors that determine the costs and, consequently, prices. By specifying the construction time, the client is able to influence costs, and thus pricing. Unduly short construction times (with maximum values for the productive use of equipment and/or labor exceeded) lead to losses of productivity and thus to higher costs. In some cases, the thresholds for losses of productivity are not yet universally known or still controversial.

If normal construction times are specified, contractors will try to combine their production factors in the best possible way so that they can perform the work owed under the contract at the lowest cost for the earmarked budget.

If there is a high degree of simultaneous work, which is necessary with short construction periods, the coordination effort increases because several resources are provided at the same time. This can subsequently lead to a reduction in productivity. For the client, the effect on the project costs results from higher bid prices and, potentially, a larger number of additional cost claims by the contractor. On the one hand, the reduction in the flexibility on the market and, consequently, productivity is caused by using a large number of permanently employed workers or resorting to temporary or auxiliary staff.

A short construction time also tends to prompt contractors to move towards excessively low wages or social dumping because the works, or portions thereof, can no longer be performed economically by their in-house workforce. This is how contractors try to avoid imminent losses on the construction site caused by bid prices that are far too low. Illegal wage and social dumping is thus committed in order to increase the profit potential, but this is obviously no excuse for this activity.

8.5.3 Significance for Quality

Short construction times generally have an adverse effect on quality because the production factors will have to be combined very quickly or, in many cases, even too quickly. This situation can result in problems due to underrunning technological limits, thus leading to quality losses. Certain materials need time (for instance for setting or drying) to develop their required characteristics such as strength, bond, or bearing capacity. In addition, in the event of additional staff requirements, contractors will have to resort to partially unskilled temporary or auxiliary staff who do not possess the same level of training as the permanently employed workers. As a result, processing errors can occur more frequently.

8.5.4 Significance for Construction Process Disruptions

In the event of disruptions to the construction process, the contractual distribution of chances and risks must be assumed with regard to the consequences (additional costs and/or longer construction period). It is necessary for each of the parties to the contract to deal with the causes from their own perspective. The contractor will usually submit additional cost claims to the client. Even where such claims are justified, there is usually disagreement about their monetary scope.

If the contractor substantiates their losses of productivity by their records, such as daily work reports, the client will often suspect that, for instance, the effects of pricing errors, insufficient process planning, or the selection of inappropriate construction methods (for which the contractor bears responsibility) are also hidden in such claims. The contractors themselves and/or their experts will try to prove the price adequacy of their claims by means of model calculations and by referring to methods discussed in the literature.

8.5.5 Legal Significance

Depending on the specific provisions in the contract, the client's right of instruction is limited to the work owed under the contract. If the need arises subsequently to change the work in order to achieve the envisaged outcome, the work thus changed may become either more expensive or cheaper. In addition, the work can be performed either more slowly or more quickly. Depending on these variables, the individual parties to the contract will be entitled to either an adjusted monetary compensation or to a change in the specified construction time. In this context, it is also conceivable that the work may be more costly, but can be performed more quickly. In this case, the contractor would, for example, be entitled to a price adjustment, whereas the client would be in a position to justify a time adjustment of the contract.

Case law also establishes a relationship between construction time and productivity. This can be deduced from the following court judgment: In the case of a construction contract, the presumption of a balance between work performance and valuable consideration does not apply irrespective of the period agreed upon for work performance because the latter will usually have an influence on the agreement on the amount of the contractor's remuneration (see BGH, ruling of April 15, 2008 X ZR 129/06, NZBau 2008, 505 = ZfBR 2008, 614).[16)]

In this respect, we refer to the requirement of OENORM B 2061[17)], which stipulates that the cost consumption based on the predicted type and form of the combination of production factors must correspond to the objectively and economically justified value input. As an essential parameter for work performance, the construction time will in any case have a major influence on the loss of value.

In addition, OENORM A 2050[18)] specifies that all circumstances (e.g. local or temporal circumstances or special requirements with regard to the way the work is to be performed) that are important for the execution of the work, and thus for the preparation of the bid, must be listed and specific exemptions or impediments be described. This standard also provides for the construction time to be regarded as an essential criterion for the performance of the work.

---

16) Cf. Gölles/Link (2011), p. 220

17) OENORM B 2061:2020, p. 7

18) OENORM A 2050:2006. p. 16

8.6 Specification of Construction Time

The construction time is usually specified by the client. Contractors use this information to derive the utilization of production factors required to fulfill their contractual duties. Labor consumption rates serve as a baseline indicator for all labor-intensive activities; they are subsequently used as key input variables to calculate performance values. Bidders use labor consumption rates as an important basis for their costing and final pricing.

The way in which the type, number, and utilization of production factors evolve over the construction period is of crucial importance for productivity. On the one hand, this essential question arises for the client in the project preparation phase. On the other hand, this question is also of relevance for bidders, or subsequent contractors, for executing, pricing, and invoicing the construction work. The client is well-advised to thoroughly consider the sensible use of production factors in order not to jeopardize its project targets in terms of quality, completion dates, costs, etc. As a matter of course, clients are free to neglect this issue and to rely on a comprehensively worded construction contract. However, this approach might be associated with major disadvantages in terms of the project outcome (including cost overruns, quality issues, and construction delays) because, obviously, any building or structure will always be constructed not by virtue of a mere contract but by an appropriate combination of production factors. It is the purpose of the construction contract to provide an appropriate framework for performing and billing construction work as well as rules for resolving potential conflicts. It would thus also be advisable for the client to deal with the change in productivity and production factors over time. Bidders and contractors are bound to do this anyway because they would otherwise not be in a position to prepare a source-based costing exercise and to write the "optimal screenplay" for executing construction works (including spatial, temporal, and intensity-related progress of work steps). In this context, the following question needs to be answered: Which trend should be assumed with regard to productivity and production factors during the total construction period?

There will always be variations in productivity because it will not remain constant over the entire period to project completion.

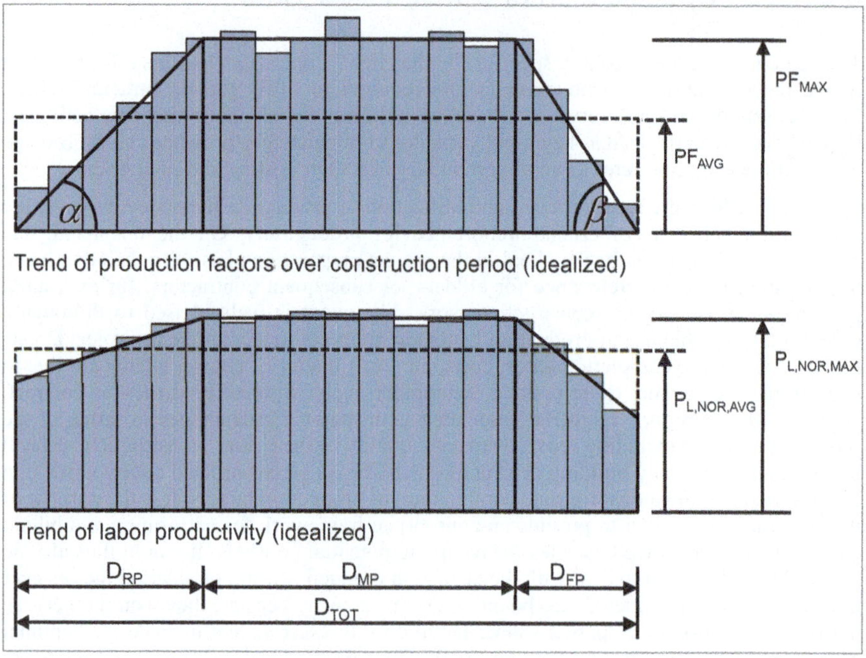

Fig. 8.8 Development of production factors and productivity over the construction period[19)]

When considering productivity, the various construction phases in the execution of the work must be taken into account (see Fig. 8.8). We basically need to differentiate between:

- Productivity during the ramp-up phase (D_{RP} = duration of ramp-up phase)
- Productivity during the main construction phase (D_{MP} = duration of main construction phase)
- Productivity during the final phase (D_{FP} = duration of final phase)

Disruptions to the construction process have a particularly detrimental effect during the ramp-up and final phases because site operations are not yet, or no longer, able to rely on an optimal coordination of resources.

8.6.1 Significance of Construction Time for the Construction Contract

By specifying the construction time, the client exerts a direct influence on costs, and thus on construction prices. Construction times that are too short lead to losses of productivity and to higher costs. In the short term, higher costs are incurred, for instance, if the planning-ahead process does not work as intended or if other circumstances related to work performance (within the scope of the client) have changed.

---

19) Cf. Hofstadler (2014), p. 14

At a later stage, this situation will lead to issues caused by poorer structural quality as well as medium to long-term obstructions of use due to changed maintenance and repair scenarios.

The thresholds for losses of productivity are sometimes not universally known or are discussed controversially in the literature (including threshold values, depth of effect, width of effect, contractual relevance, among other factors).

Provided the client specifies normal construction times, contractors are in a position to optimally combine their production factors so as to perform their work owed under the contract at the lowest possible construction cost while adhering to the allocated budget (see Fig. 8.11). There is thus a lower risk of non-compliance with the specified construction time and cost as well as the agreed-upon quality standard.

When determining construction time, the client usually considers limiting boundary conditions with an influence on site operations and management only to an insufficient extent, or not at all. These limiting factors result, for instance, from structural and operating conditions and have a major influence on the parameters that determine work performance, such as the maximum number of available and deployable workers or the maximum number of usable pieces of equipment.

A major issue in the context of calculating construction time is the linear nature of currently applied methods. In most cases, linear construction management-related relationships are used for the calculation process although these do not reflect reality at all. Any over- or underrun of thresholds or limits will lead to losses of productivity when capacities increase or decrease; these losses have a significant influence on calculation results. To accurately determine construction time, the various influences of productivity, such as limited available space for construction equipment, on production factors must always be considered.

Hruschka stresses that any calculation or optimization of construction time must consider the mutual obstruction of the existing potential[20)] to prevent unrealistically short construction times. The construction time of a building project cannot be shorter than the working capacity of the maximum installable transport equipment allows.[21)]

Blecken notes in his comments that the assumption of a deterministic production process would oversimplify the production model. Linear methods are chosen although most relationships of the combination of production factors do not follow a linear pattern. He considers the additional application of stochastic methods to be a key tool for achieving significantly improved outcomes.[22)]

8.6.1.1 Determination of Construction Time

In the macro-analysis, the construction time is calculated on the basis of the indicative quantity and the type of work determining the construction time. In building construction, for example, the calculation approach initially uses the gross cubic volume and then proceeds to using the cubic meter of concrete as the governing unit for the construction of the building shell. From each calculation we get exactly one value for the duration. When applying the deterministic calculation method, no conclusions can be drawn with respect to the chance/risk ratio linked to the respective calculated value. On the micro-level, the

20) Potential refers to the production factors.

21) Cf. Hruschka (1968), p. 165

22) Cf. Blecken (1976), p. 208

construction time results from interlinking processes through establishing logical sequencing relationships relative to the critical path. Realistic quantities and performance estimates are required in order to calculate the individual process times. As with any determination of values, the input data is subject to uncertainties. In most cases, linear dependencies between the production factors are also assumed, but this does not always correspond to the reality in construction and site management. Fig. 8.9 shows a qualitative representation of the relationship between productivity and construction time. It also schematically illustrates how the construction costs change when moving outside the range of the normal construction time. In this normal construction time range, proportional relationships between the production factors can be assumed since losses of productivity should not be expected yet within this range. Outside this area, losses of productivity will occur when combining the production factors. This means that nonlinearities will emerge, and the uncertainties of the values will continue to increase.

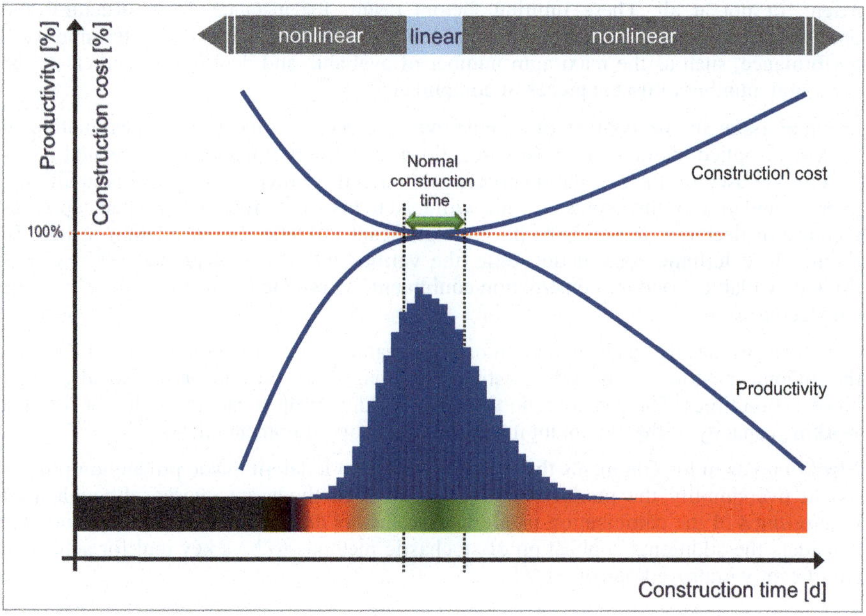

Fig. 8.9 Modeling of the construction time – Linear relationships between the production factors prevail in the range of the normal construction time (the construction time is in the "green area")[23]

Site operations will respond to any changes in the production factors in a linearly elastic (proportional) pattern until moving outside the range of normal productivity. The behavior of the construction process will then change to a plastic (non-proportional) mode. This will cause losses of productivity and thus higher costs. This is justified by the fact that a change in the number of resources will usually not result in a proportional change in performance. Instead, performance values will decrease owing to losses of productivity.

[23] Cf. Hofstadler (2014), p. 39

These uncertainties should be grouped into the aleatoric (inherent, physical) and epistemic (model-specific or statistical) categories. Epistemic uncertainties result from a lack of knowledge, inadequate or inaccurate modeling, or limited information and/or too few data sets, but such uncertainties can be reduced by providing additional data or information. More precise calculation methods can be identified if the works are specified in sufficient detail. Calculation of the construction time is usually based on a model that incorporates equations or spreadsheets.

A frequent epistemic error in the modeling of equations for calculating the construction time is that permanent linear relationships between the influential parameters are assumed. However, real on-site activities and processes provide a completely different picture. The interactions between the production factors are linear only within narrow ranges (where the construction time lies in the "green area", see Fig. 8.9). In general, nonlinearities between the production factors must be taken into account in order to arrive at accurate calculation models. The in-depth, contextual knowledge of the relationships between the internal and external production conditions makes a major contribution to reducing epistemic uncertainties.

In this respect, the type and combination (in terms of time, space, and intensity) of production factors is crucial. These factors have a direct influence on the labor consumption and performance rates, and thus on the productivity level.

Productivity is a key performance indicator in construction management and economics. It constitutes a crucial parameter for planning, costing, and implementing construction projects. It is used for evaluating the effectiveness of individual work steps or of the entire production or economic process and is expressed by the output/input ratio. Likewise, productivity plays a key role in forecasting costs and times. The shorter the specified construction times, the more workers and equipment that manipulate (install or remove) large quantities of materials simultaneously within this short period of time have to be deployed simultaneously on the construction site. As a result, the available working space for the production factors will become increasingly smaller, and coordination between workers and pieces of equipment will be moving outside the optimal range due to a productivity loss. In addition, the coordination and interaction between elementary and discretionary production factors will no longer be economical.

Besides the elementary production factors (i.e. labor, equipment, and materials), total productivity is also influenced by discretionary production factors, including planning, management, and control. These parameters also determine if the "normal" productivity level can be reached. The available resources must be utilized to a greater extent in order to compensate for the generated losses of productivity if the specific production factor limits are not adhered to. Due to existing interactions, we should adopt a systemic view of the production factors in order not to optimize only one or a few of them at the expense of others.

If the contractor is able to utilize its production factors at a "normal" productivity level on-site, we can assume that its pricing assumptions developed for this purpose will be associated with a high probability of occurrence.

8.6.2 Definitions of Construction Time

The concept of construction time designates the calculated construction time prior to considering any buffer (assumed by the contractor). After determining construction time and considering the boundary conditions in terms of construction management and economics, the client should additionally account for a buffer to be able to compensate for any potential delays within its scope of responsibility.

8.6.2.1 Differentiation of Construction Times

Construction times are categorized while integrating bottleneck equipment. In building construction, cranes belong in this category. The differentiation into construction time categories is based on the maximum number of workers who can be served by a single crane without loss of productivity.

According to *Hofstadler*, construction times can generally be differentiated as follows:[24)]

Extremely short construction time

Construction time is determined such that the number of workers and pieces of equipment to be utilized productively exceed the corresponding maximum values (i.e. thresholds to losses of productivity) by 20%.

Very short construction time

Construction time is determined such that the number of workers and pieces of equipment to be utilized productively exceed the corresponding maximum values by 10%.

Short construction time

Construction time is determined such that the number of workers and pieces of equipment to be utilized productively represent the corresponding maximum values. Any disruption to the construction process may immediately lead to losses of productivity when adhering to the construction time target.

Normal construction time

Construction time is determined such that the number of workers and the number of pieces of equipment to be utilized productively underrun the relevant maximum values by 10%.

Long construction time

Construction time is determined such that the number of workers and the number of pieces of equipment to be utilized productively underrun the relevant maximum values by 25%.

8.6.2.2 Calculation of Normal Construction Time – Example

In socio-technical systems, the achievable productivity in construction is determined by the optimal coordination of the workforce with the equipment. We consider the combination of the number of workers with the number of cranes to illustrate the concept of normal construction time. To highlight the interaction between the number of pieces of equipment and the number of workers, we apply the crane proportionality factor indicating the maximum number of workers (for formwork, reinforcing, and concreting operations) who can be served by a single crane without any loss of productivity due to insufficient crane capacity (see Table 8.1).

---

24) Cf. Hofstadler (2014), p. 57

Concreting method [-]	Crane and crane bucket for all structural components	Crane only for walls and columns	Crane only for columns	No crane
Number of workers per top-slewing crane	13	16	18	20

Table 8.1 Crane proportionality factors – Maximum usable number of workers per crane without incurring losses of productivity – Use of a top-slewing crane[25)]

Fig. 8.10 shows the number of workers per crane in combination with the defined construction time categories. For the maximum value, we assumed that concreting takes place without using a crane (i.e. 20 workers/crane). Thus, the 10% reduction applied to the maximum value (18 workers/crane = 20 workers/crane * 0.9) represents the case in which a normal construction time is possible.

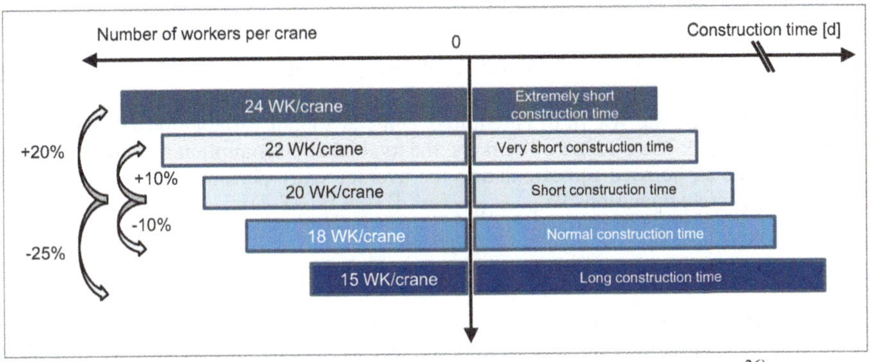

Fig. 8.10 Relationship between the proportional factor for cranes and construction time[26)]

It is necessary to determine the number of cranes that can be installed for site logistics in the area of the building in order to calculate the normal construction time from the bidder's/contractor's perspective. Thus, the maximum number of cranes that can be installed is very important for the achievable productivity level because this type of equipment constitutes bottleneck equipment, which is why losses of productivity will occur if the crane capacity is too low.

This calculated number of cranes can then be used to derive the maximum number of workers who can be deployed during the main construction phase (see Fig. 8.8). It is essential to ensure that this maximum value is not used for further calculation of the average performance value. The mean should be used instead, which can be considerably lower. The ratio of the average to the maximum number of workers usually lies between 0.65 and 0.85 and essentially depends on the duration of the ramp-up and final phases.[27)] The shorter the ramp-up and final phases, the longer the main construction phase (for the same total construction time) and the higher the proportional factor for workers f_{WK} [-] will be (see Eq. (8-2)).

25) Cf. Hofstadler (2014), p. 231

26) Cf. Hofstadler/Kummer (2014), p. 62

27) Cf. Hofstadler (2010), p. 219

$$f_{WK} = \frac{WK_{RCO,AVG}}{WK_{RCO,MAX}} \tag{8-2}$$

Taking into account the available data and information on the project under study, an expected ratio is selected and multiplied by the maximum value $WK_{RCO,MAX}$ [wh/h]. Besides the labor consumption rate, the resulting average value $WK_{RCO,AVG}$ [wh/h] is an essential prerequisite for calculating the average performance, which is related to the indicative quantity. The indicative quantity refers to the main construction material to be installed. For example, the cubic meter of reinforced concrete is used as the indicative quantity for the construction of the building shell. The amount of labor required per cubic meter of reinforced concrete comprises the proportions for formwork, reinforcing, and concreting operations. This figure is calculated using the total labor consumption rate LCR_{RCO} [wh/m³] while also considering the formwork ratio fr [m²/m³] and the reinforcement ratio rr [t/m³] (see Eq. (8-3)).

$$LCR_{RCO} = LCR_{FO} \cdot fr + LCR_{RO} \cdot rr + LCR_{CO} \tag{8-3}$$

The daily working time WT [h/d] is required as the last input variable for calculating the performance value. The number of workers, the total labor consumption rate, and the daily working time can be used for calculating the average daily performance relative to the reinforced concrete operations $PV_{RCO,AVG}$ [m³/d] (see Eq. (8-4)).

$$PV_{RCO,AVG} = \frac{WT \cdot WK_{RCO,AVG}}{LCR_{RCO}} \tag{8-4}$$

Dividing the total reinforced concrete volume C_V [m³] by the daily performance results in the normal construction time for reinforced concrete operations D_{RCO} [d] (without buffer):

$$D_{RCO} = \frac{C_V}{PV_{RCO,AVG}} \tag{8-5}$$

Depending on the extent of the assumed project uncertainties, this period can be regarded as the normal construction time. Alternatively, a buffer of, for example, 5 to 10% can be added.

This construction time calculated by the bidder/contractor can then be compared with the construction time specified by the client. If both time calculations arrive at the same or roughly the same result, we can assume that the production factors can be used with normal productivity based on the current state of knowledge.

If the specified construction time is shorter or even extremely short compared to the normal construction time, this increases the risk for the bidder/contractor not being able to adhere to this construction time in the execution phase, or for high losses of productivity to occur because the production factors can no longer be utilized economically. There will also be subsequent cost increases unless the bidder/contractor has based their pricing on this excessively short construction period. In this context, the bidder/contractor should consider different scenarios with regard to pricing.

8.6.2.3 Scenarios if Construction Time is Too Short

If an exceedingly short construction time is specified (shorter than the "normal construction time"), this can result in the following scenarios for the bid or contract pricing exercise:[28)]

Scenario 1

Bidders recognize that the specified short construction period means that they can no longer make optimum use of their production factors and increase the unit prices for the directly and indirectly affected work items.

Scenario 2

Bidders recognize that the specified short construction period means that they can no longer make optimum use of their production factors and partially increase the unit prices for the directly and indirectly affected work items. The remaining portion of the expected losses of productivity is taken into account in the risk premium.

Scenario 3

Bidders recognize that the specified short construction period means that they can no longer make optimum use of their production factors and partially increase the unit prices for the directly and indirectly affected work items. A certain portion of the expected losses of productivity is not taken into account in the pricing, and the remaining part is included in the risk premium.

Scenario 4

Bidders recognize that the specified short construction period means that they can no longer make optimum use of their production factors, but they do not increase their unit prices (for instance owing to the current competitive environment and market situation).

Scenario 5

Bidders recognize that the specified short construction period means that they can no longer make optimum use of their production factors, but they do not increase their unit prices (unit prices prior to accounting for potential losses of productivity). The bidders recognize that the specification is incomplete and are determined to have the deficits compensated for by submitting additional claims.

Scenario 6

Bidders do not recognize that the specified short construction period means that they can no longer make optimum use of their production factors, and they do not increase their unit prices (unit prices prior to accounting for potential losses of productivity).

8.6.3 Consequences for the Client

If the client specifies a shorter construction period than determined by the bidder/contractor as the normal construction time, they should expect consequences for the prices quoted and subsequently invoiced. In addition, adverse effects on quality should be expected, which will subsequently influence the use or reuse of the building.

[28)] Cf. Hofstadler (2014), p. 57f.

Fig. 8.11 shows the construction time specified by the client in comparison to the normal construction time specified by the bidder/contractor. It is shorter than the normal construction time and is thus associated with a higher cost minimum. The bidder/contractor attempts to optimally coordinate their production factors for the specified construction time so as to achieve the local cost minimum for this specified period by combining the production factors.

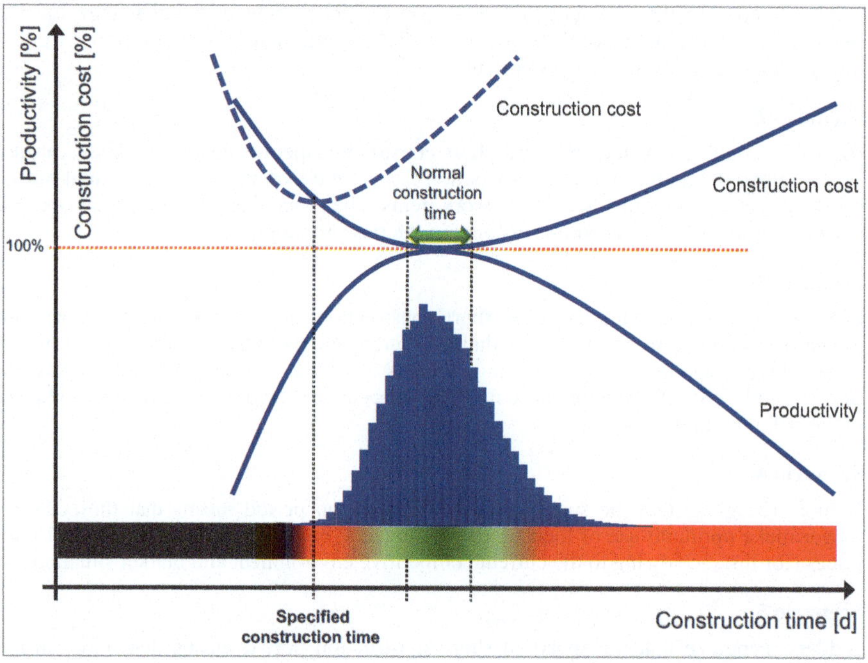

Fig. 8.11 Construction time specified by the client compared to the bidder's/contractor's normal construction time[29)]

The conditions specified by the client for performing work, including the highly significant construction period, should be defined in such a way that the contractors do not have to start at the productivity threshold right from the beginning and that they are not bound to continually exceed it in subsequent phases. If projects already start at the productivity limit (see selected construction time [i.e. specified construction time] in Fig. 8.11), their resilience will be weakened from the outset. Such production systems reach their efficiency limit right away. If, on the other hand, "normal production conditions" (essentially determined by a "normal construction time") are specified, it will be easier to adapt to changes or disruptions to the construction process. Resilient production systems are characterized by their flexibility, diversity, redundancy, and/or free capacities. These are precisely the characteristics that are reduced by the drive for efficiency.

[29)] Cf. Hofstadler (2014), p. 39

8.6.3.1 In the Contract Award Phase

For a normal construction time, the bidder can determine the combination of production factors with normal productivity. It can thus be assumed in this pricing phase that no losses of productivity will occur due to an exceedingly short construction time. In addition, the normal construction time is also expected to incur the lowest construction cost. This circumstance can be factored into the bidder's pricing strategy.

However, if the construction time specified by the client is shorter than the normal construction time calculated by the bidder/contractor (which is usually the case), the latter can no longer assume a normal combination of production factors for the purpose of its pricing. For this reason, losses of productivity need to be taken into account when determining the costs (zero-base costing), which is expected to lead to a local higher cost minimum. The client would have to expect a higher bid price if the bidder were able to factor these circumstances into the pricing.

According to the principle of the balance between work performance and consideration, this would also lead to higher bid prices if all bidders were to identify the situation correctly, taking into account the corresponding effects on the construction management and economics levels in their costing and pricing processes. If bidders engage in wage and social dumping in their competitive environment, this effect will obviously be distorted because such bidders will always be in a position to quote cheaper prices than construction companies performing their costing and pricing in compliance with the applicable legal requirements.

8.6.3.2 In the Construction Phase

If the specified construction time is shorter than the normal construction time, greater losses of productivity should be expected in the case of deviations from the local minimum than in the case of deviations from the minimum in the normal construction time. The bend is more pronounced in the curve representing the specified construction time because site operations will respond more sensitively to changes and disruptions (see Fig. 8.11). This is due to the fact that site operations are bound to hit the productivity limit in line with an unfavorable construction time with regard to the combination of production factors.

If the client causes deviations from the contractually agreed work, it would have to account for higher additional costs both in terms of accelerating work and extending the construction time.

8.6.3.2.1 Work Acceleration

The work would have to be accelerated or intensified while retaining the same construction time if the production potential were to be increased due to deviations from the contractually agreed work, or if the estimate of the amount of labor were too low as a result of applying excessively low labor consumption rates. This will result in higher costs when combining the production factors. Correspondingly, these cost variations will be more pronounced due to the stronger bend of the curve (see Fig. 8.11). In the case of work acceleration, attention must be paid to the number of production factors that the construction site can tolerate while maintaining its productivity. In practice, the costs of work acceleration or intensification will usually vary depending on which party initiated the work intensification process. If the client orders the acceleration, the costs will be higher than if the contractor has to bear the related responsibility. This largely depends on the party by which the costs should be borne. If the client is at fault, they will also have to cover the costs. If the reasons for acceleration are attributable to the contractor, the related costs will fall within their sphere of responsibility. If the contractor has to carry the costs themselves, it is only natural that they will make an even greater effort to achieve the most economical combination of production factors through optimization in order to keep the costs incurred as low as possible.

8.6.3.2.2 Extension of Construction Time

If the construction time is extended on the basis of the originally specified, excessively short construction period, site operations will become more expensive because we should assume that site operations had already been disrupted from the outset, which leads to more significant cost effects even with a longer construction period and due to the more pronounced bend of the curve. Since, initially, an excessively short construction time was specified, the cost level that could have been realized in a normal construction time can no longer be achieved by a subsequent extension of the construction period.

8.7 Credible Construction Time

To be able to assess construction process disruptions and additional cost claims, it is essential to ascertain if a plausible target construction schedule based on an equally plausible process schedule exists that represents a feasible operational basis for performing the contractually agreed construction works. The required standard of workmanship is generally specified in the construction contract. If the bidder was also required to submit a construction schedule with its bid, this will constitute the target construction schedule that would also have to be checked, in the course of the bid analysis, with a view to the plausibility of the process times and sequencing relationships contained in this document.

If an expert is consulted in the dispute between the client and the contractor, such expert would have to check whether it is at all possible to perform the agreed works according to the target construction schedule. If this is not feasible in terms of site operations and technology, the expert is to amend the target construction schedule by introducing corrections with respect to ensuring buildability. In construction practice, this often poses a serious problem because the essential pricing parameters agreed in conjunction with the unit prices, including labor consumption rates and performance values, do not correspond to the (actual) durations of the individual processes. Labor consumption rates are usually too low, whereas performance values are too high. If these values are logically related to the construction schedule submitted by the bidder, one often arrives at the conclusion that the assumed process times and sequencing relationships are not feasible. Some essential characteristics must be taken into account when preparing and reviewing construction schedules.

A credible construction schedule will essentially meet the following requirements (non-hierarchical ranking; see Fig. 8.12):

- Plausible
- Reliable
- Complete

- Calculable
- Binding

The most important basis for a credible construction schedule is a credible process schedule.

Fig. 8.12 shows key features of the above five essential attributes that characterize a credible construction schedule.

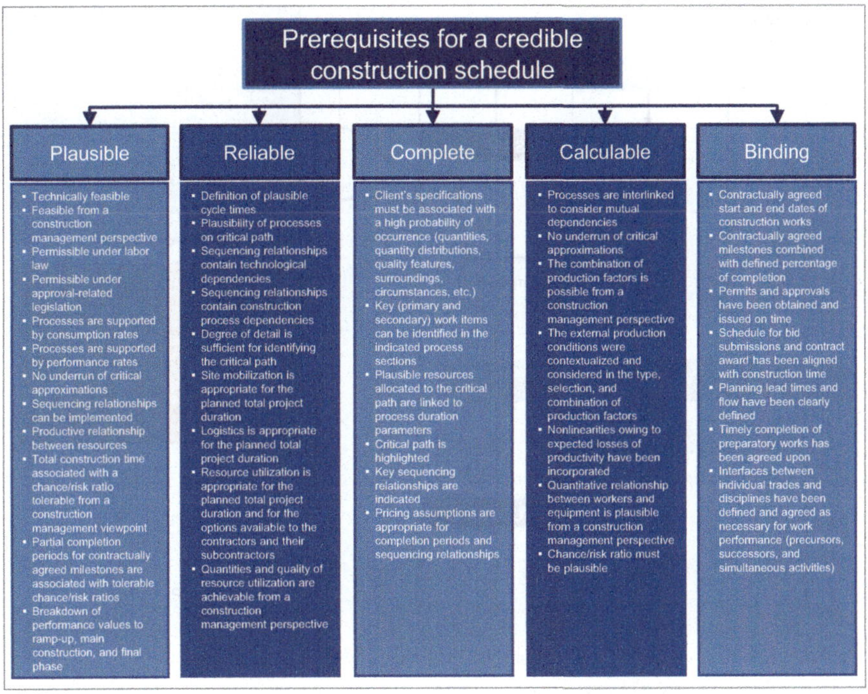

Fig. 8.12 The five essential characteristics of a credible construction schedule

These requirements also apply to the "should-be" and actual construction schedules. For instance, credibility depends on whether the beam lengths and sequencing relationships shown in the construction schedule are buildable at all (the laws of productivity of the internal production system and the underlying conditions of the external production system correspond to the total and/or partial construction times). In addition, we need to check if the combination of production factors shown in conjunction with the processes is credible in technological, operational, and environmental terms. It is also important whether critical approximations representing the earliest start of a process or of several processes have been underrun. Plausible sequencing relationships will result from considering external and internal dependencies. Fig. 8.13 shows three target processes and gives examples of possible dependencies that must be considered or verified.

The most frequent error sources in construction schedules are inappropriate, exceedingly short process times, incorrect sequencing relationships, and violations of the critical approximation. The critical approximation is the result of the external and internal dependencies. For example, with respect to internal dependencies, the contractor would not start the subsequent work if this were to result in losses of productivity owing to mutual obstructions and insufficient crane capacity.

As an example of a technological dependency, we refer to the rule that a climbing formwork may only be moved to the next concreting section once a previously defined minimum compressive strength of the concrete (usually 15 N/mm²) has been reached.

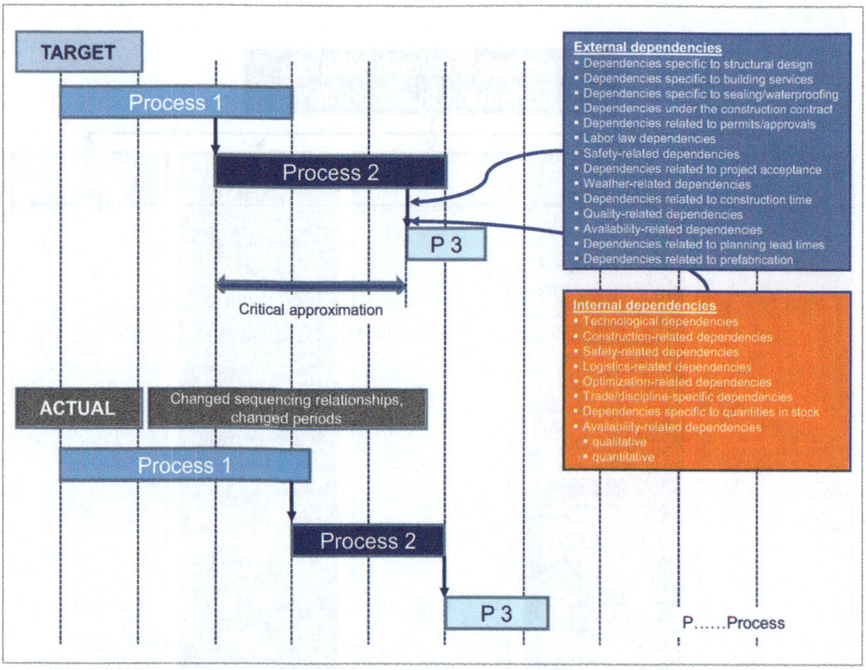

Fig. 8.13 Effects on sequencing relationships between processes

The aspect of meaningfulness will depend on the processes on the critical path being plausible and on the presence of a productive relationship between resources and processes. The construction site and logistics concept is sufficiently plausible to also render plausible the total construction time or its portions.

Yet the client must also make a major contribution to arriving at a plausible construction time by providing relevant information and adhering to their duty to cooperate. Its information regarding quantities, qualities, the site environment, etc. must have a high probability of occurrence. The construction schedule must be checked particularly in respect of the critical path in order to ensure that the main works have been captured completely and that the essential ancillary works are also included. The resources are linked to the processes along the critical path so as to verify if the utilization of workers and equipment is consistent with the duration of these processes. For the target process schedule, the pricing parameters are linked to the planned use of resources and the networked construction process.

To ensure calculability, the processes included in the construction schedule are interlinked with sequencing relationships and resources as well as performance values and quantities. Critical approximations are not underrun because, otherwise, the construction schedule or parts thereof would become unrealistically short. The combination of the production factors must be possible from a site operations perspective and the influences stemming from the external production system are integrated into the calculation. There are plausible quantitative and qualitative relationships between the workers as well as between the individual pieces of equipment. When determining the construction time, we should also consider the extent to which quantity variations (±20% in Austria; ±10% in Germany) to be accepted by the contractor will potentially affect the construction process.

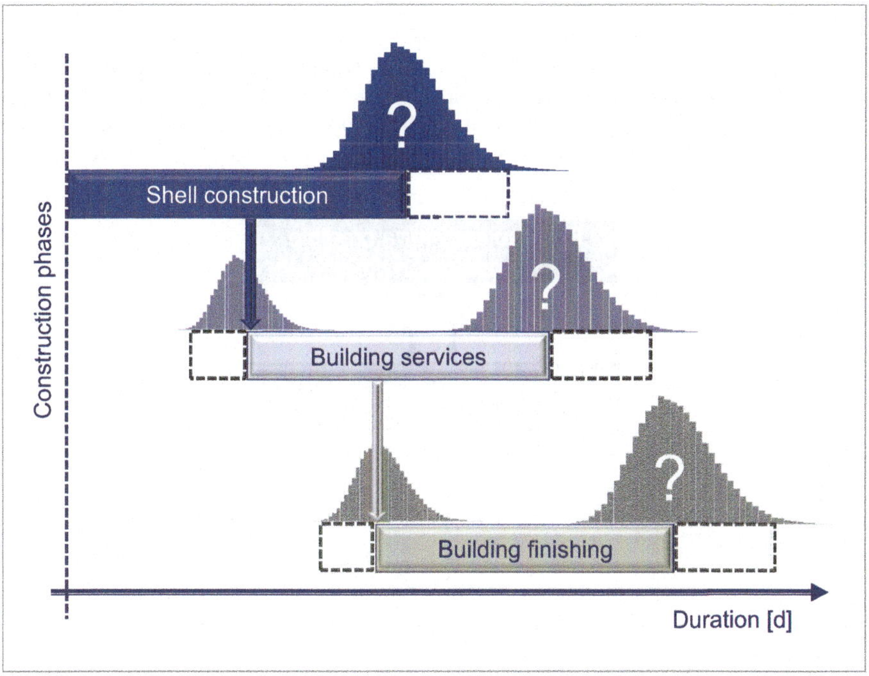

Fig. 8.14 Sequencing relationships between the shell, finishing, and building services phases – Systematic consideration of uncertainties during and between the different construction phases[30)]

The start and end date as well as important milestones are contractually agreed for the construction schedule, and permits and approvals as well as the planning-ahead process are aligned with it and are binding. The planning flow and the procedure for planning changes are also precisely stipulated in the contract, and the timely completion of advance works by the client or other contractors of the client is specified. The interfaces to the trades closely related to the performance of the work have also been defined and agreed. Besides the detailed sequencing relationships, a successful project outcome will also heavily depend on the networking between the shell construction, finishing, and building services phases (see Fig. 8.14). The finishing and building services phases must not start

[30)] Cf. Hofstadler (2014), p. 40

too early (violating the critical approximation) in order to prevent mutual obstructions. Compared to the shell construction phase, the finishing and building services phases require consideration of many more trades and disciplines on the construction site. If interfaces between the trades and higher-level logistics are not explicitly agreed beforehand, disruptions occurring in the execution phase can amplify significantly owing to the large number of stakeholders involved in the process. Parameter ranges should be applied to choose the right start dates for the finishing and building services phases. The stochastic Monte Carlo simulation method is a very effective tool for systematically mapping the associated uncertainties. A histogram can be generated for the respective durations and the offset between two phases. Ultimately, a chance/risk ratio can be determined for each selected value. If a higher risk is taken, for instance due to a selected longer offset period, the preceding coordination between the trades must be planned and organized even more accurately.

The most important basis for a credible, and thus buildable, construction schedule is a credible process schedule. This credible process schedule takes into account all requirements with regard to identifying and defining the sequencing relationships while ensuring that critical approximations will not be underrun.

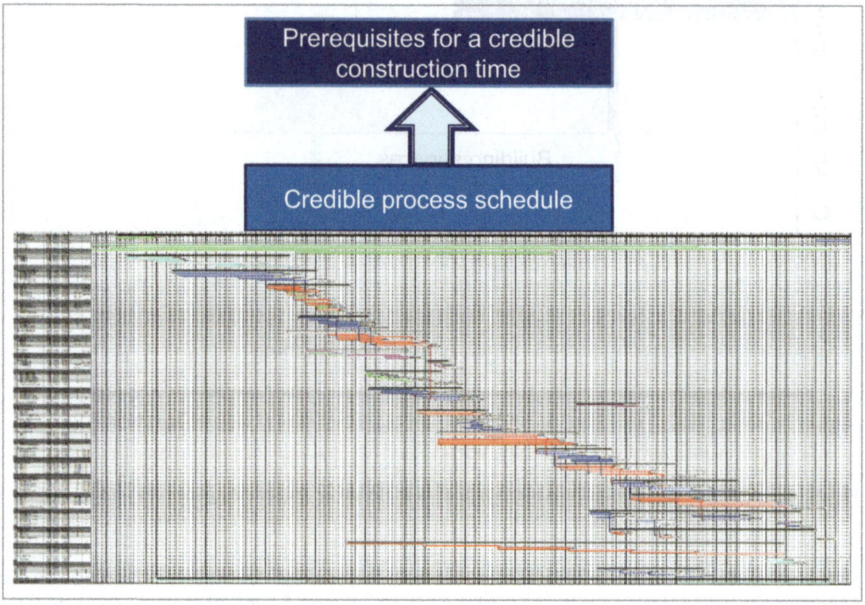

Fig. 8.15 From a credible process schedule to a credible construction time

Fig. 8.15 includes a schematic representation of a credible construction process schedule that contains the essential processes determining the construction period while identifying the critical path. This networked process schedule is linked to a range of resources (not shown in this figure) that indicates the number of workers and the number of pieces of bottleneck equipment. Linking the processes to the resources and quantities identifies the types of work required for achieving the related performance values. Provided these performance values also lie within a plausible range, all requirements for a credible construction schedule will have been met. The range of resources and the building and construction site conditions make it relatively easy to verify if this construction project can actually tolerate the indicated number and distribution of resources in terms of site operations, technology, and production.

8.8 Influence of Construction Time on Productivity

The qualitative relationship between the construction time and productivity has already been described in previous sections of this book. We also refer to the pertinent literature on construction management (see Section 8.8.2). However, the quantitative relationship between the productivity and construction time will also be of great interest. It is important for all parties involved in the project to find out how productivity changes in the event of deviations from the "normal construction time". In the bid phase, the bidder will essentially be interested in identifying the chance/risk ratio associated with the final bid price. In the construction phase, the magnitude of the losses of productivity resulting from deviations from the agreed construction time should be determined.

8.8.1 Construction Time and Losses of Productivity

How can reliable data be derived that quantitatively describe the relationship between construction time and productivity in a generic manner (i.e. independent of specific projects)? Any *ex-post* analysis of completed projects will be difficult to establish because the type and circumstances of the work performed will have changed dramatically, and additional work will have been commissioned compared to the original contract. The originally agreed-upon construction project does not correspond to the project actually completed. It is almost impossible to correctly differentiate between individual losses of productivity according to their causes, namely changes in the construction time and/or in the type of work, the circumstances under which work is performed, or additional work performed.

For this reason, subsequent analyses started from an *ex-ante* consideration as a feasible alternative.

To gather valid data, *Hofstadler* conducted an expert survey at Graz University of Technology in order to identify the quantitative relationship between the construction time and the labor consumption rate, and thus productivity. For this purpose, a total of 35 experts from construction contractors were asked to participate in this survey (from construction trades and industry; survey period from August 2012 to April 2013). Respondents had an average work experience of 17 years (minimum: five years; maximum: 43 years).

The experts were asked to indicate the increases in the labor consumption rate or losses of productivity to be expected if deviations from the optimal construction time occur from a site operations and construction economics viewpoint (first stage: completion of the questionnaire by the expert; second stage: interview with the expert for validation purposes). The "normal" construction time during which the production factors can be utilized without incurring losses of productivity was taken as the reference base.

In the first step, an exploratory data analysis was carried out in order to obtain an overview of the collected data. This analysis is necessary for checking plausibility, verifying distributions, and identifying outliers. Box plots of the variables were prepared as part of the exploratory data analysis to graphically represent distributions. A box plot describes the location and spread of a distribution and provides information on any outliers included in the data. Outliers and extremes may significantly distort arithmetic means depending on the distance from the box plot whisker. *Huber's* M-estimator method can be applied to prevent such an outcome. Depending on the specific case, a lower weighting can be assigned to the existing outliers/extremes so as to move them closer to the "main set" of data. In the following figures, the robust M-estimators are used as a basis for creating diagrams and illustrating the trends of labor consumption rate increases and losses of productivity (see Figs. 8.16 to 8.21).

The differences in the increases in labor consumption rates for shorter or longer construction times are reflected in the diagram shown in Fig. 8.16 in order to consider the influence of construction time at the costing stage or in the process planning phase. The abscissa shows the extension/reduction of the construction time relative to the normal construction time [%] while the ordinate indicates the increase in the labor consumption rate of reinforced concrete operations [%].

The two curves representing the labor consumption rate increase are almost identical up to a 10% change in construction time. Only from the 15% threshold do the two curves deviate from each other. For instance, a 50% change corresponds to a difference of about 27.6 percentage points. At a 50% shortening of construction time, the labor consumption rate increase amounts to about 67.4% and is thus 69.3% greater than in the case of construction time extension [= 67.4% * 100% / 39.8% - 100%].

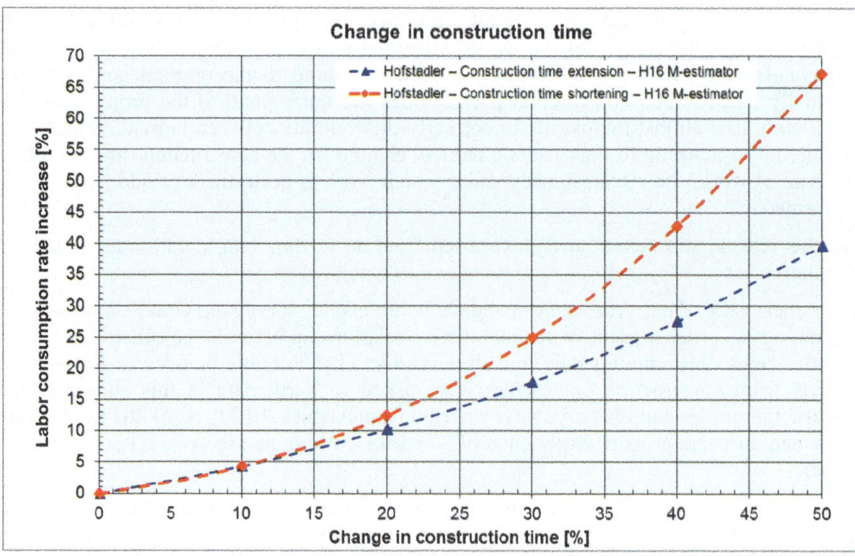

Fig. 8.16 Trend of consumption rate increases – Comparison between extension and shortening of construction time – Reinforced concrete operations[31)]

This comparison clearly demonstrates that a shortening in construction time is associated with greater labor consumption rate increases compared to an extended construction

[31)] Cf. Hofstadler (2014), p. 455

period. It is well-known that a reduction in construction time will lead to a superposition of several disruptions with an adverse effect on productivity. In this context, it should be noted, in particular, that the minimum working space and crane capacity are underrun.

The diagram in Fig. 8.17 is expanded to also show the loss of productivity directly related to the increase in the labor consumption rate.

The effect on productivity (loss of productivity ΔPL_D [%]) can be quantified by inserting the labor consumption rate increase ΔLCR_D [%] in Eq. (8-6).

$$\Delta PL_D = \left(\frac{\Delta LCR_D}{100\% + \Delta LCR_D}\right) \cdot 100\% \quad (8\text{-}6)$$

The abscissa shows the extension/reduction of the construction time relative to the normal construction time [%] while the ordinate indicates the increase in the labor consumption rate and the loss of productivity, respectively, for reinforced concrete operations [%]. The central portion of the diagram shows the normal construction time at 0%, where no loss of productivity or consumption rate increase occurs.

Based on this global view of the relationship between construction time and productivity, the question arises as to which and how many influential factors can lead to these losses. The type and nature of these influences will depend on the specific construction project. The number of these factors was determined in an expert survey. Compared to an extended construction time, a shortened construction time tends to cause a larger number of different losses of productivity at the same time.

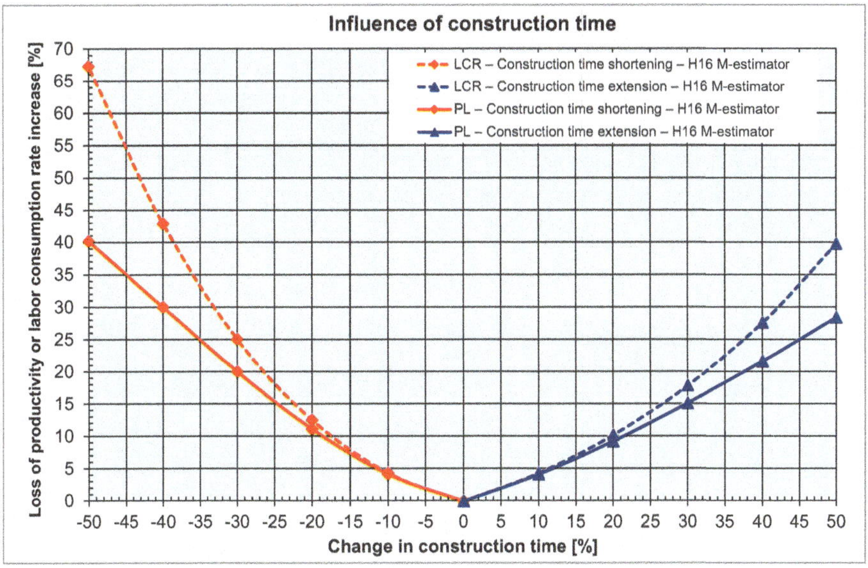

Fig. 8.17 Trend of labor consumption rate increases and losses of productivity – Comparison between extension and shortening of construction time – Reinforced concrete operations[32)]

[32)] Cf. Hofstadler (2014), p. 454f.

In the expert survey conducted by *Kummer/Hofstadler* (2015), the experts (a sample of 29 respondents) stated that if the construction period is too short, three losses of productivity usually occur simultaneously. This result is based on the mode (most frequent mention) obtained in the course of the statistical analysis. The median was also identified to cause three different losses (see Fig. 8.18).

No.	Parameter	Construction time too short	Construction time too long
0	A	B	C
1	n	27.00	26.00
2	Median	3.00	2.00
3	Mode	3.00	2.00
4	Minimum	2.00	1.00
5	Maximum	> 5.00	3.50

Fig. 8.18 Descriptive representation of data on the number of productivity losses due to excessively short and long construction times in building construction

For the excessively long construction period, two losses of productivity were identified both as the median and the mode (ten mentions). However, we should also account for the fact that, on the other hand, eight experts stated just a single loss of productivity if the construction time is too long (see Fig. 8.19). Thus, the result is not as clear-cut as in the case of the excessively short construction period (i.e. a single-peak distribution with eleven mentions and three productivity losses). The expert survey clearly shows that an unduly short construction time tends to result in a larger number of productivity losses compared to an exceedingly long construction period.

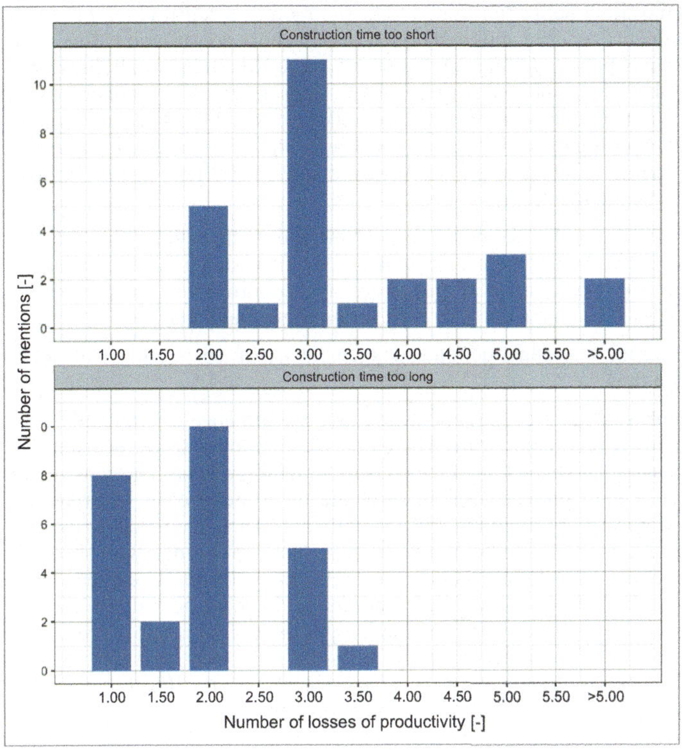

Fig. 8.19 Average number of losses of productivity occurring simultaneously in building construction – Exceedingly short and long construction times[33)]

8.8.2 Comparison with Pertinent Literature

In the following sections, we compare the qualitative and quantitative survey results with current approaches taken from the literature.

8.8.2.1 Qualitative Comparison

Hürlimann/Bucher indicate a qualitative range for the construction time within which the costs are minimal. A construction period that differs only slightly from this optimal duration will only lead to a moderate cost increase. A sudden cost "leap" is also possible if the construction time is shortened too much. This sharp increase can be caused, for example, by the need for multi-shift operation in the case of a significantly shortened construction period.[34)]

33) Cf. Hofstadler/Kummer (2017a), p. 344

34) Cf. Hürlimann/Bucher (2010), p. 10f.

Schach/Otto relate the relationship between construction time and construction cost particularly to the site mobilization aspect. The duration of a construction project has a considerable influence on site mobilization costs. A short construction time will cause site mobilization expenses to increase because more equipment will be required over a relatively short construction period, which tends to result in higher costs due to the associated transport, assembly, and dismantling costs. Likewise, a longer construction period will incur higher costs because the on-site equipment needs to be kept available for a longer period, and idle costs are incurred due to the non-optimal utilization of the facilities and equipment.[35)]

Karasek/Duve argue that the construction time should be determined taking into account the specific conditions of the project under study. In so doing, all circumstances should be considered that would arise if they had also occurred for comparable projects and contractors, such as a long delivery time for building materials.[36)]

Karasek stresses that construction management expertise is crucial in determining the construction time.[37)]

Kropik emphasizes that construction time and performance are mutually dependent. We can thus apply a very simple rule: The larger the amount of work (or the number of work steps) to be completed, the longer the construction time will be. The construction time will also influence total project costs because the time-bound costs incurred in the on-site production process are anything but insignificant. An unrealistically short construction time leads to higher construction costs because losses of productivity should be expected compared to an optimal construction time. This phenomenon is caused, for instance, by overtime work, a large number of workers to be coordinated, or non-adherence to minimum work areas. A longer construction period translates into a prolonged accrual of time-bound costs. The bidder must also consider the execution time provided for in the specifications in its pricing exercise.[38)] *Kropik* also states the following in relation to specified times: "If time constraints become too tight, this will often result in unnecessary remedial work in later phases, as well as in lost time and money and, even more importantly, dissatisfaction among the project stakeholders. Sufficient time should thus be allotted to the individual project phases."[39)]

Roquette/Viering/Leupertz argue that a detailed planning approach is necessary for adhering to the calculated construction time. Before awarding the contract, the technical design must have reached a sufficiently detailed status. The same applies to the detailed scheduling of work completion dates and milestones.[40)] The authors also emphasize that the agreed construction time is particularly important for the contractor to determine the corresponding costs. In this context, is is important to not only define the overall duration of the project but to also allocate the individual construction phases to suitable seasons of the year. This is not only about time-bound costs but also about the way in which the production factors can be combined. The authors also establish the link between the construction time and the standard of workmanship achieved for the building. Defects will occur if the planning time is too short and/or the construction process is too fast.[41)]

---

35) Cf. Schach/Otto (2011), p. 355

36) Cf. Karasek/Duve (2009), p. 15

37) Cf. Karasek (2016), p. 261

38) Cf. Kropik (2014), p. 659f.

39) Cf. Kropik (2014), p. 140 (translation by Steffen Walter)

40) Cf. Roquette/Viering/Leupertz (2013), p. VII

41) Cf. Roquette/Viering/Leupertz (2013), p. 2

Hoffmann/Motzko/Corsten stress that sufficient time must be made available by the client for the preparation of bids, for the preparation of the construction process, and for the execution of construction work. Factors to be considered when determining the construction time include seasonal effects and specific site conditions as well as other influential factors and difficulties. Exceedingly short construction periods and building under unfavorable seasonal conditions are problematic and can adversely affect the project.[42)]

Wolkerstorfer/Lang argue that bid prices can only be determined in a cost-covering manner if the specification prepared by the client includes very detailed and precise quantities and times for the performance of the work. The information provided by the client must be associated with a high probability of occurrence.[43)]

In summary, the statements of the authors cited above show a clear relationship between construction time and construction cost on the one hand and between construction time and building quality on the other. Yet this is not the only reason why the client should take a close look at the influences on the construction time so as to define a normal construction time for their project. This is how the client will lay the foundation for a fair and equitable handling of the project and the most favorable project outcome.

8.8.2.2 Quantitative Comparison with the Survey conducted by Hofstadler

The diagram in Fig. 8.20 compares the quantitative results from the literature with the results of the expert survey (*Hofstadler* 2012/13). This diagram represents the effects of shortened and extended construction times and the resulting losses of productivity. *Oberndorfer*[44)], *Petzschmann*[45)], and *Reister*[46)] subsequently cite blanket limits or ranges for plausible losses of productivity. Prior to using these limit values, care must be taken to ensure that the concept of productivity losses as interpreted by these authors is fully understood, as well as the calculation methods applied by them.

The comparison with the limit defined by *Oberndorfer* shows that this limit is equivalent to the losses of productivity that would occur at a construction time shortening by about 22% and a construction time extension by about 27% as determined in the study conducted by *Hofstadler*.

In the event of simultaneous disruptions to the construction process, *Petzschmann* indicates a potential loss of productivity that ranges from 15% to 30%. The 15% productivity reduction according to *Petzschmann* corresponds approximately to a 24% reduction or a 30% extension of the construction time. When considering the upper limit of 30%, the construction time is reduced by 40% and extended by more than 50% (the two curves do not intersect in this case).

Reister includes ranges that cover the entire area of losses of productivity due to construction time extension or shortening. When considering a 45% extension of the construction period, it is striking that, according to the classification of *Reister*, the 25% loss of productivity coincides exactly with the end of the average and the beginning of the high productivity losses. The productivity loss of 40% identified for a 50% reduction in

42) Cf. Hoffmann/Motzko/Corsten (2012), p. 29

43) Cf. Wolkerstorfer/Lang (2014), p. 221

44) Cf. Oberndorfer (2010), p. 127

45) Cf. Petzschmann (2000), p. 138

46) Cf. Reister (2007), p. 504

the construction time coincides exactly with the end of the high and the beginning of the exceptionally high productivity losses according to *Reister*.

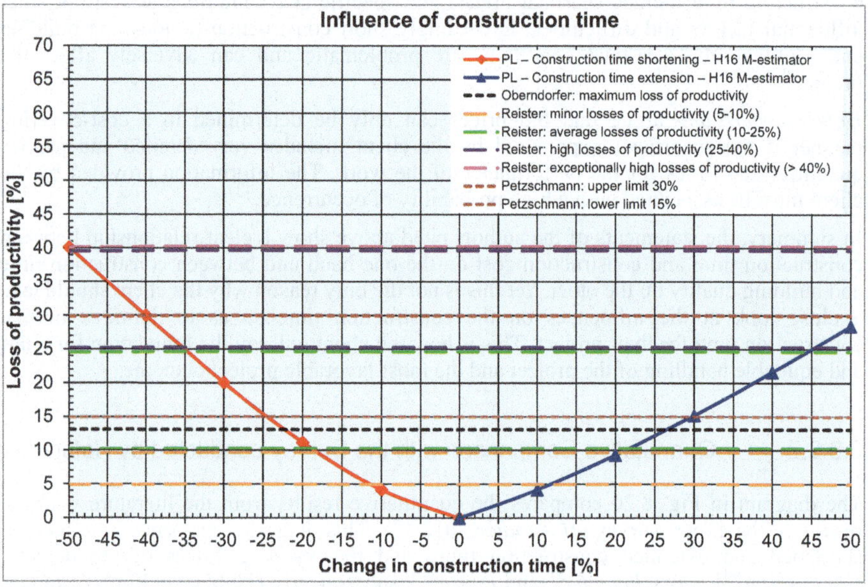

Fig. 8.20 Losses of productivity – Comparison of the approaches of *Oberndorfer*, *Petzschmann*, and *Reister* with the findings of the expert survey conducted by *Hofstadler*[47)]

Comparing the results of the expert survey with the *Oberndorfer* limit shows that the limit does not necessarily apply and that higher losses of productivity can also occur. The limits stated by *Petzschmann* appear to lie within a plausible range. The ranges defined by *Reister* appear to be fairly plausible from a construction management perspective, depending on the extent of changes in construction time.

8.8.2.3 Quantitative Comparison with the Survey conducted by *Kummer/Hofstadler*

From December 2014 to October 2015, *Kummer/Hofstadler* conducted another expert survey on the subject of uncertainties and productivity losses in reinforced concrete (in-situ concrete) and masonry operations. One of the questions dealt with how construction costs change in the case of a deviation from the normal construction time. The experts graphically represented the changes in construction cost for an extension and a shortening of the construction period. A total of 29 experts were interviewed in the course of a primary survey. A standardized written questionnaire was used, and the responses were subsequently discussed in interviews conducted with the respondents. Depending on the extent of the change in construction time, between 19 and 29 experts indicated the cost variations as percentages (see Tables 8.2 and 8.3, Row 1).

[47)] Cf. Hofstadler (2014), p. 479

The resulting curves were transferred to a drawing program (AutoCAD) and scaled accordingly in order to generate quantitative values for the cost trend as a function of the specified construction time. For this purpose, the derived values for an exceedingly short and long construction time were rounded to integer values (see Fig. 8.21).

No.	Parameter	Construction time too short (reduction to normal construction time)					
		50%	60%	70%	75%	80%	90%
0	A	B	C	D	E	F	G
1	n	19.00	24.00	28.00	28.00	29.00	29.00
2	Mean	182.00	163.88	143.29	131.86	124.83	110.03
3	Standard error of mean	5.51	5.52	4.87	4.45	4.43	2.76
4	Standard deviation	24.04	27.03	25.77	23.56	23.84	14.87
5	Coefficient of variation	13.21%	16.50%	17.98%	17.87%	19.10%	13.51%
6	Median	190.00	163.00	138.00	123.50	116.00	105.00
7	Mean absolute deviation	11.00	19.50	11.50	7.50	5.00	2.00
8	Robust coefficient of variation (based on median)	5.79%	11.96%	8.33%	6.07%	4.31%	1.90%
9	Mode	200.00	163.00	140.00	118.00	112.00	102.00
10	Minimum	124.00	116.00	110.00	107.00	105.00	102.00
11	Maximum	213.00	222.00	218.00	200.00	199.00	171.00
12	Spread	89.00	106.00	108.00	93.00	94.00	69.00
13	Skewness	-1.24	0.11	1.32	1.92	2.01	3.05
14	M-estimator (H12 (80%))	185.31	163.51	139.33	127.1	118.48	105.43
15	Standard deviation (H12 (80%))	20.37	29.80	20.37	14.85	11.70	3.83
16	Robust coefficient of variation (based on M-estimator)	10.99%	18.23%	14.62%	11.68%	9.87%	3.63%
17	Normal distribution (Shapiro-Wilk, $p < 0.05$)	No	Yes	No	No	No	No

Table 8.2 Descriptive representation of data to illustrate the relationship between construction cost and construction time – Excessively short construction time[48)]

A descriptive data analysis was subsequently carried out on the basis of the derived values. As part of this analysis, mean values and robust M-estimators were formed to account for potential outliers and extremes, and the means and M-estimators were used for trend analyses (see Tables 8.2 and 8.3).

No.	Parameter	Construction time too long (increase to normal construction time)					
		110%	120%	125%	130%	140%	150%
0	A	B	C	D	E	F	G
1	n	29.00	29.00	29.00	29.00	28.00	26.00
2	Mean	103.83	109.83	113.66	118.52	129.57	139.19
3	Standard error of mean	0.86	1.51	1.82	2.20	3.33	3.06
4	Standard deviation	4.64	8.13	9.82	11.82	17.63	15.62
5	Coefficient of variation	4.47%	7.40%	8.64%	9.98%	13.60%	11.22%
6	Median	103.00	110.00	114.00	117.00	127.00	139.00
7	Mean absolute deviation	2.00	4.00	5.00	6.00	7.50	10.00
8	Robust coefficient of variation (based on median)	1.94%	3.64%	4.39%	5.13%	5.91%	7.19%
9	Mode	101.00	107.00	114.00	117.00	127.00	139.00
10	Minimum	96.00	94.00	95.00	96.00	102.00	112.00
11	Maximum	122.00	140.00	150.00	158.00	183.00	177.00
12	Spread	26.00	46.00	55.00	62.00	81.00	65.00
13	Skewness	2.47	1.81	1.78	1.31	1.43	0.44
14	M-estimator (H12 (80%))	103.13	108.96	112.68	117.23	126.84	138.49
15	Standard deviation (H12 (80%))	2.48	5.51	6.90	9.13	12.63	15.89
16	Robust coefficient of variation (based on M-estimator)	2.40%	5.06%	6.12%	7.79%	9.96%	11.47%
17	Normal distribution (Shapiro-Wilk, $p < 0.05$)	No	No	No	No	No	Yes

Table 8.3 Descriptive representation of data to illustrate the relationship between construction cost and construction time – Excessively long construction time[49)]

Fig. 8.21 illustrates the change in construction time as a percentage on the abscissa. The left-hand ordinate shows the changes in construction costs while the right-hand ordinate demonstrates the effects on the increase in the consumption rate as percentage values.

48) Cf. Hofstadler/Kummer (2017a), p. 348

49) Cf. Hofstadler/Kummer (2017a), p. 348

To assess the changes in the wage proportion, four different distributions between the wages and the "Other" components are applied to the construction costs in relation to the trend. The "Other" item comprises the costs of equipment and materials.

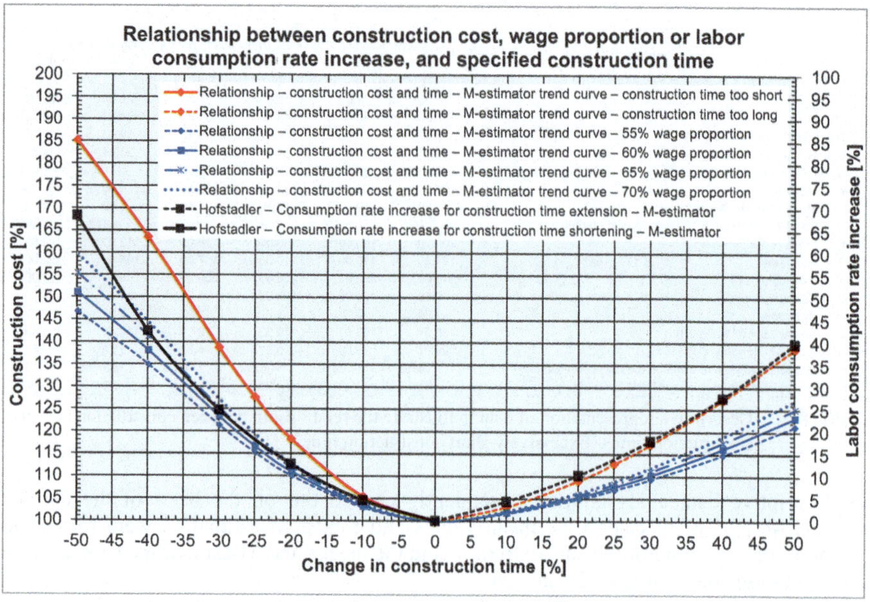

Fig. 8.21 Construction cost and labor consumption rate increase[50] – Comparison of the findings of the expert survey of *Hofstadler* (2012/13)[51] with the expert survey of *Kummer/Hofstadler* (2015)

The following wage proportions were used as a basis for the diagram so as to provide options for construction projects with different labor intensities:

- Wage proportion = 55%
- Wage proportion = 60%
- Wage proportion = 65%
- Wage proportion = 70%

For the selected 55% wage proportion, for example, the values of the curve representing the "Relationship – construction cost and time – M-estimator trend curve – construction time too short" are multiplied by 0.55 for the area of the short construction time, and the curve labeled "Relationship – construction cost and time – M-estimator trend curve – 55% wage proportion" is derived. For the longer construction times, the values of the curve representing the "Relationship – construction cost and time – M-estimator trend curve – construction time too long" are also multiplied by 0.55, resulting in the "Relationship – construction cost and time – M-estimator trend curve – 55% wage proportion" curve for the range from 0 to 50% on the abscissa.

[50] Cf. ibid., p. 349

[51] Cf. Hofstadler (2014), p. 455ff.

The lowest curve in the diagram represents a wage proportion of 55%. The curves derived for the different wage proportions for the exceedingly short and long construction times can be compared directly with those from the *Hofstadler* survey (2012/13) for the labor consumption rate increases. At the short end, a good consistency is found up to a 40% time reduction.

For extended construction times, the curve representing the consumption rate increase consistently runs at a higher level than the curves for the different wage proportions.

8.9 Summary

It is not only the analysis of "chaotic" projects that proves the great importance and necessity of integrating the aspects of project lead time and construction time more thoroughly and systematically into organizational, management, planning, and construction processes. As a result, optimization potentials will emerge even for successful projects. We will be able to pave the road to success by implementing a predictive, smart chance and risk policy that defines the objectives and methods of the management of chances and risks.

The relationship between the construction time and the labor consumption rate as well as productivity was presented qualitatively on the basis of literature references and quantitatively with the aid of studies conducted at Graz University of Technology. We demonstrated the effects on productivity and labor consumption rates to be expected if construction time is shortened or extended compared to the "normal" period. Furthermore, the effects of deviations from the "normal" construction time on construction cost were evaluated.

We can assume a "normal" construction time if the type, number, and combination of production factors are suitable for implementing on-site operations according to plan without exceeding the thresholds to losses of productivity. A notional shortening of the construction time (see the introduction to this book for an explanation of the concept of notional shortening) will occur in the bid phase only if the client departs from the "normal" construction time calculated by the bidder. Irrespective of the specified construction period, the bidder will calculate its construction time considered to be "normal" from a construction management perspective as an internal benchmark for assessing the existing level of productivity.

The baseline data for presenting the interactions was gathered directly from construction practice in the course of expert surveys. In addition to the arithmetic means, robust indicators were calculated using *Huber's* M-estimator method. This approach ensures that a lower weighting is consistently allocated to any existing outliers and extremes in the statistical analyses, but these values are not excluded from the analysis.

The relationships shown in the diagrams are thus very reliable. In addition, statistical parameters such as the standard deviation can be used so that ranges can be adjusted to the specific underlying conditions of the project under study. Users of the diagrams and/or values are thus able to account for presumed smaller or larger disruptions to the construction project by using the full range of the standard deviation in both the downward and the upward direction. This setting provides a statistically sound flexibility in applying the presented values and diagrams and allows for the adaptation to the specific project conditions.

Any construction time that leaves sufficient room for considering all relevant construction management aspects ("normal" construction time) will be beneficial for each of the parties to the contract. On the one hand, we can expect a construction process with very few disruptions if the client specifies a "normal" construction period. On the other hand, bidders or subsequent contractors can be highly confident in being able to utilize their production factors at their planned "normal productivity" level if a normal construction time is specified. Any disruptions can be overcome more easily compared to a construction process that had been disrupted already when work commenced due to an exceedingly short construction time. The client must also ensure that the project documentation is of a high quality – depending on sufficient project lead time – even before the contract is awarded.

In any case, a credible construction time will create the basis for the client, the contractor, and the other project stakeholders to achieve their project targets together.

8.10 References

Blecken, Udo (1976). Die Produktions- und Kostentheorie im instationären Baubetrieb. In: Der Baubetriebsberater, Heft 4, 1/1976. 199-210. Wiesbaden, Berlin. Bau-Verlag.

Gölles, Hans; Link, Doris (2011). OENORM-Bauvertrag – Praxiskommentar – OENORM B 2110, OENORM B 2118 und die BVergG-Vertragsbestimmungen. Vienna. Austrian Standards plus Publishing. (ISBN 978-3854022428)

Hoffmann, Friedrich; Motzko, Christoph; Corsten, Bernhard (2012). Aufwand und Kosten zeitgemäßer Schalverfahren – 3rd ed. Neu-Isenburg. Zeittechnik-Verlag GmbH. (ISBN 978 3- 939216-11-7)

Hofstadler (2010). Calculation of Construction Time for Building Projects – Application of the Monte Carlo Method to Determine the Period Required for Shell Construction Works. In: Full Paper Proceedings – CIB 2010 World Congress | May 10-13, 2010, Salford Quays – United Kingdom. Eds.: Barrett, Peter et al. International Council for Building (CIB). (ISBN 978-1-905732-91-3)

Hofstadler, Christian (2014). Produktivität im Baubetrieb – Bauablaufstörungen und Produktivitätsverluste. Berlin, Heidelberg. Springer Verlag (ISBN 978-3-642-41632-3)

Hofstadler, Christian (2014b). Methoden zur Ermittlung von Lebenszykluskosten. In: Nachhaltig Bauen mit Beton: Werkstoff und Konstruktion, Beton Graz'14 – 2. Grazer Betonkolloquium; September 25-26, 2014. Eds.: Nguyen, Viet Tue; Maydl, Peter; Freytag, Bernhard; Santner, Gerhard. 139-152. Graz. Verlag der Technischen Universität Graz. (ISBN 978-3-85125-360-3)

Hofstadler, Christian; Franzl, Gerald (2011). Bewehrungsarbeiten im Baubetrieb. Ed.: Verband Österreichischer Biege- und Verlegetechnik (Austrian Association for Bending and Installation of Concrete Reinforcement) (ISBN 978-3-200-02149-5)

Hofstadler, Christian; Kummer, Markus (2014). Systematischer Umgang mit Produktivitätsrisiken in der Auftragskalkulation. In: Tagungsband – 12. Grazer Baubetriebs- und Bauwirtschaftssymposium – Risiken im Bauvertrag – Baubetriebliche, bauwirtschaftliche und rechtliche Aspekte. Eds.: Heck, Detlef; Mauerhofer, Gottfried; Hofstadler, Christian. 53-90. Graz. Verlag der Technischen Universität Graz. (ISBN 978-3-85125-312-2)

Hofstadler, Christian; Kummer, Markus (2015a). Risiken und Chancen der Mischpreiskalkulation. In: Tagungsband – 13. Grazer Baubetriebs- und Bauwirtschaftssymposium – Preisermittlung und Vergabe in der Bauwirtschaft – Baubetriebliche, bauwirtschaftliche und rechtliche Aspekte. Eds.: Heck, Detlef; Mauerhofer, Gottfried; Hofstadler, Christian. 49-92. Graz. Verlag der Technischen Universität Graz. (ISBN 978-3-85125-385-9)

Hofstadler, Christian; Kummer, Markus (2017a). Chancen- und Risikomanagement in der Bauwirtschaft – Für Auftraggeber und Auftragnehmer in Projektmanagement, Baubetrieb und Bauwirtschaft. Berlin, Heidelberg. Springer-Verlag. (ISBN 978-3-662-54318-4)

Hruschka, Johann (1968). Die Anwendung des Normprodukts in der Baubetriebsplanung, Erläutert am Beispiel der Rohbauarbeiten im Hochbau. Doctoral thesis. Munich. Fakultät für Bauwesen der Technischen Hochschule München.

Hürlimann, Roland; Bucher, Jörg (2010). Bauzeit und Bauablauf im schweizerischen Baurecht. In: bau aktuell, 1. Jahrgang, Heft Nr. 1, January 2010. 10-15. Vienna. Linde-Verlag Ges.m.b.H. (ISSN 2077-4737)

Karasek, Georg (2016). Kommentar zur OENORM B 2110 – 3rd ed. Vienna. Manz'sche Verlags- und Universitätsbuchhandlung GmbH. (ISBN 978-3-214-13575-1)

Karasek, Georg; Duve, Helmuth (2009). Die Bauzeit im Bauvertrag – die baurechtliche und bauwirtschaftliche Betrachtung. In: Tagungsband – 2. Grazer Baubetriebs- und Baurechtsseminar – Spezielle Probleme des Bauvertrages und die neue OENORM B 2110. Eds.: Lechner, Hans; Heck, Detlef. 13-61. Graz. Verlag der Technischen Universität Graz. (ISBN 978-3-85125-030-5)

Kropik, Andreas (2014). Bauvertrags- und Nachtragsmanagement. Perchtoldsdorf. Eigenverlag. (ISBN 978-3-200-03502-7)

Kummer, Markus (2015a). Aggregierte Berücksichtigung von Produktivitätsverlusten bei der Ermittlung von Baukosten und Bauzeiten – Deterministische und probabilistische Betrachtungen. Doctoral thesis. Graz. Institute of Construction Management and Economics, Graz University of Technology.

Motzko, Christoph; Boska, Erik (2013). Ermittlung von Aufwandswerten bei Schalungsarbeiten im Sichtbetonbereich. In: BauPortal – Fachzeitschrift der Berufsgenossenschaft der Bauwirtschaft, 3/2013. 18-24. Berlin. Erich Schmidt Verlag GmbH & Co. KG. (ISSN 1866-0207)

Oberndorfer, Wolfgang (2010). Claim Management und alternative Streitbeilegung im Bau- und Anlagenvertrag – Teil 1: Grundlagen und Methoden – 2nd ed. Vienna. MANZ Verlag. (ISBN 978-3-2140-0546-7)

Oberndorfer, Wolfgang; Jodl, Hans Georg et al. (2001). Handwörterbuch der Bauwirtschaft – ON V 208 – 2nd ed. Vienna. ON Österreichisches Normungsinstitut. (ISBN 3-85402-072-4)

OENORM B 2061 (2020-05-01 edition) Preisermittlung für Bauleistungen – Verfahrensnorm.

OENORM B 2050 (2006-11-01 edition) Vergabe von Aufträgen über Leistungen – Ausschreibung, Angebot, Zuschlag – Verfahrensnorm.

Petzschmann, Eberhard (2000). Handbuch für Bauingenieure – Bauverfahrenstechnik und Baumaschineneinsatz. 5. Jahrgang, Heft 6. Ed.: Petzschmann, Eberhard. Brandenburgische Technische Universität Cottbus. (ISSN 1431-3243)

Reister, Dirk (2007). Nachträge beim Bauvertrag: Erkennen – Dokumentieren – Aufbau. Neuwied. Werner Verlag. (ISBN 978-3-8041-4647-1)

Roquette, Andreas; Viering, Markus; Leupertz, Stefan (2013). Handbuch Bauzeit – 2nd ed. Cologne. Werner Verlag. (ISBN 978-3-8041-4656-3)

Schach, Rainer; Otto, Jens (2011). Baustelleneinrichtung – Grundlagen – Planung – Praxishinweise – Vorschriften und Regeln – 2nd ed. Wiesbaden. Vieweg+Teubner Verlag. (ISBN 978-3-8348-1399-2)

Spranz, Dieter (1975). Arbeitszeiten im Baubetrieb – Ermittlung und Anwendung bei Kalkulation, Arbeitsvorbereitung und Leistungsentlohnung – Eine Genauigkeits- und Wirtschaftlichkeitsbetrachtung. Schriftenreihe des Instituts für Baubetriebslehre der Universität Stuttgart (TH) – Nr. 13. Ed.: Drees, Gerhard. Wiesbaden, Berlin. Bauverlag GmbH. (ISBN 3-7625-0570-5)

Wolkerstorfer, Herbert; Lang, Christian (2014). Praktische Baukalkulation – Erstellen der Kalkulationsformblätter gemäß OENORM B 2061 – 4th ed. Vienna. Linde Verlag. (ISBN 978-3-7073-1971-2)

9 Worked Examples – Client

This chapter describes worked examples from the client's perspective. However, the methodology and approaches outlined in this chapter can also be applied by contractors or other project stakeholders as far as their underlying structure is concerned.

Worked examples are categorized into project phases and broken down into fundamental aspects, or basics (if and when required), the actual task or initial situation, the modeling exercise, and the calculation and interpretation of results.

All calculations and/or simulations were performed using MS Excel in combination with the @Risk add-in. Manual recalculation may result in rounding differences compared to the figures stated in this chapter.

Unless otherwise indicated, simulation results are based on 50,000 iterations using the Latin Hypercube sampling method.

9.1 Project Phase 1 (PPH 1)

In the first project phase, the client performs developer or project development calculations as well as initial projections of the building's life-cycle costs. We then demonstrate the value that selected use cases of the Monte Carlo simulation method can add for the client with regard to their decision-making and project specification processes.

9.1.1 Developer Costing

A developer or project development costing exercise initially involves a rough cost estimate for the construction or extension of a specific building. This estimate is then continuously refined on the basis of increasingly specific planning and design briefs. In the second step, initial cost efficiency or commercial viability analyses can be performed for the project.[1)]

In the early project phases in particular, input parameters are fraught with uncertainties due to the early planning and design stage involved. The related calculations are thus

C. Hofstadler and M. Kummer, *Chances and Risks in Construction Management and Economics*, Springer Tracts in Civil Engineering,
https://doi.org/10.1007/978-3-030-64014-9_9

forecasts that need to be performed on the basis of experiential evidence and/or values stated in the literature.

By way of example[2], the application of the Monte Carlo simulation method is outlined for a simple developer analysis (calculation of total capital expenditure). This example should be allocated to Project Phase 1 (see Fig. 9.1). It has its own specific cost structure even though it largely replicates the system specified in DIN 276 or OENORM B 1801-1. This system requires adjustment to the circumstances and conditions prevailing in the specific situation, which particularly refers to the question of identifying the detailed cost items included in the individual cost components.

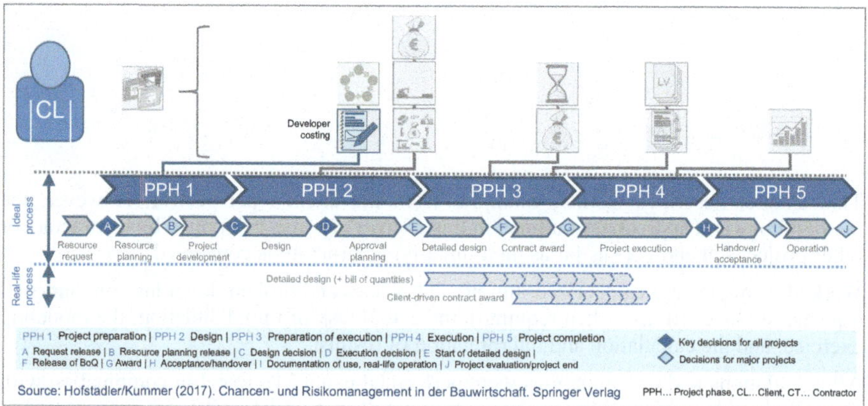

Fig. 9.1 Allocation of example in the project sequence

9.1.1.1 Basics

The simple developer analysis aims to clearly structure and break down the total costs of investing in a construction project. This example exclusively considers costs, including ancillary items, associated with land purchase, construction, and financing. Accordingly, this analysis does not include the costs of operation, maintenance, and dismantling/demolition.

Overall, six sub-areas of the developer analysis are examined and added up in order to arrive at the total cost:

- Land acquisition costs
- Site preparation costs
- Construction costs[3]
- Ancillary construction costs
- Project management fees
- Financing costs

---

[1] Cf. Glatte (2014), p. 203ff.
Cf. Diederichs (2006), p. 74ff.

[2] Based on an example by Lechner (2009), p. 134ff.

[3] In this context, the concept of construction costs refers to the costs of the building or structure as specified in OENORM B 1801-1:2015, p. 28. They comprise the shell construction, services and utilities, and finishing components.

9.1.1.2 Task

The total capital expenditure is calculated for a building construction project in Phase 1. The applied parameters are fraught with uncertainties, which is why they are factored into the calculation as distribution functions. Applying Monte Carlo simulations permits conclusions regarding potential capital expenditure ranges as well as the chance/risk ratio taken on the basis of a defined budget.

This calculation should make it possible to identify the range within which the proposed expenditure or investment is likely to lie, and whether the associated chance/risk ratio is appropriate. These findings can be used as a basis for defining project budgets that also include sufficiently large contingencies. All costs stated in this example are net amounts.

9.1.1.3 Modeling

Cost components are calculated independently of one another. For each of these components, the calculation sequence or flow is also shown separately in order to ensure effective guidance and to enhance the plausibility of calculation steps. Uncertain input parameters are represented by distribution functions, whereas deterministic inputs appear as arrows. In the figures, both intermediate and final results are shown as histograms.

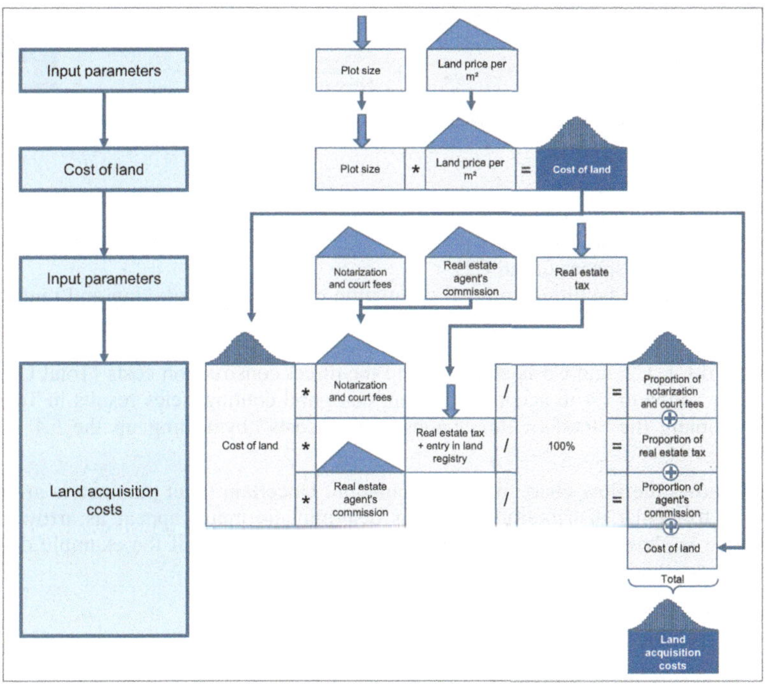

Fig. 9.2 Calculation flow chart – Land acquisition costs

Land acquisition costs

Land acquisition costs include the cost of the land itself, the cost of notarization, the clerk's fee, real estate tax, land registry fees, and the real estate agent's commission. For the purpose of simulating this example, all input parameters whose costs lie within a specific range are represented by triangular distributions. The only deterministic parameters are the plot size or area and associated fixed markups (see Fig. 9.2).

Site preparation costs

Additional site preparation costs are incurred if the plot to be purchased is covered with buildings or contaminated. Likewise, costs may be incurred by creating external access via driveways or by establishing utility connections, as well by the transfer of rights or title or other items. These costs must be considered in the developer analysis of site preparation costs.

In this example, only the costs of demolishing an existing building and of transferring an easement are considered. These two cost components are modeled by triangular distributions (see Fig. 9.3).

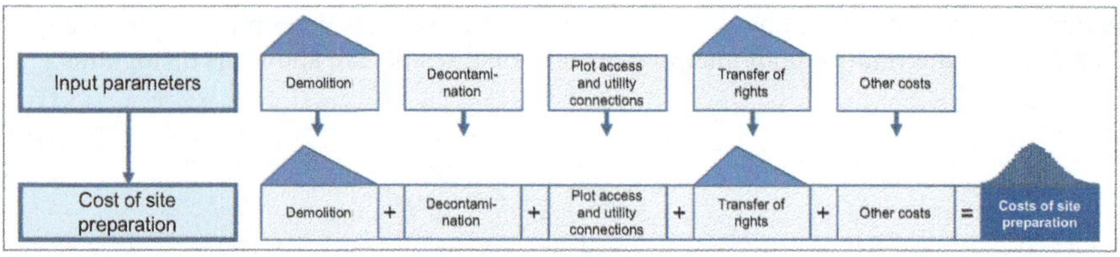

Fig. 9.3 Calculation flow chart – Site preparation costs

Construction costs

Direct construction costs include the following cost components:

- Costs of usable areas (Total C1)
- Costs of parking spaces (Total C2)
- Costs of outdoor facilities, internal circulation and access, marketing and rental costs (Total C3)[4)]

The total of C1, C2, and C3 is equivalent to the direct construction costs (Total C4). The percentage added to C4 to account for other costs and contingencies results in Total C5. We then obtain the "total of direct construction costs" by adding up the C4 and C5 portions.

Fig. 9.4 shows the flow chart for this calculation. Uncertain input parameters are represented by triangular distributions, whereas deterministic inputs appear as arrows.[5)] No distribution is shown in the flow chart (such as "Other areas") if the example does not specify costs of individual parameters.

4) The costs of internal access and circulation and marketing costs are not considered in this example.

5) The amount payable for the real estate agent's commission was previously agreed to be equivalent to 2.5 monthly rents.

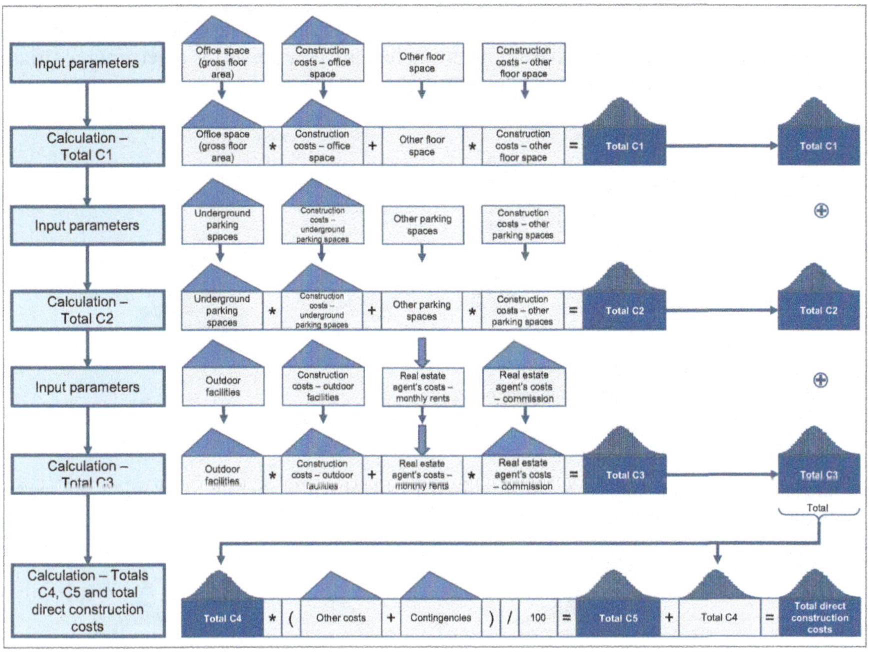

Fig. 9.4 Flow chart – Direct construction costs

Ancillary construction costs

Besides direct construction costs, ancillary construction costs must also be considered in the developer analysis. These include, for example, the costs of insurance, expert opinions, or surveying.

In this example, ancillary construction costs are determined as a percentage of direct construction costs. Adding up direct and ancillary construction costs results in the total cost of construction (see Fig. 9.5).

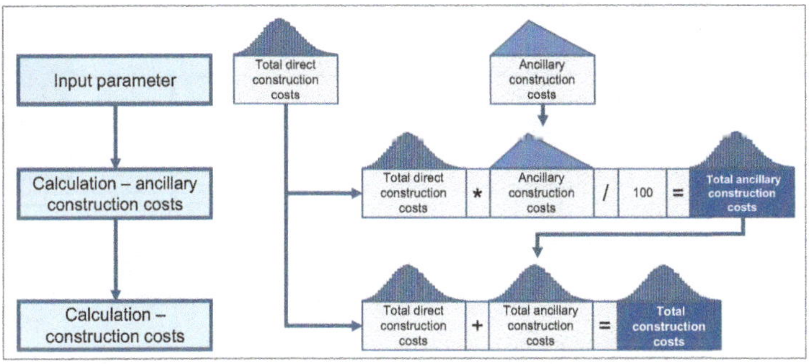

Fig. 9.5 Flow chart – Ancillary construction costs and total cost of construction

Project management fees

Project management fees are another component to be added to land acquisition and site preparation costs. In the worked example, these fees are determined as a percentage (triangular distribution) of the total cost of construction (see Fig. 9.6).

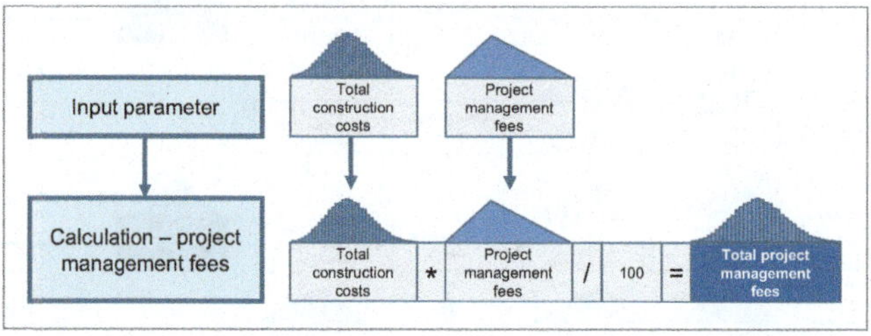

Fig. 9.6 Flow chart – Project management fees

Financing costs

Additional debt financing costs are incurred if the client is unable to provide the entire construction budget from their own funds.

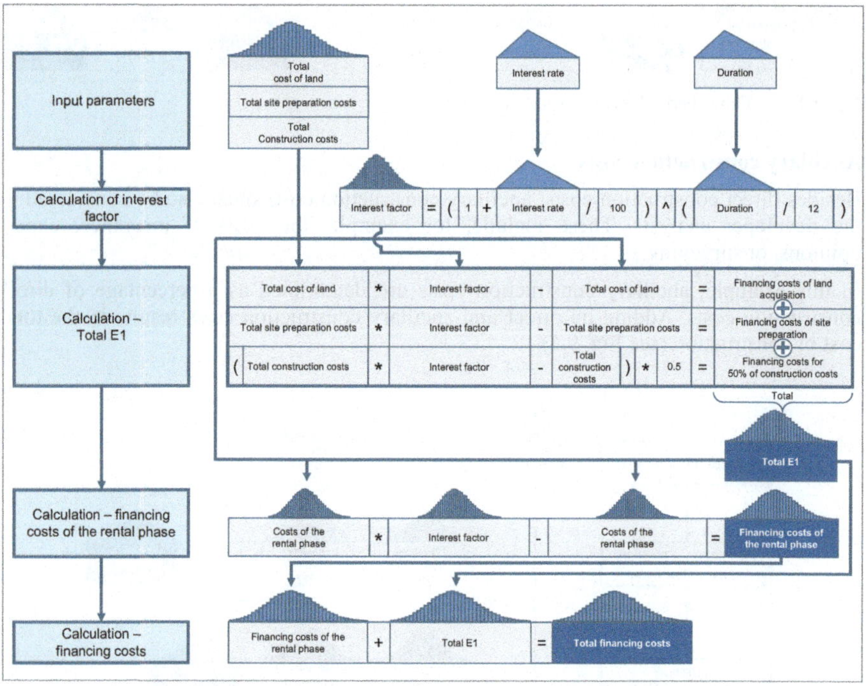

Fig. 9.7 Calculation flow chart – Financing costs

In this example, we assume the costs of land acquisition and site preparation, as well as the costs of the rental phase, to be fully covered by external funds. However, only half of the construction costs are to be debt-financed.[6)]

The interest rates shown as triangular distributions are taken to be identical for all four financing cost components, but the maturities of these four options (stated in months) differ. Fig. 9.7 shows the flow chart for this calculation.

Total capital expenditure

Total capital expenditure includes the following cost components:

- Land acquisition costs
- Site preparation – connection of utilities
- Construction costs
- Ancillary construction costs
- Project management fees
- Financing costs

Since the proportion of contingency costs was already included in the construction costs, with percentages of fees and ancillary costs added, the category of contingencies (reserves) is not explicitly referred to at this stage.

Fig. 9.8 shows the flow chart for this calculation. The histograms previously determined for the individual cost components are used as input parameters.

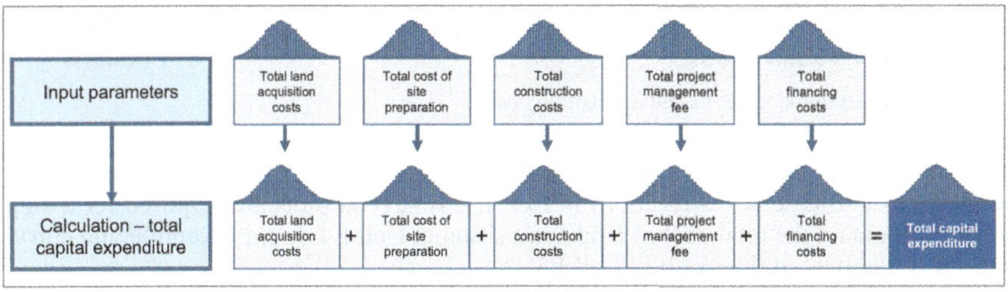

Fig. 9.8 Calculation flow chart – Total capital expenditure

9.1.1.4 Calculation and Interpretation of Results

Total capital expenditure is calculated separately for the individual categories. This section explains and interprets the calculation tables, including the associated input parameters, and the results in the form of histograms. At the end of this section, total capital expenditure is shown in a tornado diagram in order to evaluate the influence of the input parameters on the output.

Land acquisition costs

Table 9.1 determines the land acquisition costs for the proposed construction project. In this step, the plot size is applied as a deterministic value because a specific plot of land in an inner-city location was identified for the project. However, the price per square meter is

6) This is a general assumption although construction costs are associated with uncertainties, which is why the amount to be financed will change in each iteration. Alternatively, available equity could be incorporated into the calculation as a fixed value.

still fraught with uncertainties; it is estimated between €1,150 and €1,400/m². The most likely price amounts to €1,200/m². The land purchase price (Row 2) is defined as a triangular distribution in Cell E2.

Land acquisition costs					
No.	Parameter	Amount/value			
		MIN	MOST LIKELY	MAX	Distribution
0	A	B	C	D	E
1	Plot size		1,400.00 m²		1,400.00 m²
2	Land price per m²	€1,150.00/m²	€1,200.00/m²	€1,400.00/m²	€1,250.00/m²
3	Cost of land				€1,750,000.00
4	**Total A1**				**€1,750,000.00**
5	Percentage – Notarization and court fees	1.20%	1.50%	1.80%	1.50%
6	Notarization and court fees				€26,250.00
7	Percentage – Real estate tax (3.50%) + entry in land registry (1.10%)		4.60%		4.60%
8	Real estate tax				€80,500.00
9	Percentage – Real estate agent's commission	2.40%	3.00%	3.60%	3.00%
10	Real estate agent's commission				€52,500.00
11	**Total A2**				**€159,250.00**
12	**Total land acquisition costs**				**€1,909,250.00**

Table 9.1 Calculation table – Land acquisition costs

Land acquisition costs are then determined by multiplying the plot size with the land price (modeled as a triangular distribution) in Cell E3. If several plots are acquired for a single project, we obtain the total cost of land acquisition in Cell E4. In this example, the entry in Cell E4 is identical to the result in Cell E3.

Beyond the cost of the land as such, other cost components must also be considered as a percentage of the land cost in the context of acquiring landed property, including notarization and clerk fees, real estate tax, and real estate agents' commissions. Rows 5 and 9 show the percentages of notarization and court fees and of the real estate agent's commission as triangular distributions. The percentage of real estate tax is prescribed by law, which is why it is stated as a deterministic value at 3.50%, whereas the fee for entry in the land registry is assumed to be 1.10% (Row 7). Each of the cost components associated with the land purchase transaction is calculated on the basis of the cost of the land (Cell E4). Cell E11 shows the total of the costs incurred over and above the land costs as such. Total land acquisition costs are shown as a histogram in Cell E12 as the sum of Cells E4 and E11 (see Fig. 9.9).

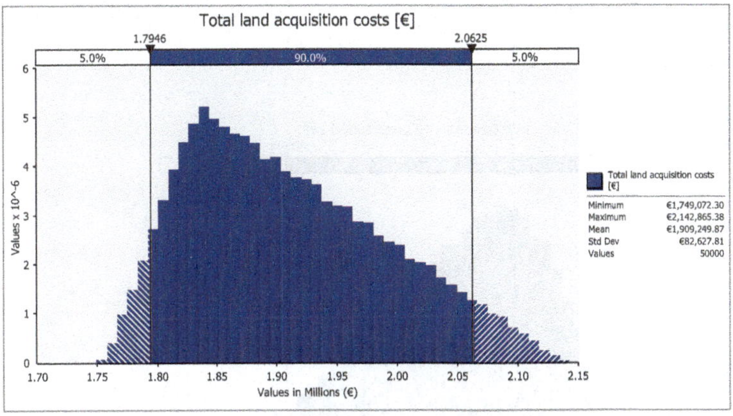

Fig. 9.9 Histogram – Land acquisition costs

The curve of the histogram of land acquisition costs is essentially determined by the triangular distribution of the land price per square meter or, ultimately, by the cost of the land. The proportion of the costs associated with land acquisition is relatively small compared to the cost of land as such, and thus hardly influences the curve of the distribution. Land acquisition costs lie between approx. €1.75 million and €2.14 million. The mean of the histogram amounts to approx. €1.91 million.

Site preparation – connection of utilities

The second component of total capital expenditure is associated with the site preparation stage. This component includes the following cost items (see also Table 9.2):

- Demolition of existing buildings
- Decontamination (e.g. due to soil contamination)
- Plot access and utility connections
- Transfer of rights (easements)
- Other costs

In this example, only the costs of demolishing an existing building (Row 1) and of transferring an easement (Row 4) are considered. In an early project phase, it is not yet possible to state exact amounts (i.e. deterministic values) for these two cost components. Rather, they are associated with uncertainties, and are thus represented by triangular distributions.

Site preparation					
No.	Parameter	Amount/value			
		MIN	MOST LIKELY	MAX	Distribution
0	A	B	C	D	E
1	Demolition	€40,000.00	€50,000.00	€80,000.00	€56,666.67
2	Decontamination				€0.00
3	Plot access and utility connections				€0.00
4	Transfer of rights	€15,000.00	€20,000.00	€26,000.00	€20,333.33
5	Other costs				€0.00
6	**Total costs of site preparation**				**€77,000.00**

Table 9.2 Calculation table – Site preparation costs

We obtain the total cost of site preparation as a histogram in Cell E6 of Table 9.2. This histogram is shown in Fig. 9.10. The mean amounts to approximately €77,000, and 90% of all cases lie roughly between €64,240 and €93,012.

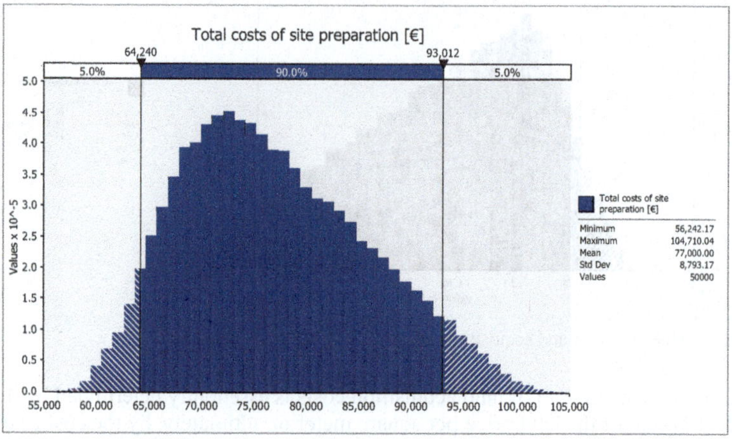

Fig. 9.10 Histogram – Site preparation costs

Direct construction costs

Table 9.3 is used for calculating the components of direct construction costs with the aid of Monte Carlo simulations. Columns B, C, and D show the minimum, most likely, and maximum values (as far as these are available) for each calculation parameter. The defined triangular distributions of input parameters, or the intermediate and final results, are entered in Column E.

The office space (usable floor space) relates to the gross floor area; it is shown in Row 1 of Table 9.3. The corresponding construction costs of the office space can either be derived from experiential evidence from completed projects, or values stated in pertinent literature, such as BKI, can be referred to.[7)]

Rows 3 and 4 provide the option of including additional areas, such as residential or warehousing, in the costing exercise. Further rows can be inserted in the table for this purpose.

---

7) Baukosteninformationszentrum Deutscher Architektenkammern (BKI; Construction Cost Information Center of the German Chambers of Architects). An own related example is included in Section 9.2.3.

Construction costs					
No.	Parameter	Amount/value			
		MIN	MOST LIKELY	MAX	Distribution
0	A	B	C	D	E
1	Office space (gross floor area)	6,450.00 m²	7,000.00 m²	7,350.00 m²	6,933.33 m²
2	Construction costs – office space	€950.00/m²	€1,100.00/m²	€1,400.00/m²	€1,150.00/m²
3	Other floor space				€0.00
4	Construction costs – other floor space				€0.00
5	**Total C1**				**€7,973,333.33**
6	Underground parking spaces	50.00 pcs	55.00 pcs	60.00 pcs	55.00 pcs
7	Construction costs – underground parking spaces	€11,000.00/pc	€15,000.00/pc	€21,000.00/pc	€15,666.67/pc
8	Other parking spaces				€0.00
9	Construction costs – other parking spaces				€0.00
10	**Total C2**				**€861,666.67**
11	Outdoor facilities	300.00 m²	360.00 m²	450.00 m²	370.00 m²
12	Construction costs – outdoor facilities	€90.00/m²	€100.00/m²	€120.00/m²	€103.33/m²
13	Real estate agent's costs – monthly rents		2.50 mo		2.50 mo
14	Real estate agent's costs – commission	€16,800.00/mo	€17,500.00/mo	€18,000.00/mo	€17,433.33/mo
15	**Total C3**				**€81,816.67**
16	**Total C4 (C1 + C2 + C3)**				**€8,916,816.67**
17	Other costs	1.80%	2.00%	2.40%	2.07%
18	Contingencies	2.50%	3.00%	5.00%	3.50%
19	**Total C5**				**€496,369.46**
20	**Total direct construction costs**				**€9,413,186.13**

Table 9.3 Calculation table – Direct construction costs

Cell E5 calculates the costs for the usable areas (Total C1). For this purpose, the individual areas are multiplied with the construction cost parameters and added up.

Only underground car parking is specified for the building construction project referred to in this example (no other parking spaces are included in the brief). In this early phase, the exact number of parking lots cannot be specified because the design exercise has not yet been completed; it is thus associated with uncertainties (Row 6). Consequently, the construction costs incurred by the underground car parking facility are also uncertain. This is why a triangular distribution is chosen to incorporate them into the calculation (Row 7).

Further rows can be inserted in the table if needed for other types of facilities, such as garages or outdoor parking, or for transferring existing parking spaces.

Cell E10 shows the total cost of parking spaces (Total C2) as a result of multiplying the number of parking spaces with the associated portions of the construction costs.

Total C3 (Cell E15) is composed of the costs of outdoor facilities and brokerage costs (real estate agents' commissions). In addition, it would be plausible to include the costs of internal access and circulation, marketing, etc. The outdoor facilities include an area of 300 to 450 m² (most likely 360 m²) and incur estimated costs between €90 and €120/m² (most likely €100/m²). The agent's commission has been estimated at 2.5 monthly rents (i.e. a deterministic value entered in Row 13), yet rental revenues are also associated with uncertainties owing to the interplay of supply and demand and thus realizable rents while also considering a selected default risk (Row 14).

Direct construction costs (Total C4) are composed of the cost components calculated above (Total C1, Total C2, and Total C3). In addition, for the purpose of determining construction costs, percentage markups are applied to other costs, such as land tax over the service life of the building determined by the project developer, costs of waste disposal, sewage, municipal fees, and contingencies. The percentages for other costs (1.80 | 2.00 | 2.40%) and contingency reserves (2.50 | 3.00 | 5.00%) are entered as triangular distributions in Rows 17 and 18. The sum of these markups (relative to Total C4) is given in Cell E19 (Total C5).

Finally, Cell E20 shows the histogram of the total of direct construction costs (Total C4 + Total C5). The histogram is shown in Fig. 9.11. At 50,000 iterations, it includes a range from €7.43 million to €11.81 million. The mean amounts to approximately €9.41 million, whereas the standard deviation roughly amounts to €732,000.

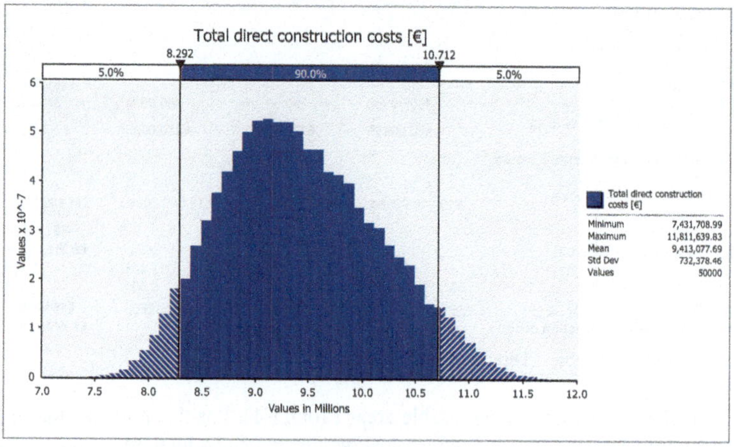

Fig. 9.11 Histogram – Total direct construction costs

Ancillary construction costs

In this example, ancillary construction costs are expressed as percentages (10 | 15 | 20%) of the total of direct costs (Cell E1 in Table 9.4 – this is the histogram shown in Fig. 9.11). These percentages are considered in the modeling exercise.

Ancillary construction costs					
No.	Parameter	Amount/value			
		MIN	MOST LIKELY	MAX	Distribution
0	A	B	C	D	E
1	Total direct construction costs				€9,413,186.13
2	Ancillary construction costs	10.00%	15.00%	20.00%	15.00%
3	**Total ancillary construction costs**				**€1,411,977.92**
4	**Total construction costs**				**€10,825,164.05**

Table 9.4 Calculation table – Total ancillary construction costs and total construction costs

Cell E3 shows the ancillary construction costs as a histogram (see Fig. 9.12). The mean amounts to approximately €1.41 million whereas the standard deviation roughly amounts to €221,400. In about 5% of simulated cases, an amount of €1.06 million is underrun and an amount of €1.79 million is exceeded.

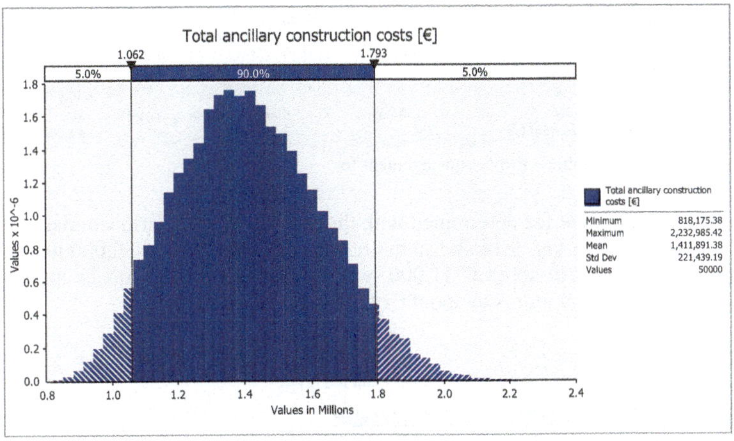

Fig. 9.12 Histogram – Total ancillary construction costs

Calculating the total of ancillary construction costs (Cell E3) and direct construction costs (Cell E1) results in the total cost of construction shown in Cell E4. Fig. 9.13 shows a histogram of this total cost, which forms the basis for determining the project management fee. As demonstrated in Fig. 9.13, an estimated budget of €11.0 million is sufficient to cover about 60% of simulated construction cost cases. For risk mitigation purposes, the construction cost budget would have to be increased further until the envisaged chance/risk ratio is reached.

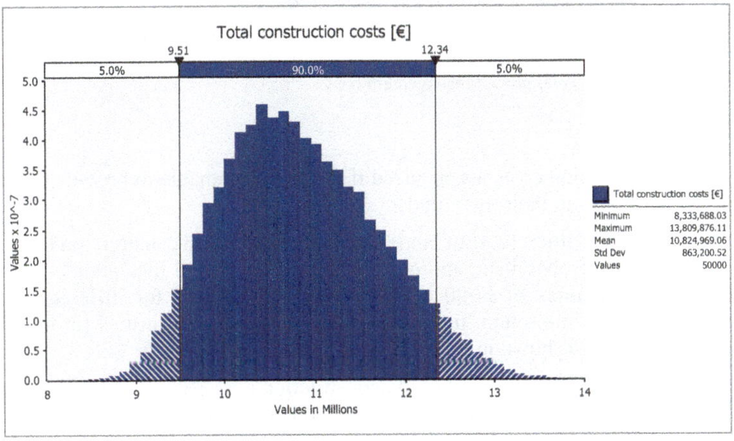

Fig. 9.13 Histogram – Total construction costs

Project management fee

In this example, the project management fee is also modeled in the form of percentages (8 | 10 | 12%) (see Row 2 of Table 9.5). It refers to the total construction cost shown as a histogram in Fig. 9.13.

No.	Parameter	Amount/value			
		MIN	MOST LIKELY	MAX	Distribution
0	A	B	C	D	E
1	Total construction costs				€10,825,164.05
2	Project management fee	8.00%	10.00%	12.00%	10.00%
3	**Total project management fee**				**€1,082,516.40**

Table 9.5 Calculation table – Project management fee

The project management fee determined with the aid of a Monte Carlo simulation is given in Cell E3 of Table 9.5. Fig. 9.14 shows the related histogram. After 50,000 iterations, the fee minimum amounts to about €711,000, whereas the maximum equals approximately €1,557,000. The mean amounts to about €1,082,000.

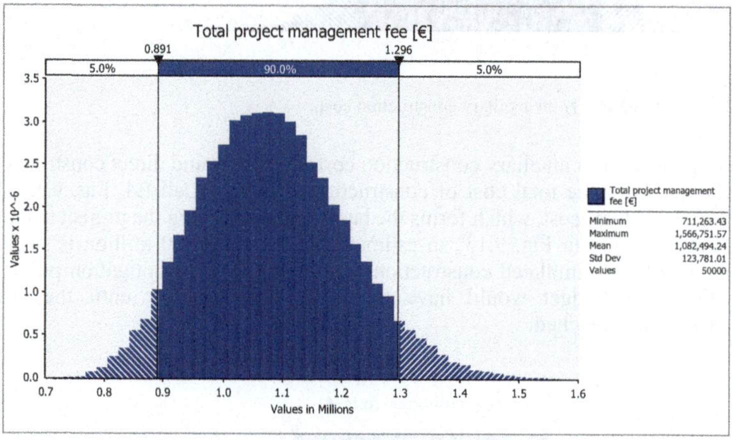

Fig. 9.14 Histogram – Total project management fee

Financing costs

Additional debt financing costs are incurred if the client is unable to provide the required construction budget from their own funds.

The costs of land acquisition (cost of land), site preparation, and construction have already been calculated in a probabilistic approach, resulting in related histograms. These costs carry interest at the rates of (2.00 | 3.00 | 4.50%) specified for different maturities (compounding[8)]). The minimum, most likely, and maximum maturities (in months) and related interest rates are shown in Columns B, C, and D of Table 9.6.

It should also be noted that only half of the construction cost is covered by debt financing. Thus, the calculated cost of financing the cost of construction is reduced by 50%.

The financing costs of land acquisition, site preparation, and 50% of the construction costs are shown in Cells E4, E8, and E12 of Table 9.6, and added up to the Total E1 in Cell E13.

---

8) The calculation of financing costs is based on a deterministic example for the values taken from Rows 1 to 4 of Table 9.6: Financing costs of land acquisition:
€98,345.78 = €1,909,250.00 * (1 + 3.17% / 100%)^(19.33 mo / 12 mo) - €1,909,250.00
Minor deviations occur during manual recalculation owing to the fact that the interest rate and the duration include more than two decimal places.

No.	Parameter	Financing costs			
		Amount/value			
		MIN	MOST LIKELY	MAX	Distribution
0	A	B	C	D	E
1	Total cost of land				€1,909,250.00
2	Duration	16.00 mo	18.00 mo	24.00 mo	19.33 mo
3	Interest rate	2.00%	3.00%	4.50%	3.17%
4	**Financing costs of land acquisition**				**€98,345.78**
5	Total site preparation				€77,000.00
6	Duration	16.00 mo	18.00 mo	24.00 mo	19.33 mo
7	Interest rate				3.17%
8	**Financing costs of site preparation**				**€3,966.28**
9	Total construction costs				€10,825,164.05
10	Duration	12.00 mo	14.00 mo	18.00 mo	14.67 mo
11	Interest rate				3.17%
12	**Financing costs for 50% of construction costs**				**€210,218.09**
13	**Total E1**				**€312,530.15**
14	Costs of the rental phase				€13,123,944.20
15	Duration	5.00 mo	6.00 mo	10.00 mo	7.00 mo
16	Interest rate				3.17%
17	**Financing costs of the rental phase**				**€240,852.53**
18	**Total financing costs**				**€553,382.68**

Table 9.6 Calculation table – Financing costs

The costs of the rental phase (Cell E14) include the land acquisition costs (Cell E1), the costs of site preparation (Cell E5), construction costs (Cell E9), and the Total E1 (Cell E13). The duration of financing the rental phase is indicated as a triangular distribution between 5 and 10 months (most likely value: 6 months). The interest rate is left unchanged compared to the calculation of the other financing cost items.[9)]

Adding up the financing costs of the rental phase (Cell E17) and Total E1 (Cell E13) results in the histogram of the total financing costs shown in Cell E18 (see Fig. 9.15).

The mean amounts to approximately €554,000 whereas the standard deviation roughly amounts to €106,000. In 90% of simulated cases, financing costs roughly lie between €0.39 and €0.74 million.

[9)] The calculation of financing costs for the rental phase is based on a deterministic example for the values taken from Rows 14 to 17 of Table 9.6:
Financing costs of the rental phase:
€240,852.53 = €13,123,944.20 * (1 + 3.17% / 100%)^(7 mo / 12 mo) - €13,123,944.20
Minor deviations occur during manual recalculation owing to the fact that the interest rate and the duration include more than two decimal places.

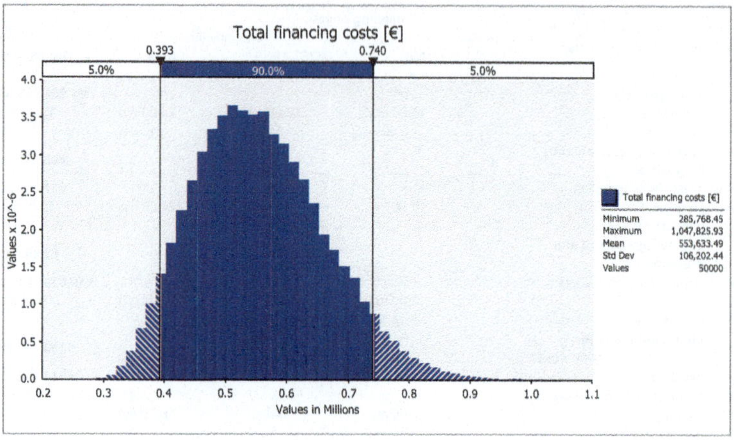

Fig. 9.15 Histogram – Total financing costs

Total capital expenditure

Table 9.7 is used for calculating total capital expenditure on the basis of the histograms previously derived for the individual cost components. Columns B, C, and D are empty because no new input parameters are introduced in this calculation step. Total capital expenditure is shown in Cell E6 (see the histogram in Fig. 9.16).

Total capital expenditure					
No.	Parameter	Amount/value			
		MIN	MOST LIKELY	MAX	Distribution
0	A	B	C	D	E
1	Total land acquisition costs				€1,909,250.00
2	Total cost of site preparation				€77,000.00
3	Total construction costs				€10,825,164.05
4	Total project management fee				€1,082,516.40
5	Total financing costs				€553,382.68
6	**Total capital expenditure**				**€14,447,313.13**

Table 9.7 Calculation table – Total capital expenditure

The histogram shows a range between approximately €11.64 million and €17.86 million. This relatively large spread is due to the uncertainties that are inevitably associated with the input parameters in an early project phase (PPH 1) because of the lack of project detailing.

The mean shown in the histogram amounts to approximately €14.45 million, whereas the standard deviation roughly amounts to €1.00 million. In 90% of simulated cases, total capital expenditure roughly lies between €12.93 and €16.20 million.

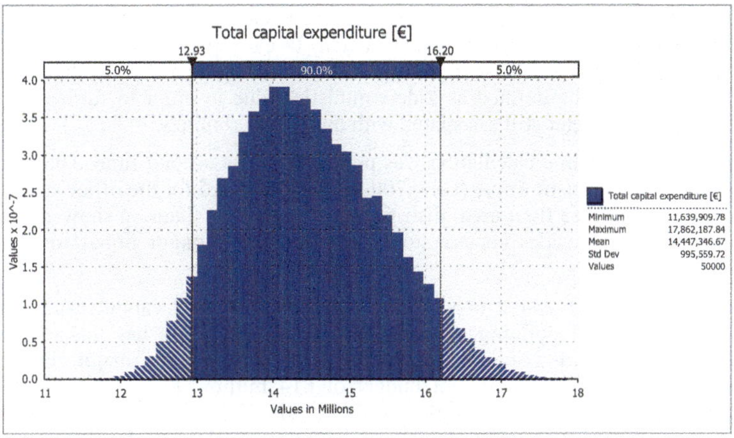

Fig. 9.16 Histogram – Total capital expenditure

Besides showing the spread and the probabilities of occurrence of total capital expenditure amounts, it is also of particular interest to identify the input parameters with the most significant influence on the output (i.e. total capital expenditure).

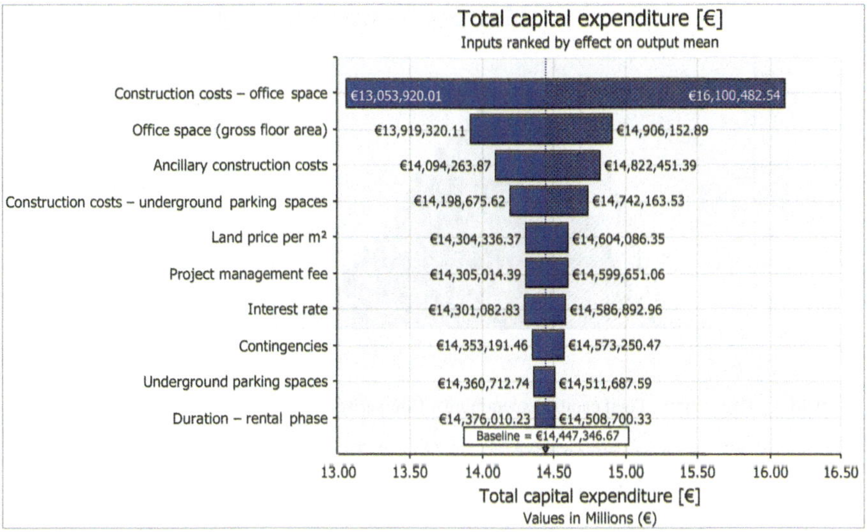

Fig. 9.17 Tornado diagram – Total capital expenditure

Fig. 9.17 includes a tornado diagram for total capital expenditure (see Section 7.4.2 for more information on how to interpret tornado diagrams). The two inputs for the office space construction costs (Row 2 in Table 9.3) and for the office space as such (Row 1 in Table 9.3) have the greatest influence on the spread of total capital expenditure. It is thus recommended to narrow the range of construction costs in order to implement cost savings or to limit the total capital expenditure spread. This is achieved by defining qualitative and quantitative factors, but would be possible only at a stage when the construction project has been specified and designed in greater detail. In early project phases, however, it is already possible to define fixed land prices or debt financing interest rates. In addition, the office space can also be defined as a deterministic value in order to further specify an important input parameter still associated with major uncertainties.

For comparison, the same calculation was performed for a second time. The only difference is that a deterministic value (i.e. 6,700 m²) was assumed for the office space instead of an office space (gross floor area) distribution. The tornado diagram showed that it was precisely this input, besides construction costs, that had a major impact on the office space.

Fig. 9.18 illustrates the extent to which the histogram of total capital expenditure has changed as a result of defining a fixed office space. Not only has this resulted in a reduction in the mean and standard deviation; it has also led to a major change in the chance/risk ratio. For an investment budget of €14 million, the chance/risk ratio has shifted from about 36 to 64% to almost a 50/50 ratio. This means that the trend of the chance/risk ratio was reversed simply by defining the office space while keeping the budget constant. As a result of defining a specific office space, the probability of under-running the budget of €14 million has risen to about 50% and thus become roughly identical to the risk of exceeding this budget.

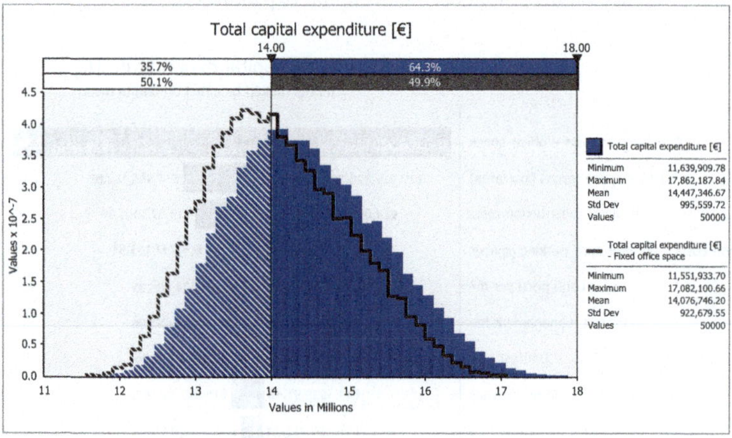

Fig. 9.18 Histogram – Total capital expenditure – Comparison

9.1.1.5 Benefit for the Client

In early project phases in particular, clients still exert a significant influence and have ample room to maneuver in their projects. Conversely, however, this also means that the uncertainties in terms of costs and deadlines are also very high. Uncertain parameters are thus modeled using distribution functions in order to consider these uncertainties in the related calculations, and to derive recommendations for actions. In this model, the sequence of computation steps corresponds to a deterministic calculation. However, the Monte Carlo simulation method (which delivers histograms) is applied for determining capital expenditure in a probabilistic approach.

On the one hand, this type of representation comes with the advantage of raising awareness of inevitable spreads and ranges. On the other hand, chance/risk ratios can be shown for specific deterministic values.

In addition, an in-depth analysis of total capital expenditure (for instance by using tornado diagrams) can be performed in order to identify those input parameters that have a particularly large influence on the output.

For the example outlined in this section, one specific recommendation was identified, i.e. to define a fixed office space as early as possible in the project. This simple response has a direct effect on the spread of future capital expenditure amounts and on the chance/risk ratio of individual deterministic values.

We recommend to update and further refine the developer costing exercise in subsequent project phases in the light of new results. Besides mere costing, the developer analysis is also suitable for estimating revenues (such as from rental income). This step is useful to determine the initial yield or the refinancing factor for the project. However, care should be taken to also consider potential vacancy rates or maintenance and rehabilitation costs in the course of this calculation. Such an analysis paves the way to life-cycle cost analyses.

9.1.2 Life-Cycle Cost Considerations

Initial life-cycle analyses are performed as early as in Project Phase 1 (see Fig. 9.19) and are then part of all project phases. Yet life-cycle costs are influenced not only by capital expenditure or investment cost (corresponding to construction costs in building projects) but, even more importantly, by the costs of building operation and maintenance.

Due to the applied interest rate, it is crucial to determine when and to what extent expenses are made or revenues received over the life cycle of a construction project.

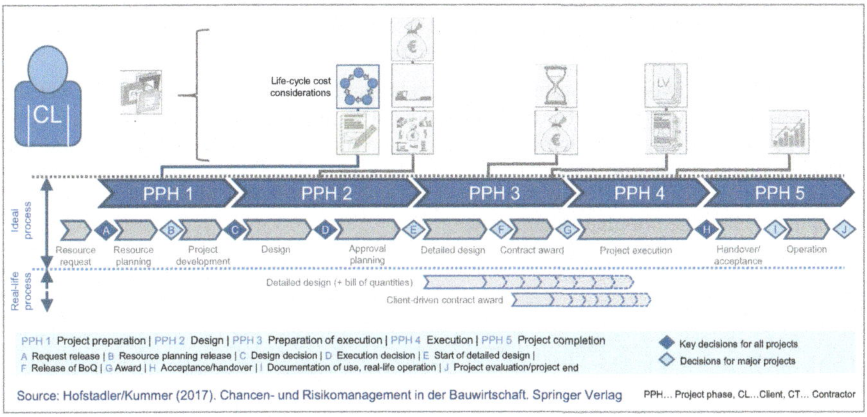

Fig. 9.19 Allocation of example in the project sequence

9.1.2.1 Basics

In the construction sector, life-cycle costs are essentially determined by a number of factors, including the type of building or structure, the characteristics of discretionary and elementary production factors, the quality of the combination and processing of these factors, the type of use, maintenance and care, and, last but not least, the applied calculation methods and their input parameters.[10)]

Calculating life-cycle costs forms part of investment calculus and aims to evaluate investment decisions with regard to their beneficial nature, or profitability.

Specific methods of investment calculus were developed because investment decisions are one of the most important decision categories in business and management, which is why such decisions should never be based on pure intuition. Generally speaking, the wide range of economic and commercial criteria for investment alternatives is condensed to just a single metric that should be as reliable as possible. Fig. 9.20 shows different methods of investment calculus focusing on both micro- and macroeconomic criteria. Besides this categorization, it is also possible to make a distinction in terms of expectations, which is not explicitly referred to at this stage. Methods of reliable, or certain, expectation are employed if sound and plausible input parameters are available. If data is fraught with uncertainties (which is almost always the case for forward-looking analyses), it can be varied using methods of unreliable, or uncertain, expectation.[11)]

The basic principle of this approach also referred to as the conventional method is to relate the economic or commercial criteria that apply to the entire service life to an average period, which is usually an average year of use for which annual averages are used. This is why the temporal aspect of payments, i.e. the exact times or dates of payments, is neglected.[12)]

In terms of the calculation, it thus makes no difference if scheduled major repair and maintenance works are carried out in the fourth or fifth year of use, for instance, as long as total maintenance costs are equal across the entire service life in both cases (and thus the average maintenance cost per year of use). This method has been strongly criticized not least because of its neglect of exact payment times.[13)]

Dynamic methods avoid averaging; instead, exact payment times are considered over the entire service life or specified period. Consequently, the totality of cash flows must be known for the building's service life, i.e. the actual income generated (cash inflows) and expenses made (cash outflows) in the individual periods. These cash flows are then compounded or discounted employing the compounding method known from financial mathematics and the imputed interest rate i_i.[14)]

10) Cf. Hofstadler (2014c), p. 139

11) Cf. Veit (2012), p. 106ff.

12) Cf. Bauer (2009), p. 7-19ff.

13) Cf. Hofstadler (2014c), p. 141

14) Cf. Bauer (2009), p. 7-32ff.

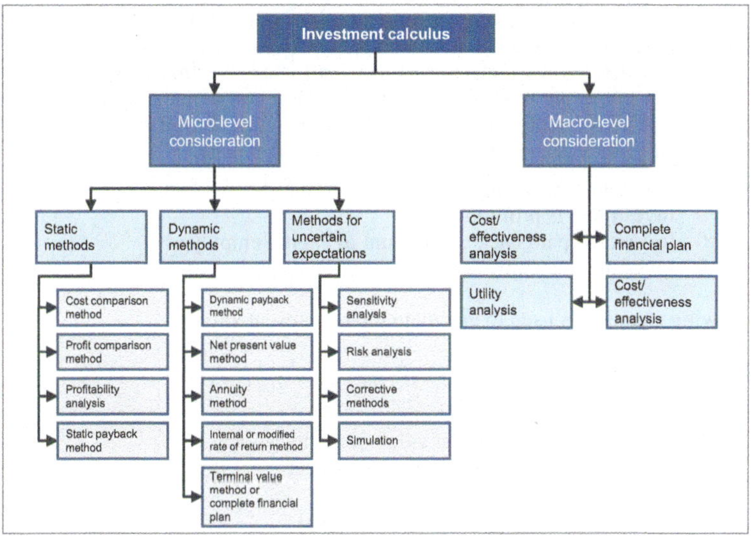

Fig. 9.20 Methods of investment calculus[15)]

Generally speaking, life-cycle analyses involve the determination of costs in order to allocate a value to cash flows (inflows and outflows) in relation to their amounts and points in time when they are incurred. Dynamic methods are predominantly applied to comparatively long periods. Discounting makes these cash flows comparable to the interest rate, which enables a phase-specific identification and analysis of life-cycle costs.[16)]

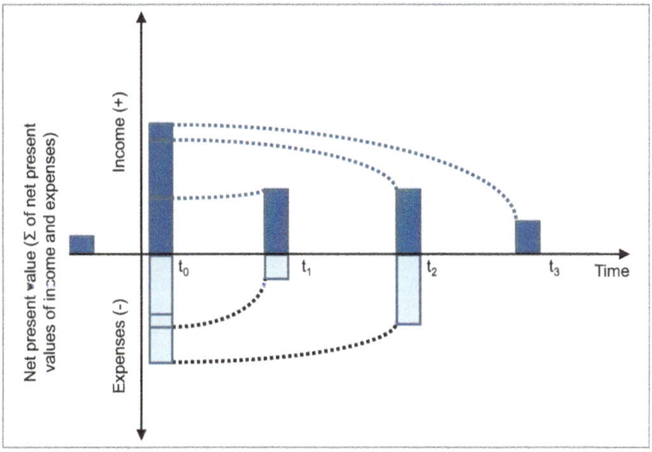

Fig. 9.21 Methodology for determining the net present value

[15)] Cf. Bichler (2014), p. 36

[16)] Cf. Hofstadler (2014c), p. 142

In this example, we apply the net present value method for this purpose (for the methodology, see Fig. 9.21). The net present value method starts from the assumption that the net present value of any and all future expenses (costs) and revenues (income) plus construction cost permits a conclusion as to whether an investment is profitable or not. The net present value is the total of the cash flows discounted by the imputed interest rate i, adjusted for the investment amount at a given point in time (t_0). It represents the amount that the investor generates over and above the capital employed.

This results in the following decision-making criteria:[17)]

- $NPV > 0$ Investment is profitable
- $NPV = 0$ Investor is paid back the amount of capital employed
- $NPV < 0$ Investment is unprofitable

The following equation is used to calculate the net present value:

$$NPV = -I_0 + \sum_{t=1}^{T} (I_t - O_t) \cdot (1 + i_I)^{-t} + D_G \cdot (1 + i_i)^{-T} \qquad (9\text{-}1)$$

NPV .... Net present value of an investment
I_0 .... Investment amount or cost
I_t .... Cash inflows (income)
O_t .... Cash outflows (expenses)
T .... Reporting period, most recent period
t .... Continuous index in period (e.g. years)
i_i .... Imputed interest rate
D_G .... Divestment gain or dismantling/demolition costs

9.1.2.2 Task

We use a simple example to illustrate the sequence of a life-cycle cost analysis based on the net present value method. In this example, an investment I_0 is made at the time t_0, and nominally identical income is assumed at the times t_1, t_2, and t_3 (see Fig. 9.22).

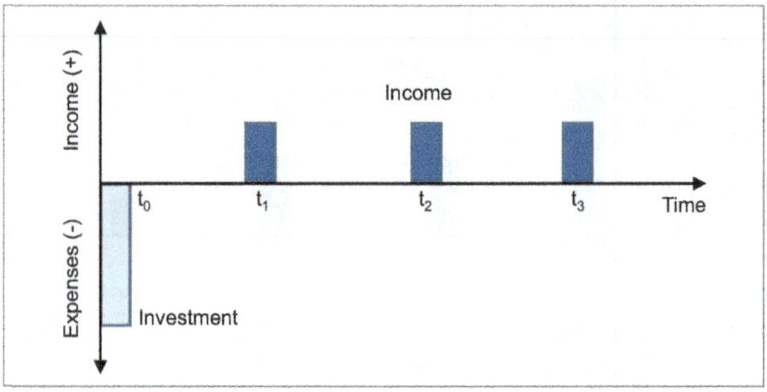

Fig. 9.22 Task

17) This particularly applies to investments that are assumed to generate income. Construction projects often do not generate income, with the exception of rental income in real estate or road toll income in the field of infrastructure. In this case, however, the net present values of different options can be compared with each other nonetheless. For this purpose, expenses are also often stated as positive amounts in order not to use negative parameters throughout the process.

The input parameters are represented by triangular distributions in order to link the net present value method to the Monte Carlo simulation. This applies to both the investment and the income side, as well as to the interest rate on capital.

9.1.2.3 Modeling

Fig. 9.23 shows the flow chart for this calculation; it refers to the net present value equation (see Eq. (9-1)). However, no additional expenses (with the exception of the investment made at the time t_0), disposal gains, or demolition costs are considered in this example.

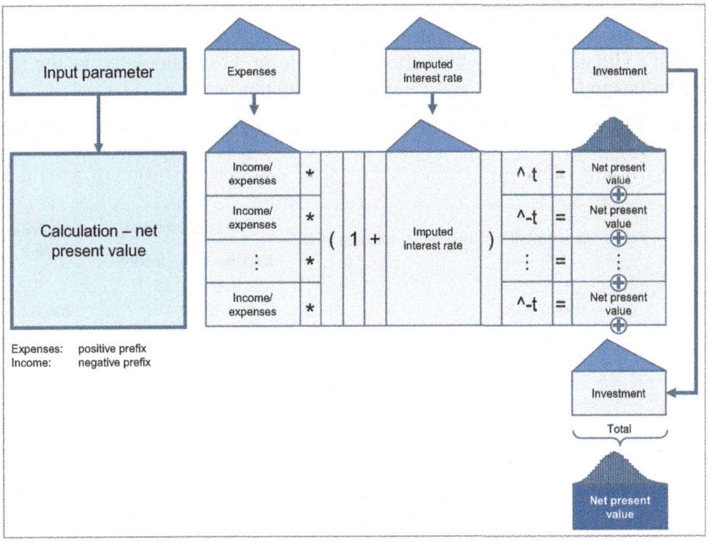

Fig. 9.23 Flow chart – Net present value

9.1.2.4 Calculation and Interpretation of Results

Table 9.8 lists the input parameters for calculating the net present value. It should be noted, though, that both the investment amount and the annual income are still uncertain because the project has not yet been completed (see Rows 1 and 2).

No.	Item	MIN	MOST LIKELY	MAX	Distribution
0	A	B	C	D	E
1	Investment	-4,500.00	-5,000.00	-5,500.00	-€5,000.00
2	Income	1,500.00	1,900.00	2,300.00	€1,900.00/a
3	Imputed interest rate	2.00	3.40	4.80	3.40%
4	Period considered				3.00 years

Table 9.8 Calculation table – Total capital expenditure

One of the crucial factors for calculating the net present value is the imputed interest rate, which is also reflected by a triangular distribution (Row 3). In this example, the period under consideration comprises three years (Row 4).

Table 9.9 calculates the net present value separately for each year. The years are entered in Column A, beginning with Year "Zero". Column B shows the only investment made in Year Zero, whereas annual income figures for Years 1, 2, and 3 are entered in Column C. In each case, the triangular distribution defined in Table 9.8 is applied. This means that both the investment and the income vary in each iteration.[18)]

The interest factor [= $(1 + i_i)^t$] (see Column D) changes year after year and also in each iteration due to modeling the imputed interest rate as a triangular distribution. In Year Zero, the interest factor is 1 [= $(1 + i_i)^0$], which means that no interest accrues on the investment shown in Cell E1.

Income figures are discounted depending on the points in time when they were generated (Column F).[19)] The net present value shown in Column G is calculated as the total of the net present values in Columns E and F up to the relevant year.

For a period of three years, the net present value is finally shown as a histogram in Cell G4.

No.	Year	Investment	Income	Interest factor	Net present value		Net present value
					Expenses	Income	
0	A	B	C	D	E	F	G
1	0	-5,000.00		1.00	-5,000.00		-5,000.00
2	1		1,900.00	1.03		1,837.52	-3,162.48
3	2		1,900.00	1.07		1,777.10	-1,385.37
4	3		1,900.00	1.11		1,718.67	**333.29**

Table 9.9 Calculation table – Net present value

Fig. 9.24 includes the net present value histogram for a period of three years. The mean amounts to €333.84 and is thus positive. This would lead to a recommendation to invest according to the basic philosophy of the net present value method. However, this isolated deterministic value does not permit conclusions regarding the chance/risk ratio that would be associated with this investment. After 50,000 iterations, the histogram ranges from -€1,188.62 to €1,862.45. The standard deviation amounts to approximately €506.02.

18) Since all income items refer to the same cell in Excel, income per iteration is identical in each year. If we were to define three input distributions (in three separate cells), this would result in different income values per iteration for the three years.

19) The deterministic calculation of the discounted income in the second year proceeds as follows: €1,900.00 / 1.07 = €1,777.10 (calculation with decimal places in MS Excel).

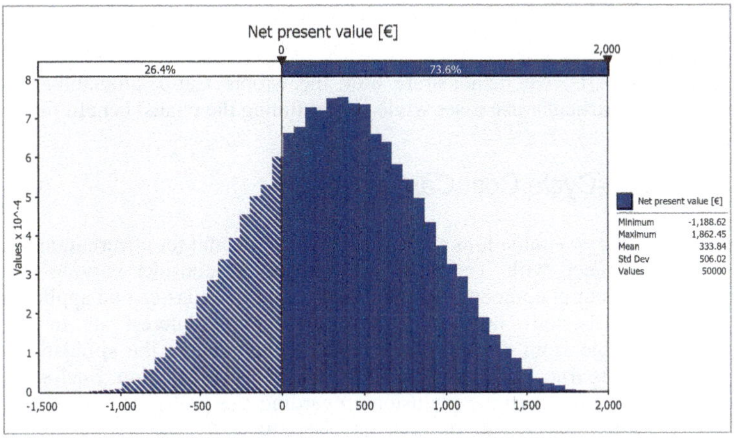

Fig. 9.24 Histogram – Net present value

In Fig. 9.24, the slider has been set to the limit for a profitable investment (net present value = €0) at the top of the diagram. Only this division of the histogram at a net present value of €0 reveals the chance/risk ratio of an investment. The chance of reaching a positive net present value, and thus to make a profitable investment, amounts to about 73.6%. The risk taken with this investment equals about 26.4%.

9.1.2.5 Benefit for the Client

Incorporating the uncertainties of the input parameters makes it possible to indicate the net present value range. If several investment opportunities exist, it is possible to compare not only the associated means and ranges, but also their chance/risk ratios to each other.

Sensitivity analyses with varying income and expense levels or capital interest rates are carried out automatically by selecting the distribution functions for the inputs and by performing a Monte Carlo simulation.

The client's awareness of the effects of income and expenses and of the points in time when they occur during the life cycle of a building or structure can be fed directly into the planning process of construction projects. The later the time and the lower the frequency in which expenses (such as for replacement, repair, maintenance) are required during the life cycle, the more positive the effect will be on the net present value and thus on life-cycle costs. Life-cycle costs are predominantly influenced by the selected materials (i.e. their quality standards and sustainability aspects), but also by the type of use and maintenance and care.

9.2 Project Phase 2 (PPH 2)

Project Phase 2 includes the design of the project. Functional requirements as well as qualities and quantities are thus specified at this stage. In this phase, the client performs life-cycle cost analyses and cost estimates or calculations. In addition, the plot of land to be developed will have been identified by this time. Phase 2 also includes a soil survey or assessment because the subsoil presents a major risk factor for both the client and the contractor.

In the following parts, we demonstrate how the Monte Carlo simulation method is employed in these particular use cases while also outlining the related benefit for the client.

9.2.1 Life-Cycle Cost Calculations

The use of Monte Carlo simulations can systematically account for uncertainties that input variables are associated with. This method allows us to consider varying anticipated service lives of structural components, and distribution functions can be applied to costs, rehabilitation intervals, and – last but not least – the imputed interest rate. In the field of infrastructure, income from road tolls can also be included in the simulation of PPP models. The same is true for rental income in building construction. Such simulations result in histograms that permit conclusions regarding the probability of occurrence of specific net present values on the basis of input parameters. This approach allows more effective allocation of deterministic values and their assessment with respect to the chance/risk ratio. It delivers a larger amount of information while also increasing the level of decision confidence.

This example outlines a possible approach to the opportunity- and risk-based life-cycle costing of bridges. Life-cycle cost analyses belong in the planning phase (Project Phase 2) (see Fig. 9.25), but they are refined in the following phases because the basis on which such analyses can be performed becomes increasingly detailed as a result of a more specific design brief including materials, use(r) requirements, etc.

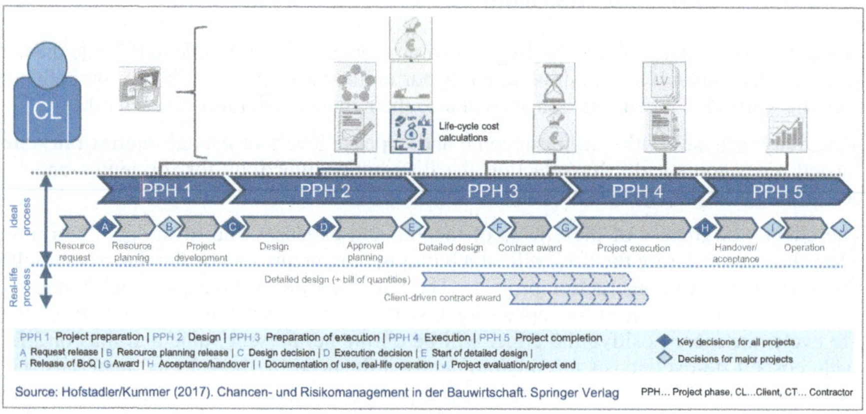

Fig. 9.25 Allocation of example in the project sequence

9.2.1.1 Basics

Any calculation of life-cycle costs includes elements that stem from costing and investment calculus. What is crucial for arriving at sound comparisons of various options is the difference in the number and type of included categories of costs of use, as well as the quality and completeness of the data on which the calculation is based.

The results of life-cycle costing models permit conclusions as to the commercial viability or profitability of a given project. Sensitivity analyses make it possible to identify those calculation parameters that have the most significant influence on the final results. Life-cycle models are developed and analyzed on the basis of the user requirements, design briefs, or rehabilitation scenarios. They create the basis for decision making and/or certification.

The basic formula shown in Eq. (3-1) provides the starting point for calculating the life-cycle costs. These costs include the costs of a property (such as a production line or a building) and its components that are incurred over the entire life cycle during which they fulfill relevant use(r) requirements. Life-cycle costs (LCC) are composed of construction costs (CC), operating costs (OC), maintenance costs (MC), and demolition costs (DC).

Methods of investment calculus (see Fig. 9.20) are employed for calculating life-cycle costs. One of the methods frequently used in this context is the net present value or terminal value method. We refer to Section 9.1.2.1 for an outline of the basics of net present value calculation.[20)]

The application of the net present value method provides a deterministic calculation where each computation step results in only a single net present value that is associated with many uncertainties while permitting no conclusion regarding the opportunities seized or risks taken. In fact, this single value represents only one of many possible combinations of input parameters that vary within ranges. This situation is unsatisfactory for decision makers, but it can be rectified by applying probabilistic computation methods.

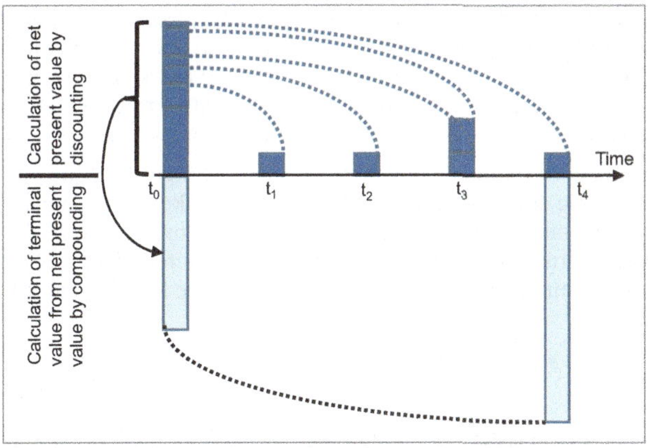

Fig. 9.26 Relationship between net present value and terminal value

Compounding makes it possible to directly convert the net present value to a terminal value. The latter represents the value of cash inflows and outflows at the end of the period under consideration. Fig. 9.26 shows the qualitative relationship between the net present value and the terminal value in a deterministic calculation. The upper portion of the diagram includes bars that symbolically represent individual expenses. At the time t_0, an investment is made (e.g. construction costs), followed by an annual expense in the same nominal amount at the times t_1, t_2, t_3, and t_4. At the time t_3, an additional expense is

[20)] In addition to the imputed interest rate, inflation indices dependent on cost groups can also be considered in the net present value calculation. This is not shown in the example.

incurred due to the repair of specific structural components. All expenses made after t_0 must be discounted by entering them in Eq. (9-1), resulting in the net present value relative to t_0 (t_r = reference point in time).

Accordingly, the terminal value of expenses is calculated by compounding this net present value (see Eq. (9-2)) up to the time t_4 (i.e. the end of the period under consideration). Net present values and terminal values can be calculated for any point in time by compounding and discounting; their graphical representation is a curve over time.

$$\text{Terminal value} = \text{NPV}_{t_r} \cdot (1 + i_i)^{(T - t_r)} \tag{9-2}$$

T ....Period under consideration

t_r ....Reference point in time

9.2.1.2 Task

Of all types of structures, modern bridges are one of the most durable, but they are subject to rehabilitation and upgrade at more or less regular intervals to meet the specified load-bearing and serviceability requirements over their exceedingly long service lives. For life spans of 100 years or more, it is particularly difficult to calculate life-cycle costs or to arrive at a general conclusion regarding the commercial benefits or disadvantages of certain construction methods and processes.

Any comparison of different construction methods and processes will no longer rely on just a single, possibly extremely uncertain figure, but on histograms that provide a much more reliable basis. If the net present values or terminal values of two scenarios overlap, a safety margin can be determined for any of the two scenarios relative to the other, which is similar to the procedure commonly applied in structural engineering.

This worked example[21)] should demonstrate the application of the Monte Carlo method to calculate the life-cycle costs of a bridge. Basic data was taken from an example by *Jodl*[22)] and refers to the Wientalbrücke. The substructure of this bridge was built from concrete; the skeleton (superstructure) consists of reinforced concrete. The values expected for construction costs, maintenance costs, and service lives are listed in Table 9.10.

No.	Component	Service life [years]	Construction cost [€]	Maintenance cost [% of construction cost]	Demolition and admin [% of construction cost]
0	A	B	C	D	E
1	Substructure	129	3,100,000.00	0.41	22.00
2	Skeleton	77	4,900,000.00	0.72	22.00
3	Equipment 1	20	600,000.00	1.50	22.00
4	Equipment 2	30	1,400,000.00	1.20	22.00

Table 9.10 Expected input values for the life-cycle costing exercise

Jodl used a constant interest rate of 4.00% to compound and discount expenses over the years. To calculate total construction costs, construction costs in the stricter sense are multiplied by 1.10 to take administration costs into account. Life-cycle costs (terminal value) are calculated for a service life of 70 years. During this period, pieces of equipment

21) Cf. Hofstadler/Kummer (2015b), p. 751ff.

22) Jodl (2010a) and Jodl (2010b)

are repeatedly removed and replaced (see service lives of components in Column B of Table 9.10). The costs of demolition/removal, including administration costs, amount to 22% (deterministic value) of relevant construction costs. In addition, varying maintenance costs are incurred each year for the individual structural components, which are shown as percentages of the relevant construction costs in Column D.

9.2.1.3 Modeling

The deterministic values taken from Table 9.10 are associated with major uncertainties; they do not provide any information on the probability the calculated life-cycle costs are associated with. To incorporate these uncertainties in the calculation process, (symmetrical) triangular distributions are applied to the input parameters instead of deterministic values. Any available historical data or specific distributions that are known for certain input parameters (such as from analyses of expert surveys) can be used for the simulations.

No.	Component	Service life [years]				Coefficient – administration costs	Construction cost [€]			
		MIN	MOST LIKELY	MAX	Distribution		MIN	MOST LIKELY	MAX	Distribution
0	A	B	C	D	E	F	G	H	I	J
1	Substructure	110	129	148	129	1.1	3,190,000.00	3,410,000.00	3,630,000.00	3,410,000.00
2	Skeleton	70	77	84	77	1.1	4,950,000.00	5,390,000.00	5,830,000.00	5,390,000.00
3	Equipment 1	18	20	22	20	1.1	605,000.00	660,000.00	715,000.00	660,000.00
4	Equipment 2	28	30	32	30	1.1	1,320,000.00	1,540,000.00	1,760,000.00	1,540,000.00

Table 9.11 Input values – Construction costs and expected service life

Tables 9.11 and 9.12 state the minimum, most likely, and maximum values that define the triangular distributions of the calculation parameters for construction costs, expected service lives, and maintenance cost percentages. Costs of demolition and administration associated with the individual components amount to a constant share of 22% (deterministic value) of construction costs. A triangular distribution is also used for the imputed interest rate (3.00% | 4.00% | 5.00%).

No.	Component	Maintenance cost [% of construction cost]			
		MIN	MOST LIKELY	MAX	Distribution
0	A	B	C	D	E
1	Substructure	0.35	0.41	0.47	0.41
2	Skeleton	0.68	0.72	0.76	0.72
3	Equipment 1	1.00	1.50	2.00	1.50
4	Equipment 2	0.80	1.20	1.60	1.20

Table 9.12 Input values – Maintenance costs

9.2.1.4 Calculation and Interpretation of Results

For the purpose of calculating life-cycle costs, maintenance costs need to be considered that are incurred each year; these are determined as a percentage of construction costs. For each iterative step, the use of distribution functions on the input side results in random maintenance costs that lie within the defined ranges. For each iteration, a service life within the defined ranges is randomly selected for the maintenance and repair (demolition and replacement) of equipment. Stated costs of demolition, administration, and replacement are then incurred in this randomly determined year. Eq. (9-1) allows for the calculation of a net present value for reference point t_0 for any period under consideration. Net present value totals can be converted to terminal values in a compounding step (for each of the years in the considered period t). Revenues are not included in this example, but they can be integrated in the calculation very easily by stating annual revenues, for example (note the inverse prefixes: expenses are positive and revenues are negative in this example). Nor are inflation effects (inflation proper as well as price increases, such as for energy or services) taken into account for simplicity reasons.

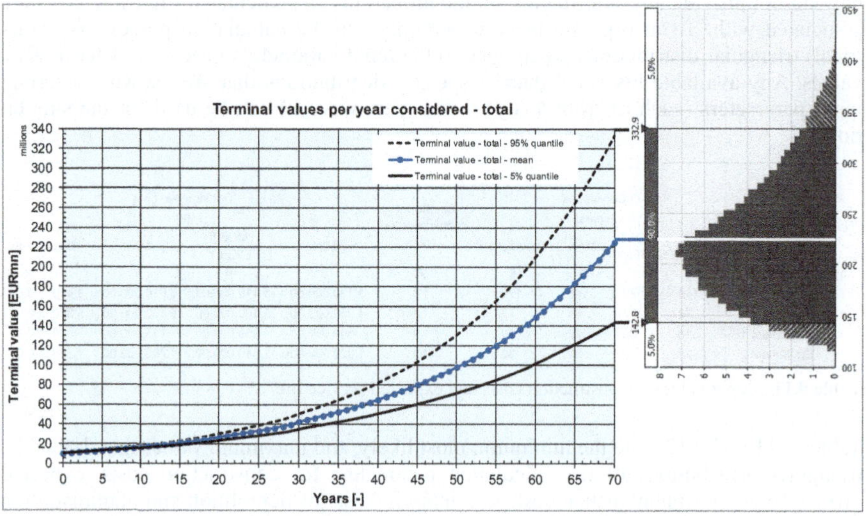

Fig. 9.27 Terminal values for the entire project for a 70-year period, including a terminal value histogram at the end of this period

Fig. 9.27 shows the terminal value curves during the considered period of 70 years. The middle curve represents the mean terminal values. The other two curves define the 5th and 95th quantiles of the terminal values determined in a probabilistic calculation applying the Monte Carlo simulation method.

Despite the fact that symmetrical distributions were used for all input variables, terminal value distributions become increasingly skewed over the length of the considered period, resulting in the distribution turned about a 90 degree angle on the right-hand side of Fig. 9.27. This is due to the fact that the distribution for the interest factor also shows an increased skewness for longer periods due to the mathematical relationship applied $[= (1 + i_i)^{(T - tr)}]$. Spreads of simulation results, mean values, and other statistical variables can be compared to one another when considering other project options in order to arrive at a conclusion regarding the opportunities and risks associated with the individual options.

9.2.1.5 Benefit for the Client

When performing life-cycle analyses, it is crucial to consider the system boundaries in terms of underlying costs and related cost groups. To analyze different investment decisions, the defined system boundaries and selected methods need to be thoroughly evaluated to enable this comparability in the first place.

The probabilistic calculation of life-cycle costs employing Monte Carlo simulations and applying the net present value or terminal value method is associated with the advantage of not only calculating a deterministic value but also providing the option of histogram outputs that permit estimates of the probabilities of occurrence of specific deterministic values.

The use of Monte Carlo simulations can systematically account for uncertainties that calculation parameters (particularly service lives, costs, and interest rates, but also revenues, such as from road toll schemes or rents) are associated with on the basis of distribution functions. Only a probabilistic calculation will permit conclusions as to how uncertain life-cycle costs stated as single figures are for exceedingly long periods. The spread of probabilistically determined terminal values increases continuously in line with the length of the considered period. Application of probabilistic methods thus improves results while increasing the level of decision confidence in the case of several project options with varying maintenance costs and service lives. Repairs are distributed randomly within defined ranges at recurring intervals depending on the expected service lives of structural components.

Incorporating correlations would make it possible to also consider dependencies in repair and maintenance (if, for instance, elements of Equipment 1 need to be replaced, there is also an increased probability of having to replace elements of Equipment 2).

A higher standard of workmanship will influence the anticipated service lives of components and can be integrated into the simulation within certain ranges. The computation model is suitable for reflecting both periodic and non-periodic costs, such as incurred by repair or damage.

Cost drivers and main factors influencing the net present value or terminal value can be identified by applying appropriate analytical methods, including spider and tornado diagrams.

9.2.2 Soil Survey

In construction projects, the site's subsoil presents a major risk factor for both the client and the contractor. A soil survey is performed as part of project preparation, establishing its basis, and design and tendering. This survey creates the basis for specifications and requests for quotations and thus the pricing on the part of bidders. At the same time, this phase also determines the future chance/risk potential for both the client and the contractor.

The soil survey is allocated to Project Phase 2 (see Fig. 9.28) because the plot to be developed will be known at this stage and initial soil investigations are carried out, or have already been performed as part of the land acquisition process.

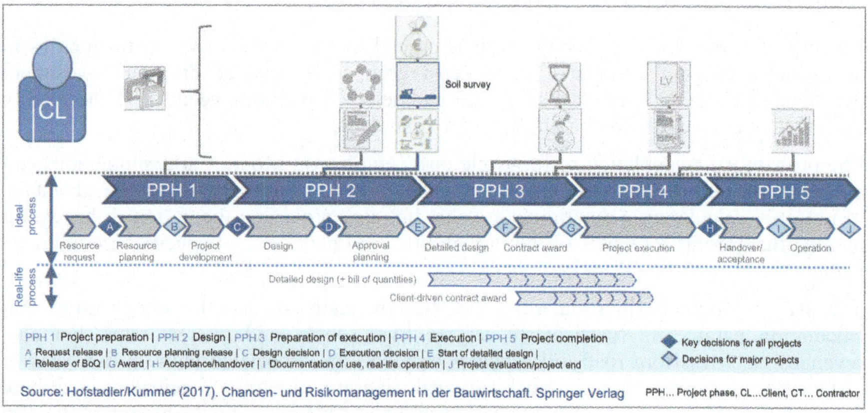

Fig. 9.28 Allocation of example in the project sequence

9.2.2.1 Basics

The greater the degree of detail in the request for quotations or the quantities specified in the tender documents, the more accurate the pricing by the bidders can be, thus minimizing the potential for contract addenda. Even though the additional cost incurred by such surveys will translate into a higher contract volume, or total amount, it will also reduce additional costs that may potentially be incurred by a subsoil risk that has materialized.

Fig. 9.29 illustrates two related cases. In Case A, the fact that the scope of the soil survey is limited and no risk assessment is performed initially leads to an advantage at the contract award stage. In the course of the project, however, additional expenses (costs) may be incurred by a subsoil risk that has actually materialized, which result in an increased overall risk.

In Case B, the contract volume is higher owing to the fact that an appropriate scope was chosen for the soil survey, combined with an adequate risk assessment. On the other hand, additional costs will be much lower than in Case A. The difference between Case A and Case B can be referred to as an avoidable risk.

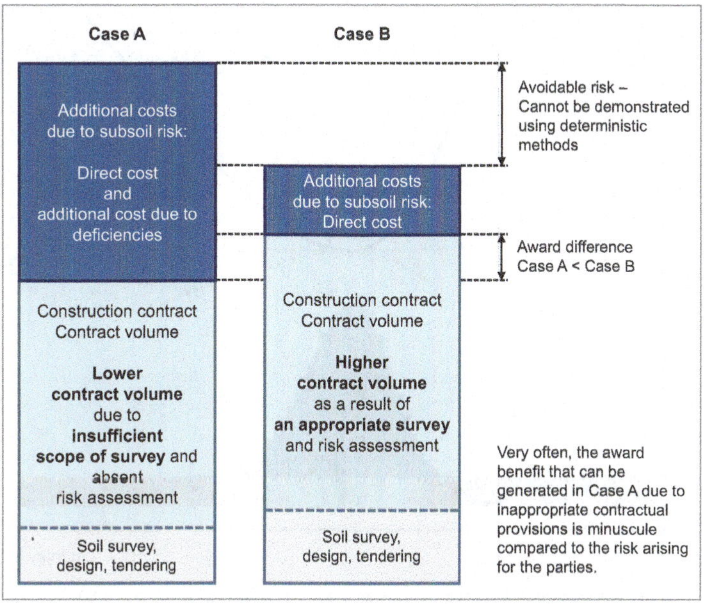

Fig. 9.29 Risk mitigation for the parties to the contract – Case A: Limited survey – Case B: Adequate survey

If the individual quantity proportions of a combined pricing item are associated with uncertainties, the range within which stable unit prices can be assumed will be relatively narrow. If quantities are exceeded or underrun, the positions on the unit price curve will inevitably differ between the client and the contractor (see Fig. 9.30).

The unit price upon which the parties ultimately agree (by claiming additional costs or reducing cost claims) is associated with a number of uncertainties for the client and the contractor since the related effects are known, but the probability of occurrence is not.

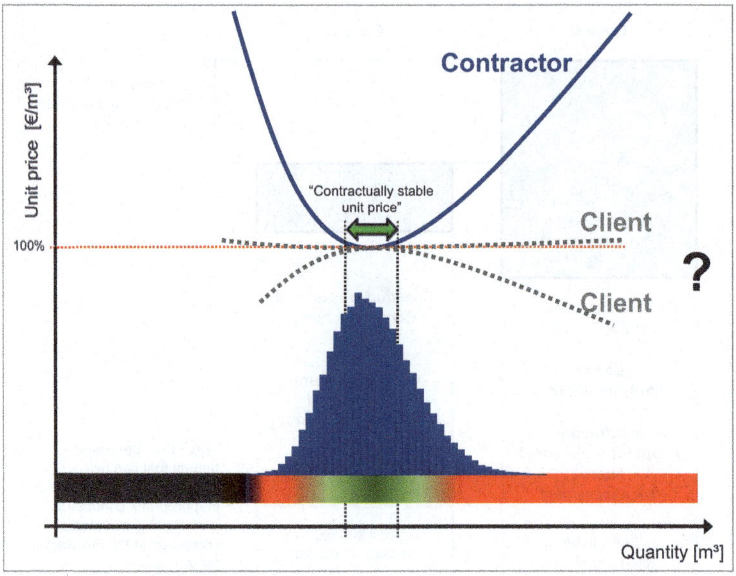

Fig. 9.30 Link between quantity uncertainties and stability of unit prices – Qualitative representation

9.2.2.2 Task

A calculated example demonstrates the effects that additional costs incurred by a soil survey will have on the uncertainties in specifying soil and contamination classes, and how these uncertainties will subsequently affect the item price.

The total excavated material amounts to 10,000 m³ (corresponding to the volume of compact soil without applying a decompaction factor), where the values expected for soil and contamination classes are structured as follows:

- Soil classes 3-5: 5,000 m³
- Soil class 6: 2,500 m³
- Soil class 7: 2,500 m³

- Contamination class I: 5,000 m³
- Contamination class II: 2,500 m³
- Contamination class III: 1,500 m³
- Contamination class IV: 1,000 m³

The client is interested in the chances and risks taken by performing excavation works after soil surveys with a varied scope. Three survey options are available, each at a different price. The more expensive the survey method, the more accurately (in the sense of a narrower range) the quantity proportions of the individual soil and contamination classes can be stated. The greater the inaccuracy of the survey, and the more uncertain the quantity proportions of the individual classes, the higher the unit prices of excavation works, transport, and landfill of the material will be.

The client intends to earmark a budget of €500,000 to excavation works and wants to know more about the chances and risks it would take when performing the different surveys.

Probabilistic calculations and a comparison of three cases with different survey activities demonstrate how the client can take into account uncertainties with regard to the subsoil risk in order to prepare a decision on the type of survey to be carried out within the constraints of a defined budget while indicating the related implications by highlighting the chances and risks to be expected. The results of the probabilistic calculations (i.e. three cases) are compared to each other.

We can also demonstrate the effect an additional financial buffer would have on the chance/risk ratio in each of the considered cases.

9.2.2.3 Modeling

The probabilistic calculation uses the Monte Carlo simulation method to vary both specified quantities and unit prices within the expected ranges. Fig. 9.31 shows the flow chart for this calculation.

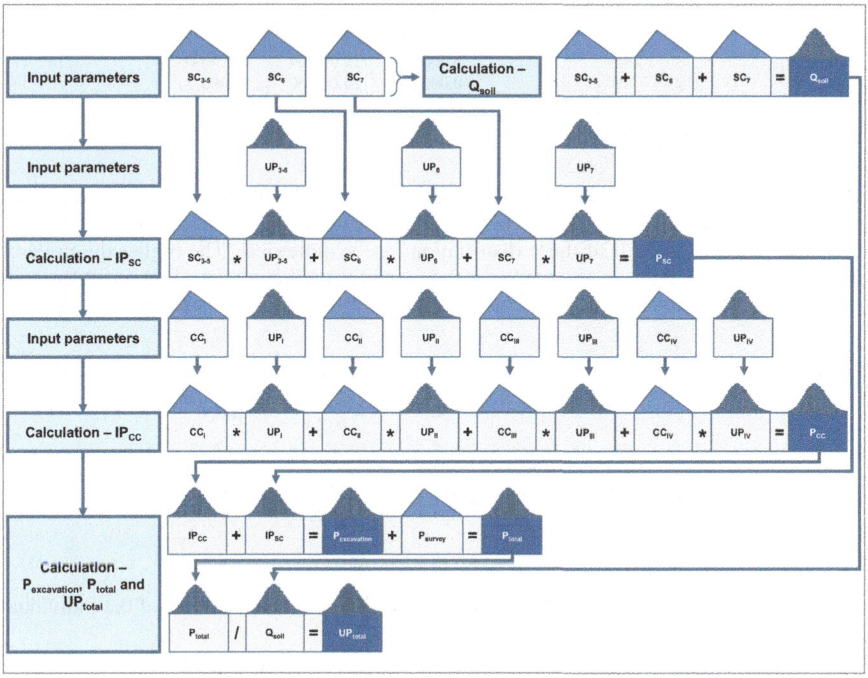

Fig. 9.31 Calculation flow chart for determining the unit price of excavated material

The quantity proportions of soil classes SC and contamination classes CC are directly translated into triangular distributions. The unit prices UP are obtained from the triangular distributions of the basic unit prices (not shown in Fig. 9.31) plus the percentage markups (not included in Fig. 9.31 either) that are also modeled as triangular distributions. This is why the unit prices UP in Fig. 9.31 are symbolically represented by histograms rather than triangular distributions.

The results of the Monte Carlo simulation are shown as histograms. Thus, they also permit conclusions regarding the ranges and probabilities of specific deterministic values.

9.2.2.4 Calculation and Interpretation of Results

Three cases with different ranges of quantity deviations and percentage markups on unit prices, as well as different soil survey prices, are evaluated for the purpose of the calculations.

Soil survey prices are represented by triangular distributions as indicated in Table 9.13. Column C shows the most likely values, whereas Columns B and D state the minimum and maximum soil survey prices. In each case, the range around the most likely value reaches ±10%.

No.	Case	Survey prices [€]		
		MIN	MOST LIKELY	MAX
0	A	B	C	D
1	Case 1	9,000.00	10,000.00	11,000.00
2	Case 2	18,000.00	20,000.00	22,000.00
3	Case 3	36,000.00	40,000.00	44,000.00

Table 9.13 Prices of soil survey

The different soil survey methods should also be associated with different deviations from the most likely quantities. Quantity deviations are also represented by triangular distributions (see Table 9.14). All of these distributions are symmetrical. The higher the price of the survey method per case, the narrower the range for predicting per-class quantities will be. Quantity deviations per case refer to both soil and contamination classes.

No.	Case	Quantity deviations [%]		
		MIN	MOST LIKELY	MAX
0	A	B	C	D
1	Case 1	30.00	40.00	50.00
2	Case 2	25.00	30.00	35.00
3	Case 3	15.00	20.00	25.00

Table 9.14 Quantity deviations

The basic unit price information for excavation works and for the landfill of the individual soil and contamination classes is shown in Table 9.15.

This table includes the basic unit price information for the three cases considered (modeled as triangular distributions), but also the percentage markups (also modeled as triangular distributions). The basic unit prices show a certain range associated with the full disclosure of the quantities for the individual soil and contamination classes. The percentage increases in these unit prices are justified by the uncertainties prevailing in terms of the composition of the quantity structure for different types of soil survey.

No.	Soil or contamination class	Unit prices (basic data) [€/m³]		
		MIN	MOST LIKELY	MAX
0	A	B	C	D
1	3 - 5	2.00	3.00	5.00
2	6	15.00	17.00	21.00
3	7	25.00	27.00	32.00
4	I	10.00	11.00	13.00
5	II	40.00	43.00	52.00
6	III	60.00	63.00	71.00
7	IV	90.00	95.00	110.00
No.	**Case**	**Percentage markups on basic unit price [%]**		
		MIN	MOST LIKELY	MAX
8	Case 1	10.00	30.00	45.00
9	Case 2	2.00	10.00	20.00
10	Case 3	0.00	2.00	4.00

Table 9.15 Basic information and percentage markups on unit prices per soil and contamination class and/or considered case

Fig. 9.32 uses triangular distributions to indicate the ranges defined for unit prices, quantity deviations, and markups on basic unit prices for Cases 1 to 3.

In Case 1, a narrow range should be expected for soil survey prices, but the associated quantity deviations and price markups reveal very wide spreads in the upper value range.

Case 2 can be referred to as an interim solution because the ranges of quantity deviations and price markups are narrower (and located in the medium value range) despite the higher survey price.

Case 3 includes the highest soil survey price but also the smallest expected quantity deviations. It also applies very small markups on the basic unit prices.

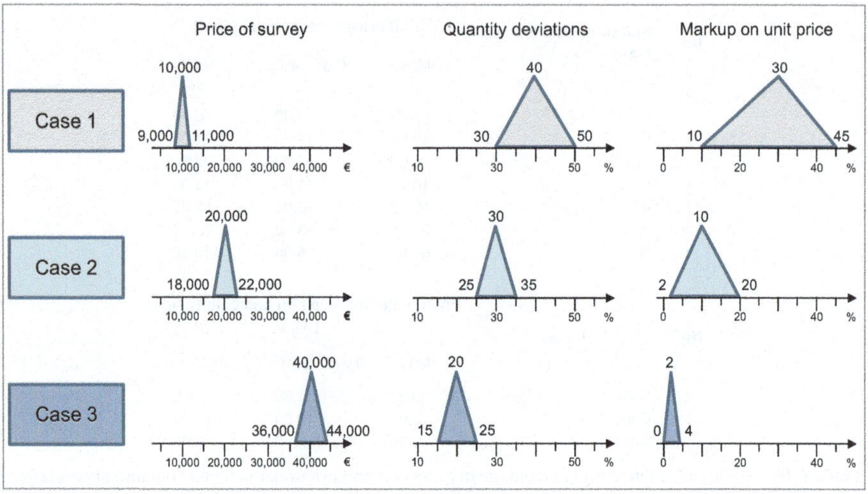

Fig. 9.32 Triangular distributions of unit prices, quantity deviations and markups on basic unit prices – Cases 1 to 3

Tables 9.16 to 9.18 show the calculation tables separately for each of the Cases 1 to 3. The flow chart for this calculation (see Fig. 9.31) is identical for each of these cases. Only input parameters and spreads vary.

Columns B to D state the minimum, most likely, and maximum volumes per soil or contamination class. In each of these cases, the sum of the most likely values is equivalent to the volume of 10,000 m³ determined for the total excavated (compact) material. Upward and downward deviations are calculated using the triangular distributions of the expected quantity deviations (see Table 9.14); they vary in Columns B and D in each iterative step. The distributions of the per-class quantities are shown in Columns E, and lead to a total that deviates from the target volume (i.e. 10,000 m³) in each iterative step.

This would distort simulation results because each iterative step would use a different total soil volume to derive the total price. To prevent this from happening, these quantity proportions are normalized to a total of 10,000 m³ in an additional step (see Column F).

No.	Soil classes	Quantities [m^3] MIN	MOST LIKELY	MAX	Quantity [m^3]	Normalized quantities [m^3]	Distributions of increased unit prices [€/m^3]	Price per class [€]
0	A	B	C	D	E	F	G	H
1	3 - 5	3,000.00	5,000.00	7,000.00	5,000.00	5,000.00	4.28	21,388.89
2	6	1,500.00	2,500.00	3,500.00	2,500.00	2,500.00	22.67	56,680.56
3	7	1,500.00	2,500.00	3,500.00	2,500.00	2,500.00	35.93	89,833.33
4	**Total**		**10,000.00**		**10,000.00**	**10,000.00**		**167,902.78**
No.	**Contamination classes**	**Quantities [m^3]** MIN	MOST LIKELY	MAX	**Quantity [m^3]**	**Normalized quantities [m^3]**	**Distributions of increased unit prices [€/m^3]**	**Price per class [€]**
5	I	3,000.00	5,000.00	7,000.00	5,000.00	5,000.00	14.54	72,722.22
6	II	1,500.00	2,500.00	3,500.00	2,500.00	2,500.00	57.75	144,375.00
7	III	900.00	1,500.00	2,100.00	1,500.00	1,500.00	82.99	124,483.33
8	IV	600.00	1,000.00	1,400.00	1,000.00	1,000.00	126.19	126,194.44
9	**Total**		**10,000.00**		**10,000.00**	**10,000.00**		**467,775.00**
10	**Price – Case 1**							**635,677.78**
11	Price of survey							10,000.00
12	**Total price – Case 1**							**645,677.78**
13	**Unit price – Case 1**							**64.57**

Table 9.16 Calculation table – Case 1

A percentage markup is added to the basic unit prices for the individual soil and contamination classes (see Table 9.15, Rows 1 to 7) depending on the specific case (see Table 9.15, Rows 8 to 10).[23] The marked-up unit prices are shown in Columns G in Tables 9.16 to 9.18. These price increases are introduced to reflect the quantity uncertainties also on the monetary level.

No.	Soil classes	Quantities [m^3] MIN	MOST LIKELY	MAX	Quantity [m^3]	Normalized quantities [m^3]	Distributions of increased unit prices [€/m^3]	Price per class [€]
0	A	B	C	D	E	F	G	H
1	3 - 5	3,500.00	5,000.00	6,500.00	5,000.00	5,000.00	3.69	18,444.44
2	6	1,750.00	2,500.00	3,250.00	2,500.00	2,500.00	19.55	48,877.78
3	7	1,750.00	2,500.00	3,250.00	2,500.00	2,500.00	30.99	77,466.67
4	**Total**		**10,000.00**		**10,000.00**	**10,000.00**		**144,788.89**
No.	**Contamination classes**	**Quantities [m^3]** MIN	MOST LIKELY	MAX	**Quantity [m^3]**	**Normalized quantities [m^3]**	**Distributions of increased unit prices [€/m^3]**	**Price per class [€]**
5	I	3,500.00	5,000.00	6,500.00	5,000.00	5,000.00	12.54	62,711.11
6	II	1,750.00	2,500.00	3,250.00	2,500.00	2,500.00	49.80	124,500.00
7	III	1,050.00	1,500.00	1,950.00	1,500.00	1,500.00	71.56	107,346.67
8	IV	700.00	1,000.00	1,300.00	1,000.00	1,000.00	108.82	108,822.22
9	**Total**		**10,000.00**		**10,000.00**	**10,000.00**		**403,380.00**
10	**Price – Case 2**							**548,168.89**
11	Price of survey							20,000.00
12	**Total price – Case 2**							**568,168.89**
13	**Unit price – Case 2**							**56.82**

Table 9.17 Calculation table – Case 2

23) The calculation is explained using a deterministic example relative to the most likely values:
In relation to the most likely basic unit price of soil class 6, this is associated, in Case 1 (also relative to the most likely value), with an increase from €17.00/m^3 (Cell C2 of Table 9.15) to €22.10/m^3 [= €17.00/m^3 * (100% + 30%) / 100%)].
The mean values of the respective distributions are entered in Column G of Tables 9.16 to 9.18.

Column H in Tables 9.16 to 9.18 multiplies the normalized quantities with the marked-up unit prices to get the prices per class. Cells H4 or H9 calculate the price totals of excavating the different soil classes or putting the individual contamination classes to landfill.

Cells H10 show the total prices (excluding the soil survey price) for the three cases. In Cells H12, the applicable soil survey price is added to obtain the total price per case as a histogram.

No.	Soil classes	Quantities [m³] MIN	MOST LIKELY	MAX	Quantity [m³]	Normalized quantities [m³]	Distributions of increased unit prices [€/m³]	Price per class [€]
0	A	B	C	D	E	F	G	H
1	3 - 5	4,000.00	5,000.00	6,000.00	5,000.00	5,000.00	3.40	17,000.00
2	6	2,000.00	2,500.00	3,000.00	2,500.00	2,500.00	18.02	45,050.00
3	7	2,000.00	2,500.00	3,000.00	2,500.00	2,500.00	28.56	71,400.00
4	**Total**		**10,000.00**		**10,000.00**	**10,000.00**		**133,450.00**
No.	**Contamination classes**	**Quantities [m³]** MIN	MOST LIKELY	MAX	**Quantity [m³]**	**Normalized quantities [m³]**	**Distributions of increased unit prices [€/m³]**	**Price per class [€]**
5	I	4,000.00	5,000.00	6,000.00	5,000.00	5,000.00	11.56	57,800.00
6	II	2,000.00	2,500.00	3,000.00	2,500.00	2,500.00	45.90	114,750.00
7	III	1,200.00	1,500.00	1,800.00	1,500.00	1,500.00	65.96	98,940.00
8	IV	800.00	1,000.00	1,200.00	1,000.00	1,000.00	100.30	100,300.00
9	**Total**		**10,000.00**		**10,000.00**	**10,000.00**		**371,790.00**
10	**Price – Case 3**							**505,240.00**
11	Price of survey							40,000.00
12	**Total price – Case 3**							**545,240.00**
13	**Unit price – Case 3**							**54.52**

Table 9.18 Calculation table – Case 3

Fig. 9.33 includes the three histograms for the total prices (including survey) of the three cases. It becomes abundantly clear that a limited scope of the soil survey (Case 1) should result in a significantly wider total price range. The higher the amount invested in survey activities, the narrower the spread of results will be (Case 3). In all three histograms, the budget of €550,000 earmarked by the client is entered as a slider at the top of the diagram.

We find that, for this budgeted amount, higher or lower chances and risks are taken depending on the survey method because uncertain quantities are also associated with a higher unit price level.

The ratio of underrunning or exceeding the budget is thus different for each of these cases. The client can mitigate this risk by including an additional financial buffer in the budget. The effect of one and the same buffer amount on the chance/risk ratio will vary depending on the survey method.

In this example, Case 3 would be the preferred scenario from the client's perspective because the range of calculated total prices is the narrowest, thus reaching the most favorable chance/risk ratio with the earmarked budget of €550,000.

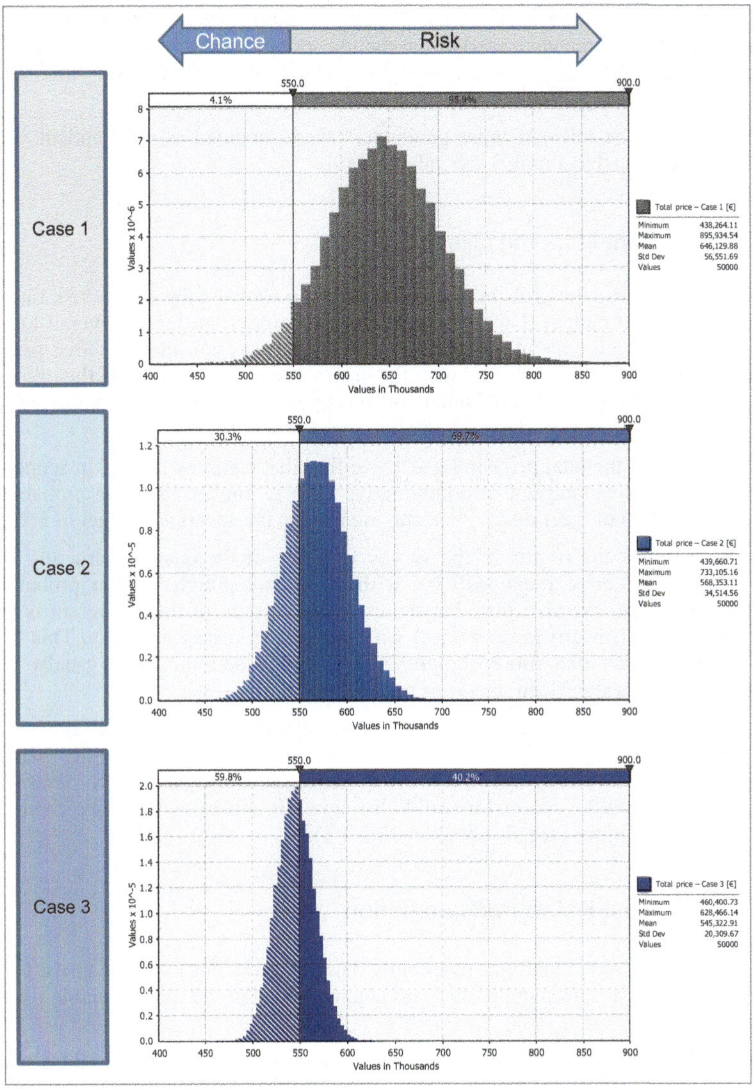

Fig. 9.33 Total price histograms for the three cases[24)]

For an in-depth analysis, the survey prices, percentages of quantity deviations, and percentage markups on unit prices could be correlated to each other for each of the cases considered. A higher survey price is likely to result in a narrower range of quantity deviations and unit price markups (i.e. negative correlations). A 3-by-3 matrix can be chosen as

24) Theoretical considerations regarding the selected input parameter ranges result in minimum and maximum total prices that lie above or below the extremes of the simulation. This difference is due to the normalization of quantities; it was described in detail in Section 6.2.5.2.

the corresponding correlation matrix for the three parameters. In this case, we would have to define three correlation coefficients. The relationship between the quantity deviation percentage and the percentage markup on the unit prices should be positive because they are dependent on the survey method, and thus show a similar pattern (of simultaneously taking higher or lower values).

In this example, the relevant input parameters were modeled as independent variables (without correlations) to simplify the calculation.

9.2.2.5 Benefit for the Client

The client can refer to the probabilistic calculations to evaluate the chance/risk ratio taken with the earmarked budget. This information is very important for the overall budgeting exercise and cost certainty of the project. The three cases considered above permit the conclusion that Case 3 is associated with the most favorable conditions in the given setup. Although the client has to accept higher soil survey prices in this scenario, the probability of meeting the earmarked budget of €550,000 increases.

When considering the total price in Case 1, we find that a survey limited in scope should result in a very wide range. Correspondingly, there is only a very low probability (of approximately 4%) of meeting (or even staying below) the specified budget of €550,000.

The more accurate the results of the survey in terms of the quantity structure are, the narrower the total price range will be, with a correspondingly greater probability of meeting the budget. We also find that the center of gravity of the histogram is moving farther to the left (toward lower prices) with increasing survey accuracy. This effect is justified by the fact that more comprehensive survey activities will usually also be associated with less significant markups on the basic unit prices.

Generally speaking, the itemized specification of soil and contamination classes should be preferred over specifying combined pricing items. Any shift in the quantity proportions could thus be considered by appropriately adjusting the unit price of each item. Alternatively, the client and the contractor could also agree on mutually disclosing their costing and pricing assumptions (see also Section 10.3.1.1).

9.2.3 Cost Estimate/Calculation

Project-related decision making is essentially determined by the expected costs. Costs are directly correlated to selected qualities and quantities and to the time available for project execution.

In Project Phase 2 (see Fig. 9.34), cost estimates or calculations are based on relatively rough approximations to the building dimensions (building-based methods of estimating costs) or on a design-based project structure in the case of more accurate cost calculations.

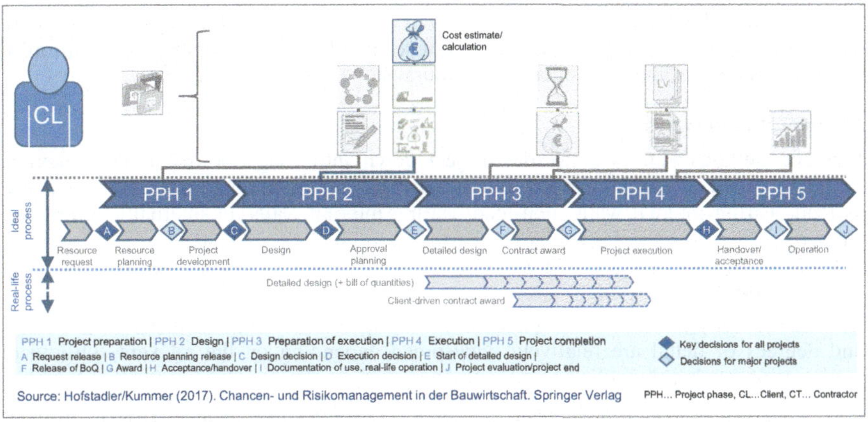

Fig. 9.34 Allocation of example in the project sequence

9.2.3.1 Basics – Budgeting using Cost Parameters

Depending on the degree of design detail and project progress, construction projects will require cost indications (and planning) in different levels of detailing. A number of methods are available for this purpose.

Use-based methods

Use-based methods are primarily applied in early project phases and deliver a first cost indication. In the costing exercise, project-specific units are used, such as the euro amount incurred per underground car parking space, per hospital bed, or per workplace. More generally, such values can be expressed in euros per unit of use.[25)]

Building-based methods

Building-based methods rely on existing project planning information. Geometrical dimensions of the building to be constructed are utilized and multiplied with cost parameters. These include, for example, the gross cubic volume (GCV) and gross floor area (GFA) determined in accordance with OENORM B 1800:2013 or DIN 277:2016.[26)]

The *Baukosteninformationszentrum Deutscher Architektenkammern GmbH* (BKI; Construction Cost Information Center of the German Chambers of Architects) is known to be a provider of reference cost indicators. The center maintains a database that contains construction costs from "similar" projects that can be retrieved and adjusted to the project at hand. This approach is closely linked to the issue of ensuring comparability because each building or structure is unique in its own right.[27)]

25) Cf. Lechner/Liebenau (2010), p. 23

26) Cf. Blecken et al. (2007), p. 106f.

27) Cf. Lechner/Liebenau (2010), p. 23

Execution-based methods

Execution-based methods divide the building or structure into work groups (WG) or work subgroups (WSG). Costing is thus carried out separately for, say, concrete and reinforced concrete operations, masonry operations, natural stone operations, etc.[28)]

Element methods

Element methods aim to break the project down into rough or overarching element categories, such as internal walls, and into individual elements on a further sublevel, such as load-bearing internal walls, non-load-bearing internal walls, or internal columns and supports.[29)]

Mixed methods

In construction practice, further methods involving a wide range of different structures and degrees of detail are relatively widespread depending on the relevant subfield of construction, such as tunneling or infrastructure.

The *Baukosteninformationszentrum Deutscher Architektenkammern* (BKI) provides indicative values for costing based on various assumptions (see Fig. 9.35). These figures represent statistically analyzed data compiled for several benchmark projects that were analyzed in a defined workflow after completion.

Depending on the level of available information, these cost indicators can be applied to derive initial cost conclusions even in the absence of a preliminary design. It should be noted, however, that using mean values alone will not ensure a sufficiently accurate costing exercise.[30)]

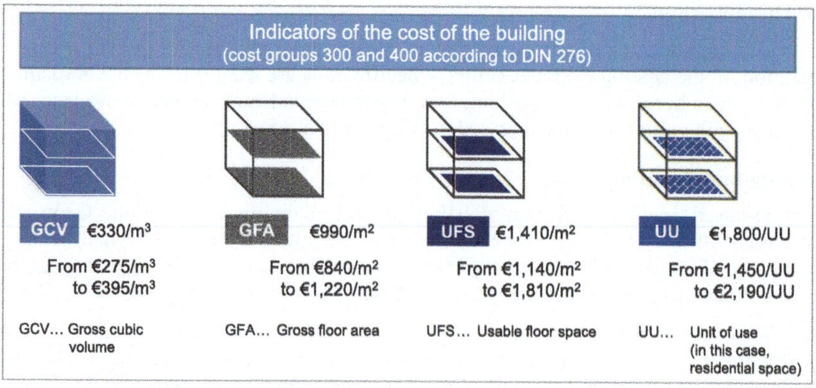

Fig. 9.35 Indicators for determining the cost of a building (cost groups 300 and 400 according to DIN 276) – Multi-family homes with more than 20 dwelling units, medium standard[31)]

The projects on which the determination of these cost indicators is based are also stated with their complete spread from minimum to maximum cost values, as well as with a statistically determined range. Fig. 9.36 shows a diagram where these values are represented

28) Cf. ibid., p. 24

29) Cf. Blecken et al. (2007), p. 75 and 108f.

30) Cf. Kalusche et al. (2016), p. 39f.

31) Cf. ibid., p. 512 – Costs: Data as of the first quarter of 2016, federal average including VAT at 19%.

by colored bars, or arrows. Each of the points represents a cost value of one of the 16 analyzed benchmark properties (i.e. multi-family homes with more than 20 dwelling units, medium standard).[32)]

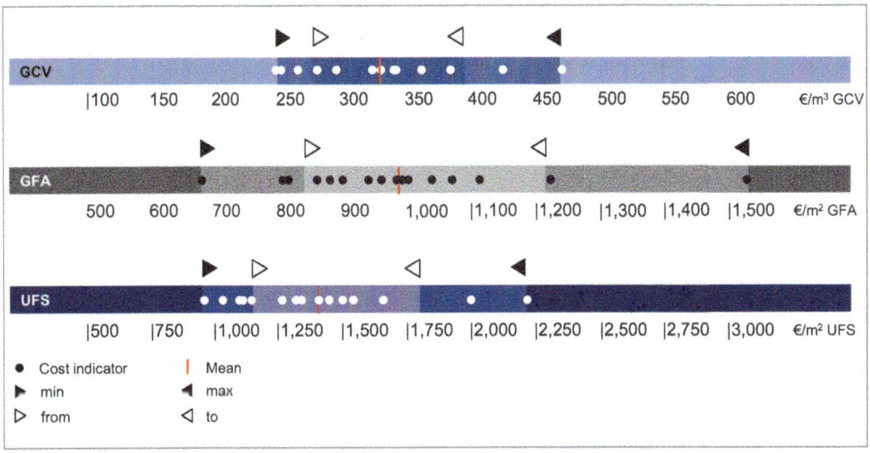

Fig. 9.36 Indicators for determining the cost of a building (cost groups 300 and 400 according to DIN 276) – Multi-family homes with more than 20 dwelling units, medium standard – 16 benchmark properties[33)]

The BKI provides not just cost indicators for building-based methods, such as GCV, GFA, or usable floor space (UFS) but also ranges of cost indicators structured in accordance with the areas of work or service provided in the *Standardleistungsbuch* (StLB; Standard Services Book) (see Table 9.19).

The "from-to" values stated for both rough approximations (e.g. GCV and detailed analyses of work areas) define a range of the cost indicators. Any determination of this range on the basis of historical data accounts for the fact that upward deviations are more likely than downward deviations in the construction industry.[34)] This is why the values above the mean are analyzed separately from the values below the mean. Excessively expensive buildings thus have no influence on the statistical values below the mean. Hence, the mean does not necessarily appear in the middle of the indicated spread (e.g. 28 | 44 | 56). The tables highlight whether there is only a single value (e.g. - | 5 | -) or whether there are several values that, however, do not yet permit a calculation of the spread (e.g. 27 | 27 | 27). This approach should reflect the reality in the construction industry more closely.[35)]

Table 9.19 also specifies the percentage shares of the individual cost groups in order to be able to carry out additional plausibility checks on the basis of the cost proportions.

32) Superposition of points of identical cost indicators is possible.

33) Cf. Kalusche et al. (2016), p. 512 – Costs: Data as of the first quarter of 2016, federal average including VAT at 19%.

34) Cf. ibid., p. 9

35) Cf. ibid., p. 9f.
According to the BKI, it is not possible to replicate the calculation method by standard Excel routines because new means would be derived for each of the partial areas (above and below the mean).

Although a costing exercise performed on the basis of the cost indicators (in the sense of budgeting) cannot replace the itemized breakdown of costs in the individual work areas, it is well suited for plausibility checks or rough cost approximations, for instance for the purpose of submitting a loan or mortgage request.[36)]

No.	WA	Work or service areas	€/m² GFA			% of 300+400		
			from	Mean	to	from	Mean	to
0	A	B	C	D	E	F	G	H
1	000	Safety and security, site facilities incl. 001	28	**44**	56	2.9	**4.4**	5.7
2	002	Earthworks	12	**24**	36	1.2	**2.4**	3.7
3	006	Custom civil works incl. 005	4	**12**	21	0.4	**1.2**	2.1
4	009	Drainage channels incl. 011	2	**5**	5	0.2	**0.5**	0.5
5	010	Drainage works	-	**-**	-	-	**-**	-
6	012	Masonry works	5	**56**	108	0.5	**5.7**	10.9
7	013	Concrete works	173	**210**	210	17.5	**21.2**	21.2
8	014	Natural and cast stone works	13	**13**	18	1.3	**1.3**	1.8
9	016	Carpentry and timber construction	2	**12**	22	0.2	**1.2**	2.2
10	017	Steel construction works	0	**40**	80	0.0	**4.0**	8.1
11	018	Sealing works	0	**3**	6	0.0	**0.3**	0.6
12	020	Roofing	2	**10**	18	0.2	**1.0**	1.8
13	021	Roof sealing	6	**15**	15	0.6	**1.5**	1.5
14	022	Plumbing works	6	**11**	16	0.6	**1.1**	1.6
15		**Shell construction**	454	**454**	478	45.9	**45.9**	48.4
16	023	Render and stucco, heat insulation systems	39	**56**	77	3.9	**5.7**	7.8
17	024	Tiling	16	**24**	32	1.6	**2.4**	3.2
18	025	Screed laying	17	**21**	25	1.7	**2.1**	2.5
19	026	Windows, external doors incl. 029, 032	48	**48**	70	4.9	**4.9**	7.1
20	027	Joinery	35	**50**	50	3.6	**5.1**	5.1
21	028	Parquet and wood block floorings	-	**5**	-	-	**0.5**	-
22	030	Window shutters	6	**12**	19	0.6	**1.2**	1.9
23	031	Metal works incl. 035	22	**41**	57	2.2	**4.2**	5.8
24	034	Paint/decoration works incl. 037	19	**32**	45	1.9	**3.2**	4.5
25	036	Floor coverings	10	**14**	18	1.0	**1.5**	1.9
26	038	Ventilated curtain façades	-	**-**	-	-	**-**	-
27	039	Dry walls	30	**34**	37	3.1	**3.4**	3.8
28		**Finishing**	340	**340**	358	34.4	**34.4**	36.2
29	040	Heat supply systems - utility systems/units incl. 041	29	**42**	56	2.9	**4.2**	5.6
30	042	Gas and water supply systems, works incl. 043	23	**36**	36	2.4	**3.6**	3.6
31	044	Wastewater system installation, piping	22	**22**	29	2.2	**2.2**	2.9
32	045	Gas, water, wastewater fixtures and fittings incl. 046	0	**8**	17	0.0	**0.8**	1.7
33	047	Insulation of utility systems/units	0	**5**	10	0.0	**0.5**	1.0
34	049	Fire-extinguishing devices and systems	-	**1**	-	-	**0.1**	-
35	050	Lightning protection and earthing systems	-	**1**	2	-	**0.1**	0.2
36	052	Medium-voltage systems	-	**-**	-	-	**-**	-
37	053	Low-voltage systems incl. 054	34	**40**	45	3.4	**4.0**	4.5
38	055	Backup power supply	-	**-**	-	-	**-**	-
39	057	Building systems equipment	-	**-**	-	-	**-**	-
40	058	Lights and luminaires incl. 059	0	**3**	6	0.0	**0.3**	0.6
41	060	Electroacoustic systems, intercoms	0	**1**	1	0.0	**0.1**	0.1
42	061	Communications networks incl. 062	0	**1**	3	0.0	**0.1**	0.3
43	063	Alarm systems	0	**0**	1	0.0	**0.0**	0.1
44	069	Elevators	0	**28**	56	0.0	**2.8**	5.7
45	070	Building automation	-	**-**	-	-	**-**	-
46	075	Ventilation and air-conditioning systems	0	**4**	10	0.0	**0.5**	1.0
47		**Services and utilities**	160	**191**	226	16.1	**19.3**	22.8

Table 9.19 Cost indicators for work or service areas as defined in the *Standardleistungsbuch* (StLB; Standard Services Book – building costs according to DIN 276) – Multi-family homes with more than 20 dwelling units, medium standard[37)]

36) Cf. ibid., p. 41

37) Cf. Kalusche et al. (2016), p. 514 – Costs: Data as of the first quarter of 2016, federal average including VAT at 19%.

As part of the costing exercise performed for a new construction project, a cost indicator must be assumed for each of the specified cost groups, for instance at the work or service area level. This selection is based on choosing corresponding reference projects and the associated degree of workmanship and quality standards. If the related considerations are consistently based on mean values, this may cause a significant underestimation of costs. Nor can any conclusions be drawn with respect to possible ranges or the chance/risk ratio.

The BKI values refer to Germany, which is why they include value-added tax at a rate of 19%. Application of this system to other countries would require either deduction of this tax component in order to arrive at net values or adjustment of the tax rate to the specific country.

Regional factors should be considered in the calculations in order to compensate for the influence of regional differences within Germany or Europe. Fig. 9.37 shows the 2016 regional factors for Europe.

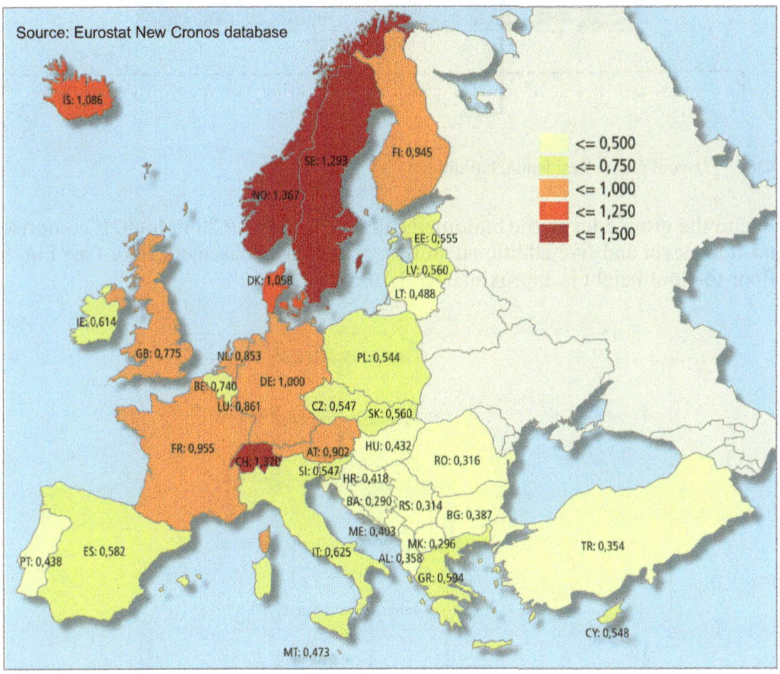

Fig. 9.37 2016 regional factors for Europe[38)]

9.2.3.2 Task

A cost estimate or costing exercise is to be performed for the construction of a residential building with six stories, 36 dwelling units and an underground car park with 45 spaces,

[38)]
BKI BAUKOSTEN-Poster 2016; source: Eurostat New Cronos database.
Please note that the English notation would use decimal points instead of commas.

basement units, cycle stands, and utility rooms. The plot extends over an area of about 5,800 m², the gross floor area of the ground-floor level amounts to approximately 430 m². The gross floor area of the basement is about 1,590 m²; the basement level is thus about 3.6 times larger than the ground floor. For the residential building, we obtain a total gross floor area of about 4,050 m². Apartments are accessed via a gallery. Fig. 9.38 shows the layout plan of the new building.

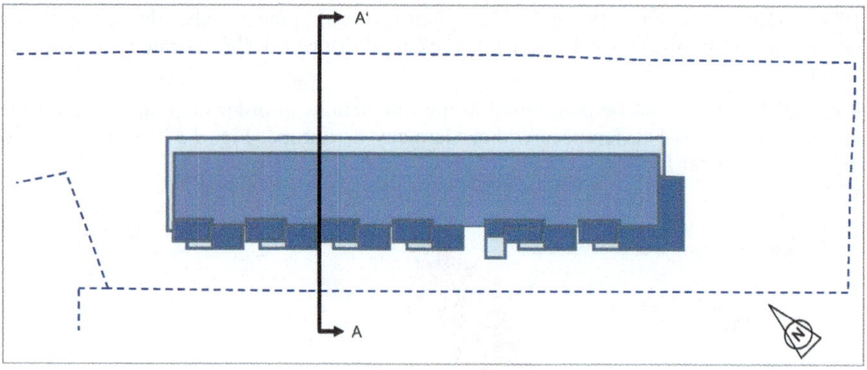

Fig. 9.38 Layout plan – Residential building

Relative to the ground level, the building is approximately 19.20 m high. It comprises the ground-floor level and five additional stories, as well as a basement floor (see Fig. 9.39). The floor-to-floor height is approximately 3.10 m.

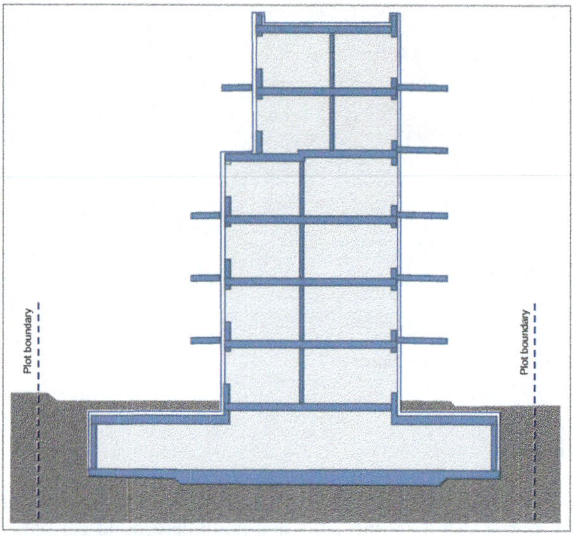

Fig. 9.39 A-A′ section – Residential building

9.2.3.3 Modeling

Since the BKI sources specify costs per work area only as mean values and "from-to" values (i.e. spreads or ranges), any probabilistic analysis will require their conversion into a distribution function. The specified ranges, however, do not permit direct conclusions regarding the respective probabilities above and below the means.

Pertinent literature and previous analyses identified right-skewed distributions for labor consumption rates, quantities, and cost values of construction projects (see Sections 6.4, 7.2.1, and 7.2.2).

In addition, the BKI user instructions state that "[...] in the construction industry, upward deviations are more probable than downward deviations [...]"[39].

Based on the above considerations, it appears plausible to use right-skewed distributions for cost parameters.

Since the complete database of reference projects is either not available or needs to be requested separately, we demonstrate an option of combining the three specified values with theoretical considerations in order to develop a distribution function of cost parameters, which can then be used for probabilistic calculations.

We investigated the possibility of converting the BKI cost parameters into PERT distributions in order to keep the input process as simple as possible. This distribution is defined by a minimum (MIN), most likely, and maximum (MAX) value.

We obtain the mean of the distribution by applying Eq. (9-3).

$$\text{MV} = \frac{\text{MIN} + 4 \cdot \text{MOST LIKELY} + \text{MAX}}{6} \tag{9-3}$$

However, the BKI parameters include a mean value (MV) instead of a most likely value, which is why this mean needs to be converted to a most likely value (see Eq. (9-4)).

$$\text{MOST LIKELY} = \frac{6 \cdot \text{MV} - \text{MIN} - \text{MAX}}{4} \tag{9-4}$$

Yet this conversion is not possible if six times the mean value is smaller than the sum of the minimum and maximum values, which is the case if the interval between the mean and the maximum value is too large. In this case, we obtain a negative most likely value that does not define a PERT distribution if the minimum and maximum values are positive.

We thus apply the more generic beta general distribution in order to be able to generate a distribution function nonetheless. The PERT distribution is a specific type of beta distribution.

This distribution is defined by four parameters:

- Minimum
- Maximum
- α_1
- α_2

[39] Cf. Kalusche et al. (2016), p. 9 (translation by Steffen Walter)

α_1 and α_2 are curve parameters that can be calculated by applying the minimum, mean, and maximum values to Eq. (9-5) and Eq. (9-6).[40)]

$$\alpha_1 = 6 \cdot \left(\frac{MV - MIN}{MAX - MIN}\right) \tag{9-5}$$

$$\alpha_2 = 6 \cdot \left(\frac{MAX - MV}{MAX - MIN}\right) \tag{9-6}$$

The beta general distribution provides the advantage of eliminating the need for converting the mean value, thus reflecting the full interval between the mean and the maximum.

The beta general and PERT distributions are identical for input values with a defined PERT distribution. We therefore suggest using the beta general distribution as a more generic approach.

Since the specified BKI parameters indicate a range rather than an absolute minimum and maximum of observed values, we provide the option of specifically adjusting the lower and/or upper cost parameter limit in order to arrive at a more realistic representation.

A percentage adjustment of the ranges specified in the BKI sources makes it possible to increase or decrease the applied spread for the individual cost parameters if and when required for the project under consideration.

The obtained cost parameter distributions are then multiplied by the gross floor area.

This worked example includes both a rough and a detailed analysis. The former covers the shell construction and finishing stages as well as the installation of services and utilities. The latter uses the work areas listed as items 000 to 075 in Table 9.19 for calculating the costs.[41)]

Fig. 9.40 shows the flow chart for this calculation. In this setting, beta general distributions represent the cost parameters, whereas a triangular distribution symbolizes the gross floor area.

---

40) The two curve parameters must be greater than zero for defining a beta general distribution. The beta general distribution is not defined if this is not the case. Identical minimum and mean values are one example of this non-defined state. The PERT distribution is applied to such cost indicators.

41) In the example, this calculation is performed for the sake of comparison, but is unusual in actual practice. The preferred approach is to apply the element-based method (in Austria, the specifications for building construction projects are used for this purpose).

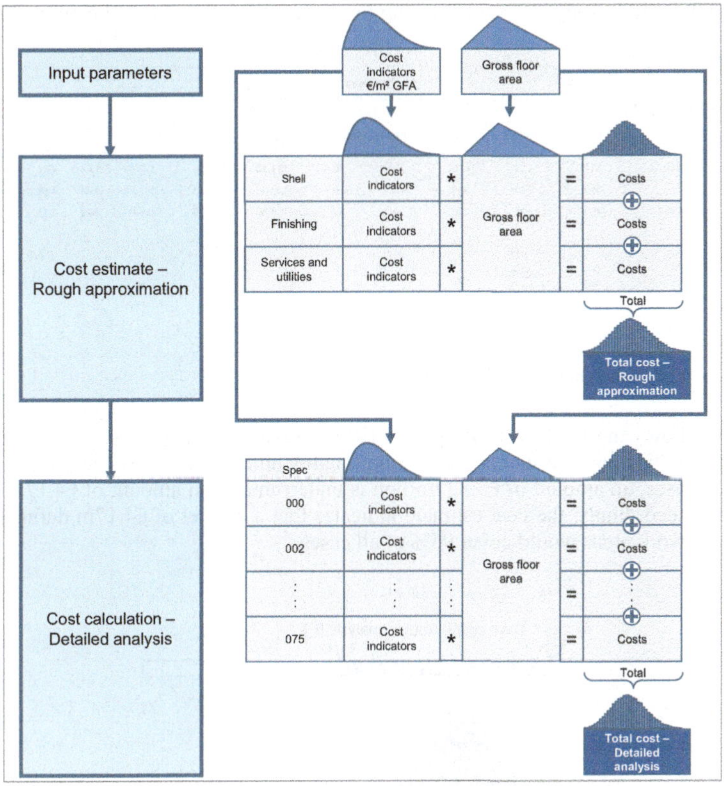

Fig. 9.40 Flow chart – Cost estimate and calculation – Rough and detailed analysis

9.2.3.4 Calculation and Interpretation of Results

The cost estimate (rough analysis) and the cost calculation (detailed analysis) are performed for the considered residential project, covering the shell construction, finishing, and services and utilities work areas. The BKI cost parameters are used for this purpose. Percentages are applied for implementing range adjustments so that it is possible to reduce the BKI "from" values and to increase the BKI "to" values.

Rough cost estimate

Columns B to D (Rows 1 to 3) of Table 9.20 state the spreads and the means of the BKI values for the shell construction, finishing, and services and utilities work areas for multi-family homes with over 20 dwelling units (medium standard of fit-out and finish). Columns E to F permit adjustments of the ranges of these values to specific project requirements. For the project outlined in this section, each of the BKI "from" and "to" values was lowered by 5% and increased by 10%, respectively. Columns G, H, and I of Table 9.20 specify the resulting adjusted cost parameters. According to this approach, we obtain a minimum shell construction cost of €431.30 per square meter of gross floor area in Cell G1 (€454.00/m² of GFA * 0.95). The gross floor area is represented by a triangular distribution (4,030 | 4,050 | 4,070 m² GFA) (see Row 5). Cell L6 shows the total cost determined in the rough analysis by multiplying the costs per work area by the gross floor area.

No.	Work or service areas	€/m² GFA			Percentage adjustment of spread [%]		€/m² GFA - Adjusted values			BetaGeneral distribution		
		from	Mean	to	Reduction	Increase	MIN	Mean	MAX	α_1	α_2	Distribution
0	A	B	C	D	E	F	G	H	I	J	K	L
1	Shell construction	454	454	478	5	10	431.30	454.00	525.80	1.44	4.56	454.00
2	Finishing	340	340	358	5	10	323.00	340.00	393.80	1.44	4.56	340.00
3	Services and utilities	160	191	226	5	10	152.00	191.00	248.60	2.42	3.58	191.00
4	Parameter	m² GFA										Distribution
		MIN	MOST LIKELY	MAX								
5	Gross floor area	4,030	4,050	4,070								4,050.00
6	Total cost – Rough analysis											3,989,250.00

Table 9.20 Calculation table – Cost estimate – Rough analysis

Fig. 9.41 shows the total cost histogram. The mean shown in the histogram amounts to €3.99 million whereas the standard deviation roughly amounts to €106,200. In about 5% of simulated cases, an amount of €3.82 million is underrun and an amount of €4.17 million is exceeded. Accordingly, the cost estimate indicates that a budget of €4.17m defined for the considered work areas would cover 95% of all cases.

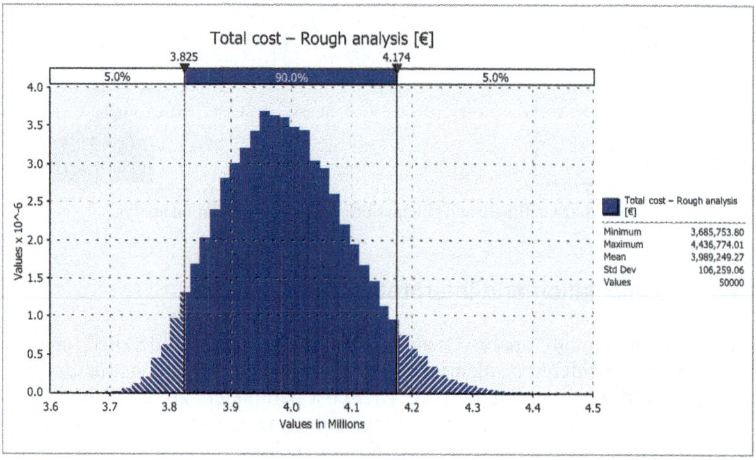

Fig. 9.41 Histogram – Total cost – Rough analysis[42)]

Cost calculation (detailed analysis)

The costs of the categories resulting from the work area breakdown (see Table 9.19) are used for the cost calculation at the detailed analysis stage. To be able to estimate the ranges of cost parameters for the individual work areas, costs must be calculated to a greater degree of detail in relation to the construction project, especially with regard to qualities. In this example, we assume that works have been scheduled for all areas. This

42) Calculations include VAT at 19%, but they do not consider regional factors.

setting also permits the addition or exclusion of individual work areas in accordance with specific project needs.

Following the same approach as in the cost estimate, the BKI from-to and mean values are applied (see Columns C to E of Table 9.21). Again, limits can be adjusted on a project-specific basis by applying percentages (see Columns F and G). Compared to the rough analysis, smaller values are applied for the percentage adjustment of the relevant ranges. This is due to the greater degree of detail in determining the costs, as well as to the more detailed specification of qualities. In this example, applying equal percentages to all work areas listed in Table 9.21 provides a welcome simplification. Individual percentage values can be used for this purpose.

No.	WA	Work or service areas	€/m² GFA			Percentage adjustment of spread [%]		€/m² GFA - Adjusted values			BetaGeneral distribution		
			from	Mean	to	Reduction	Increase	MIN	Mean	MAX	α_1	α_2	Distribution
0	A	B	C	D	E	F	G	H	I	J	K	L	M
1	000	Safety and security, site facilities incl. 001	28	**44**	56	2	5	27.44	44.00	58.80	3.17	2.83	44.00
2	002	Earthworks	12	**24**	36	2	5	11.76	24.00	37.80	2.82	3.18	24.00
3	006	Custom civil works incl. 005	4	**12**	21	2	5	3.92	12.00	22.05	2.67	3.33	12.00
4	009	Drainage channels incl. 011	2	**5**	5	2	5	1.96	5.00	5.25	5.54	0.46	5.00
5	010	Drainage works	-	-	-	-	-	-	-	-	-	-	-
6	012	Masonry works	5	**56**	108	2	5	4.90	56.00	113.40	2.83	3.17	56.00
7	013	Concrete works	173	**210**	210	2	5	169.54	210.00	220.50	4.76	1.24	210.00
8	014	Natural and cast stone works	13	**13**	18	2	5	12.74	13.00	18.90	0.25	5.75	13.00
9	016	Carpentry and timber construction	2	**12**	22	2	5	1.96	12.00	23.10	2.85	3.15	12.00
10	017	Steel construction works	0	**40**	80	2	5	0.00	40.00	84.00	2.86	3.14	40.00
11	018	Sealing works	0	**3**	6	2	5	0.00	3.00	6.30	2.86	3.14	3.00
12	020	Roofing	2	**10**	18	2	5	1.96	10.00	18.90	2.85	3.15	10.00
13	021	Roof sealing	6	**15**	15	2	5	5.88	15.00	15.75	5.54	0.46	15.00
14	022	Plumbing works	6	**11**	16	2	5	5.88	11.00	16.80	2.81	3.19	11.00
15		**Shell construction**											
16	023	Render and stucco, heat insulation systems	39	**56**	77	2	5	38.22	56.00	80.85	2.50	3.50	56.00
17	024	Tiling	16	**24**	32	2	5	15.68	24.00	33.60	2.79	3.21	24.00
18	025	Screed laying	17	**21**	25	2	5	16.66	21.00	26.25	2.72	3.28	21.00
19	026	Windows, external doors incl. 029, 032	48	**48**	70	2	5	47.04	48.00	73.50	0.22	5.78	48.00
20	027	Joinery	35	**50**	50	2	5	34.30	50.00	52.50	5.18	0.82	50.00
21	028	Parquet and wood block floorings	-	**5**	-	2	5	4.90	5.00	5.25	1.71	4.29	5.00
22	030	Window shutters	6	**12**	19	2	5	5.88	12.00	19.95	2.61	3.39	12.00
23	031	Metal works incl. 035	22	**41**	57	2	5	21.56	41.00	59.85	3.05	2.95	41.00
24	034	Paint/decoration works incl. 037	19	**32**	45	2	5	18.62	32.00	47.25	2.80	3.20	32.00
25	036	Floor coverings	10	**14**	18	2	5	9.80	14.00	18.90	2.77	3.23	14.00
26	038	Ventilated curtain façades	-	-	-	-	-	-	-	-	-	-	-
27	039	Dry walls	30	**34**	37	2	5	29.40	34.00	38.85	2.92	3.08	34.00
28		**Finishing**											
29	040	Heat supply systems - utility systems/units incl. 041	29	**42**	56	2	5	28.42	42.00	58.80	2.68	3.32	42.00
30	042	Gas and water supply systems, works incl. 043	23	**36**	36	2	5	22.54	36.00	37.80	5.29	0.71	36.00
31	044	Wastewater system installation, piping	22	**22**	29	2	5	21.56	22.00	30.45	0.30	5.70	22.00
32	045	Gas, water, wastewater fixtures and fittings incl. 046	0	**8**	17	2	5	0.00	8.00	17.85	2.69	3.31	8.00
33	047	Insulation of utility systems/units	0	**5**	10	2	5	0.00	5.00	10.50	2.86	3.14	5.00
34	049	Fire-extinguishing devices and systems	-	**1**	-	2	5	0.98	1.00	1.05	1.71	4.29	1.00
35	050	Lightning protection and earthing systems	-	**1**	2	2	5	0.98	1.00	2.10	0.11	5.89	1.00
36	052	Medium-voltage systems	-	-	-	-	-	-	-	-	-	-	-
37	053	Low-voltage systems incl. 054	34	**40**	45	2	5	33.32	40.00	47.25	2.88	3.12	40.00
38	055	Backup power supply	-	-	-	-	-	-	-	-	-	-	-
39	057	Building systems equipment	-	-	-	-	-	-	-	-	-	-	-
40	058	Lights and luminaires incl. 059	0	**3**	6	2	5	0.00	3.00	6.30	2.86	3.14	3.00
41	060	Electroacoustic systems, intercoms	0	**1**	1	2	5	0.00	1.00	1.05	5.71	0.29	1.00
42	061	Communications networks incl. 062	0	**1**	3	2	5	0.00	1.00	3.15	1.00	4.10	1.00
43	063	Alarm systems	0	**0**	1	2	5	0.00	0.00	1.05	-	-	0.18
44	069	Elevators	0	**28**	56	2	5	0.00	28.00	58.80	2.86	3.14	28.00
45	070	Building automation	-	-	-	-	-	-	-	-	-	-	-
46	075	Ventilation and air-conditioning systems	0	**4**	10	2	5	0.00	4.00	10.50	2.29	3.71	4.00
				m² GFA									
47		**Parameter**	MIN	MOST LIKELY	MAX								**Distribution**
48		Gross floor area	4,030	4,050	4,070								4,050.00
49		**Total cost – Detailed analysis**											**3,985,908.75**

Table 9.21 Calculation table – Cost calculation – Detailed analysis

No separate calculations were performed for work areas without available cost parameters, such as WA 010, drainage works, because no such works are scheduled as part of the project. Own assumptions need to be made if works within these areas are expected. Alternatively, other structural categories according to the BKI system may be chosen.

For cost parameters with only a mean value stated in the BKI table, such as WA 028, parquet and wood block floorings, and WA 049, fire-extinguishing devices and systems (Rows 21 and 34), percentage adjustments were made to calculate a minimum and a maximum value (on the basis of the specified mean) so as to generate a beta general distribution.

For work areas for which no "from" value is available, but mean and "to" values are specified (WA 050, lightning protection and earthing systems – Row 35), the mean is also lowered by the specified percentage whereas the "to" value is increased accordingly. This step again leads to the three values required for defining the beta general distribution.

Such a procedure is not possible in Work Area 063, alarm systems (Row 43), because both the mean and the "from" value equal zero. A PERT distribution (0 | 0 | €1/m² GFA) is thus applied to this cost parameter.

Similarly to the rough analysis, the gross floor area is represented by a triangular distribution in the total cost calculation (see Row 48).[43] The total costs obtained in the detailed analysis are shown as a histogram (see Fig. 9.42) in Cell M49.

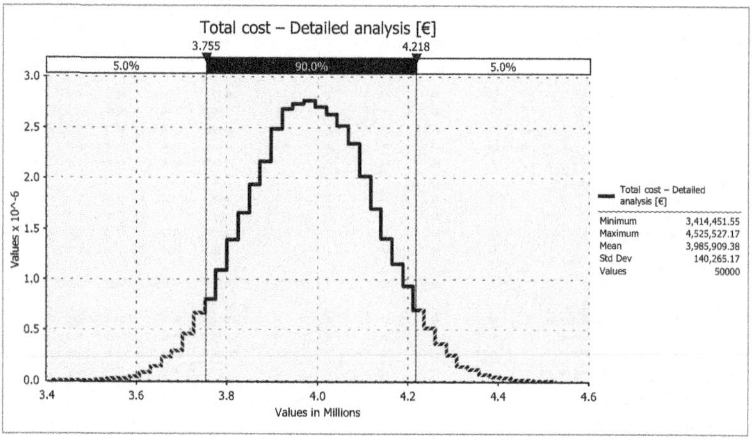

Fig. 9.42 Histogram – Total cost – Detailed analysis[44]

The total cost distribution generated from 50,000 iterations ranges from approximately €3.41 million to €4.53 million, where 90% of the individual results lie between €3.76 million and €4.22 million. The mean of the histogram amounts to approx. €3.99 million.

Fig. 9.43 compares the detailed analysis with the rough analysis. We find that the total cost range is wider in the detailed analysis (see also the standard deviations indicating the spread) although the percentage adjustments to cost parameter limits assumed in the detailed analysis were consistently smaller than in the rough analysis.

This is due to the fact that, although the sum of the means obtained for the individual work

[43] The gross floor area is assumed to be associated with uncertainties also at the detailed analysis stage.

[44] Calculations include VAT at 19%, but they do not consider regional factors.

areas in the detailed analysis is approximately equal to that of the means in the rough analysis in the specified building category (multi-family houses with more than 20 dwelling units, medium standard of fit-out and finish), the sum of the "from-to" values reflects a larger spread of possible values in the detailed analysis.

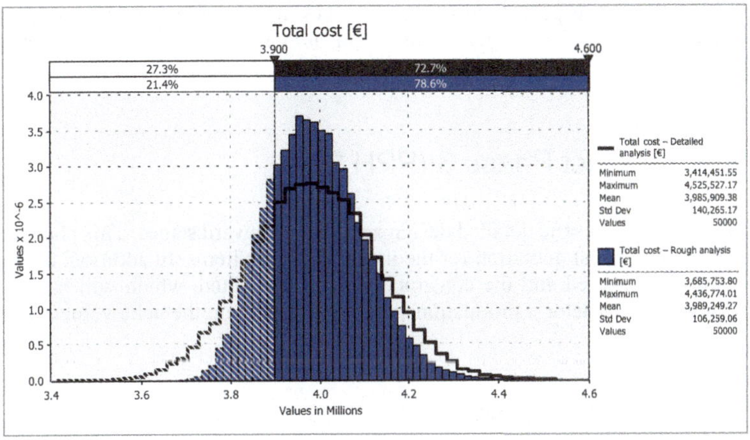

Fig. 9.43 Histograms – Total cost – Rough and detailed analysis

We conclude that the rough analysis is an appropriate means to arrive at a relatively accurate initial estimate. However, the results also show that merely applying the BKI values and performing a more detailed analysis need not necessarily lead to a sufficiently narrow total cost range. Only when the qualities are specified properly, thus enabling a sufficient restriction of the cost range, will we be able to narrow down the ranges.

In the detailed analysis, the applied input parameters or selected ranges of cost parameters result in a lower risk potential (72.7%) than in the rough analysis (78.6%) for an assumed budget of €3.90 million for the shell construction, finishing, and services and utilities work areas. A reduction in the qualities and/or quantities should be considered to increase the likelihood of adhering to the budget of €3.90 million. Qualities will have to be detailed to a greater degree if ranges are narrowed further. In addition, more detailed analyses can be performed, such as on the cost group or element/item levels.

9.2.3.5 Benefit for the Client

From the client's point of view, the determination of costs is a particularly important part of their planning process because costs are directly associated with the specified qualities (and quantities). An iterative process makes it possible to continuously detail and specify the requirements for the building or structure with respect to the costs to be expected.

The system for considering the associated uncertainties by applying beta general or PERT distributions frees the client from having to choose a deterministic value for each cost parameter. Instead, the client can specify related ranges. Only for the final results (total cost) is a deterministic value chosen if the specific chance/risk ratio is to be calculated for a given budget.

The client is able to incorporate the uncertainties inevitably associated with the individual items into the results, and to base any further considerations, analyses, and planning activities on them. The BKI cost parameters create a database or a set of indicative values that can be adjusted if and when required for the purpose of the project. Alternatively, these values can be replaced with own cost parameters.

Depending on the degree of design detail, cost calculations should be adjusted and continuously updated with respect to the expected ranges in order to be able to derive accurate conclusions regarding the expected costs and associated uncertainties (expressed as ranges).

9.3 Project Phase 3 (PPH 3)

Project Phase 3 includes the detailed design and contract award stages. This phase requires the preparation of a cost quotation for the individual work items. In addition, the building specification is analyzed and the construction time calculated, which will have a major influence on the contractor's production system as well as on the achievement of project targets.

9.3.1 Cost Quotation

Preparing the cost quotation is part of Project Phase 3. It should ideally be based on a sufficiently advanced detailed design (see Fig. 9.44).

The cost quotation's purpose is to obtain even more accurate knowledge of the costs or bid prices to be expected. It is prepared on the level of individual work items.

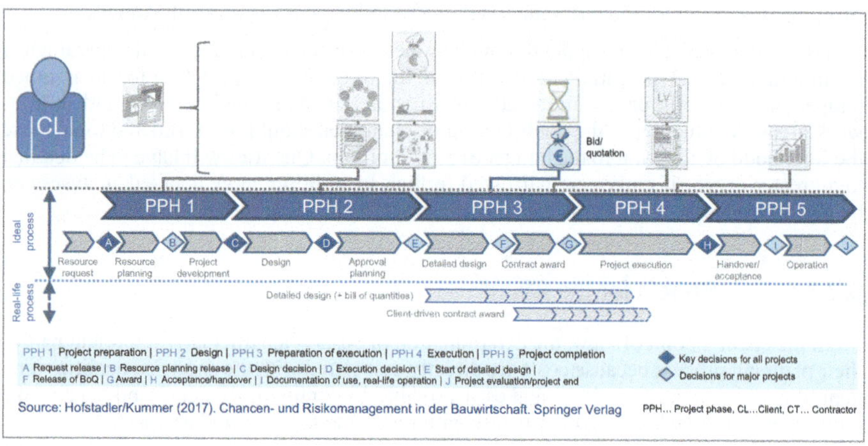

Fig. 9.44 Allocation of example in the project sequence

9.3.1.1 Basics

The cost quotation involves a costing exercise prior to releasing the bills of quantities. It is prepared on the basis of the detailed design, the building specification, and the work execution schedule.

The cost quotation has an itemized structure defined by the individual work steps; it reflects the detailed design and contract award preparation stages. The cost quotation is usually prepared by the contracted designer or preparer of the bills of quantities for the request for proposals by entering market prices for the individual items (own pricing) in order to be able to make comparisons with previous costing exercises (budget of the award package) and to enable the client/project manager to implement corrective actions if and when required.[45)]

It ultimately helps define the type of tender procedure (threshold values in the case of public clients). In addition, the prices quoted by the bidders can be checked for plausibility in an initial step after the bid opening phase.

9.3.1.2 Task

After preparation of the bill of quantities, prices are entered for their individual items. At this stage, the client seeks to identify reliable cost and/or pricing assumptions for the individual items. Pricing ranges will depend on the type of work and the circumstances and conditions under which it is performed, as well as on the current market situation. The client is able to refer to information published in pertinent literature, such as statistical cost indicators for work items[46)], as well as to their own experience and analyses, and requests sent to construction contractors. However, the client's assumptions will always be associated with uncertainties.

The plausibility of the cost quotation will essentially be influenced by quantity uncertainties associated with the individual work items, but also by the uncertainties associated with unit prices, such as in the proportions of wages and other cost items.

Using distribution functions for quantities and prices makes it possible to express these uncertainties as ranges and probabilities for individual items, work groups, work subgroups, or for the entire construction project.

To keep the associated cost and effort at a reasonable level, uncertainties should be analyzed predominantly for those items with the most significant influence on total cost (A items[47)] and essential items).

If possible, combined pricing items should be avoided (as a minimum, the expected quantity structure should be disclosed to the bidders) in order to reduce the uncertainties referred to above. This step eliminates the need for bidders to make their own assumptions (with regard to details usually provided by the client), thus preventing additional cost claims in the event of deviations from these assumptions.

---

45) Cf. Lechner/Heck (2011), p. 71f.

46) Cf. Fetzer et al. (2016)

47) In analogy to the Pareto principle, these are the 20% of all items that account for about 80% of the total price.

9.3.1.3 Modeling

In the case of reinforced concrete and masonry operations, we can base the modeling exercise to be performed for quantity uncertainties on the results of the expert survey (see Section 7.2.2).

The same system as used for the cost estimate or cost calculation in Section 9.2.3.3 can be applied to reflect the uncertainties associated with itemized prices.

Fig. 9.45 shows the flow chart for this calculation. In addition, the proportions of wages and other components can be stated for the prices of the individual items, which permits a more detailed modeling exercise.

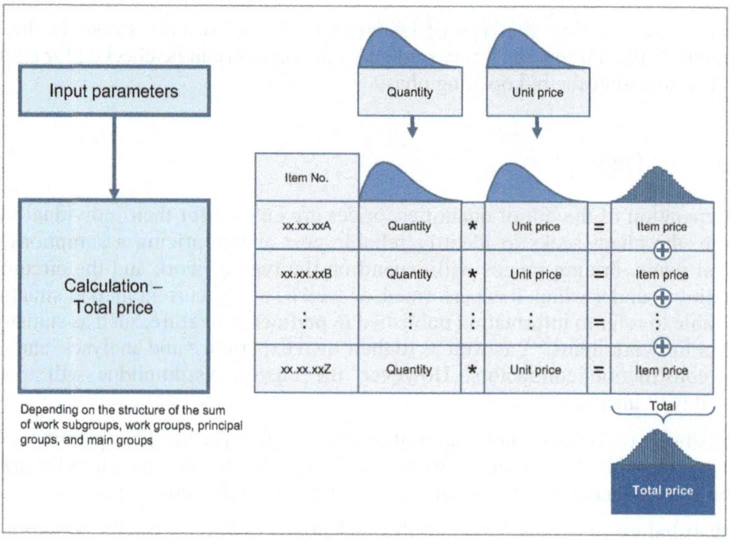

Fig. 9.45 Calculation flow chart – Cost quotation

9.3.1.4 Calculation and Interpretation of Results

No separate example is given in this section because the calculation process is more or less identical to that used for the cost estimate and/or cost calculation (see Section 9.2.3).

9.3.1.5 Benefit for the Client

The consistent consideration of uncertainties in preparing the cost quotation and the application of Monte Carlo simulations enable the client to get an overview of possible bid price ranges even before the tender documents are published. As a result of thoroughly considering the possible ranges of itemized prices, the client has the opportunity to correct the bill of quantities if required, for instance by breaking down combined pricing items into separate sub-items.

In addition, the pricing histograms can be used for checking the bids for plausibility or deriving the chance/risk ratios for the prices quoted by the individual bidders both on the level of individual work items and for work groups or work subgroups.

9.3.2 Determination of Construction Time

We discussed the importance of construction time and the relationship between the specified construction time and construction cost in Chapter 8. Construction time is one of the essential specifications provided by the client; it has a major influence on the utilization of the contractor's production factors. Various methods and indicators are available for determining the construction time. The BKI also provides related indicative values.

Other parameters on which this calculation can be based include the number of workers or the number of essential pieces of equipment on-site. In building construction, for example, one key indicator is the number of cranes installed on the construction site. In this regard, it is crucial to consider the ramp-up, main, and final construction phases for the purpose of determining resource utilization. Section 10.3.5 includes an example of how to calculate construction time on the basis of the number of available cranes or the number of workers.

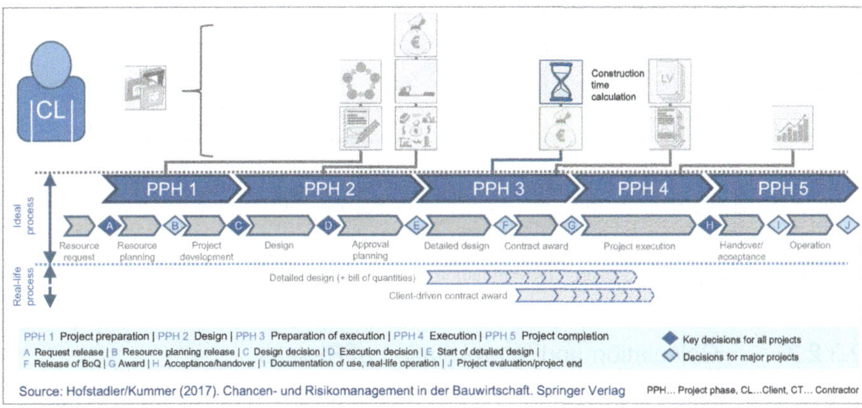

Fig. 9.46 Allocation of example in the project sequence

Determining the construction time from the client's perspective is part of Project Phase 3 (see Fig. 9.46). It is an essential prerequisite for preparing the tender and contractual documents.

9.3.2.1 Task

For a building construction project (Linz Opera House), the time required for construction works is calculated on the basis of the expected gross cubic volume (GCV) [m³], the average number of workers WK_{AVG} [wh/h], the daily working time WT [h/d], and the average labor consumption rate of shell construction works $LCR_{SC,AVG}$ [wh/m³].

Indicative values for the labor consumption rate can be derived from pertinent literature and from analyzing previously completed projects, as well as on the basis of expert responses to surveys.

In the field of building construction, one possible method of estimating or calculating the number of workers is to start from the number of cranes to be installed on the job site as required for the specified works (see also Section 10.3.5).

9.3.2.2 Modeling

Fig. 9.47 shows the flow chart for calculating the duration of shell construction works D_{SC} [d]. In this example, all input parameters are represented by triangular distributions. In the first step, we obtain the average performance value $PV_{SC,AVG}$ [m³/d] from the average number of workers[48] WK_{AVG} [wh/h], the daily working time WT [h/d], and the average labor consumption rate of shell construction works $LCR_{SC,AVG}$ [wh/m³]. We divide the gross cubic volume (GCV) [m³] by this calculated performance value to obtain the duration of shell construction works D_{SC} [d].

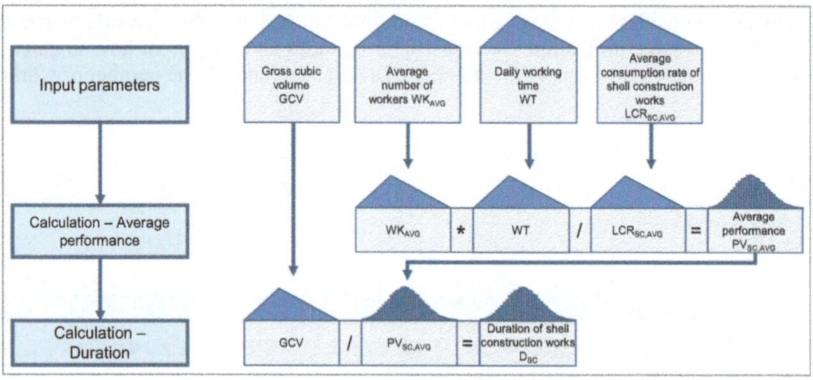

Fig. 9.47 Calculation flow chart – Determination of construction time

9.3.2.3 Calculation and Interpretation of Results

Table 9.22 illustrates the calculation of the time to completion of shell construction works. Columns C, D, and E state the minimum, most likely, and maximum values for the uncertain input parameters in order to define the related triangular distributions. Column G includes the distributions as such, or the related interim and final results.

The expected gross cubic volume GCV of this building construction project amounts to 289,860 m³. Depending on project progress in terms of its design and detailing, the gross cubic volume is either a deterministic (fixed) value, or it is still fraught with uncertainties. In this example, we assume a certain (narrow) range within which the gross cubic volume may still vary (minimum: 289,750 m³; maximum: 290,000 m³ – see Row 1 in Table 9.22).

The average number of workers WK_{AVG} was also defined as a triangular distribution between 65 and 80 workers (most likely value: 70 workers) (see Row 2). Likewise, the daily working time WT (Row 3) can range from 8.00 to 10.00 h/d (most likely value: 9.00 h/d).

[48] In building construction, the number of workers (maximum) essentially depends on the number of cranes used.

No.	Calculation parameter		Unit	MIN	MOST LIKELY	MAX	Distribution
0	A	B	C	D	E	F	G
1	Gross cubic volume	GCV	m³	289,750.00	289,860.00	290,000.00	289,870.00
2	Average number of workers	WK_{AVG}	wh/h	65.00	70.00	80.00	71.67
3	Daily working time	WT	h/d	8.00	9.00	10.00	9.00
4	Average labor consumption rate of shell construction works	$LCR_{SC,AVG}$	wh/m³	0.70	0.80	1.00	0.83
5	Average performance value	$PV_{SC,AVG}$	m³/d				774.00
6	Duration of shell construction works	D_{SC}	d				374.51

Table 9.22 Calculation table – Determination of construction time

In this example, the average labor consumption rate of shell construction works $LCR_{SC,AVG}$ (Row 4) is also represented by a triangular distribution. However, any data or information available from completed, clearly delineated projects (e.g. broken down to individual types of buildings or structures) or results of an expert survey can be subjected to data fitting in order to convert them into a theoretical distribution function to be applied to the subsequent calculation steps.

In Cell G5, the average performance of shell construction works $PV_{SC,AVG}$ is obtained by multiplying the working time by the average number of workers and dividing the product by the labor consumption rate of shell construction works.

In the last step, we obtain the duration of shell construction works D_{SC} by dividing the gross cubic volume by the average performance. Cell G6 shows the related histogram (see Fig. 9.48).

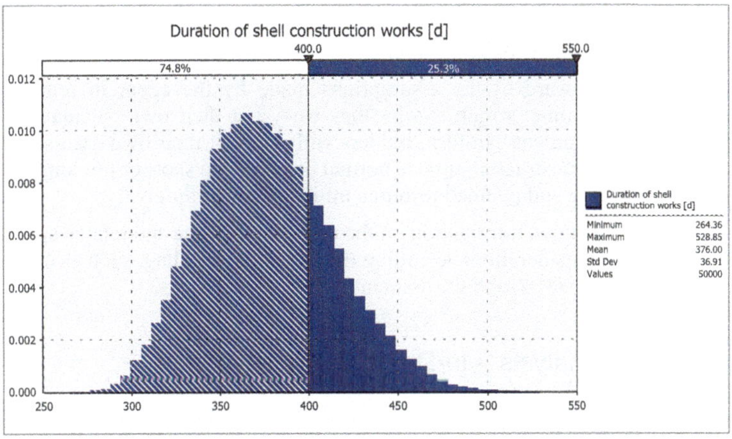

Fig. 9.48 Histogram – Duration of shell construction works

In this example, the duration of shell construction works ranges from 264 to 529 days after 50,000 iterations. The mean value amounts to about 376 days, whereas the standard deviation is approximately 37 days.

When specifying a certain construction time, the client can use the result of the probabilistic calculation in order to arrive at direct conclusions regarding the associated chance/risk ratio for their assumptions made for the purpose of the calculation. In the example, a construction time of 400 days would correspond to a chance/risk ratio of about 75% (chance) to 25% (risk).

The client should thus aim to specify a construction time during which the contractors are able to fully utilize their production factors. In this scenario, the potential for required acceleration measures or additional cost claims is reduced significantly.

9.3.2.4 Benefit for the Client

The probabilistic calculation of the construction time provides the client with a comprehensive representation (in the form of a histogram) of the uncertainties associated with the specification of a given construction time. The client would be well-advised not to define a time to completion near the lower end of the calculated range because this would be associated with a higher risk of non-adherence to the schedule.

Starting from the gross floor area to calculate construction time is merely a rough approximation, which is why it can be useful to add a certain percentage as a buffer to the calculated construction time.

The latter is obviously directly related to the assumptions made by the client regarding the number of workers, daily working time and, particularly, the labor consumption rate [wh/m^3]. If relatively optimistic ranges are applied in the modeling exercise, this will result in a histogram with a construction time range near the lower end. The applied input parameter distributions should thus be based on a solid database or information provided by experts.

Monte Carlo simulations make it possible to systematically consider the associated uncertainties in the calculation process and to visualize the possible spreads as part of the interpretation of results. This step creates a basis for decision making in order to specify a deterministic value for the construction time.

The bidders are not aware of the assumptions made by the client in relation to the specified construction time, which is why they carry out their own calculations on the basis of the tender documents. In turn, bidders will arrive at their own assessments as to whether the specified construction time is normal or either too short or too long. They will then adjust their pricing and planned resource utilization accordingly.

Clients can use the obtained information on the construction time and related spreads to be expected for further considerations regarding the use of the building, such as in relation to letting or shop opening dates, and the associated chances and risks.

9.3.3 Bid Analysis – In-Depth Bid Assessment

For the bidder (and future contractor), the bill of quantities forming part of the structural building specification is the most important basis upon which to proceed with costing and pricing, although it should always be relied upon in conjunction with all other tender documents. The client must describe all circumstances and conditions for works to be performed so as to enable the bidder to prepare their proposal in accordance with the objective explanatory content of the bid documents.

Conversely, it is the contractor's responsibility to justify all assumptions regarding pricing and work execution made on the basis of the tender documents (costing assumptions or cost risk), as well as all discretionary decisions on the part of the contractor and the risks associated with the suppliers and subcontractors it selects.

Thus, this means that the client usually prepares the building specification, whereas the bidder performs the costing exercise for their quotation and defines the related assumptions. If, for instance, the client merges several types of work or service for different structural components, trades, soil types, rebar diameters etc. into a single item, and if the costing assumptions for such combined items are not disclosed, the bidder will be forced to make their own free costing assumptions in order to be able to perform their costing exercise at all.

Any change in the proportions of the works included in the merged item compared to the bidder's original pricing assumptions will lead to a changed pricing structure associated with potentially higher or lower costs compared to the bid price.

The structural building specification constitutes the main basis for the contractor to perform their costing and pricing exercise. In relation to the planned construction works, the bidder should thus justify or interpret their assumptions by their understanding of the explanatory content of the tender documents on the basis of an objective assessment of the prevailing circumstances. Accordingly, the bidder or cost surveyor can be confident with regard to the documents provided, and will consider their content combined with the communicated or expected conditions and circumstances under which the work is to be performed for the purpose of their costing and pricing exercise.

Clients need to specify, for instance, a mean rebar diameter in their building specification if they wish to prevent the bidder from determining their own mean rebar diameter for the concrete steel reinforcement and using this diameter as a basis for their costing exercise. The same applies to the distribution of bend shapes and reinforcement ratios of individual structural components. Although the client can reasonably expect the bidder to not blindly trust the details contained in the bill of quantities, the bidder is bound to only identify, and warn the client of, blatantly obvious inaccuracies or incomplete specifications.

In a unit price contract, itemized prices are stated alongside unit prices and quantities. These itemized prices must add up to the total price. Besides a significantly lower total price pointing to an element of speculation, higher bids can also include speculative considerations. The following section demonstrates how unit prices can be evaluated on a systematic basis if there is a necessity of performing an in-depth bid assessment for a unit price contract.[49)]

[49)] Cf. Hofstadler (2018), p. 232ff.

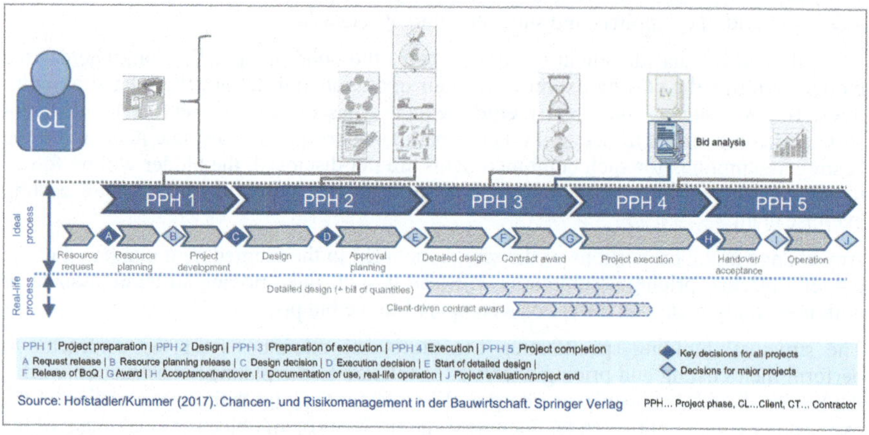

Fig. 9.49 Allocation of example in the project sequence

The bid analysis or in-depth bid assessment is allocated to the final stage of Project Phase 3. It usually constitutes the last step prior to awarding the contract (see Fig. 9.49).

9.3.3.1 Basics

In their "*Handwörterbuch der Bauwirtschaft*" [Construction Industry Dictionary], *Oberndorfer/Jodl* consider a price to be fair and reasonable if "[...] it lies within a range that ensures full cost coverage. Thus, a fair and reasonable price lies within a range between an unduly low price and an excessively high price."[50)]

In their "*Handkommentar zur VOB*" [Commentary on the VOB], *Heiermann/Riedl/Rusam et al.* postulate that a reasonable price level is reached if and when "[...] an impeccable standard of workmanship can be expected, including liability for defects, while ensuring cost-efficient construction management and economical business operations."[51)]

It should be noted that bidders are generally free in their approach to pricing. However, the components of the total price should be transparent and plausible. Cost estimates may significantly deviate from one another owing to different performance values or productivity estimates, or the application of innovative construction methods by the bidders.

Even bid prices that fail to cover the full cost could be justified by market entry considerations or the bidder's pronounced interest in being awarded landmark projects or ensuring full capacity utilization within their business. Such an extraordinarily low price, however, must never put the liquidity of the business at risk because this would contradict the principle of the economic justifiability of prices. Consequently, even bids without full cost coverage may include a plausible composition of the total price, and should thus not necessarily be excluded at the outset.

50) Cf. Oberndorfer/Jodl et al. (2001), p. 121 (translation by Steffen Walter)

51) Cf. Heiermann/Riedl/Rusam et al. (2003), p. 67 (translation by Steffen Walter)

Besides a significantly lower total price pointing to an element of speculation, higher bids can also include speculative considerations. Care must be taken, in this regard, to bear in mind that high and low unit prices can offset each other in the total price.

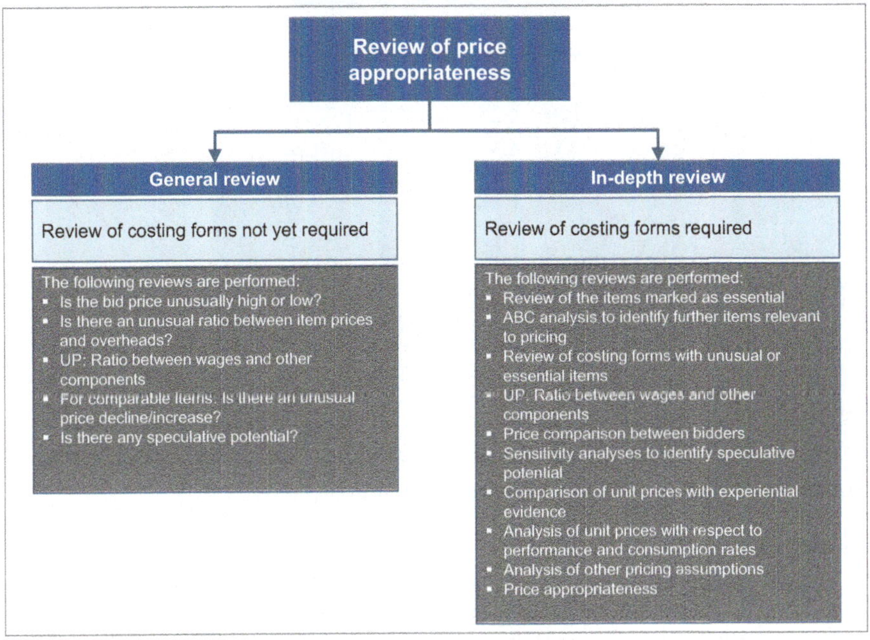

Fig. 9.50 Review of price appropriateness for public-sector clients

As a minimum, the primary costs usually recorded as expenses should be covered, including salaries and wages, materials, equipment repairs, and third-party services. The major part of these costs is variable and would be incurred at any rate. This proportion of the cost cannot be earned by other works or services or construction sites. An excessively low price will always be lower than the total variable cost. Fixed costs such as depreciation and interest on equipment, overheads, or risk and profit, can be earned through other work items or job sites.

A basic distinction should be made between a general and an in-depth bid assessment (see Fig. 9.50).

Costing forms are analyzed as part of assessing price components and bases (see examples in Fig. 9.51).

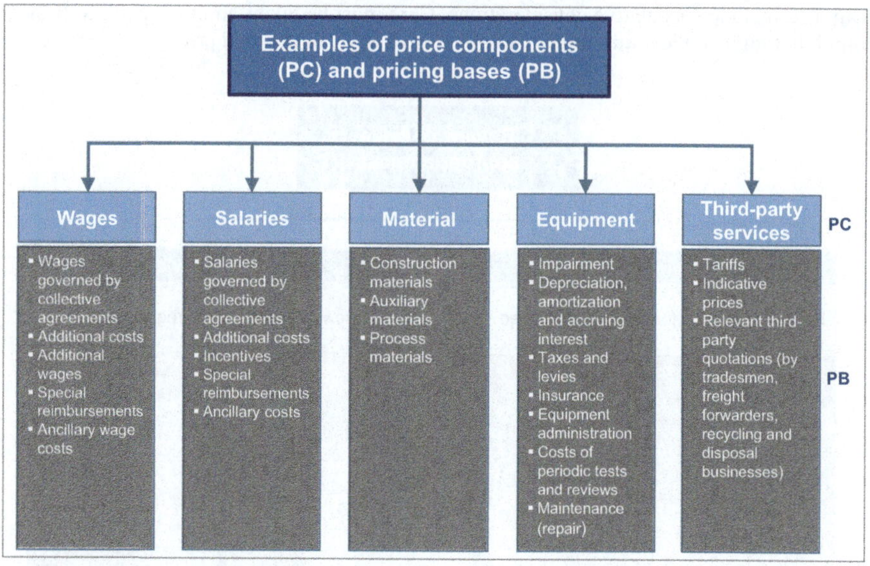

Fig. 9.51 Distinction between pricing bases (PB) and price components (PC)

Costing forms

In Austria, costing forms ("C forms") are used to standardize costing exercises. Using these forms is mandatory for public-sector projects. The C forms are an essential prerequisite for determining the basis for pricing. Appropriate preparation of these forms simplifies the presentation of the underlying pricing assumptions in the context of the bid assessment, and subsequently in the event of additional cost claims. The client and the contractor are well-advised to agree on these basic assumptions at the outset of the project.

We use flow charts to describe the general bid assessment sequence for unit price contracts, followed by more detailed descriptions and explanations for the in-depth-assessment of specific items.

Flow chart for bid assessment

Fig. 9.52 shows the flow chart representing the general bid assessment sequence for unit price contracts.

The first step after submission of the quotations by the bidders and expiration of the bid submission period is to exclude those bids that arrived after the final submission date. If a bid bond was requested, proof of its receipt must be provided at the time of bid opening. The bid must be rejected if this is not the case.

In the next step, the suitability of bidders, the formal accuracy and correctness of the calculations, and the completeness and freedom from errors of bids must be verified. Other reasons for rejecting a bid include unethical behavior or participation in agreements in contravention of anti-trust law on the part of the bidder.

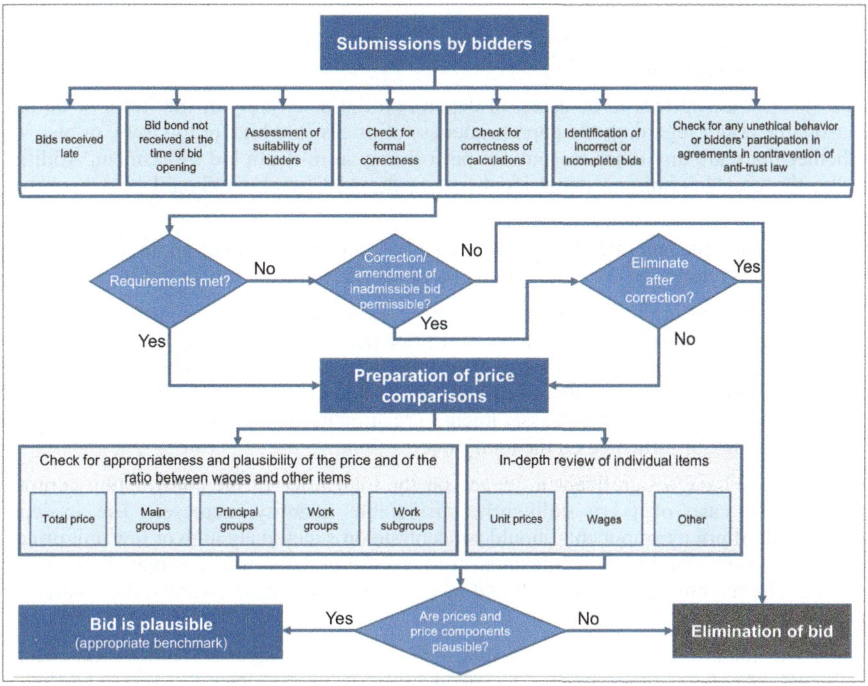

Fig. 9.52 General bid assessment workflow (in Austria)

If these basic requirements are met, the client can proceed with preparing a pricing table and comparing and further assessing the individual bids. If bidders fail to fulfill these requirements, specific corrections and amendments are permissible under certain circumstances, particularly in the case of incorrect calculations contained in the bid. However, such bids may still be rejected directly after such corrections have been made.[52)]

Further steps include the assessment of the appropriateness and plausibility of prices, as well as of the ratio between wages and other components, on a project-specific basis on different levels of the bill of quantities. These levels are defined by the structure of the bill of quantities. The assessment procedure ranges from a high-level review to a detailed check, i.e. from the total price to individual work subgroups.

The in-depth assessment of individual work items constitutes the most detailed bid assessment level, starting from the verification of unit prices and then checking the proportions of wages and other components for appropriateness and plausibility.

If all prices and price components are found to be appropriate and plausible (after additional information provided by the bidder, if required), the bids can be considered to be comparable and plausible so that they are eligible for being shortlisted for the contract award.

52) For instance if the total of absolute amounts of all corrections amounts to 2% or more of the original total price (excluding VAT). Cf. Bundesvergabegesetz (BVergG; Federal Act on Public Procurement in Austria).

9.3.3.1.1 Flow Chart for Unit Price Review

The flow charts below illustrate how an in-depth assessment of unit prices can be performed on a systematic basis.

The specific unit prices to be assessed depend on the materiality of the items or on any reasonable doubt regarding the appropriateness of prices (too high or too low), or else, on whether there are any other reasons for performing an in-depth bid assessment. Additionally, the cost-determining items identified in the ABC analysis should be assessed in more detail.

In the first step, the unit price is assessed as a whole with respect to its appropriateness and plausibility in relation to the described work, or whether it is too low or too high.

The prices quoted in the bill of quantities are used to consider the proportions of wages and other components or the respective unit price per item. In so doing, it is necessary to clarify whether to perform an assessment of the bid prices at all, or for which items the assessment should be performed. Immaterial or C items will usually be negligible so that they will not require an in-depth assessment in relation to the total contract. Nor will these items have a major influence on the total price.

C items may have a significant influence on the total price of the construction contract only in the case of (very unlikely) considerable quantity increases. The in-depth assessment of price components should concentrate on excessively high or low unit prices. This assessment should protect the client from awarding works at excessively high prices. Conversely, the bidder/contractor should be protected from being awarded the contract at an unduly low price.

Any assessment of unit prices should differentiate between wages and other components. In a first step, the appropriateness and plausibility of the ratio between wages and other components should be checked. If, for instance, the proportion of wages stated for labor-intensive works only represents a very low percentage of the unit price, the plausibility of this share should always be challenged and discussed in more detail with the bidder if required.

In addition, the proportions of wages (mean wage price and labor consumption rate) and other components (material and equipment prices) should be assessed more thoroughly.

If, for example, certain portions of the wage component are shifted to the "Other" component of the respective item as a result of determining the costs of subcontractor services, or if the bidder uses this argument, this will not constitute a source-based costing exercise in accordance with OENORM B 2061:2020. Thus, the relevant bid would have to be rejected.

9.3.3.1.2 Review of Wage Proportion

The review of the wage proportion (see flow chart in Fig. 9.53) is of particular importance for labor-intensive works. More specifically, disclosure of the costing form makes it possible to divide the wage proportion by the mean wage price to obtain the labor consumption rate applied by the bidder to the specific work item. If these calculated labor consumption rates result in a very small number of paid working hours by unit of quantity [wh/UQ], this is potentially indicative of excessively low wages or social dumping, or of the shift of price components to other items.

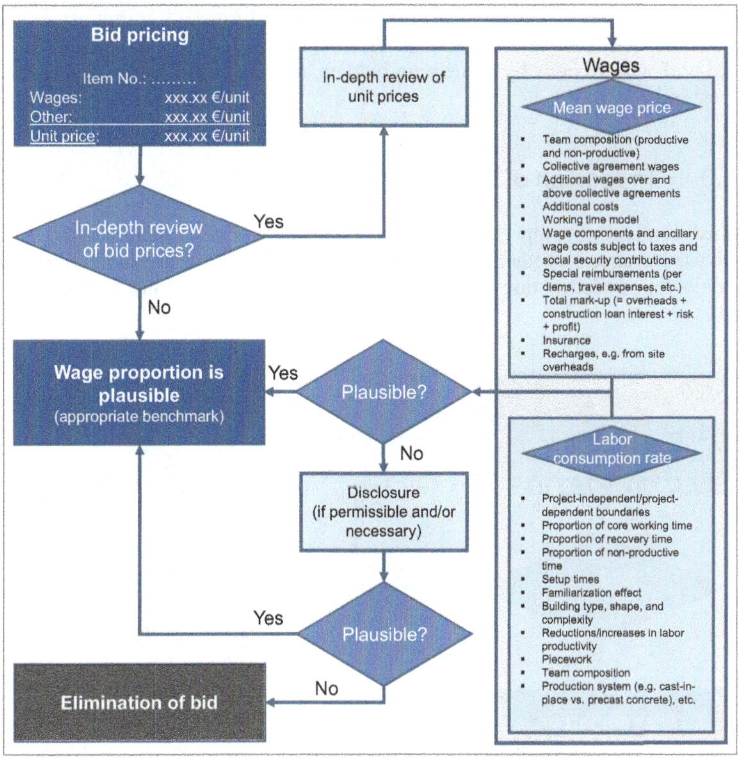

Fig. 9.53 Plausibility check of prices in unit price contracts – Wage proportion

This is justified by the assumption that the performance potential assigned to a productive construction worker appears to be exhausted. Collective agreements include uniform provisions and clear limits. Any increase in the productivity of performing construction work will be restricted by physical, labor law, construction management, technological, environmental, and manufacturing limitations. Nowadays, there is a sufficient understanding of the workload limits for employees (normal performance).

Labor consumption rates have been defined for labor-intensive types of work, and thus process-specific and technological minimum levels that cannot be underrun. Depending on the specific project, the influences of weather, the environment, quality and quantity, and the type, shape, and complexity of the building or structure will lie above these lower limits.

If consumption rates are found to be much lower than the minimum possible levels, this indicates that the mean wage price was either included incorrectly or should be checked separately for its plausibility. In addition, certain wage proportions may have also been included in the "Other" component, such as in the case of the bidder commissioning subcontractors.

If individual unit price components are found to be implausible, the bidder can be requested to provide additional explanatory information on their submissions as far as permissible and/or necessary. The bid must be rejected if such components continue to be implausible even after explanatory information has been furnished. If unit prices are found to be plausible before or after additional discussions with the bidder, such bids can be compared with submissions by competing bidders, and the contract can be awarded to the best/lowest bid.

9.3.3.1.3 Proportion of Other Items

The "Other" component comprises equipment and material prices. The flow chart in Fig. 9.54 should be used to check whether the ratio between the two main components is equivalent to a commonly found distribution or whether a bias towards one of the components can be identified.

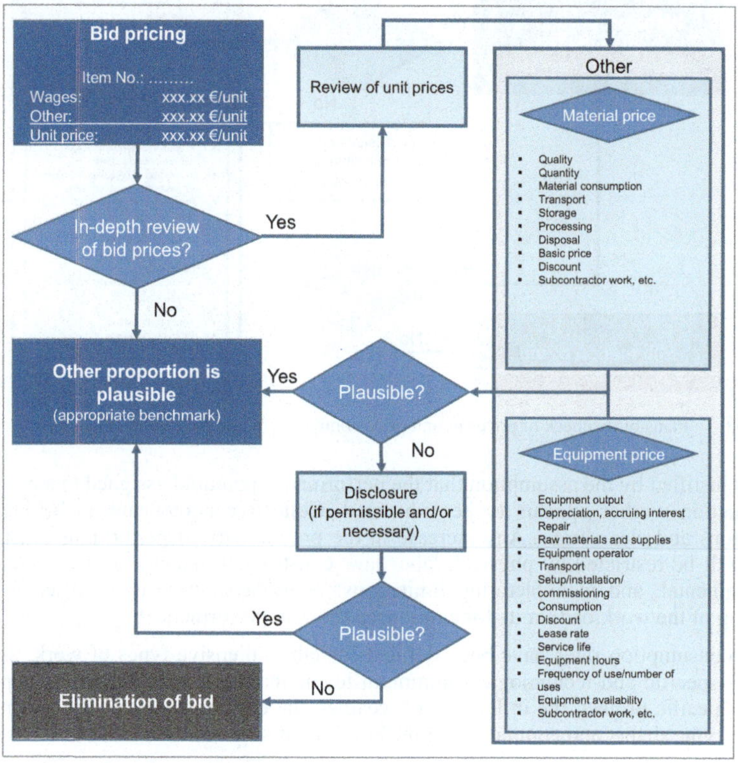

Fig. 9.54 Plausibility check of prices in unit price contracts – Other proportion

Plausibility check of equipment prices

Equipment prices are essentially determined by the costs of equipment and operation (machine operator). It should also be investigated whether one or more additional items are needed in the context of equipment use. Any equipment-intensive work will require a certain number of equipment hours. These hours are expended by individual pieces of equipment and/or chains of equipment. Technical output levels have been defined for equipment-intensive types of work, and thus process-specific and technological minimum levels that cannot be underrun. Depending on the specific project, the influences of weather, the environment, quality and quantity, and the type, shape, and complexity of the building or structure will lie above these lower limits.

If OEBGL or BGL[53] values are used for checking the plausibility of depreciation and interest on equipment, these values should be reduced accordingly because they represent excessively high average values that need not necessarily be applicable to the case at hand.

This notion is underpinned by the following citation of what *Kropik/Krammer* write on reducing the values for depreciation and interest: "In costing practice, the OEBGL values for depreciation and interest are normally reduced. This reduction results from the difference between the economic depreciation and the tax depreciation while also considering that the service life of the equipment is usually longer in practice than stipulated by the OEBGL values. The reduction factor for OEBGL values normally lies roughly between 0.40 and 0.80."[54]

Plausibility check of material prices

Any check of material prices will include the verification whether the material, transport, storage, handling, and monitoring price components were actually considered. In addition, any applied discounts will be checked for their arm's-length nature.

9.3.3.2 Task

As part of the in-depth bid assessment, the unit price quoted by the best bidder should be checked for plausibility and appropriateness. Fig. 9.55 shows the extract from the bill of quantities for reinforcing operations (in a simplified and anonymized format).

---

[53] (OE)BGL = (Österreichische) Baugeräteliste [(Austrian) List of Construction Equipment]

[54] Kropik/Krammer (1999), p. 239f. (translation by Steffen Walter)

Item No.	Item text Quantity Unit / Price components / Item price			
01 xx xx	Reinforcement Standing preliminary remarks 1. Technical Conditions of Contract The Technical Conditions of Contract RVS 08.06.02 must be complied with.			
01 xx xx xx	Deliver, cut to size, bend, and install reinforcing steel of grade Bx for untensioned reinforcement. The unit price applies regardless of different diameters or shapes and to all lengths specified in the design, as well as to round steel of grade S235 if required. This work also includes: • appropriate storage • any required interim transport including loading and unloading • the final transport to the installation site • delivery of binding wire and of spacers to the formwork • welding of the supports of tension members and expenses for the increased accuracy of such supports A separate compensation is paid for: • the additional costs of rebar in lengths greater than 14 m The following items are taken into account: • the theoretical weight of the reinforcement, of the support or spacer (Z plate, spacer strips or similar), and bracings as specified in the approved plans without offcuts			
01 xx xx xxA	**Reinforcing steel B550B** LT 1 reinforcing steel B550B conforming to OENORM B 4707.			
		Wages:	€160.00/t	
		Other:	€540.00/t	
	3,500.00 t	Unit price:	€700.00/t	€2,450,000.00

Fig. 9.55 Extract from the bill of quantities of the bid to be assessed

The unit price of €700 per ton comprises the wage and other components. The project relates to an engineering structure (civil engineering category) for which only rebar was specified as the type of reinforcement. The combined price applies independently of the type of structural component because no separate items were defined for the individual groups of components.

The quotation (K-3 form) also indicates a mean wage of €40/wh.

9.3.3.3 Modeling

The extract from the bill of quantities shows that a combined price model was used. At first glance, the unit price appears to be unusually low. We can then check the proportions of wages and other components more thoroughly on the basis of the provided breakdown. We obtain the wage proportion [€/t] by multiplying the labor consumption rate [wh/t] with the mean wage price [€/wh]. Dividing the wage proportion [€/t] by the mean wage of €40 per working hour, on which the quotation was based, results in a corresponding consumption rate of 4 wh/t (see Fig. 9.56).

In the next step, we need to answer the question if this blended labor consumption rate would actually make it possible to perform reinforcing operations from a construction management point of view. We thus need to establish a practicable range to determine if the calculated consumption rate is at all achievable.

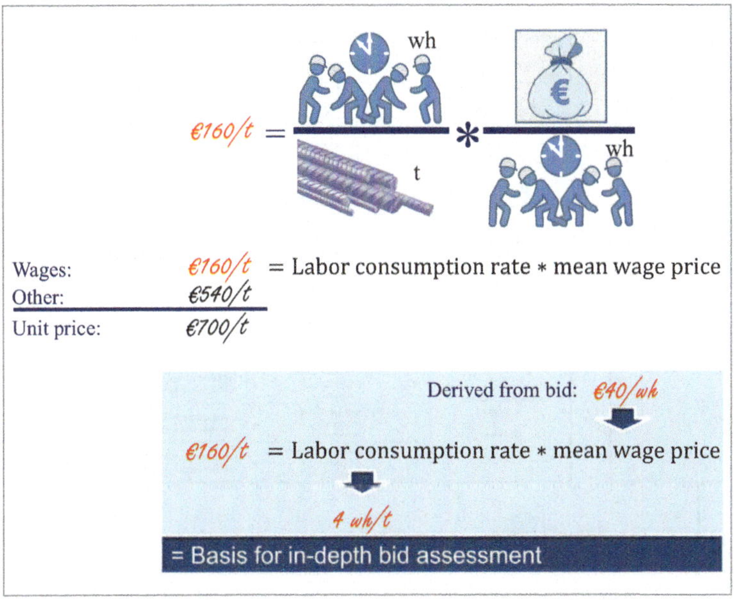

Fig. 9.56 Calculation of the labor consumption rate from the wage proportion[55)]

Generally accepted values stated in the literature (or reference values taken from site surveys or final costing exercises) should be used to arrive at a representative reference base for the in-depth assessment. Since this project example belongs in the civil engineering category, the specific diameter and bend shape distributions as well as consumption rates are applied for this purpose.

These deterministic values are combined with ranges for performing the analysis in order to obtain a robust result for the spread of the reference consumption rates.

The quantity proportions of rebar diameters and bend shapes as well as the consumption rates per diameter and bend shape are defined as uncertain input parameters. Fig. 9.57 includes the flow chart illustrating these calculation steps and a qualitative representation of the corresponding distribution functions. In the computation model, the mean wage price is included as a deterministic value following from the quoted amount of €40 per working hour.

The quantity and bend shape proportions need to be normalized to 100% in order not to exceed or underrun the specified reinforcement ratio in the simulation (the quantity for the in-depth bid assessment is equal to the quantity of 3,500 tons specified in the bill of quantities).

55) Cf. Hofstadler (2018), p. 236

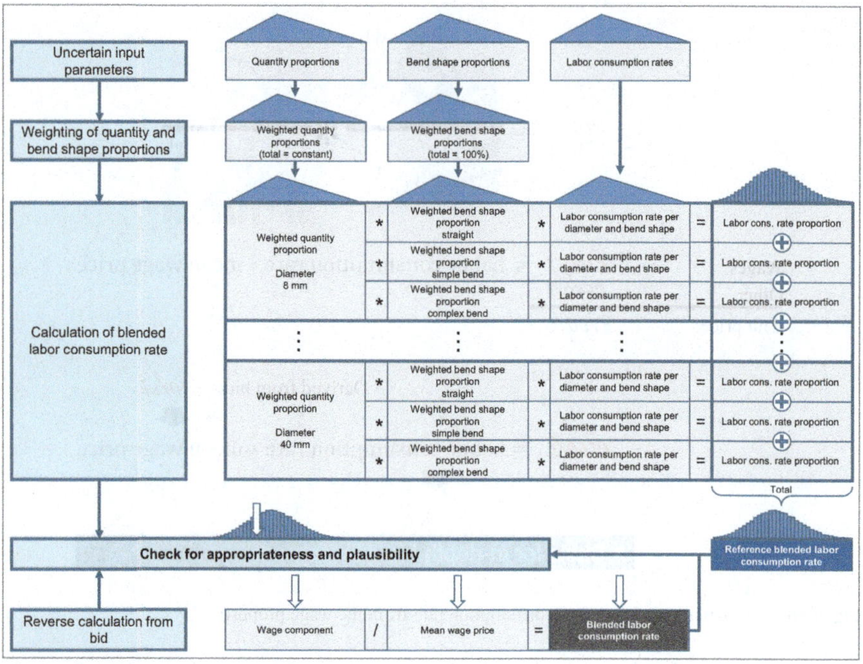

Fig. 9.57 Flow chart – Modeling

To calculate the blended labor consumption rate, each rebar diameter quantity proportion is then multiplied with the weighted bend shape proportions (i.e. straight, simple, and complex bend) as well as with the distribution functions representing the individual consumption rates. We thus obtain proportional consumption rates in the form of histograms for each diameter and bend shape (i.e. a total of 30 = 10 diameters times three bend shapes). In the final step, the proportional consumption rates derived from these results are added up to form the histogram representing the blended labor consumption rate. This histogram serves as a reference for assessing the plausibility of the consumption rate contained in the bid pricing. To establish this comparison, the blended labor consumption rate quoted in the bid also needs to be back-calculated from the unit price for the wage proportion. This is done by dividing the wage proportion [€/t] by the mean wage price. This step determines the blended labor consumption rate as a deterministic value that can be used to determine its appropriateness and plausibility.

9.3.3.4 Calculation and Interpretation of Results

The question thus arises how prices quoted in bids can be assessed systematically with respect to their viability and plausibility while ensuring a sound basis from a construction management and economics as well as legal point of view.

The authors expressly stipulate that this verification or falsification will be successful only if the chance/risk ratio taken by the bidders is analyzed.

The flow chart shown in Fig. 9.57 is thus used for creating the calculation matrix as included in Table 9.23. The quantities are varied for the individual rebar diameters relative to the normalized quantity of 3,500 tons.

Table 9.23 illustrates the calculation of the blended labor consumption rate. The uncertain parameters (quantity and bend shape proportions as well as consumption rates) are modeled as triangular distributions. For this purpose, Columns B, C, D, I, J, K, N, O, and P include the minimum (MIN), most likely, and maximum (MAX) values defining the triangular distributions. In this setting, the most likely values are equivalent to the qualified information provided by the VÖBV for the category of engineering structures.[56)]

A 10% quantity underrun and a 20% quantity overrun were assumed for each diameter. Column E includes the triangular distributions of the quantities, whereas the table lists the individual mean values of the distributions. In the simulation, these values may vary within the specified ranges and can represent random figures in each iterative step.

However, the individual quantity proportions need to be weighted because the specification includes a fixed total rebar quantity of 3,500 metric tons. Based on this weighting, each iteration involves the proportional calculation of the random quantity composition so that the quantity proportions always (i.e. in each iteration) add up to 3,500 tons. Consequently, the proportions of individual diameters in the total quantity will change during the simulation, but the total quantity as such will always remain constant. Column G then includes a normalization of the weighted quantity proportions for calculating the reference blended labor consumption rate. These quantity proportions represent the percentage quantity ratios for each rebar diameter. The percentages must add up to 100%.

56) Cf. VÖBV (2010), p. 2ff.
Cf. Hofstadler/Franzl (2011), p. 25ff.

No.	Diameter	Quantities						Proportion of quantities for bend shapes						Labor consumption rates				Labor consumption rate
		MIN	MOST LIKELY	MAX	Distribution	Weighting	Standardi-zation	Bend shape	MIN	MOST LIKELY	MAX	Distribution	Weighting	MIN	MOST LIKELY	MAX	Distribution	
	[mm]	[t]	[t]	[t]	[t]	[t]	[%]	[-]	[%]	[%]	[%]	[%]	[%]	[wh/t]	[wh/t]	[wh/t]	[wh/t]	[wh/t]
0	A	B	C	D	E	F	G	H	I	J	K	L	M	N	O	P	Q	R
					=f(B, C, D)	=E/E11*C11	=F/F11*100					=f(I,J,K)	=(L/ΣL per diameter)*100				=f(N,O,P)	=G*M*Q/10,000
								straight	45.09	56.36	67.63	56.36	56.36	14.29	15.88	19.06	16.41	0.05
1	8	17.0	18.9	22.7	19.5	18.9	0.54	simple	19.32	24.15	28.98	24.15	24.15	31.44	34.93	41.92	36.09	0.05
								complex	15.59	19.49	23.39	19.49	19.49	51.59	57.32	68.78	59.23	0.06
								straight	45.09	56.36	67.63	56.36	56.36	10.98	12.20	14.64	12.61	0.60
2	10	266.8	296.5	355.7	306.3	296.5	8.47	simple	19.32	24.15	28.98	24.15	24.15	23.45	26.05	31.26	26.92	0.55
								complex	15.59	19.49	23.39	19.49	19.49	40.15	44.61	53.53	46.10	0.76
								straight	45.09	56.36	67.63	56.36	56.36	8.41	9.34	11.21	9.65	0.78
3	12	453.9	504.4	605.2	521.2	504.4	14.41	simple	19.32	24.15	28.98	24.15	24.15	16.74	18.60	22.32	19.22	0.67
								complex	15.59	19.49	23.39	19.49	19.49	29.08	32.31	38.77	33.39	0.94
								straight	45.09	56.36	67.63	56.36	56.36	6.68	7.42	8.90	7.67	0.74
4	14	542.4	602.7	723.2	622.8	602.7	17.22	simple	19.32	24.15	28.98	24.15	24.15	13.02	14.47	17.36	14.95	0.62
								complex	15.59	19.49	23.39	19.49	19.49	20.76	23.07	27.68	23.84	0.80
								straight	45.09	56.36	67.63	56.36	56.36	5.47	6.08	7.30	6.28	0.54
5	16	484.2	538.0	645.5	555.9	538.0	15.37	simple	19.32	24.15	28.98	24.15	24.15	9.79	10.88	13.06	11.24	0.42
								complex	15.59	19.49	23.39	19.49	19.49	16.40	18.22	21.86	18.83	0.56
								straight	45.09	56.36	67.63	56.36	56.36	3.80	4.22	5.06	4.36	0.44
6	20	565.7	628.6	754.3	649.6	628.6	17.96	simple	19.32	24.15	28.98	24.15	24.15	6.79	7.54	9.05	7.79	0.34
								complex	15.59	19.49	23.39	19.49	19.49	11.56	12.84	15.41	13.27	0.46
								straight	45.09	56.36	67.63	56.36	56.36	3.18	3.53	4.24	3.65	0.37
7	26	569.5	632.8	759.4	653.9	632.8	18.08	simple	19.32	24.15	28.98	24.15	24.15	5.72	6.35	7.62	6.56	0.29
								complex	15.59	19.49	23.39	19.49	19.49	10.09	11.21	13.45	11.58	0.41
								straight	45.09	56.36	67.63	56.36	56.36	3.18	3.53	4.24	3.65	0.12
8	30	190.9	212.1	254.5	219.2	212.1	6.06	simple	19.32	24.15	28.98	24.15	24.15	5.54	6.15	7.38	6.36	0.09
								complex	15.59	19.49	23.39	19.49	19.49	9.43	10.48	12.58	10.83	0.13
								straight	45.09	56.36	67.63	56.36	56.36	3.18	3.53	4.24	3.65	0.02
9	36	26.5	29.4	35.3	30.4	29.4	0.84	simple	19.32	24.15	28.98	24.15	24.15	5.36	5.95	7.14	6.15	0.01
								complex	15.59	19.49	23.39	19.49	19.49	9.25	10.28	12.34	10.62	0.02
								straight	45.09	56.36	67.63	56.36	56.36	3.18	3.53	4.24	3.65	0.02
10	40	33.1	36.8	44.1	38.0	36.8	1.05	simple	19.32	24.15	28.98	24.15	24.15	5.36	5.95	7.14	6.15	0.02
								complex	15.59	19.49	23.39	19.49	19.49	9.07	10.08	12.10	10.42	0.02
11	**Sum**		**3,500.0**		3,616.7	3,500.0	100.00											**10.92**

Table 9.23 Calculation of the reference blended labor consumption rate

Yet there is another parameter besides the rebar quantity distributions that is fraught with uncertainties, namely the distribution of bend shapes (i.e. straight, simple, and complex bends). Again, triangular distributions (Column L) are applied for the Monte Carlo simulation. Columns I to K specify the minimum, most likely, and maximum percentages of the individual bend shape distributions. The most likely values are equivalent to the details provided by the VÖBV for the category of engineering structures. In this context, it should be noted that the bend shape distributions are assumed to be equal for each diameter, and further that the triangular distributions are identical for each diameter. This means that the randomly selected proportions of bend shapes are identical for the individual diameters in each iterative step. Accordingly, in each calculation sequence, the proportion of straight 8 mm rebar will be identical to the proportions of straight 10 to 40 mm rebar. The same applies to the proportions of rebar with simple and complex bends.

Since triangular distributions are also applied to individual bend shapes, these must also be weighted accordingly in the simulation in order to ensure that the sum of bend shape proportions equals 100% in each iterative step. This conversion is completed in Table 9.23, Column M.[57] The labor consumption rates expected for the individual rebar diameters and bend shapes (Column O) also correspond to the details provided in the VÖBV for the purpose of determining the reference blended labor consumption rate. In so doing, we assume a 10% downward and a 20% upward spread of consumption rates (Columns N and P). The triangular distributions are entered in Column Q.

In each row, the weighted quantity and bend shape proportions (Columns G and M) are multiplied with the consumption rate distributions (Column Q) to calculate the proportions of consumption rates for each diameter and bend shape (Column R). The sum of these consumption rate proportions calculated in Cell R11 in the final step represents the reference blended labor consumption rate. Fig. 9.58 shows this value as a histogram, which is used for checking the appropriateness and plausibility of the bid price.

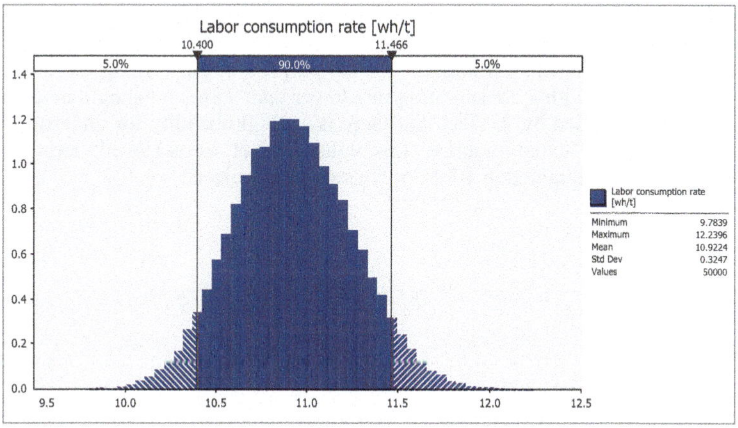

Fig. 9.58 Histogram of the reference blended labor consumption rate

[57] Column L of Table 9.23 shows the mean values of the distributions. These add up to 100% (= 56.36% + 24.15% + 19.49%), which is why no weighting of the proportions is necessary in this case. For this reason, the values entered in Column M for this iterative step are identical to those in Column L.

We obtain a mean blended labor consumption rate of 10.92 wh/t with a standard deviation of approximately 0.32 wh/t. The 5% and 95% quantiles amount to 10.40 wh/t and 11.47 wh/t, respectively.

We can use the histogram established from historical values to define a minimum that describes the transition point beyond which the project would no longer be viable. The authors recommend to reject the bid if the current value is, for instance, 10% lower than the minimum. This minimum represents the value for which we can safely assume that it is almost impossible, and thus highly improbable, to reach or underrun it. We demonstrate the probabilistic simulation approach in order to strengthen the transparency and reliability of this method. Probabilistic calculation methods make it possible to systematically integrate existing uncertainties with the aid of ranges. In addition, selecting the type of distribution function enables us to determine the location of the most likely value as well as to assign a higher probability of occurrence to the range around this value.

At the in-depth bid assessment stage, chance/risk ratios need to be verified for back-calculated consumption rates. The 5% quantile is applied as an assessment measure indicating that 5% of the values lie below this parameter while 95% of the values lie above. If, for example, the labor consumption rate derived (back-calculated) from the bid is 10% lower than the 5% quantile value, this consumption rate would no longer pass the plausibility check.

Fig. 9.59 shows the histogram resulting from a Monte Carlo simulation used for calculating the blended labor consumption rate for reinforcing operations in an engineering structure (based on the histogram shown in Fig. 9.58). We apply a ±20% range to the calculations to take into account uncertainties in the bend shape distributions. A -10% to +20% range below and above the most likely values was defined for both the diameter distributions and labor consumption rates. A median of 10.91 wh/t is applied to the histogram while the 5% and 95% quantiles amount to 10.40 wh/t and 11.47 wh/t, respectively. The 5% quantile of 10.40 wh/t is used as a basis from which 10% are subtracted to determine the minimum labor consumption rate for the construction project under assessment. The minimum labor consumption rate of 9.36 wh/t resulting from this calculation constitutes the absolute minimum for the purpose of performing the in-depth bid assessment. Any bids with a consumption rate lower than 9.36 wh/t should be eliminated. This approach is justified by the fact that there is a 0% probability for underrunning this consumption rate. In actual practice, this value cannot be achieved because higher consumption rates are attained in 100% of construction projects.

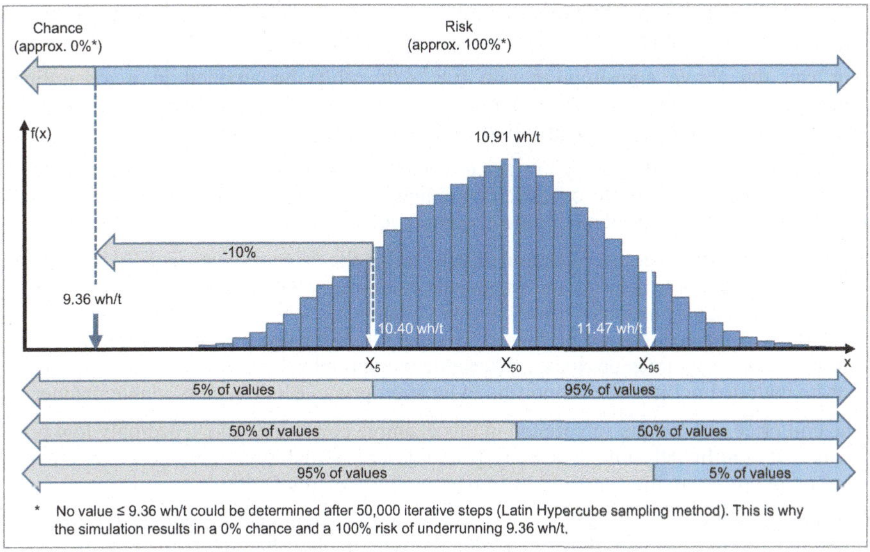

Fig. 9.59 Distribution resulting from a Monte Carlo simulation illustrating how to determine a minimum labor consumption rate for reinforcing operations[58)]

In the next step, the consumption rate of 4 wh/t on which the bid is based is compared with the range of justifiable consumption rates. Fig. 9.60 also clearly visualizes that this value is far below the possible range. There is a difference of about 57% to the minimum [= (9.36 wh/t - 4 wh/t) / 9.36 wh/t * 100%). In other words, the minimum plausible consumption rate lies approximately 234% (= 9.36 wh/t / 4 wh/t * 100%) above the consumption rate of 4 wh/t stated in the bid.

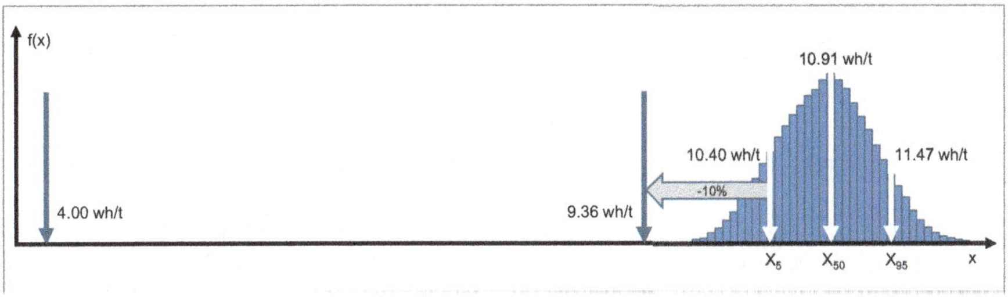

Fig. 9.60 Labor consumption rate quoted in the bid compared to the range of plausible labor consumption rates[59)]

58) For the cited example, we obtain a minimum labor consumption rate of 9.36 wh/t with a chance/risk ratio of 0 to 100%. This clearly shows that the underrun probability tends to move towards zero. Consequently, all labor consumption rates below this minimum should be considered extremely questionable with respect to the plausibility of the bid price. Cf. Hofstadler (2018), p. 238

59) Cf. ibid., p. 239

This example clearly illustrates that the consumption rates, and thus the wage proportion, lie clearly outside the possible range of full cost coverage. This price is thus inappropriate because it is considerably below the minimum acceptable level.

Nor can the above consumption rate (i.e. 4.00 wh/t) be attained in a streamlined construction environment and economical management, which is why it prevents an impeccable standard of workmanship, including any liability for defects.

In this case, this extraordinarily low price would put the liquidity of the business at risk if the wages paid to the workforce indeed complied with legal requirements. At any rate, this clearly contradicts the principle of the economic justifiability of prices.

Applying this principle, prices should at least cover the primary costs usually recognized as expenses, including salaries and wages, materials, equipment repairs, and/or third-party services. The major portion of these costs is variable and would be incurred anyway. This proportion of the cost cannot be earned by other works or services, or construction sites. Labor consumption rates obviously determine the variable share of the costs; they should thus be included in the wage proportion in an appropriate, economically viable manner.

The unit price of €700 per ton quoted in this example constitutes an exceedingly low price because it significantly underruns any plausible or justifiable variable cost level, which even more applies to the wage proportion amounting to only €160/t.

Nor is it possible to explain this gap between the bid price and the unduly low price in additional discussions with the bidder, which is why the bid should be eliminated.

9.3.3.5 Bid Assessment for Lump-Sum Contracts

Unlike unit-price contracts, global lump-sum contracts do not include an itemized list of works and services in terms of quantities and unit prices.

9.3.3.5.1 Bid Assessment for Itemized Lump-Sum Contracts

The in-depth bid assessment of itemized lump-sum contracts can proceed in the same way as previously discussed for unit-price contracts as long as prices are assigned to the individual work items.

9.3.3.5.2 Bid Assessment for Global (Conventional) Lump-Sum Contracts

The detailed assessment of bids for global lump-sum contracts without specified bills of quantities or unit prices is usually an excessively difficult and time-consuming exercise. This type of contract does not specify unit prices, which is why the only basis for assessing its plausibility is to start from the totals of the wage and other components from which the bid price is derived. This check concentrates on the stated number of paid working hours. Total paid working hours are back-calculated on the basis of the wage proportion applying the mean wage price. Dividing the total paid working hours proportion [€] by the mean wage price [€/wh] results in the sum of paid working hours [wh], for instance for the shell construction phase. We then divide this paid working hours total either by the building's gross cubic volume [m^3] or total gross floor area [m^2] and obtain the associated ratios. These values can be compared with similar projects or details stated in the literature. If, for instance, the current value lies more than 10% below the minimum of similar, previously completed projects, this can be an indicator of excessively low wages or social dumping.

Checking whether there is any reason for eliminating bids for lump-sum contracts follows the same logic as for unit-price contracts. At the in-depth bid assessment stage, chance/risk ratios need to be verified for back-calculated consumption rates and output values. If, for example, the value derived from the bid is 10% lower than the 5% quantile, the bid price would no longer pass the plausibility check, and the bid should be eliminated.

9.3.3.6 Assessment of Additional Cost Claims

Assessing additional cost claims essentially depends on the ability to determine the costs of works to be performed. The same 5% quantile approach is applied to verifying whether or not it is possible to determine such costs. From a legal point of view, the limit beyond which it would no longer be possible to do so is defined by assessing the ratio between the work owed under the contract and the related agreed-upon compensation.

9.3.3.7 Benefit for the Client

From the client's perspective, the greatest benefit is to arrive at a representative, plausible benchmark for verifying the labor consumption rates back-calculated from the unit prices specified in the bid.

Performing the in-depth bid assessment will thus put the client in a position to ensure that they have embarked upon a consistent approach that is also sound from a construction management and economics as well as legal point of view.

Each type of work will be associated with minimum labor consumption rates specific to the type of building or structure as well as to the specifications and requirements. There is no way to work more quickly or productively because there are limits to the workload that can be imposed on employees. The same applies to the other production factors, such as raw materials and supplies, which are also subject to certain technological and manufacturing constraints. In addition, element shapes and dimensions as well as the site environment and weather conditions will add to the influence and increase consumption rates. This will inevitably lead to consumption rates (far) above the stated minimum depending on the specific project.

If the calculated consumption rates exceed the commercially viable minimum determined by the above method (this minimum includes primary and secondary activities, waiting and recovery periods, non-productive and distribution times, as well as off-peak hours), the work items thus calculated can be considered plausible while taking into account the specific requirements for execution and workmanship.

Excessively low wages and/or social dumping can be identified when applying the above verification system (i.e. consumption rate = wage proportion divided by the mean wage price; subsequent comparison with the minimum determined in a probabilistic calculation) to all essential labor-intensive items as part of the bid assessment.

The wage proportion and the "Other" component need to be considered individually when checking the agreed compensation amount. This differentiation is necessary in order to separate labor-intensive works from works that require large amounts of equipment or materials. It will thus prevent, or at least reduce, the intermingling of different pricing assumptions. Analyzing the wage proportion makes it possible to determine the paid working hours on which the bid is based for the entire building or structure, work groups, or individual items, depending on the objective of the assessment. This approach can be likened to the situation in track and field athletics, where, for instance, possible (plausible) achievable times in the 100 m sprint race are subject to certain physical, biomechanical, and medical limits – there just isn't any other way to be faster. The same is true in construction management[60)] since any type of work will inevitably require a certain minimum cost and time to be completed. It is impossible to reduce paid working hours below this limit even if only ordinary fitness for purpose is to be assumed.

In conclusion, we find that only bidders quoting plausible consumption rates should be considered for construction contracts in order to prevent exceedingly low wages and/or social dumping. In contrast, bids with unduly low prices and consumption rates below the minimum determined for a specific type of work need to be rejected.

In so doing, we would, on the one hand, protect the construction industry against illegal practices in the country in which the project is to be completed. On the other hand, legally operating companies sending employees to the construction site could be granted markups for projects in the respective country.

9.3.4 Contract Award Decision

The bid price is particularly important at the stage of awarding construction contracts, which applies to both the lowest and the best bidder principle. In a structural specification, bidders specify unit prices for individual items. Combined with the individual quantity proportions, these will result in prices of individual items, and thus bid prices.

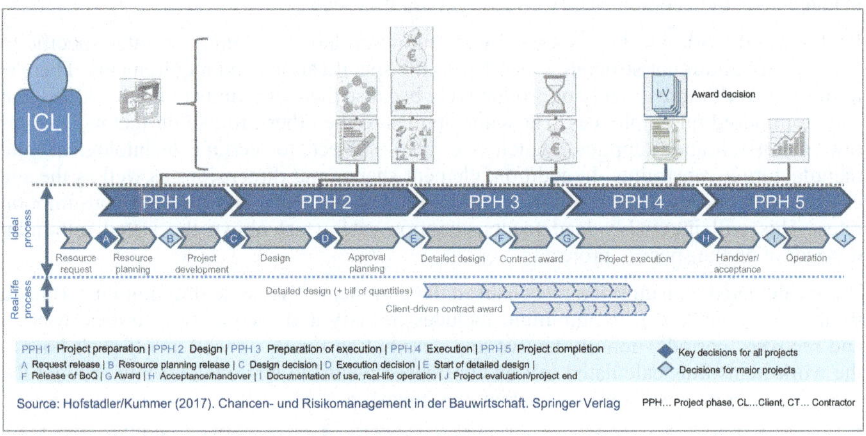

Fig. 9.61 Allocation of example in the project sequence

The client will be interested not only in the deterministic bid prices, but also in the effects of any quantity changes on them. In this regard, the crucial question is whether quantity changes may trigger the selection of a different favored bidder, and how probable such a

[60)] Construction management is characterized by the combination of elementary production factors. Construction work is thus performed not while walking fast but while walking slowly with frequent interruptions in order to perform manual work in specific locations on-site.

change in the selection is. The contract award decision made by the client belongs in the final part of Project Phase 3 (see Fig. 9.61).

9.3.4.1 Task

The prices quoted by the bidders are usually summarized in an itemized pricing table. Columns E to L of Table 9.24 show the unit prices quoted by the eight bidders. After appropriate editing and processing, the pricing table will deliver information on the range of unit prices and their deviations from the respective lowest unit price or total bid price.

However, the quantities assigned to the individual items (see Column D in Table 9.24) are usually fraught with uncertainties, which may change the pricing structure and thus the ranking of bidders (quoted prices). Triangular distributions are allocated to the quantities, which are then multiplied by the deterministic unit prices quoted by the bidders in order to integrate these spreads into the considerations.

To illustrate this approach, we do not consider a complete bill of quantities but merely an individual work group (07 Concrete and Reinforced Concrete Operations – WG 07) that forms part of a structural specification prepared for a building construction project.

The OENORM B 2110 Austrian standard stipulates that quoted unit prices are binding for quantities that exceed or underrun specified levels by a maximum of 20%:

"If the contractual quantity of a unit price item is exceeded or underrun by more than 20%, any party to the contract may request a new unit price to be agreed upon for the quantity actually completed whilst considering additional costs or saved costs if such change is solely due to a quantity variation (inaccurate quantity specifications without the work deviating from the agreed standard). Any such request must preferably be submitted in conjunction with appropriate evidence proving its merits."[61)]

In this example, the ranges of bid prices are shown in a diagram that also considers quantity uncertainties, and selected unit price histograms are compared to one another.

[61)] Cf. OENORM B 2110:2013, p. 28f. (translation by Steffen Walter)

No.	Item No.		Item	Unit	MOST LIKELY	Bidder A	Bidder B	Bidder C	Bidder D	Bidder E	Bidder F	Bidder G	Bidder H
									[€/unit]				
0	A		B	C	D	E	F	G	H	I	J	K	L
1	07.02.02A	Z	External concrete sandwich wall, 4.00 m high, C30/37 35 cm	m^3	270.45	159.18	104.62	250.98	153.42	125.24	108.44	258.77	96.55
2	07.02.02B	Z	Internal concrete wall, 5.00 m high, C30/37, >30-40 cm thick	m^3	12.69	90.14	104.62	91.92	106.79	100.47	93.01	87.23	100.01
3	07.02.02C	Z	Internal concrete wall, 5.00 m high, C30/37, >20-30 cm thick	m^3	3.83	92.16	125.55	93.54	109.69	107.51	93.01	88.90	103.56
4	07.02.02D	Z	Internal concrete wall, 5.00 m high, C30/37, >40-65 cm thick	m^3	37.22	92.14	102.74	91.92	106.79	100.47	93.01	85.79	100.01
5	07.02.02S	Z	Formwork, 5.00 m high concrete wall	m^2	232.06	19.40	23.72	28.26	43.83	30.34	28.13	28.87	23.63
6	07.02.02T	Z	Formwork, 4.00 m high concrete sandwich wall	m^2	1,499.68	18.39	28.48	38.99	77.15	44.39	71.64	30.43	32.19
7	07.02.02U	Z	Markup for exposed concrete formwork	m^2	614.75	20.51	4.49	2.45	20.14	6.63	8.09	14.54	10.71
8	07.02.02V	Z	Concrete wall rebar reinforcement	kg	22,317.20	1.34	1.16	1.21	1.27	1.50	1.32	1.34	1.38
9	07.02.02W	Z	Concrete wall mesh reinforcement	kg	16,227.00	1.39	1.16	1.26	1.27	1.40	1.35	1.30	1.57
10	07.02.96I	Z	Wall connecting element, two rows 10/150mm	m	160.00	11.52	21.02	27.53	25.11	19.24	16.12	23.49	22.56
11	07.03.02D		Concrete floor/cantilever slab, 5.00 m high, C30/37, 25 cm wide	m^3	50.63	90.25	103.65	92.64	111.02	95.78	103.85	92.66	97.39
12	07.03.02S		Formwork, soffit of 5.00 m high floor/cantilever slab	m^2	168.75	19.39	23.04	23.79	33.80	29.47	26.19	29.70	35.70
13	07.0302T		Formwork, 5.00 m high floor/cantilever slab, grids, perimeter	m^2	78.14	19.40	37.87	54.82	33.80	32.35	37.80	40.48	34.02
14	07.03.02U	Z	Markup for exposed concrete formwork	m^2	116.36	7.47	4.49	0.35	14.21	4.97	6.15	14.83	10.71
15	07.03.02V		Rebar reinforcement, 5.00 m high floor/cantilever slab	kg	7,000.00	1.34	1.16	1.21	1.27	1.40	1.32	1.34	1.38
16	07.03.48A		Formwork, 0.5 m^2 opening of floor/slab	m^2	3.57	24.34	50.13	63.61	67.60	65.98	64.90	79.31	75.61

Table 9.24 Pricing table – WG 07

9.3.4.2 Modeling

Fig. 9.62 shows the flow chart for this calculation. In this exercise, the unit prices quoted by the bidders are integrated into the calculation as deterministic values (represented by arrows) while quantity deviations are represented by triangular distributions (see the corresponding symbol assigned to the respective item quantity).

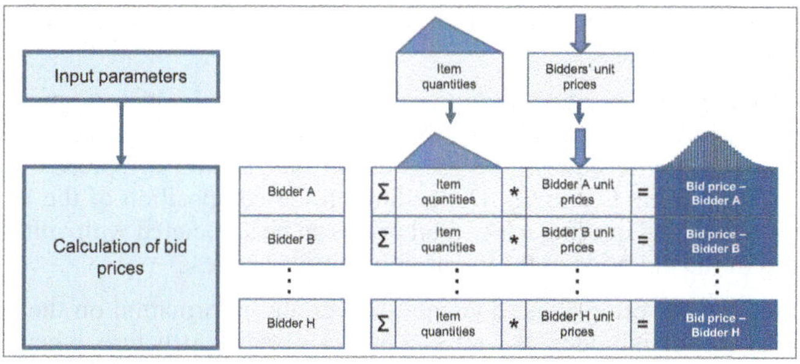

Fig. 9.62 Calculation flow chart

In this example, quantity uncertainties are modeled using symmetrical triangular distributions where the minimum and maximum equal 80% and 120% of the specified quantity, respectively. The most likely value is equivalent to the specified quantity as stated in the bill of quantities. Symmetrical distributions are assigned to the quantities of all items.

It is also possible to only include the essential or A items in an initial analysis for specific practical use cases with several hundred items contained in a single bill of quantities as part of the overall assessment. It should also be noted that items need not necessarily be associated with uncertainties. If the client is in a position to reliably assume that the quantities specified for certain items will not be exceeded or underrun, such quantities should be incorporated into the calculations as deterministic values.

Prices quoted by the bidders will then show different spreads resulting from the consideration of quantity uncertainties; these spreads can be compared to one another.

9.3.4.3 Calculation and Interpretation of Results

Table 9.25 can be used as a calculation table for determining the prices submitted by the bidders. Initially, the most likely quantities taken from the bill of quantities are increased or reduced by 20% in order to take the quantity uncertainties into account. For each item, this step results in the minimum and maximum quantities (Columns D and F) defining the related triangular distributions. The specified quantity per item (Column E) is used as the most likely value for the triangular distributions. The distributions as such are entered in Column G.

The deterministic unit prices of the individual bidders shown in Table 9.24 are multiplied with the triangular distributions assigned to the quantities (Columns H to O) in order to calculate the item prices. Finally, the sum of the item prices per bidder results in the bid prices shown in Row 17. The stated values correspond to the total means of the item prices. In Row 18, these values are applied to indicate the ranking from the lowest (1) to the highest bid (8).

However, these bid prices are not deterministic values (individual figures) but 50,000 individual results of a Monte Carlo simulation (50,000 iterations). These individual results can be shown as histograms and analyzed statistically. The minimum, maximum, mean, standard deviation, and spread (Rows 18 to 22) were processed for the purpose of establishing the histograms of the prices quoted by each bidder.

The minimum and maximum bid prices do not correspond to the extremes that would occur if the quantities were simultaneously exceeded or underrun by 20% in each item. We identify the lowest and highest bid prices obtained in a Monte Carlo simulation involving 50,000 iterations.

Row 24 shows the ranking of spreads (narrowest spread = 1 | widest spread = 8). This initial analysis already demonstrates that the ranking of the bid prices (i.e. the total of item price means) need not necessarily be identical to the ranking of spreads (compare the prices quoted by Bidders C and G). Depending on the composition of the unit price and quantity structures, a bid quoting a low total price can be associated with either a wide or a narrow spread of the bid price in the event of quantity changes.

The amount of the bid price alone does not deliver any information on the stability of a price quoted by a certain bidder, as well as a bidder's pricing structure, when impacted by quantity changes. Excessively wide spreads indicate a low level of price stability of the bid and carry a higher risk for the client having to tolerate greater price increases if quantities change.

Narrow bid price spreads provide the client with the advantage that the total price will be more stable if quantities change within an assumed range of ±20%.

No.	Item No.		Item	Unit	MIN	MOST LIKELY	MAX	Distribution	Bidder A	Bidder B	Bidder C	Bidder D	Bidder E	Bidder F	Bidder G	Bidder H
												[€]				
0	A		B	C	D	E	F	G	H	I	J	K	L	M	N	O
1	07.02.02A	Z	External concrete sandwich wall, 4.00 m high, C30/37 35 cm	m³	216.36	270.45	324.54	270.45	43,050.23	23,294.48	67,877.54	41,492.44	33,871.16	29,327.60	69,984.35	26,111.95
2	07.02.02B	Z	Internal concrete wall, 5.00 m high, C30/37, >30-40 cm thick	m³	10.15	12.69	15.23	12.69	1,143.88	1,327.63	1,166.46	1,355.17	1,274.96	1,180.30	1,106.95	1,269.13
3	07.02.02C	Z	Internal concrete wall, 5.00 m high, C30/37, >20-30 cm thick	m³	3.06	3.83	4.60	3.83	352.97	480.86	358.26	420.11	411.76	356.23	340.49	396.63
4	07.02.02D	Z	Internal concrete wall, 5.00 m high, C30/37, >40-65 cm thick	m³	29.78	37.22	44.66	37.22	3,429.45	3,823.98	3,421.26	3,974.72	3,739.49	3,461.83	3,193.10	3,722.37
5	07.02.02S	Z	Formwork, 5.00 m high concrete wall	m²	185.65	232.06	278.47	232.06	4,501.96	5,504.46	6,558.02	10,171.19	7,040.70	6,527.85	6,699.57	5,483.58
6	07.02.02T	Z	Formwork, 4.00 m high concrete sandwich wall	m²	1,199.74	1,499.68	1,799.62	1,499.68	27,579.12	42,710.89	58,472.52	115,700.31	66,570.80	107,437.08	45,635.26	48,274.70
7	07.02.02U	Z	Markup for exposed concrete formwork	m²	491.80	614.75	737.70	614.75	12,608.52	2,760.23	1,506.14	12,381.07	4,075.79	4,973.33	8,938.47	6,583.97
8	07.02.02V	Z	Concrete wall rebar reinforcement	kg	17,853.76	22,317.20	26,780.64	22,317.20	29,905.05	25,887.95	27,003.81	28,342.84	33,475.80	29,458.70	29,905.05	30,797.74
9	07.02.02W	Z	Concrete wall mesh reinforcement	kg	12,981.60	16,227.00	19,472.40	16,227.00	22,555.53	18,823.32	20,446.02	20,608.29	22,717.80	21,906.45	21,095.10	25,476.39
10	07.02.96I	Z	Wall connecting element, two rows 10/150mm	m	128.00	160.00	192.00	160.00	1,843.20	3,363.20	4,404.80	4,017.60	3,078.40	2,579.20	3,758.40	3,609.60
11	07.03.02D		Concrete floor/cantilever slab, 5.00 m high, C30/37, 25 cm wide	m³	40.50	50.63	60.76	50.63	4,569.36	5,247.80	4,690.36	5,620.94	4,849.34	5,257.93	4,691.38	4,930.86
12	07.03.02S		Formwork, soffit of 5.00 m high floor/cantilever slab	m²	135.00	168.75	202.50	168.75	3,272.06	3,888.00	4,014.56	5,703.75	4,973.06	4,419.56	5,011.88	6,024.38
13	07.0302T		Formwork, 5.00 m high floor/cantilever slab, grids, perimeter	m²	62.51	78.14	93.77	78.14	1,515.92	2,959.16	4,283.63	2,641.13	2,527.83	2,953.69	3,163.11	2,658.32
14	07.03.02U	Z	Markup for exposed concrete formwork	m²	93.09	116.36	139.63	116.36	869.21	522.46	40.73	1,653.48	578.31	715.61	1,725.62	1,246.22
15	07.03.02V		Rebar reinforcement, 5.00 m high floor/cantilever slab	kg	5,600.00	7,000.00	8,400.00	7,000.00	9,380.00	8,120.00	8,470.00	8,890.00	9,800.00	9,240.00	9,380.00	9,660.00
16	07.03.48A		Formwork, 0.5 m² opening of floor/slab	m²	2.86	3.57	4.28	3.57	86.89	178.96	227.09	241.33	235.55	231.69	283.14	269.93
17			**Sum**	€					**166,663.35**	**153,893.38**	**212,941.21**	**263,214.37**	**199,220.76**	**230,027.05**	**214,911.85**	**176,515.75**
18			Ranking of the bid prices						2	1	5	8	4	7	6	3
19			Minimum	€					146,672.42	134,905.52	183,932.01	228,909.25	174,141.61	199,314.37	188,419.41	156,341.00
20			Maximum	€					187,898.71	173,358.23	241,747.28	299,680.28	225,438.95	262,412.66	243,263.47	198,452.21
21			Mean	€					166,663.35	153,893.38	212,941.21	263,214.37	199,220.76	230,027.05	214,911.85	176,515.75
22			Standard deviation	€					5,389.93	5,061.76	7,932.46	10,604.70	7,062.88	9,667.59	7,604.22	5,707.55
23			**Spread**	€					**41,226.29**	**38,452.71**	**57,815.27**	**70,771.02**	**51,297.34**	**63,098.29**	**54,844.06**	**42,111.21**
24			Ranking of spreads						2	1	6	8	4	7	5	3

Table 9.25 Calculation table – Bid prices for uncertain quantities

Fig. 9.63 includes the spreads of simulated bid prices per bidder together with the mean values. We find that the submissions by Bidders D and F show the widest spread while Bidders B and A show the narrowest spread in terms of the total bid price. The submission by Bidder B is also associated with the lowest mean value. Bidder A submitted the second-best bid.

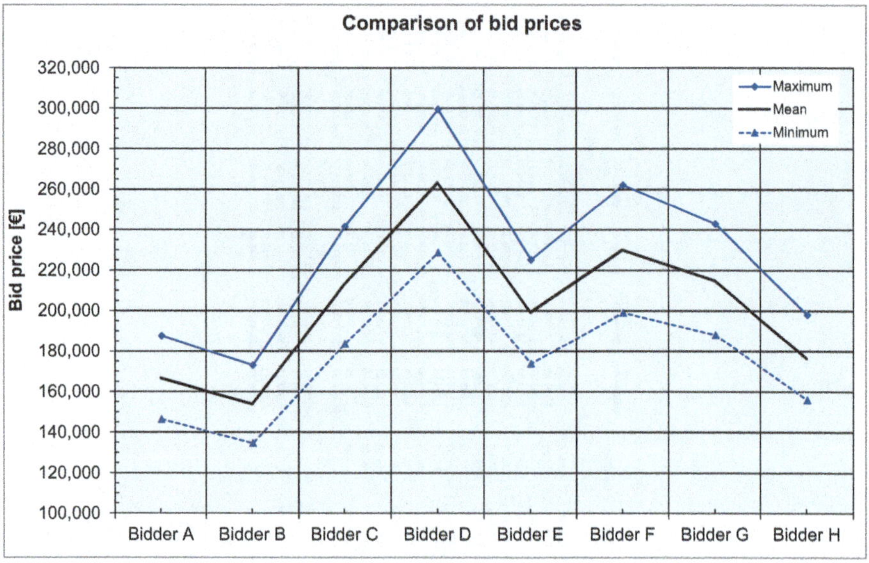

Fig. 9.63 Spreads of bid prices considering quantity uncertainties

We should thus investigate any cases in which Bidder A submitted a more favorable bid than Bidder B where quantity changes could shift the contract award to Bidder B. For this purpose, the total bid price of Bidder B is subtracted from that of Bidder A. The difference will be negative if the total quoted by Bidder A is lower than that of Bidder B. Fig. 9.64 shows the difference as a histogram. All of the values are positive, which means that the price quoted by Bidder B is lower than that of Bidder A in all simulated cases.

Fig. 9.65 shows a similar comparison for Bidders C and G, where the bid price quoted by Bidder C was subtracted from the price submitted by Bidder G. In this case, the price of bid G is lower than that of bid C in about 5.9% of simulated cases. If the bidders analyzed in this exercise were the two lowest bidders, we would be able to exclude a bidder change owing to quantity variations by ±20% in approximately 94.1% of cases. However, a risk of about 5.9% remains for such a case to occur.

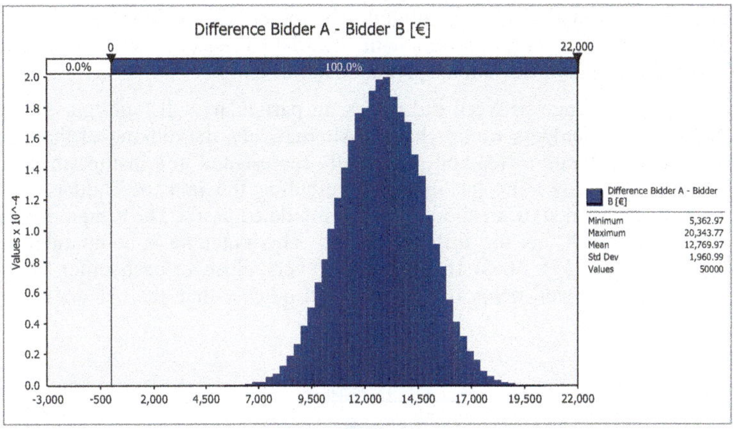

Fig. 9.64 Histogram – Difference of bid prices – Bidder A – Bidder B

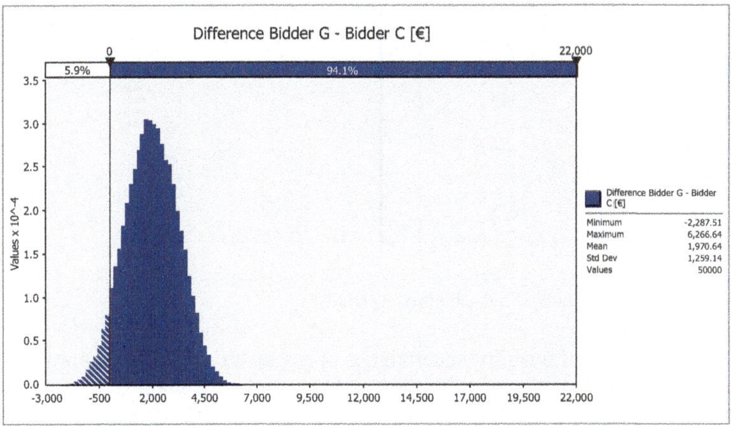

Fig. 9.65 Histogram – Difference of bid prices – Bidder G – Bidder C

9.3.4.4 Benefit for the Client

When preparing specifications, clients are bound to determine specific quantities for the individual items of a structural building specification. This exercise is made even more complex because the planning phase will usually not have been completed at this stage, which also brings about related uncertainties. The bidders will submit unit prices that are converted into item prices by multiplying them with the specified quantities. This exercise will ultimately result in the bid price.

This price is a fundamental component informing the contract award decision ultimately made by the client. Accordingly, it is particularly interesting to take a closer look at the stability of quoted prices in the event of quantity variations. The method discussed in this example considers the uncertainties associated with quantities and highlights their influence on the spread of the prices submitted by the bidders. We thus obtain histograms of bid prices that can be investigated with regard to the statistical values (minimum, maximum, mean, standard deviation, spread, quantiles, etc.).

Calculating the difference between bid prices, in particular, will highlight any cases in which the ranking of bidders would change. Alternatively, the ranking of the bid can be determined in each iterative step and graphically represented in a histogram. By way of example, Fig. 9.66 shows the histogram representing the rank of Bidder C. The bid submitted by C came in sixth in about 5.9% of simulated cases. The remaining iterations result in C ranked fifth, i.e. the fifth lowest bid. This outcome is comparable with the percentages shown in Fig. 9.65. If the bids are very close to each other, the ranking exercise can be interpreted much more easily and quickly than the calculation of differences.

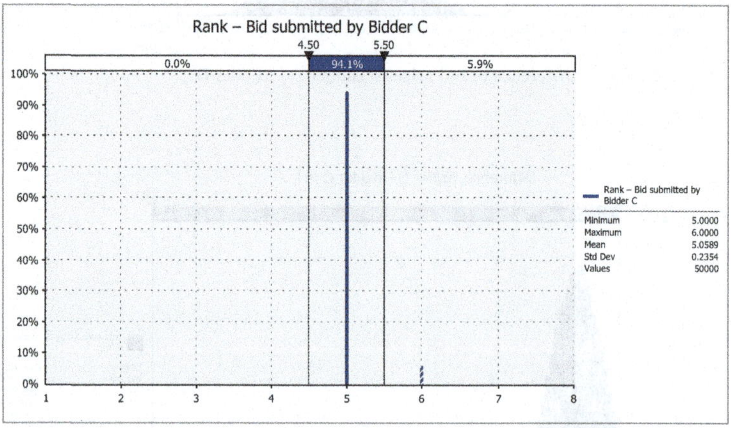

Fig. 9.66 Histogram – Rank of bid submitted by Bidder C

Overall, the inclusion of quantity uncertainties opens up an additional opportunity for the client to assess the bids with respect to their structure and price stability. Bid price stability could also be included as effectively influencing the contract award decision in accordance with the best bidder principle.

9.4 Project Phase 4 (PPH 4)

In Project Phase 4, contractors carry out the commissioned works. At this stage, the client or their representatives or agents perform coordination, monitoring, and management tasks. The next section deals with the fundamentals of conducting trend analyses as a tool for forecasting future developments.

9.4.1 Trend Analysis

Historical data or project progress information can be used to forecast future developments with the aid of mathematical/statistical methods. Trend analyses are a suitable tool for the client to monitor and control project progress during the execution phase of the construction project (PPH 4 – see Fig. 9.67), to forecast the future course of events, and to intervene if adverse trends become foreseeable.

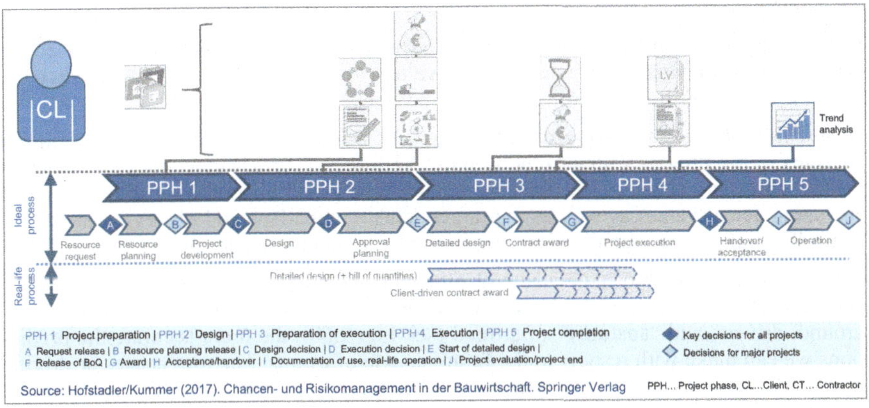

Fig. 9.67 Allocation of example in the project sequence

9.4.1.1 Basics

The notion of trend stands for:

- A particularly pervasive and long-term development (in the field of sociology)
- A development in the fields of mathematics, statistics, metrology, and business/economics (statistics)
- Price trend – a development of share prices (chart analysis)
- Market trend – a projected development of demand in a defined market

A trend thus describes a certain development (tendency) or a process of change that can be observed over a certain period and measured statistically. It is a component of a time series that we assume to have a longer-term, sustainable effect. A trend is often considered to be a linear function of time expressing the basic direction into which a time series is developing, which is also usually referred to as a deterministic trend. Its counterpart is a stochastic trend ("random walk").[62)]

In the field of statistics, available data or information is subjected to regression analysis in order to derive a future development hypothesis.

Commonly used methods include linear but also polygonal and logarithmic regression on the basis of existing observation points (point clouds). All these methods can be grouped under the heading of deterministic trend analyses.

[62)] Cf. http://wirtschaftslexikon.gabler.de/Definition/trend.html. Accessed December 22, 2020.

If we include uncertainties in the trend analysis, this can also be referred to as a time series analysis. Typical time series include share prices, pre-electoral surveys, and weather observations. Time series analyses apply inferential statistics, a statistical method arriving at conclusions regarding the statistical population based on sampling, to time series and predict their future development (trends). Time series analysis is a special type of regression analysis.

Trend analyses usually follow mathematical principles. They investigate historical data and/or events. The resulting graphical representations are used to derive forecasts of future developments.

In the field of construction management and economics, trend analyses are based on target-actual comparisons in which actual values are projected in order to predict the progress of the construction project in future periods.[63)]

Girmscheid/Motzko state that project management would be impossible if we limited our analyses to the present without relying on trend forecasts or accurate analyses of actual values. Thus, any sound project management will have to be based on a combination of analyzing actual values as well as the target-actual comparison and observing these two controlling outcomes over a longer period in order to identify tendencies and trends.[64)]

Any change in a variable associated with a benchmark can be attributed to certain causes, and such changes will translate into a trend. The better our understanding of the background, drivers, and "anatomy" of a trend (see Fig. 9.68), the more accurate the predictions we can make with respect to the future course of events. The root causes or driving forces (drivers) need to be investigated and analyzed in order to establish an understanding of the trend.[65)]

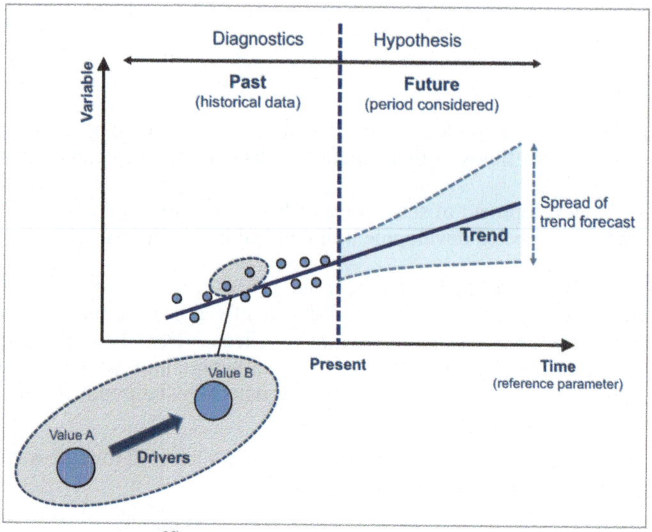

Fig. 9.68 Anatomy of a trend[66)]

[63)] Cf. Girmscheid/Motzko (2013), p. 421
[64)] Cf. ibid.
[65)] Cf. Pillkahn (2007), p. 131f.
[66)] Cf. ibid., p. 131

Trend analyses deliver a forecast projecting the further development of the project by clearly illustrating how the assessment of target values changes from one reporting date to the next.

Such analyses can be performed to compare, for instance, quantities, costs, times, or paid working hours.

The easiest way to identify a trend within a series of measurements is to establish its graphical representation as a time-dependent point diagram (alternatively, depending on the problem to be investigated, based on the measurement point). The greater the similarity of the point cloud to a horizontal line (or a round point cloud), the smaller a visible trend will be. Linear regression is usually the next analytical step.

This method is a special type of regression analysis in which a linear relationship between an influential parameter and a target variable is established. A straight line runs through a point cloud in such a way that the difference between the regression line and the measured values is as low as possible (least squares method or least square error method).

Eq. (9-7) is the straight-line equation where k is the straight line's slope and d represents the y axis intersection point ("offset").

$$y = k \cdot x + d \tag{9-7}$$

The slope of a regression line that does not intersect with the coordinate origin (see Fig. 9.69, left) is calculated by dividing the covariance of x and y by the variance of x (see Eq. (9-8)).

In this example, the covariance is a non-standardized measure describing a monotonic relationship between two random variables. The value of this measure permits conclusions as to whether a high value of one of the random variables tends to be associated with either a high or a low value of the other variable.

$$k = \frac{Cov(X;Y)}{Var(X)} = \frac{\sum_{i=1}^{n}(x_i - \bar{x}) \cdot (y_i - \bar{y})}{\sum_{i=1}^{n}(x_i - \bar{x})^2} \tag{9-8}$$

We then define the y axis intersection by applying the means of x and y to Eq. (9-9).

$$d = \bar{y} - k \cdot \bar{x} \tag{9-9}$$

A linear regression analysis involving a zero intersection (see Fig. 9.69, right) considers the condition that the regression line must run through the coordinate origin (d = 0). Depending on the specific task, this approach can be useful if the influential parameter takes a zero value, which would automatically result in the target variable also becoming zero. In this case, we calculate the slope using Eq. (9-10).

$$k = \frac{\sum_{i=1}^{n} x_i \cdot y_i}{\sum_{i=1}^{n} x_i^2} \tag{9-10}$$

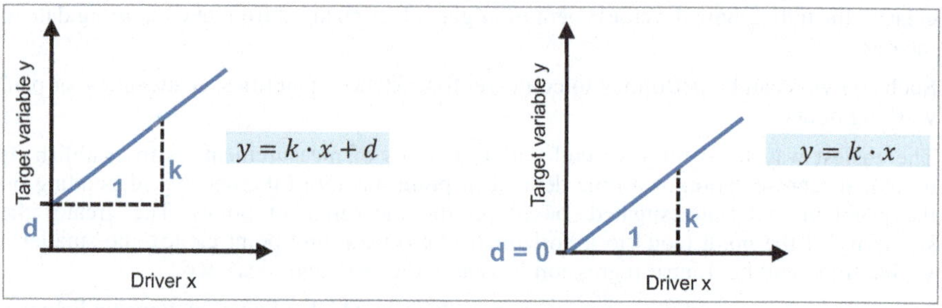

Fig. 9.69 Straight-line equation without (left) and with zero intersection (right)

The linear regression line provides a simplified method to forecast the future trend of a series of measurements. However, any such conclusion will inevitably be associated with spreads and, in some cases, major uncertainties. This is why the results should be critically reviewed at the stage of their interpretation.

9.4.1.2 Task

The client will be particularly interested in adhering to the specified construction timeline or schedule as well as to the agreed construction cost, or contract price. Different metrics can be applied to the continuous monitoring of project progress. We consider the variables of construction progress relative to an indicative quantity (such as the amount of reinforced concrete for project or site management purposes), working hours (taken from daily work reports), and costs (partial invoices) to be particularly helpful in this regard.

9.4.1.3 Modeling

Documenting work progress at the shortest possible intervals leads to an interval-driven trend of progress data over time, which can be used as a database for forecasting future developments.

For this purpose, the differences between individual progress values can be calculated, and these differences can then be converted into a distribution function (e.g. through data fitting). In the next step, we apply this distribution to forecast work progress for future time intervals. We use a Monte Carlo simulation to select random progress values based on the range, or spread, of the selected distribution in order to simulate a possible project progress scenario.

If this process is repeated (for instance in 50,000 iterations), this will result in a forecast of the progress trend over time that also permits conclusions regarding the probability of exceeding or underrunning values.

9.4.1.4 Calculation and Interpretation of Results

Section 10.4.10 includes an example of a trend analysis performed to determine the progress of reinforced concrete operations. This example comprises both deterministic and probabilistic trend analyses. The results are displayed in diagrams and interpreted.

9.4.1.5 Benefit for the Client

It is relatively easy to show target-actual comparisons in both bar and line diagrams. In contrast, it is relatively difficult to reflect changes in target values over a longer period in such diagrams. This is why trend analyses are used in such cases because they clearly reveal any future bottlenecks or tendencies.

A trend analysis makes it possible to determine the magnitude of deviations from a baseline schedule (structured by cost, deadline, content, and scope parameters) caused by applying data taken from previous progress reports.

Likewise, we can project the magnitude of the deviation of a single project parameter at any given point in the future if no unforeseeable changes are implemented in the project and if the previously established documents and records are used for forecasting future developments.

Forecasts can be shown not only as isolated deterministic values but also as histograms if ranges are included in the analysis. We are thus able to reflect possible ranges and the chances and risks associated with a specific value.

9.5 Project Phase 5 (PPH 5)

No examples are currently included in this book that would be grouped under Project Phase 5. However, Monte Carlo simulations appear to be suitable for operational and strategic facility management applications. In addition, in real estate rental and operations, lost rental income and its repercussions on the profitability of the property can be assessed together with the associated uncertainties.

9.6 Portfolio Management

Rather than assigning portfolio management directly to a specific project phase, we should consider it a management tool to be applied across all phases (see Fig. 9.70).

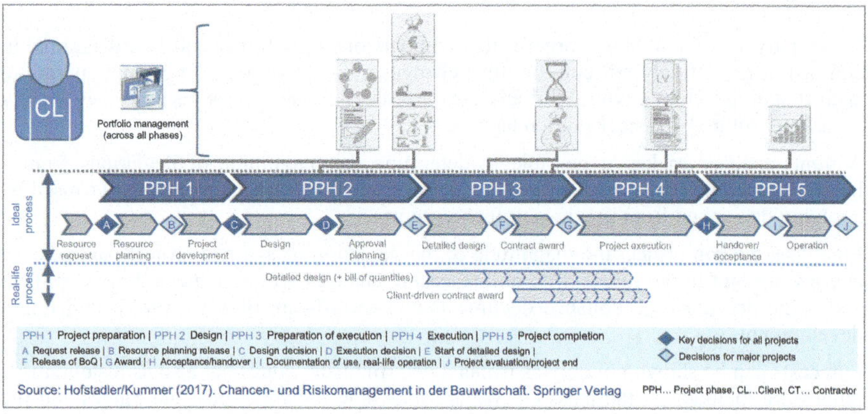

Fig. 9.70 Allocation of example in the project sequence

We assess several projects in terms of their costs and the budget made available by the client. Our aim is to obtain an overview of the chance/risk ratio for the entire portfolio and, conversely, to demonstrate and evaluate the influence of specific projects on the portfolio.

This application is particularly interesting for large (public-sector) clients with several projects under construction at the same time, each with a different size and progress level.

9.6.1 Basics

The portfolio theory was developed by *Markowitz*[67]. In its original form, it represents a subfield of capital market theory. Its starting point is a defined available investment amount distributed across several asset classes, such as equities or bonds, by its spreading or diversification. This diversification aims to reduce the associated risk relative to the variance and percentage of the return compared to an investment in only a single asset class. In relation to the capital market, this raises the question which securities to aggregate in a portfolio at which proportions. On the practical level, such an aggregation of investment opportunities happens through the establishment and marketing of funds, such as equity funds.[68]

In practice, theoretical considerations on how to establish an optimal portfolio are compromised by uncertainties of data and information on available investment alternatives as well as by market volatility.

Clients are faced with a similar situation in the construction industry. They can use the available budget to implement a number of different projects. If the completed projects do not generate a direct return, as is the case, for example, for infrastructural projects, the portfolio analysis will concentrate on the expenditure side (expenses or costs). In so doing, the client should primarily ensure that project expenses do not exceed the available budget. However, the client will be able to achieve this goal only with a certain probability

[67] Cf. Markowitz (1952), p. 77ff.

[68] Cf. http://wirtschaftslexikon.gabler.de/Definition/portfolio-selection.html. Accessed December 22, 2020.

owing to the uncertainties inherent in future developments. Clients utilize the selection of projects to be completed or the narrowing of the cost ranges of individual projects (for instance by entering into lump-sum and/or fixed-price agreements) as management instruments. Depending on the type of client (public vs. private), there is a greater or lesser degree of flexibility in terms of project selection.

Unlike financial mathematics, which, for instance, enables the purchase of any number of shares, the project portfolio management approach does not enable the "cutting into pieces" of individual projects. Any investment in a project is either effected in full or not at all.

9.6.2 Task

Clients complete construction projects in order to meet a certain demand and to generate a specific benefit. However, meeting this demand needs to be contrasted with a limited available budget, which can be spread across several projects within a certain department or field. Such projects usually widely differ in their size (volume) and progress rates. This is why they are also associated with different chances and risks.

Depending on the progress of planning at the time of the tender or contract award, more or less significant quantity and/or scope of work changes may occur in the course of the project and result in a cost increase. Any incomplete or inaccurate planning and a large number of changes implemented in the construction phase will usually have an adverse effect on the chance/risk ratio relative to budget compliance.

Portfolio management makes it possible to merge specific projects and their chance/risk ratios and to compare them with the available budget. Likewise, we can analyze the effects of a new project on the portfolio so that these effects inform the decision-making process in favor or against the project considered.

9.6.3 Modeling

In the modeling exercise, the cost uncertainties of the individual portfolio projects need to be reflected in the form of distributions in order to determine the chance/risk ratio for meeting a specified budget. Any probabilistic costing exercises performed for the projects can be used directly for the purpose of the portfolio analysis. The probabilities of exceeding or underrunning the individual project budgets and the ranges of expected costs will have to be estimated if no such probabilistic calculations exist.

The sum of these cost distributions can then be fed into a histogram representing the total cost of the project, which we can compare with the overall budget. This comparison results in a chance/risk ratio for budget compliance for the entire portfolio. Likewise, we can also add new projects and evaluate them in respect of their influence on the portfolio and on budget compliance.

It is crucial that the cost distributions established for the individual projects be continuously updated in line with project progress. This is the only way to consider current trends and developments in the individual projects in the context of the full portfolio.

9.6.4 Calculation and Interpretation of Results

Section 10.6 includes a portfolio management example. This example shows, from the contractor's point of view, how individual projects can be merged in a project portfolio and how portfolio management can be used as a management and controlling tool.

9.6.5 Benefit for the Client

The system outlined above makes it possible to effectively aggregate projects at varying stages of progress, with different contract volumes and chance/risk ratios in a single portfolio, which then permits conclusions regarding individual new projects.

Analyzing the complete set of projects in a portfolio provides the advantage of being able to proactively manage budget certainty in an iterative process by selecting new projects if the database is adjusted and updated continuously. In this process, the chance/risk ratio determined for the portfolio is used as a key management tool.

If the selection of a new project alters the chance/risk ratio of the entire portfolio to such an extent that there is no sufficient probability of meeting the budget, it appears to be useful to postpone the proposed project and to complete it at a later point in time (in a different portfolio constellation).

Section 10.6 discusses the practical application of portfolio management from the contractor's perspective because the underlying considerations are very similar.

Project portfolio management should be considered to be a management tool involving controlling and steering features. In this process, input data needs to be continuously updated in an iterative process to be able to draw conclusions regarding the current chance/risk ratios.

9.7 Summary

This chapter outlined examples of Monte Carlo simulations from the client's vantage point. The examples discussed in this chapter primarily relate to Project Phases 1 to 4. However, the described methods and tools can also be applied to other phases. They can be used by bidders, or subsequent contractors, if required.

The worked examples were outlined in a plausible manner; they follow a clear, consistent structure. Calculation tables were explained in the relevant sections and complemented by graphical representations of the calculation workflows. We put particular emphasis on the interpretation of the simulation results and on demonstrating their benefits for the user.

9.8 References

Bauer, Ulrich (2009). Enzyklopädie Betriebswirtschaftslehre – Grundlagen der Betriebswirtschaftslehre – Betriebswirtschaftslehre Bau. Lecture script. Graz. Graz University of Technology.

Bichler, Daniel (2014). Lebenszykluskostenrechnung in der Bauwirtschaft. Master's thesis. Graz. Institute of Construction Management and Economics, Graz University of Technology.

BKI BAUKOSTEN-Poster 2016; Source: Eurostat New Cronos database.

Blecken, Udo et al. (2007). Kosten im Hochbau – Praxis-Handbuch und Kommentar zur DIN 276. Eds.: Blecken, Udo; Hasselmann, Willi. Cologne. Verlagsgesellschaft Rudolf Müller GmbH & Co. KG. (ISBN 978-3-481-02245-7)

Diederichs, Claus Jürgen (2006). Immobilienmanagement im Lebenszyklus – Projektentwicklung, Projektmanagement, Facility Management, Immobilienbewertung – 2nd ed. Berlin, Heidelberg, New York. Springer-Verlag. (ISBN-10 3-540-25509-5)

DIN 276-1 (2008-12 edition). Kosten im Bauwesen – Teil 1: Hochbau.

DIN 277-1 (2016-01 edition). Grundflächen und Rauminhalte im Bauwesen.

Fetzer, Robert et al. (2016). BKI Baukosten 2016 – Neubau – Teil 3 – Statistische Kostenkennwerte für Positionen. Ed.: BKI Baukosteninformationszentrum. Stuttgart. Baukosteninformationszentrum Deutscher Architektenkammern GmbH. (ISBN 978-3-945-649-15-2)

Girmscheid, Gerhard; Motzko, Christoph (2013). Kalkulation, Preisbildung und Controlling in der Bauwirtschaft – Produktionsprozessorientierte Kostenberechnung und Kostensteuerung – 2nd ed. Berlin, Heidelberg. Springer-Verlag. (ISBN 978-3-642-36637-6)

Glatte, Thomas (2014). Entwicklung betrieblicher Immobilien – Beschaffung und Verwertung von Immobilien im Corporate Real Estate Management. Eds.: Berner, Fritz; Kochendörfer, Bernd. Wiesbaden. Springer-Verlag. (ISBN 978-3-658-05687-2)

Heiermann, Wolfgang; Riedl, Richard; Rusam, Martin et al. (2003). Handkommentar zur VOB – Teile A und B, Rechtsschutz im Vergabeverfahren – 10th ed. Wiesbaden. Vieweg & Sohn Verlag/GWV Fachverlage GmbH. (ISBN 3-528-11715-X)

Hofstadler, Christian (2014c). Methoden zur Ermittlung von Lebenszykluskosten. In: Nachhaltig Bauen mit Beton: Werkstoff und Konstruktion, Beton Graz'14 – 2. Grazer Betonkolloquium; September 25-26, 2014. Eds.: Nguyen, Viet Tue; Maydl, Peter; Freytag, Bernhard; Santner, Gerhard. 139-152. Graz. Verlag der Technischen Universität Graz. (ISBN 978-3-85125-360-3)

Hofstadler, Christian (2018). In-Depth Bid Assessment for Unit-Priced Contracts. In: The Tenth International Conference on Construction in the 21st Century (CITC-10). Eds.: Ahmed, Syed M.; Shah, Attaullah; Azhar, Salman; Smith, Norma A.; Campbell, Shaunna C.; Mahaffy, Kelly; Saul, Amelia. 232-240. Greenville, North Carolina, USA. (ISBN 978-1-7324416-1-3)

Hofstadler, Christian; Franzl, Gerald (2011). Bewehrungsarbeiten im Baubetrieb. Ed.: Verband Österreichischer Biege- und Verlegetechnik (Austrian Association for Bending and Installation of Concrete Reinforcement) (ISBN 978-3-200-02149-5)

Hofstadler, Christian; Kummer, Markus (2017a). Chancen- und Risikomanagement in der Bauwirtschaft – Für Auftraggeber und Auftragnehmer in Projektmanagement, Baubetrieb und Bauwirtschaft. Berlin, Heidelberg. Springer-Verlag. (ISBN 978-3-662-54318-4)

Hofstadler, Christian; Kummer, Markus (2015b). Application of the Monte-Carlo Method to Calculate the Life-Cycle Costs of Bridges. In: Multi-Span Large Bridges. Eds.: Pacheco, Pedro; Magalhães, Filipe. London. 751-758. CRC Press – Taylor & Francis Group. (ISBN 978-1-138-02757-2)

Jodl, Hans (2010a). Lebenszykluskosten von Brücken – Teil 1 – Berechnungsmodell LZKB. Bauingenieur 85, May 2010. 221-230. Düsseldorf. Springer-VDI-Verlag. (ISSN 0005-6650)

Jodl, Hans (2010b). Lebenszykluskosten von Brücken – Teil 2 – Software LZKB. Bauingenieur 85, May 2010. 231-240. Düsseldorf. Springer-VDI-Verlag. (ISSN 0005-6650)

Kalusche, Wolfdietrich et al. (2016). BKI Baukosten 2016 Neubau – Teil 1 – Statistische Kostenkennwerte für Gebäude. Ed.: BKI Baukosteninformationszentrum. Stuttgart. Baukosteninformationszentrum Deutscher Architektenkammern GmbH. (ISBN 978-3-945-649-13-8)

Kropik, Andreas; Krammer, Peter (1999). Mehrkostenforderungen beim Bauvertrag – Ansprüche aus Leistungsänderungen, ihre Geltendmachung und Abwehr. Vienna. Österreichischer Wirtschaftsverlag. (ISBN 3-85212-105-1)

Lechner, Hans (2009). Projektentwicklung. Lecture script. Graz. Institute of Construction Management and Economics, Graz University of Technology.

Lechner, Hans; Heck, Detlef [Eds.] (2011). Wörterbuch (dt./engl.) – BauProjektManagement – 3rd ed. Graz. Verlag der Technischen Universität Graz. (ISBN 978-85125-148-7)

Lechner, Hans; Liebenau, Sabine (2010). PM Teil 7a – Kostenplanung. Graz. Verlag der Technischen Universität Graz.

Markowitz, Harry (1952). Portfolio Selection. In: The Journal of Finance. Vol. 7, No. 1. March 1952. 77-91. American Finance Association. (ISSN 1540-6261)

Oberndorfer, Wolfgang; Jodl, Hans Georg et al. (2001). Handwörterbuch der Bauwirtschaft – ON V 208 – 2nd ed. Vienna. ON Österreichisches Normungsinstitut. (ISBN 3-85402-072-4)

OENORM B 1800 (2013-08-01 edition). Ermittlung von Flächen und Rauminhalten von Bauwerken und zugehörigen Außenanlagen.

OENORM B 1801-1 (2015-12-01 edition). Bauprojekt- und Objektmanagement – Teil 1: Objekterrichtung.

OENORM B 2061 (2020-05-01 edition) Preisermittlung für Bauleistungen – Verfahrensnorm.

OENORM B 2110 (2013-03-15 edition) Allgemeine Vertragsbestimmungen für Bauleistungen – Werkvertragsnorm.

Pillkahn, Ulf (2007). Trends und Szenarien als Werkzeuge zur Strategieentwicklung – Wie Sie die unternehmerische und gesellschaftliche Zukunft planen und gestalten. Erlangen. Publicis Corporate Publishing. (ISBN 978-3-89578-286-2)

Springer Gabler Verlag (Ed.), Gabler Wirtschaftslexikon, Keyword: Trend. Available online at http://wirtschaftslexikon.gabler.de/Definition/trend.html. Accessed December 22, 2020.

Springer Gabler Verlag (Ed.), Gabler Wirtschaftslexikon, Keyword: Portfolio Selection. Available online at http://wirtschaftslexikon.gabler.de/Definition/portfolio-selection.html. Accessed December 22, 2020.

Veit, Peter (2012). Betriebswirtschaftslehre – Bau – Kostenrechnung, Investition. Lecture script. Graz. Graz University of Technology.

VÖBV – Verband Österreichische Biege- und Verlegetechnik (2010). Wirtschaftliches Bewehren. Available online at http://www.voebv.at/pdf/VOEBV%20Wirtschaftliches%20Bewehren.pdf. Accessed January 8, 2020.

10 Worked Examples – Contractor

This chapter discusses worked examples from the bidder's and contractor's vantage points. However, the outlined methodology and approaches can also be applied by clients or by other project stakeholders as far as their underlying structure is concerned.

The examples described in this chapter are allocated to the individual project phases and divided into the Basics (if appropriate), Task/Initial Situation, Modeling, Calculation and Interpretation of Results, and Benefit subsections.

All calculations and/or simulations were performed using MS Excel in combination with the @Risk add-in. Manual recalculation may result in rounding differences compared to the figures stated in this chapter.

Unless otherwise indicated, simulation results are based on 50,000 iterations using the Latin Hypercube sampling method.

10.1 Project Phase 1 (PPH 1)

Contractors are usually not yet involved in this phase, with the exception of the projects they manage and supervise themselves or, for example, PPP projects. If and when the contractor also assumes client responsibilities, the methods and tools available to the client can be applied (see Section 9.1).

10.2 Project Phase 2 (PPH 2)

As above, contractors are usually not yet involved in this phase, with the exception of the projects they manage and supervise themselves or, for example, PPP projects. If and when the contractor also assumes client responsibilities, the methods and tools available to the client can be applied (see Section 9.2).

C. Hofstadler and M. Kummer, *Chances and Risks in Construction Management and Economics*, Springer Tracts in Civil Engineering,
https://doi.org/10.1007/978-3-030-64014-9_10

10.3 Project Phase 3 (PPH 3)

In PPH 3, bidders calculate the expected costs and derive the bid price from these figures. The following sections include specific cases of applying Monte Carlo simulations in order to demonstrate the value added to the bidder's decision-making process.

10.3.1 Quantity Uncertainties/Combined Pricing

Specifiers require combined pricing if works or services for different structural components, construction trades, ground conditions, rebar diameters, etc. requiring a variable mode and combination of production factors are merged into a single item. Combined pricing is necessary whenever a unit price is derived from different cost estimates with respect to the utilization of production factors.

We refer to the specifications of reinforcing operations in an engineering structure in order to discuss the uncertainties associated with free pricing assumptions becoming necessary due to undisclosed pricing bases. By resorting to a combined pricing item, the client fails to provide detailed information on the rebar diameter and/or bend shape distributions.

This example exclusively considers the wage cost incurred by installing rebar reinforcement. Equipment and material costs as well as wage costs incurred by cutting and bending are not taken into account. Within the overall project sequence, this example is allocated to PPH 3 (see Fig. 10.1).

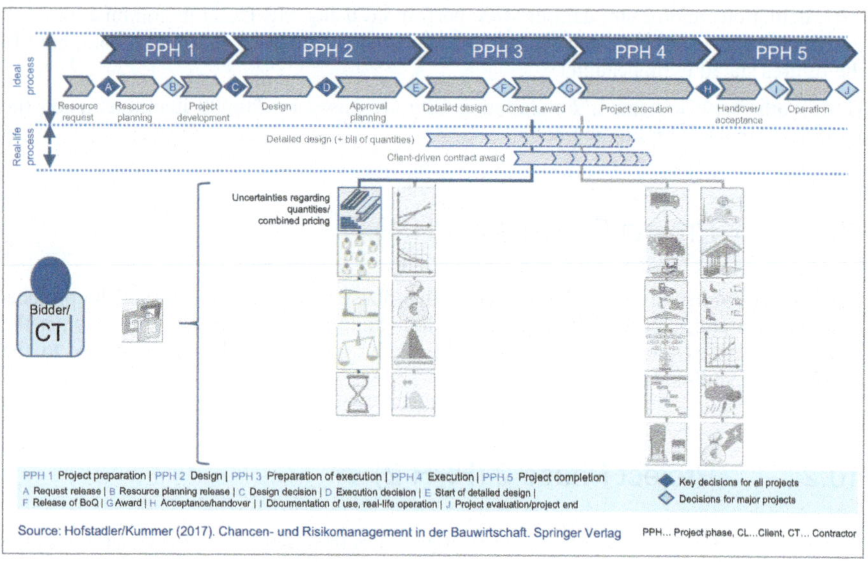

Fig. 10.1 Allocation of example in the project sequence

10.3.1.1 Initial Situation

In its specification for reinforcing operations, the client included only a global rebar item and the associated total quantity of reinforcing steel. In contrast, the specification does not detail the individual rebar diameters and the bend shape distribution (quantity distribution). Thus, the bidder will be forced to determine a blended consumption rate and, consequently, a combined price for the purpose of preparing their bid (see Fig. 10.2).

In the case of reinforcing operations, bidders will have to resort to combined pricing models in the absence of any differentiation in terms of rebar diameters and/or lacking detail on the complexity of bend shapes, or on the allocation of the reinforcement to individual structural components.

In its pricing exercise, the bidder is bound to start from those reinforcement distributions and quantities that the client specifies in their bill of quantities. The bidder has to define free pricing assumptions if the specification does not provide information on the allocation of the reinforcement to the individual groups of structural components, or on the mean rebar diameter and bend shapes.

Item No.	Item text *Quantity* *Unit* *Price components* *Item price*
01 xx xx	Reinforcement Standing preliminary remarks 1. Technical Conditions of Contract The Technical Conditions of Contract RVS 08.06.02 must be complied with.
01 xx xx xx	Deliver, cut to size, bend, and install reinforcing steel of grade Bx for untensioned reinforcement. The unit price applies regardless of different diameters or shapes and to all lengths specified in the design, as well as to round steel of grade S235 if required. This work also includes: • appropriate storage • any required interim transport including loading and unloading • the final transport to the installation site • delivery of binding wire and of spacers to the formwork • welding of the supports of tension members and expenses for the increased accuracy of such supports A separate compensation is paid for: • the additional costs of rebar in lengths greater than 14 m The following items are taken into account: • the theoretical weight of the reinforcement, of the support or spacer (Z plate, spacer strips or similar), and bracings as specified in the approved plans without offcuts
01 xx xx xxA	**Reinforcing steel B550B** LT 1 reinforcing steel B550B conforming to OENORM B 4707. Wages: Other: 3,500.00 t Unit price:

Fig. 10.2 Extract from the bill of quantities[1)]

If the client partially or fully fails to disclose their pricing bases or specifications, this will force the bidder, or subsequent contractor, to come up with their own assumptions in order to be able to perform the pricing exercise in the first place. If the bidder also fails to disclose their assumptions in the submission process, this is likely to result in disputes in the event of quantity changes or changes in the distribution of combined pricing components (see Fig. 10.3).

1) Based on a real bill of quantities anonymized by the authors for this example.

In this case, the contractor cannot know or derive the quantity structure that the client used for their specification or request for bids. Conversely, the client is not in possession of any information with respect to the assumptions made by the bidder for the purpose of costing and pricing (if the latter failed to disclose their pricing assumptions).

Yet another distinction needs to be made regarding the bidder's disclosure, namely if it is based on their own assumptions or assumptions made by a subcontractor, which may result in different mean wage prices, for example. If the bidder discloses their assumptions, the client will know the bases of pricing.

With these pricing assumptions remaining undisclosed, however, only the pricing components will be obvious (with the unit price derived from wages and other items), but they cannot be used as a sound basis for verifying the most appropriate pricing model. In the event of a dispute, the contractor's subsequent pricing disclosures need to be checked for plausibility, including a critical comparison with relevant information provided in the literature and, as far as reasonably possible, by back-calculating the bid price.

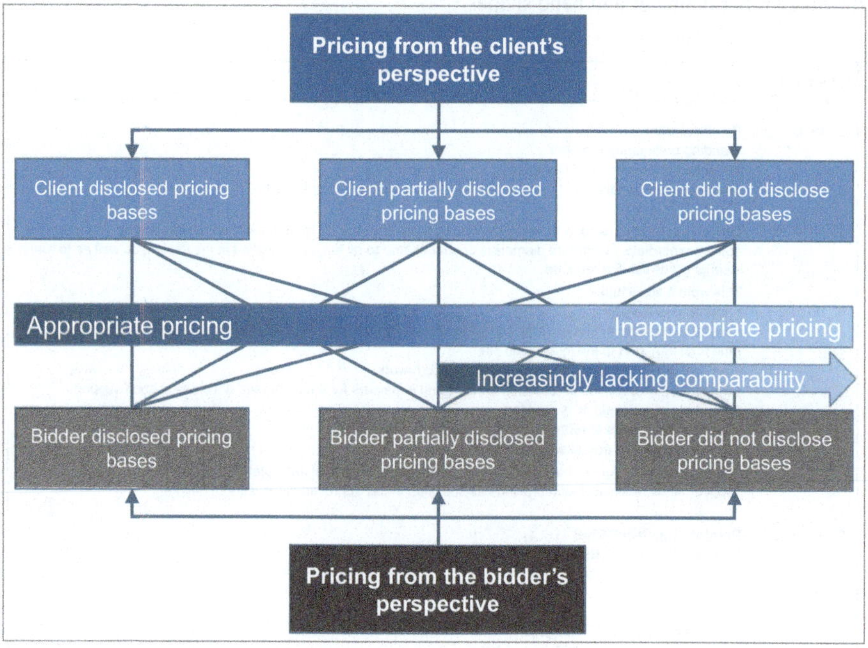

Fig. 10.3 Pricing bases from the client's and bidder's perspective

Generally speaking, the contractor is forced to determine free assumptions regarding quantities and their distribution if the client fails to provide detailed information in their specification. However, such free assumptions must not be unrealistic and must also comply with legal requirements, such as minimum wages in order to prevent excessively low pay or social dumping.

In this example, the bidder used the VÖBV information[2)] as a basis for their pricing assumptions. This document provides details on bend shape and diameter distributions commonly found in specific building or structure categories.

10.3.1.2 Modeling

The bidder had to determine a combined price for all diameters (8 to 40 mm) and bend shapes (straight, simple, and complex bend). The pricing exercise was thus fraught with major information gaps, which is why the bidder was bound to resort to free pricing assumptions.

It is generally the bidder's responsibility to verify the sources from which they derive their project-specific assumptions. At any rate, these assumptions should be plausible in relation to the works or services defined in the specification.

For the purpose of determining the wage proportion, the initial step includes making assumptions on the distribution of diameters and bend shapes and subsequently weighting their proportions in conjunction with the corresponding consumption rates. The resulting blended labor consumption rate can then be multiplied by the mean wage price, from which we derive the wage proportion of the unit price (see Fig. 10.4).

The quantity proportions of rebar diameters and bend shapes as well as the consumption rates per diameter and bend shape are defined as uncertain input parameters. Fig. 10.4 includes the flow chart illustrating these calculation steps and a qualitative representation of the corresponding distribution functions. The mean wage price is integrated into the computation model as a deterministic value.

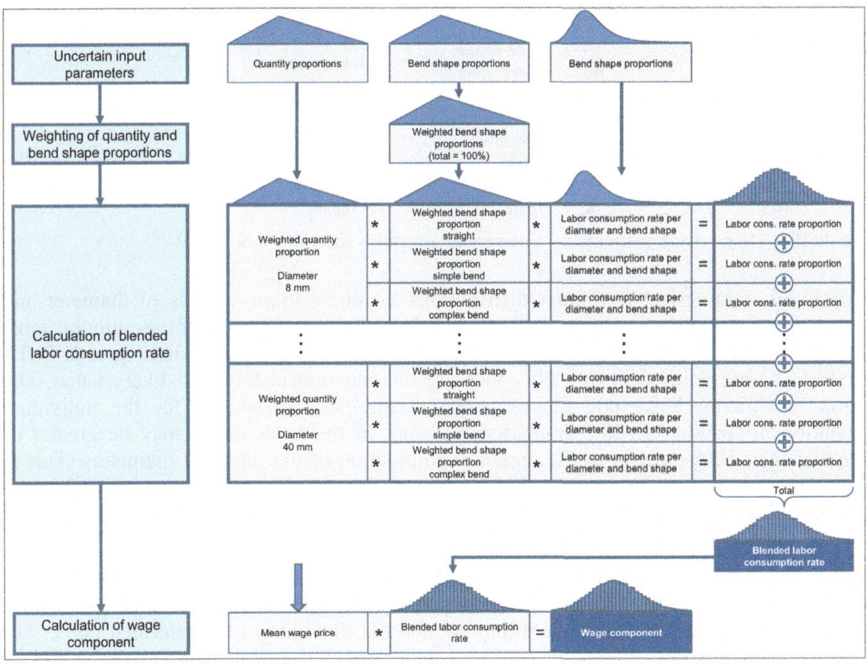

Fig. 10.4 Calculation flow chart

[2] Verband Österreichische Biege- und Verlegetechnik (Austrian Association for Bending and Installation of Concrete Reinforcement)

The VÖBV states diameter and bend shape distributions commonly found in engineering structures. Tables 10.1 and 10.2 list these percentages for the individual diameters and bend shapes.

No.	Diameter [mm]	Structural engineering [%]
0	A	B
1	8	0.54
2	10	8.47
3	12	14.41
4	14	17.22
5	16	15.37
6	20	17.96
7	26	18.08
8	30	6.06
9	36	0.84
10	40	1.05
11	**Total**	**100.00**

Table 10.1 Diameter distribution – Engineering structures according to the VÖBV[3)]

Since the VÖBV values are averages associated with uncertainties and/or spreads, we assume, for the probabilistic calculation, that the individual proportions may be up to 20% higher or lower than the stated values.[4)]

No.	Bend shape distribution [-]	Structural engineering [%]
0	A	B
1	straight	56.36
2	simple	24.15
3	complex	19.49
4	**Total**	**100.00**

Table 10.2 Bend shape distribution – Engineering structures according to the VÖBV[5)]

We apply symmetrical triangular distributions to model these spreads of diameter and bend shape proportions. Columns B, C, and D of Table 10.3 show the minimum, most likely, and maximum quantity proportions for each diameter. The related spreads (MIN and MAX) are equivalent to a 20% increase or reduction in the most likely value. The rows in Column E include the respective triangular distributions for the individual quantity proportions. In the simulation, the sum of the percentages may be greater or smaller than 100% owing to the random combination of the different diameters. This is why these proportions are weighted in Column F. For simply bent 20 mm rebar (Row 6), we thus obtain a weighted quantity proportion of 19.635% (= 19.494% / 99.281% * 100%) for the random values stated in Table 10.3.

Thus, this simulation also considers the fact that the specified quantity need not necessarily be identical to the actually installed quantity. Borderline cases in which all quantity proportions are equal to both the minimum and the maximum of the defined ranges are extremely improbable. No such cases were generated in the simulation involving 50,000 iterations.

3) Cf. VÖBV (2010), p. 2

4) Assumption based on the 20% provision contained in OENORM B 2110:2013, p. 28f.

5) Cf. VÖBV (2010), p. 6

At this stage, it would be possible to normalize the quantity proportions for each iterative step if the specified quantity were not associated with uncertainties (see Section 9.3.3.4).

Structural engineering						
		Quantity distribution				
No.	Diameter	MIN	MOST LIKELY	MAX	Distribution	Weighted proportion
	[mm]	[%]	[%]	[%]	[%]	[%]
0	A	B	C	D	E	F
1	8	0.432	0.540	0.648	0.470	0.473
2	10	6.776	8.470	10.164	8.633	8.695
3	12	11.528	14.410	17.292	12.843	12.936
4	14	13.776	17.220	20.664	18.215	18.347
5	16	12.296	15.370	18.444	15.878	15.993
6	20	14.368	17.960	21.552	19.494	19.635
7	26	14.464	18.080	21.696	15.748	15.862
8	30	4.848	6.060	7.272	6.095	6.139
9	36	0.672	0.840	1.008	0.924	0.931
10	40	0.840	1.050	1.260	0.982	0.989
11	**Total**	**80.000**	**100.000**	**120.000**	**99.281**	**100.000**

Table 10.3 Diameter distribution – Engineering structures – Ranges

Table 10.4 shows the modeling of the bend shape distribution according to the same principle as used for the diameter distribution. Columns B to D also state the minimum, most likely, and maximum quantity proportions per bend shape. A spread of ±20% is also applied to the bend shapes. The random quantity proportions listed in Column E lie within the defined ranges and demonstrate that the sum of the bend shape proportions can also be greater or smaller than 100%. This is why these proportions are weighted in Column F. For simply bent rebar (Row 2), we thus obtain a weighted quantity proportion of 26.62% (= 27.43% / 103.05% * 100%) for the random values stated in Table 10.4.

Structural engineering						
		Bend shape distribution				
No.	Bend shape	MIN	MOST LIKELY	MAX	Distribution	Weighted proportion
	[-]	[%]	[%]	[%]	[%]	[%]
0	A	B	C	D	E	F
1	straight	45.09	56.36	67.63	58.10	56.38
2	simple	19.32	24.15	28.98	27.43	26.62
3	complex	15.59	19.49	23.39	17.51	16.99
4	**Total**	**80.00**	**100.00**	**120.00**	**103.05**	**100.00**

Table 10.4 Bend shape distribution – Engineering structures – Ranges

In the next step, labor consumption rates must be applied for each diameter and bend shape in order to calculate the rebar wage proportion. For this purpose, the VÖBV also provides a table listing mean consumption rates (see Table 10.5).

No.	Diameter	Labor consumption rates		
		Bend shape		
		straight	simple bend	complex bend
	[mm]	[wh/t]	[wh/t]	[wh/t]
0	A	B	C	D
1	8	15.88	34.93	57.32
2	10	12.20	26.05	44.61
3	12	9.34	18.60	32.31
4	14	7.42	14.47	23.07
5	16	6.08	10.88	18.22
6	20	4.22	7.54	12.84
7	26	3.53	6.35	11.21
8	30	3.53	6.15	10.48
9	36	3.53	5.95	10.28
10	40	3.53	5.95	10.08

Table 10.5 Labor consumption rates for rebar installation – Differentiation according to bend shapes and diameter[6)]

Labor consumption rates are lower for straight bars compared to rebar with simple or complex bends. They generally decrease as the rebar diameter increases (see Fig. 10.5). Particularly small diameters with complex bends incur the largest number of paid working hours per ton of reinforcement.

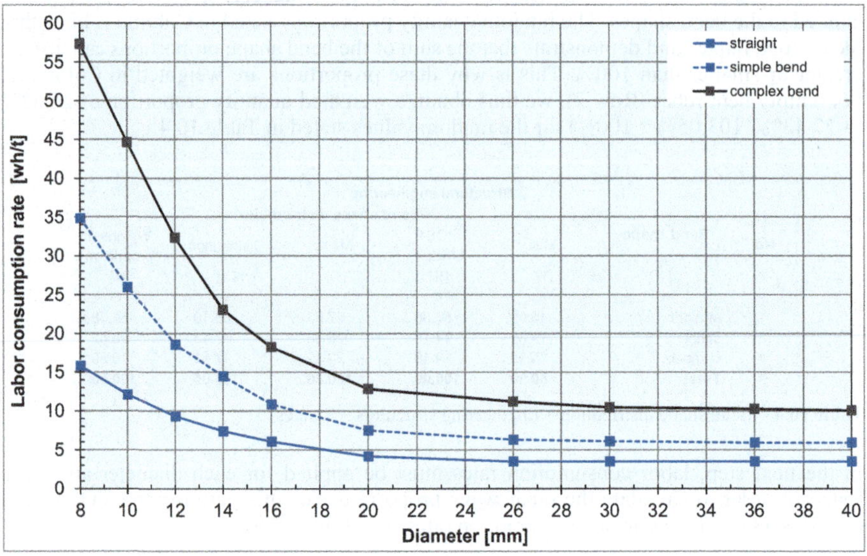

Fig. 10.5 Labor consumption rates for rebar installation, broken down to bend shapes according to the VÖBV[7)]

Log-logistic-like distributions are defined for the consumption rates in order to also account for the uncertainties associated with consumption rate data. In so doing, we apply

6) Cf. Hofstadler/Franzl (2011), p. 174

7) Cf. ibid.

the same methodology as for the labor consumption rates of formwork operations for floor slabs (see Section 7.2.1). Since the distribution curve for reinforcing operations does not provide a corresponding database for a separate data fitting step, we assume that the distribution curve determined for the labor consumption rate of formwork operations can also be applied to reinforcing operations.[8)] Hence, we only apply the distribution curve determined for formwork operations while using project-specific consumption rates for reinforcing operations (minimum and most likely values).

Besides the most likely values (see Table 10.5), we also need to specify minimum values in order to transform the polygonal curve of the log-logistic-like distribution of the consumption rates of formwork operations into the range of values for reinforcing operations. Each of the minimum values is said to be 5% lower than the corresponding most likely consumption rate.

For 20 mm straight rebar, we thus obtain a most likely labor consumption rate of 4.22 wh/t (see Cell B6 in Table 10.5) and a minimum labor consumption rate of 4.01 wh/t (= 4.22 wh/t * 0.95). We start from these two values and convert the coordinates of the polygonal approximation of the consumption rate for formwork operations (see Table 7.3) in order to determine the consumption rate distribution for 20 mm rebar (see Fig. 10.6). The minimum value of the distribution function is found to be identical to the calculated minimum. In addition, the most likely value of 4.22 wh/t is associated with the highest probability (mode). The consumption rate maximum of roughly 5.02 wh/t is derived from the distribution curve and the defined spread between the minimum and the most likely value.

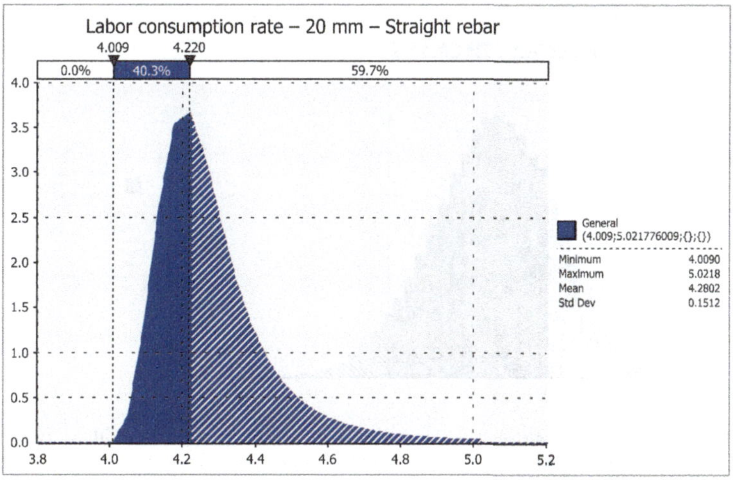

Fig. 10.6 Log-logistic-like distribution – Labor consumption rate of rebar installation – Diameter 20 mm – Straight rebar

8) This assumption is realistic because there is also a lower limit for the labor consumption rates of reinforcing operations, and a wider spread is possible in the upward direction. In addition, we start from the minimum principle to assume that comparatively lower consumption rates are associated with a higher probability of occurrence than comparatively higher values. We also state that the mean values specified by the VÖBV should be considered to be most likely values. Our theoretical considerations and characteristics thus result in a continuous, two-sided closed, unimodal, right-skewed distribution modeled by applying the log-logistic-like distribution determined for formwork operations. For the sake of simplicity, it would also be possible to apply PERT or triangular distributions, but we would have to define not only a minimum but also a maximum consumption rate for this purpose.

The same principle is applied to the remaining rebar diameters and bend shapes in order to define log-logistic-like distributions for the corresponding consumption rates, which are used as input parameters for the Monte Carlo simulation.

10.3.1.3 Calculation and Interpretation of Results

In the first step, determining the wage proportion for rebar reinforcement requires the calculation of the blended labor consumption rate of the individual diameters and bend shape distributions. For this purpose, the quantity proportions of each diameter are multiplied by the corresponding weighted quantity proportions per bend shape and the consumption rates. We allocate distribution functions determined in the modeling exercise to each of these parameters in order to consider the existing uncertainties and spreads. Multiplication of the distributions then results in 30 histograms for the consumption rate proportions (three bend shape distributions multiplied by ten diameters).

By way of example, we obtain the proportion of the labor consumption rate for 20 mm straight rebar by multiplying the distributions for the quantity proportions of this diameter (see Cell E6 of Table 10.3) and for the weighted quantity proportion of the bend shape (see Cell F1 of Table 10.4) by the log-logistic-like distribution allocated to the consumption rate (Fig. 10.6). For 20 mm straight rebar, the Monte Carlo simulation results in the histogram shown in Fig. 10.7 (see Cell B6 of Table 10.6).

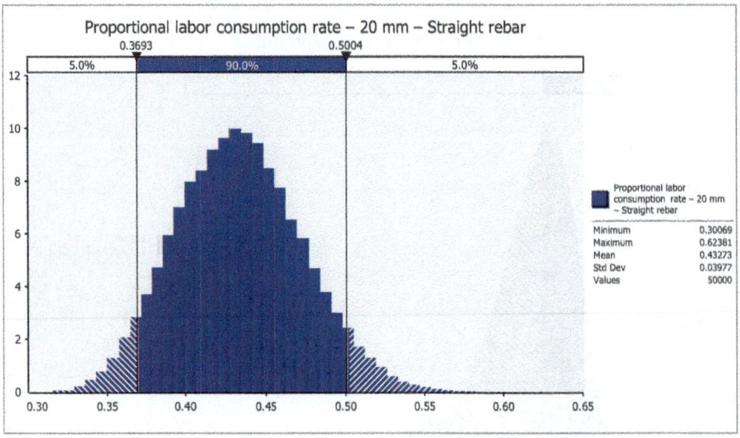

Fig. 10.7 Log-logistic-like distribution – Labor consumption rate of rebar installation – Diameter 20 mm – Straight rebar

This approach is applied to calculate the proportions of consumption rates in the form of a histogram for each of the three bend shapes (straight, simple, and complex bend) of the individual diameters (8 to 40 mm). These values are shown in Table 10.6 in Columns B to D (Rows 1 to 10). Column E then calculates the proportions of consumption rates per diameter, and Cell E11 determines the blended consumption rate histogram for rebar installation.

No.	Diameter	Blended labor consumption rate Bend shape straight	simple bend	complex bend	Proportional consumption rate	Mean wage price	Wage proportion
	[mm]	[wh/t]	[wh/t]	[wh/t]	[wh/t]	[€/wh]	[€/t]
0	A	B	C	D	E	F	G
1	8	0.05	0.05	0.06	0.16		
2	10	0.59	0.54	0.75	1.88		
3	12	0.77	0.66	0.92	2.35		
4	14	0.73	0.61	0.79	2.13		
5	16	0.53	0.41	0.55	1.50		
6	20	0.43	0.33	0.46	1.22		
7	26	0.36	0.28	0.40	1.05		
8	30	0.12	0.09	0.13	0.34		
9	36	0.02	0.01	0.02	0.05		
10	40	0.02	0.02	0.02	0.06		
11	**Total**				**10.71**	**40.00**	**428.58**

Table 10.6 Calculation of the blended labor consumption rate

Based on the uncertainties integrated into the simulation, the histogram derived for the blended labor consumption rate (see Fig. 10.8) shows the range of realistic consumption rates established for rebar installation. Running 50,000 iterations results in a range between about 9.59 and 11.93 wh/t. We obtain a mean blended labor consumption rate of 10.72 wh/t with a standard deviation of 0.31 wh/t. In addition, we find that 90% of the simulation values range from 10.23 wh/t to 11.24 wh/t. There is a 5% probability for obtaining values above or below these limits.

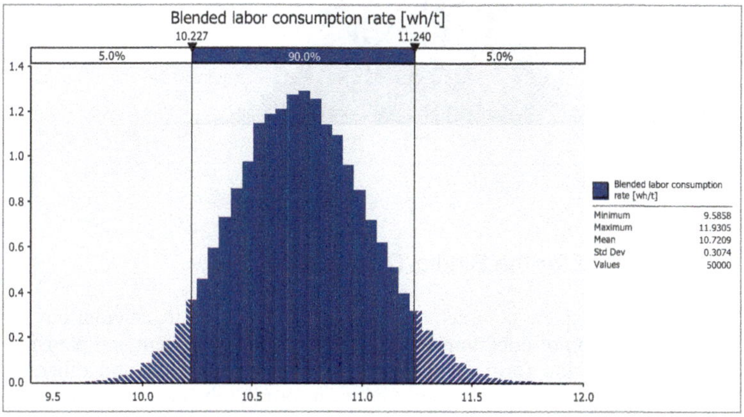

Fig. 10.8 Blended labor consumption rate histogram

Multiplication of the consumption rate by the mean wage price (see Cell F11 of Table 10.6) results in the wage proportion for rebar stated in euros per ton (see Cell G11 of Table 10.6). Based on this calculation, we are again able to generate a wage proportion histogram to be used for further analyses and interpretations (see Fig. 10.9).

We obtain a distribution mean of approximately €429/t while the 5% and 95% wage cost quantiles amount to about €409/t and €450/t, respectively. The standard deviation equals approximately €12.30/t.

In the course of the pricing exercise, the bidder is confronted with the crucial question of which value to select from within the calculated range for the purpose of the bid. This decision is informed by the risk-averse, risk-neutral, or risk-taking attitude of the cost surveyor and, ultimately, of the decision maker as well as by the chance/risk policy of the company and, last but not least, by the overall market situation.

The median of the derived histogram should be used for the wage proportion if a balanced chance/risk ratio (50% chance versus 50% risk) is to be achieved. If an increased risk is taken for this item, it is possible to go down from the mean by one standard deviation (€428.84/t - €12.30/t = €416.54/t). This adjustment would lead to a chance/risk ratio of about 16% (chance) to 84% (risk).

In exceptional cases, it would also be possible to go down to the 5% quantile if there is a disproportionately high urgency for the bidder to win the contract, provided the increased risk can be compensated by other items or projects. This approach would still be plausible while being associated with an excessively high risk (see Section 9.3.3).

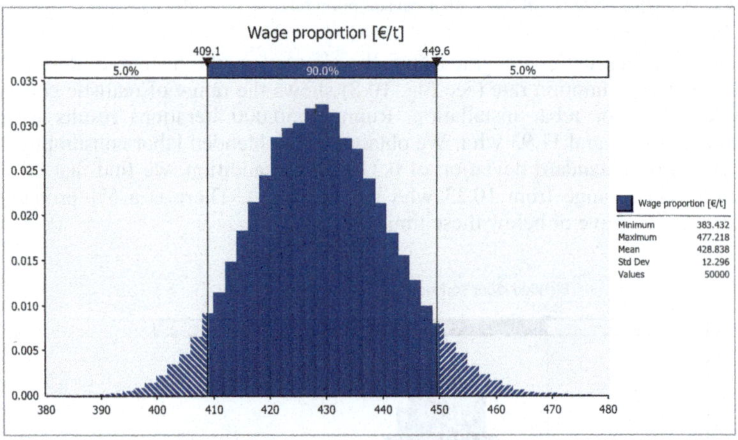

Fig. 10.9 Wage proportion histogram

10.3.1.4 Benefit for the Bidder/Contractor

For the bidder, or subsequent contractor, the benefit of probabilistic calculations lies in the systematic consideration of uncertainties as well as in the transparent and plausible representation of results in the form of histograms. Ranges, quantiles, and other statistical variables can be determined quickly and used for informing the decision-making process.

A specific value is defined on the basis of the corresponding preferred chance/risk ratio despite the fact that empirical values and project-specific factors are integrated into the model or into the selection of distribution functions and the simulation.

Implausible values that lie significantly outside the determined ranges will be excluded at the decision-making stage because the results are represented in a histogram. If values outside the range are selected nonetheless, this decision needs to be justified or deliberately made with a view to the associated chance/risk ratio.

The simulation results and defined input parameters can also enhance the communication and mutual understanding regarding the uncertainties or chances and risks in the course of contractual negotiations with the client.

If the diameter and/or bend shape distribution changes or is detailed further in the project execution phase, the applied labor consumption rates can be used to derive a new histogram to be compared with the original calculation. Retaining the same chance/risk ratio for target and actual figures adds value because it subsequently enables the contractor to claim plausible additional costs (see Section 10.4.12).

10.3.2 Macro-Analysis of Number of Workers

The number of workers is a key variable within the contractor's production system. It influences the number of usable pieces of equipment and has a direct impact on site logistics and mobilization. We perform a macro-analysis of a building construction project (Linz Opera House) in order to demonstrate how to calculate the average and maximum number of workers for reinforced concrete operations. This macro-analysis forms part of PPH 3 (see Fig. 10.10).

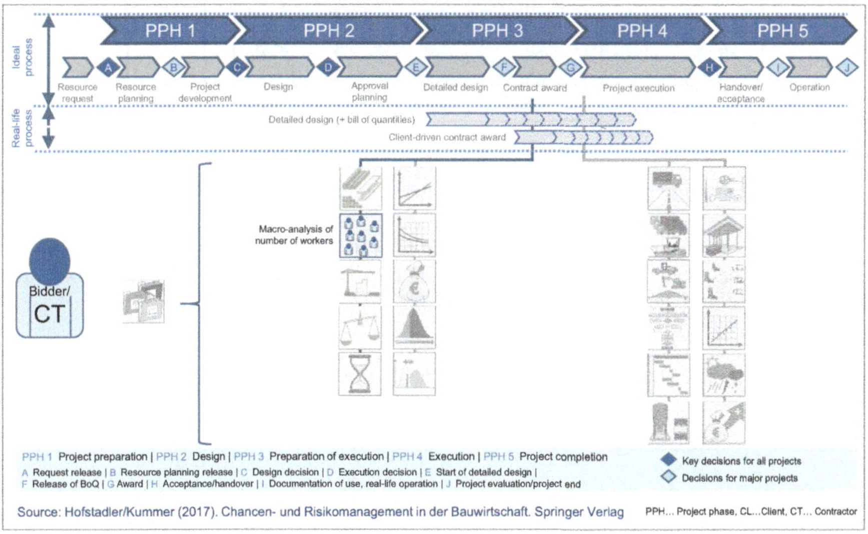

Fig. 10.10 Allocation of example in the project sequence

10.3.2.1 Basics

The number of workers required for reinforced concrete operations primarily depends on the specified concrete volume C_V [m³] and the construction time available for this type of work D_{RCO} [d]. These two variables are used in Eq. (10-1) for calculating the average daily performance value of reinforced concrete operations PV_{RCO} [m³/d].

$$PV_{RCO} = \frac{C_V}{D_{RCO}} \tag{10-1}$$

We determine the work performance by the daily working time WT [h/d], the mean total labor consumption rate of reinforced concrete operations LCR_{RCO} [wh/m³], and the average number of workers WK_{AVG} [wh/h]. Transforming the equation results in the average number of workers according to Eq. (10-2).

$$WK_{AVG} = \frac{PV_{RCO} \cdot LCR_{RCO}}{WT} \tag{10-2}$$

The maximum rather than the average number of workers should be considered as the determining factor particularly when it comes to designing and dimensioning site facilities and calculating the number of key pieces of equipment, such as cranes.

We apply a trapezoidal model for converting the average to the maximum (see Fig. 10.11). The proportional factor for workers f_{WK} [-] represents the relationship between the average and the maximum number of workers.

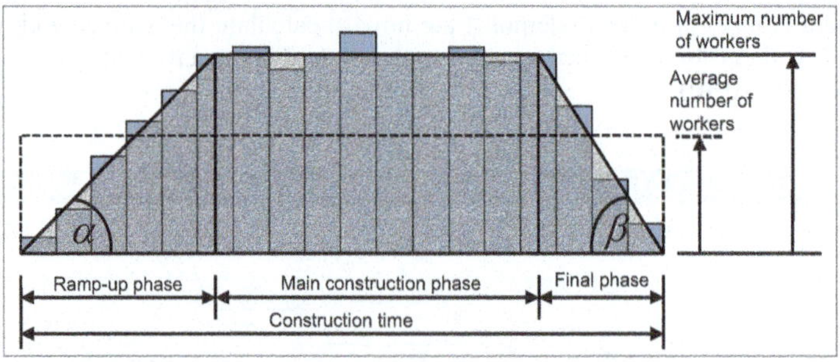

Fig. 10.11 Trapezoidal distribution of workers over the construction period

The proportional factor for workers is applied in Eq. (10-3) to calculate the maximum number of workers. The greater the approximation of the trapezoidal model shown in Fig. 10.11 to a rectangle, the closer the factor f_{WK} will be to 1.00.

$$WK_{MAX} = \frac{WK_{AVG}}{f_{WK}} \tag{10-3}$$

10.3.2.2 Modeling

Fig. 10.12 shows the flow chart for calculating the average or maximum number of workers. The equations included in Section 10.3.2.1 are used for this purpose.

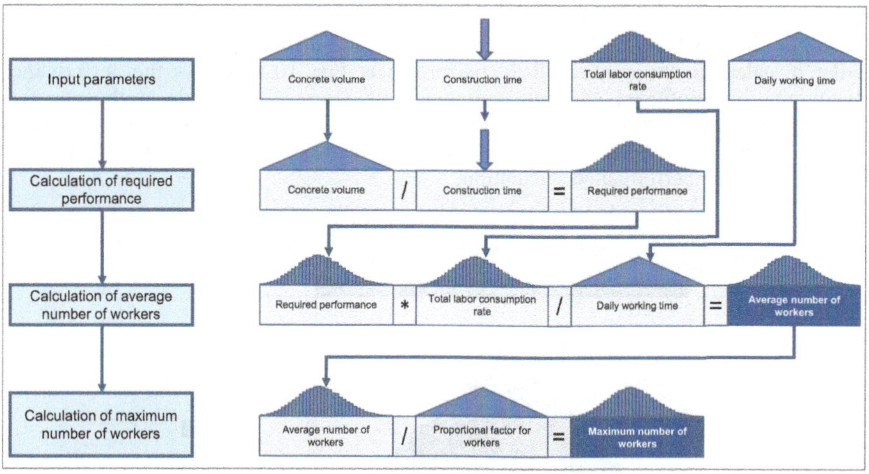

Fig. 10.12 Calculation flow chart

This example refers to the Musiktheater Linz (Linz Opera House), for which a construction time of 400 days was specified as a deterministic value for reinforced concrete operations. A triangular distribution is used for the concrete volume, and a ±10% spread relative to the value most likely for the concrete volume (i.e. 38,275 m³) is identified (see Table 10.7). Daily working time lies between 8.00 and 10.00 h/d, with the most likely value equaling 9.00 h/d (triangular distribution).

We take the total labor consumption rate from the calculation for correlated parameters explained in Section 7.3.3. It is then directly applied in the simulation, as the result of a probabilistic calculation, to determine the number of workers (the distribution is entered in Cell G4 of Table 10.7).[9)]

The proportional factor for workers indicates the relationship between the average and maximum number of workers; it is also represented by a triangular distribution (0.75 | 0.80 | 0.85) (see Row 5 of Table 10.7).

No.	Calculation parameter		Unit	MIN	MOST LIKELY	MAX	Distribution
0	A	B	C	D	E	F	G
1	Concrete volume	C_V	m³	34,447.50	38,275.00	42,102.50	38,275.00
2	Construction time	D_{RCO}	d		400.00		400.00
3	Daily working time	WT	h/d	8.00	9.00	10.00	9.00
4	Total labor consumption rate	LCR_{RCO}	wh/m³				6.10
5	Proportional factor for workers	f_{WK}	-	0.75	0.80	0.85	0.80

Table 10.7 Calculation parameters

[9)] Information on construction time, daily working time, and labor consumption rates provided by the authors.

10.3.2.3 Calculation and Interpretation of Results

We apply Eq. (10-1) and enter the concrete volume and the specified construction time in order to calculate the required daily performance. Fig. 10.13 shows the performance histogram for reinforced concrete operations. Its shape corresponds to the concrete volume (symmetrical triangular distribution) because the construction time was applied as a deterministic value in order to prevent any additional uncertain parameters from being considered in the performance calculation.

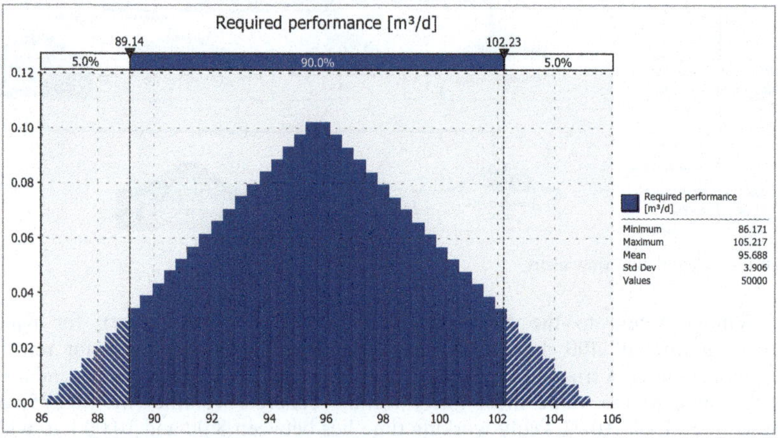

Fig. 10.13 Required daily performance histogram

In the next step, we calculate the average number of workers using Eq. (10-2) by incorporating the total labor consumption rate of reinforced concrete operations as well as the daily working time. Fig. 10.14 shows the result as a histogram. The average amounts to about 69 workers, the standard deviation equals approximately eight workers.

However, during the main construction phase, the maximum rather than the average number of workers constitutes the determining factor for designing and dimensioning site facilities and key pieces of equipment, such as the number of cranes.

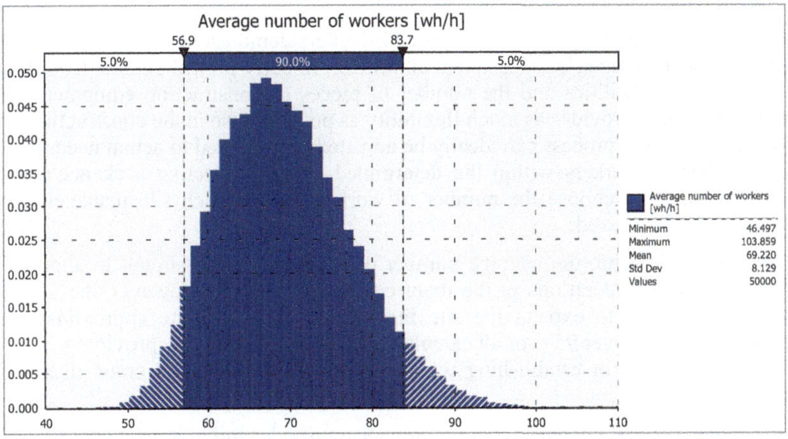

Fig. 10.14 Average number of workers histogram

Dividing the average number of workers by the proportional factor for workers (see Eq. (10-3)) yields the maximum number of workers from the simulation. This result is also shown as a histogram (see Fig. 10.15). The maximum number of workers amounts to about 87 with a standard deviation of approximately 10 workers.

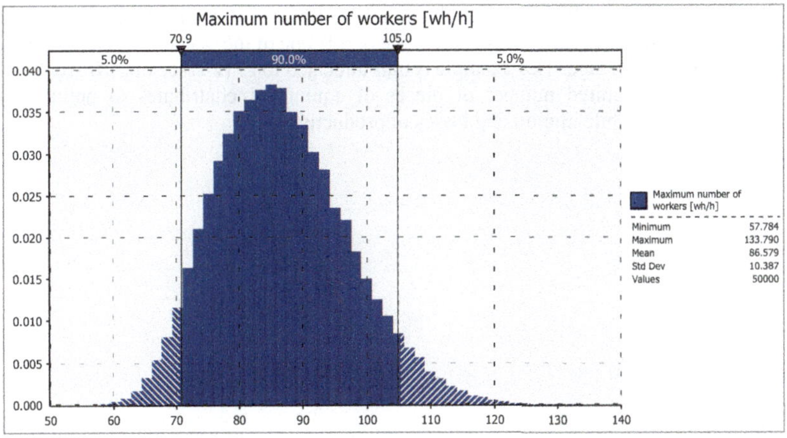

Fig. 10.15 Maximum number of workers histogram

The maximum number of workers ranges from 58 to 134 owing to the uncertainties associated with the concrete volume, the daily working time, the total labor consumption rate of reinforced concrete operations, and the proportional factor for workers.

Any limitation of the ranges of individual input parameters or their inclusion as deterministic values into the calculation will have a direct influence on the spread of results obtained for the maximum number of workers.

10.3.2.4 Benefit for the Bidder/Contractor

The probabilistic calculation of the number of workers demonstrates how large the uncertainties are for the bidder, or subsequent contractor, in early project phases. Coordinating and planning site facilities and the number of pieces of construction equipment should follow a model that provides as much flexibility as possible also in the construction phase. The site mobilization process can ideally be adjusted or upgraded to actual needs in terms of the number of workers within the determined ranges. Selecting a chance/risk ratio makes it possible to choose the number of workers upon which site management and operation should be based.

In the example at hand, the average number of workers would amount to about 87 for reinforced concrete operations in the main construction phase. However, the contractor should also be able to expand the site facilities to accommodate approximately 105 workers in order to cover 95% of all calculated cases, for instance by providing spaces for additional containers or establishing the footprint for an additional crane if and when required.

The macro-analysis of the number of workers also provides initial indications of losses of productivity that may potentially occur owing to insufficient crane capacity or by underrunning the minimum working area.

10.3.3 Calculation of Number of Required Pieces of Construction Equipment

In construction projects, selecting the "right" number of pieces of equipment plays a major part in ensuring a smooth, efficient construction process. It is one of the crucial tasks to be completed at the process planning stage. This example is thus allocated to PPH 3 (see Fig. 10.16). Selecting and providing the required number of pieces of equipment contributes to optimizing the construction process while minimizing losses of productivity.

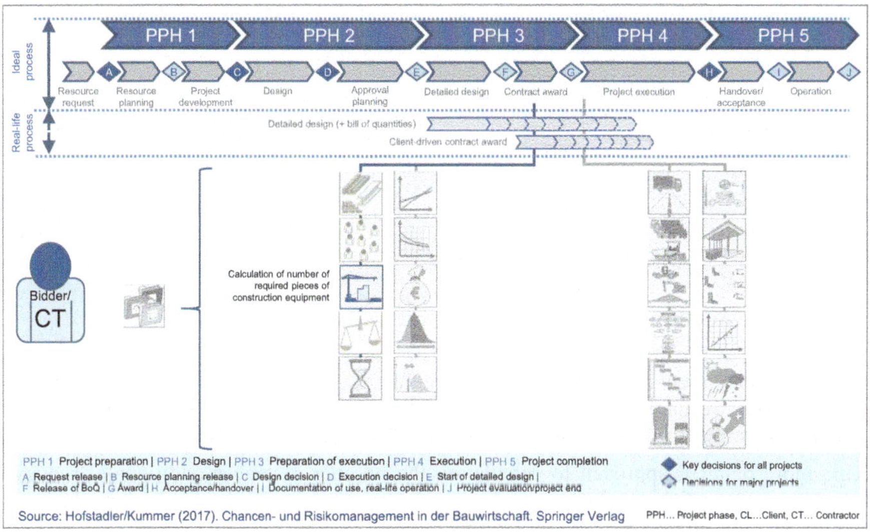

Fig. 10.16 Allocation of example in the project sequence

In building construction, cranes are considered to be key pieces of on-site equipment. They constitute crucial elements of site logistics. We find interdependencies between the specified construction time, the number of workers, and the number of cranes. A comparatively short construction time translates to more workers being present on-site. A larger number of workers usually also requires a larger number of cranes or other lifting equipment.

This example compares two methods for determining the number of cranes. On the one hand, we obtain the number of cranes through the maximum number of workers (based on the example in Section 10.3.2). On the other hand, we perform a more detailed analysis on the basis of crane utilization values. This example refers to reinforced concrete operations in a building construction project.

10.3.3.1 Basics

In the first step, we need to distinguish between the two applied calculation methods in order to identify the basis for determining the number of required cranes. We perform this calculation based on the maximum number of workers and on crane utilization values.[10)]

Deriving the number of cranes from the number of workers

We apply the proportional factor for cranes $f_{C,WK}$ [workers/crane] for calculating the number of required cranes on the basis of the maximum number of workers. This factor indicates the number of workers that can be served by a single crane during reinforced concrete operations. The proportionality factor for cranes depends on the type of crane (top-slewing versus fast-erecting) and the type of concrete placement (i.e. using a crane and skip or pouring the concrete without a crane). Table 10.8 shows the maximum number

[10)] Other methods for determining the number of required cranes start from the building's gross cubic volume, the weight of construction materials, or the application of queueing models.

of workers for each crane type and concrete pouring method. All stated values are averages taken from expert survey data. No unusual non-productive or distribution times should occur if these maximum values are not exceeded.[11)]

Concreting method [-]	Crane and skip for all structural components	Crane only for walls and columns	Crane only for columns	No crane
Number of workers per top-slewing crane	13	16	18	20
Number of workers per fast-erecting crane	10	12	14	15

Table 10.8 Maximum number of workers per crane[12)]

Deriving the number of cranes from crane utilization values

Another option for determining the number of cranes to a greater degree of detail is to base the calculation on crane utilization values. Crane utilization values indicate the required basic available time of a crane per unit of quantity. Table 10.9 lists minimum and maximum crane utilization values for reinforced concrete operations.

This more detailed approach to calculating the number of cranes additionally requires known or estimated quantities per group of structural components, broken down to formwork, reinforcing, and concreting operations.

No.	Type of work/item	Crane utilization values MIN [h/UQ]	MAX [h/UQ]
0	A	B	C
1	**Formwork operations**	[h/m²]	[h/m²]
2	Conventional slab formwork	0.020	0.090
3	Slab tables	0.020	0.030
4	Columns	0.020	0.045
5	Extensive wall units	0.040	0.080
6	Conventional wall formwork	0.020	0.045
7	Beams on tables	0.025	0.035
8	Beams on support frames	0.030	0.040
9	Ribbed floors	0.050	0.090
10	Foundations	0.010	0.020
11	**Reinforcing operations**	[h/t]	[h/t]
12	Mesh	0.300	0.550
13	Rebar	0.200	0.350
14	Total reinforcement	0.240	0.400
15	**Concreting operations**	[h/m³]	[h/m³]
16	Floor slabs	0.060	0.120
17	Foundations	0.050	0.090
18	Walls	0.080	0.150
19	Columns	0.120	0.260

Table 10.9 Maximum and minimum crane utilization values for reinforced concrete operations[13)]

11) Cf. Hofstadler (2014), p. 227ff.

12) Cf. ibid.

The ranges identified for crane utilization values and the corresponding quantities are used to calculate the basic equipment utilization time t_{BE} [h]. Furthermore, the additional utilization time t_{AU} [h] is taken into account in relation to the available equipment time t_E [h] by adding a contingency of 10 to 15%. This contingency permits the inclusion of initially unpredictable transport processes with inaccurate pricing, such as loading and unloading or relocating components for subsequent installation.

An idle time proportion t_I [h] of 30 to 35% relative to the available equipment time should be considered according to empirical evidence because it is usually impossible to prepare an exact schedule for the sequence of individual transport processes, as well as for minimizing waiting periods for workers.[14)]

We obtain the required available equipment time t_E [h] using Eq. (10-4).

$$t_E = t_{BE} + t_{AU} + t_I \tag{10-4}$$

The available equipment time t_E represents the total number of crane hours required for the building or structure for reinforced concrete operations (see the distribution in Fig. 10.17). We then derive the number of required cranes from the scheduled shell construction time and the daily working time.

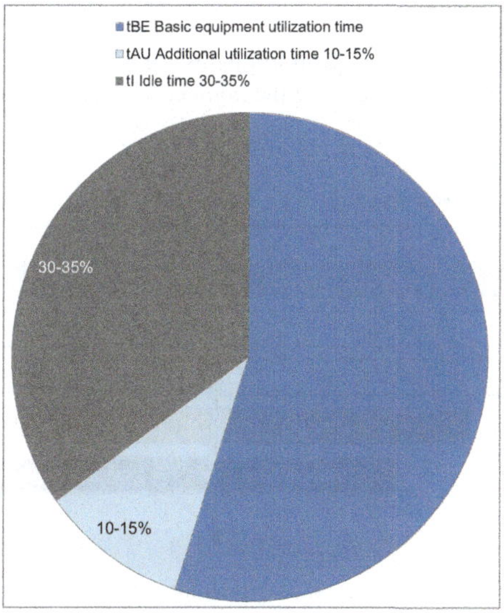

Fig. 10.17 Components of available equipment time t_E

10.3.3.2 Task

Probabilistic calculations are performed for the two selected methods of calculating the number of required cranes, and the results are compared to one another. The calculations

13) Cf. Spranz (2003), p. 113

14) Cf. ibid., p. 114

refer to the Musiktheater Linz (Linz Opera House) building construction project. We use the results of the calculations described in Section 10.3.2 to derive the number of cranes from the number of workers.

10.3.3.3 Modeling

Similarly to our approach chosen in the Basics section, we also make a distinction between the two described methods for calculating the number of cranes (i.e. using the maximum number of workers versus crane utilization values).

Deriving the number of cranes from the number of workers

In this example, we assume the use of top-slewing cranes and concrete pouring without a crane or skip.

Fig. 10.18 shows the usable number of workers per crane in combination with the construction time. In the case of a normal construction time, the values are 10% lower than the maximum number of workers and of the pieces of equipment available for productive use.

We apply a 10% reduction to the maximum proportionality factor for cranes (20 workers/crane) in order not to run the risk of generating losses of productivity owing to insufficient crane capacity.

This is why we use a crane proportionality factor of 18 workers per crane (20 workers/crane * 0.90) for calculating the number of required cranes.

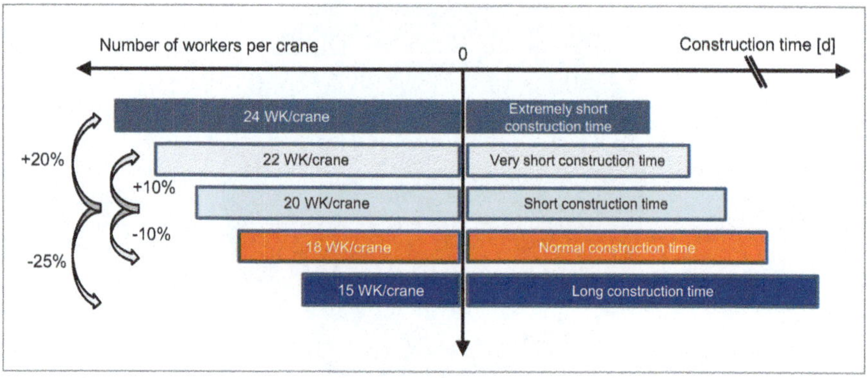

Fig. 10.18 Relationship between the proportional factor for cranes and construction time

Fig. 10.20 shows the flow chart for this calculation. This calculation primarily involves the division of the maximum number of workers calculated probabilistically in Section 10.3.2 by the proportionality factor for cranes (i.e. 18 workers/crane). The result is transformed into a histogram to be used for interpreting the results.

The trapezoidal model (see Fig. 10.19) should be applied in order to arrive at a realistic number of cranes. For this purpose, we first determine the average number of workers required per working day WK_{AVG} [wh/h] and apply the proportionality factor for workers to obtain the maximum number of required workers WK_{MAX} [wh/h].

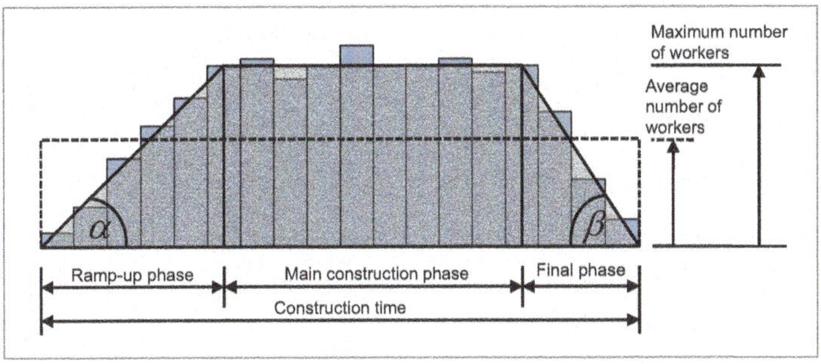

Fig. 10.19 Trapezoidal distribution of workers over the construction period

The individual simulation results can also be rounded to integers and represented in a bar diagram in order to permit a direct conclusion regarding the chance/risk ratio for the integer multiple of the number of cranes.

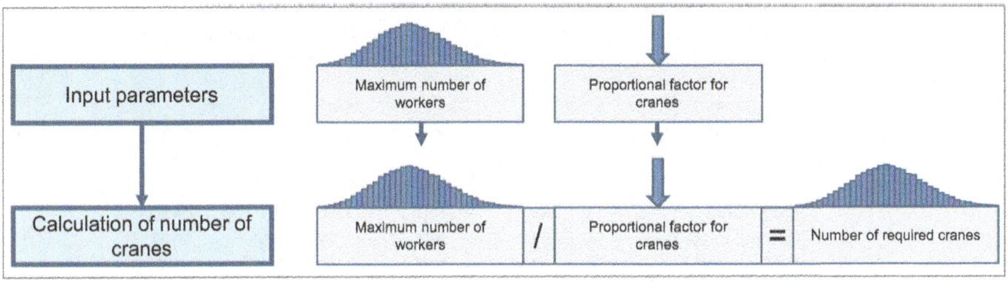

Fig. 10.20 Calculation flow chart – Number of cranes derived from number of workers

Deriving the number of cranes from crane utilization values

The crane utilization times are calculated separately for formwork, reinforcement, and concrete by multiplying the quantities with the corresponding crane utilization values. Their sum is referred to as the basic equipment utilization time t_{BE} [h]. In the next step, the proportions of the additional utilization time t_{AU} [h] and idle time t_I [h] are added to the basic equipment utilization time. This addition results in the available equipment time t_E [h], which can be used to determine the number of required cranes. Other parameters needed to perform this operation include construction time, daily working time, and a factor to consider the trapezoidal distribution of resources over the construction period.

No.	Structural components	Formwork area			
		MIN	MOST LIKELY	MAX	Distribution
		[m²]	[m²]	[m²]	[m²]
0	A	B	C	D	E
1	Foundation slab	1,070	1,150	1,231	1,150
2	Columns	3,441	3,700	3,959	3,700
3	Walls	75,330	81,000	86,670	81,000
4	Floor slabs	27,900	30,000	32,100	30,000
5	Beams, joists	5,952	6,400	6,848	6,400
6	Other slabs	1,581	1,700	1,819	1,700
7	Stairs	465	500	535	500
8	Parapets	2,604	2,800	2,996	2,800
9	**Total**	**118,343**	**127,250**	**136,158**	**127,250**

Table 10.10 Quantities – Formwork

No.	Structural components	Reinforcement quantity			
		MIN	MOST LIKELY	MAX	Distribution
		[t]	[t]	[t]	[t]
0	A	B	C	D	E
1	Foundation slab	1,395	1,500	1,620	1,500
2	Columns	112	120	130	120
3	Walls	1,395	1,500	1,620	1,500
4	Floor slabs	1,349	1,450	1,566	1,450
5	Beams, joists	151	162	175	162
6	Other slabs	93	100	108	100
7	Stairs	19	20	22	20
8	Parapets	74	80	86	80
9	**Total**	**4,587**	**4,932**	**5,327**	**4,932**

Table 10.11 Quantities – Reinforcement

No.	Structural components	Concrete volume			
		MIN	MOST LIKELY	MAX	Distribution
		[m³]	[m³]	[m³]	[m³]
0	A	B	C	D	E
1	Foundation slab	9,024	9,600	10,272	9,600
2	Columns	376	400	428	400
3	Walls	14,570	15,500	16,585	15,500
4	Floor slabs	9,776	10,400	11,128	10,400
5	Beams, joists	1,297	1,380	1,477	1,380
6	Other slabs	376	400	428	400
7	Stairs	85	90	96	90
8	Parapets	475	505	540	505
9	**Total**	**35,979**	**38,275**	**40,954**	**38,275**

Table 10.12 Quantities – Concrete

Tables 10.10 to 10.12 list the expected quantities for reinforced concrete components separately for formwork, reinforcement, and concrete for the individual groups of structural components. Columns B, C, and D show the minimum, most likely, and maximum quantities of formwork, reinforcement, and concrete. The minimum and maximum values are derived from the quantity deviations taken from an expert survey (relative to the M-estimators for private-sector clients; see Section 7.2.2). To represent the quantity uncertainties, we apply a simplified model using triangular distributions. These distributions are entered in Column E (the tables show the respective most likely value). Fig. 10.22 shows the flow chart for this calculation.

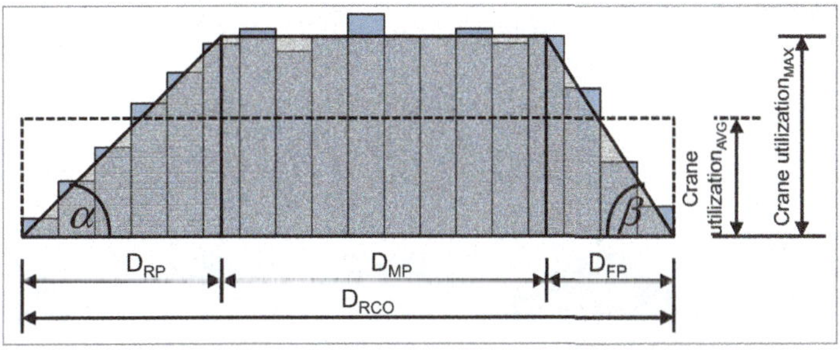

Fig. 10.21 Trapezoidal distribution of crane utilization over the construction period

The trapezoidal model of the distribution of required crane hours (see Fig. 10.21) should be applied in order to arrive at a realistic number of cranes. For this purpose, we first determine the average total number of crane hours required per working day and then consider the proportion of the main construction phase in order to obtain the maximum number of crane hours per working day.

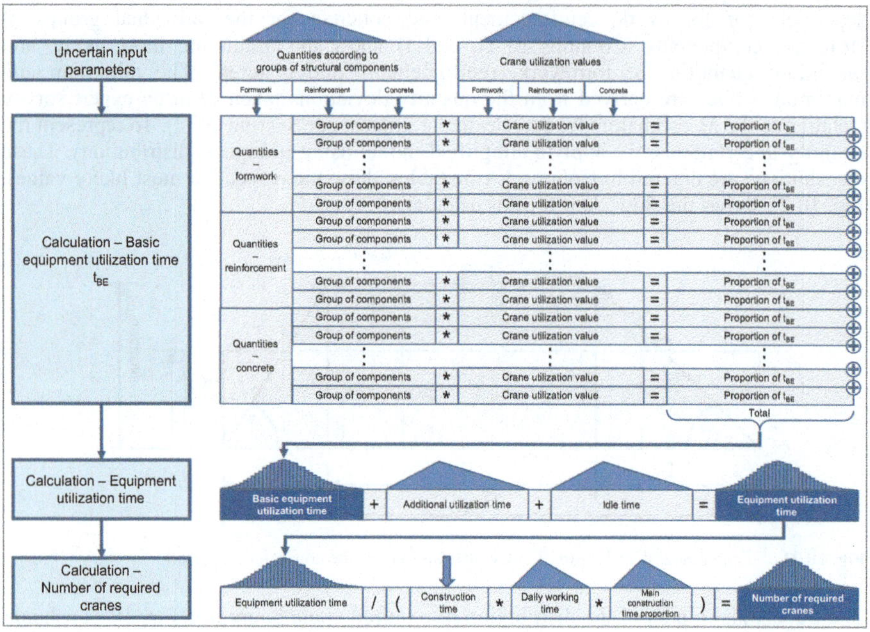

Fig. 10.22 Calculation flow chart – Number of cranes derived from crane utilization values

10.3.3.4 Calculation and Interpretation of Results

The calculations and interpretations of results are shown separately for each of the two applied methods.

Deriving the number of cranes from the number of workers

To determine the number of required cranes, we divide the probabilistically calculated maximum number of workers (see Fig. 10.15) by the proportionality factor for cranes (maximum reduced by 10%: 20 workers/crane * 0.90 = 18 workers/crane). Fig. 10.23 shows the result of this calculation as a histogram. This graphical representation proves that this calculation results not only in even numbers. If we set the slider at the top of the diagram to five cranes, for instance, this scenario would cover approximately 65.5% of all simulated cases.

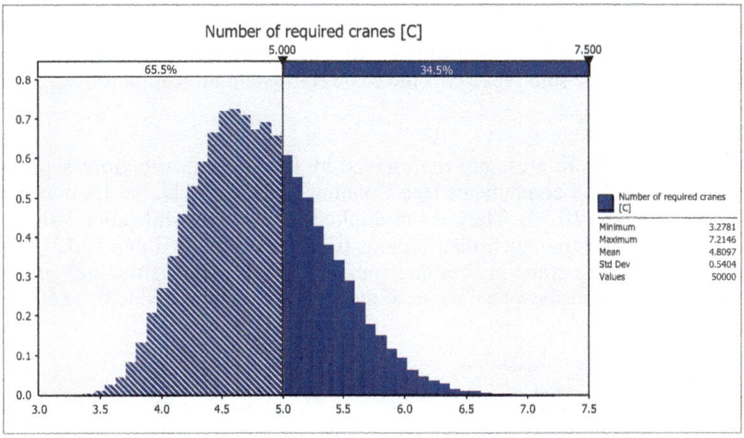

Fig. 10.23 Number of required cranes histogram – Derived from number of workers

Each of the individual results is rounded to an integer value after each iteration and represented in a bar chart (see Fig. 10.24) in order to merge the decimals obtained in the calculation to integer multiples of the number of cranes. We find that, in this representation, five cranes would even cover more than 88% of the calculated cases. This is due to the fact that values between 5.00 and 5.49 are also rounded down to five cranes in the integer (discretized) representation. Overall, the value range from 4.50 to 5.49 (rounded to 5.00 in each case) is found to be occurring most frequently. This result refers to the relatively rough estimate of the number of cranes derived from the number of workers.

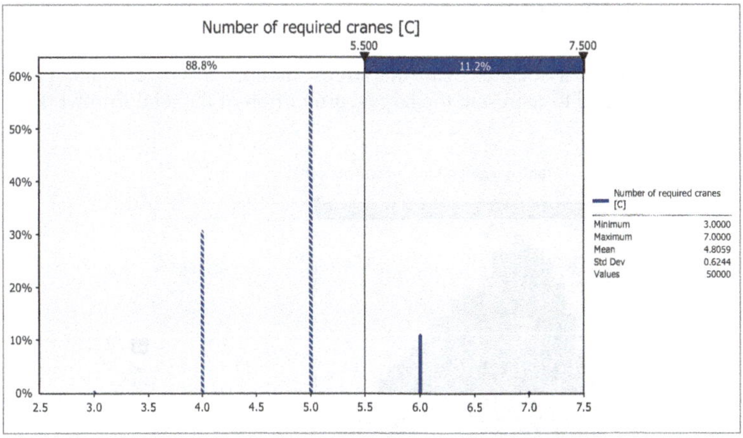

Fig. 10.24 Bar diagram for number of required cranes – Derived from number of workers

Deriving the number of cranes from crane utilization values

The crane utilization times are calculated and added up separately for formwork, reinforcement, and concrete by multiplying the quantities with the corresponding crane utilization values. Their sum represents the basic equipment utilization time t_{BE} [h].

Formwork

The individual formwork areas are represented by triangular distributions separately for each group of structural components (see Columns B to E in Table 10.13; corresponding to the values in Table 10.10). They are multiplied by the crane utilization values, which are also assumed to be triangular distributions (Columns F to I in Table 10.13). Column J is used to calculate the crane utilization times for each group of structural components. Each cell in this column shows the mean of the distribution allocated to this cell.

No.	Structural components	Formwork area				Crane utilization values				Crane utilization time
		MIN	MOST LIKELY	MAX	Distribution	MIN	MOST LIKELY	MAX	Distribution	Distribution
		[m²]	[m²]	[m²]	[m²]	[h/m²]	[h/m²]	[h/m²]	[h/m²]	[h]
0	A	B	C	D	E	F	G	H	I	J
1	Foundation slab	1,070	1,150	1,231	1,150	0.010	0.015	0.020	0.015	17.25
2	Columns	3,441	3,700	3,959	3,700	0.020	0.030	0.045	0.030	111.00
3	Walls	75,330	81,000	86,670	81,000	0.040	0.050	0.080	0.050	4,050.00
4	Floor slabs	27,900	30,000	32,100	30,000	0.020	0.025	0.030	0.025	750.00
5	Beams, joists	5,952	6,400	6,848	6,400	0.020	0.030	0.045	0.030	192.00
6	Other slabs	1,581	1,700	1,819	1,700	0.020	0.060	0.090	0.060	102.00
7	Stairs	465	500	535	500	0.050	0.070	0.090	0.070	35.00
8	Parapets	2,604	2,800	2,996	2,800	0.020	0.030	0.050	0.030	84.00
9	**Total**	**118,343**	**127,250**	**136,158**	**127,250**					**5,341.25**

Table 10.13 Calculation of crane utilization time – Formwork

Cell J9 includes the sum of crane utilization times for formwork as a histogram; see Fig. 10.25. There is a 90% probability for the crane utilization time for formwork to lie between 4,887 and 7,186 hours. The probability of occurrence of a crane utilization time of more than 7,186 hours equals 5%; the same 5% probability was identified for a crane utilization time of fewer than 4,887 hours.

Formwork for walls is associated with the largest number of crane hours. This type of formwork is also found to represent the largest proportion of the total formwork quantity.

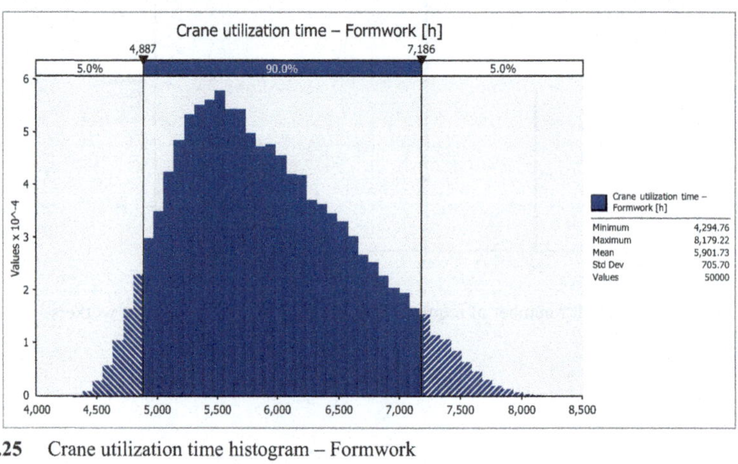

Fig. 10.25 Crane utilization time histogram – Formwork

Reinforcement

The individual reinforcement quantities are represented by triangular distributions separately for each group of structural components (see Columns B to E in Table 10.14; corresponding to the values in Table 10.11). They are multiplied by the crane utilization values, which are also assumed to be triangular distributions (Columns F to I in Table 10.14).

No.	Structural components	Reinforcement quantity				Crane utilization values				Crane utilization time
		MIN	MOST LIKELY	MAX	Distribution	MIN	MOST LIKELY	MAX	Distribution	Distribution
		[t]	[t]	[t]	[t]	[h/t]	[h/t]	[h/t]	[h/t]	[h]
0	A	B	C	D	E	F	G	H	I	J
1	Foundation slab	1,395	1,500	1,620	1,500	0.200	0.400	0.550	0.400	600.00
2	Columns	112	120	130	120	0.200	0.250	0.350	0.250	30.00
3	Walls	1,395	1,500	1,620	1,500	0.250	0.300	0.400	0.300	450.00
4	Floor slabs	1,349	1,450	1,566	1,450	0.300	0.400	0.550	0.400	580.00
5	Beams, joists	151	162	175	162	0.200	0.300	0.350	0.300	48.60
6	Other slabs	93	100	108	100	0.300	0.350	0.550	0.350	35.00
7	Stairs	19	20	22	20	0.200	0.300	0.350	0.300	6.00
8	Parapets	74	80	86	80	0.200	0.250	0.350	0.250	20.00
9	Total	4,587	4,932	5,327	4,932					1,769.60

Table 10.14 Calculation of crane utilization time – Reinforcement

Cell J9 includes the sum of crane utilization times for reinforcement as a histogram, which is shown in Fig. 10.26. There is a 90% probability for the crane utilization time for reinforcement to lie between 1,568 and 2,039 hours. The probability of occurrence of a crane utilization time of more than 2,039 hours equals 5%; the same 5% probability was identified for a crane utilization time of fewer than 1,568 hours.

The largest number of crane hours for reinforcement is associated with the foundation slab, followed by floor slabs and walls.

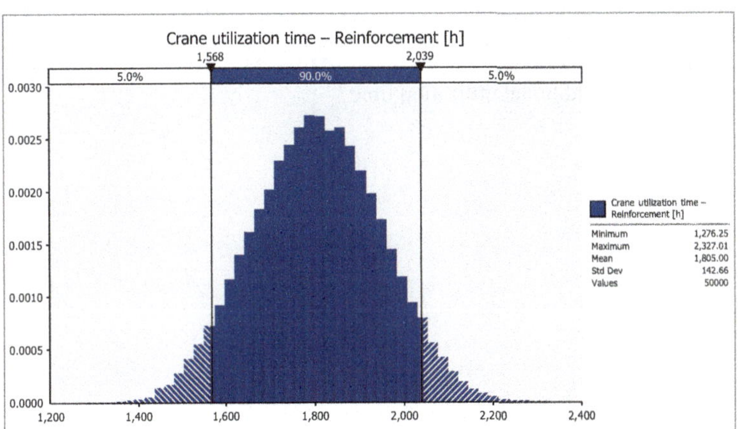

Fig. 10.26 Crane utilization time histogram – Reinforcement

Concrete

The individual concrete volumes are represented by triangular distributions separately for each group of structural components (see Columns B to E in Table 10.15; corresponding to the values in Table 10.12). They are multiplied by the crane utilization values, which are also assumed to be triangular distributions (Columns F to I in Table 10.15). In addition, Column J shows a deterministic value indicating the percentage of the respective concrete volume per structural component group poured with the aid of a crane and skip. In this example, we assume that no concrete is placed via a crane or skip.

In the next step, the crane utilization times per structural component group are calculated in Column K by multiplying the concrete volume with the crane utilization value and the percentage of the concrete volume poured through a skip.

Since all of the percentages shown in Column J are set to zero, we obtain no crane utilization time for concreting operations (transport for vibrators etc. belongs in the additional utilization time category).

No.	Structural components	Concrete volume				Crane utilization values				Volume concreted with crane and skip	Crane utilization time
		MIN	MOST LIKELY	MAX	Distribution	MIN	MOST LIKELY	MAX	Distribution	det. value	Distribution
		[m³]	[m³]	[m³]	[m³]	[h/m³]	[h/m³]	[h/m³]	[h/m³]	[%]	[h]
0	A	B	C	D	E	F	G	H	I	J	K
1	Foundation slab	9,024	9,600	10,272	9,600	0.050	0.060	0.090	0.060	0.00	0.00
2	Columns	376	400	428	400	0.120	0.200	0.260	0.200	0.00	0.00
3	Walls	14,570	15,500	16,585	15,500	0.080	0.100	0.150	0.100	0.00	0.00
4	Floor slabs	9,776	10,400	11,128	10,400	0.060	0.080	0.120	0.080	0.00	0.00
5	Beams, joists	1,297	1,380	1,477	1,380	0.080	0.120	0.150	0.120	0.00	0.00
6	Other slabs	376	400	428	400	0.060	0.090	0.120	0.090	0.00	0.00
7	Stairs	85	90	96	90	0.070	0.100	0.140	0.100	0.00	0.00
8	Parapets	475	505	540	505	0.080	0.110	0.150	0.110	0.00	0.00
9	**Total**	**35,979**	**38,275**	**40,954**	**38,275**						**0.00**

Table 10.15 Calculation of crane utilization time – Concrete

The sum of the calculated crane utilization times for formwork, reinforcement, and concrete constitutes the basic equipment utilization time t_{BE} [h] (see Fig. 10.27) to which additional utilization times t_{AU} [h] and idle times t_I [h] are added to obtain the available equipment time t_E [h]. *Spranz*[15)] states a range of 30 to 35% for idle time and a proportion of 10 to 15% for the additional utilization time.

[15)] Cf. Spranz (2003), p. 114

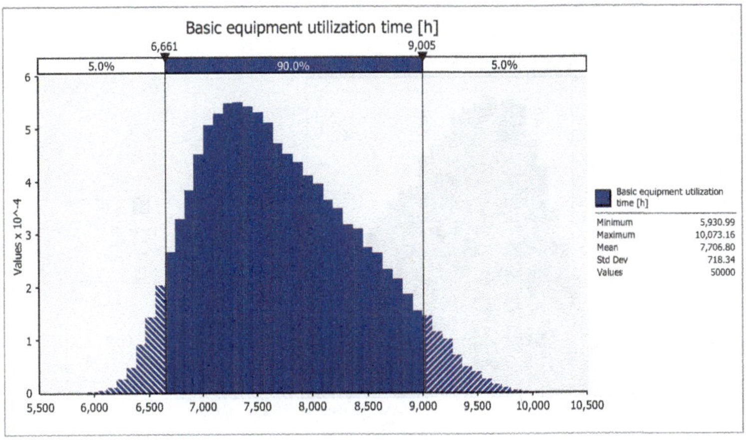

Fig. 10.27 Basic equipment utilization time histogram

Triangular distributions are defined for these two equipment time components (see Table 10.16, Columns C, D, E, and F). We obtain the proportion of the basic equipment utilization time t_{BE} by performing a back-calculation to 100% for the available equipment time t_E. If, for instance, a proportion of 12.50% and a proportion of 32.50% are randomly assumed in an iterative step of the simulation for the additional utilization time and idle time components (corresponding to the most likely values), respectively, the share of the basic equipment utilization time would equal 55% (100% - 12.5% - 32.5%). We subsequently calculate an additional utilization time of 1,616.10 h (7,110.85 h * 12.50% / 55.00%) and an idle time of 4,201.87 h (7,110.85 h * 32.50% / 55.00%).

No.	Time components		Time proportion MIN	MOST LIKELY	MAX	Distribution	Distribution
	[-]	[-]	[%]	[%]	[%]	[%]	[h]
0		B	C	D	E	F	G
1	Basic equipment utilization time	t_{BE}				55.00	7,110.85
2	Additional utilization time	t_{AU}	10.00	12.50	15.00	12.50	1,616.10
3	Idle time	t_I	30.00	32.50	35.00	32.50	4,201.87
4	**Available equipment time (= $t_{AU} + t_I + t_{BE}$)**	t_E				**100.00**	**12,928.82**

Table 10.16 Calculation of available equipment time

The sum of the basic equipment utilization time, additional utilization time, and idle time components constitutes the available equipment time t_E [h] that is the key input parameter for calculating the number of required cranes (Cell G4 in Table 10.16). Fig. 10.28 shows the histogram of the available equipment time.

There is a 90% probability for the available equipment time to lie between 12,031 and 16,480 hours. The probability of occurrence of an available equipment time of more than 16,480 hours equals 5%; the same 5% probability was identified for an available equipment time of fewer than 12,031 hours.

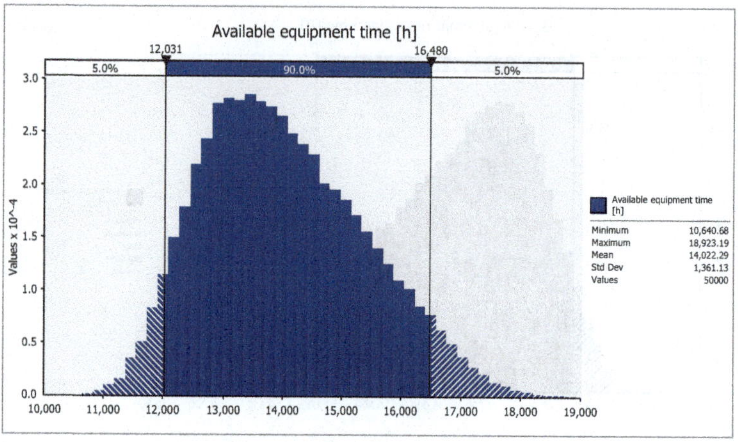

Fig. 10.28 Available equipment time histogram

The construction time for reinforced concrete operations and the daily working time must be known in order to derive the number of required cranes from the available equipment time. In addition, the share of the main construction time in the total construction time must be identified. This approach considers a trapezoidal distribution of resources over the construction period.

A construction time of 400 days is given as a deterministic value. We define the daily working time and the proportion of the main construction time as triangular distributions (see Rows 3 and 4 in Table 10.17).

The basic equipment utilization time is divided by the product of construction time, daily working time, and the main construction time proportion in order to determine the number of required cranes. This calculation is performed in Cell F5 of Table 10.17. Fig. 10.29 shows the related histogram.

No.	Item	Unit	MIN	MOST LIKELY	MAX	Distribution
0	A	B	C	D	E	F
1	Available equipment time t_E	[h]				12,928.82
2	Construction time	[d]				400.00
3	Daily working time	[h/d]	8.00	9.00	10.00	9.00
4	Proportion of main construction time	[%]	75.00	80.00	85.00	80.00
5	**Number of required cranes**	**[C]**				**4.49**

Table 10.17 Calculation of number of required cranes

The histogram displayed in Fig. 10.29 shows no clear result for the number of required cranes, but the output is found to be similar to deriving the number of cranes from the number of workers. If we set the slider at the top of the diagram to five cranes, this scenario would cover approximately 61% of all simulated cases.

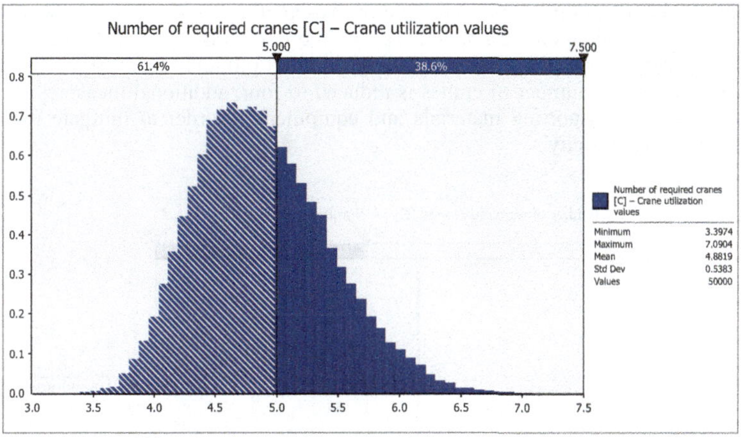

Fig. 10.29 Number of required cranes histogram

Each of the individual results is rounded to an integer multiple after each iteration and represented in a bar chart (see Fig. 10.30) in order to merge the decimals obtained in the calculation to integer multiples of the number of cranes. We find that, in this representation, five cranes would cover more than 86% of the calculated cases (including the rounding down of values of up to 5.49). The value range from 4.50 to 5.49 (rounded to 5.00 in each case) is found to be occurring most frequently. This result refers to the detailed calculation using crane utilization values and assuming that no concrete is poured with the aid of a crane and skip.

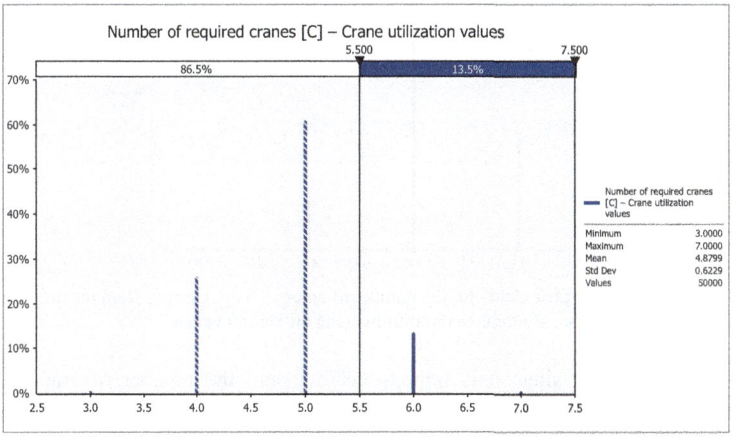

Fig. 10.30 Bar diagram for number of required cranes – Derived from crane utilization values

In this example, comparing the two methods (i.e. starting from the maximum number of workers versus crane utilization values) reveals a very strong consistency between the two approaches (see Fig. 10.31). For a chance/risk ratio of about 86% (chance) to 14% (risk), five top-slewing cranes should be used for reinforced concrete operations in the building construction project considered in this chapter. It appears advisable to provide a certain footprint for the temporary use of a mobile or fast-erecting crane in the course of the site mobilization (depending on the available space) in order to also cover the remaining cases and to be prepared for peaks while preventing losses of productivity owing to insufficient crane capacity. If the number of cranes is reduced to four, additional measures should be implemented for transporting materials and equipment in order to mitigate the risk of insufficient crane capacity.

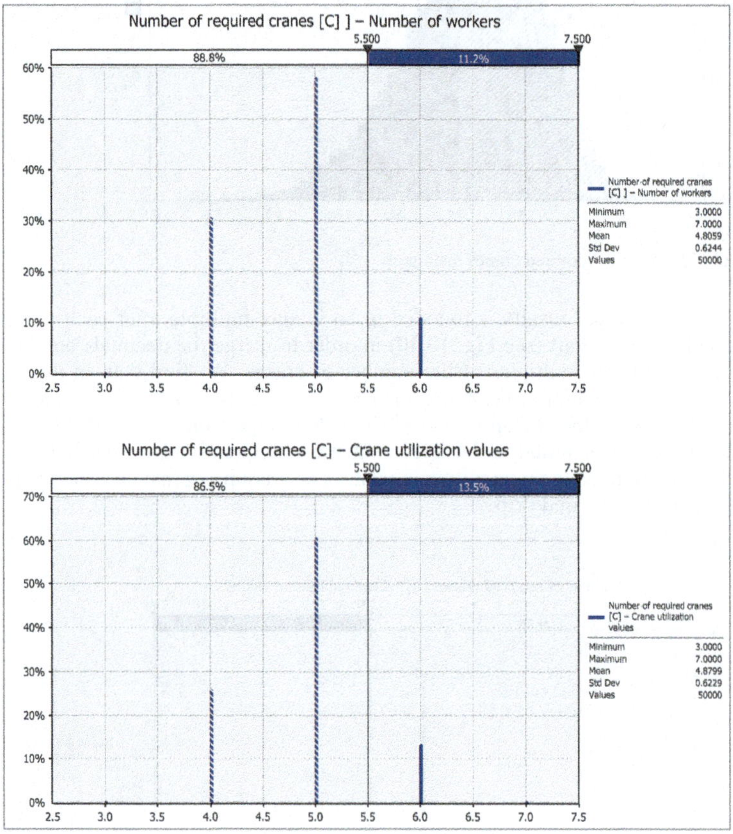

Fig. 10.31 Comparison of bar charts for the number of cranes – Top: Derived from the maximum number of workers – Bottom: Derived from crane utilization values

On many construction sites, it is impossible to install the number of cranes actually required from a construction management point of view. This shortage may be due to the shape and dimensions of the plot as well as to a lack of available space. In the case at hand, installation of only four cranes would increase the risk of losses of productivity and of non-adherence to the scheduled time to completion. The modified chance/risk ratio for using only four cranes is shown in Fig. 10.32 for the calculation method relying on crane utilization values. The two histograms refer to the same simulation output, but they are shown as a continuous distribution (top) and as a discrete distribution (bottom).

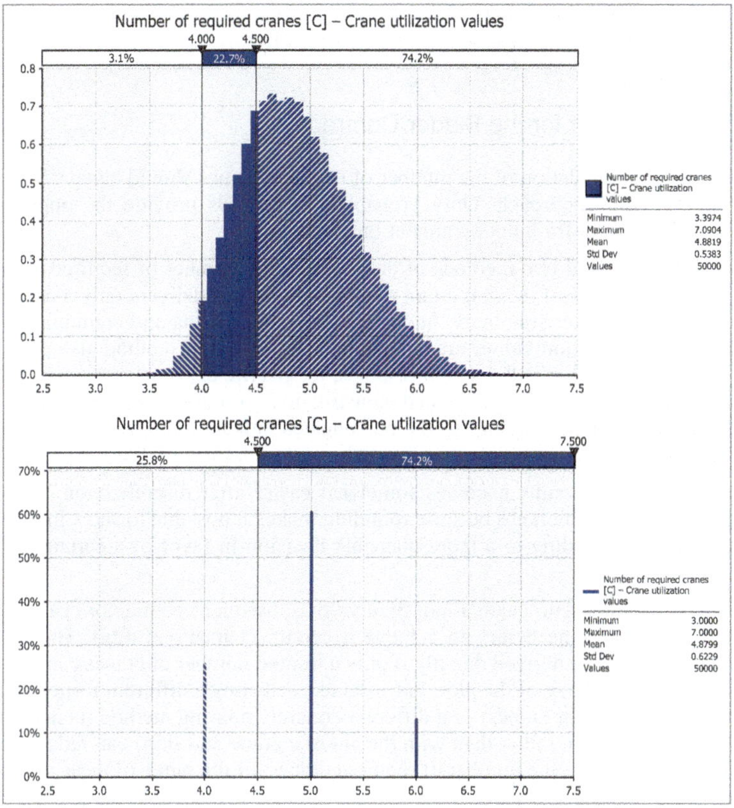

Fig. 10.32 Number of required cranes histogram – Chance/risk ratio for four cranes – Derived from crane utilization values

The continuous histogram reveals that the chance and the risk amount to 3% and 97% (22.7% + 74.2%), respectively, for four cranes (see Fig. 10.32, top). The discrete distribution indicates a chance of about 26% for requiring four or fewer cranes, while the risk of requiring more than four cranes approximates 74% (see Fig. 10.32, bottom).

This difference in the chance/risk ratios is due to the fact that values between 4.00 and 4.49 are rounded down in the discrete histogram (bottom), and their frequency is added to the bar representing four cranes. If we set the slider to 4.5 in the upper histogram, we also obtain a chance/risk ratio of approximately 26% (3.1% + 22.7%) to 74% for the continuous distribution.

Generally speaking, comparing the two histograms demonstrates that the discretization of continuous distributions facilitates the process and proves to be beneficial for the interpretation and communication of results in the case of parameters for which the decision must be made in favor of an integer value because of their indivisibility, including the number of cranes. The frequencies of the individual simulation results captured as decimals are merged for the corresponding integer values. This step is performed by the "ROUND" command available in Excel.

Overall, the result of the calculation shows that the use of only four cranes is associated with a considerably higher risk (74%) of insufficient crane capacity (see Fig. 10.32) compared to the five-crane scenario (13.5%; see Fig. 10.30).

10.3.3.5 Benefit for the Bidder/Contractor

The probabilistic calculation of the number of required cranes should always be preferred over a deterministic approach. Only probabilistic methods provide the opportunity to systematically account for input parameter uncertainties.

This example described two methods of determining the number of required cranes. The calculation on the basis of the maximum number of workers delivers only a rough approximation but also greater simplicity. In contrast, more input data and computing power is needed if crane utilization values are used. However, the latter method also provides the benefit of a more flexible and detailed response to specific circumstances encountered in the construction phase. The scheduled construction process can be reflected in the simulation very accurately thanks to the percentages indicated for the proportions of the individual groups of structural components to be cast with the aid of a crane and skip.

The interpretation of results becomes somewhat easier after discretization of the histograms for the number of cranes because rounding makes it possible to show integer values in a bar chart, thus leading to a more plausible decision in favor of a certain number of cranes.

Based on the probabilistic calculation, bidders or subsequent contractors can determine the number of cranes necessary to achieve a certain chance/risk ratio. The associated chance/risk ratio can be derived directly if only a limited number of cranes can be installed owing to the shape or size of the plot. For instance, selecting a different formwork system (eliminating the need for cranes) or a different concrete pouring method (concreting via a pump and placing boom rather than with the aid of a crane and skip) can reduce the crane utilization time and thus the number of required cranes, or the range of these numbers.

10.3.4 Register of Chances and Risks

Particularly in conventional risk management, the aim of assessing a risk (see Chapter 4 for the description of the chance/risk management process) is to obtain an overview of possible cases of damage (i.e. the consequences of risks after they materialize) and to allocate appropriate funds to cover these expected risks or to implement risk mitigation measures (prevention, reduction, diversification, and/or transfer). This approach thus allows for an estimate of the required risk capital or of the expected investment amount.

The example discussed below demonstrates how to assess individual chances and risks applying the so-called practitioner method while performing Monte Carlo simulations on the macro-level (PPH 3; see Fig. 10.33).

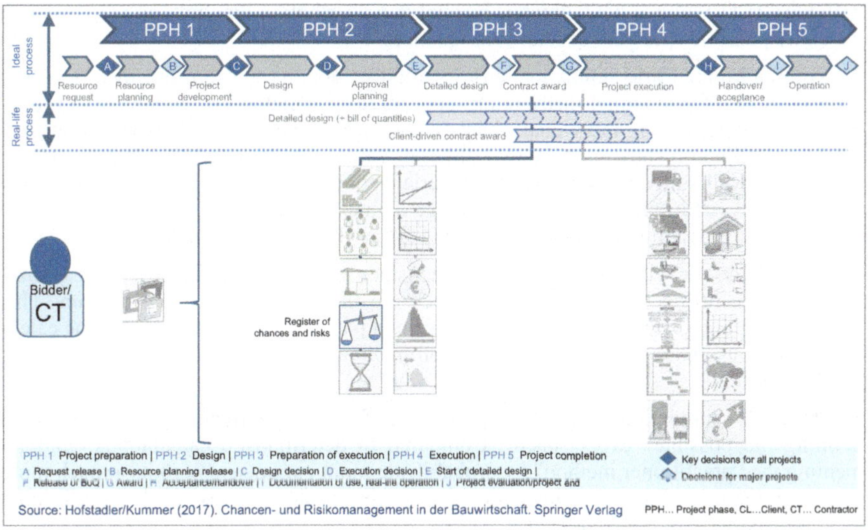

Fig. 10.33 Allocation of example in the project sequence

10.3.4.1 Initial Situation

It is often impossible to carry out a detailed calculation of risks (or chances) because macro-analyses are usually performed in early project phases. We thus generally resort to the "practitioner method", where the expected probability of occurrence is multiplied by the assumed consequences of a defined risk (see Eq. (10-5)).

$$\text{Risk} = \text{Probability of occurrence} \cdot \text{Consequence} \tag{10-5}$$

This equation can be modified accordingly for analyzing chances:

$$\text{Chance} = \text{Probability of occurrence} \cdot \text{Benefit} \tag{10-6}$$

Particularly at an early project stage or when performing an initial analysis of a new project, we are confronted with the difficulty of having to define individual (deterministic) figures for probabilities of occurrence and the related consequences or benefits. In this regard, we refer to Section 2.1.2 and more specifically to Fig. 2.2.

When performing a deterministic assessment of chances and risks applying the "practitioner method", we are faced with the problem that risks with a low probability but severe consequence exhibit the same risk value as risks associated with only a minor consequence but high probability of occurrence (see Fig. 2.2). This means that we arrive at an identical assessment for two risks whose structure and characteristics are completely different. However, risks associated with a severe consequence may require different corrective or remedial actions compared to risks associated with a high probability of occurrence but minor consequence.[16)]

Any probabilistic analysis involves the option of representing the consequences of individual risks within a defined range by applying distribution functions so as to consider the entire possible spectrum of total consequences.

As part of a Monte Carlo simulation, a Bernoulli distribution (or a zero-one distribution as a special case of a binomial distribution) appears to be suitable for modeling the probability of occurrence of individual chances and risks. The Bernoulli distribution is a discrete distribution showing only two different values, i.e. 0 or 1. This characteristic is particularly well-suited to modeling the materialization of a risk on the basis of a defined probability of occurrence. Stating a percentage defines the probability of occurrence (= 1) or the counterprobability of non-occurrence (= 0) for the considered chance or risk.[17)]

Any simulation run will make it possible to consider several thousands of possible combinations of occurring chances or risks, which are then merged into a histogram. This approach will integrate the inevitably existing input parameter uncertainties into the calculation. We thus create the basis for a decision-making process that starts from a specific chance/risk ratio.

Monte Carlo simulations provide significant additional information with a view to defining the possible "event space". Compared to deterministic calculations, complementing the "practitioner method" by probabilistic input parameters permits a much more realistic reflection of the assessment situation in early project phases or when performing an initial analysis for a new project.

For the assessment (macro-analysis) of chances and risks, we use the concepts of benefit or consequence to refer to a positive or negative effect expressed in monetary units that will occur when the chance or risk materializes.

10.3.4.2 Modeling

The probability of occurrence is modeled as a Bernoulli distribution for adding a probabilistic component to the "practitioner method". This distribution can only take two values, i.e. 0 or 1. A value of zero indicates that the risk or chance will not materialize. In contrast, a value of one represents the occurrence of the chance or risk in the considered iteration. Depending on the specific probability of occurrence, the simulation will select either zero (probability of occurrence of less than 50%) or one (probability of occurrence greater than 50%) in the majority of cases. At a probability of occurrence of exactly 50%, the values of one and zero will occur randomly, but equally frequently in the simulation.

Modeling probabilities of occurrence using Bernoulli distributions thus makes it possible to derive histograms reflecting the number of simultaneously occurring chances and risks.

Triangular distributions are applied in order to assess individual chances and risks for their benefits and consequences. We use these distributions as input parameters for the following example. We then calculate the risk or chance value per each individual risk or chance by multiplying the random value taken from the relevant Bernoulli distribution (0 or 1) by a random consequence or benefit within the defined ranges.

Chance and risk values are then transformed into histograms taking a value of either zero (for all cases in which the chance or risk has not materialized) or being equivalent to the defined benefit or consequence.

[16)] In this context, the assessment of chances is associated with similar problems.

[17)] Cf. Kummer (2014), p. 178

The sums of the chance or risk values can also be shown as histograms available for further interpretation and decision-making processes. Fig. 10.34 shows the flow chart for this calculation.

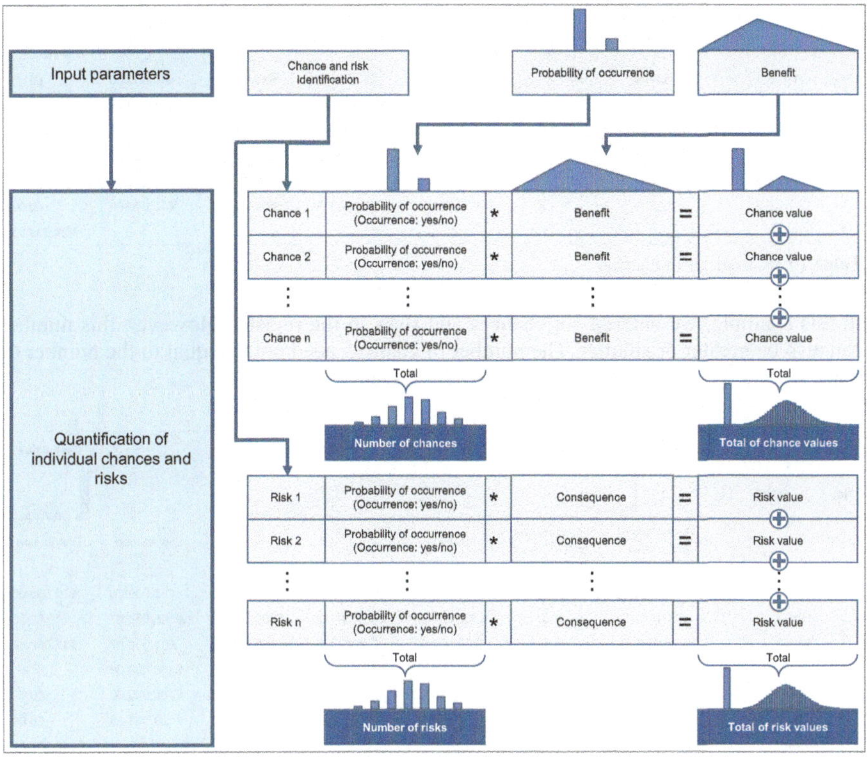

Fig. 10.34 Calculation flow chart – Register of chances and risks

The individual chances and risks should be listed in Column A of Table 10.18 or Table 10.19 (as a result of the chance/risk identification; see Section 4.5.1.1) for performing a probabilistic assessment of chances and risks on the basis of a macro-analysis supported by a Monte Carlo simulation. The probability of occurrence is entered as a percentage in Column B; it describes the probability with which a chance or risk will materialize in the simulation (or take a value of 1 in the Bernoulli distribution).

No.	Chances	Bernoulli distribution		Benefit				Chance value
		Probability of occurrence	Occurrence: yes=1 / no=0	MIN	MOST LIKELY	MAX	Distribution	Distribution
0	A	B	C	D	E	F	G	H
1	Chance 1	15.00%	0	€25,000.00	€30,000.00	€35,000.00	€30,000.00	€0.00
2	Chance 2	20.00%	0	€150,000.00	€200,000.00	€210,000.00	€186,666.67	€0.00
3	Chance 3	60.00%	1	€20,000.00	€22,000.00	€40,000.00	€27,333.33	€27,333.33
4	Chance 4	10.00%	0	€55,000.00	€60,000.00	€75,000.00	€63,333.33	€0.00
5	Chance 5	5.00%	0	€250,000.00	€280,000.00	€300,000.00	€276,666.67	€0.00
6	Chance 6	25.00%	0	€30,000.00	€40,000.00	€50,000.00	€40,000.00	€0.00
7	**Total**		**1**					**€27,333.33**

Table 10.18 Register of chances

In this example, we entered six chances and risks in the register. However, this number can also be greater or smaller. The number of chances need not be equal to the number of risks.

No.	Risks	Bernoulli distribution		Consequence				Risk value
		Probability of occurrence	Occurrence: yes=1 / no=0	MIN	MOST LIKELY	MAX	Distribution	Distribution
0	A	B	C	D	E	F	G	H
1	Risk 1	80.00%	1	€10,000.00	€15,000.00	€20,000.00	€15,000.00	€15,000.00
2	Risk 2	20.00%	0	€500,000.00	€600,000.00	€900,000.00	€666,666.67	€0.00
3	Risk 3	70.00%	1	€20,000.00	€35,000.00	€40,000.00	€31,666.67	€31,666.67
4	Risk 4	45.00%	0	€110,000.00	€140,000.00	€200,000.00	€150,000.00	€0.00
5	Risk 5	15.00%	0	€340,000.00	€360,000.00	€500,000.00	€400,000.00	€0.00
6	Risk 6	30.00%	0	€80,000.00	€120,000.00	€185,000.00	€128,333.33	€0.00
7	**Total**		**2**					**€46,666.67**

Table 10.19 Register of risks

By way of example, Risk 1 has a probability of occurrence of 80% as indicated in Row 1 of Table 10.19. More specifically, this means that the bar for the value of one (representing an 80% probability of occurrence) included in the applied Bernoulli distribution (see Fig. 10.35) is four times as high as the bar representing the value of zero (i.e. the 20% counter-probability). In this example, Risk 1 will materialize with an 80% probability in the simulation run. Columns D to F include detailed definitions of the consequences associated with such an occurrence of a chance or risk.

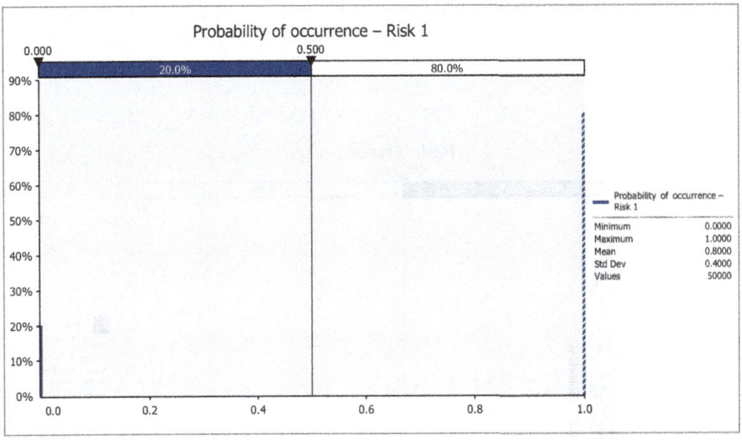

Fig. 10.35 Probability of occurrence – Risk 1 – Bernoulli distribution

For the sake of simplicity, the benefit or the consequence expressed in monetary units are defined as triangular distributions.[18)] This approach requires defining a minimum, most likely, and maximum value (Columns D to F in Tables 10.18 and 10.19). The triangular distributions are entered in Column G; the table lists mean values of the individual distributions.

Column H calculates the product of the occurrence (0 or 1; Column C) and the benefit/consequence (a random value within the defined range) for each chance or risk. The smaller the proportion of the probability of occurrence, the more frequently the result of this multiplication will be 0 because the chance or risk does not materialize and thus will not lead to any benefit or consequence.

In this example, we assumed the considered chances and risks to be independent of one another. However, possible correlations between individual risks or between risks and chances could be considered by applying a correlation matrix.

10.3.4.3 Calculation and Interpretation of Results

Columns G and H in Tables 10.18 and 10.19 show the mean values of the relevant distributions. The sum of the risk or chance values is calculated in Cells H7. The Monte Carlo simulation involves 50,000 iterations. In each of these steps, a random value is selected for the uncertain input parameters and used for calculating results for chance and risk values. These results can be represented as histograms in order to draw conclusions regarding the probabilities associated with specific deterministic values. Figs. 10.36 and 10.37 include the histograms for the sum of the chance and risk values [€].

Fig. 10.36 shows two different slider settings at the top of the histogram. The top image reveals that in about 96.0% of cases the chance will not materialize at an amount greater than approximately €300,000. The bottom image includes the same histogram showing

[18)] *Busch*, for example, selected BetaPERT distributions for the consequences. These distributions are also defined by minimum, most likely, and maximum values, but they generally tend to assign a lower probability to comparatively higher values. Cf. Busch (2004), p. 60f.

that in about 72.9% of cases the sum of chances will merely take a value between 0 and €100,000. The probability that no chance will be realized at all amounts to about 17.6% (see the slider in Fig. 10.38, top). This involves iterations in which none of the defined chances materialized.

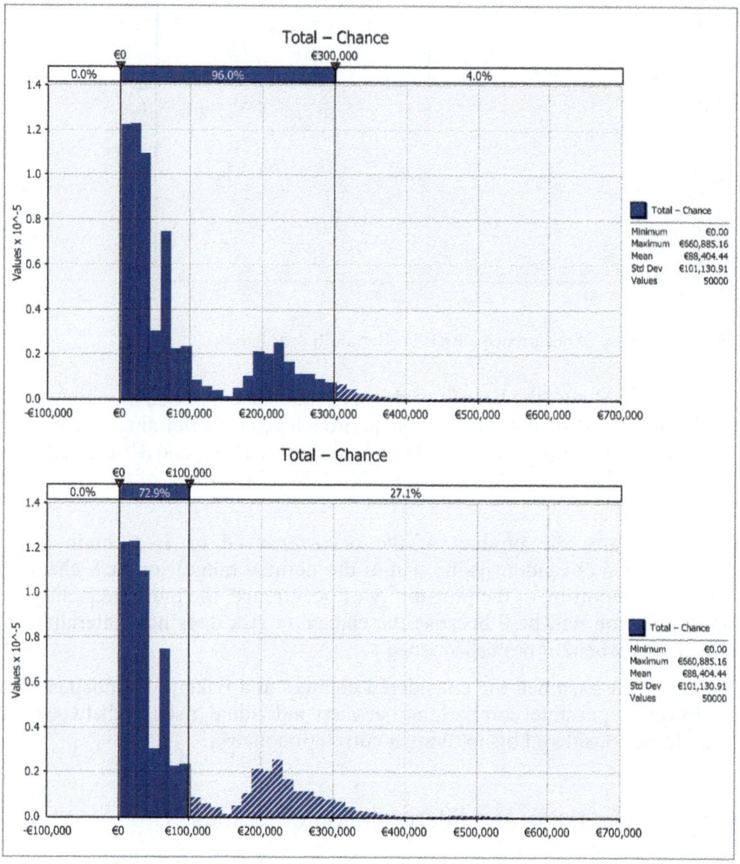

Fig. 10.36 Histogram – Sum of chance values – Different slider settings

Fig. 10.37 also shows two different slider settings at the top edge of the histogram. The top image reveals that in about 96.0% of cases the consequence of a risk to materialize would be smaller than €1 million. In addition, the same histogram shown in the bottom image indicates that in about 72.6% of cases the sum of the risks will take a value between 0 and €500,000. The probability that none of the identified risks occurs amounts to about 1.5% (see the slider in Fig. 10.38, bottom). These are the iterations in which none of the identified risks materialized. In this case, all entries in Column C of Table 10.19 are 0.

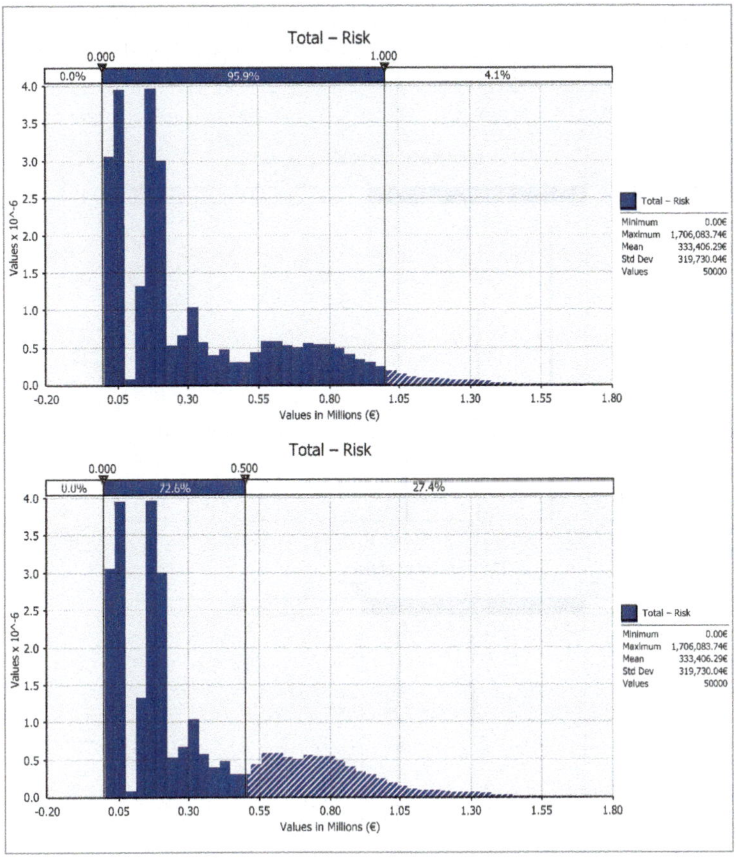

Fig. 10.37 Histogram – Sum of risk values – Different slider settings

Figs. 10.36 and 10.37 reveal relatively wide ranges and multimodal trends resulting from the consideration of the individual chances and risks.

Any chance-only or risk-only evaluation is suitable for making a decision in favor of a deterministic value on the basis of a selected chance/risk ratio using the histograms shown in Figs. 10.36 and 10.37. If, for instance, we were to cover 80% of the potential risks by a capital buffer or reserve, we would have to provide an amount of about €643,000 in the case at hand.

We can conduct an additional analysis with respect to the number of simultaneously occurring chances and risks. The sum of the chance/risk occurrences is calculated in Cells C7 of Table 10.18 or Table 10.19. The fact that different combinations of chances and risks may occur in each iterative step of the simulation also results in different frequencies for the number of simultaneously occurring chances and risks.

Fig. 10.38 shows the frequencies of occurrence of several chances and risks that are expected on the basis of the probability of occurrence determined for each of these chances and risks.

The probability that none of the chances materializes is relatively high at 17.6% for the number of simultaneously occurring chances (see Fig. 10.38, top). There is an extremely low probability for more than three chances to materialize (1.4%). The simultaneous occurrence of one or two chances can be considered to be most frequent.

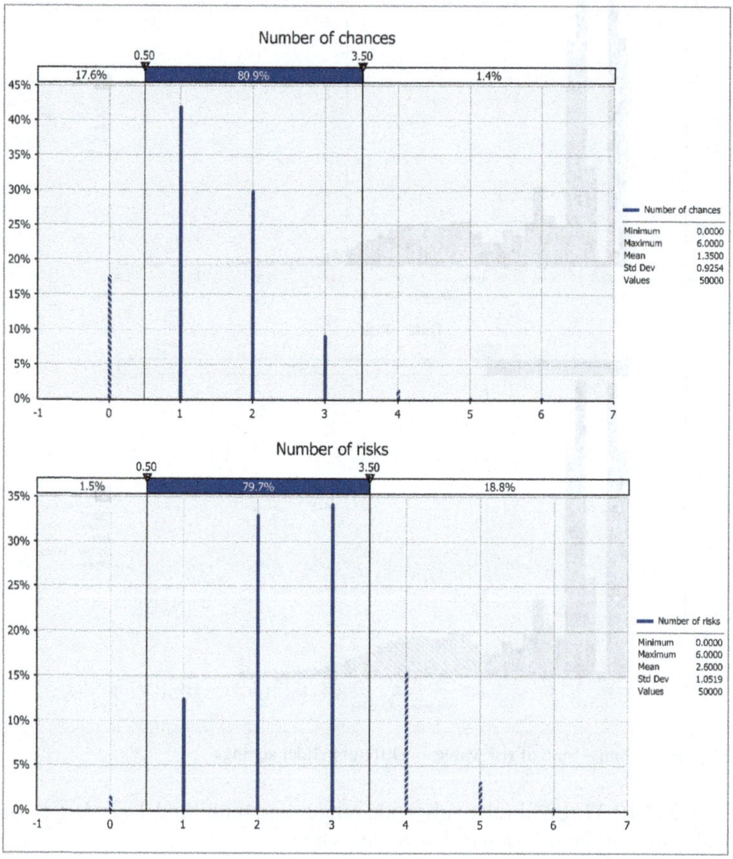

Fig. 10.38 Number of simultaneously occurring chances (top) and risks (bottom)

Fig. 10.38 (bottom) indicates that the probability for none of the risks occurring amounts to about 1.5%. The probability for four or more risks occurring simultaneously equals about 18.8%. Two or three risks occur most frequently depending on the selected input parameters.

Subtracting the sum of the risk values from that of the chance values (difference = chance values - risk values) results in the histogram shown in Fig. 10.39. This diagram also includes negative values, which represent results of scenarios in which the consequences of occurred risks are greater than the benefit of the chances that materialized in the simulation. The effects of chances can be expected to outweigh the effects of risks in about 23% of all cases if no risk mitigation measures are implemented. This scenario does not consider non-identified or non-identifiable chances and risks.

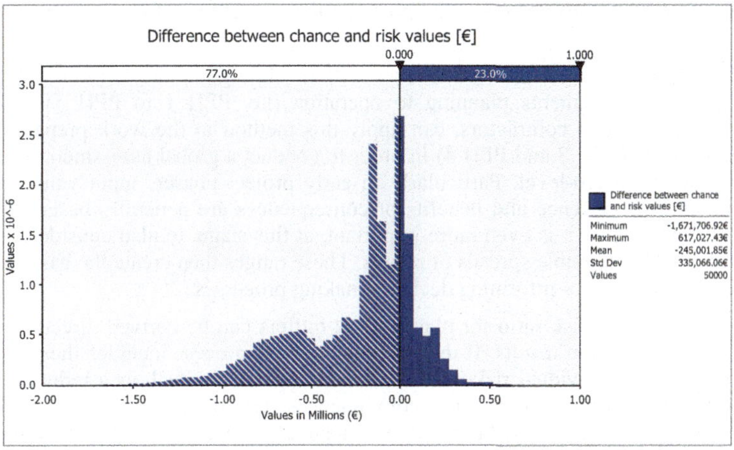

Fig. 10.39 Difference of chance and risk values

Any increase in the chance/risk ratio in favor of budget certainty or of chances to materialize requires the selection of a risk buffer that delivers the intended ratio on the basis of the derived histogram.

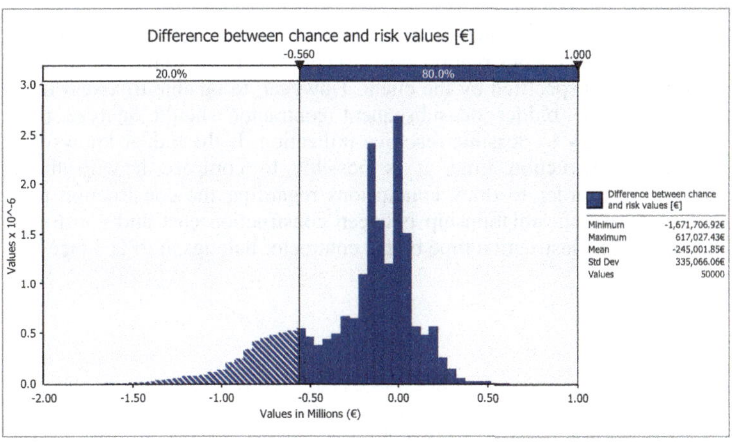

Fig. 10.40 Difference of chance and risk values – Risk buffer for 80% of cases

If, for instance, an additional buffer of about €560,000 were provided, this would cover about 80% of all cases when considering chances and risks together (see Fig. 10.40).

10.3.4.4 Benefit for the Bidder/Contractor

When compiling lists of chances and risks, a system or categorization into different groups appears to be useful, such as a breakdown into different types of production factors, stakeholders, or project phases. This grouping into categories should take place as early as at the chance/risk identification stage that precedes the assessment of chances and risks (see also Section 4.5).

Clients can apply the enhanced "practitioner method" throughout the course of the project, reaching from requirements planning to operation (i.e. PPH 1 to PPH 5). Likewise, bidders, or subsequent contractors, can apply this method in the work preparation and execution phases (PPH 3 and PPH 4) in order to conduct a global assessment of chances and risks at the macro-level. Particularly in early project phases, input values such as probabilities of occurrence and benefits or consequences are generally based on expert estimates. This is why it is even more important, at this stage, to also consider ranges in order to reflect the possible spreads of results. These ranges then create the basis for more detailed analyses and for informing decision-making processes.[19)]

The modified chance/risk ratio for planned risk buffers can be derived directly from the probabilistic calculation results. If the risk management process includes the prevention, mitigation, etc. of individual risks, the corresponding effects, such as a reduction in the probability of occurrence and/or the consequences, can be entered in the register of chances and risks to be assessed in relation to the total chance/risk ratio of the project.

We recommend to proactively maintain and update the register of chances and risks throughout the project. This approach permits an up-to-date overview of the expected individual chances and risks of the project.

10.3.5 Determination of Construction Time

Chapter 8 already included a detailed discussion of the significance of construction time and its relationship with construction costs. The construction time, or time to project completion, is usually specified by the client. However, to be able to assess the specified construction time, the bidder or subsequent contractor should analyze the time to completion with a view to possible resource utilization. If the bidder knows the project-specific normal construction time, it is possible to compare it with the specified construction time in order to draw conclusions regarding the construction process, the chance/risk ratio, and the relationship between construction cost and construction time. Determination of the construction time by the contractor belongs in PPH 3 (see Fig. 10.41).

19) Cf. Hofstadler/Kummer (2019), p. 739ff.

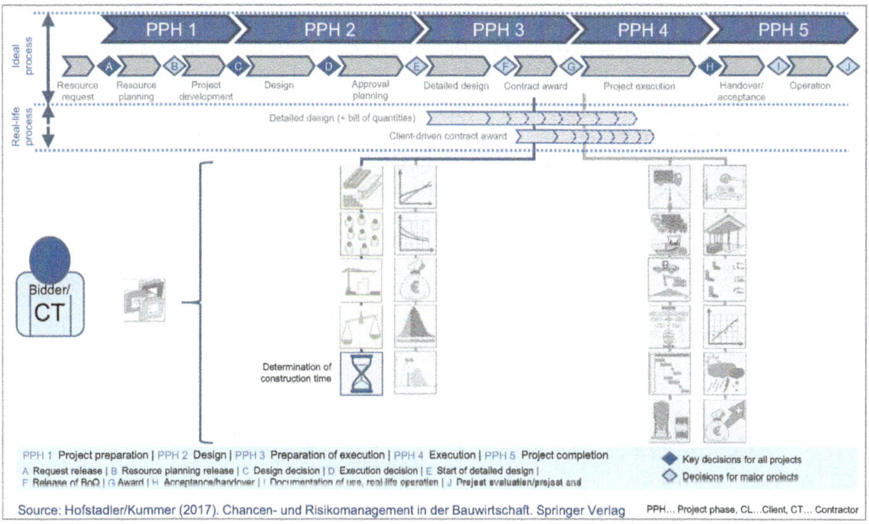

Fig. 10.41 Allocation of example in the project sequence

10.3.5.1 Task

For a building construction project, the number of available cranes is used for calculating the normal construction time (excluding losses of productivity) for reinforced concrete operations.

We take the total labor consumption rate of reinforced concrete operations from the calculation for correlated parameters discussed in Section 7.3.3. This rate is then directly applied in the simulation, as the result of a probabilistic calculation, in order to determine normal construction time.

10.3.5.2 Modeling

In building construction projects, cranes are considered to be the key elements of site logistics. Depending on the method selected for concrete pouring (e.g. using cranes and skips for all structural components) and the type of crane (top-slewing versus fast-erecting), the proportionality factor[20)] for cranes indicates the number of workers that the available cranes can serve.

We assume that only top-slewing cranes are used and that no cranes are deployed for concrete placement. Applying the trapezoidal model[21)] of resource utilization, these considerations make it possible to calculate not only the maximum but also the average number of workers.

20) See Table 10.8

21) See Fig. 10.19

This variable is combined with the scheduled daily working time and the total labor consumption rate of reinforced concrete operations in order to calculate the average daily performance. We obtain the normal construction time of reinforced concrete operations by relating this performance figure to the volume of reinforced concrete to be poured.

Fig. 10.42 shows the flow chart of the process of calculating normal construction time. Deterministic input parameters are represented by arrows. Parameters fraught with uncertainties are modeled using triangular distributions. Only the total labor consumption rate of reinforced concrete operations is used as an output value of a preceding calculation (see Section 7.3.3) and fed into the simulation as a histogram.[22)]

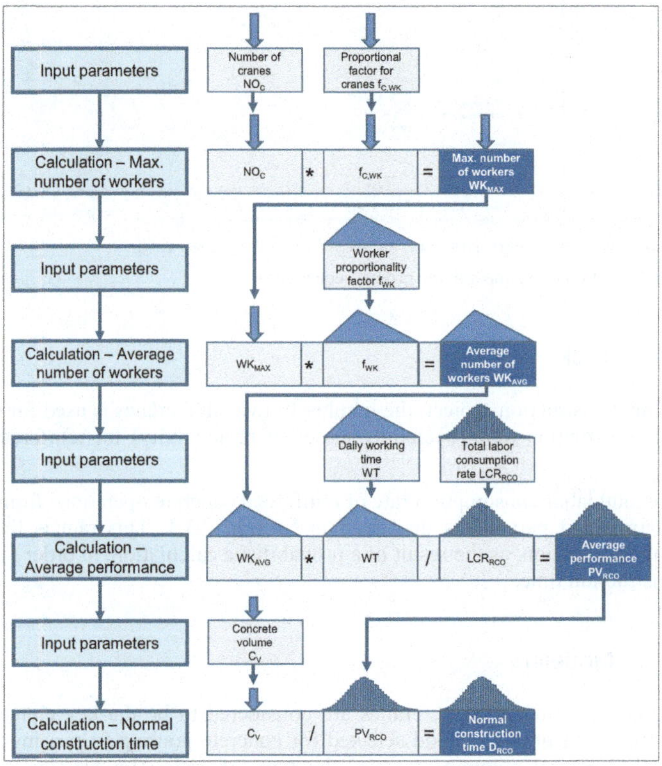

Fig. 10.42 Calculation flow chart – Calculation of normal construction time (working days)

Using distribution functions for the input parameters makes it possible to also derive histograms for the intermediate results, such as the average performance. Likewise, we are able to derive conclusions regarding probabilities and to use these histograms for plausibility checks.

22) We apply the histogram determined on the basis of correlated input parameters (see Fig. 7.19).

10.3.5.3 Calculation and Interpretation of Results

Table 10.20 shows the calculation of normal construction time. The volume of reinforced concrete C_V [m³], the number of cranes NO_C [cranes] available for site operations, and the proportionality factor for cranes $f_{C,WK}$ [workers/crane] are applied as deterministic values (see Rows 1 to 3).

For top-slewing cranes, the proportionality factor for cranes amounts to 20 workers per crane if no cranes are used for concrete pouring. We apply a 10% reduction to this value (18 WK/crane = 20 WK/crane * 0.90) in order to calculate the normal construction time (see Fig. 10.18).

Thus, we also obtain the maximum number of workers WK_{MAX} [wh/h] of 90 (= 5 cranes * 18 WK/crane) as a deterministic value (Row 4). The proportional factor for workers indicates the relationship between the average and maximum number of workers; it is also represented by a triangular distribution (0.75 | 0.80 | 0.85) (see Row 5 of Table 10.20). Multiplying the maximum number of workers by the worker proportionality factor results in the average number of workers WK_{AVG} [wh/h] (Row 6).

No.	Calculation parameter		Unit	MIN	MOST LIKELY	MAX	Distribution
0	A	B	C	D	E	F	G
1	Concrete volume	C_V	m³		38,275.00		38,275.00
2	Number of cranes	NO_C	crane		5.00		5.00
3	Proportional factor for cranes	$f_{C,WK}$	WK/crane		18.00		18.00
4	Maximum number of workers	WK_{MAX}	wh/h				90.00
5	Worker proportionality factor	f_{WK}	-	0.75	0.80	0.85	0.80
6	Average number of workers	WK_{AVG}	wh/h				72.00
7	Daily working time	WT	h/d	8.00	9.00	10.00	9.00
8	Total labor consumption rate	LCR_{RCO}	wh/m³				6.10
9	Average performance	PV_{RCO}	m³/d				106.32
10	Normal construction time	D_{RCO}	d				360.01

Table 10.20 Calculation table – Normal construction time

We then calculate the average daily performance PV_{RCO} [m³/d] (Cell G9) by applying this value in combination with the daily working time (triangular distribution; Row 7), the total labor consumption rate of reinforced concrete operations LCR_{RCO} [wh/m³] (Row 8), and the average number of workers.

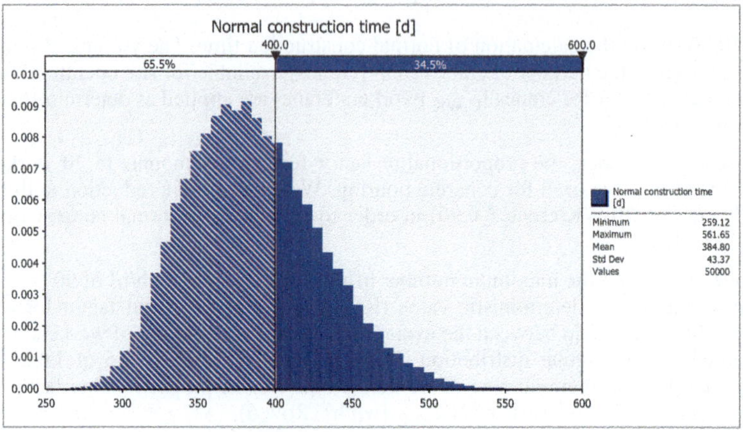

Fig. 10.43 Histogram – Normal construction time

Division of the reinforcement quantity by the daily performance results in the histogram (see Fig. 10.43) for the normal construction time in Cell G10.[23)]

The mean of the normal construction time equals about 385 days, with a standard deviation of approximately 43 days. If the contractor were in a position to define the construction time, it would go in the direction of about 400 days with a view to the chance/risk ratio. This time to completion would be associated with higher chances (approx. 65.5%) than risks (approx. 34.5%). Thus, the probability of adhering to or even shortening the specified construction time would be almost twice as high as the risk of exceeding this period.

10.3.5.4 Benefit for the Bidder/Contractor

The contractor will plan their resource utilization so as to ensure that there is still a certain buffer for contingencies or disruptions caused by it. In the example considered in this section, we take this buffer into account by reducing the maximum crane proportionality factor by 10% (from 20 WK/crane to 18 WK/crane).

We use distribution functions to incorporate the uncertainties associated with the input parameters, which is why we obtain a spread also for the calculated normal construction time. The bidder or subsequent contractor can rely on this distribution for evaluating and assessing construction times specified by the client with respect to the associated chance/risk ratio.

The chance of the bidder/contractor will drop to just 1.1% if the client specifies a construction time of only 300 days (see Fig. 10.44). In this case, the bidder/contractor should check if an additional crane or more workers should be used. In addition, there is also the option of accelerating certain work steps by prefabrication, such as of reinforcement cages, or of extending the daily working time or determining whether shift work is necessary at all. If the client specifies a particularly short construction time, the

[23)] The normal construction time specified in Table 10.20 is derived from the input parameter means. This value will not necessarily be identical to the mean value shown in the histogram for normal construction time (see Fig. 10.43).

associated effects on site management and the construction process, on the standard of workmanship, and on issues arising with respect to construction economics should definitely be assessed and priced prior to submitting the bid (see Section 8.5).

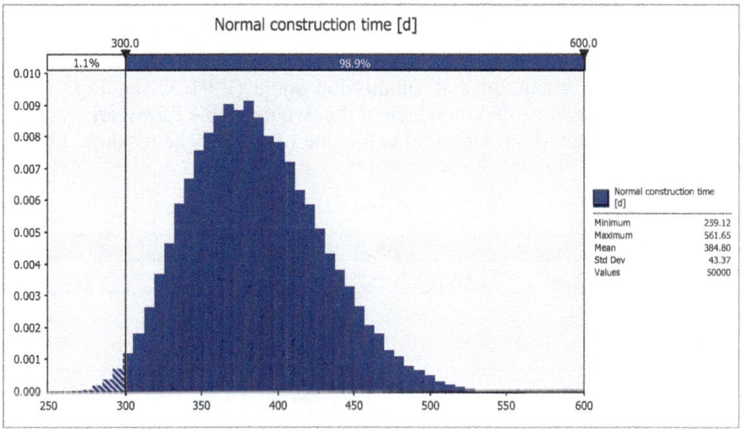

Fig. 10.44 Histogram – Normal construction time

It should also be noted that an excessively short specified construction time will require increased resource utilization and may thus lead to losses of productivity and higher construction costs.

The contractor applies probabilistic methods to take these input parameter uncertainties into account. They are thus in a position to assess the risk or chance to be taken in a project with a construction time specified by the client in order to adhere to this predetermined period. In the case of a high probability for using a large number of workers and cranes, the contractor should perform an in-depth analysis in order to assess whether this number of workers can at all be used on the construction site while ensuring a normal productivity level, or how many cranes can actually be installed on-site. If the contractor is unable to provide sufficient crane capacity, the use of alternative lifting equipment such as platforms, different construction methods (including precast elements), or process modifications (for instance using a formwork system requiring fewer cranes) should be considered prior to commencing construction work.

In this example, the contractor was not in possession of any information indicating how the client arrived at the specified construction time. Nor were any details available regarding the associated chance/risk ratio. Irrespective of this fact, the contractor had to prepare their schedule for resource utilization (with the related uncertainties) on the basis of the construction time specified by the client.

If a particularly short construction time is specified, effects on site management and the construction process, on the standard of workmanship, and on issues arising with respect to construction economics should be expected (see Section 8.5).

The results of the Monte Carlo simulation should also raise awareness of this situation and create the basis for entering into a communication process regarding chances and risks. Deterministic values will not deliver satisfactory results because they only provide the option of evaluating ranges or performing isolated sensitivity analyses. In contrast, probabilistic calculation methods yield several thousand individual results in a single simulation run and translate them systematically into histograms. Only if we apply such methods will we be able to assess and estimate the chances and risks associated with specific decisions.

10.3.6 Rough Process Comparison

We intend to identify the most efficient construction processes by performing related comparisons. As early as at the bid submission stage (PPH 3, see Fig. 10.45), it is important for the contractor to know which of the two available formwork systems comes at a lower price or from which marginal value one of these systems should be preferred over the other.

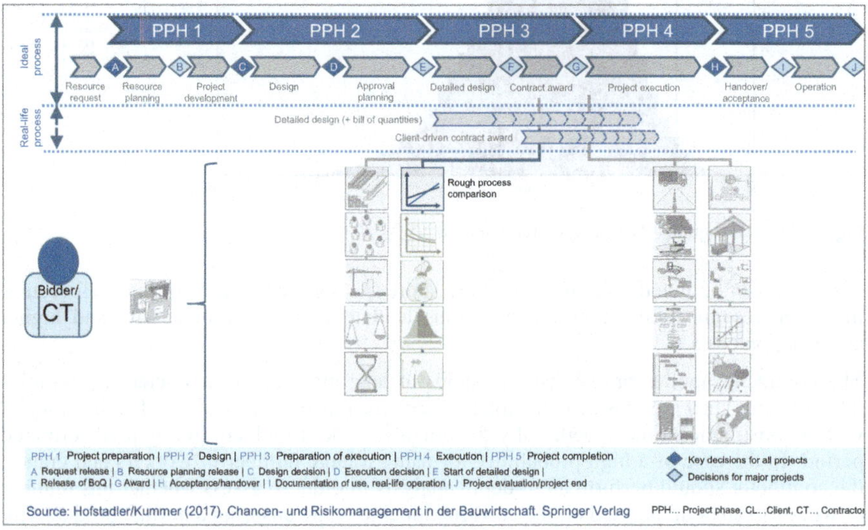

Fig. 10.45 Allocation of example in the project sequence

10.3.6.1 Basics

Different construction processes or formwork systems are usually available for performing the individual works necessary to complete a building or structure while achieving the same specified standard of workmanship. However, selecting the right process or system has a major influence on the degree of target achievement in terms of cost, performance, quality, time, etc. Processes and systems differ, for instance, in terms of the technology, cost effectiveness, safety, and site operations. We distinguish between cost-based and differentiated process comparisons depending on the number of different influential factors considered when selecting the relevant process.[24)]

Cost-based process comparisons start from costing exercises performed for the processes or systems to be assessed.

[24)] Cf. Hofstadler (2008), p. 339

The differentiated comparison of formwork systems (process comparison) also considers other factors besides cost or price criteria. Thus, the latter type of comparison is more comprehensive than the former. One option of conducting such a comparison involves the application of a decision matrix (see Section 10.4.4).

10.3.6.2 Task

We perform a costing exercise for a building construction project in order to determine if the use of girder formwork or of slab tables consisting of girder formwork elements is more useful from a cost point of view when installing flat floor slabs. The formwork area of a standard floor amounts to 1,500 m². The project involves the construction of six floors divided into three work sections (see Fig. 10.46). We thus obtain a total formwork area of 9,000 m². The calculation tables in Section 10.3.6.4 list additional input parameters for both girder formwork and slab table systems.

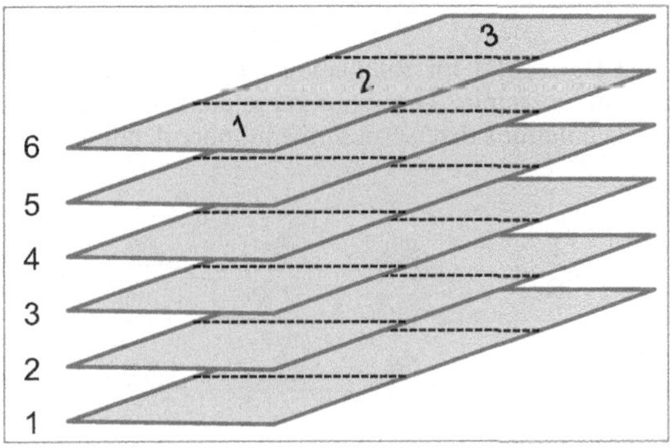

Fig. 10.46 Schematic representation of floor slabs (six floors with three work sections each)

10.3.6.3 Modeling

The cost curves representing the compared formwork systems depend on the total formwork area and are associated with uncertainties. The cost-based comparison considers wage costs as well as equipment and material costs and compares their sums as direct costs of the two processes.

When comparing the formwork systems including and excluding assembly and disassembly, we need to identify the formwork area or number of use cycles from which the respective process/system will incur the lowest cost. For a formwork system excluding assembly and disassembly, we obtain the direct cost $C_{F,DC,A}$ from Eq. (10-7). In the parentheses, the unit wage, equipment, and material costs are calculated and added up. We then multiply these costs per unit by the corresponding formwork area FA [m²].

$$C_{F,DC,A} = FA \cdot (LCR_{FO,A} \cdot MW + C_{F,EQ,A} + C_{F,MA,A}) \quad (10\text{-}7)$$

The costs $C_{F,DC,B}$ of a formwork system including assembly and disassembly are calculated according to Eq. (10-8). The first term in parentheses multiplies the labor consumption rate for formwork assembly and disassembly $LCR_{F,AD}$ [wh/m²] with the specific formwork area FA_{AD} [m²] for which assembly and disassembly takes place. The resulting product is divided by the total formwork area FA [m²].

$$C_{F,DC,B} = FA \cdot \left[\left(\frac{LCR_{F,AD} \cdot FA_{AD}}{FA} + LCR_{FO,B}\right) \cdot MW + C_{F,EQ,B} + C_{F,MA,B}\right] \tag{10-8}$$

When evaluating two formwork systems, we are particularly interested in the formwork area from which a process or system incurs lower direct costs compared to the other. We calculate the break-even formwork area FA_{BE} [m²] using Eq. (10-9) (by equating Eq. (10-7) with Eq. (10-8)).

$$FA_{BE} = \frac{LCR_{F,AD} \cdot FA_{AD}}{\frac{LCR_{FO,A} \cdot MW + C_{F,EQ,A} + C_{F,M,A} - C_{F,EQ,B} - C_{F,MA,B}}{MW} - LCR_{FO,B}} \tag{10-9}$$

Provided the total formwork area of the building under consideration is greater than that calculated in Eq. (10-9), the incurred direct costs will be lower when using formwork system B (slab tables).

Fig. 10.47 shows the flow chart for calculating the direct costs for the two formwork systems analyzed in this example. The input parameters represented by histograms are taken from previous calculations that we describe in more detail in the following section.

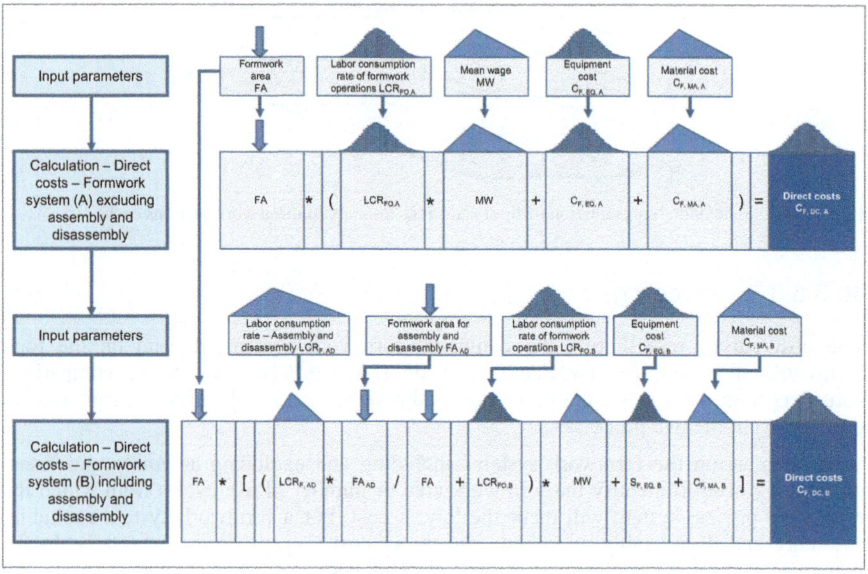

Fig. 10.47 Calculation flow chart – Direct costs

10.3.6.4 Calculation and Interpretation of Results

The direct costs of the two formwork systems are calculated while considering the associated uncertainties with the aid of a Monte Carlo simulation. Those parameters for which we are unable to determine fixed values are modeled as triangular distributions. The minimum, most likely, and maximum values per parameter are shown in Columns B, C, and D of Tables 10.21 and 10.22. Column E includes the histograms representing the deterministic values, the distributions, and/or the intermediate or final results.

In the first step, we determine the total formwork area required for the floor slabs of the building (Row 4). This figure is a deterministic value obtained by multiplying the formwork area per standard floor (Row 1) by the number of floors (Row 2) and the number of work sections to be completed (Row 3). We thus obtain a total area of 9,000 m². This formwork area applies to both formwork systems.

Formwork system A – Girder formwork

The mean labor consumption rate for girder formwork $LCR_{FO,A}$ [wh/m²] (Cell E17) is composed of the consumption rates for formwork installation and stripping, relocation, and other work items (Rows 5 to 8).

We calculate the average daily formwork performance [m²/d] (Row 18) applying the number of workers (Row 13), the daily working time (Row 14), and the labor consumption rate. However, this calculation does not include the proportion of the consumption rates allocated to the other items shown in Row 8. Multiplying the average daily formwork performance by the number of working days (Row 15) results in the average monthly formwork performance (Cell E19). This parameter is necessary for calculating the equipment cost of crane utilization (Cell E20) together with the crane cost per month (Row 11) and the average crane utilization for girder formwork (Row 12).[25]

No.	Parameter		MIN	MOST LIKELY	MAX	Distribution
0	A		B	C	D	E
1	Formwork area of standard floor			1,500.00 m²		1,500.00 m²
2	Number of floors			6.00		6.00
3	Number of work sections			3.00		3.00
4	Formwork area – Ceilings	FA				9,000.00 m²
5	Mean labor consumption rate – Formwork installation		0.18 wh/m²	0.20 wh/m²	0.25 wh/m²	0.20 wh/m²
6	Mean labor consumption rate – Formwork stripping		0.12 wh/m²	0.15 wh/m²	0.20 wh/m²	0.15 wh/m²
7	Mean labor consumption rate – Relocation		0.02 wh/m²	0.02 wh/m²	0.05 wh/m²	0.02 wh/m²
8	Mean labor consumption rate – Other		0.05 wh/m²	0.05 wh/m²	0.10 wh/m²	0.05 wh/m²
9	Mean equipment cost		€3.00/m²	€3.50/m²	€5.00/m²	€3.50/m²
10	Mean material cost	$C_{F,MA,A}$	€0.40/m²	€0.50/m²	€0.80/m²	€0.50/m²
11	Crane cost per month		€7,000.00/mo	€7,500.00/mo	€8,200.00/mo	€7,500.00/mo
12	Average crane utilization for girder formwork		12.00%/mo	15.00%/mo	20.00%/mo	15.00%/mo
13	Number of workers		5.00 wh/h	5.00 wh/h	6.00 wh/h	5.00 wh/h
14	Daily working time		8.00 h/d	9.00 h/d	10.00 h/d	9.00 h/d
15	Number of working days per month		19.00 d/mo	20.00 d/mo	21.00 d/mo	20.00 d/mo
16	Mean wage cost	MW	€32.00/wh	€35.00/wh	€37.00/wh	€35.00/wh
17	Mean labor consumption rate – Girder formwork	$LCR_{FO,A}$				0.42 wh/m²
18	Average daily performance (excl. LCR, Other)					121.62 m²/d
19	Average monthly performance					2,432.43 m²/mo
20	Equipment cost of crane utilization					€0.46/m²
21	Equipment cost – Total for girder formwork	$C_{F,EQ,A}$				€3.96/m²
22	**Direct costs – Girder formwork**	$C_{F,DC,A}$				**€172,462.50**
23	**Unit costs – Girder formwork**					**€19.16/m²**

Table 10.21 Calculation table – Direct costs – Formwork system A (girder formwork)[26]

[25] Deterministic calculation example based on most likely values:
(€7,500.00/mo * 15.00% / 100%) / 2,432.43 m²/mo = €0.46/m²

The total equipment cost of girder formwork (Cell E21) includes the mean equipment cost (Row 9) and the equipment cost of crane utilization. We enter the parameters for which abbreviations are listed in Table 10.21 in the equation used for calculating direct costs (Eq. (10-7)) in order to obtain the histogram shown in Fig. 10.48.

The mean shown in the histogram (see Fig. 10.48) equals approximately €188,600 whereas the standard deviation amounts to about €10,140. In about 5% of simulated cases, costs of approximately €172,300 are underrun while costs of about €205,700 are exceeded.

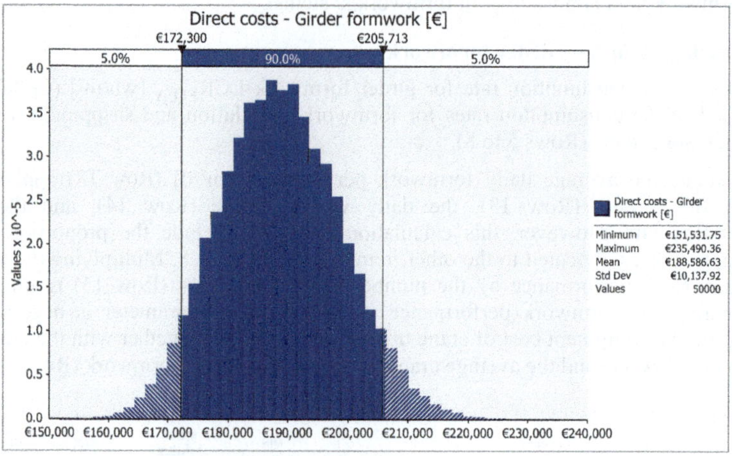

Fig. 10.48 Histogram – Direct costs – Girder formwork

Formwork system B – Slab tables

The direct costs of the second formwork system (slab tables) refer to the same formwork area as determined for the girder formwork (i.e. 9,000 m²; see Row 4 in Table 10.22).

Similarly to the girder formwork, we obtain the mean labor consumption rate for slab tables $LCR_{FO,B}$ [wh/m²] (see Cell E20) from the proportions of formwork installation and stripping, relocation, and other items (Rows 5 to 8).

We apply the same principle as to the girder formwork for calculating the equipment cost of crane utilization (Cell E23). However, the crane utilization for slab tables (between 32 and 40%/month; see Row 15) is significantly higher than the percentage determined for the girder formwork.

In addition, the assembly and disassembly of the slab tables should be incorporated in the direct cost figure. The consumption rate for formwork assembly and disassembly $LCR_{F,AD}$ [wh/m²] is stated as a triangular distribution in Row 10 of Table 10.22. Cell E25 calculates the formwork area for assembly and disassembly FA_{AD} [m²].

In this example, we assumed the provision of formwork in a quantity sufficient for two work sections (Row 13) and slab tables being suitable for 80 to 92% of the total floor area (see Row 12). Relative to the most likely value, we thus obtain a formwork area of 900 m² (= 1,500 m² / 3 * 0.90 * 2) for the assembly and disassembly steps.

26) The direct costs indicated in Table 10.21 for girder formwork result from a deterministic calculation applying the most likely input parameter values. The fact that right-skewed distributions were predominantly used means that this value is not identical to the mean and is located in the lower value range of the histogram shown in Fig. 10.48.

Entering the parameters for which abbreviations are listed in Table 10.22 in Eq. (10-8) results in the direct cost of slab tables expressed as a histogram (see Fig. 10.49).

No.	Parameter		MIN	MOST LIKELY	MAX	Distribution
0	A		B	C	D	E
1	Formwork area of standard floor			1,500.00 m²		1,500.00 m²
2	Number of floors			6.00		6.00
3	Number of work sections			3.00		3.00
4	Formwork area – Ceilings	FA				9,000.00 m²
5	Mean labor consumption rate – Formwork installation		0.07 wh/m²	0.08 wh/m²	0.09 wh/m²	0.08 wh/m²
6	Mean labor consumption rate – Formwork stripping		0.04 wh/m²	0.05 wh/m²	0.08 wh/m²	0.05 wh/m²
7	Mean labor consumption rate – Relocation		0.02 wh/m²	0.02 wh/m²	0.05 wh/m²	0.02 wh/m²
8	Mean labor consumption rate – Other		0.05 wh/m²	0.05 wh/m²	0.10 wh/m²	0.05 wh/m²
9	Mean labor consumption rate – Assembly and disassembly	$LCR_{F,AD}$	0.90 wh/m²	1.00 wh/m²	1.30 wh/m²	1.00 wh/m²
10	Mean equipment cost		€4.00/m²	€4.50/m²	€5.50/m²	€4.50/m²
11	Mean material cost	$C_{F,MA,B}$	€0.65/m²	€0.75/m²	€0.95/m²	€0.75/m²
12	Proportion of formwork area per floor for which slab tables can be used		80.00%	90.00%	92.00%	90.00%
13	Number of work sections for which slab tables are provided			2.00		2.00
14	Crane cost per month		€7,000.00/mo	€7,500.00/mo	€8,200.00/mo	€7,500.00/mo
15	Average crane utilization for slab tables		32.00%/mo	35.00%/mo	40.00%/mo	35.00%/mo
16	Number of workers		5.00 wh/h	5.00 wh/h	6.00 wh/h	5.00 wh/h
17	Daily working time		8.00 h/d	9.00 h/d	10.00 h/d	9.00 h/d
18	Number of working days per month		19.00 d/mo	20.00 d/mo	21.00 d/mo	20.00 d/mo
19	Mean wage cost	MW	€32.00/wh	€35.00/wh	€37.00/wh	€35.00/wh
20	Mean labor consumption rate – Slab tables	$LCR_{FO,B}$				0.20 wh/m²
21	Average daily formwork performance (excl. LCR, Other)					300.00 m²/d
22	Average monthly formwork performance					6,000.00 m²/mo
23	Equipment cost of crane utilization					€0.44/m²
24	Equipment cost – Total for slab tables	$C_{F,MA,B}$				€4.94/m²
25	Formwork area for assembly and disassembly	FA_{AD}				900.00 m²
26	**Direct costs – Slab tables**	$C_{F,DC,B}$				**€145,687.50**
27	**Unit costs – Slab tables**					**€16.19/m²**

Table 10.22 Calculation table – Direct costs – Formwork system B (slab tables)[27)]

The mean direct cost shown in the histogram for slab tables equals approximately €158,400 whereas the standard deviation amounts to about €7,400. In about 5% of simulated cases, costs of approximately €146,600 are underrun while costs of about €170,900 are exceeded.

[27)] The direct costs indicated in Table 10.22 for slab tables result from a deterministic calculation applying the most likely input parameter values. Since right-skewed distributions were predominantly used, this value is not identical to the mean and is located in the lower value range of the histogram shown in Fig. 10.49.

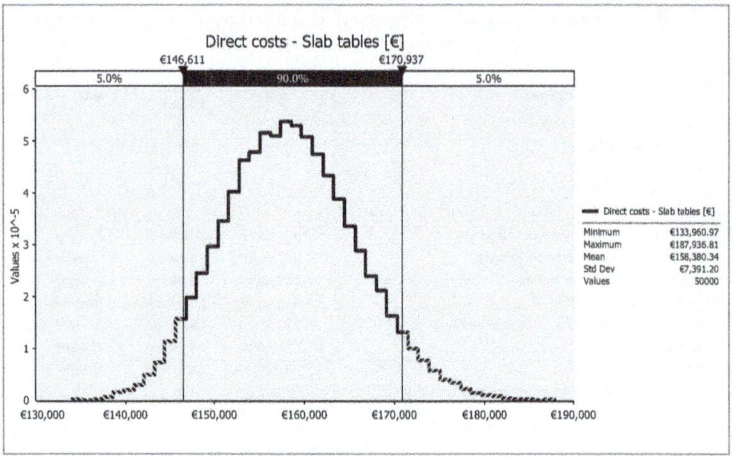

Fig. 10.49 Histogram – Direct costs – Slab tables

Fig. 10.50 shows the comparison of the direct cost histograms for the two selected formwork systems. The cost of slab tables is lower than the cost of girder formwork, and the spread of the former is narrower than that of the latter. The cost of girder formwork lies below the 95% quantile derived for the slab tables only in 3.7% of all simulated cases.

For the project considered in this example, we thus define a clear preference for formwork system B (slab tables) in terms of cost. However, the use of girder formwork may also become necessary from a construction management point of view if the underlying conditions, including crane capacity, are not aligned with this formwork system. For a differentiated process comparison, we can apply a decision matrix, among other methods. Section 10.4.4 discusses a related example.

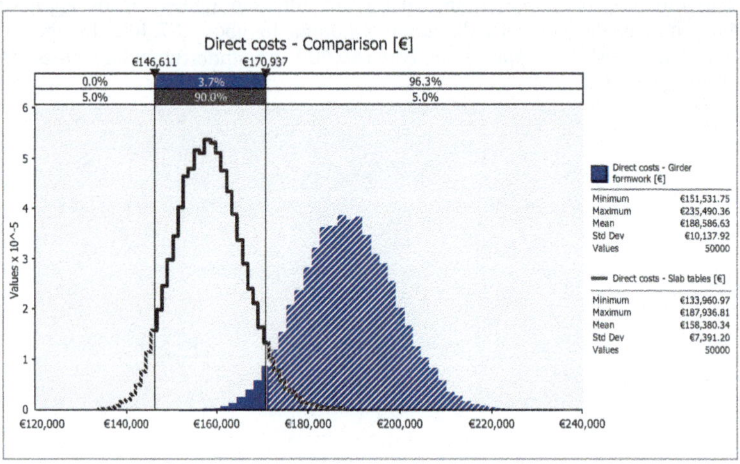

Fig. 10.50 Histogram – Direct costs – Comparison

Break-even formwork area

Besides establishing a head-to-head comparison of direct costs, bidders and subsequent contractors will be particularly interested in determining the break-even formwork area, which is defined as the area from which the use of slab tables would be more viable or cost-efficient compared to girder formwork. Like direct costs, the break-even formwork area also depends on the same uncertain parameters. Any probabilistic calculation will thus also result in a histogram rather than a specific figure.

The direct costs were calculated for different formwork areas FA [m²] and entered in a diagram (see Fig. 10.51) in order to illustrate the cost trends for girder formwork and slab tables. The fact that the costs were calculated as histograms necessitates five curves for each formwork system:

- Mean
- Mean - standard deviation
- Mean + standard deviation
- 1% quantile
- 99% quantile

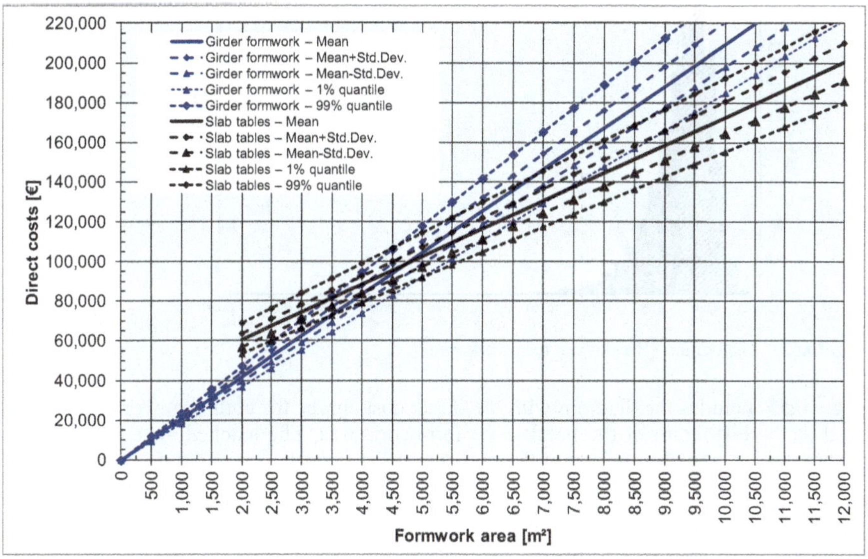

Fig. 10.51 Cost trend comparison – Girder formwork vs slab tables

We calculated the direct costs of slab tables starting from a formwork area of 2,000 m² and assumed an identical formwork area for assembly and disassembly for all calculations performed as part of this rough comparison.[28)]

[28)] Since the formwork area for assembly and disassembly was taken to equal about 900 m², a formwork area of approximately 2,000 m² is roughly equivalent to two slab table use cycles. In this case, it appears implausible to include a smaller area in the comparison. If smaller work sections are chosen for formwork areas below 2,000 m², it can be useful to perform a separate comparison by recalculating the assembly and disassembly steps also in this range (i.e. up to 2,000 m²).

The two sets of curves show a relatively long taper of the overlapping area. Even with a formwork area of 12,000 m², we are still unable to identify an intersection of the 1% quantile of the girder formwork with the 99% quantile of the slab tables. Accordingly, we expect a wide range for the histogram representing the break-even formwork area.

The parameters listed in Tables 10.21 and 10.22 are entered in Eq. (10-9) to calculate the break-even formwork area. Fig. 10.52 shows the histogram for the break-even formwork area. In this histogram, we recognize a long branch towards high break-even areas owing to the uncertainties associated with the input parameters. This means that the probability for the girder formwork to incur lower costs than the slab tables will still be (very) low for an area of 12,000 m² or more.

Correspondingly, for an area of about 5,000 m², we obtain a probability of about 62.7% for the slab tables incurring a lower cost than the girder formwork under the defined conditions and applying the defined input parameters.

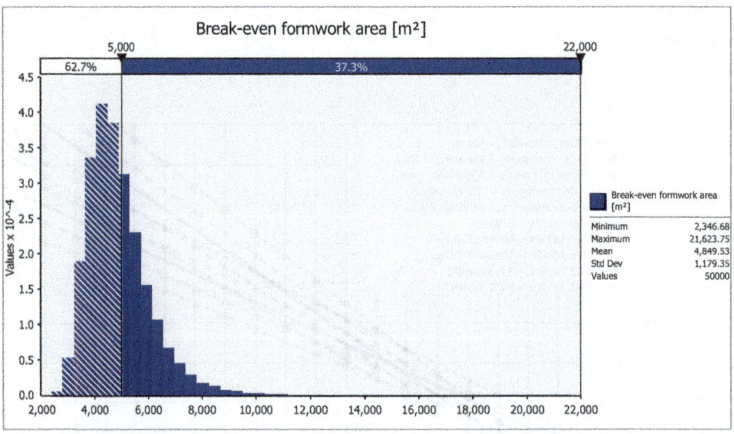

Fig. 10.52 Histogram – Break-even formwork area

Fig. 10.53 includes the diagram with the direct cost curves for both formwork systems as well as the histogram of the break-even formwork area. The hatched area shown in the diagram represents the overlap of the two curves between the 1% and 99% quantiles. The long taper of the overlapping area is also reflected in the shape of the histogram.

In most cases, the calculated break-even formwork areas roughly lie between 4,000 and 5,000 m². If the specified formwork area lies within this range, it is advisable to not only perform a cost-based comparison but also a differentiated process comparison.

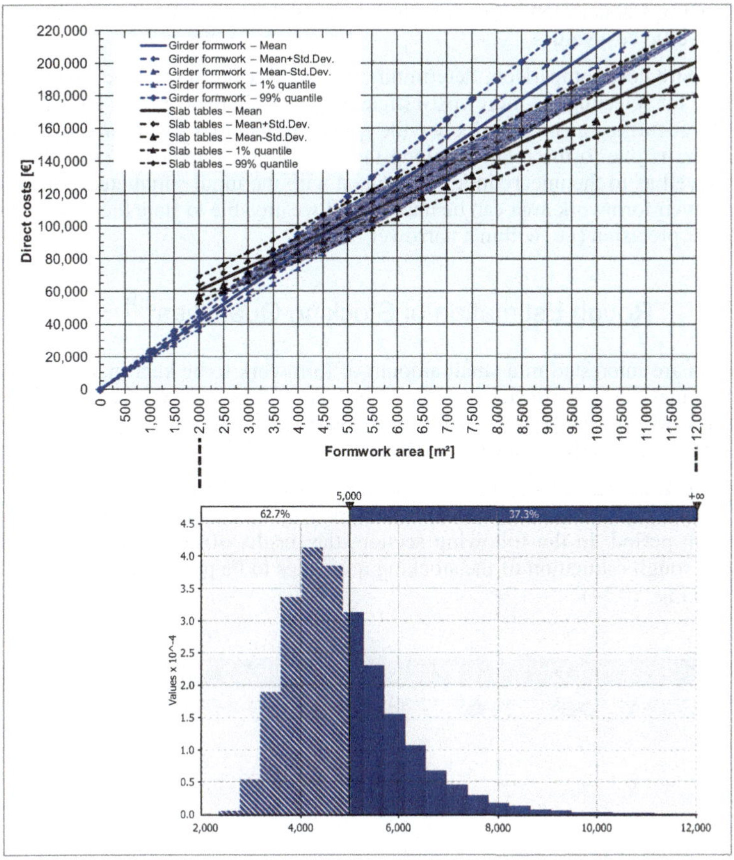

Fig. 10.53 Comparison – Diagram and histogram of break-even formwork area

10.3.6.5 Benefit for the Bidder/Contractor

The rough process comparison performed for different formwork systems only refers to a comparison of direct costs, including wages, equipment, and materials. A pure cost comparison will be helpful whenever the circumstances under which the work is to be performed as well as the external conditions prevailing on the construction site allow for the selection of one of the available processes.

However, if there are any limitations, for instance with respect to crane availability, a differentiated process comparison should also be conducted (see Section 10.4.4). This type of comparison makes it possible to assess and incorporate criteria in the process that cover commercial, site management, safety, and environmental aspects.

At any rate, the probabilistic calculation of the direct costs of formwork systems will be highly beneficial for the bidder or subsequent contractor because this method reveals the range of possible costs, which can then be analyzed and interpreted accordingly. The histograms derived for different processes can be combined in a single diagram, thus permitting direct conclusions regarding the spreads or probabilities of exceeding or underrunning specific cost values.

Besides a pure cost comparison, determining the break-even formwork area is of crucial importance for selecting the more cost-efficient process. In the example considered in this section, we identified a comparatively wide spread of the break-even formwork area. This phenomenon is due to the relatively long taper of the overlapping area of the two cost curves as well as to the uncertainties associated with the input parameters. The spread of the break-even formwork area can be narrowed if we are able to state the input parameters with greater precision (i.e. within a narrower range).

10.3.7 Rough Estimation of Stocking Quantities[29)]

Contractors are interested in a small amount of formwork to be kept in stock and thus in short idle times. This parameter is strongly influenced by the selected dimensions of the respective work sections and the daily formwork installation rate. The latter is primarily dependent on the specified construction time.

Hofstadler conducted a survey among experts in order to determine how large the required stocking quantities of wall and slab formwork had to be in relation to the specified construction period. In the following section, the results of this survey will be used to carry out a rough estimation of the stocking quantities to be provided by the contractor in PPH 3 (see Fig. 10.54).

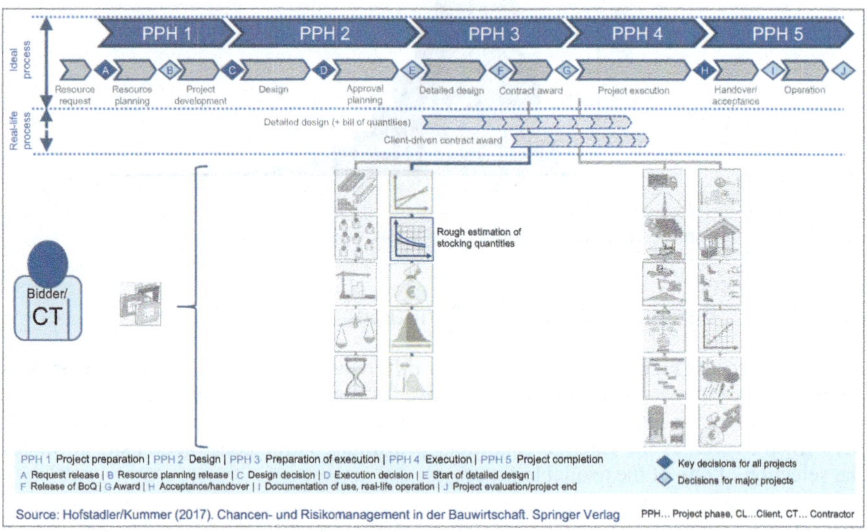

Fig. 10.54 Allocation of example in the project sequence

[29)] Stock available and deployable on-site

10.3.7.1 Basics

A questionnaire for an expert survey was developed using the Delphi method with the aim of describing the functional relationship between construction times and stocking quantities.

To generate reliable data, *Hofstadler* conducted an expert survey at Graz University of Technology between August 2012 and April 2013 in order to ascertain the quantitative relationship between construction times and stocking quantities. For this purpose, a total of 35 experts from construction contractors were asked to participate in this survey (from both construction trades and industry). Respondents had an average work experience of 17 years (minimum: five years; maximum: 43 years).

The experts were asked to indicate the stocking quantity percentages required for different construction periods (first stage: independent completion of the questionnaire by the expert; second stage: discussion of the questionnaire with the expert). The "normal" construction time during which the production factors can be utilized without incurring losses of productivity was taken as the reference.

Prior to a multivariate analysis of survey data, an exploratory analysis was conducted to obtain an overview of gathered data. This analysis is necessary for checking plausibility, verifying distributions, and identifying outliers. As part of this analysis, box plots of the variables were established to graphically represent distributions. A box plot describes the location and spread of a distribution and provides information on any outliers included in the data. Outliers and extremes (defined on the basis of their distance from the box plot whisker) may significantly distort arithmetic means. This distortion can be prevented by applying *Huber's* robust M-estimator method in which the existing outliers/extremes are not excluded. Instead, they are taken into account by a minutely adjustable lower weighting. Arithmetic means and, as and when required in the given situation, the above-described M-estimators were used as a basis for plotting diagrams and trend curves for increases in labor consumption rates and losses of productivity.

A distinction was made between extremely short, very short, short, normal, and long construction times (compare the definitions provided in Section 8.6.2.1). In the survey, a distinction was made between the stocking quantities for walls and floor slabs. In so doing, the stocking quantity is related to the specific formwork area of the building or structure considered.

Wall formwork

The experts were asked to assess the influence of construction time on the formwork quantity to be kept in stock. They specifically concentrated on the formwork required for walls. Table 10.23 lists the statistical values identified in the exploratory data analysis.

The shorter the construction period, the greater the average quantity of formwork to be kept in stock will be. This percentage value refers to the proportion of the total wall formwork area. The standard errors are somewhat larger for the shorter construction times, which can be explained by the higher standard deviation at a constant sample size. Overall, however, we can consider these estimates to be relatively precise. Since all of the coefficients of variation lie in the somewhat higher range, the experts' opinions on the amount of formwork stocking quantity diverge in some cases. However, this may also be due to the individual outliers or extremes, which weigh more heavily for this indicator owing to the way the standard deviation is calculated. All skewness values are positive, which is why the distributions are only right-skewed with the largest number of mentions in the lower value range. A normal data distribution could not be confirmed for any of the variables (Shapiro-Wilk, $p < 0.05$).

The basic data also includes outliers and extremes for all construction time estimates, all of which lie in the upper value range. The M-estimator method was used in order to limit the influence of these values on the averaging results and to prevent an upward bias. The H12 M-estimator applies full weighting to 80% of the data and thus makes it possible to generate plausible and realistic values recommended for construction practice.

Construction time [-]	Extremely short	Very short	Short	Normal	Long
n	30.00	30.00	30.00	30.00	30.00
Mean	31.89	25.16	18.65	11.64	9.51
Standard error of mean	4.69	3.75	2.83	1.80	1.28
Standard deviation	25.67	20.56	15.50	9.85	7.03
Coefficient of variation	80.49%	81.72%	83.09%	84.62%	73.97%
Median	25.00	20.00	15.00	10.00	8.00
Mean absolute deviation	13.00	9.50	6.92	2.75	3.00
Robust coefficient of variation	52.00%	47.50%	46.10%	27.50%	37.50%
Minimum	3.00	3.00	2.00	2.00	1.00
Maximum	100.00	85.00	75.00	50.00	33.00
Spread	97.00	82.00	73.00	48.00	32.00
Skewness	1.44	1.60	2.05	2.52	1.79
M-estimator (H12)	27.15	21.22	15.81	9.45	8.09
Standard deviation (H12)	19.38	14.85	10.54	4.90	4.46
Robust coefficient of variation	71.39%	69.96%	66.65%	51.84%	55.11%
Normal distribution (Shapiro-Wilk)	No	No	No	No	No

Table 10.23 Descriptive statistics – Stocking quantity – Wall formwork[30)]

Slab formwork

The experts were asked to indicate the influence of construction time on the formwork quantity to be kept in stock (in relation to the total floor slab area). This question referred to the amount of slab formwork. Table 10.24 includes the related key descriptive indicators.

The shorter the construction period, the greater the percentage of the formwork stocking quantity will be in relation to the total slab formwork area (see the mean values). However, the standard errors of mean are somewhat higher for the shorter construction periods, which makes the estimates based on the collected data slightly less accurate (larger spreads in the population are more probable owing to the higher mean values). The coefficients of variation indicate that the experts' opinions diverge in some cases. This indicator expresses the average spread relative to the mean, which allows for direct comparisons. Unlike with the wall formwork discussed above, however, the experts appear to agree to a somewhat greater extent. All skewness values are positive, which is why the distributions are only right-skewed with the largest number of mentions in the lower value range. A normal data distribution could not be confirmed for any of the variables (Shapiro-Wilk, $p < 0.05$).

For each of the considered construction periods, the basic data also include outliers and extremes in the upper value range. The M-estimator method was used in order to limit the influence of these values on the averaging results. The H12 M-estimator applies full weighting to 80% of the data and thus makes it possible to generate plausible and realistic values.

30) Of the 35 experts surveyed, 30 responded to this question.

Construction time [-]	Extremely short	Very short	Short	Normal	Long
n	30.00	30.00	30.00	30.00	30.00
Mean	42.48	33.90	24.36	15.53	12.60
Standard error of mean	5.09	4.39	3.03	2.04	1.65
Standard deviation	27.90	24.05	16.60	11.16	9.06
Coefficient of variation	65.67%	70.94%	68.18%	71.86%	71.92%
Median	31.50	25.00	20.00	10.00	10.00
Mean absolute deviation	12.50	10.00	5.00	2.00	2.00
Robust coefficient of variation	39.68%	40.00%	25.00%	20.00%	20.00%
Minimum	8.00	6.00	5.00	4.00	2.00
Maximum	100.00	100.00	75.00	50.00	50.00
Spread	92.00	94.00	70.00	46.00	48.00
Skewness	1.13	1.50	1.77	2.24	2.75
M-estimator (H12)	36.92	29.01	20.52	12.63	10.48
Standard deviation (H12)	21.05	16.59	9.80	5.05	3.78
Robust coefficient of variation	57.02%	57.19%	47.73%	39.99%	36.05%
Normal distribution (Shapiro-Wilk)	No	No	No	No	No

Table 10.24 Descriptive statistics – Stocking quantity – Slab formwork[31)]

Comparison

An exponential trend adjustment of the M-estimators is performed on the basis of the expert data in order to obtain a continuous curve for the relationship between the specified construction time and the stocking quantity. Fig. 10.55 includes a diagram showing these trend curves and the M-estimators for slab and wall formwork, proving the very close approximation of the trend curves to the M-estimators.

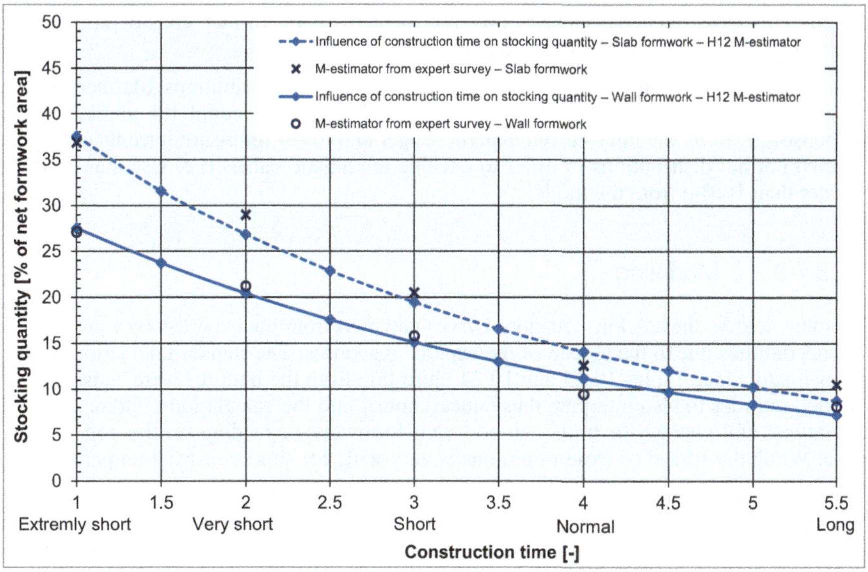

Fig. 10.55 Percentage stocking quantity relative to the specified construction time – Result of an expert survey – M-estimator and exponential trend adjustment

31) Of the 35 experts surveyed, 30 responded to this question.

The value of the percentage stocking quantity SQ_{WF} [%] determined by an exponential trend analysis is a function of the construction time CT [-] for the wall formwork using Eq. (10-10) below.

$$SQ_{WF} = 37.230329 \cdot e^{-0.299509 \cdot CT} \tag{10-10}$$

This applies to a range of $1 \leq CT \leq 5.5$.

The value of the percentage stocking quantity SQ_{SF} [%] determined by an exponential trend analysis is a function of the construction time CT [-] for the slab formwork using Eq. (10-11) below.

$$SQ_{SF} = 51.525680 \cdot e^{-0.323307 \cdot CT} \tag{10-11}$$

This applies to a range of $1 \leq CT \leq 5.5$.

10.3.7.2 Task

We refer to an example to show how the uncertainties of the functional relationship between the stocking quantity and the specified construction time, as derived from the expert survey, can be integrated into a Monte Carlo simulation, as well as the uncertainties associated with assessing the specified construction time.

In qualitative terms, the specified construction time needs to be classified into being too short, normal, or too long, particularly in an early phase of the contractor's project preparation. This assessment is subsequently represented in the further considerations as a corresponding unitless numerical value between 1 and 5.5.

The uncertainties of the survey are simplified by normal distributions (defined by the mean value and the standard deviation) while taking into account the spread of the responses given by the interviewed experts. Upper and lower limits are introduced for the applied normal distributions in order to exclude unrealistic values (i.e. less than 0% and greater than 100%) from the model.

10.3.7.3 Modeling

To some degree, the stocking quantity curves gathered from the expert survey are subject to uncertainties due to the spread of the experts' responses. The standard deviations of the M-estimators (see Tables 10.23 and 10.24, third line from the bottom) were subsequently applied in order to also integrate these uncertainties into the calculations. These standard deviations will change for both slab and wall formwork depending on the construction time. While the spread of expert opinions is very wide for short construction periods, the standard deviations decrease with longer specified construction times.

The standard deviations of the M-estimators and the minima and maxima of the expert data are transformed into exponential trend curves. This step is necessary in order to be able to specify not only the M-estimator, but also the standard deviation and the upper and lower limits as a function of the construction time when defining the distribution functions representing the required stocking quantity.

Wall formwork

The standard deviation was also converted into an exponential trend curve (see Eq. (10-12)) to calculate the curve of the standard deviation for each value of the construction time (between 1 and 5.5) and to base a distribution on it.

$$\sigma_{WF} = 28.240029 \cdot e^{-0.361428 \cdot CT} \tag{10-12}$$

This applies to a range of $1 \leq CT \leq 5.5$.

Although the descriptive statistics of the expert interviews fail to confirm normal distributions of the responses, a normal distribution is defined for the stocking quantity and subsequently truncated in order to exclude unrealistic values (such as negative stocking quantities or quantities over 100%) as early as during the modeling of the distribution function. This truncation also leads to a right-skewed distribution, which accurately reflects the experts' responses.

For the wall formwork, we obtain the exponential trend curve of the minima (see Table 10.23) from the following equation:

$$MIN_{WF} = 4.345321 \cdot e^{-0.242708 \cdot CT} \tag{10-13}$$

This applies to a range of $1 \leq CT \leq 5.5$.

The trend curve of the maximum stocking quantity required for wall formwork is defined by Eq. (10-14). For the purpose of truncating the distribution, we assume the largest permissible stocking quantity to be 100%.

$$MAX_{WF} = 139.302797 \cdot e^{-0.252220 \cdot CT} \tag{10-14}$$

This applies to a range of $1 \leq CT \leq 5.5$.

For a short to very short construction time (CT = 2.5), insertion into Eq. (10-10) and Eq. (10-12) results in a stocking quantity of 17.61% (= μ) and a standard deviation of 11.44% (= σ). These two values are then used for defining a normal distribution (μ = 17.61% | σ = 11.44%). The distribution boundaries of 2.37% and 74.15% result from the trend curves determined for the applied construction time (CT = 2.5). Fig. 10.56 shows the corresponding normal distribution with these boundaries.

Even though the maximum of the distribution was set to 74.15%, this value has such a low probability for the defined distribution that it was not reached in the simulation with 50,000 iterations. The maximum of the distribution determined in the simulation is approximately 67.46% for CT = 2.5.

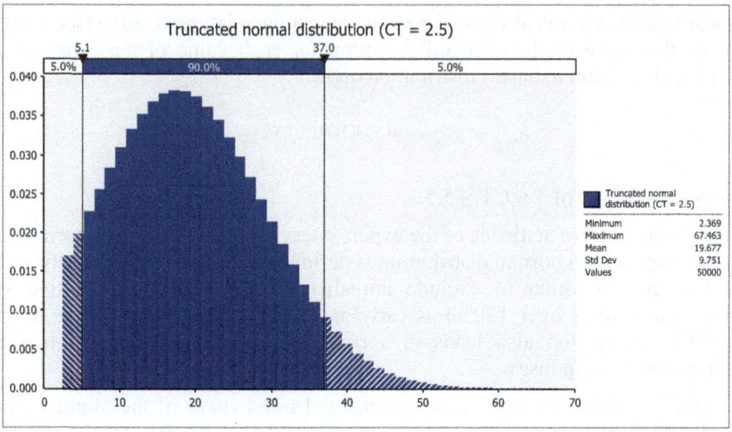

Fig. 10.56 Truncated normal distribution of the percentage stocking quantity for a short construction time (CT = 2.5) – Wall formwork

Slab formwork

The curve of the standard deviation with which the stocking quantity distributions can be defined is an exponential trend curve (see Eq. (10-15)).

$$\sigma_{SF} = 33.293242 \cdot e^{-0.415474 \cdot CT} \tag{10-15}$$

This applies to a range of $1 \leq CT \leq 5.5$.

Although the descriptive statistics of the expert interviews fail to confirm normal distributions of the responses, a normal distribution is defined for the stocking quantity and subsequently truncated in order to exclude unrealistic values (such as negative stocking quantities or quantities over 100%) as early as during the modeling of the distribution function. This truncation also leads to a right-skewed distribution, which accurately reflects the experts' responses.

For the slab formwork, we obtain the exponential trend curve of the minima (see Table 10.24) from Eq. (10-16) below.

$$MIN_{SF} = 11.286316 \cdot e^{-0.294056 \cdot CT} \tag{10-16}$$

This applies to a range of $1 \leq CT \leq 5.5$.

The trend curve for the maximum slab formwork stocking quantity is defined by Eq. (10-17). For the purpose of truncating the distribution, we assume the largest permissible stocking quantity to be 100%.

$$MAX_{SF} = 127.012550 \cdot e^{-0.185133 \cdot CT} \tag{10-17}$$

This applies to a range of $1 \leq CT \leq 5.5$.

For a short to very short construction time (CT = 2.5), insertion into Eq. (10-11) and Eq. (10-15) results in a stocking quantity of 22.96% (= μ) and a standard deviation of 11.78% (= σ). These two values are then used for defining a normal distribution (μ = 22.96% | σ = 11.78%). The distribution limits of 5.41% and 79.95% result from the trend curves determined for the applied construction time (CT = 2.5). Fig. 10.57 shows the corresponding normal distribution with these boundaries.

Even though the maximum of the distribution was set at 79.95%, this value has such a low probability for the defined distribution that it was not reached in the simulation with 50,000 iterations. The maximum of the distribution determined in the simulation is approximately 75.01% for CT = 2.5.

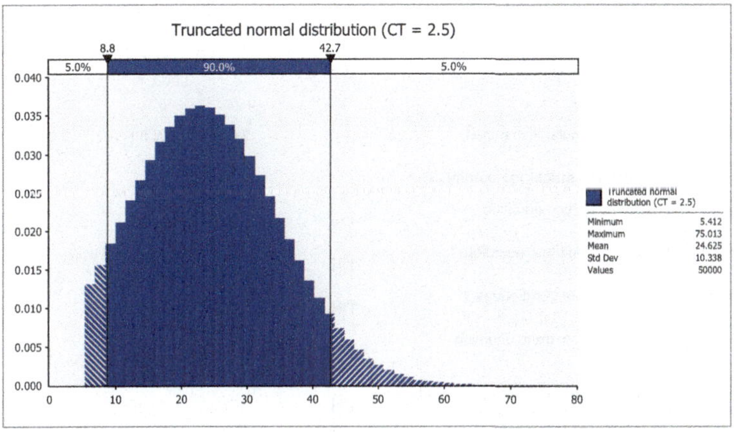

Fig. 10.57 Truncated normal distribution of the percentage stocking quantity for a short construction time (CT = 2.5) – Slab formwork

If the specified construction time is not defined as a single value, such as 2.5, but also as a continuous distribution (for instance, a triangular distribution), a new value for the construction time that lies within the defined limits is selected for each iteration. This value is used as an input parameter for the determined trend curves, and a new truncated normal distribution is defined in each iteration from which a random value is selected each time. Taken together, all stocking quantities expressed as percentages that are calculated as described above produce a histogram taking into account both the uncertainties arising from the functional relationship between construction time and stocking quantity and from the uncertainty associated with the definition of construction time.

10.3.7.4 Calculation and Interpretation of Results

In this specific example, we assume that the initial, rough analysis of a building construction project resulted in a short (CT = 3) to very short (CT = 2) construction time. We apply a triangular distribution (2.0 | 2.5 | 3.0) to the construction time in order to determine the percentage of wall and slab formwork to be kept in stock based on the experts' responses while taking the related uncertainties into account.

For the simulation, we use the system for defining a truncated normal distribution (see Section 10.3.7.3).

Wall formwork

The total net formwork area of the vertical structural members (walls) amounts to approximately 10,000 m².

Table 10.25 is the calculation table for determining the wall formwork stocking quantity. Row 1 states the unitless values representing the expected construction time (2.0 | 2.5 | 3.0). The corresponding triangular distribution is stored in Cell F1.

Cells F3 to F6 each contain the calculation of the mean, standard deviation, minimum, and maximum of the subsequently defined truncated normal distribution. Each of these four cells includes a reference to Cell F1, in which a different value (within the defined range) is randomly selected for each iteration. The equations (exponential trend curves) for calculating the parameters in Cells F3 to F6 are shown in Rows 3 to 6.

No.	Parameter	Unit	MIN	MOST LIKELY	MAX	Distribution
0	A	B	C	D	E	F
1	Construction time – triangular distribution	[-]	2.00	2.50	3.00	2.50
2	**Calculation of required stocking quantity**					
3	Mean – calculated from trend equation – $y = 37.230329e^{-0.299509x}$	[%]				17.61
4	Std.dev. – calculated from trend equation – $y = 28.240029e^{-0.361428x}$	[%]				11.44
5	Minimum – calculated from trend equation – $y = 4.345321e^{-0.242708x}$	[%]				2.37
6	Maximum – calculated from trend equation – $y = 139.302797e^{-0.252220x}$	[%]				74.15
7	**Distribution for the required stocking quantity – Wall formwork [% of net formwork area]**	**[%]**				**19.68**
8	Net formwork area	[m²]				10,000.00
9	**Stocking quantity**	**[m²]**				**1,967.68**

Table 10.25 Calculation table – Stocking quantity – Wall formwork

The truncated normal distribution is defined in Cell F7 and gives the stocking quantity as a percentage of the net formwork area. A new distribution is thus generated and a random value selected in each iteration of the Monte Carlo simulation. This random value is then incorporated in the final result obtained for the stocking quantity.

Thus, the simulation results in a histogram representing the wall formwork stocking quantity as a percentage, which takes into account both the uncertainties associated with the experts' responses and those of the estimates of the construction time (see Fig. 10.58).

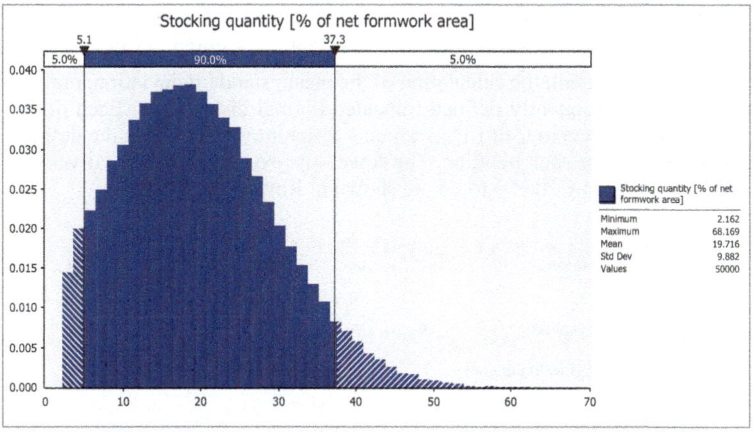

Fig. 10.58 Percentage stocking quantity – Wall formwork

Likewise, multiplying this histogram by the total net formwork area (= 10,000 m²) will result in a histogram of the wall formwork stocking quantity (see Fig. 10.59). This histogram shows a mean value of about 1,972 m² and a standard deviation of about 988 m². This range represents a relatively large spread of the results attributable to the variations in the figures stated by the experts. However, this wide spread is also plausible on the basis of the rough analysis and shows, for example, that a spread of approximately 2,000 m² would cover over 50% of all cases.

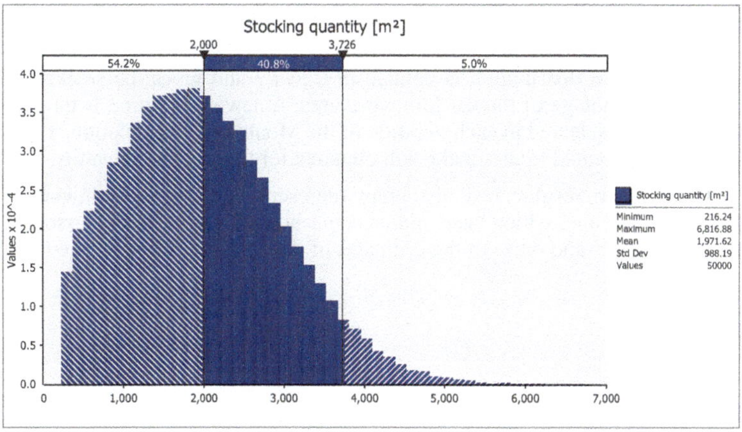

Fig. 10.59 Stocking quantity – Wall formwork

Slab formwork

The total net formwork area of the horizontal structural members (floor slabs) amounts to approximately 8,000 m². The figure indicating the ratio between the formwork areas of the horizontal and vertical structural components[32)] thus amounts to 0.80 (= 8,000 m² / 10,000 m²).

Table 10.26 is the calculation table for determining the slab formwork stocking quantity. Row 1 states the unitless values representing the expected construction time (2.0 | 2.5 | 3.0). The corresponding triangular distribution is stored in Cell F1.

Cells F3 to F6 each contain the calculation of the mean, standard deviation, minimum, and maximum of the subsequently defined truncated normal distribution. Each of these four cells includes a reference to Cell F1, in which a different value (within the defined range) is randomly selected for each iteration. The equations (exponential trend curves) for calculating the parameters in Cells F3 to F6 are shown in Rows 3 to 6.

No.	Parameter	Unit	MIN	MOST LIKELY	MAX	Distribution
0	A	B	C	D	E	F
1	Construction time – triangular distribution	[-]	2.00	2.50	3.00	2.50
2	**Calculation of required stocking quantity**					
3	Mean – calculated from trend equation – $y = 51.525680e^{-0.323307x}$	[%]				22.96
4	Std.dev. – calculated from trend equation – $y = 33.293242e^{-0.415474x}$	[%]				11.78
5	Minimum – calculated from trend equation – $y = 11.286316e^{-0.294056x}$	[%]				5.41
6	Maximum – calculated from trend equation – $y = 127.012550e^{-0.185133x}$	[%]				79.95
7	**Distribution for the required stocking quantity – Slab formwork [% of net formwork area]**	**[%]**				**24.63**
8	Net formwork area	[m²]				8,000.00
9	**Stocking quantity**	**[m²]**				**1,970.02**

Table 10.26 Calculation table – Stocking quantity – Slab formwork

The truncated normal distribution is defined in Cell F7 and gives the stocking quantity expressed as a percentage of the net formwork area. A new distribution is thus generated and a random value selected in each iteration of the Monte Carlo simulation. This random value is then incorporated in the final result obtained for the stocking quantity.

Thus, the simulation results in a histogram representing the slab formwork stocking quantity as a percentage, which takes into account both the uncertainties associated with the experts' responses and those of the estimates of the construction time (see Fig. 10.60).

[32] Cf. Hofstadler (2008), p. 461f.

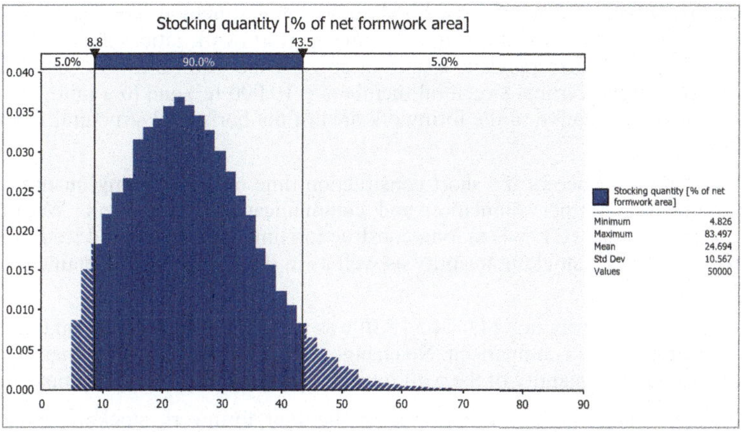

Fig. 10.60 Percentage stocking quantity – Slab formwork

Likewise, multiplying this histogram by the total net formwork area (= 8,000 m²) will result in a histogram of the slab formwork stocking quantity (see Fig. 10.61). This histogram shows a mean value of about 1,975 m² and a standard deviation of approximately 845 m². This range represents a relatively large spread of the results attributable to the variations in the figures stated by the experts. However, this wide spread is also plausible on the basis of the rough analysis and shows, for example, that a spread of approximately 2,000 m² would cover over 50% of all cases.

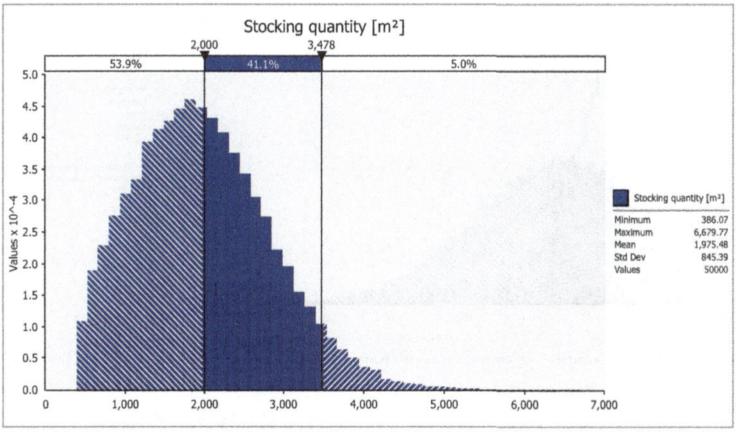

Fig. 10.61 Stocking quantity – Slab formwork

10.3.7.5 Benefit for the Bidder/Contractor

Although this determination of the formwork stocking quantity is a very rough estimate, the example given shows that about 2,000 m² of formwork must be kept in stock for both the slab and wall formwork in order to cover about 54% of all simulated cases. It should be noted that the net formwork area for the horizontal structural members is 2,000 m² smaller than that needed for the vertical members (formwork ratio = 0.80). A stocking quantity of 2,000 m² corresponds to a ratio of 20% for the wall formwork (relative to the formwork area of the vertical structural members = 10,000 m²) and to a ratio of 25% for the slab formwork (relative to the formwork area of the horizontal structural members = 8,000 m²).

In addition, the influence of the short construction time on the stocking quantity can be demonstrated in a further simulation and communicated to the client. We find, for instance, that a normal (CT = 4) to long construction time (CT = 5) would lead to a significant reduction in the stocking quantity as well as in the spread of the obtained stocking quantity figures.

Again, a triangular distribution (4.0 | 4.5 | 5.0) was applied to the construction time for the purpose of establishing a comparison. No changes were introduced to the remaining calculation steps. Only the quantity of the wall formwork is considered in this comparison.

Fig. 10.62 compares the two histograms of the wall formwork stocking quantity for different specified construction times. We find that for normal or long construction periods, the stocking quantity can be stated in a much narrower range than for short or very short construction periods. To reach the same chance/risk ratio (i.e. approximately 54% chance to 46% risk), about 2,000 m² of formwork need to be kept in stock for walls if a short to very short construction time is specified, whereas this figure amounts to only about 1,070 m² for a normal to long construction period.

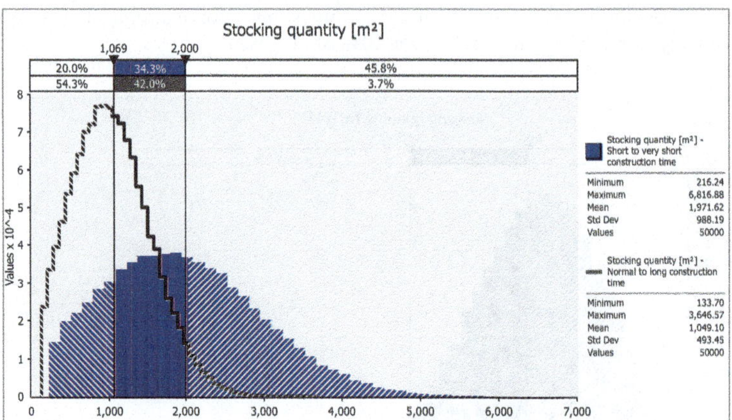

Fig. 10.62 Stocking quantity – Comparison of short and long construction time – Wall formwork

Although the rough determination of the stocking quantity is not directly applicable to costing or process planning, it does require an initial assessment of the circumstances in which the work is to be performed (i.e. assessing the specified construction time) while indicating the ratio of the stocking quantities required for slab and wall formwork.

Particularly if a very short or extremely short construction time is specified, the described calculation method serves as a communication tool for internal decision making on the one hand and as a line of argument in discussions with the client on the other.

10.3.8 Costing (Zero-Base Approach)

Costing creates the basis for pricing and thus for the preparation of the bids to be submitted by the bidders (PPH 3, see Fig. 10.63). Only if the bidders are aware of the costs to be expected for the specified works will they be able to set prices on a sound basis.

Unlike in pricing, the costing exercise (applying a zero-base approach) merely determines prices that cover the full costs, excluding speculative elements, of construction works to be performed at a later stage.[33)]

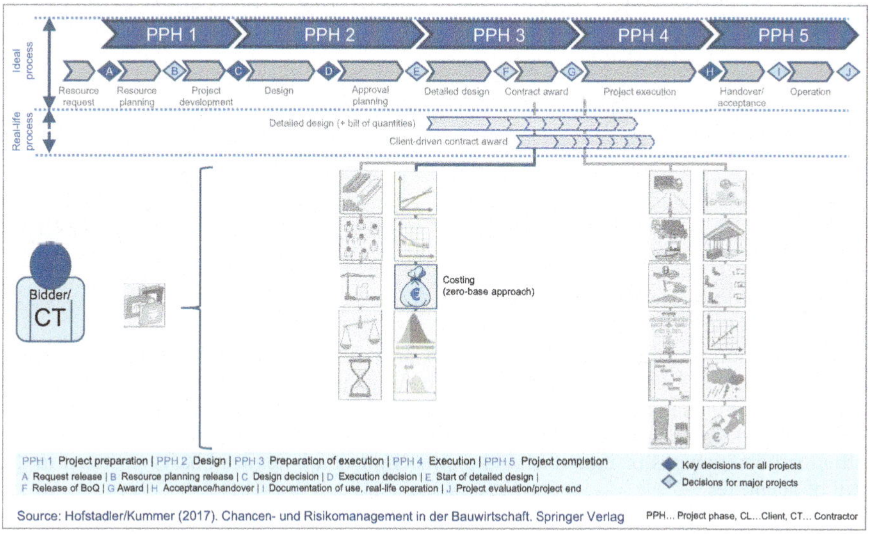

Fig. 10.63 Allocation of example in the project sequence

10.3.8.1 Task

Determining the costs of construction works to be performed in the future is obviously fraught with uncertainties. These uncertainties can be taken into account either non-systematically through percentage markups or systematically by referring to ranges of input parameters.

For the systematic approach, the calculation parameters are modeled using distribution functions instead of individual (deterministic) numerical values. The probabilistic calculation with the aid of a Monte Carlo simulation subsequently provides a cost histogram that can be assessed in the pricing phase with regard to the associated chances and risks.

As an illustrative example, the direct costs of reinforced concrete operations are determined in a Monte Carlo simulation, and the result is interpreted in a realistic manner.

33) Cf. Oberndorfer/Jodl et al. (2001), p. 36

10.3.8.2 Modeling

The direct costs of reinforced concrete operations C_{RCO} [€] (see Eq. (6-3) and Fig. 10.64) are calculated by multiplying the mean wage MW [€/wh] by the total labor consumption rate of reinforced concrete operations LCR_{RCO} [wh/m³] plus the equipment costs C_{EQ} [€/m³] and material costs C_{MA} [€/m³]. This sum of the unit costs is then multiplied by the reinforced concrete volume C_V [m³].

This example assumes a range of possible values (minimum and maximum) and a most likely value (highest probability) for each of the input parameters. These three values make it possible to define a triangular distribution serving as an input for the simulation (see Fig. 10.64).

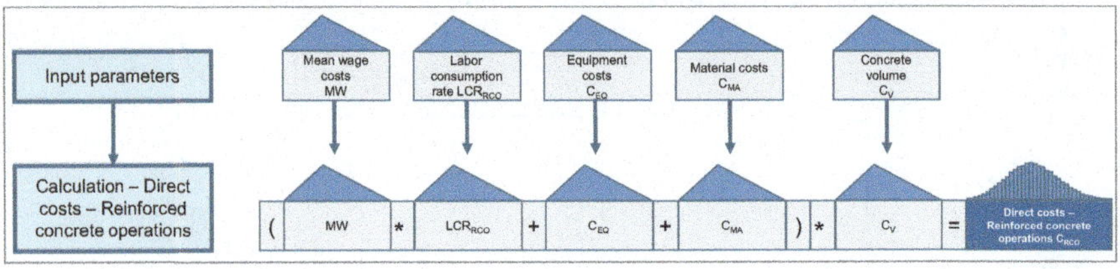

Fig. 10.64 Calculation flow chart

The calculation process shown in Fig. 10.64 provides a rough approximation for determining the direct costs of reinforced concrete operations. Depending on the degree of detail, the individual input parameters can be calculated in detail (for instance separately for formwork, reinforcing, and concreting operations and/or separately for groups of structural components or individual components) or on an itemized basis as stated in the bill of quantities.

10.3.8.3 Calculation and Interpretation of Results

The calculation of the direct costs of reinforced concrete operations has already been discussed in Section 6.2.5.3 both with deterministic values and triangular input parameter distributions while applying the Monte Carlo simulation method.

Fig. 10.65 shows the histogram determined in the simulation for the direct costs of reinforced concrete operations. The shape of the histogram roughly replicates a normal distribution, but it includes both a defined minimum (approx. €15.09 million) and a defined maximum (approx. €19.02 million). The mean of the direct costs amounts to approximately €17.0 million whereas the standard deviation roughly amounts to €545,100. About 90% of the total 50,000 iteration results lie between €16.12 million and €17.92 million (see the slider at the top of Fig. 10.65). Costs are below or above this range in 5% of the simulated cases.

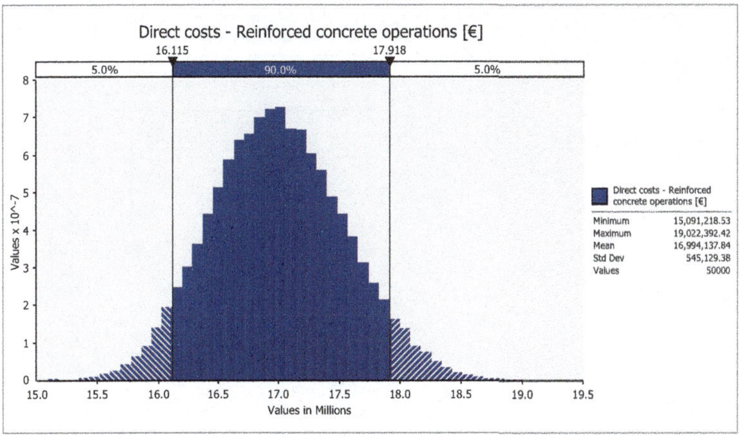

Fig. 10.65 Histogram – Direct costs of reinforced concrete operations

10.3.8.4 Benefit for the Bidder/Contractor

The probabilistic zero-base costing is associated with the advantage that the "true" cost values are shown, i.e. excluding speculative elements while considering inevitable uncertainties.

The cost surveyor does not have to choose a value within a range for each input parameter, but can directly and systematically incorporate experience and findings from studying the tender documents in the form of ranges (minimum and maximum) and, for example, the specification of an expert (most likely) value. The results of the zero-base costing exercise are represented by histograms, such as for the entire building, individual structural components, or specific work groups or items, depending on the applied classification system. Histograms deliver transparent information about the ranges and probabilities to be anticipated for specific deterministic values.

Only at the pricing stage will a single value (price) and the associated chance/risk ratio be determined. The probabilistic calculation of costs creates the basis for any conclusions regarding chances and risks associated with specific prices.

10.3.9 Assessment of Costing Risk

After completion of the probabilistic cost calculation (zero-base costing; see Section 10.3.8), the question arises during the pricing process which calculation risk the bidder takes, or can/should take, when submitting their bid. The related considerations concentrate on selecting an appropriate chance/risk ratio as well as on the influence of markups and discounts on this ratio. This example is part of PPH 3 (see Fig. 10.66).

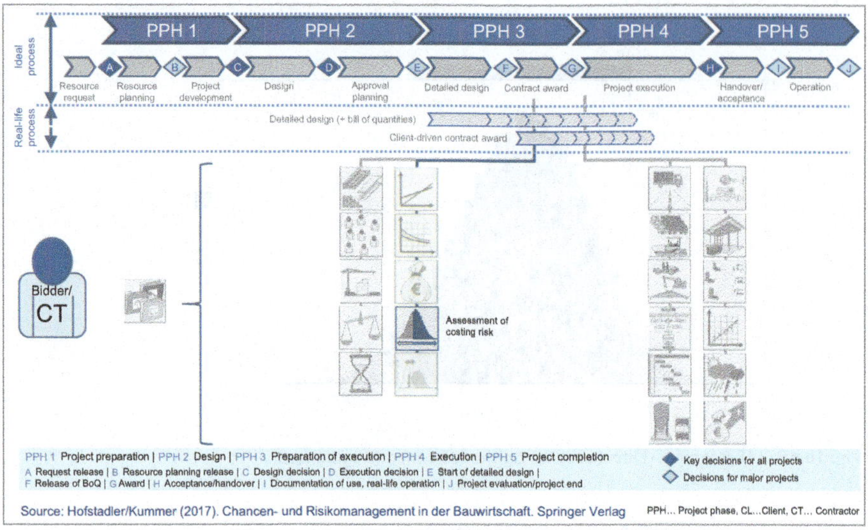

Fig. 10.66 Allocation of example in the project sequence

10.3.9.1 Task

In this example, the histogram determined in Section 10.3.8 (see Fig. 10.65) for the direct costs of reinforced concrete operations is analyzed with regard to the associated chance/risk ratio.

Assessing the probability of specific deterministic values being exceeded or underrun permits conclusions with respect to the effects of percentage markups or discounts. Decision making becomes more transparent and can be understood and interpreted more easily thanks to the graphical representation of the chance/risk ratio.

10.3.9.2 Modeling

We previously discussed the modeling for the probabilistic calculation of the direct costs of reinforced concrete operations in Section 10.3.8.2. The assessment of the costing risk is based on the results of this probabilistic cost calculation.

10.3.9.3 Calculation and Interpretation of Results

The probabilistic cost calculation provides the basis for selecting a deterministic value within the determined range and thus for defining the chance/risk ratio. There are various possible reasons for selecting a certain chance/risk ratio. For example, a bidder may accept or offer contracts with a comparatively high risk potential in order to ensure full resource utilization of their operations. In addition, the market situation or strategic considerations (such as entering a new market segment or acquiring landmark projects) may lead to the choice of a low potential chance. If the bidder chooses their basis (i.e. the deterministic cost value) in the middle (at the median) of the histogram in Fig. 10.67, the probability of exceeding or underrunning it is 50% in each case. There is thus a balance between chances and risks.

If a particularly high cost value is chosen (e.g. the 5% quantile at €17.92 million), there is a high chance that the bidder or subsequent contractor will incur lower costs in the execution phase, provided they are awarded the contract at all given the high bid price (see also Fig. 7.26).

Bidders tend to offer low prices owing to the fierce competition on the market. The selected basis will thus be lower than the median. With an offered value of €16.12 million, the bidder or contractor has only a chance of approximately 5% that the costs can be underrun. On the other hand, the risk of exceeding these costs is much higher at about 95%.

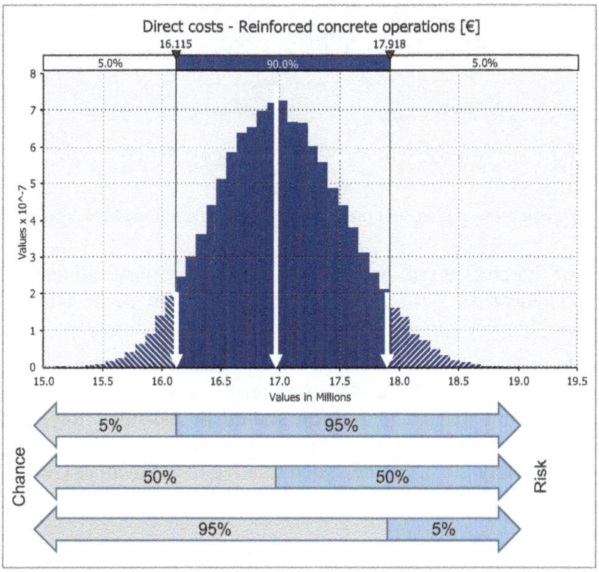

Fig. 10.67 Chance/risk ratios – Direct costs of reinforced concrete operations

There is a 50/50 chance/risk ratio if the bidder enters into contractual negotiations with the client on the basis of a deterministic value corresponding to the median. If, in the course of the negotiations, the bidder decides to grant a blanket 2% discount on the contract amount, it will be in the bidder's best interest to identify the effect this discount would have on the chance/risk ratio compared to their costing/pricing. In this context, knowledge of the expected spread of costs is essential in order to explore the price limits in the negotiations.

Fig. 10.68 shows a graphical representation of this situation. Granting the 2% discount (i.e. reducing the amount from about €16.98 million to about €16.64 million) increases the bidder's/contractor's risk from 50% to over 73%. Conversely, the chance of incurring lower costs in the execution phase decreases from 50% to about 27%.

The effects of a percentage markup can be assessed in the same way. As in the previous example, such an increase in the baseline will lead to an increase in the potential for chances while reducing the risk potential.

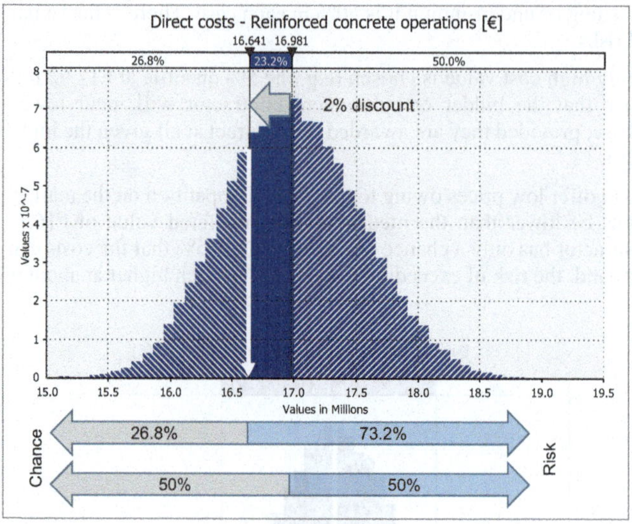

Fig. 10.68 Chance/risk ratios – Discount on the direct costs of reinforced concrete operations

If, on the basis of the chance/risk policy or of the project portfolio analysis (see Section 10.6), a certain chance/risk ratio is sought, the maximum permissible discount or the minimum permissible price can also be stated in the reverse direction.

For this example, we assume that, based on a management decision, a ratio of 20% (chance) to 80% (risk) may not be underrun in terms of the chance potential. This means that a minimum amount of €16.53 million should be assumed for the direct costs of reinforced concrete operations based on the derived histogram. A maximum permissible discount of approximately 2.6% can be stated relative to the median (see Fig. 10.69).

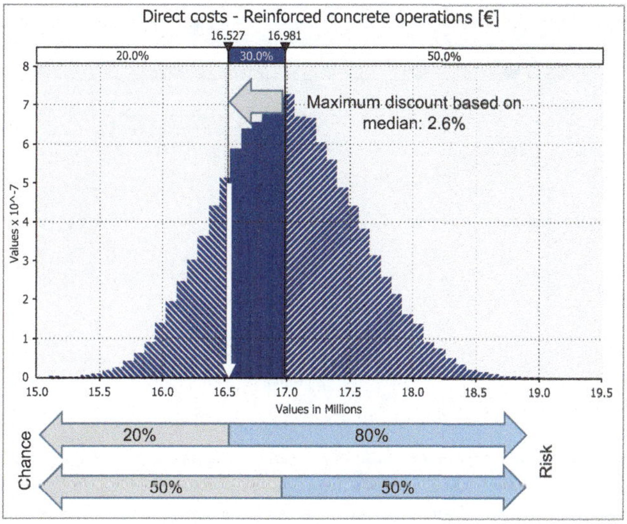

Fig. 10.69 Chance/risk ratios – Maximum discount on the direct costs of reinforced concrete operations

10.3.9.4 Benefit for the Bidder/Contractor

Bidders, or subsequent contractors, can only estimate the risk they are taking with their bids if they know the range of possible costs. Combined with the corresponding probabilities for the values considered, this range provides the result of a probabilistic cost calculation performed with the aid of a Monte Carlo simulation.

Based on the distribution, the ratio of chances and risks can be derived directly and displayed graphically using histograms. Histograms are useful both for internal communications on uncertainties and for external communications with the client or with other contractors in order to highlight the possible event space in a compelling manner.

The analysis of cost histograms with regard to the changed probability of exceeding or underrunning individual deterministic values is particularly suitable for evaluating blanket discounts or markups while taking the chance/risk ratio into account.

10.3.10 Bid Pricing

Pricing is one of the most important tasks for businesses in order to get orders in the first place and to be able to compete in the free market. Particularly in the case of public contracts, price is one of the main award criteria and thus requires specific, in-depth consideration.

Pricing follows the costing exercise and is assigned to the final part of PPH 3 (see Fig. 10.70).

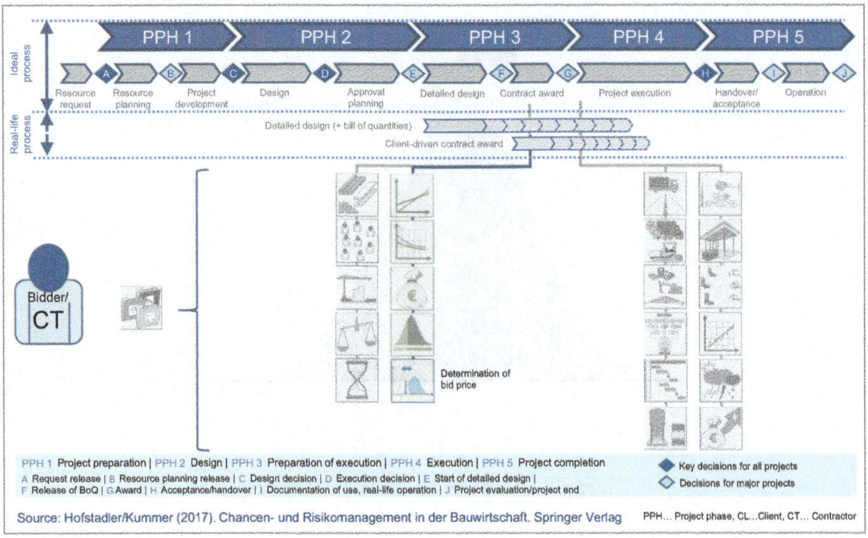

Fig. 10.70 Allocation of example in the project sequence

10.3.10.1 Basics

In general, a financial loss is imminent if the quotient of construction cost and construction price is greater than 1 (see Eq. (10-18)). The risk of this to occur is referred to as construction price risk.[34)]

$$\frac{\text{Construction cost}}{\text{Construction price}} \tag{10-18}$$

It makes sense to apply probabilistic calculation methods with which histograms for construction costs can be determined separately for items, cost groups, building components, work groups, principal groups, trades, or the entire building, because costs will always be subject to some uncertainties, which applies particularly to the costs of works to be performed in the future. Any agreement on specific prices will thus be based on a cost forecast.[35)]

The concept of construction price risk that *Raaber* defines as the quotient of construction cost and construction price can also be illustrated graphically by presenting the building costs as a histogram or distribution function and the construction price as a deterministic value (see Fig. 10.71).

34) Cf. Raaber (1979), p. 1

35) Cf. ibid.

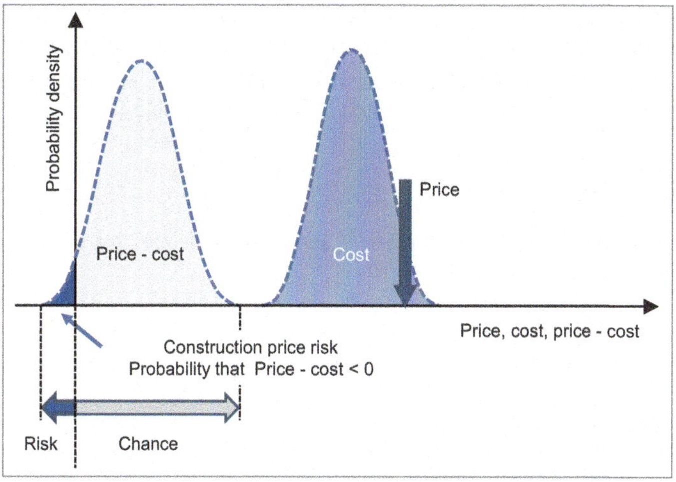

Fig. 10.71 Construction price risk

The construction price risk will have materialized if the costs exceed the bid price, i.e. if the difference between price and cost lies below zero. In all other cases where the costs are lower than the bid price the contractor has the chance to generate a profit.[36)]

No.	Cost groups	Unforeseeable influences	
		market-related	performance-related
0	A	B	C
1	Construction management costs	▪ Salaries ▪ Wages ▪ Preparation: land, connection of utilities	▪ Number of required staff ▪ Technical and organizational requirements
2	Equipment costs	▪ Cost of purchase ▪ Book values ▪ Cost of spares	▪ Performance ▪ Susceptibility to repair ▪ Suitability
3	Material costs	▪ Purchase price ▪ Transport costs ▪ Handling	▪ Quantity required for contractual performance ▪ Handling
4	Wage costs	▪ Pay under collective agreements and beyond ▪ Travel and accommodation ▪ Daily allowances ▪ Social security contributions	▪ Skills and qualifications ▪ Number of working hours required for contractual performance
5	Ancillary costs (materials and wages)	See Nos. 2-4	See Nos. 2-4
6	Fixed costs	General overheads	Allocation to sales

Table 10.27 Unforeseeable influences on different cost groups – Market and performance-related[37)]

36) Provided full costs (direct costs and overheads) were taken into account.

37) Cf. Raaber (1979), Appendix, Sheet 1

The tense situation in the construction industry and the fierce competition for contracts force bidders to quote prices near the lower end of the cost distribution range. Any increase in the construction price risk will lead to correspondingly reduced chances.

Raaber summarizes the uncertainties that affect different costs in tabular form, divided into market-related and performance-related influences (see Table 10.27).

10.3.10.2 Task

This example was developed by *Raaber*[38)] and refers to factors for the individual cost groups, rather than to absolute cost values.

It includes two analyses, and the base case assumes a certain composition of the total costs and unforeseeable (market and performance-related) influences by means of percentages. This initial consideration results in a spread of the total costs, which can be displayed as a histogram.

The second analysis refers to the same construction project. However, we assume that the composition of the cost group proportions has changed as a result of the exercise of influence. In addition, the performance-related uncertainties have also been reduced, for instance by optimizing the construction process, such that their range has narrowed. In this case, too, the total costs are subject to a certain spread, which, however, differs from that documented in the first case in that it is lower.

The aim is to compare the two histograms and to reflect and interpret the effects of a percentage risk premium on the chance/risk ratios of the two cases.

10.3.10.3 Modeling

Raaber's model reflecting the costs per cost group is based on normal distributions and simultaneously gives percentage values for the upper and lower cost limits. This reveals a contradiction since, in statistical terms, we define a normal distribution as being two-sided open without boundaries. For the subsequent calculation steps, we thus assume that the given "limit values" correspond to the 1% and 99% quantiles.

Fig. 10.72 shows the flow chart for this total cost calculation.

[38)] Cf. ibid.

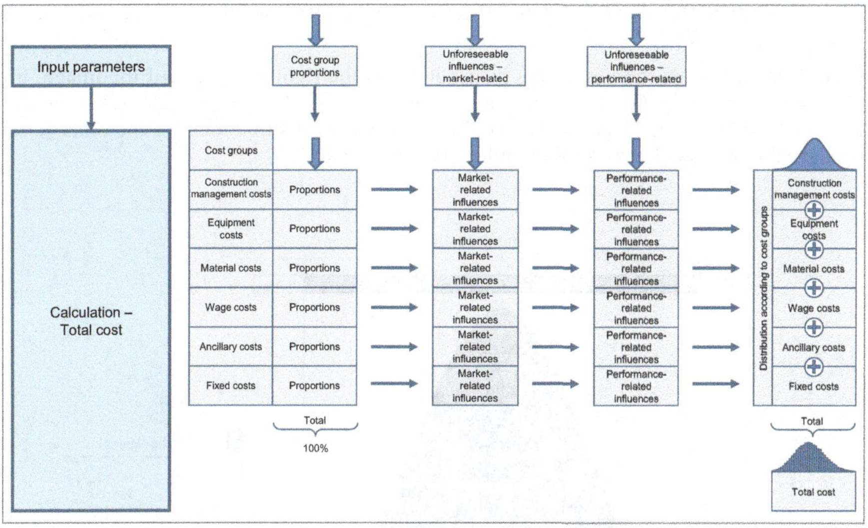

Fig. 10.72 Calculation flow chart

For a calculation example, the costs per cost group are not stated as absolute values, but as factors that assume the total value of 1.00, or 100% (see Table 10.28). Wage costs account for the largest share at 50% (see Cell B4), followed by material costs at 32% (Cell B3).

Raaber describes this cost breakdown as conventional or old-fashioned since the wage proportion was set high while the costs of equipment and construction management were assumed to be very low.[39]

No.	Cost groups		Unforeseeable influences	
			market-related	performance-related
0	A	B	C	D
1	Construction management costs	0.03	2.00%	5.00%
2	Equipment costs	0.04	2.00%	6.00%
3	Material costs	0.32	2.00%	3.00%
4	Wage costs	0.50	6.00%	15.00%
5	Ancillary costs	0.03	4.00%	6.00%
6	Fixed costs	0.08	0.00%	2.00%
7	Total cost	1.00		

Table 10.28 Calculations by the contractor

The percentages given in Columns C and D of Table 10.28 indicate uncertainties in the cost estimates due to market- and performance-related effects and should be understood in both the positive and the negative direction. To model the entire spread (within the 1% and 99% quantiles), the percentages of market- and performance-related effects must be added up. For example, a share of 4% (a factor of 0.04) in the total cost is determined for equipment costs (Row 2). The corresponding range is ±8% (±2% market-related and ±6% performance-related). Fig. 10.73 illustrates the distribution function (normal distribution)

[39] Cf. Raaber (1979), p. 13

of the equipment costs. The mean is equivalent to the equipment cost factor (i.e. 0.04). The 1% quantile was determined as 0.0368 (= 0.04 * 0.92), and the 99% quantile equals 0.0432 (= 0.04 * 1.08).[40)]

As can be seen from the key of the distribution on the right-hand side of the image, the minimum and maximum of the distribution are not defined as numerical values because the distribution extends from minus to plus infinity. The distributions for the other listed cost groups are modeled according to the same approach.

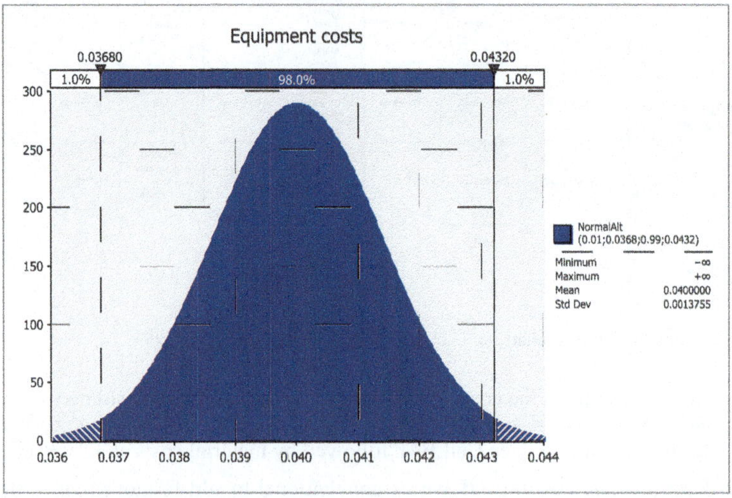

Fig. 10.73 Equipment costs – Normal distribution

10.3.10.4 Calculation and Interpretation of Results

The costs modeled as normal distributions of the six cost groups are added up using the Monte Carlo simulation method. This calculation results in a histogram (see Fig. 10.74) that permits conclusions regarding the probability of specific values being over- or underrun.

Although the individual distributions of the cost groups were set as normal distributions with two-sided open boundaries (from minus to plus infinity), the total cost histogram shows a specific minimum as well as a specific maximum.

This can be explained by the fact that random values within the ranges (from minus to plus infinity) were selected and added up to arrive at the total cost. With a finite number of iterations (in this case, 50,000) and presentation of the entirety of the results as a histogram, we get a closed range of the total costs as shown in Fig. 10.74.

If we assume the mean of the histogram (= 1.00), a cost overrun of more than 13% appears to be possible. Yet the selected input parameters would also allow for a cost underrun by over 13%.

40) Feature for defining the distribution function in @Risk: =RiskNormalAlt(1%;0.0368;99%;0.0432)

There is only a 49% probability that costs will vary by less than ±3% (which could be a plausible approach to the risk premium in the positive direction) around the mean value. The probability that the mean will not be exceeded by more than 3% is approximately 74.4% (= 48.9% + 25.5%).

In this case, this means that a 3% risk premium still leaves an (uncovered) risk of about 26%.

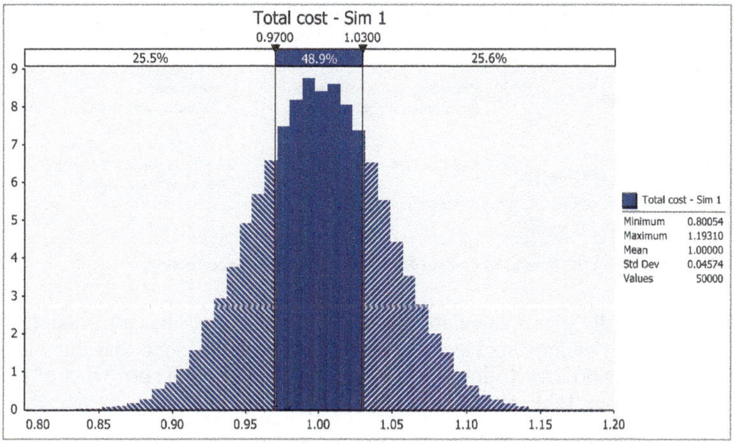

Fig. 10.74 Histogram – Total cost – Simulation 1

Measures can be implemented to control and determine the workflow and to counter human inadequacies in order to narrow the total cost range so as to organize manageable, maximally predictable, and reproducible site operations. Such measures must be prepared and planned in the process planning phase. A detailed and comprehensive process planning exercise will reduce not only the costs in the construction phase, but also the spreads caused by uncertainties.[41)]

Raaber cites a poorly constructed site road as an example. This road will provide a transport capacity of 100 t/h under normal (average) weather conditions. In particularly good weather it will allow for 120 t/h. In bad weather, such as during long periods of rain, it will become unusable. Should cost surveyors apply the long-term weather average or should they assume a less probable, but extremely severe bad weather case in their costing exercise? In other words, how high should the risk premium be? Under certain conditions, a road upgrade might appear to be the better option. This will lead to higher costs of the "site road" item, but reduce the total risk associated with the costing exercise. The remaining part of the construction process, which is dependent on transport, will also be organized more efficiently, which will help to narrow down the ranges further. The overall cost forecast will become more accurate in terms of precision and correctness.[42)]

Raaber lists general options or starting points for influencing the spread of costs per cost group in tabular form, separated according to market- and performance-related effects (see Table 10.29).

[41)] Cf. Raaber (1979), p. 13f.

[42)] Cf. ibid., p. 14

No.	Cost groups	Exerting influence on spread	
		market-related	performance-related
0	A	B	C
1	Construction management costs	Recruiting the required staff	**Process planning:** • Planning of workflows • Time/work studies • Time schedule (network techniques) • Site mobilization plan • Equipment use plan • Traffic routes • Accessibility of workplaces • Transport and storage • Preliminary tests • Consumption and wear • Protection against harmful environmental effects
2	Equipment costs	• Sourcing the required equipment • Condition (overhauled)	
3	Material costs	Supply contracts	
4	Wage costs	Recruiting the required staff (permanent employment)	
5	Ancillary costs (materials and wages)	See Nos. 2-4	
6	Fixed costs		

Table 10.29 Influencing the spread of costs – Market- and performance-related[43)]

With reference to the given calculation example, we assume that all possible ways of influencing the cost parameters have been identified and used, and that this has led to a shift in the cost proportions (Column B) on the one hand and to a narrowing of the ranges on the other hand (see Table 10.30 compared to Table 10.28).

We find that there has been a shift in the cost proportions between the groups. More specifically, the wage cost proportion was reduced from 0.50 to 0.40. Conversely, the shares of construction management and equipment costs were increased. The uncertainties resulting from market-related influences (Column C) were left unchanged. However, the cost ranges were narrowed significantly due to the performance-related effects (Column D).

No.	Cost groups		Unforeseeable influences	
			market-related	performance-related
0	A	B	C	D
1	Construction management costs	0.06	2.00%	3.00%
2	Equipment costs	0.10	2.00%	4.00%
3	Material costs	0.32	2.00%	2.00%
4	Wage costs	0.40	6.00%	5.00%
5	Ancillary costs	0.04	4.00%	3.00%
6	Fixed costs	0.08	0.00%	2.00%
7	Total cost	1.00		

Table 10.30 Contractor's calculation after utilizing available options

For the total cost calculation, we again assume that the costs per cost group are defined as normal distributions. The "limits" defined by the sum of the percentages refer to the 1% and 99% quantiles. After 50,000 iterations applying the Latin Hypercube sampling method, the Monte Carlo simulation delivers the total cost histogram shown in Fig. 10.75.

43) Cf. Raaber (1979), Appendix, Sheet 5

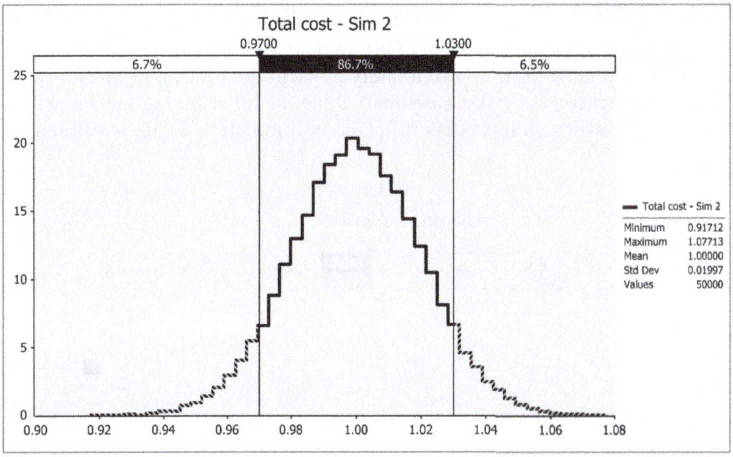

Fig. 10.75 Histogram – Total cost – Simulation 2

The spread has become significantly smaller for the same average expected costs (= 1.00). At this stage, approximately 86.7% of the simulation results lie within a range of 3% on either side of the mean (compared to approximately 48.9% in Simulation 1). Overall, we thus succeeded in narrowing the cost spread.

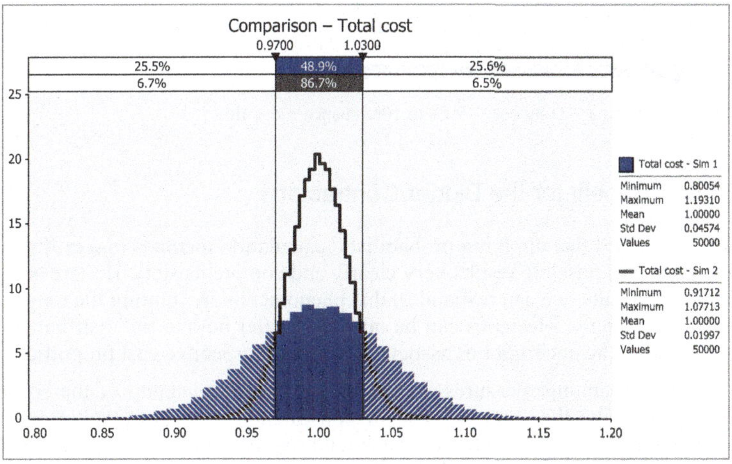

Fig. 10.76 Comparison – Total cost – ±3% from mean

A direct comparison of the two histograms derived from the first and second simulation makes the narrowing of the spread even clearer for the total cost (see Fig. 10.76). While in the first simulation the total spread of costs is in a range of approximately ±19% (see the minimum and maximum of the histogram of the first simulation in Fig. 10.76), in the second simulation this range was narrowed to approximately ±8% (see the minimum and maximum of the histogram of the second simulation in Fig. 10.76). In the second case, a

3% risk premium could cover about 93.4% (= 86.7% + 6.7%) of the calculated cases. Accordingly, the construction price risk would amount to about 7% (counterprobability).

If we aim to achieve a chance/risk ratio of 90% (chance) to 10% (risk) (see Fig. 10.77), the risk premium would have to be approximately 5.9% in the first case, but only about 2.6% in the second. For reasons of risk assessment alone, or for retaining the same chance/risk ratio, the planned construction process reflected in Simulation 2 can be offered at a lower price.

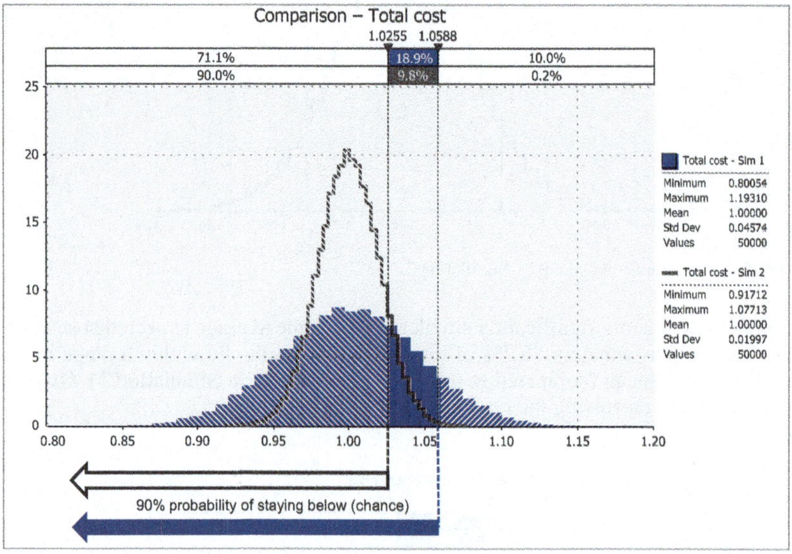

Fig. 10.77 Comparison – Total cost – 90% to 10% chance/risk ratio

10.3.10.5 Benefit for the Bidder/Contractor

In summary, we find that applying probabilistic calculation methods makes it possible to present the range of possible results very clearly and comprehensibly. If there is too much variation in the results, we can respond to this phenomenon by limiting the range of input parameters accordingly. Measures can be taken that refer both to the distribution of cost proportions and to the uncertainties associated with the respective cost proportions.

Specific process planning measures as well as the precise planning of the construction process help to make the future course of construction more predictable and thus to narrow the possible ranges. In addition, such a narrowing will have a direct effect on the achievable chance/risk ratio – applying the same percentage risk premium –, as the example demonstrates.

The considerations regarding the definition of the distribution functions for the input parameters of the simulation help raise awareness among cost surveyors to refrain from consistently using the lowest values found in the internal database or in the literature. Dealing with probabilistic calculations thus also promotes the critical examination of final costing values and of theoretically and empirically sound references in the literature. Limits are thus set to an overly optimistic approach to determining costs and, subsequently, prices. Even if the market tends to move in the direction of lower prices, the methodology outlined above is suitable for consistently quantifying the risk taken or the chance remaining for the bidder or subsequent contractor.

Raaber urges construction practitioners to conscientiously implement the tasks of internal data collection and processing in order to create a sound basis for future costing. Its purpose is not limited to just collecting mean values; it also includes the targeted collection and illustration of statistically determined variations.[44)]

Nowadays, commercially available software programs require only little mathematical and probabilistic knowledge to carry out a fitting of a wealth of data, which allows for a graphical representation of the ranges and weightings of specific costing parameters.

10.4 Project Phase 4 (PPH 4)

In Project Phase 4, contractors perform the commissioned works. At this stage, they are strongly interested in optimizing their production system. In so doing, the decision regarding the type, number, and combination of production factors is of particular importance. The following sections highlight the value that selected applications of the Monte Carlo simulation can add for contractors if they use it in the construction phase.

10.4.1 Logistics/Number of Transport Cycles

The number of transport cycles is a key variable for site logistics planning. For example, an uninterrupted supply of concrete to the site must be ensured during concreting operations. This challenge should not be underestimated, particularly on inner-city construction sites where traffic congestions can occur at peak times ("rush hour"). If the daily number of transport cycles is known in advance, this data can be used to better investigate the effects on the transport network or the logistics concept developed for the construction site. The calculation of the number of transport cycles belongs in PPH 4 (see Fig. 10.78).

Transports in conjunction with reinforced concrete operations play a major role in the shell construction phase. We thus consider only this type of transport for a specific building construction project (i.e. the Opera House in Linz, Austria). The calculations are carried out separately for the transport cycles required for materials (i.e. the concrete and the reinforcing steel) and those needed for equipment (i.e. formwork).

44) Cf. Raaber (1979), p. 5

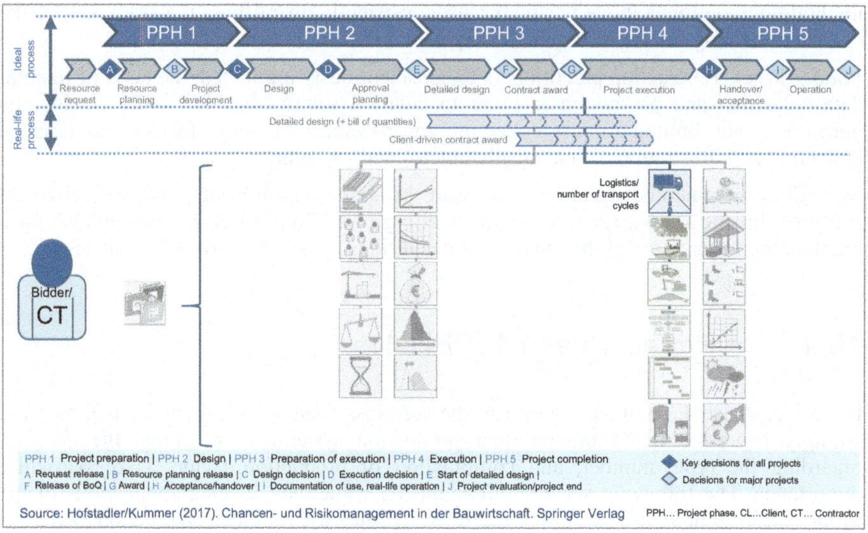

Fig. 10.78 Allocation of example in the project sequence

10.4.1.1 Basics

Construction logistics can be divided into procurement, production, and disposal logistics. Depending on the location of the construction site (such as in an inner-city environment), site logistics is of major importance with regard to an economically effective and efficient utilization of the production factors. While the total number of transport cycles is significantly influenced by the construction method and the type of construction project, the transport density (number of transports [TR] per unit of time) depends on the available construction time. Accordingly, short construction times are more demanding in terms of logistics.

For example, relatively precise equations are available for earthworks for the purpose of calculating the total number of transports. The number of transport cycles can be derived from the net technical performance levels of the excavation equipment in combination with the capacity of the transport vehicles. In earthworks, care must be taken to ensure that a loosening coefficient is factored in for the excavated material.[45)]

A set of performance indicators is used for calculating the number of transport cycles required for reinforced concrete operations. More specifically, these include transport coefficients t_c [1/UQ] for the formwork, the reinforcement, and the concrete. Multiplying these values by the corresponding quantities will give the number of required transport cycles.

[45)] Cf. Hofstadler (2009), p. 146

Transport coefficient of formwork

The transport coefficient of formwork $t_{c,f}$ [1/m²] is calculated from the formwork stocking quantity SQ_F [m²], the total formwork area of the structure FA_{STR} [m²] and the average formwork area FA_{TR} [m²] delivered to the construction site per transport cycle (see Eq. (10-19)).[46)]

$$t_{c,f} = \frac{SQ_F}{FA_{STR} \cdot FA_{TR}} \tag{10-19}$$

The transport coefficient will essentially be influenced by the type of formwork. Prefabricated formwork usually requires more space on a vehicle than disassembled formwork transported in containers or pallets. A range from 0.00083 to 0.00125 [1/m²] is specified for the transport coefficient of formwork operations. However, these values may be lower for small construction projects such as single-family homes, resulting in fewer transport cycles.[47)]

Transport coefficient of reinforcement

The transport coefficient of reinforcement $t_{c,r}$ [1/t] is calculated using the quantity of reinforcement $R_{Q,TR}$ [t] delivered to the site per transport cycle.

$$t_{c,r} = \frac{1}{R_{Q,\,TR}} \tag{10-20}$$

For example, if 20 tons can be delivered per trip, the transport coefficient of the reinforcement is 0.05 [1/t]. With only 10 tons per delivery, the transport coefficient rises to 0.10 [1/t]. The figure determined for the transport coefficient of the reinforcement is mainly influenced by external circumstances. These can include, for example, weight restrictions for bridges to be crossed on the way to the construction site. If semitrailer trucks cannot access the construction site due to space restrictions, this will also affect the transport coefficient.[48)]

Transport coefficient of concrete

The transport coefficient of concrete $t_{c,c}$ [1/m³] is calculated using the standard volume of concrete $C_{V,TR}$ [m³] delivered to the site per transport cycle.

$$t_{c,c} = \frac{1}{C_{V,\,TR}} \tag{10-21}$$

For example, if 12 m³ can be delivered per trip, the transport coefficient of the concrete is 0.08 [1/m³]. With only 7 m³ per delivery, the transport coefficient rises to 0.14 [1/m³]. The figure determined for the transport coefficient of the concrete is mainly influenced by external circumstances, including weight restrictions, the condition of access roads, or the number and size of vehicles available at the mixing plant.[49)]

46) Cf. ibid., p. 150

47) Cf. ibid., p. 464

48) Cf. ibid., p. 151

49) Cf. ibid.

10.4.1.2 Task

In this example, the number of transport cycles required for the construction of the building shell (reinforced concrete operations) is calculated using transport coefficients. The total number of transport cycles comprises the proportions of the formwork, the reinforcement, and the concrete.

This example can be assigned to procurement logistics (transports to the construction site), which is essentially concerned with supplying the site with equipment and materials. The number of transport cycles calculated in this example must be differentiated from production logistics (intra-site transport) and disposal logistics (transports away from the construction site).

10.4.1.3 Modeling

The triangular distributions for the formwork areas, the reinforcement quantities, and the concrete volumes per structural component group of the Linz Opera House project are used for modeling the probabilistic calculation. These were already given in Section 10.3.3 in Tables 10.10 to 10.12, each with minimum, most likely, and maximum values.

Fig. 10.79 shows the flow chart for calculating the number of transport cycles or the transport density per day.

The quantities of each structural component group are added up to obtain the total quantities of the formwork, reinforcement, and concrete, and multiplied by the transport coefficients, which are also indicated by triangular distributions. These calculations result in histograms for the number of transport cycles (divided into formwork, reinforcement, and concrete). Another histogram indicates the total number of transport cycles required for reinforced concrete operations.

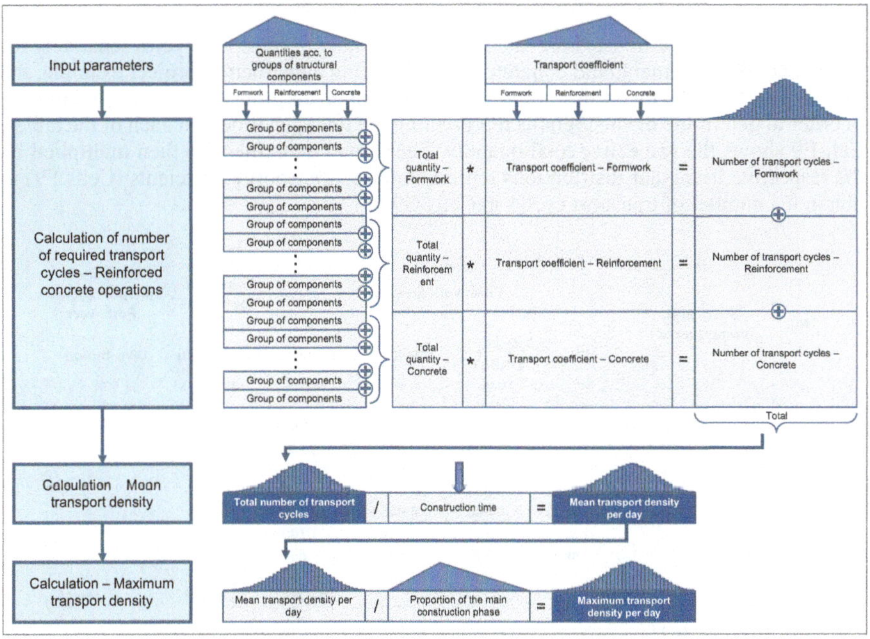

Fig. 10.79 Calculation flow chart

The mean transport density per day is calculated by dividing the total number of transport cycles by the specified construction time (deterministic value = 400 days). However, transport peaks cannot be derived from the mean transport density because the resource utilization follows a trapezoidal pattern. For this purpose, the mean still has to be divided by the proportion of the main construction phase (triangular distribution: 0.75 | 0.80 | 0.85). This calculation step results in the histogram representing the maximum transport density per day during the main construction phase.

Table 10.31 indicates the transport coefficients used in this example. The minimum, most likely, and maximum values are shown in Columns B, C, and D. Column E contains the triangular distributions, and the most likely value of each of these distributions is shown in Table 10.31.

No.	Components	Transport coefficients			
		MIN	MOST LIKELY	MAX	Distribution
		[1/UQ]	[1/UQ]	[1/UQ]	[1/UQ]
0	A	B	C	D	E
1	Formwork	0.00083	0.00110	0.00125	0.00110
2	Reinforcement	0.05000	0.08500	0.10000	0.08500
3	Concrete	0.08000	0.13000	0.14000	0.13000

Table 10.31 Input values for transport coefficients – Formwork, reinforcement, and concrete

10.4.1.4 Calculation and Interpretation of Results

Tables 10.32 to 10.34 are used for calculating the number of transport cycles separately for formwork, reinforcement, and concrete. In the building construction project example, the first step adds up the quantities of the structural component groups for the entire building in order to determine the histograms representing the total quantities. In each of the tables, Cell E9 shows the respective total quantity. These total quantities are then multiplied by the respective triangular distributions representing the transport coefficients (Cells F9) to obtain the number of transport cycles in Cells G9.

No.	Structural components	Formwork area				Transport coefficient	Number of transport cycles – Formwork
		MIN	MOST LIKELY	MAX	Distribution	Distribution	Distribution
		[m²]	[m²]	[m²]	[m²]	[1/m²]	[TR]
0	A	B	C	D	E	F	G
1	Foundation slab	1,070	1,150	1,231	1,150		
2	Columns	3,441	3,700	3,959	3,700		
3	Walls	75,330	81,000	86,670	81,000		
4	Floor slabs	27,900	30,000	32,100	30,000		
5	Beams, joists	5,952	6,400	6,848	6,400		
6	Other slabs	1,581	1,700	1,819	1,700		
7	Stairs	465	500	535	500		
8	Parapets	2,604	2,800	2,996	2,800		
9	**Total**				**127,250**	**0.00110**	**139.98**

Table 10.32 Calculation of number of required transport cycles – Formwork

The number of transport cycles is thus represented by separate histograms for the formwork, the reinforcement, and the concrete. We find that the supply of concrete requires the largest number of transport cycles. On the other end of the scale, the formwork triggers the smallest number of transport cycles. This is due to the fact that the formwork is reused several times, which is why only the formwork stocking quantity has to be delivered to the site.

No.	Structural components	Reinforcement quantity				Transport coefficient	Number of transport cycles – Reinforcement
		MIN	MOST LIKELY	MAX	Distribution	Distribution	Distribution
		[t]	[t]	[t]	[t]	[1/t]	[TR]
0	A	B	C	D	E	F	G
1	Foundation slab	1,395	1,500	1,620	1,500		
2	Columns	112	120	130	120		
3	Walls	1,395	1,500	1,620	1,500		
4	Floor slabs	1,349	1,450	1,566	1,450		
5	Beams, joists	151	162	175	162		
6	Other slabs	93	100	108	100		
7	Stairs	19	20	22	20		
8	Parapets	74	80	86	80		
9	**Total**				**4,932**	**0.08500**	**419.22**

Table 10.33 Calculation of number of required transport cycles – Reinforcement

Due to the large number of transport cycles required for the concrete, an increased transport density should also be expected for those structural components that account for the largest share of the total concrete volume (walls, ceilings, and the foundation slab) in the construction phase. There is a high likelihood for the number of transport cycles to reach a peak for the foundation slab, which constitutes a coherent component as opposed to individual walls or floor slabs. In this case, a large amount of concrete needs to be delivered to the site within a correspondingly short period owing to the limitations of concrete technology. Thus, the transport density will be particularly high for this component.

We recommend a more detailed analysis. Such a detailed analysis of the transport density will require determination of the construction method (individual work sections vs. monolithic) as well as of the time required for the concreting process or individual process steps.

No.	Structural components	Concrete volume				Transport coefficient	Number of transport cycles – Concrete
		MIN	MOST LIKELY	MAX	Distribution	Distribution	Distribution
		[m³]	[m³]	[m³]	[m³]	[1/m³]	[TR]
0	A	B	C	D	E	F	G
1	Foundation slab	9,024	9,600	10,272	9,600		
2	Columns	376	400	428	400		
3	Walls	14,570	15,500	16,585	15,500		
4	Floor slabs	9,776	10,400	11,128	10,400		
5	Beams, joists	1,297	1,380	1,477	1,380		
6	Other slabs	376	400	428	400		
7	Stairs	85	90	96	90		
8	Parapets	475	505	540	505		
9	**Total**				**38,275**	**0.13000**	**4,975.75**

Table 10.34 Calculation of number of required transport cycles – Concrete

Fig. 10.80 shows the three histograms for the number of transport cycles required for the formwork (top), the reinforcement (middle), and the concrete (bottom). When comparing the histograms, it should be noted that the scales of the abscissae (and of the ordinates) differ significantly.

All of the histograms represent left-skewed (negatively skewed) distributions. This result is due to the distributions selected for the transport coefficients. In the modeling exercise, we assumed that rather few units of quantity can be delivered per transport.

The mean shown in the histogram for the formwork equals approximately 135 transport cycles whereas the standard deviation amounts to about 11 cycles. The spread (minimum and maximum) of the number of transport cycles lies between approximately 102 and 165 transport cycles.

The number of transport cycles required for the reinforcement is generally greater than that for the formwork. The mean amounts to approximately 388 cycles. There is a 90% probability for the number of transport cycles to lie between approximately 293 and 465.

By far the largest share of transports in the total number is required for the concrete deliveries needed for reinforced concrete operations. These works are associated with a mean value of about 4,480 and a standard deviation of approximately 508 transport cycles.

If a plant for supplying the construction site can already be selected in the process planning phase, and if the loading capacities of the individual truck mixers are known, it is possible to specify this transport coefficient as a deterministic value or at least to set a narrower range. Thus, the number of transport cycles would be determined mainly by the uncertainties in the concrete quantities.

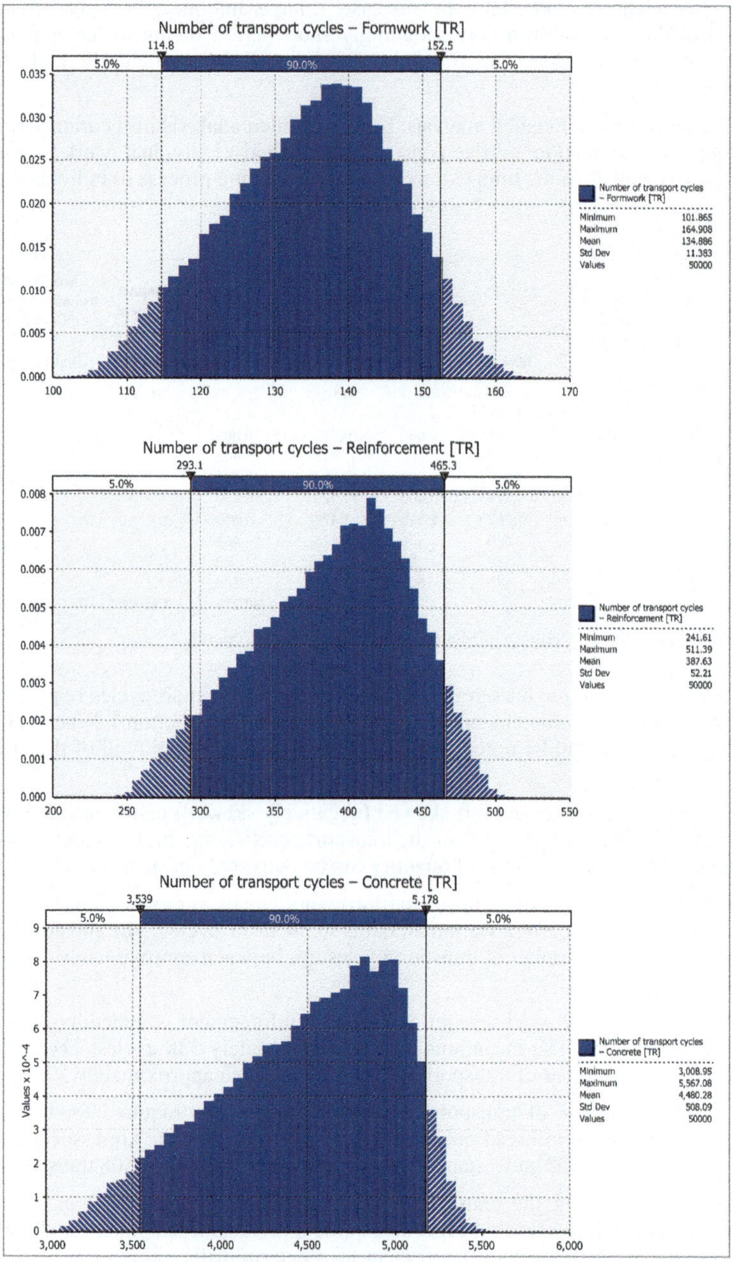

Fig. 10.80 Histograms for the number of transport cycles – Formwork, reinforcement, and concrete

Adding up the transport cycles gives the total number of transports that Fig. 10.81 shows as a histogram. The curve of the histogram is dominated by the number of concrete deliveries since these account for the largest share in the total number. Overall, a mean of about 5,003 transport cycles should be assumed for reinforced concrete operations over the entire construction period. The 5% quantile and the 95% quantile amount to 4,060 and 5,707 transport cycles, respectively.

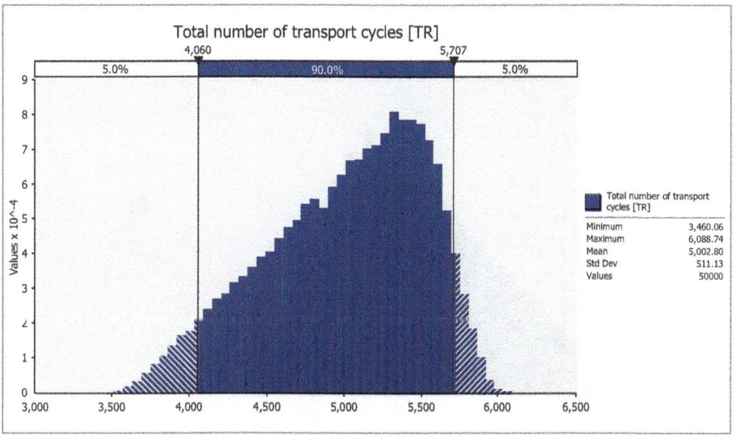

Fig. 10.81 Histogram for the total number of transport cycles

If we divide the total number of transport cycles by the specified construction time of 400 days (deterministic value), we obtain the mean transport density per day shown as a histogram in Fig. 10.82. The mean transport density should be understood as an average consideration over the entire construction period. Accordingly, we should expect an average of about 9 to 15 transport cycles per day over the entire 400-day period.

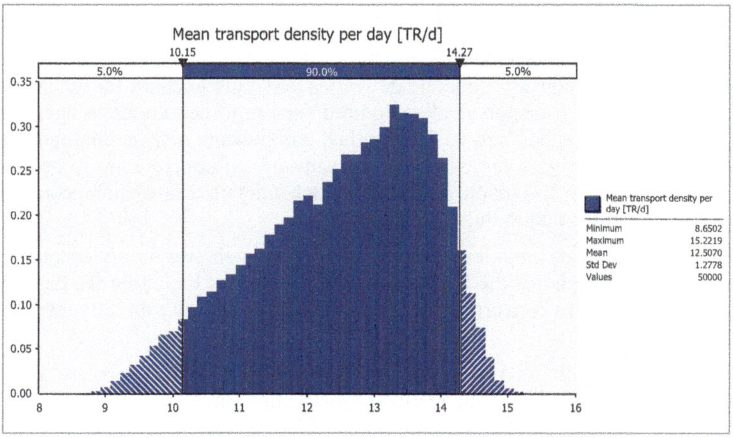

Fig. 10.82 Histogram for the mean transport density per day

The demand for materials required on-site will usually increase in the main construction phase associated with maximum resource utilization. Accordingly, we can expect an increase in transport density to occur at this project stage. The maximum transport density per day is obtained by dividing the mean transport density by the proportion of the main construction phase (assumed as a triangular distribution: 0.75 | 0.80 | 0.85). Fig. 10.83 shows the histogram representing the maximum transport density.

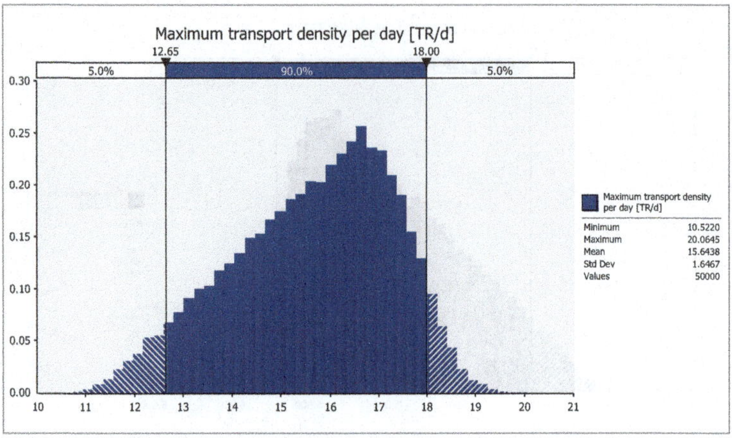

Fig. 10.83 Histogram for the maximum transport density per day

Thus, 11 to 20 transport cycles per day should be expected during the main construction phase, with a mean of about 16 cycles. Assuming a daily working time of 8 hours, this would correspond to a frequency of about two transport cycles per hour.

10.4.1.5 Benefit for the Contractor

The probabilistic calculation of the number of transport cycles provides ranges of the expected total number of transport cycles and – depending on the construction time – of the expected transport density. Concrete deliveries generally account for the largest share in the total number of transport cycles required for reinforced concrete operations. We recommend a more detailed analysis if individual components such as the foundation slab or work sections requiring a large amount of ready-mixed concrete are to be completed within a short time. The systematic approach to such a detailed assessment corresponds to the procedure demonstrated in this example.

The calculated (mean and maximum) transport intervals should be understood as direct quantitative requirements for the logistics planning exercise. Including the daily working time makes it possible to convert the determined transport density directly into a required transport interval [TR/h].

If the expected transport interval cannot be adhered to, either an increase in the daily working time or an extension of the relevant construction times should be anticipated.

For particularly short transport intervals, time slots must be allocated to deliveries or waiting zones must be installed in the vicinity of the construction site. From these zones, the individual transports can then be ordered to the construction site in order to better coordinate the loading and unloading processes. The risk that access roads to the construction site and/or unloading or storage areas are blocked by waiting trucks can thus be reduced.

10.4.2 Coordination of Equipment Chains

It is not only in labor-intensive activities, such as reinforced concrete operations, that the best possible combination of production factors is crucial in order to achieve the highest possible productivity and thus to incur the lowest possible costs. Even for equipment-intensive work, the achievable performance will depend on the combination of production factors. Only if the equipment chain is optimally coordinated can the specified performance values and the related project targets be achieved.

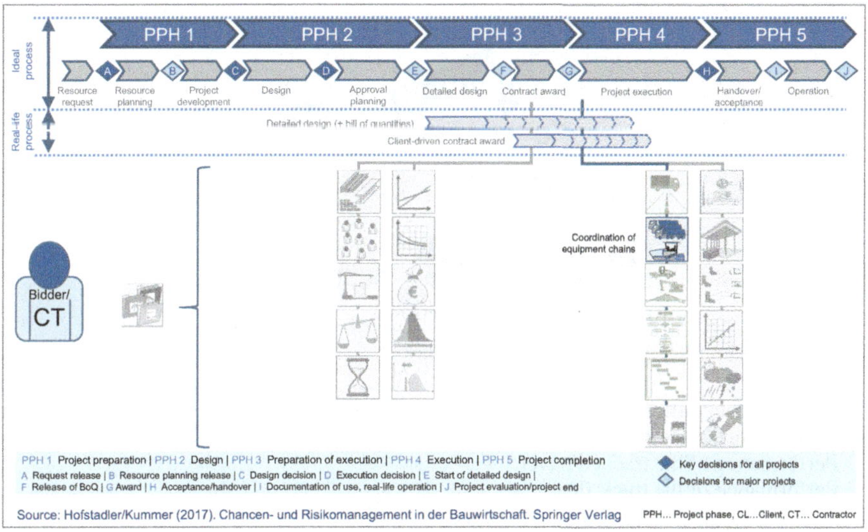

Fig. 10.84 Allocation of example in the project sequence

This example shows the coordination of the equipment chain for asphalt paving. It belongs in PPH 4 (see Fig. 10.84). More specifically, we compare the performance of the asphalt paver with the transport performance of the truck fleet and relate these parameters to the specified construction time. This comparison also determines the bottleneck capacity, which, in turn, will determine the shortest possible construction time.

10.4.2.1 Basics

A tracked paver is used to install an asphalt surface course. Asphalt is supplied by dump trucks which can dock onto the asphalt paver and are emptied during operation (see the schematic sequence in Fig. 10.85). The paver will not be able to deliver its full performance if no or too little material is available.

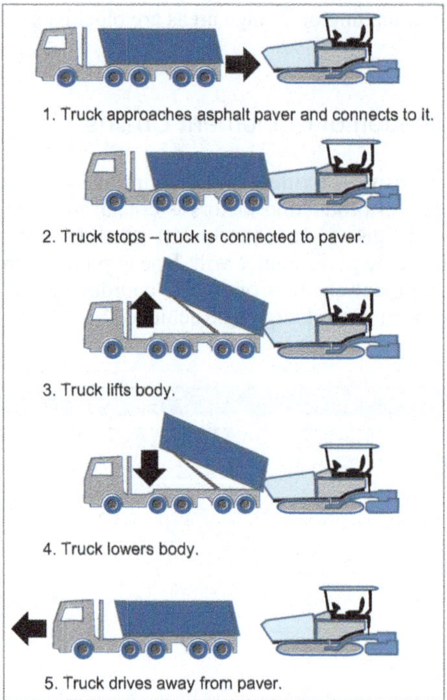

Fig. 10.85 Schematic feed process using a dump truck[50)]

Essentially, four different performance values will determine the installation process of the asphalt surface course:

- Performance of the tracked paver
- Performance of the truck fleet
- Performance of the truck unloading/feed cycle (directly at the paver)
- Performance of the asphalt mixing plant

In this example, we assume that the supply from the mixing plant is sufficient and therefore non-critical for achieving the overall performance of the equipment chain. The truck cycle involving switching between empty and loaded dumpers directly at the asphalt paver is not considered critical either.[51)]

This example thus only compares the performance of the truck fleet with that of the paver. Only a seamlessly working truck chain will provide the basis for the paver to reach its envisaged technical performance level.

---

50) Cf. Joseph Vögele AG – Product Data Sheet – Vögele tracked paver SUPER 1900-3i – Highway Class

51) For a comprehensive analysis, it would also be possible to evaluate these two performance values in detail, and to compare them with the other performance values.

10.4.2.2 Task

A three-course asphalt pavement with an expected width of 9.50 m and an expected total depth of 18.5 cm is to be installed. The road section to be completed extends over a length of 10 km. A construction time of 25 days was defined for the asphalt pavement. Table 10.35 lists the parameters determining the paving process as deterministic values.

Based on the dimensions of the asphalt pavement, the asphalt volume (Row 4 in Table 10.35) and the asphalt mass (Row 6) are calculated. The scheduled construction time (Row 7) is used for calculating the required performance per day (Row 8) (1,722.35 t/d = 43,058.75 t / 25 d). We then include the daily working time (Row 9) to calculate the required hourly performance and obtain a value of 172.24 t/h (= 1,722.35 t/d / 10.00 h/d).

The average required performance per working day or per working hour thus depends on the asphalt quantity to be installed, the specified construction time, and the effective daily working time[52).

No.	Parameter	Value
0	A	D
1	Width of asphalt pavement	9.50 m
2	Total asphalt depth (all 3 courses)	0.185 m
3	Length of road section	10,000.00 m
4	Asphalt volume	17,575.00 m³
5	Asphalt density	2.45 t/m³
6	Asphalt mass	43,058.75 t
7	Scheduled construction time	25.00 d
8	Required performance per day	1,722.35 t/d
9	Daily working time	10.00 h/d
10	**Required performance per hour**	**172.24 t/h**

Table 10.35 Calculation table – Required performance

The results can be displayed as curves in a diagram (see Fig. 10.86) if the deterministic calculation is carried out using different input values. The abscissa shows the specified (planned) construction time in days, whereas the ordinate shows the required performance [t/h]. Table 10.36 also lists the required performance values as a function of the construction time and the daily working time.

52) In the truck chain, legal regulations governing maximum permissible driving times may affect the net technical performance.

No.	Scheduled construction time	Required performance per hour [t/h]				
		Daily working time				
	[d]	8 h/d	9 h/d	10 h/d	11 h/d	12 h/d
0	A	B	C	D	E	F
1	10	538.23	478.43	430.59	391.44	358.82
2	11	489.30	434.94	391.44	355.86	326.20
3	12	448.53	398.69	358.82	326.20	299.02
4	13	414.03	368.02	331.22	301.11	276.02
5	14	384.45	341.74	307.56	279.60	256.30
1	15	358.82	318.95	287.06	260.96	239.22
2	16	336.40	299.02	269.12	244.65	224.26
3	17	316.61	281.43	253.29	230.26	211.07
4	18	299.02	265.79	239.22	217.47	199.35
5	19	283.28	251.81	226.63	206.02	188.85
6	20	269.12	239.22	215.29	195.72	179.41
7	21	256.30	227.82	205.04	186.40	170.87
8	22	244.65	217.47	195.72	177.93	163.10
9	23	234.01	208.01	187.21	170.19	156.01
10	24	224.26	199.35	179.41	163.10	149.51
11	25	215.29	191.37	172.24	156.58	143.53
12	26	207.01	184.01	165.61	150.56	138.01
13	27	199.35	177.20	159.48	144.98	132.90
14	28	192.23	170.87	153.78	139.80	128.15
15	29	185.60	164.98	148.48	134.98	123.73
16	30	179.41	159.48	143.53	130.48	119.61
17	31	173.62	154.33	138.90	126.27	115.75
18	32	168.20	149.51	134.56	122.33	112.13
19	33	163.10	144.98	130.48	118.62	108.73
20	34	158.30	140.71	126.64	115.13	105.54
21	35	153.78	136.69	123.03	111.84	102.52

Table 10.36 Required performance per hour depending on the daily working time and the scheduled construction time

The curves illustrate the relationship between these two variables, with each curve representing a different effective daily working time (from 8 h/d to 12 h/d). For this calculation, the asphalt mass was kept constant at 43,058.75 t. Approximately 43,000 metric tons of asphalt are equivalent to the total quantity required for the three courses of a directional roadway (highway) without taking the central reservation and ramps into account.

If, for example, the work is to be completed within 16 days and if a daily working time of 10 hours is scheduled, an hourly performance of approximately 269 t/h would have to be achieved. On the other hand, assuming the same daily working time and a target construction time of 25 days, a performance value of “only” approximately 172.2 t/h would be required.

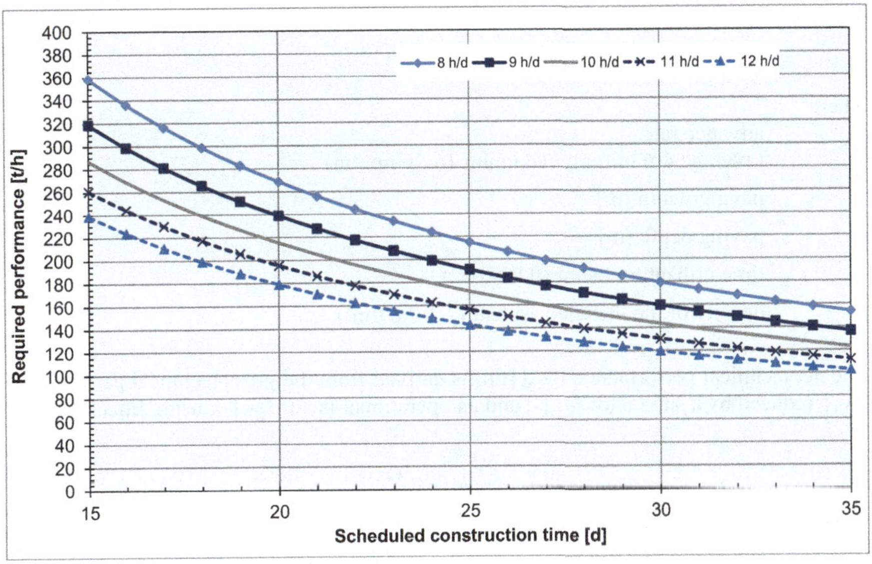

Fig. 10.86 Relationship – Scheduled construction time, daily working time, and required hourly performance

As such, the deterministic calculation cannot deliver an answer to the question of whether the risk of adhering to a certain construction period is acceptable to the contractor or not. Nor does it permit any conclusions as to whether the performance of the paver or of the truck fleet is sufficient for the envisaged construction time. Thus, the application of probabilistic methods is necessary in order to adequately analyze these key project components.

10.4.2.3 Modeling

As part of the modeling exercise, the input parameters and equations for calculating the performance of the tracked paver and the transport performance (truck fleet) are presented and explained.

For the probabilistic calculations, the paving width of the carriageway and the installation depth of the asphalt pavement are each taken as a triangular distribution in order to account for uncertainties in the routing and the subgrade in the simulation.

Tracked paver performance – Complete asphalt pavement

Performance values for tracked pavers provided by the respective manufacturers must not be applied without due reflection because the conditions under which these values were determined are unknown (ideal conditions and maximum possible paving depth, etc.). These are merely theoretical values that are difficult to achieve in a real-life construction environment. In this example, the gross technical performance and the net technical performance of the asphalt paver are determined in a probabilistic approach.

We calculate the gross technical performance PV_G [t/h] using the following equation:[53)]

$$PV_G = 60 \cdot v_A \cdot w \cdot d \cdot f_T \cdot \rho \quad (10\text{-}22)$$

where

v_A	advance rate (average 4-6 m/min; maximum 16-20 m/min)
w	paving width [m]
d	paving depth [m]
f_T	time utilization factor (0.83-0.92)
ρ	density of the installed asphalt (2.3-2.6 t/m³)

The net technical performance PV_N [t/h] is derived from the gross technical performance PV_G, reduced by a site factor f_{E1} [-] and an operational factor f_{E2} [-], using Eq. (10-23).[54)]

$$PV_N = PV_G \cdot f_{E1} \cdot f_{E2} \quad (10\text{-}23)$$

where

f_{E1}	site factor (0.5-1.0)
f_{E2}	operational factor (see Table 10.37)

Process planning Age and condition of machine	Skillset and dedication of operators			
	Very good	Good	Average	Poor
Very good	1.00	0.90	0.80	0.70
Good	0.90	0.81	0.72	0.63
Average	0.80	0.72	0.64	0.56
Poor	0.70	0.63	0.56	0.49

Table 10.37 Operational factor[55)]

Fig. 10.87 shows the flow chart for the calculation of the net technical performance of the asphalt paver. In addition, the input parameters and their respective characteristics (triangular distribution or deterministic value) are shown qualitatively in the diagram.

[53)] Cf. Scheuer (1997), p. 5.8ff.

[54)] Cf. ibid.

[55)] Cf. Jurecka (1975), p. 109

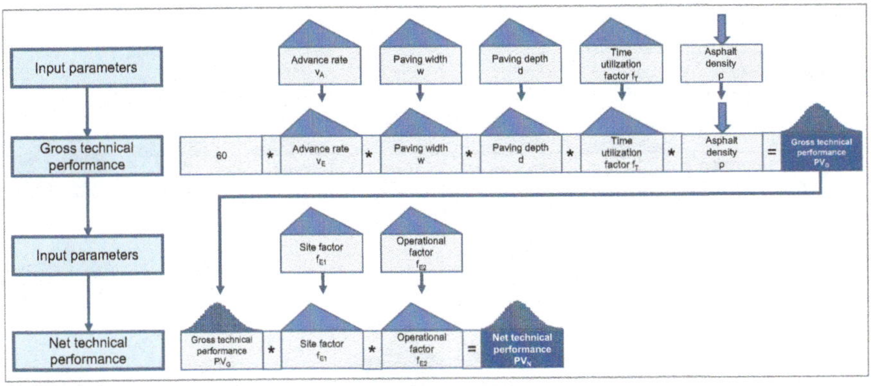

Fig. 10.87 Calculation flow chart – Net technical performance of asphalt paver

Truck fleet performance – Transport performance

To calculate the transport performance of the truck fleet, we first determine the transport performance for a single vehicle and then extrapolate it to the entire fleet. For this purpose, we record and add up the time spent at the mixing plant (loading), the net driving time (transport), and the time required for unloading. Relating this time to the loading capacity of a single truck gives the average transport performance per truck [t/h]. Multiplication by the total number of trucks then results in the hourly transport performance [t/h] for the entire fleet.

Fig. 10.88 shows the flow chart for this calculation, including the qualitatively represented inputs and outputs.

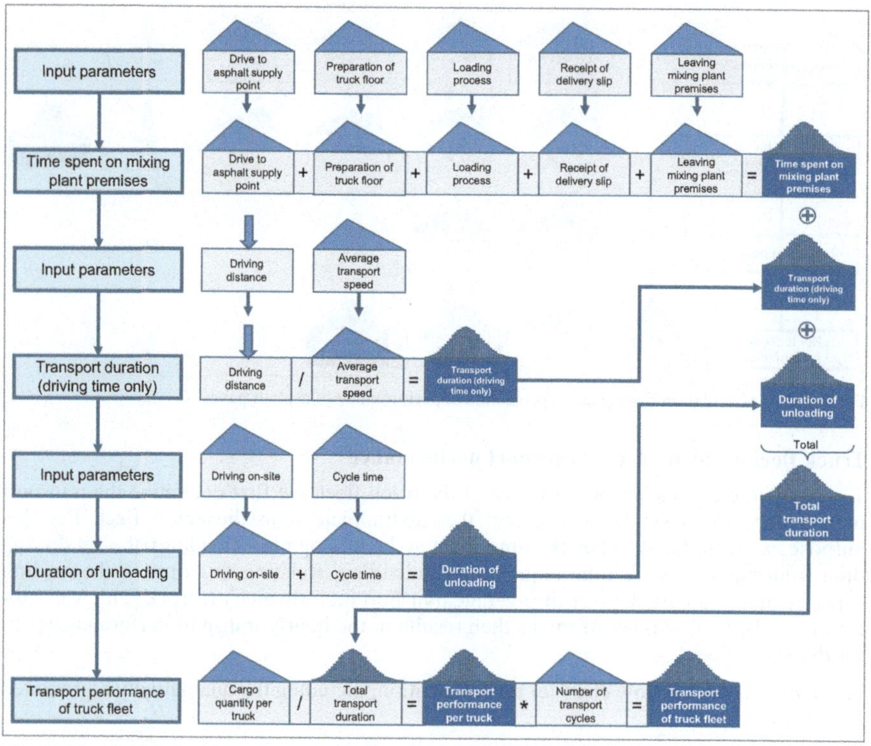

Fig. 10.88 Calculation flow chart – Truck fleet transport performance

10.4.2.4 Calculation and Interpretation of Results

The calculation tables for determining the performance of the tracked paver and of the truck fleet are given below, and the simulation results are presented and interpreted in the form of histograms.

Tracked paver performance – Complete asphalt pavement

The corresponding calculation parameters are entered in Eq. (10-22) and Eq. (10-23) in order to calculate the net technical performance of the asphalt paver. In the simulation, parameters fraught with uncertainties are represented by triangular distributions. Table 10.38 shows the minimum, most likely, and maximum values of the uncertain parameters in Columns B, C, and D. The distributions (or the deterministic value for the density of the paved asphalt) are entered in Column E.

No.	Parameter	MIN	MOST LIKELY	MAX	Distribution
0	A	B	C	D	E
1	Paving width	9.400	9.500	9.600	9.500 m
2	Paving depth	0.180	0.185	0.190	0.185 m
3	Time utilization factor	0.830	0.875	0.920	0.875 -
4	Density of installed asphalt				2.450 t/m³
5	Average advance rate	1.100	1.150	1.200	1.150 m/min
6	Gross technical performance				259.967 t/h
7	Site factor	0.750	0.850	0.900	0.833 -
8	Operational factor	0.720	0.810	0.950	0.827 -
9	**Net technical performance**				**179.089 t/h**

Table 10.38 Calculation table – Asphalt paver performance

We obtain the gross technical performance PV_G [t/h] by applying the calculation parameters in Eq. (10-22) as a histogram (see Fig. 10.89) or in Cell E6 of Table 10.38. The mean of the gross technical performance amounts to approximately 260 t/h whereas the standard deviation roughly amounts to 7.7 t/h.

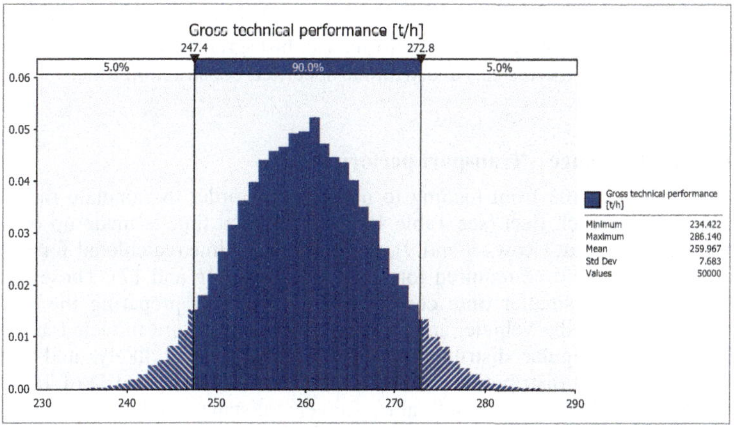

Fig. 10.89 Histogram – Gross technical performance – Asphalt paver

To determine the net technical performance PV_N [t/h] (Cell E9 in Table 10.38), the gross technical performance must subsequently be reduced by a site factor and an operational factor. Both factors are fraught with uncertainties, which is why they are incorporated into the simulation as triangular distributions. We carry out the calculation by inserting the values in Eq. (10-23). Fig. 10.90 shows the histogram of the net technical performance PV_N [t/h]. This value lies between approximately 158 and 202 t/h in 90% of the simulated cases. The mean amounts to approximately 179 t/h whereas the standard deviation roughly amounts to 13.3 t/h.

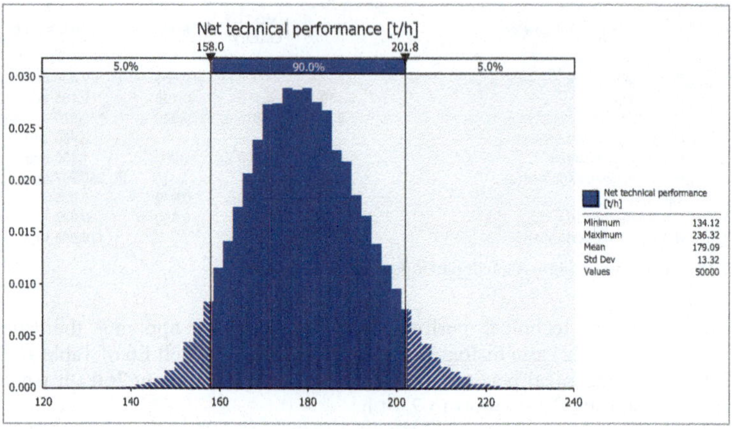

Fig. 10.90 Histogram – Net technical performance – Asphalt paver

We can then use the net technical performance histogram for comparison with the transport performance (truck fleet) and with the specified construction time.

Truck fleet performance – Transport performance

We add up the total time from loading to unloading in order to calculate the transport performance of the truck fleet (see Table 10.39). This total time is made up of the time spent at the mixing plant (Rows 8 and 9), the net driving time (calculated for two directions; Row 13) and the time required for unloading (Rows 16 and 17). These three time components comprise smaller time components, such as for preparing the load floor, checking the load and the vehicle, and driving on-site, which are modeled as uncertain parameters using triangular distributions. The minimum, most likely, and maximum values of the triangular distributions are shown in Columns B, C, and D of Table 10.39. The distributions, deterministic input values (traveled distance) and the intermediate and final results of the calculations are entered in Column E of Table 10.39.

If we divide the cargo quantity (defined as a triangular distribution in Row 1 of Table 10.39) by the total transport duration of a truck (shown in Cell E18), we obtain a histogram of the transport performance of a single truck given in Cell E19 (see Fig. 10.91).

No.	Parameter	MIN	MOST LIKELY	MAX	Distribution
0	A	B	C	D	E
1	Cargo quantity	24.80	25.00	25.20	25.00 t/truck
2	Drive to asphalt supply point	0.80	1.00	1.20	1.00 min
3	Preparation of truck floor	0.80	1.00	1.20	1.00 min
4	Loading process	8.00	8.50	9.00	8.50 min
5	Receipt of delivery slip	2.50	3.00	3.50	3.00 min
6	Checking cargo and vehicle	4.50	5.00	5.50	5.00 min
7	Leaving mixing plant premises	0.80	1.00	1.20	1.00 min
8	Time spent on mixing plant premises				19.50 min
9	**Time spent on mixing plant premises**				**0.33 h**
10	Distance				120.00 km
11	Total traveled distance				240.00 km
12	Average transport speed	50.00	60.00	70.00	60.00 km/h
13	**Transport duration (driving time only)**				**4.00 h**
14	Driving on-site	8.00	9.00	10.00	9.00 min
15	Cycle time	8.00	8.50	9.00	8.50 min
16	Total duration of unloading				17.50 min
17	**Total duration of unloading**				**0.29 h**
18	Transport duration (loading - driving distance - unloading - return)				4.62 h
19	**Transport performance per truck**				**5.42 t/h**
20	Number of trucks	35.00	40.00	45.00	40.00 -
21	**Transport performance of truck fleet**				**216.61 t/h**

Table 10.39 Calculation table Truck fleet transport performance

The mean transport performance of a single truck amounts to approximately 5.41 t/h. After 50,000 iterations, we get a maximum of about 6.20 t/h and a minimum of about 4.60 t/h. The histogram representing the transport performance of a single truck corresponds approximately to a triangular distribution because the cargo quantity was assumed to be a triangular distribution with a relatively narrow range. In this case, this distribution thus determines the curve of the quotient.

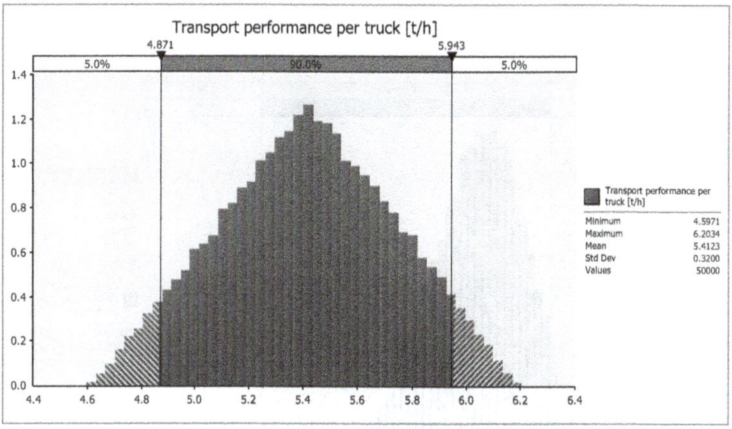

Fig. 10.91 Histogram – Transport performance of a single truck

To calculate the transport performance of the entire truck fleet (see Row 21 of Table 10.39), the transport performance of a single truck (Cell E19) is multiplied by the number of trucks used (triangular distribution; see Row 20). Fig. 10.92 shows a histogram of the transport performance of the entire fleet [t/h].

The mean amounts to approximately 216.49 t/h whereas the standard deviation roughly amounts to 16.90 t/h. In 90% of all cases, a transport performance in an approximate range from 189 to 245 t/h is achieved. The histogram representing the transport performance is then used for comparing it with the performance of the asphalt paver.

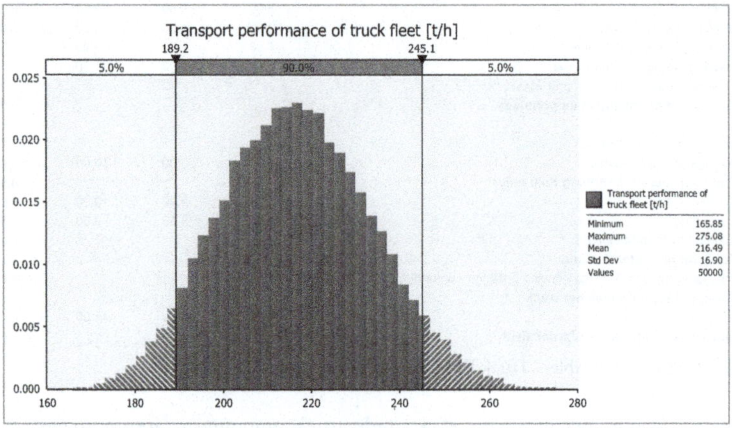

Fig. 10.92 Histogram – Truck fleet transport performance

Comparison of performance values

A comparison of the two histograms of the performance of the asphalt paver and transport performance of the truck fleet shows that, in most cases, the performance of the truck fleet exceeds the performance of the asphalt paver (see Fig. 10.93).

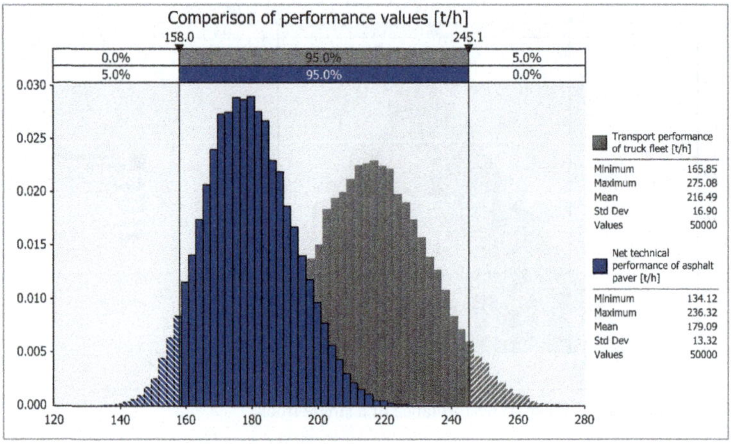

Fig. 10.93 Histograms – Comparison of performance values

We obtain the histogram shown in Fig. 10.94 by calculating the difference between the two histograms in Fig. 10.93 (i.e. subtracting the paver performance from the truck fleet transport performance). We find that the transport performance of the truck fleet is lower than the paver performance in only 4.1% of all cases. We can thus assume, under the defined boundary conditions, that the supply to the asphalt paver will be sufficient in over 95% of all cases. Accordingly, in the example considered, the performance of the asphalt paver as a bottleneck piece of equipment will be crucial for adhering to the specified construction time.

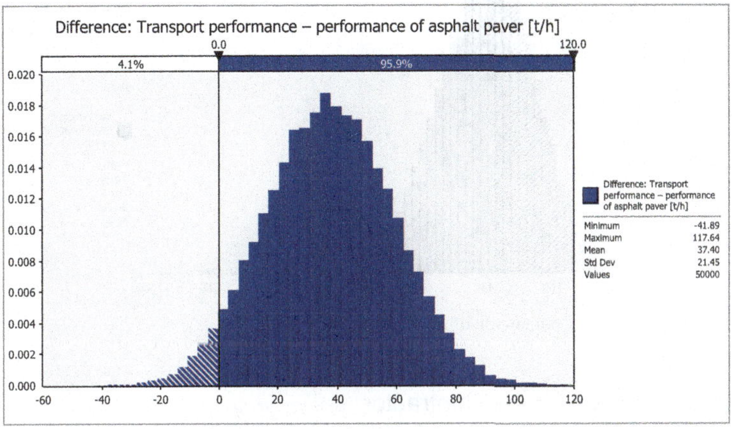

Fig. 10.94 Histograms – Difference of performance values

In the next step, the chance/risk ratios for the installation of the asphalt pavement can be evaluated. Installation of the asphalt pavement within 25 days and during a daily working time of 10 hours would require a daily performance of approximately 172.2 t/h (see Cell D11 in Table 10.36). If, on the other hand, the pavement is to be completed in only 20 days (assuming the same daily working time), an average performance of about 215.29 t/h would be required (see Cell D6 in Table 10.36).

Fig. 10.95 again compares the two histograms representing the asphalt paver performance and the transport performance of the truck fleet. At the top of the diagram, however, the sliders have been shifted to such an extent that the two performance values required for a construction time of 20 and 25 days (at a daily working time of 10 hours) are shown. Accordingly, the asphalt paver cannot deliver the mean performance for a construction time of 25 days in 31.9% of all cases. In 99.9% of cases, however, the transport performance is sufficiently high for adhering to a construction time of 25 days.

For a construction time of only 20 days, only about 0.4% of the asphalt paver's performance values exceed the required performance of 215.29 t/h (daily working time = 10 h/d). The truck fleet could still achieve or exceed this performance in approximately 52% of all cases.

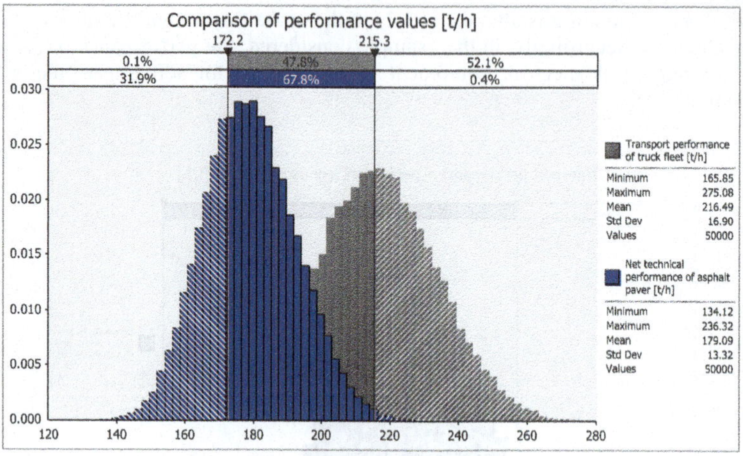

Fig. 10.95 Histograms – Comparison of performance values

10.4.2.5 Benefit for the Contractor

For evaluating and analyzing equipment chains, the performance values of their individual components should be examined and represented in a calculation model while taking into account the respective uncertainty characteristics. The individual performance values must then be normalized for ensuring their comparability (in this example, performance values are measured in t/h) in order to juxtapose them.

The component determining the performance of the entire equipment chain as a "bottleneck device" can be analyzed directly on the basis of the probabilistic calculation. Calculating performance differences also permits conclusions regarding the probability for one component delivering a lower performance than another (see Fig. 10.94).

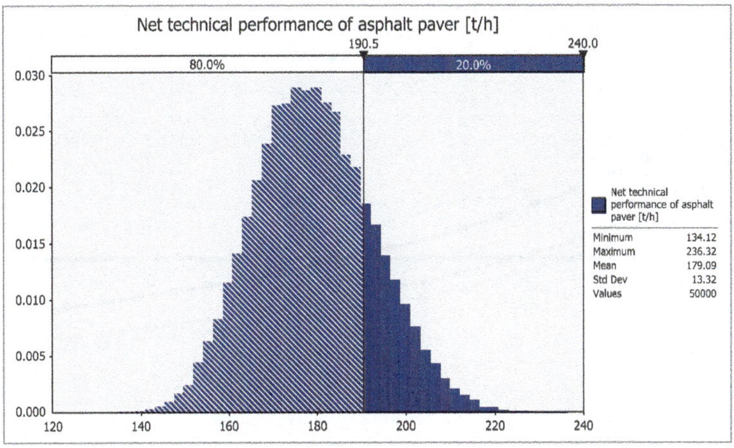

Fig. 10.96 Histogram – Asphalt paver performance – 80% to 20% chance/risk ratio

Determining the chance/risk ratio for the equipment chain depends on the bottleneck equipment on the one hand and on the scheduled construction time or daily working time on the other.

Since the asphalt paver has been identified as the bottleneck equipment in this example, a chance/risk ratio that the contractor would be prepared to accept can be selected for the performance histogram. For example, if the contractor is prepared to take a 20% risk, the asphalt paver's maximum performance is approximately 190.5 t/h (see Fig. 10.96).

If we enter this limit of 190.5 t/h as a horizontal line in the diagram for the required performance, all combinations of daily working time and scheduled construction time can be directly identified for which the selected performance limit is less than or equal to the required performance (chance). The contractor would take a higher risk in cases where the construction time or the daily working time is too short and the required performance would thus exceed the threshold of 190.5 t/h (see Fig. 10.97).

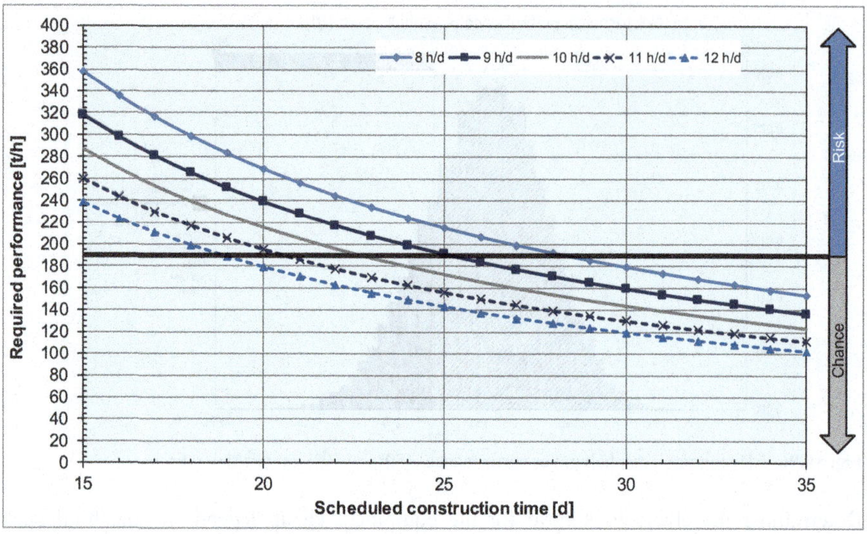

Fig. 10.97 Relationship – Scheduled construction time, daily working time, and required hourly performance – Chance and risk

If the work is performed in the holiday season, for example (involving an increased risk of traffic congestions and other construction sites en route between the mixing plant and the job site), the contractor may demand a higher safety margin for the transport performance. One option for achieving this level of safety is to use a larger number of trucks for transport (new triangular distribution: 45 | 50 | 55 trucks). The additional transport vehicles shift the performance of the truck fleet further to the right (see Fig. 10.98 compared to Fig. 10.95). Although the higher transport performance may lead to waiting times for the dump trucks, an undersupply to the bottleneck equipment (i.e. the asphalt paver) can basically be excluded.

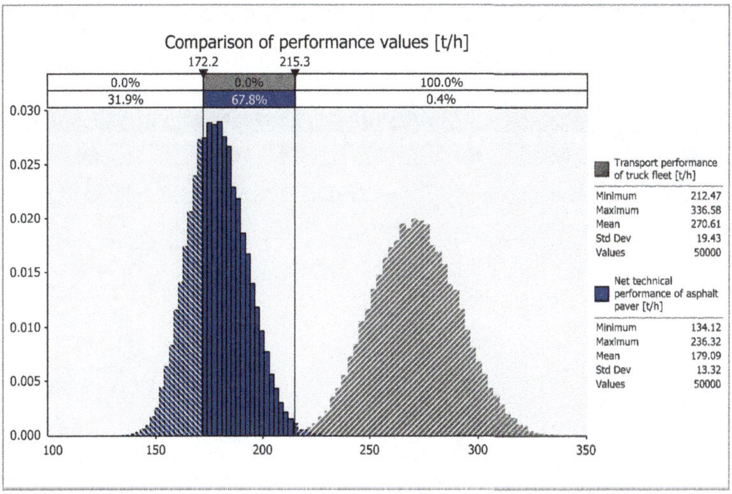

Fig. 10.98 Histograms – Comparison of performance values – Increased number of trucks

Contractors derive great benefit from probabilistic calculations carried out with the aim of coordinating equipment chains. The uncertainties associated with each component can first be investigated separately, followed by a comparison. This approach makes it easier to identify the bottleneck performance within the entire equipment chain, including the possibility of a graphical representation based on the histograms. The relationships between the pieces of equipment can be analyzed much more effectively by comparing the performance values (calculation of differences) and indicating the related chance/risk ratios, rather than by carrying out deterministic analyses.

The system outlined in this example can already be applied in PPH 3 in the bid preparation phase or at the contract award stage. If the contract is to be awarded according to the best bidder principle, for example, and if the construction time constitutes one of the assessment criteria, the probabilistic calculation makes it possible to check the plausibility of quoted short construction times, or to refute them accordingly. The respective chance/risk ratio is applied in order to carry out a clear and comprehensible assessment of the performance values determined for the equipment chain.

10.4.3 Compaction Performance in Earthworks

Determining the compaction performance in the construction of an embankment is part of the bid phase (PPH 3) and/or the construction phase (PPH 4) (see Fig. 10.99).

In the following sections, we describe the planned construction workflow for the contractor for constructing an embankment and illustrate it by schematic diagrams. We also calculate the performance values required for the transport of the fill material to the job site, for its distribution and precompaction using an excavator, and for its final compaction by means of a roller.

Taking into account the existing uncertainties, the performance calculations are carried out with the aid of Monte Carlo simulations. The input parameters for which no single numerical value can be determined are modeled as triangular distributions and are integrated as such into the simulations. The schematic sequences of the performance calculations are each represented in flow charts.

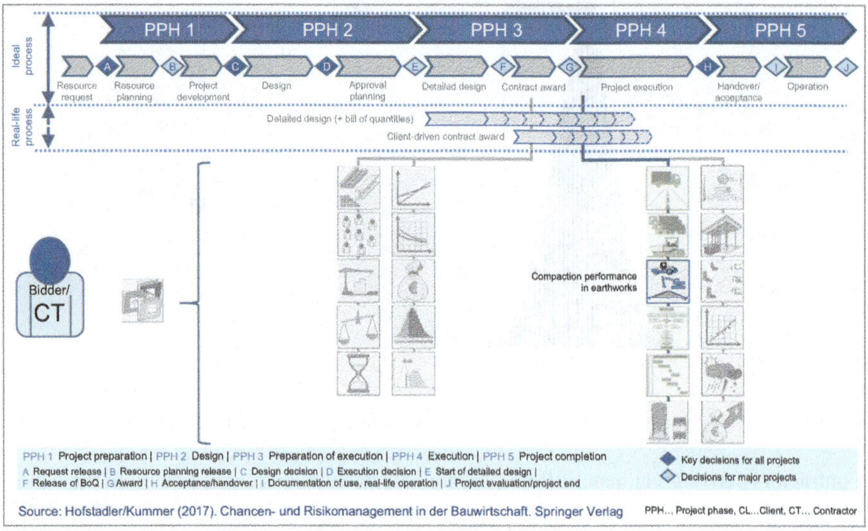

Fig. 10.99 Allocation of example in the project sequence

The results of the simulations are converted into histograms allowing to specify an over- or underrun probability for individual (deterministic) numerical values. This approach also illustrates the chance/risk ratio that the contractor has taken with their planned process flow.

For a direct comparison, the performance values of the equipment chain components are shown in a diagram and interpreted in a realistic way.

10.4.3.1 Task

The embankment to be constructed has a slope gradient of 1:2.75 on the water side and of 1:2.50 on the air side (see Fig. 10.100). Installation is carried out in layers with thicknesses of up to 30 cm. The outer sides of the embankment are to be covered to a width of about 0.50 m. After completion of an embankment section, the excess width must be removed again in accordance with the profile.

Excess material should be removed only after ground settlement has subsided. Depending on the extent of this settlement, an additional crest layer is applied and compacted if necessary using the removed excess material. This rework is non-critical for installation purposes and is thus disregarded in the calculations.

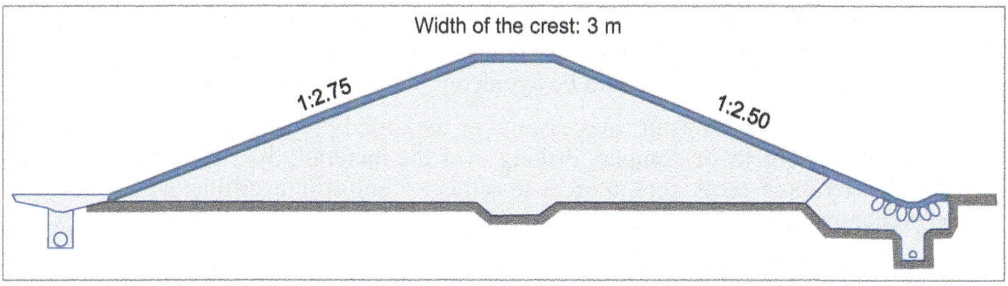

Fig. 10.100 Embankment profile – Qualitative representation

The embankment is about 870.0 m long. A total soil volume of approximately 51,000 m³ should be installed. This results in a fill material volume to be installed of about 58.62 m³ (= 51,000 m³ / 870 m) per meter of embankment.

For the installation of the fill material, dump trucks customized to earthworks (for material that is removed on-site) and four-axle trucks (for additionally supplied fill material) are used, which transport the material directly to the installation site and distribute it across a large area.

The fill material is installed and compacted over the entire embankment length in order to ensure that the load on the subsoil is as uniform as possible. Fig. 10.101 shows the flow chart for installing the individual layers of the embankment.

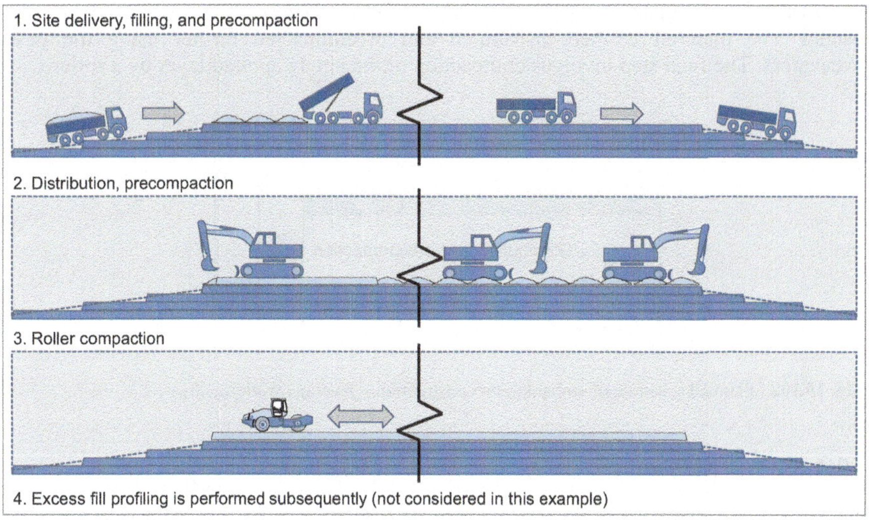

Fig. 10.101 Flow chart for constructing the embankment

Initially, trucks or dumpers are used to transport the fill material (which is either material stored on-site or material supplied from externally) via ramps to the respective height of the next layer to be installed. These vehicles drive onto the embankment and dump the fill material as evenly as possible onto the previously completed layer in order to install the next one. A "one-way operation" scheme is implemented so that the trucks or dumpers always have to drive over the previously placed material. This setup ensures initial precompaction and local distribution of the material. When unloading the fill material, care must be taken to ensure that it is distributed evenly over as large an area as possible. Isolated large piles of material should be avoided.

After completion of the transport process and of the initial precompaction of the new fill layer achieved by trucks or dumpers driving over the material, the fill material is distributed further by three excavators in order to achieve a uniform layer thickness. This layer is then precompacted further by the excavators.

Only two excavators have been directly allocated to the construction of the embankment. A third excavator is included in the overhead costs and is kept available for the entire construction period.

This third excavator is used during the construction of the embankment in order to maintain the paths to and on the dam. Once the fill work has been completed using trucks or dumpers, this excavator is used together with the other two excavators for distributing and precompacting the fill material.

In the final work step, the precompacted 30 cm fill material layer is compacted further using a roller.

The specified profile of the embankment is restored afterwards. The excess fill (0.50 m on either side of the embankment) required for construction is removed again if necessary due to ground settlement. If subsidence of the crest occurs, the removed excess material is placed as an additional layer on the crest in the affected areas and compacted.

Fig. 10.102 shows the planned schedule for the construction of the embankment. We find that in the first eight hours of the day the material is transported, placed, and precompacted. The material is then distributed and precompacted further using the three excavators. The final step involves compaction of the newly applied layer by a roller.

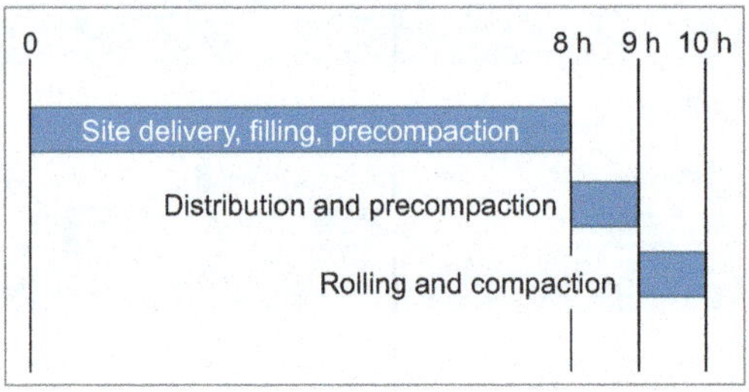

Fig. 10.102 Planned schedule of embankment construction – Average working day

10.4.3.2 Modeling

The following sections explain the individual calculation models for determining the performance values required for the delivery of the fill material, the layer-by-layer distribution and precompaction of the fill (by means of excavators), and roller compaction. The performance calculations carried out for the excavators and the roller refer to the transport performance required for the fill material. For this purpose, we compare the consolidated rather than the loosened volume of the fill material.

Input parameters subject to variations or uncertainties are modeled in the calculations as triangular distributions. The calculations are carried out using Monte Carlo simulations, each with 50,000 iterations (Latin Hypercube sampling method).

Transport performance for delivering the fill material

Fig. 10.103 shows the flow chart for calculating the transport performance. Uncertain input parameters are represented by triangular distributions, whereas deterministic inputs appear as arrows. Final and intermediate results of the simulation are shown as histograms in the flow chart.

The daily fill quantity is calculated by multiplying the number of deliveries and the average cargo quantity. We derive the average cargo quantity from the respective fill quantities (on-site or delivered) and the cargo volume of the dumpers or trucks.

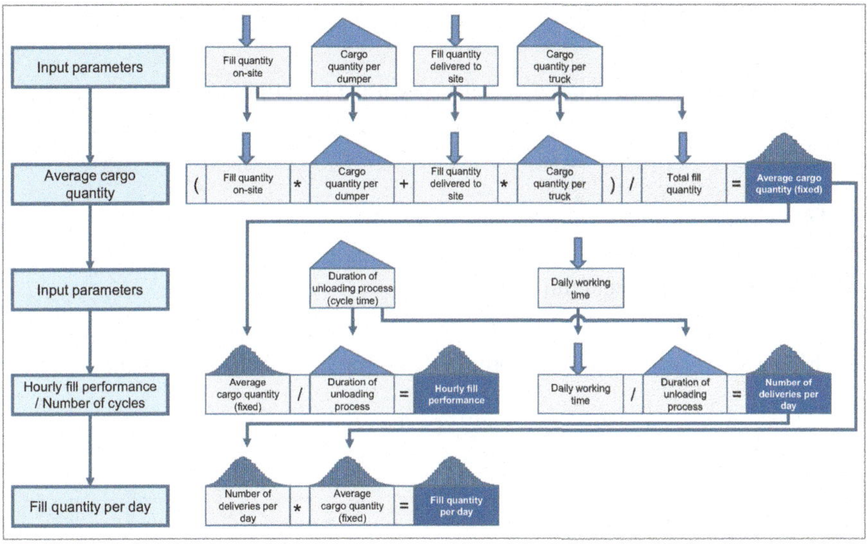

Fig. 10.103 Calculation flow chart – Transport performance

Performance of excavators for layer-by-layer distribution and precompaction of the fill material – Without GPS support

Since the trucks and dumpers will already have distributed the fill material roughly over the installation area, the excavators only need to level the loosely placed material across the entire area. The excavators already precompact the fill material by driving over the embankment and the newly applied layer.

To calculate the gross technical performance PV_G [m³/h], we apply the following equation (Eq. (10-24)):

$$PV_G = V_N \cdot n \cdot f_T \cdot f_B \cdot f_L \tag{10-24}$$

where

V_N	nominal volume – blade capacity [m³]
n	number of load cycles [1/h]
f_T	time factor – $0.83 \leq f_T \leq 0.92$ [-]
f_B	blade factor [-]
f_L	loading factor – $f_L = f_F / f_S$ [-]
f_S	loosening factor [-]
f_F	fill factor [-]

The nominal volume/blade capacity is calculated using Eq. (10-25).

$$V_N = 0.80 \cdot W \cdot H^2 \tag{10-25}$$

where

W	blade width [m]
H	mean blade height [m]

We then obtain the net technical performance PV_N [m³/h] applying Eq. (10-26).

$$PV_N = PV_G \cdot f_{E1} \cdot f_{E2} \tag{10-26}$$

where

f_{E1}	site factor [-]
f_{E2}	operational factor [-]

Fig. 10.104 shows the flow chart for calculating the excavator performance for distribution and precompaction. Uncertain input parameters are represented by triangular distributions, whereas deterministic inputs appear as arrows. Final and intermediate results of the simulation are shown as histograms in the flow chart.

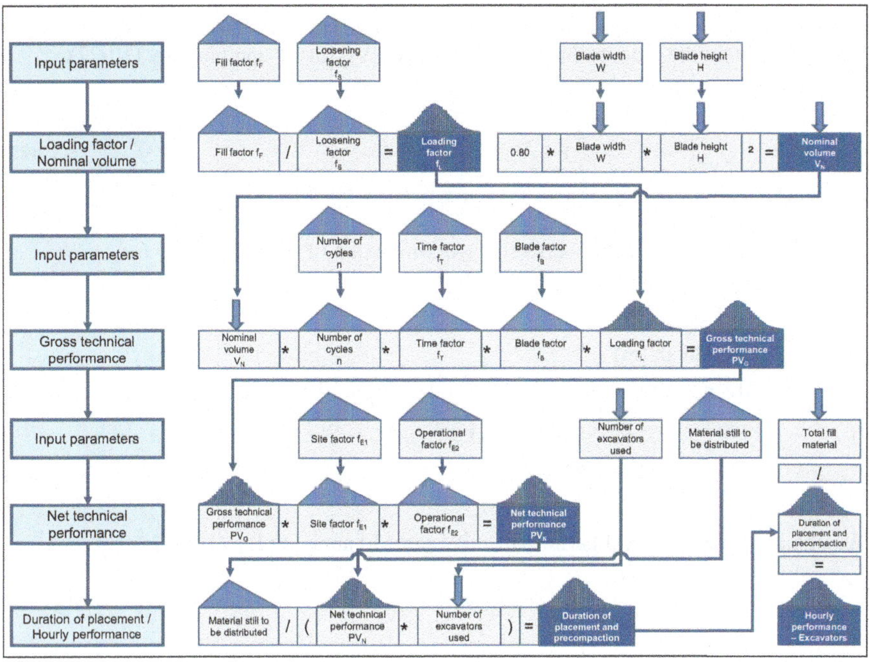

Fig. 10.104 Calculation flow chart – Excavator performance

Performance of excavators for layer-by-layer distribution and precompaction of the fill material – With GPS support

Systems have been introduced that enable automatic machine control in order to increase the productivity of construction equipment and to save costs and time as a result. However, the aim is not to replace operators but to increase the efficiency of their work. Nowadays, a large amount of additional information generated by the system helps to speed up work processes and avoid additional rework.

GPS-controlled equipment comes with the advantage that it increases the achievable accuracy while reducing the required load cycle times. The use of GPS-controlled excavators is particularly recommended for very complex tasks involving a large number of steps, as well as for subgrade construction since costly rework is largely avoided.

In a study carried out at Graz University of Technology[56)], a 14 to 23% increase in productivity was found for GPS-controlled excavators. This increase in productivity is applied as a triangular distribution (14.0 | 18.5 | 23%) directly to the hourly performance of the excavators determined without GPS support.

56) Cf. Graggober (2011), p. 118

Roller compaction performance

The compaction performance of the roller is considered to be the last link in the equipment chain. We obtain the gross technical performance PV_G [m³/h] applying Eq. (10-27).

$$PV_G = 60 \cdot \frac{w \cdot d \cdot v_A \cdot f_T}{P} \tag{10-27}$$

where

w	compaction width – effective working width of the roller w = 0.80 * W [m]
W	working width [m]
d	layer depth [m]
v_A	advance rate of the compaction roller [m/min]
f_T	time factor – $0.83 \leq f_T \leq 0.92$ [-]
P	number of roller passes [-]

We obtain the net technical performance PV_N [m³/h] applying Eq. (10-28).

$$PV_N = PV_G \cdot f_{E2} \tag{10-28}$$

where

f_{E2}	operational factor [-]

Fig. 10.105 shows the flow chart for calculating the roller performance. Uncertain input parameters are represented by triangular distributions, whereas deterministic inputs appear as arrows. The number of roller passes P is modeled as a discrete distribution with two possible values (two or three passes). Final and intermediate results of the simulation are shown as histograms in the flow chart.

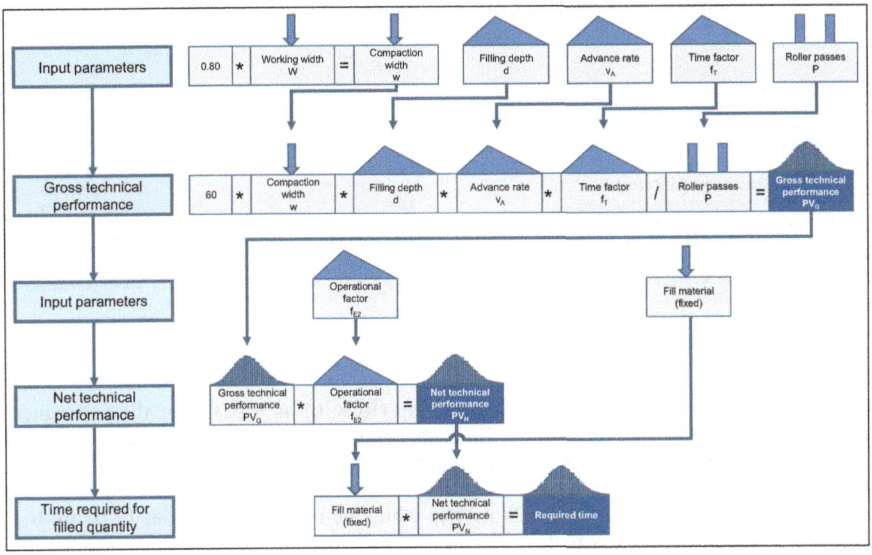

Fig. 10.105 Calculation flow chart – Roller performance

10.4.3.3 Calculation and Interpretation of Results

Calculation of the performance values required for the delivery of the fill material, the layer-by-layer distribution and precompaction of the fill (using excavators), and roller compaction is performed separately for each of these steps. A comparison of the performance histograms is given at the end of this section.

Transport performance for delivering the fill material

The total solid volume of the embankment fill amounts to approximately 51,000 m³. Of this total, about 21,000 m³ are available on-site as excavated material (see Row 1 of Table 10.40), and approximately 30,000 m³ must be obtained externally (see Row 2 of Table 10.40). Dumpers are used for transporting the site material to the point of installation. The additional material is delivered by 4-axle trucks.

The transport performance for the delivery of the fill material essentially depends on the cargo quantity per truck/dumper and on the cycle time required for the unloading process per truck/dumper.

The cargo quantity per truck or dumper is represented as a triangular distribution. The minimum, most likely, and maximum values are entered in Columns B, C, and D of Table 10.40 (see Rows 4 and 5). Column E contains the distributions or intermediate results.

Transport performance – Fill material delivery					
No.	**Parameter**	**MIN**	**MOST LIKELY**	**MAX**	**Distribution**
0	A	B	C	D	E
1	Fill material – on-site – dumpers		21,000.00		21,000.00 m³
2	Fill material – delivered – trucks		30,000.00		30,000.00 m³
3	Total fill material (fixed)				51,000.00 m³
4	Cargo quantity – dumpers (fixed)	12.00	13.50	15.00	13.50 m³/cycle
5	Cargo quantity – trucks (fixed)	9.00	9.50	10.00	9.50 m³/cycle
6	**Average cargo quantity (fixed)**				**11.15 m³/cycle**
7	Duration of unloading process (cycle time)	4.00	5.00	6.00	5.00 min/cycle
8	Duration of unloading process (cycle time)				0.08 h/cycle
9	**Hourly fill performance (fixed)**				**133.76 m³/h**
10	Daily working time for material transport		8.00		8.00 h/d
11	Number of deliveries per day				96.00 deliveries/d
12	Number of deliveries per hour				12.00 deliveries/h
13	**Fill quantity per day (fixed)**				**1,070.12 m³/d**

Table 10.40 Calculation table – Transport performance

We determine the mean cargo quantity per trip (Row 6) as a basis for the subsequent calculation steps. We obtain this result through a weighted average of the cargo quantities for dumpers and trucks based on the fill material quantities to be transported. The mean cargo quantity is shown as a histogram in Fig. 10.106. The minimum after 50,000 iterations amounts to approximately 10.28 m³/trip while the maximum is about 12.01 m³/trip. The mean amounts to approximately 11.15 m³/trip whereas the standard deviation roughly amounts to 0.28 m³/trip.

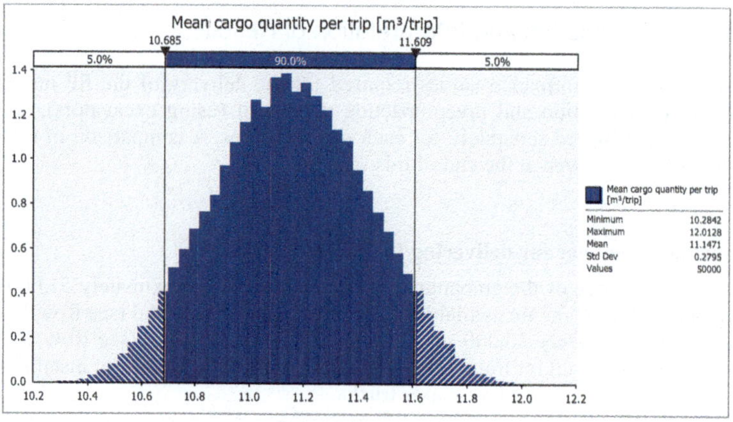

Fig. 10.106 Histogram – Mean cargo quantity per trip

The time required for a single unloading process (cycle time) is set as a triangular distribution between 4.00 and 6.00 minutes per trip, with the most likely value amounting to 5 minutes per trip (see Row 7 of Table 10.40). In Cell E8, the time required for the unloading process is converted from [min/trip] to [h/trip].

The hourly fill performance of the consolidated material is then calculated in Cell E9 of Table 10.40 by dividing the mean cargo quantity (Cell E6) by the duration of the unloading process (Cell E8). Fig. 10.107 shows the histogram representing the hourly fill performance. In 5% of all cases, 116.9 m³/h are underrun and 155.6 m³/h are exceeded. The mean amounts to about 135 m³/h.

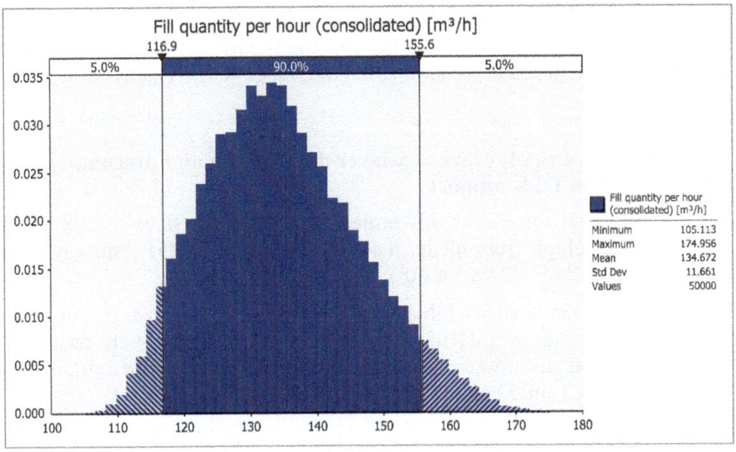

Fig. 10.107 Histogram – Fill quantity per hour (consolidated material)

The daily working time for material transport is set at 8.00 h/d (see the deterministic value in Row 10 of Table 10.40). We thus obtain the number of deliveries per day (Row 11) by dividing the working time (Row 10) by the duration of each unloading operation (see Row 8). The number of deliveries per hour is equivalent to the reciprocal of the duration of the unloading process (Cell E8) and is shown in Row 12 of Table 10.40.

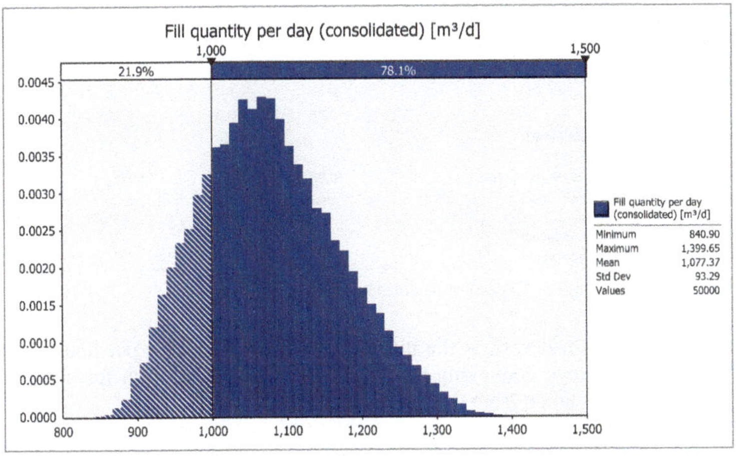

Fig. 10.108 Histogram – Fill quantity per day (consolidated material)

The delivered solid fill material quantity per day (see Cell E13) is calculated by multiplying the number of deliveries per day (Cell E11) by the mean cargo quantity (Cell E6). Fig. 10.108 shows the histogram representing the fixed daily installation performance. The mean amounts to approximately 1,077 m^3/d whereas the standard deviation roughly amounts to 93 m^3/d. The minimum after 50,000 iterations amounts to approximately 840 m^3/day while the maximum is about 1,400 m^3/day.

In the so-called embankment fill campaign, the client specified a fill volume of 500 to 1,000 m³/day. As the histogram shows, a quantity of 1,000 m³/day is exceeded in most cases (approximately 78.0%). Only in 22.0% of the iterations does the transport performance fall below 1,000 m³/day. The simulation did not generate values below 500 m³/day.

Performance of excavators for layer-by-layer distribution and precompaction of the fill material – Without GPS support

The excavators planned for use in this project example each have a 2.98 m wide and approximately 0.60 m high front blade. Insertion into Eq. (10-25) results in its nominal capacity of 0.86 m³ (= 0.80 * 2.98 * 0.60²).

The performance is calculated in Table 10.41. In the first step, we calculate the gross technical performance using Eq. (10-24). Input parameters fraught with uncertainties are modeled as triangular distributions. The minimum, most likely, and maximum values are shown in Columns B, C, and D of Table 10.41.

Performance of excavators for layer-by-layer distribution and precompaction of fill material					
No.	Parameter	MIN	MOST LIKELY	MAX	Distribution
0	A	B	C	D	E
1	Blade width				2.98 m
2	Blade height				0.60 m
3	Nominal volume V_N				0.86 m³
4	Number of cycles n	120.00	140.00	160.00	140.00 1/h
5	Time factor f_T	0.83	0.87	0.92	0.87
6	Blade factor f_B	0.70	0.80	0.85	0.78
7	Loosening factor f_S	1.20	1.25	1.30	1.25
8	Fill factor f_F	0.95	0.97	1.00	0.97
9	Loading factor f_L				0.78
10	**Gross technical performance**				**64.01 m³/h**
11	Site factor f_{E1}	0.85	1.00	1.15	1.00
12	Operational factor f_{E2}	0.88	0.94	1.00	0.94
13	**Net technical performance (fixed)**				**60.16 m³/h**
14	Fill material (fixed)				1,000.00 m³
15	Proportion of material still to be distributed	10.00	30.00	50.00	30.00%
16	Material still to be distributed				300.00 m³
17	Number of excavators used				3.00
18	Duration of filling and precompaction				1.66 h
19	**Hourly performance – Excavators**				**601.65 m³/h**

Table 10.41 Calculation table – Excavator performance

The number of cycles n describes the number of blade operations per hour [1/h] and is dependent on the average conveying distance. Table 10.42 lists data for calculating the performance of dozers taken from the literature.

Average conveying distance [m]	20.00	30.00	40.00	50.00	60.00	70.00	80.00	90.00	100.00
Number of cycles n [1/h]	100.00	78.00	63.00	50.00	42.00	36.00	31.00	27.00	24.00

Table 10.42 Number of cycles n [1/h][57)]

57) Cf. Brecheler et al. (1998), p. 207

Since the excavators are positioned directly between the roughly spread fill material and only need to distribute it evenly, the average conveying distance consistently lies below 20 m. The number of cycles is thus set between 120 and 160, with a most likely value of 140 (Row 4 of Table 10.41).

The time factor f_T [-] is selected in a range between 0.83 and 0.92 (see Row 5 of Table 10.41), while the blade factor f_B [-] ranges from 0.70 to 0.85 (see Table 10.43 or Row 6 of Table 10.41) (in this case, the lowest category for dozers was selected since no comparable values could be determined for excavators).

Blade shape	Blade factor f_B
U blade Front blade, transverse blade, standard blade Blade ends with narrow angle to front or with lateral baffle plates	1.10 - 1.25
S blade Universal blade Wide blade wings, angle to front	1.00
A blade Swivel blade Bilateral swiveling, no angular ends, large blade width, small blade height	0.70 - 0.85

Table 10.43 Blade factor[58)]

The loading factor f_L [-] captures the blade filling level and is calculated by dividing the fill factor f_F [-] by the loosening factor f_S [-].

The latter is assumed to range between 1.20 and 1.30 [-] for the fill material to be installed. This corresponds to a 20 to 30% increase in the volume of the loosened material compared to the compacted material. The fill factor f_F is given as 1.00 for topsoil and for easily removable soil types, whereas it is set to 0.95 for moderately removable soil types (see Row 8 of Table 10.41).

We insert the input parameters into Eq. (10-24) and obtain the gross technical performance PV_G [m³/h] as a histogram, as shown in Fig. 10.109.

58) Cf. ibid., p. 206f.

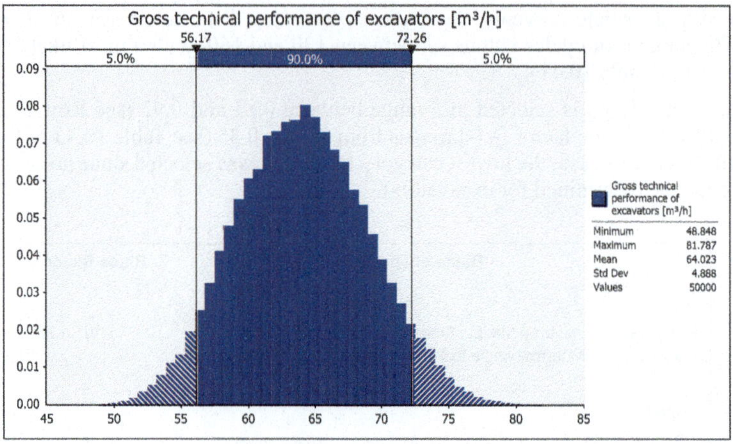

Fig. 10.109 Histogram – Gross technical performance – Excavators

Taking into account the site factor f_{E1} [-] (Row 11 of Table 10.41) and the operational factor f_{E2} [-] (Row 12 of Table 10.41), we get the net technical performance PV_N [m³/h] shown in Fig. 10.110. This value lies between 50.56 and 70.59 m³/h in 90% of the simulated cases. The mean amounts to approximately 60.18 m³/h whereas the standard deviation roughly amounts to 6.10 m³/h. These values refer to an excavator and to the actual consolidated soil volume to be distributed (i.e. moved horizontally).

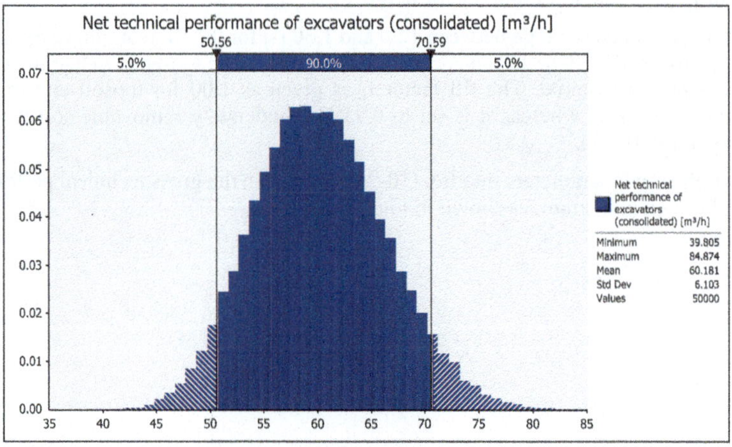

Fig. 10.110 Histogram – Net technical performance – Excavators

Cell E14 of Table 10.41 states the target volume of 1,000 m³ of consolidated material to be installed per day. The major portion of this fill material will already be in the right place due to the distribution by trucks or dumpers, and thus need not be leveled further. We assume that, ultimately, only 10 to 50% of the fill material (depending on the quality of the distribution by trucks/dumpers) would have to be distributed further by the excavators (see Row 15; triangular distribution with the most likely value set at 30%).

We thus obtain the quantity of the remaining material to be distributed in Row 16. A total of three excavators (each with identical performance) are used to distribute this quantity (see Row 17). We calculate the time required for installing as well as precompacting the material (Row 18) by dividing the remaining quantity to be distributed (Row 16) by the product of the net technical performance (Row 13) and the number of excavators (Row 17). Fig. 10.111 shows the histogram of this duration.

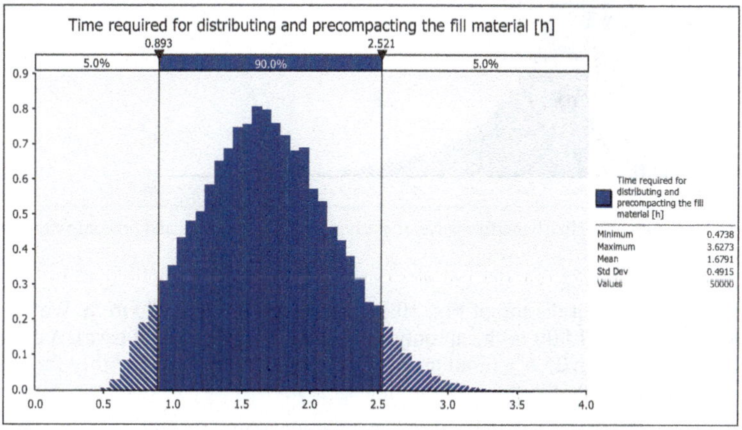

Fig. 10.111 Histogram – Time required for distributing and precompacting the fill material

If the total fill quantity is again allocated to this installation time, we get an additional histogram representing the hourly performance of the excavators for spreading and precompacting the material (see Fig. 10.112). There is an 8.5% probability for the distribution and further precompaction of 1,000 m^3 to be completed within less than an hour. Thus, there is only a small chance that this performance can be achieved within an hour in the actual construction phase, but it would still be possible to achieve this target from a construction management perspective.

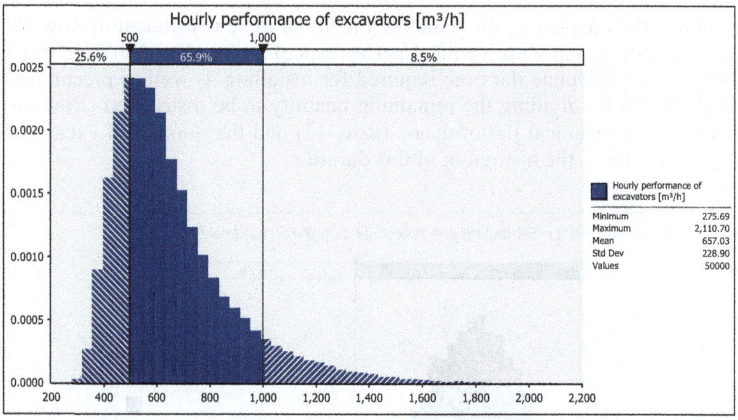

Fig. 10.112 Histogram – Hourly performance of excavators for distributing and precompacting the fill material

The sliders appearing at the top of Fig. 10.112 are set to 500 or 1,000 m^3/h. We find that a performance value of 1,000 m^3/h can only be exceeded in 8.5% of all cases. A capacity of 500 m^3/h is underrun in 25.6% of all cases, and there is a 65.9% probability for achieving a performance between 500 and 1,000 m^3/h.

Relating the quantity of 1,000 m^3 to an installation strip of 3.00 m at a depth (layer thickness) of 0.30 m results in a length of approximately 1,111 m [= 1,000 m^3 / (3.00 m * 0.30 m)]. Dividing this length by the time required for distribution and precompaction (see Fig. 10.111) then yields the average required speed of the excavators (without turning maneuvers), see Fig. 10.113.

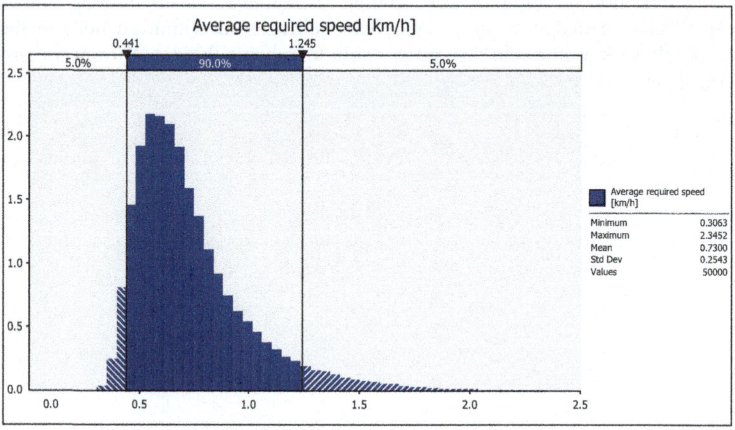

Fig. 10.113 Histogram – Average required speed of excavators for distribution and precompaction

This average speed is a theoretical value because the material is not only installed on a 3 m wide strip. In addition, overlaps between "lanes" need to be considered during precompaction. However, the average speed magnitude and range shows that the speeds required for achieving the target of 500 to 1,000 m³ per day are not unrealistic.

The excavators move back and forth on the area where they have to distribute and precompact the material. Considering an installed volume of 1,000 m³ and a layer thickness of 0.30 m, the excavators used on-site would have to jointly work a total area of 3,333 m². If three excavators are used, each of the excavators must spread and precompact the fill material over an area of approximately 1,111 m². For an embankment width of 16 m, for example, this corresponds to an embankment section per excavator of about 69 m in length (see the schematic representation in Fig. 10.114, bottom).

On the other hand, if only 500 m³ of fill material are delivered, the area to be worked by a single excavator decreases to approximately 556 m² (assuming a layer thickness of 0.30 m). Again, for an embankment width of 16 m, this corresponds to an embankment section per excavator of about 35 m (see the schematic representation in Fig. 10.114, top).

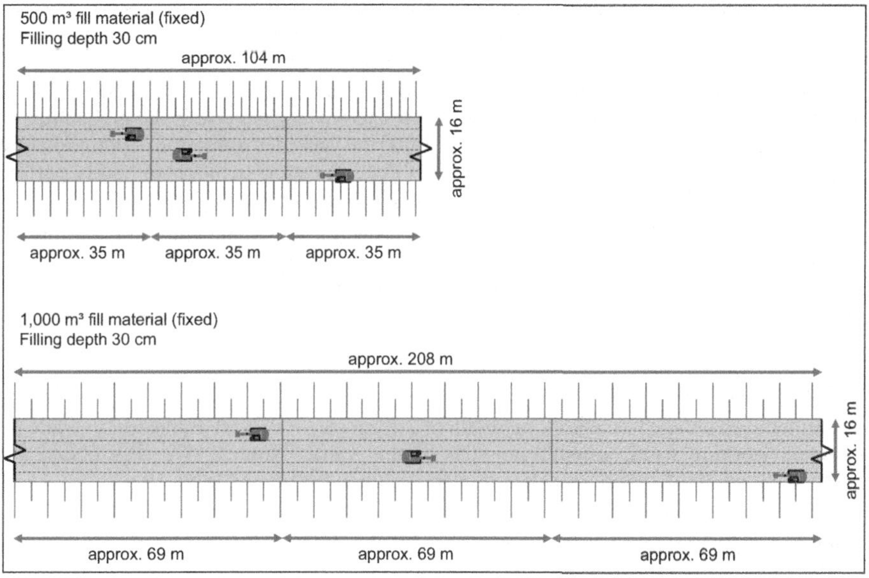

Fig. 10.114 Schematic representation – Distribution and precompaction using three excavators

Performance of excavators for layer-by-layer distribution and precompaction of the fill material – With GPS support

The increase in excavator productivity brought about by GPS support is applied as a triangular distribution (14.0 | 18.5 | 23%) directly to the hourly performance of the excavators determined without GPS support (see Fig. 10.112). Fig. 10.115 compares the histogram representing the hourly performance of the excavators with GPS support with the hourly performance achieved without GPS.

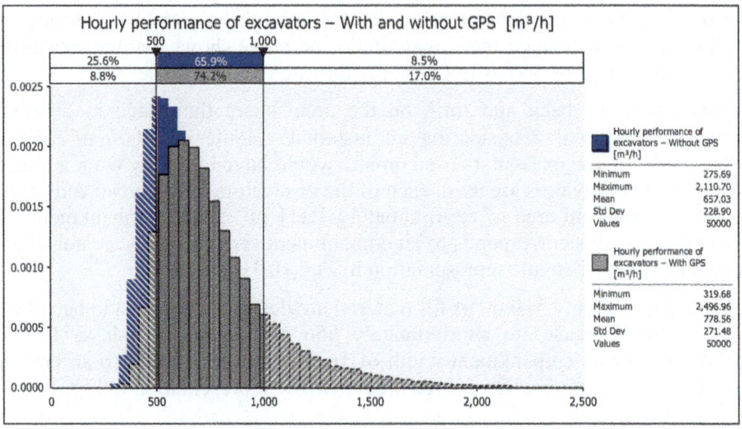

Fig. 10.115 Histograms – Comparison – Hourly performance of excavators for distribution and precompaction of the fill material – With and without GPS support

The figure shows that the probability of achieving performance values below 500 m³/h has decreased significantly from approximately 25.6% to 8.8%. Conversely, the probability of achieving performance values in excess of 1,000 m³ has increased from approximately 8.5% to about 17.0%. With GPS support, the three excavators achieve an hourly performance between 500 and 1,000 m³/h in about 74.2% of the simulated cases.

Using GPS-controlled equipment increases the probability of achieving the target installation rate of 500 to 1,000 m³/h. The risks taken by the contractor are reduced while the chances of achieving or exceeding the envisaged target increase to the same extent. With GPS support, the contractor will be in a much better position to achieve the specified performance of 1,000 m³/h.

Roller compaction performance

The roller used on-site has a working width of 2.13 m (see Row 1 of Table 10.44) while its effective working width amounts to 80% of this value (Row 2). We thus obtain the compaction width of 1.70 m as a deterministic value in Cell E3 of Table 10.44.

The embankment fill process specifies a filling depth of up to 30 cm. This is why we assume a depth (layer thickness) between 0.25 and 0.30 m as a triangular distribution (with a most likely value of 0.275 m) (Row 4).

Advance rates between 50 and 300 m/min are specified for sheep-foot rollers, whereas smooth-drum rollers usually achieve rates between 50 and 100 m/min.[59)] For the calculation, the advance rate is assumed to be between 50 and 100 m/min (with a most likely value of 75 m/min) (Row 5).

The time factor f_T [-] is defined as a triangular distribution between 0.83 and 0.92 (most likely value: 0.87), in analogy to the performance calculation carried out for the excavators.

[59)] Cf. Brecheler et al. (1998), p. 223

Compaction performance – Roller					
No.	**Parameter**	**MIN**	**MOST LIKELY**	**MAX**	**Distribution**
0	A	B	C	D	E
1	Working width		2.130		2.130 m
2	Proportion of effective working width		0.800		0.800
3	Compaction width				1.704 m
4	Filling depth	0.250	0.275	0.300	0.275 m
5	Advance rate	50.000	75.000	100.000	75.000 m/min
6	Time factor	0.830	0.870	0.920	0.873
7	Number of passes	2.000		3.000	2.000
8	**Gross technical performance**				**920.799 m³/h**
9	Operational factor	0.880	0.940	1.000	0.940
10	**Net technical performance**				**865.551 m³/h**
11	Fill material (fixed)				1,000.000 m³
12	**Time required for 1,000 m³**				**1.155 h**

Table 10.44 Calculation table – Roller performance

The number of roller passes is modeled as a discrete distribution, expecting either two or three passes. A 50% probability is assumed for each of these two options. For the simulation, this means that in 50% of all cases two passes are expected, whereas three passes are assumed in the remaining 50% of cases. Fig. 10.116 shows the discrete distribution of the roller passes.

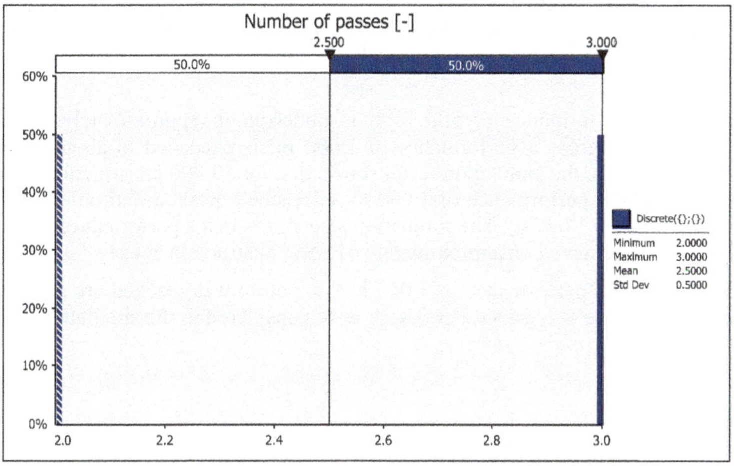

Fig. 10.116 Discrete distribution of the input parameter – Number of roller passes

We subsequently obtain the gross technical performance PV_G [m³/h] by inserting the input parameters in Eq. (10-27) as a histogram (see Fig. 10.117). The histogram is found to show a bimodal curve. This is due to the discrete input parameter for the passes (see Fig. 10.116). Higher performance values are achieved if only two roller passes are required.

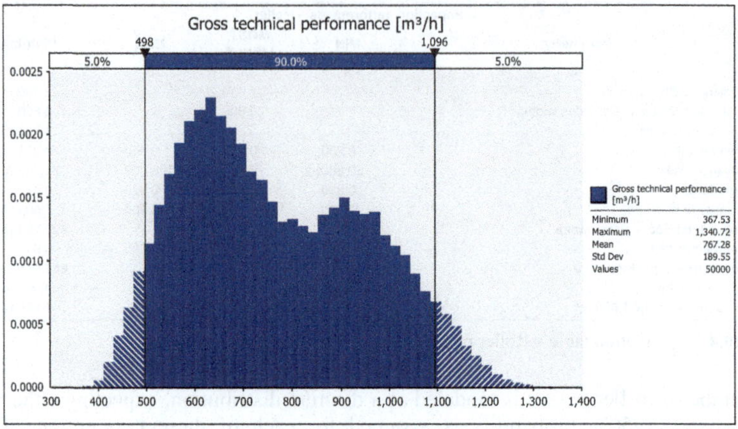

Fig. 10.117 Histogram – Gross technical performance – Compaction roller

Considering the operational factor f_{E2} [-] according to Eq. (10-28), we get the net technical performance PV_N [m³/h], which is also represented by a histogram (see Fig. 10.118). Again, we recognize a bimodal curve. The distribution mean amounts to approximately 721 m³/h whereas the standard deviation roughly amounts to 179 m³/h. The minimum after 50,000 iterations amounts to approximately 344 m³/h while the maximum is about 1,281 m³/h.

A net technical performance of 500 m³/h is underrun in approximately 9.6% of all simulated cases, whereas a performance of 1,000 m³ is exceeded in about 7.7% of all cases. In other words, the simulation result shows that for 50,000 construction sites considered, a compaction performance of 1,000 m³ is reached or exceeded on approximately 3,850 sites (= 50,000 * 0.076). The simulation also shows that a performance between 500 and 1,000 m³/h is achieved on approximately 41,350 construction sites (= 50,000 * 0.827).

All cases in which a performance of 1,000 m³/h or more was reached are attributable to those iterations where only two roller passes were considered in the simulation.

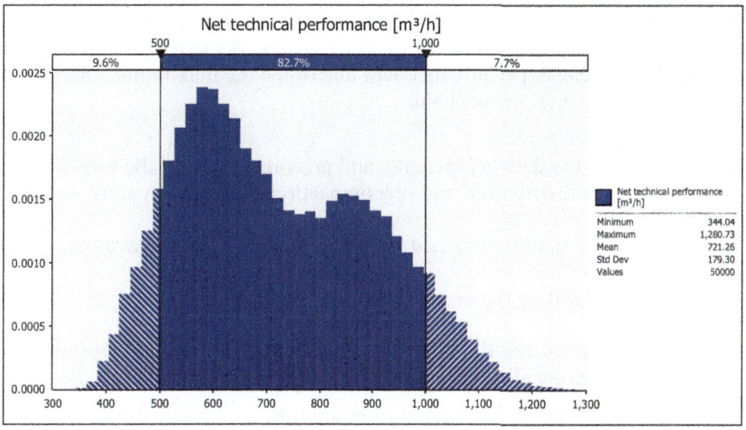

Fig. 10.118 Histogram – Net technical performance – Compaction roller

The time required for compacting a volume of 1,000 m³ (Row 11; deterministic value) is given in Row 12 of Table 10.44. The quantity is divided by the net technical performance (Cell E10). This calculation results in a histogram representing the duration of the roller compaction process (see Fig. 10.119).

This duration also exhibits a bimodal curve, which is due to the net technical performance histogram (see Fig. 10.118). Although it is relatively unlikely that a duration of one hour for compacting 1,000 m³ will be underrun (the related probability amounts to only 7.7%), this still lies within the range of what is possible from a construction management point of view.

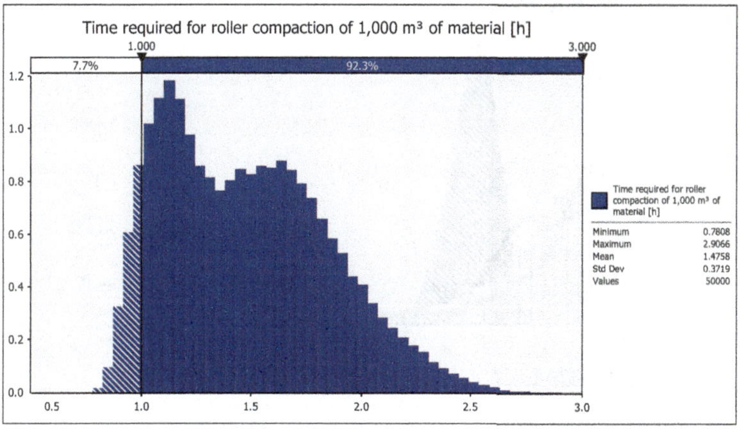

Fig. 10.119 Histogram – Time required for roller compaction of 1,000 m³ of material

Comparison of performance values

The performance values achieved by the three links in the equipment chain must be optimally coordinated in order to ensure an efficient and effective construction process. In this scenario, the bottleneck performance will determine the maximum achievable performance across the entire equipment chain.

Histograms were derived for

- the performance for the delivery, placing, and precompaction by the trucks/dumpers
- the performance for distribution and precompaction by the excavators – without GPS support
- the performance for distribution and precompaction by the excavators – with GPS support
- the performance for rolling the installed fill.

These performance values are then compared with the client's specification of 500 to 1,000 m³ expected to be installed on a single working day.

The transport performance is determined in relation to a single working day for a daily working time of 8 h/d. The excavator performance and the roller performance are each calculated for a working time of 1 h/d (see Fig. 10.120).

The abscissa of Fig. 10.120 shows the fill performance of the trucks/dumpers (transport performance) relative to the working day [m³/d]. The performance values of the excavator and the roller are each related to cubic meters per hour [m³/h]. The fact that an operating time of 1 h/d was defined for the excavator and the roller ensures comparability of the respective performance values.

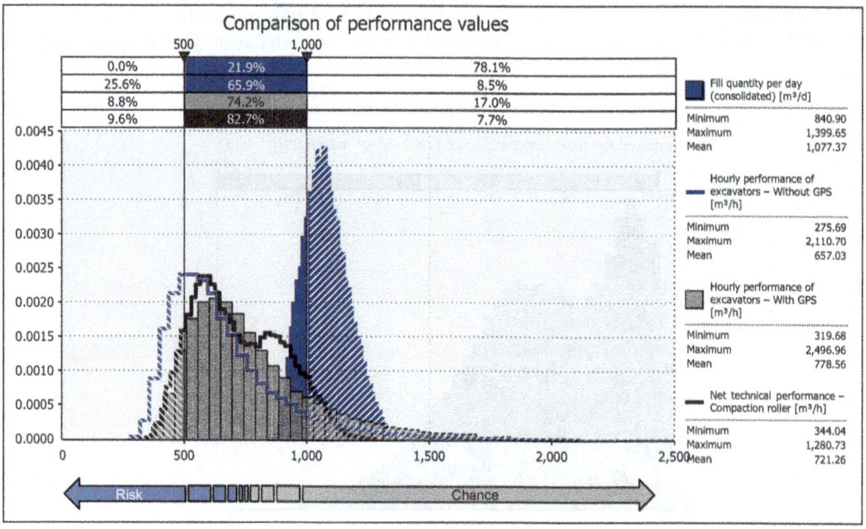

Fig. 10.120 Histogram – Comparison of performance values – Transport performance (8 h/d) – Excavator performance with and without GPS support (1 h/d) – Roller performance (1 h/d)

Due to the different cargo quantities (of the trucks/dumpers) and the varying unloading times for one working day, the transport performance of the trucks/dumpers was determined for 8.00 h/d. For a daily working time of 8 hours, the calculated histogram shows a performance value above 1,000 m³/d in approximately 78.1% of all cases (see Fig. 10.120). This means that, for the purpose of this comparison, the transport performance should not be classed as a bottleneck parameter.[60)]

Three excavators are used for one hour per day for distributing and precompacting the fill material. The performance achievable in this setting was also determined as a histogram (see Fig. 10.120; this is the hourly excavator performance).

For excavators without GPS support, there is a probability of approximately 65.9% for the performance to lie between 500 and 1,000 m³/d (assuming an operating time of 1 h/d). A lower performance is achieved in about 25.6% of all cases, whereas an even higher performance value is reached in approximately 8.5% of all cases.

If excavators with GPS support are used, the performance of the equipment increases because less rework needs to be done while achieving greater accuracy and precision overall.

With GPS-supported excavators, there is a probability of approximately 74.2% for achieving a performance between 500 and 1,000 m³/d (assuming an operating time of 1 h/d). A lower performance is achieved in about 8.8% of all cases, whereas a higher performance is reached in approximately 17.0% of all cases.

The performance values assumed by the contractor while taking into account the planned time schedule (see Fig. 10.102) appear even more realistic from a construction management perspective if GPS support is included in the scenario.

Roller compaction performance is the last equipment parameter relevant to constructing the embankment. For the purpose of this calculation, this performance is also related to a daily operating time of 1 h/d. Fig. 10.120 also shows the related histogram. We find that a roller compaction performance between 500 and 1,000 m³/d (assuming an operating time of 1 h/d) can be achieved in 82.7% of cases. This assumption applies if two roller passes are sufficient for the quantity to be compacted in 50% of cases while three passes are assumed for the remaining 50% of cases.

According to these assumptions, a roller performance below 500 m³ is achieved in 9.6% of cases while over 1,000 m³ are reached in 7.7% of all cases.

Overall, this comparison shows that the performance of the excavators and of the roller determine the total equipment chain performance and are thus on the critical path (bottleneck performance parameters). However, there are sufficient chances for both work areas that the required performance can be achieved when viewed from a construction management perspective. In any case, the above hourly performance values appear to be plausible and justifiable for the defined input parameters and processes.

10.4.3.4 Benefit for the Contractor

The daily performance specified by the client for the construction of the embankment lies between 500 and 1,000 m³/d. Both the material available on-site (approximately 21,000 m³) and material obtained externally (about 30,000 m³) are installed in order to construct the embankment. This material needs to be transported to the installation site by trucks or dumpers, and is dumped there. This process is carried out such that the truck or

60) We did not consider the performance required for loading the dumpers and the transport performance of the trucks on their way to the construction site.

dumper drivers ensure that the material is spread as evenly as possible when it is dumped, and that initial precompaction is achieved by driving over the dumped material.

Further distribution and precompaction of this fill material is performed by three excavators used together for one hour per day to produce the daily subgrade, which is subsequently compacted by a roller in a final work step.

Monte Carlo simulations were used to determine the performance of the individual components of the equipment chain. This approach offers the advantage that by using distribution functions, a large number of different combinations of input parameters can be taken into account in the simulations and subsequently displayed as histograms. The calculations show that it is realistic and plausible to install between 500 to 1,000 m^3/d of material in the specified construction process.

Comparing the histograms provides information on the chance/risk ratios of the individual links in the equipment chain. The ability to define the bottleneck equipment means that it is possible to intervene in, and manage and control, the planned process even before the actual construction phase begins. Among other conclusions, this example showed that the use of GPS-controlled excavators can lead to an increase in the potential chance for the contractor to meet the performance requirements.

Overall, the simulation proved that the installation of 500 to 1,000 m^3 of material per day for constructing the embankment is plausible and extremely probable from a construction management point of view.

10.4.4 Method Selection

In addition to economic criteria, other criteria, such as workplace safety and aesthetics, come into play in order to carry out a differentiated process comparison. This differentiated comparison is more comprehensive than the costing- or pricing-based comparison. It requires a larger amount of data, which makes it more time-consuming. However, this approach offers the advantage of a holistic view that contributes significantly to reducing the risks associated with the pricing and construction phases.[61)]

Selecting the most appropriate construction method is particularly crucial in the construction phase (PPH 4, see Fig. 10.121) and/or in the process planning phase because this is when the final decision on the use of the selected method will be made. If the selected method is found to be completely or partially unsuitable for the execution phase, this will be associated with increased costs and/or longer construction times due to reduced productivity or to the change to another method.

[61)] Cf. Hofstadler (2008), p. 345

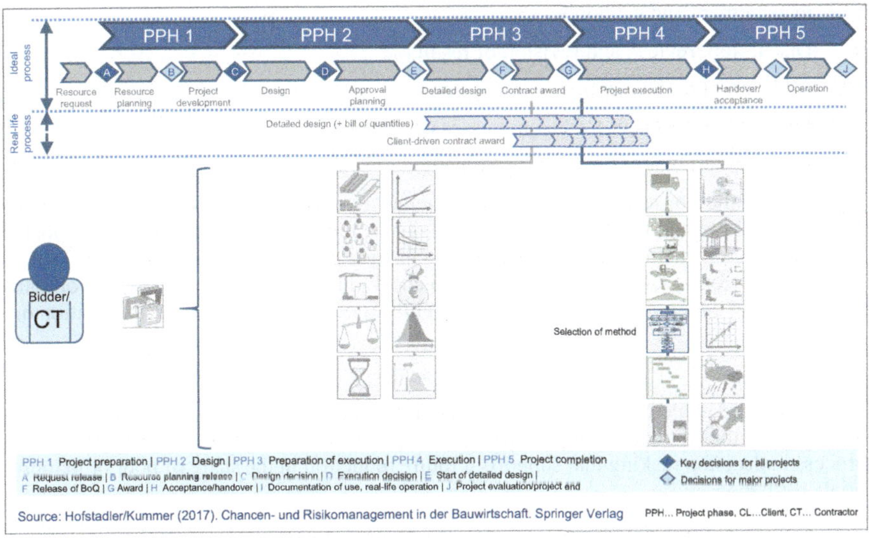

Fig. 10.121 Allocation of example in the project sequence

10.4.4.1 Basics

Fig. 10.122 provides a schematic overview of the systematic procedure for the differentiated process comparison. Its steps can be detailed as follows:[62)]

1. Definition of elimination criteria including subcriteria, taking into account the specification documents or the construction contract (depending on whether the contract has already been awarded or not):
   - Aesthetic criteria
   - Structural criteria
   - Safety criteria
   - Technical criteria
   - Criteria related to site surroundings and environment

2. Utility value analysis including a points system (elimination criteria)
3. Check if all elimination criteria were met. If some of the subcriteria were not met, we will have to determine whether specific elements of the corresponding process or system can be substituted.
4. Exclusion of all processes and systems that do not meet all of the selection criteria after substitution
5. Definition of the criteria related to construction management and economics
6. Determination of the ranges for costs and construction time with special consideration of resource utilization and logistics
7. Estimation of the chance/risk ratio of the different processes/systems with respect to the defined criteria

[62)] Cf. ibid., p. 345

8. Utility value analysis including a points system (criteria related to construction management and economics)
9. Discussion, review, and revision of results
10. Addition of points per process or system
11. Final risk assessment for the processes or systems
12. Decision in favor of a construction method or system
13. During the construction phase, target-actual comparisons are used to check if the selected method meets all requirements. Thus, the decision may have to be updated during construction, for instance due to changed circumstances under which work is to be performed. In addition, we need to check if all relevant criteria for the comparison were covered and whether their weightings corresponded to the real situation on-site.

The knowledge gained from the decision-making process and in the evaluation phase must be added to the knowledge repository. Such updates create a continuous improvement process in decision making that supports a future reduction in the share of poor or inappropriate decisions.

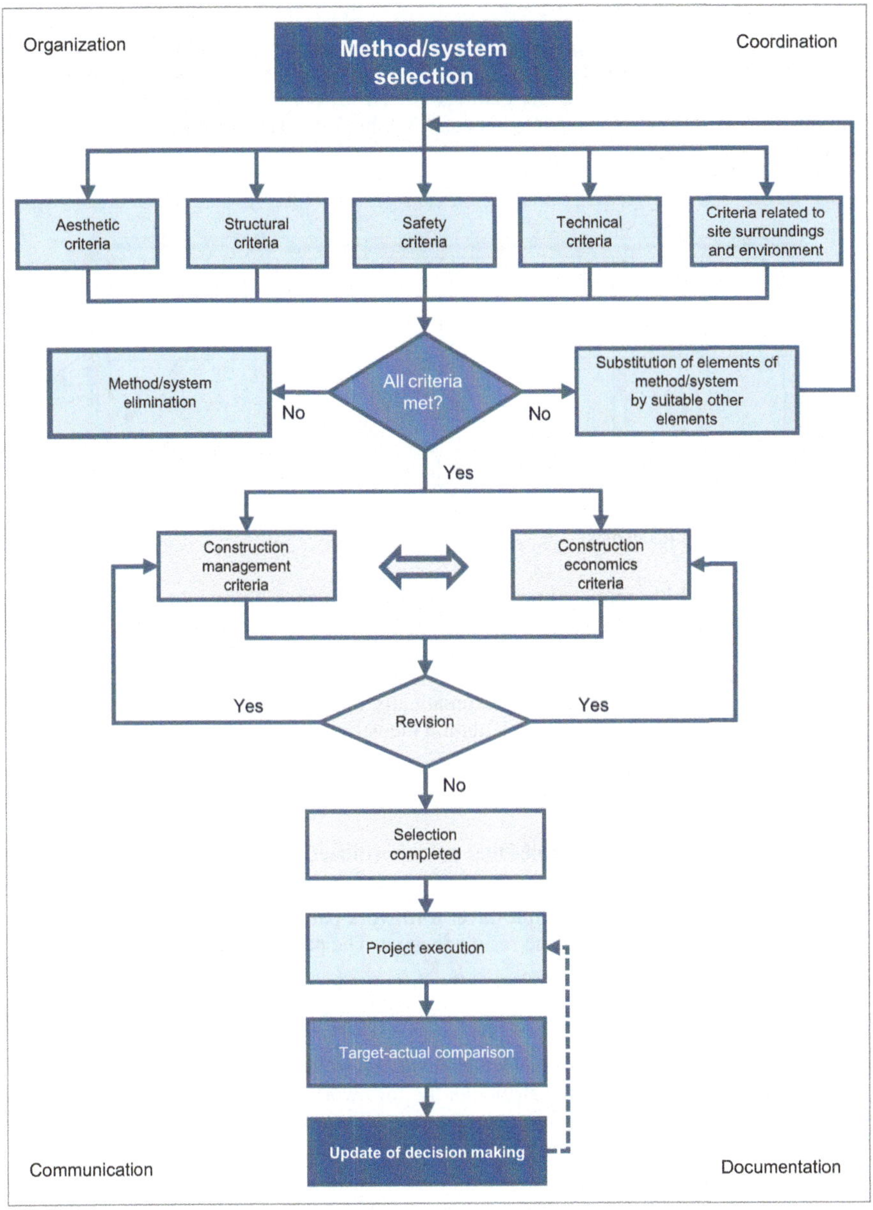

Fig. 10.122 Flow chart of the differentiated process comparison

10.4.4.2 Task

Cast-in-place concrete has been specified for the ceilings of a two-story retail park (see sectional view in Fig. 10.123). Each floor is 4 m high and the floor slab area amounts to 2,250 m² in each case. Each of the ceilings is cast in four work sections. The decision-making process should also take into account that the floor slabs must conform to exposed concrete specifications.

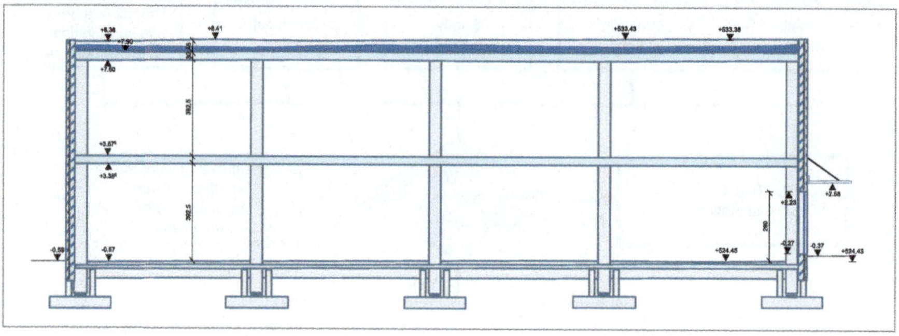

Fig. 10.123 Section – Retail park

A decision matrix (differentiated process comparison) makes it possible to evaluate three slab formwork systems and to determine a sequence of process steps for preparing final decision making. Triangular distributions are used for assessing the individual criteria, and they are weighted on the basis of ranges. The Monte Carlo simulation performed to determine the total number of points automatically combines different points and weightings by means of the probabilistic calculation (sensitivity analysis).

The following methods are compared with each other:

Method A

The first method involves slab tables that are repositioned horizontally by a table transfer unit and vertically by a crane.

The formwork facing comprises triple-layer formwork panels made of spruce wood whose surface consists of a urea-melamine resin adhesive. The panels are 21 mm thick.

The tables are 2.0 × 4.0 m, 2.5 × 4.0 m, 2.0 × 5.0 m, and 2.5 × 5.0 m in size.

Method B

The second method available for selection involves girder formwork that is repositioned using the available cranes. During stripping, the formwork elements are temporarily stored in different containers designed for crane transport if required, or moved manually directly to the next work section.

Again, the formwork facing comprises triple-layer formwork panels made of spruce wood whose surface consists of a urea-melamine resin adhesive. The panels are 21 mm thick.

Method C

The last method uses a modular or element formwork system, which can be relocated using the available cranes.

The formwork facing consists of triple-layer panels with a melamine resin coating ensuring that their edges are sealed all around. The panels consist of three 7 mm spruce board layers glued in a crosswise pattern.

Elements of 150 × 50 cm and 150 × 75 cm in size are used.

It should be noted that in Method B, the proportion of infill areas is smaller than in the case of the other two systems owing to its flexibility. However, the larger number of individual items is associated with a higher repair ratio. In contrast, transport to and from the site and on-site storage are more complex in Method A due to the larger system dimensions.

10.4.4.3 Modeling

Modeling is based on the differentiated decision matrix according to *Hofstadler*.[63] In deterministic decision-making processes, the points awarded to each criterion are multiplied by their respective weightings, and the sums for the elimination criteria are derived from them. Methods in which zero points are awarded to a criterion must be eliminated unless the substitution is successful. The respective sums are also calculated for the criteria related to construction management and economics and added to those of the elimination criteria. Depending on the specific requirements, risk factors can be applied to points in order to reduce their weighting. The method with the highest total number of points is considered to be the most appropriate for the project. Additional analyses must be performed if two methods are very close to each other in terms of their scores.

Developing the deterministic model creates the basis for the Monte Carlo simulation. The matrix is complemented by additional columns in which a minimum and a maximum value can be specified for each input parameter (weighting and scoring) in addition to the most likely value. The minimum must be smaller than or equal to the most likely value, while the maximum must be greater than or equal to the most likely value. Triangular distributions are defined on the basis of these three input values per parameter. The weighting values must be normalized in the simulation in order to correctly derive the weightings of the higher-level and individual criteria. Absent such normalization, the sum of the percentage criteria weightings resulting from the iterations can be greater or smaller than 100%. In contrast, normalization ensures that the sum of the percentage weightings is exactly 100% after each iterative step.

Tables 10.45 and 10.46 show the two parts of the decision matrix, i.e. the elimination criteria and the criteria related to construction management and economics including the awarded points and the defined weighting. Columns 7 and 12 contain the respective triangular distributions used for criteria weighting. In Columns 8 and 13, the iteration values are normalized (this step is not reflected in the tables because only the most likely values are shown in each case).

[63] Cf. Hofstadler (2008), p. 345ff.

Decision matrix																											
Criteria			Weighting										Method/system														
1	2	3	4	5	6	7	8	9	10	11	12	13	14	15	16	17	18	19	20	21	22	23	24	25	26	27	28
			Total					Individual					Distribution of points					Distribution of points					Distribution of points				
Project: Retail park			MIN	MOST LIKELY	MAX	Distr.	SIM	MIN	MOST LIKELY	MAX	Distr.	SIM	Method A					Method B					Method C				
Authors: Hofstadler/Kummer		Version: 1.0	[-]	[-]	[-]	[%]	[%]	[-]	[-]	[-]	[%]	[%]	MIN	MOST LIKELY	MAX	Distr.	SIM	MIN	MOST LIKELY	MAX	Distr.	SIM	MIN	MOST LIKELY	MAX	Distr.	SIM
Elimination criteria	Aesthetic criteria	Sealing to vertical components						5	10	15	10	10	2	3	4	3	0.060	3	4	5	4	0.080	2	3	4	3	0.060
		Actions on element edges						5	10	15	10	10	2	3	3	3	0.053	2	2	3	2	0.047	3	4	5	4	0.080
		Actions on formwork edges						5	10	15	10	10	2	3	4	3	0.060	2	2	3	2	0.047	3	4	5	4	0.080
		Evenness						10	15	20	15	15	3	4	5	4	0.120	3	4	5	4	0.120	4	5	5	5	0.140
		Element joints	15	20	25	20	20	10	15	20	15	15	2	3	4	3	0.090	4	4	5	4	0.130	4	5	5	5	0.140
		Color uniformity						5	10	15	10	10	2	3	4	3	0.060	3	4	5	4	0.080	3	4	5	4	0.080
		Surface structure						5	10	15	10	10	2	3	4	3	0.060	2	3	4	3	0.060	4	5	5	5	0.093
		Formwork joints						5	10	15	10	10	3	4	5	4	0.080	2	3	4	3	0.060	4	5	5	5	0.093
		Texture						5	10	15	10	10	3	4	5	4	0.080	2	3	4	3	0.060	3	4	5	4	0.080
	Number of points – sub-criterion								100		100	100					0.663					0.683					0.847
	Structural criteria	Number of repetitive ground plans						5	10	15	10	10	3	4	5	4	0.080	2	3	4	3	0.060	3	4	5	4	0.080
		Elevations of building						5	10	15	10	10	1	2	3	2	0.040	3	4	5	4	0.080	3	4	5	4	0.080
		Reinforcement ratio						5	10	15	10	10	4	5	5	5	0.093	4	5	5	5	0.093	4	5	5	5	0.093
		Even/curved surfaces						5	10	15	10	10	4	5	5	5	0.093	4	5	5	5	0.093	4	5	5	5	0.093
		Total height of building	15	20	25	20	20	5	10	15	10	10	1	2	3	2	0.040	3	4	5	4	0.080	3	4	5	4	0.080
		Uniform/changing ground plan						5	10	15	10	10	3	4	5	4	0.080	4	5	5	5	0.093	3	3	4	3	0.067
		Uniform/changing elevations						5	10	15	10	10	3	4	5	4	0.080	3	4	5	4	0.080	3	4	5	4	0.080
		Ground plan of building						15	20	25	20	20	3	4	5	4	0.160	4	5	5	5	0.187	3	3	4	3	0.133
		Rearrangement during cycle						5	10	15	10	10	4	4	5	4	0.087	2	3	4	3	0.060	4	5	5	5	0.093
	Number of points – sub-criterion								100		100	100					0.753					0.827					0.800
	Safety criteria	Quality and durability of elements						25	30	35	30	30	2	3	4	3	0.135	2	3	3	3	0.120	3	4	5	4	0.180
		Safe working in all phases						10	15	20	15	15	3	4	5	4	0.090	2	3	4	3	0.068	3	4	5	4	0.090
		Subjective perception of safety among workers	10	15	20	15	15	10	15	20	15	15	2	3	4	3	0.068	2	3	4	3	0.068	3	4	5	4	0.090
		Safety elements integrated into system						15	20	25	20	20	2	3	4	3	0.090	3	3	4	3	0.100	4	4	5	4	0.130
		Susceptibility to wind						15	20	25	20	20	2	3	4	3	0.090	4	5	5	5	0.140	3	4	4	4	0.110
	Number of points – sub-criterion								100		100	100					0.473					0.495					0.600
	Technical criteria	Links to infill areas						15	20	25	20	20	2	2	3	2	0.140	4	5	5	5	0.280	2	3	3	3	0.160
		Links to blockouts and fittings						5	10	15	10	10	2	3	4	3	0.090	4	5	5	5	0.140	2	2	3	2	0.070
		Links to stop-end formwork						5	10	15	10	10	2	3	4	3	0.090	3	4	4	4	0.110	2	3	3	3	0.080
		Links to joist formwork						1	2	3	2	2	2	2	3	2	0.014	3	4	5	4	0.024	2	3	4	3	0.018
		Work area for formwork stripping and repositioning						4	6	10	6	6	2	2	3	2	0.042	4	4	5	4	0.078	3	4	4	4	0.066
		Support area for load transfer (even. inclined)						2	3	4	3	3	4	5	5	5	0.042	4	5	5	5	0.042	4	5	5	5	0.042
		Type of reinforcement, reinforcement ratio	25	30	35	30	30	4	6	10	6	6	4	5	5	5	0.084	4	5	5	5	0.084	4	5	5	5	0.084
		Impermeability of element and formwork joints to cement paste spill						10	15	20	15	15	2	2	3	2	0.105	3	3	4	3	0.150	3	4	5	4	0.180
		Evenness						6	8	12	8	8	2	3	4	3	0.072	3	4	5	4	0.096	3	4	5	4	0.096
		Influence of weather on formwork facing						7	8	9	8	8	2	3	4	3	0.072	2	3	4	3	0.072	3	4	5	4	0.096
		Influence of number of use cycles on formwork facing						6	7	8	7	7	1	2	3	2	0.042	2	3	4	3	0.063	3	4	5	4	0.084
		Storage function for other materials or equipment						2	3	4	3	3	2	3	4	3	0.027	2	2	3	2	0.021	3	4	5	4	0.036
		Bearing capacity of subsoil for load transfer						1	2	3	2	2	4	5	5	5	0.028	4	5	5	5	0.028	4	5	5	5	0.028
	Number of points – sub-criterion								100		100	100					0.848					1.188					1.040
	Criteria for site surroundings and environment	Adjoining properties						5	10	15	10	10	3	4	5	4	0.060	2	3	4	3	0.045	4	5	5	5	0.070
		Development density						0	10	15	10	10	4	5	5	5	0.070	4	5	5	5	0.070	4	5	5	5	0.070
		Global transport links	10	15	20	15	15	40	50	60	50	50	3	4	5	4	0.300	4	5	5	5	0.350	4	5	5	5	0.350
		Wind conditions						25	30	35	30	30	3	4	5	4	0.180	4	5	5	5	0.210	4	5	5	5	0.210
	Number of points – sub-criterion								100		100	100					0.610					0.675					0.700
						100	100										3.347					3.868					3.987
Elimination factor (multiplicative: 1 all criteria met, 0 at least one criterion not met):																	1					1					1
Individual factor (multiplicative: between 0 and 1):																	1					1					1
Number of points for elimination criteria:																	3.35					3.87					3.99

Table 10.45 Decision matrix – Part 1

Columns 17, 22, and 27 contain the triangular distributions for the award of points per method (A, B, and C). For each row, these triangular distributions are multiplied with the weightings shown in Columns 8 and 13 in order to calculate the weighted point proportions per criterion and method in the course of the iteration steps.

The weighted point proportions are added up for each criterion as well as for all criteria taken together. The total number of points of the three methods available for selection is the last but one value shown in Columns 18, 23, and 28 of Table 10.46. The last value expresses the ranking of the methods (1 = most appropriate method).

Decision matrix																											
Criteria			Weighting										Method/system														
1	2	3	4	5	6	7	8	9	10	11	12	13	14	15	16	17	18	19	20	21	22	23	24	25	26	27	28
			Total					Individual					Distribution of points					Distribution of points					Distribution of points				
Project: Retail park			MIN	MOST LIKELY	MAX	Distr.	SIM	MIN	MOST LIKELY	MAX	Distr.	SIM	Method A					Method B					Method C				
Authors: Hofstadler/Kummer		Version: 1.0	[-]	[-]	[-]	[%]	[%]	[-]	[-]	[-]	[%]	[%]	MIN	MOST LIKELY	MAX	Distr.	SIM	MIN	MOST LIKELY	MAX	Distr.	SIM	MIN	MOST LIKELY	MAX	Distr.	SIM
Construction management and economics criteria	Construction management criteria	Distance between construction and storage focal points						4	7	11	7	7	2	3	4	3	0.105	4	5	5	5	0.163	4	5	5	5	0.163
		Requirements for workers' qualifications and skills						4	5	6	5	5	2	3	4	3	0.075	3	4	5	4	0.100	3	4	5	4	0.100
		Requirements for site mobilization						4	5	6	5	5	1	2	3	2	0.050	4	5	5	5	0.117	4	5	5	5	0.117
		Requirements for exposed concrete quality						4	5	6	5	5	1	2	3	2	0.050	2	3	4	3	0.075	2	3	4	3	0.075
		Number of workers per work group						2	3	4	3	3	2	3	4	3	0.045	2	3	4	3	0.045	4	5	5	5	0.070
		Number of usable workers per work section						4	5	6	5	5	2	3	4	3	0.075	3	4	5	4	0.100	4	5	5	5	0.117
		Number of possible uses on the same level						2	4	6	4	4	2	3	4	3	0.060	3	4	5	4	0.080	3	4	5	4	0.080
		Ratio of consumption rates, e. g. formwork installation to stripping	40	50	60	50	50	4	5	6	5	5	3	4	5	4	0.100	1	2	3	2	0.050	4	5	5	5	0.117
		Required daily formwork performance						6	8	10	8	8	3	4	4	4	0.147	3	3	4	3	0.133	3	4	4	4	0.147
		Achievable daily formwork performance						6	8	12	8	8	4	4	5	4	0.173	2	3	3	3	0.107	3	4	5	4	0.160
		Flexibility						20	25	30	25	25	1	2	3	2	0.250	4	5	5	5	0.583	3	3	4	3	0.417
		Crane availability and intensity						5	10	15	10	10	1	1	2	1	0.067	4	4	5	4	0.217	3	3	4	3	0.167
		Requirements for storage/interim storage						4	5	6	5	5	1	1	2	1	0.033	3	4	5	4	0.100	3	4	5	4	0.100
		Availability of formwork at building yard						1	2	3	2	2	2	2	2	2	0.020	2	3	3	3	0.027	2	2	2	2	0.020
		General independence of weather conditions						2	3	4	3	3	2	3	4	3	0.045	3	4	5	4	0.060	4	5	5	5	0.070
	Number of points – sub-criterion								100		100	100					1.295					1.957					1.918
	Construction economics criteria	Links to infill areas						5	10	15	10	10	1	2	3	2	0.100	4	4	5	4	0.217	2	3	4	3	0.150
		Proportion of serial components						2	3	4	3	3	2	3	3	3	0.040	4	4	5	4	0.065	4	4	5	4	0.065
		Number of individual parts						1	2	3	2	2	3	4	5	4	0.040	3	3	4	3	0.033	3	4	4	4	0.037
		Number of possible formwork use cycles						7	8	9	8	8	2	2	3	2	0.093	3	4	5	4	0.160	3	4	5	4	0.160
		Number of possible support use cycles						2	3	4	3	3	2	3	4	3	0.045	3	4	5	4	0.060	4	5	5	5	0.070
		Number of possible use cycles on the same level						5	10	15	10	10	2	3	4	3	0.150	3	4	5	4	0.200	3	4	5	4	0.200
		Labor consumption rate						5	10	15	10	10	2	3	4	3	0.150	2	3	3	3	0.133	3	4	5	4	0.200
		Discretionary labor	40	50	60	50	50	5	10	15	10	10	2	3	3	3	0.133	3	4	4	4	0.183	3	4	4	4	0.183
		Achievable daily performance						5	10	15	10	10	2	3	4	3	0.150	2	3	4	3	0.150	4	4	5	4	0.217
		Assembly and disassembly works						2	3	4	3	3	2	3	4	3	0.045	2	3	4	3	0.045	3	4	5	4	0.060
		Crane intensity						5	10	15	10	10	1	2	3	2	0.100	4	4	5	4	0.217	3	4	5	4	0.200
		Lease rate						4	5	6	5	5	2	3	4	3	0.075	2	3	4	3	0.075	2	2	3	2	0.058
		Lease/purchase ratio						0	1	2	1	1	2	2	3	2	0.012	2	3	3	3	0.013	2	3	4	3	0.015
		Productivity						4	5	6	5	5	1	2	3	2	0.050	2	3	4	3	0.075	4	5	5	5	0.117
		Stocking quantity						5	10	15	10	10	2	3	4	3	0.150	4	5	5	5	0.233	3	4	4	4	0.183
	Number of points – sub-criterion								100		100	100					1.333					1.860					1.915
						100	100										2.628					3.817					3.833
Total number of points for methods/systems:																	5.98					7.68					7.82
Risk factor (multiplicative: 1 no increased risk, 0 non-manageable risk):																	1					1					1
Final number of points:																	5.98					7.68					7.82
Decision (ranking):																	3					2					1

Table 10.46 Decision matrix – Part 2

10.4.4.4 Calculation and Interpretation of Results

The Monte Carlo simulation performed for the developed decision matrix results in a histogram calculated for the total number of points for each of the three methods (see Fig. 10.124). This diagram shows that the spread of results is similar for all methods. This is due to the fact that, on the one hand, the weighting of the criteria is identical for all methods and, on the other hand, a range of one or two points was consistently applied to the evaluation of the methods with regard to each criterion.[64)]

In principle, all three methods would be suitable for constructing the floor slabs (none of the elimination criteria attracted a zero score), but the fewest points were awarded to Method A, which is why it should be eliminated on the basis of this procedure.

Methods B and C thus remain available for the selection process. At first glance, Method C appears to show the greater mean of the total number of points. However, the two histograms for the total number of points of Methods B and C also overlap in one area. We should interpret this overlap such that the simulation also resulted in some point and weighting combinations based on the inputs where Method B received more points than Method C. The decision in favor of Method C is thus associated with some uncertainty, which needs to be quantified.

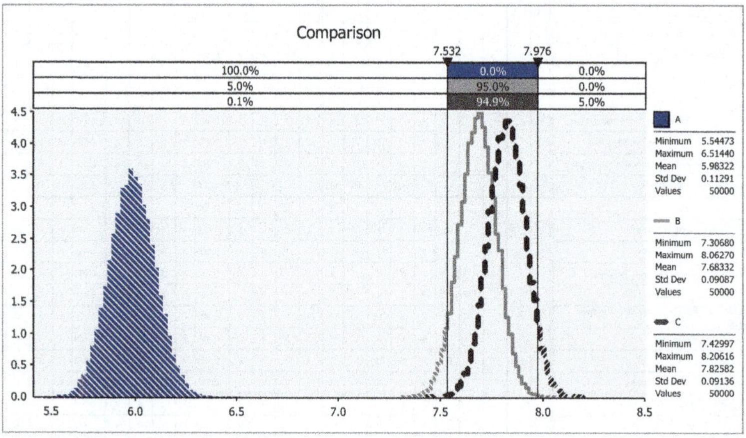

Fig. 10.124 Comparison of points awarded to the three methods

Besides the number of points per method, the ranking of the methods can also be determined automatically. The ranking follows the same principle as for sports results, but it is also possible to assign equal ranks in each iterative step if it results in a tie. However, such a case is very unlikely given the application of continuous distributions and the large number of criteria.

Fig. 10.125 shows the ranks that Method C reached in the course of the simulation (involving 50,000 iterations). In none of the calculated cases did Method C come in third, as confirmed by the initial interpretation of the histograms in Fig. 10.124.

[64)] For the "rearrangement during cycle" criterion, for example, we chose the values 4 | 4 | 5 (range = 1) for Method A and 2 | 3 | 4 (range = 2) for Method B.

In addition, Method C ranked first in about 87% of cases and second in about 13% of cases. If the contractor, at this stage, makes their decision on the basis of the differentiated probabilistic comparison of methods and opts in favor of Method C, the chance that this method will actually be the one that best meets the defined criteria amounts to approximately 87%. Thus, there is only a corresponding chance of 13% for Method B to be the best.

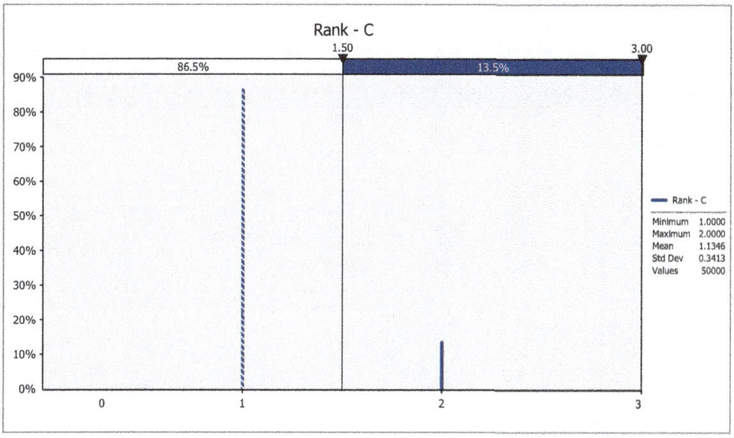

Fig. 10.125 Bar chart – Rank of Method C

10.4.4.5 Benefit for the Contractor

In the differentiated method comparison, the contractor bases their decision on the evaluation of individual criteria. This evaluation is based on both hard and soft factors associated with uncertainties.

The Monte Carlo simulation provides the benefit of introducing ranges and/or distribution functions to represent these uncertainties and to incorporate them in the calculation. In the case of a clear decision, such as assigning the third place to Method A, this clarity is again confirmed by the simulation where the result is represented by a range. However, if two methods have similar point totals, the histograms will show an overlap, and it is possible to indicate a chance/risk ratio by means of the assigned ranks. If this ratio is not significant for a decision, it may be necessary to check the allocation of points and/or the weighting again or to add further criteria.

Only probabilistic calculations will allow for an overlapping display of the histograms representing the total scores. In contrast, purely deterministic approaches will deliver only one total point value per method. This rigid reduction of the available information to a single number makes it impossible to draw conclusions regarding the spread of the number of points and the chance/risk ratio. There are thus "blind spots" within the decision.

10.4.5 Construction Process Planning

The planning of sequences of individual works or services becomes more important with the increase in the thoroughness of observation or in the degree of detail. Whereas, in early project phases, rough time estimates in relation to the entire structure or parts thereof will be sufficient, the execution phase (PPH 4, see Fig. 10.126) makes it necessary to break the scope of work down into individual structural components or work sections.

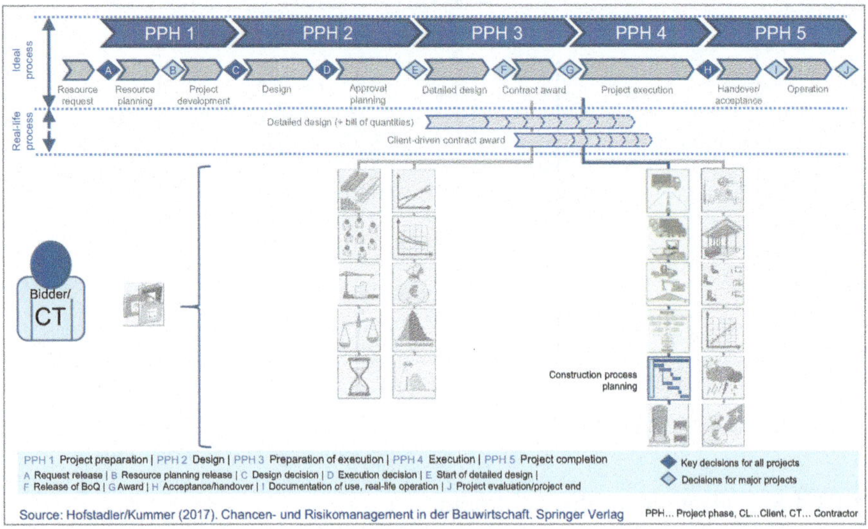

Fig. 10.126 Allocation of example in the project sequence

In building construction, networked Gantt charts are frequently used for process planning. In such a chart, each bar corresponds to an activity where the length of the bar represents the scheduled duration of the corresponding activity.

Individual times or durations are given as deterministic values that do not allow for any indication of the probability or range. If we were, for instance, to complete a floor section within five days, the corresponding Gantt chart will not permit any conclusion with regard to probability. We refer to an example in order to demonstrate how input parameter uncertainties can also be taken into account in construction process planning with the aid of Monte Carlo simulations.

10.4.5.1 Task

In a building construction project, the formwork area of a standard floor amounts to 1,500 m². The project involves the construction of six floors divided into three work sections (see Fig. 10.127). We thus obtain a total formwork area of 9,000 m².

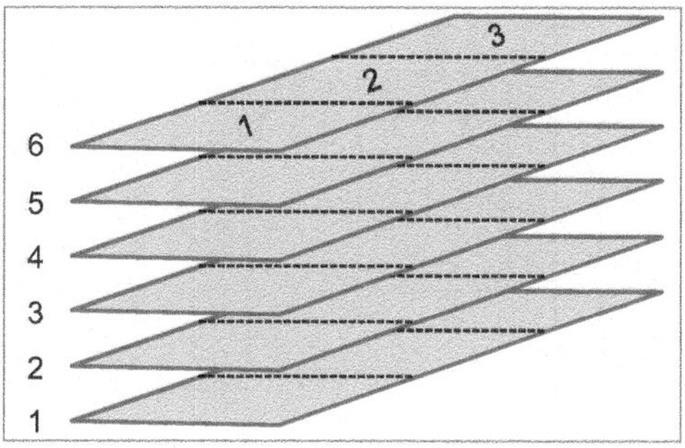

Fig. 10.127 Schematic representation of the floors to be completed

Based on the preliminary work schedule, a total period of 90 days is allotted to the construction of the floors divided into 18 work sections. This corresponds to a scheduled construction time of 5 days per work section.

We refer to the detailed schedule and take into account the existing uncertainties in order to determine the chance/risk ratio for adhering to the planned cycle (5 days per section). In so doing, we only consider works along the critical path (formwork installation, reinforcing, and concreting).

10.4.5.2 Modeling

Fig. 10.128 shows a schematic representation of the calculation process for determining the performance values or process durations for formwork installation, reinforcing, and concreting a floor slab section.

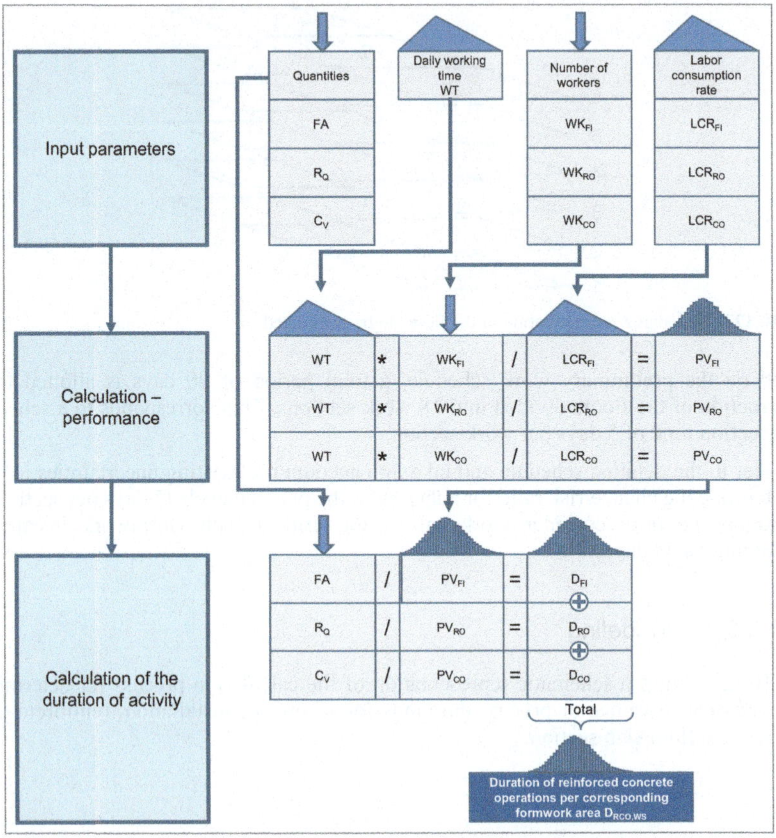

Fig. 10.128 Calculation flow chart – Continuous flow – Floor slab work section

The labor consumption rates and the daily working time are modeled as triangular distributions. All other parameters (i.e. quantities and number of workers) are incorporated in the calculation as deterministic values represented by arrows.

For flow production, the sum of the individual process durations (assuming no offset between formwork installation and reinforcing) is the duration required for completing one floor slab work section.

In this example, we only consider the processes along the critical path. In our model, we thus disregard the time needed for formwork stripping and the process of removing the formwork from the slabs.

10.4.5.3 Calculation and Interpretation of Results

Table 10.47 illustrates the calculation of the time to completion of a floor slab work section. Input parameters that are applied as triangular distributions are defined by the minimum, most likely, and maximum values in Columns B, C, and D. Column E contains the corresponding distributions for the respective rows. Only the most likely value is entered in Column C in the case of deterministic input parameters. This value is transferred to Column E for calculation purposes. Any parameters not entered in Columns C to D should be considered as intermediate results of calculations.

The building comprises a total of six floors, and the formwork area per standard floor (Row 1) amounts to 1,500 m^2. This means that, in total, 9,000 m^2 of slab formwork need to be produced, with each floor being divided into three work sections. Thus, the formwork area (FA) of the floor slabs equals 500 m^2 per work section (entered as a deterministic value in Cell E5). The floor slabs are 30 cm thick (Row 6), and their reinforcement ratio rr is 120 kg/m^3 (Row 8). These values enable us to determine the reinforced concrete volume C_V and the reinforcement quantity R_Q per work section in Cells E7 and E9.

The labor consumption rates of formwork installation, reinforcing, and concreting as well as the daily working time are fraught with uncertainties. These parameters are defined as triangular distributions in Rows 10 to 12, and 16. The number of workers varies depending on the specific activity. A group of five workers is allocated to formwork installation while reinforcing requires two groups of five workers each. Concreting also requires two groups, but with four workers each (Rows 13 to 15).

We obtain the performance values of formwork installation, reinforcing, and concreting operations in Rows 17 to 19 by inserting the calculation parameters into the equation used for the calculation of performance values (Eq. (10-29)).

$$PV = \frac{WK \cdot WT}{LCR} \tag{10-29}$$

In the next step, dividing the quantities by the performance values results in the durations of the individual activities (formwork installation, reinforcing, and concreting; Rows 20 to 22). For a fully continuous flow (without offset between the formwork installation and reinforcing processes), Cell E23 gives the duration required for completing a floor slab work section.

No.	Parameter		MIN	MOST LIKELY	MAX	Distribution
0	A		B	C	D	E
1	Formwork area of standard floor			1,500.00		1,500.00 m²
2	Number of floors			6.00		6.00
3	Number of work sections (WS)			3.00		3.00
4	Formwork area – Ceilings					9,000.00 m²
5	Formwork area per WS	FA				500.00 m²
6	Floor slab thickness	d		0.30		0.30 m
7	Reinforced concrete volume per WS	C_V				150.00 m³
8	Reinforcement ratio	rr		120.00		120.00 kg/m³
9	Reinforcement quantity per WS	R_Q				18.00 t
10	Mean labor consumption rate – Formwork installation	LCR_{FI}	0.18	0.20	0.25	0.20 h/m²
11	Mean labor consumption rate – Reinforcing	LCR_{RO}	9.00	10.00	12.00	10.00 h/t
12	Mean labor consumption rate – Concreting	LCR_{CO}	0.40	0.45	0.55	0.45 h/m³
13	Number of workers – Formwork installation	WK_{FI}		5.00		5.00 h/h
14	Number of workers – Reinforcing	WK_{RO}		10.00		10.00 h/h
15	Number of workers – Concreting	WK_{CO}		8.00		8.00 h/h
16	Daily working time	WT	8.00	9.00	10.00	9.00 h/d
17	Performance – Formwork installation	PV_{FI}				225.00 m²/d
18	Performance – Reinforcing	PV_{RO}				9.00 t/d
19	Performance – Concreting	PV_{CO}				160.00 m³/d
20	**Duration – Formwork installation**	D_{FI}				**2.22 d**
21	**Duration – Reinforcing**	D_{RO}				**2.00 d**
22	**Duration – Concreting**	D_{CO}				**0.94 d**
23	**Total of FI, RO, CO per WS**	$D_{RCO,WS}$				**5.16 d**

Table 10.47 Calculation table – Duration of formwork installation, reinforcing, and concreting a floor slab work section

We use the mean values resulting from this calculation to derive the networked Gantt chart for completing a floor slab work section (see Fig. 10.129). We find that the three activities are interlinked by start-end relationships and that the duration of formwork installation, reinforcing, and concreting operations exceeds five days (we get 5.16 d) based on the mean values. However, this example results only from a specific combination of the input parameters and does not yet provide any information about the chance/risk ratio.

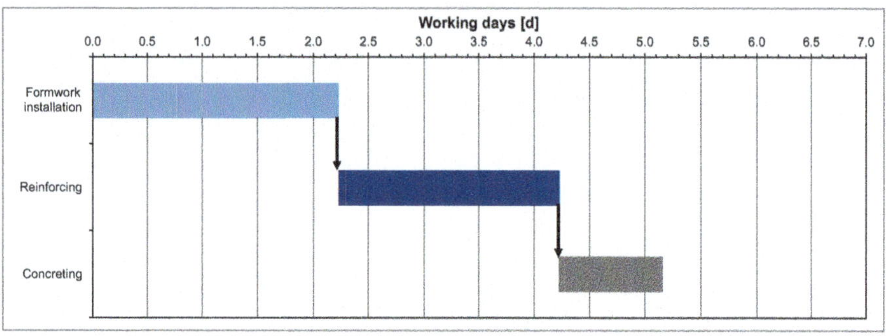

Fig. 10.129 Networked Gantt chart – Floor slab work section

In addition to the Gantt chart based on the mean durations of the activities, Fig. 10.130 also includes the histograms of the possible completion dates of the individual works. We find that the spread of the end date increases in the presence of several interlinked, uncertain processes. The larger the number of uncertain processes along the critical path that lead to a scheduled end date, the greater the range of possible completion dates will be.

These ranges can most probably be reduced again if we deliberately add buffers to the process or if we are able to restrict the ranges of the uncertain input parameters.

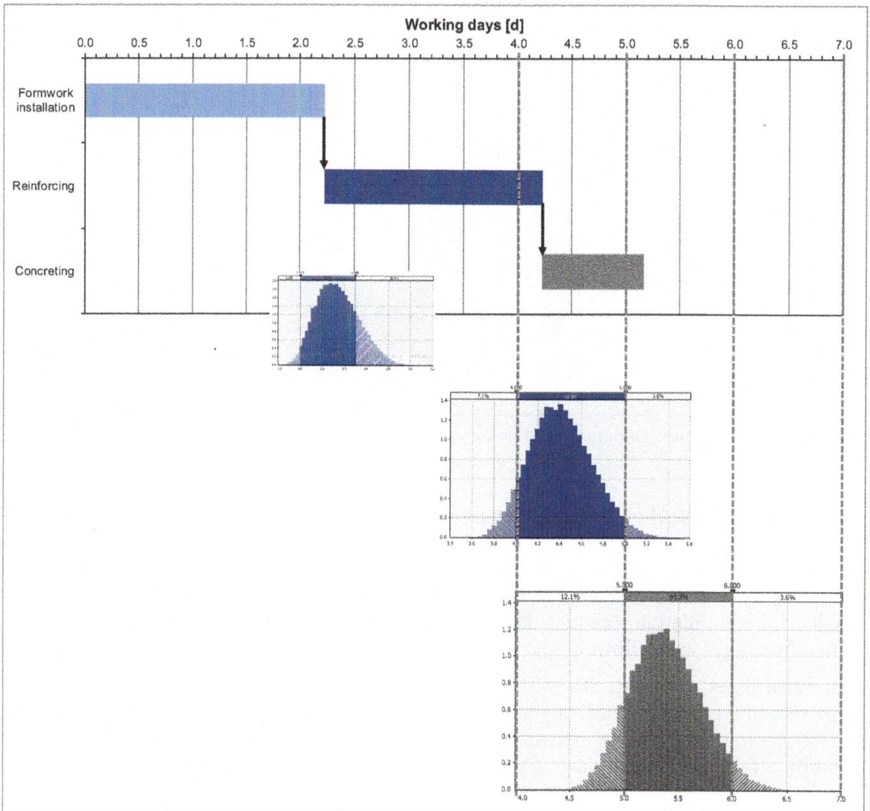

Fig. 10.130 Networked Gantt chart including histograms – Floor slab work section

In this example, we assume a construction time of five days for each of the floor slab work sections. Fig. 10.131 again shows the (enlarged) histogram of the time to completion of a floor slab section.

Based on the construction process (continuous flow) and the approaches chosen for the uncertain input parameters, we arrive at a probability of only about 12.1% of adhering to the scheduled five days per section. In about 84.3% of the simulated cases, the time to completion ranges from five to six days. In approximately 3.6% of cases, the simulation results in a duration of more than six days.

Even though it is possible, from a construction management perspective, to adhere to the construction process schedule on the basis of the completed analysis, the contractor would run a very high risk of about 88%.

To counteract this situation, it would be conceivable to overlap the two processes of formwork installation and reinforcing in order to run these processes simultaneously at least to a certain extent. In so doing, it is crucial to prevent any clash of the reinforcing with the formwork operations while allocating sufficient space to all workers.[65)]

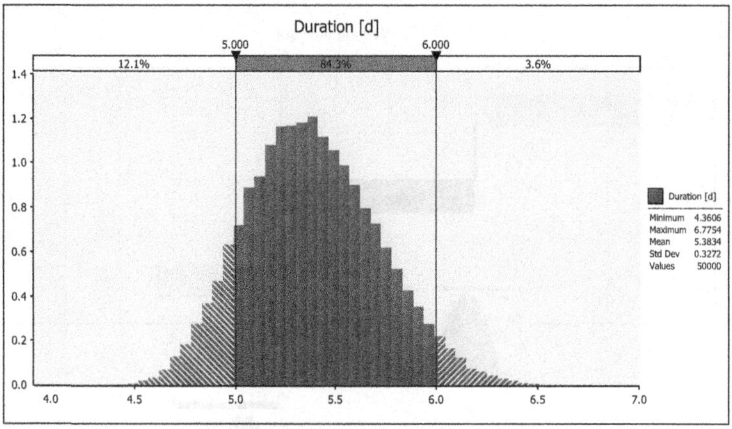

Fig. 10.131 Histogram – Duration per floor slab work section

10.4.5.4 Benefit for the Contractor

Monte Carlo simulations make it possible to consider uncertainties in the construction process planning phase and to combine different values for the input parameters. As a result of the simulation, the end dates of a sequence of interlinked processes are represented by ranges in the form of histograms.

Individual dates or durations can thus be assessed and analyzed in terms of their probability of being earlier or later, or shorter or longer, than originally planned. If the sequence or sequencing relationship between individual activities can be changed, the corresponding effects on the end date can be shown and graphically displayed by rerunning the simulation.

We find that the larger the number of uncertain processes interlinked by start-end relationships, the wider the end date or total duration range will become.

Smaller schedules with only a few processes and simple dependency relationships can be modeled in MS Excel, and input parameter uncertainties can be added using the @Risk add-in.

For larger schedules with many interdependencies, it is possible to calculate schedules created in MS Project in combination with MS Excel and @Risk while taking uncertainties into account.

[65)] Any underrun of the minimum available area per worker will result in mutual obstruction and thus losses of productivity and/or labor consumption rate increases.

10.4.6 Stocking of Construction Materials

The planning of available quantities (stocks, inventories) of construction materials is an essential element on construction sites in order to supply the elementary production factors of labor, equipment, and materials. This example is part of PPH 4 (see Fig. 10.132).

Construction material stocks need to be optimized such that the amount of materials put to interim storage is minimized as far as reasonably possible in order to avoid additional costs incurred by heating, roofing, losses, or on-site transfer. On the other hand, sufficient stocks must be held in order to ensure continuous site supply and to prevent material bottlenecks.

Determining the amount of different construction materials to be stocked on-site is one of the most important tasks in logistics. These considerations are primarily governed by construction progress, the storage space available on-site, and existing transport links to the site.[66)]

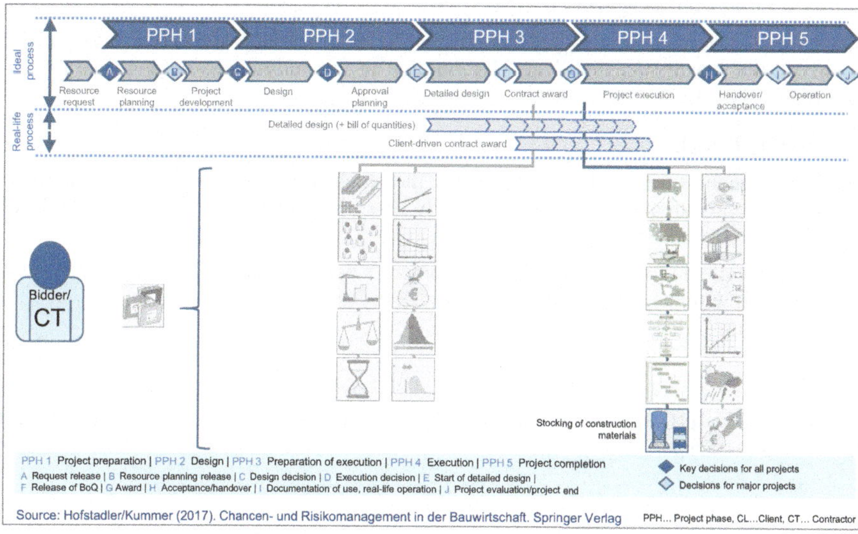

Fig. 10.132 Allocation of example in the project sequence

10.4.6.1 Basics

The following section discusses the basics for dimensioning a stockpile according to the order cycle method[67)] and outlines the individual steps of calculating the quantities to be stocked.[68)]

[66)] Cf. Hofstadler (2007), p. 176

[67)] This is where the scheduling of optimal ordering intervals (dates) is in the focus. Ordering dates are defined while the order quantity is variable and will be governed by consumption.

[68)] Cf. Hofstadler (2007), p. 176ff.

Throughput (production output)

The throughput U is the production quantity per unit of time. If we install, for example, 5 tons of reinforcement per day, this figure represents the throughput of one working day. This throughput is essentially determined by the work and/or equipment performance.

Minimum stock level (safety reserve)

The minimum stock level n_{min} [UQ], or safety or emergency reserve, is the stock level that serves to compensate for disruptions if the planned procurement interval is exceeded or if the actual consumption is higher than planned, for instance because of a higher dispatch rate.

In this example, we use the working day as the unit of time. The minimum stock level is calculated according to Eq. (10-30) from the product of the additional stocking time t_{AS} [d] and the throughput U [UQ/d].

$$n_{min} = U \cdot t_{AS} \tag{10-30}$$

The additional stocking time t_{AS} ensures that sufficient material is stocked on-site in order to exceed the procurement interval. The duration of the additional stocking time can be selected freely.

Procurement stock

The procurement stock n_P [UQ] designates the stock used up during the procurement interval. The procurement stock will normally be so large that the minimum stock level n_{min} will be retained.

We use Eq. (10-31) to calculate the procurement stock by multiplying the procurement interval t_P [d] with the throughput U.

$$n_P = U \cdot t_P \tag{10-31}$$

According to Eq. (10-32), the procurement interval t_P [d] is the sum of the scheduling time t_S [d] and the transport time t_T [d].

$$t_P = t_S + t_T \tag{10-32}$$

The supplier's scheduling time t_S is influenced, for example, by the production and procurement of the construction materials. The transport time t_T depends, for example, on delivery times, permits, customs, breakdowns, etc.

Required stock

The required stock or reorder level n_{req} [UQ] is the stock level at which a new order will be triggered. It must be defined such that the minimum stock level n_{min} is normally retained during the procurement interval (coverage period). The sum of the minimum stock level n_{min} and the procurement stock n_P is equal to the required stock n_{req} (Eq. (10-33)).

$$n_{req} = n_{min} + n_P \tag{10-33}$$

Maximum stock

The maximum possible inventory or maximum stock level n_{max} [UQ] results from the storage options and areas available on the construction site, including the number of existing bins and the staging area for reinforcement, for instance.

From a commercial point of view, the maximum stock level n_{max} is the maximum quantity that may be kept in stock while still avoiding an excessive capital absorption by stocked materials.

Effective stock

The effective stock or inventory n_{eff} [UQ] is the stock of building materials effectively available on the construction site at a given point in time. In other words, this is the stock available in the warehouse or storage facility as at the scheduling date, resulting from incoming and outgoing quantities.

The stock level n_{eff} lies between the required stock n_{req} and the maximum stock n_{max} (Eq. (10-34)).

$$n_{max} \geq n_{eff} \geq n_{req} \tag{10-34}$$

Ordering interval

The ordering interval t_O [d] is the time remaining between the effective and the required stock.

The ordering interval t_O is calculated by dividing the difference of the effective stock n_{eff} and the required stock n_{req} by the throughput U.

$$t_O = \frac{n_{eff} - n_{req}}{U} \tag{10-35}$$

10.4.6.2 Task

This example concerns the storage of bricks for the masonry construction of a residential building. The number of bricks required per day results from the scheduled daily work performance. Bricks are delivered on pallets and temporarily kept in stock in storage areas or directly adjacent to the walls under construction. We need to determine the procurement interval that corresponds to a chance/risk ratio of 20% (risk) to 80% (chance) on the basis of the uncertain input parameters.

10.4.6.3 Modeling

Fig. 10.133 shows the flow chart for this calculation. The uncertain input parameters are modeled as triangular distributions. The only known deterministic value is the effective stock (inventory) n_{eff} [m³].

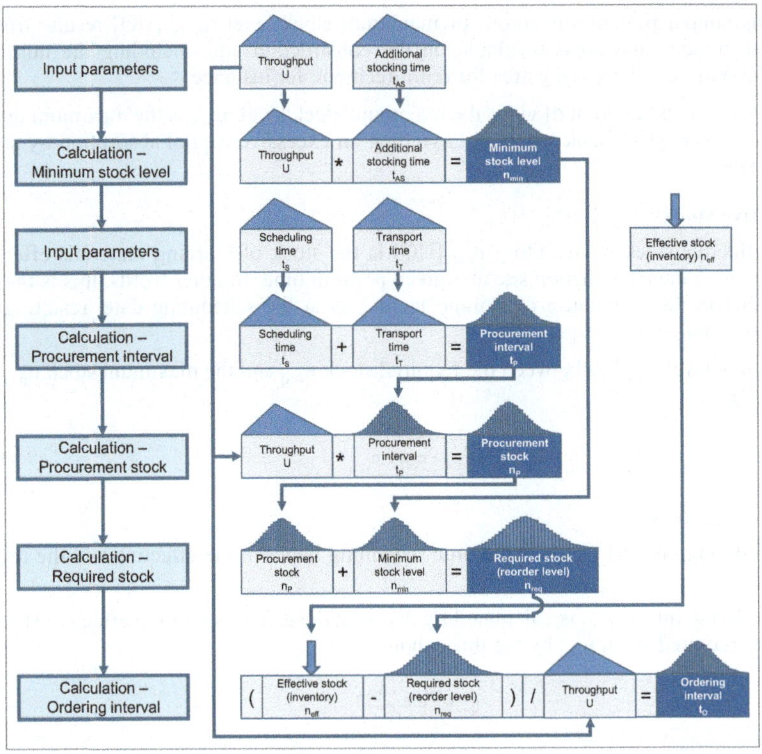

Fig. 10.133 Calculation flow chart – Stocking of construction materials

10.4.6.4 Calculation and Interpretation of Results

The uncertain input variables were modeled as triangular distributions to simulate the ordering interval. The minimum, most likely, and maximum values are shown in Columns B, C, and D of Table 10.48. Column E includes the triangular distributions and/or the histograms representing the related interim and final results.

In this example, we calculate the throughput (Row 3) by dividing the quantity of bricks per floor (Row 1) by the duration of the masonry work per floor (Row 2).

Multiplying the additional stocking time t_{AS} (Row 4) by the throughput U (Row 3) results in the minimum stock level n_{min} [m³] shown in Cell E8. We calculate the procurement interval t_P (Cell E7) as the sum of the scheduling time t_S (Row 5) and the transport time t_T (Row 6).

The product of the procurement interval t_P and the throughput U is equivalent to the procurement stock n_P (Cell E9), which together with the minimum stock level represents the required stock n_{req} [m³].

The effective stock or inventory n_{eff} [m³] was calculated as a deterministic value (without uncertainties) amounting to 100.00 m³.

Inserting the effective stock n_{eff} as well as the required stock n_{req} and the throughput U in Eq. (10-35) yields the ordering interval t_O [d] as a histogram shown in Cell E12.

No.	Parameter		MIN	MOST LIKELY	MAX	Distribution
0	A		B	C	D	E
1	Quantity of bricks per floor		35.00	40.00	45.00	40.00 m³
2	Duration of masonry works for floor		3.50	4.00	4.50	4.00 d
3	Throughput	U				10.00 m³/d
4	Additional stocking time	t_{AS}	1.00	2.00	3.00	2.00 d
5	Scheduling time	t_S	1.50	2.00	2.50	2.00 d
6	Transport time	t_T	0.30	0.50	1.00	0.50 d
7	Procurement interval	t_P				2.50 d
8	Minimum stock level	n_{min}				20.00 m³
9	Procurement stock	n_P				25.00 m³
10	Required stock	n_{req}				45.00 m³
11	Effective stock (inventory)	n_{eff}		100.00		100.00 m³
12	**Ordering interval**	t_O				**5.50 d**

Table 10.48 Calculation table – Ordering interval

The ordering interval indicates the number of days after which the next brick order must be placed in order to ensure the continuous supply of materials to the workers. The uncertainties in the calculation lead to an ordering interval range as a result of the Monte Carlo simulation (see Fig. 10.134).

The longer the order is delayed, the higher the risk of causing a disruption to the production system due to insufficient material supply will be.

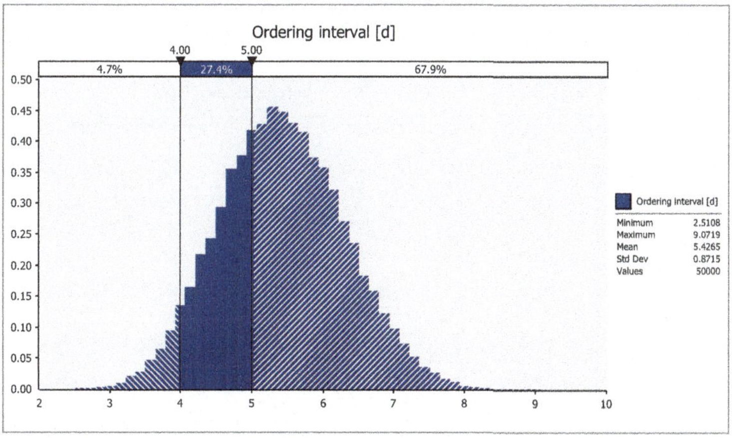

Fig. 10.134 Histogram – Ordering interval

In this example, this means that for an order after five days, there is an approximate probability of 67.9% (chance) that the order was placed at a sufficiently early point in time. Correspondingly, the late-order risk would equal approximately 31.1% in this case.

This risk can be reduced to about 4.7% by an earlier order placed after four days. In other words, combining the input parameters would result in a late order after four days only in about 5% of all cases.

10.4.6.5 Benefit for the Contractor

Since both the supply chain and the production process on a construction site are subject to certain variations, we recommend to take these uncertainties into account for the storage of construction materials. The simulation results enable the contractor to directly determine how different ordering times or intervals will influence the chance/risk ratio.

In this example, we demonstrated that shortening the ordering interval by just one day would reduce the risk from over 30% to about 4.7%.

10.4.7 Aggregation of Losses of Productivity

Required work acceleration measures are often justified by the occurrence of losses of productivity, and they frequently lead to additional cost claims, the commissioning of expert opinions, and legal disputes. Considering losses of productivity in costing/pricing is preceded by the finding that the performance will not behave linearly if the influential parameters change (utilization of the production factors) and that, conversely, the related labor consumption rates will not remain constant. Losses of productivity occur in the execution phase (PPH 4, see Fig. 10.135), but they can also be taken into account at the preceding process planning stage, provided that the productivity thresholds are expected to be exceeded or underrun. Such losses of productivity can influence and reinforce each other. In the calculation, we take this phenomenon into account by aggregating the losses of productivity.

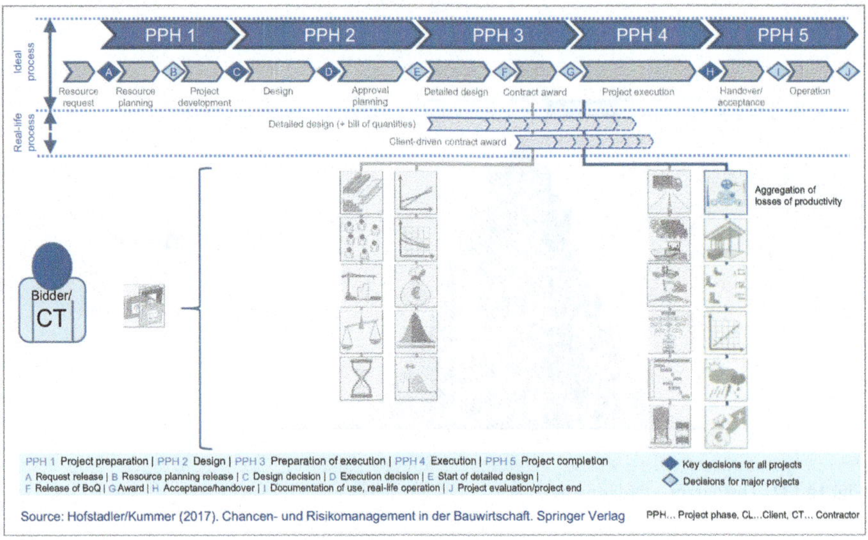

Fig. 10.135 Allocation of example in the project sequence

10.4.7.1 Basics

If the available construction time is short or very short, the daily performance has to be adjusted (increased), which will require acceleration measures. If it is not possible to spend the time necessary to achieve a normal productivity level in construction work, losses of productivity or increases in the corresponding consumption rates are likely to occur. These will materialize if specific productivity thresholds are exceeded or underrun. In such a case, we should also expect an increase in the construction cost, besides the reduced productivity level.

When evaluating losses of productivity, we find that if the same production factors are affected by losses, for example, the sum of two individual productivity losses will not be identical to the concurrent occurrence of these two losses. As a result of the predefined construction time, either a single or several losses of productivity can lead to feedback or reinforcement phenomena owing to the existing interdependencies. Since the required daily performance is predetermined by the specified construction time, the degree of resource utilization must be adjusted to make up for an increased consumption rate (for instance by increasing the number of workers). This response can reinforce the original losses or trigger new losses.

In the calculation, we consider such feedbacks by applying an incremental iteration method. Such an aggregation results in the calculated total loss of productivity being greater than in the case of a mere addition (see Fig. 10.136).

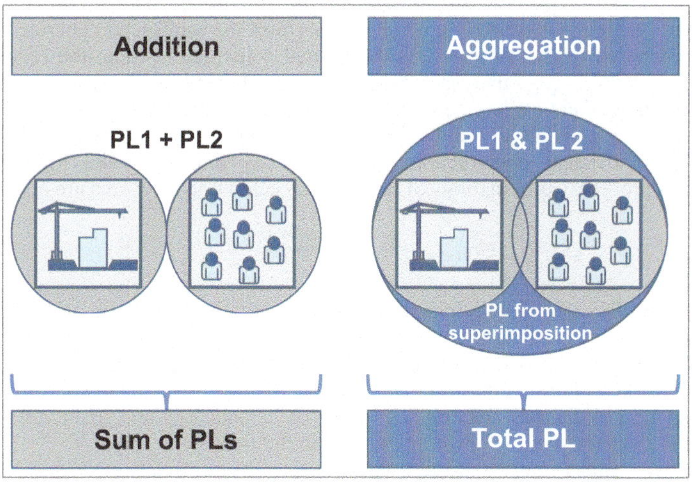

Fig. 10.136 Addition (left) and aggregation (right) of losses of productivity (PL)[69]

Addition is characterized by the commutative and associative laws and involves the pure summation of individual characteristics (including risks, chances, and losses of productivity) without considering feedback effects and interdependencies.[70]

[69] Cf. Hofstadler/Kummer (2014), p. 69

[70] Cf. Kummer (2015b), p. 172

Aggregation is the amalgamation of individual characteristics (risks, chances, and losses of productivity) into a global characteristic (bottom-up approach), taking into account interdependencies, relationships, and individual reinforcement phenomena, with the aid of iterative calculations.[71)]

In the literature, reference is usually only made to individual, non-interlinked losses of productivity or consumption rate increases. Applying an incremental iteration process makes it possible to integrate feedbacks from productivity losses or consumption rate increases into the calculations. The probabilistic aggregation of losses of productivity allows for representing the chance/risk ratio in the form of histograms that provide a sound basis for decision making.

We thus aim to determine source-based costs in a scenario that prevents the optimal utilization of resources while losses of productivity are likely to occur.

10.4.7.2 Task

In this example, we demonstrate how the mutual influence or reinforcement of two different losses of productivity in relation to reinforced concrete operations can be taken into account by incremental iterative calculations.

The calculated example refers to a building construction project in which the construction time specified for reinforced concrete operations ($C_V = 38{,}275$ m³) does not correspond to the "normal" construction time during which normal productivity can be achieved.

Resource utilization is increased to be able to perform the reinforced concrete operations within the allotted time. Compared to the normal construction time, the required daily performance ($PV_{RCO,AVG}$) increases, which can be achieved, for instance, by increasing the number of workers (WK_{AVG}) or the daily working time (WT) (see Eq. (10-36)). It would also be conceivable to reduce the labor consumption rate (LCR_{RCO}), for example through more efficient construction methods or by improving the logistics of the construction process. Combinations of the above acceleration measures are also possible.

$$PV_{RCO,AVG} = \frac{WK_{AVG} \cdot WT}{LCR_{RCO}} \tag{10-36}$$

We investigate the effects a shortened construction time would have on the total labor consumption rate of reinforced concrete operations LCR_{RCO} [wh/m³] and, subsequently, on the wage costs, if acceleration is implemented only on the basis of increasing the number of workers. For this purpose, we carry out a probabilistic analysis with the aim of integrating input parameter uncertainties into the calculations and to be able to assess chances and risks relative to consumption rates in the final step.

Approximately 400 days were determined as the normal construction time required for reinforced concrete operations. However, the specified construction time amounts to only 300 days, which is equivalent to a 25% shortening. To compensate for this excessively short construction time and to achieve the required performance, the number of workers is increased and the effects on productivity and on the basic total labor consumption rate of reinforced concrete operations (excluding losses of productivity) are determined.

[71)] Cf. Kummer (2015b), p. 172

10.4.7.3 Modeling

In this example, two specific potential losses of productivity are considered to be relevant, and integrated into the calculations. More specifically, these losses are attributed to a reduced crane capacity and to non-adherence to the minimum work area. Either of these two losses of productivity is directly determined by the number of workers employed. The larger the number of active workers, the greater the number of cranes would be that are required on-site to adequately supply all workers. In addition, the available work area per worker will become smaller with the increasing number of workers. If a certain threshold (minimum work area) is underrun, the workers will obstruct each other and thus trigger losses of productivity.

Crane capacity

If more cranes are required on the construction site than can be installed (for instance due to lack of space), this will result in losses of productivity or in increased labor consumption rates.[72] Fig. 10.137 shows the consumption rate increase as a function of the reduction in crane capacity as an M-estimator trend curve. This curve was determined in an expert survey conducted at Graz University of Technology.[73]

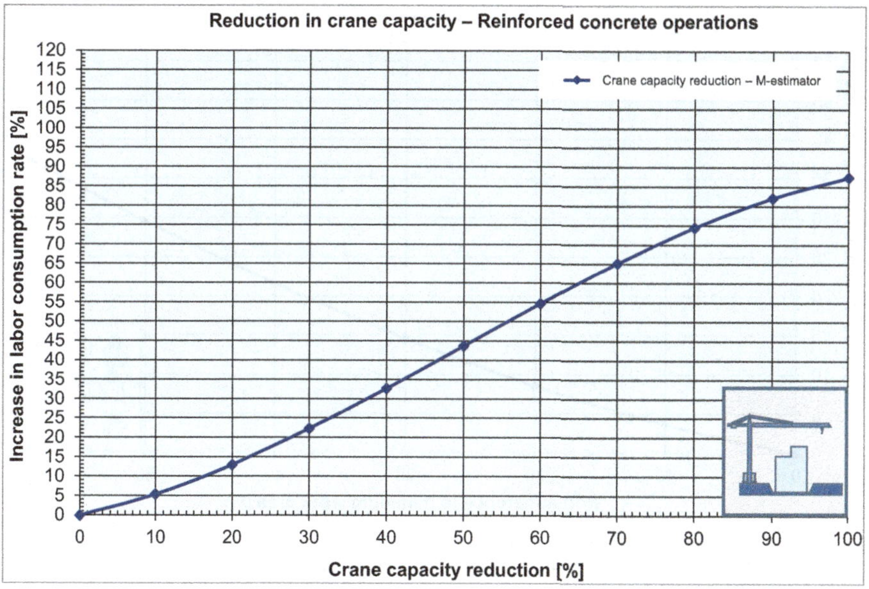

Fig. 10.137 Trend of labor consumption rate increases – Reduced crane capacity – Reinforced concrete operations[74]

Using Eq. (10-37), the crane capacity reduction CC_{RED} [%] results from the number of available cranes $NO_{C,AV}$ [cranes] and the optimum number of cranes $NO_{C,OPT}$ [cranes]. This principle applies if $NO_{C,AV}$ is smaller than $NO_{C,OPT}$.

[72] If no compensation by other measures is possible.

[73] Cf. Hofstadler (2014)

[74] Cf. Hofstadler/Kummer (2014), p. 73

$$CC_{RED} = \left(1 - \frac{NO_{C,AV}}{NO_{C,OPT}}\right) \cdot 100\% \tag{10-37}$$

The progression of the curve representing the labor consumption rate increase $\Delta LCR_{CC,INC}$ [%] is described by the following equation:

$$\Delta LCR_{CC,INC} = -0.000093 \cdot CC_{RED}^3 + 0.013934 \cdot CC_{RED}^2 + 0.410561 \cdot CC_{RED} \tag{10-38}$$

Minimum work area

On any given area (such as a floor slab section with installed formwork), only a certain number of workers can be used productively. If too many workers are employed in a confined space, they will obstruct each other and thus trigger losses of productivity. In an expert survey[75)], the magnitude of the consumption rate increase in the case of non-adherence to the minimum work area was estimated. Fig. 10.138 shows the result of this survey for extensive reinforced concrete components as a nonlinear curve.

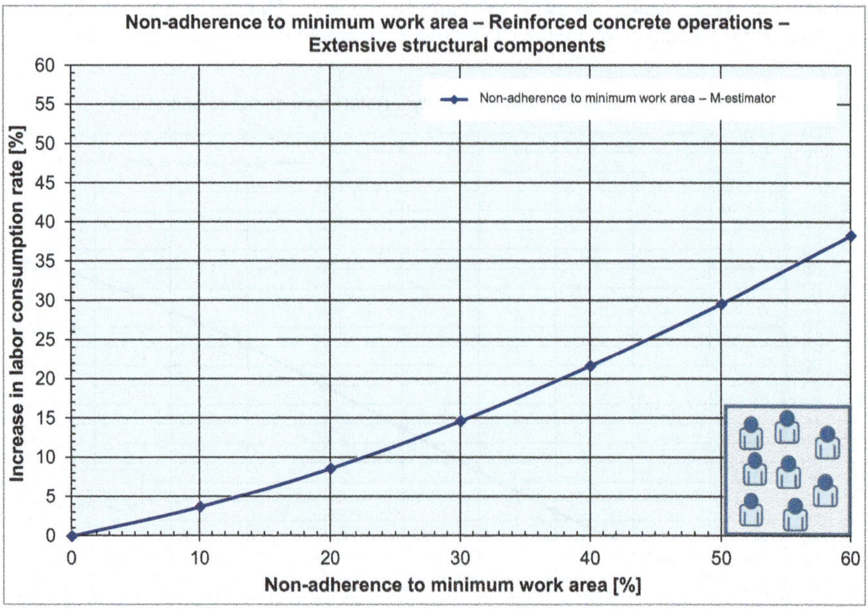

Fig. 10.138 Trend of labor consumption rate increases – Non-adherence to minimum work area – Reinforced concrete operations – Extensive structural components[76)]

Using Eq. (10-39), we obtain the work area reduction WA_{RED} [%] from the available work area WA_{AV} [m²/worker] and the minimum work area WA_{MIN} [m²/worker]. This principle applies if WA_{AV} is smaller than WA_{MIN}.

75) Cf. Hofstadler (2014)

76) Cf. Hofstadler/Kummer (2014), p. 74

$$WA_{RED} = \left(1 - \frac{WA_{AV}}{WA_{MIN}}\right) \cdot 100\% \qquad (10\text{-}39)$$

Eq. (10-40) defines the nonlinear relationship between underrunning the minimum work area and the increase in the labor consumption rate $\Delta LCR_{WA,INC}$ [%].

$$\Delta LCR_{WA,INC} = -0.000018 \cdot WA_{RED}^3 + 0.00661 \cdot WA_{RED}^2 + 0.30707 \cdot WA_{RED} \qquad (10\text{-}40)$$

10.4.7.4 Calculation and Interpretation of Results

To perform probabilistic calculations, distributions must be assigned to the input parameters. In this example, triangular distributions are used for the uncertain parameters for the sake of simplicity. Distributions of this type are defined by a minimum, a most likely, and a maximum value. Table 10.49 (Columns C, D, and E) includes the related data. Parameters for which only a most likely value is specified are included in the calculation as deterministic values.

Input parameters for the probabilistic calculation					
No.	Parameter	Unit	MIN	MOST LIKELY	MAX
0	A	B	C	D	E
1	Concrete volume	[m³]	38,000.00	38,275.00	38,700.00
2	Normal construction time	[d]		400.00	
3	Basic total labor consumption rate	[wh/m³]	5.50	6.00	6.50
4	Daily working time	[h/d]	8.00	9.00	10.00
5	Mean wage costs	[€/wh]	33.00	34.00	38.00
6	Number of cranes (possible maximum governed by site restrictions)	[cranes]		5.00	
7	Minimum work area	[m²/WK]	28.00	30.00	32.00
8	Available work area	[m²]	2,500.00	3,000.00	3,500.00
9	Maximum crane proportionality factor	[WK/crane]		20.00	

Table 10.49 Input parameters for the probabilistic calculation[77)]

Table 10.50 shows the calculation of the new labor consumption rate taking into account the two potential losses of productivity (reduced crane capacity and non-adherence to the minimum work area) for the addition (Column C) and the aggregation (Column D) of the productivity losses.

For each case, the table includes deterministic values. These are either deterministic input parameters, mean values of the applied triangular distributions, or calculation results of the corresponding computing operations using these values. We explain the calculation process on the basis of these deterministic values. In the simulation, varying values for the triangularly distributed input parameters are chosen in each repetitive iteration, and the calculations are performed using these values.

Row 2 indicates the normal construction time of reinforced concrete operations as 400 days (deterministic value). The 25% shortening of the construction time (Row 1) is thus

77) We assume that losses of productivity due to reduced crane capacity will occur only after exceeding the threshold of 20 workers per crane.

equivalent to a specified construction time of 300 days (Row 3). Row 5 gives the average performance required for the normal construction period (95.81 m³/d = 38,325 m³ / 400 d). We thus obtain a normal average number of workers of 63.88 (= 95.81 m³/d * 6.00 wh/m³ / 9.00 h/d).

For the shorter construction time of 300 days, the required average performance is 127.75 m³/d (= 38,325 m³ / 300 d).

Category	No.	Input parameter	Unit	25% construction time shortening	
	0	A	B	C	D
Baseline data	1	Construction time shortening	[%]	25.00	25.00
	2	Normal construction time	[d]	400.00	400.00
	3	Shortened construction time	[d]	300.00	300.00
	4	Concrete volume (C_V)	[m³]	38,325.00	38,325.00
	5	Required performance – Normal construction time		95.81	95.81
	6	**Required daily performance**	[m³/d]	**127.75**	**127.75**
	7	Daily working time	[h/d]	9.00	9.00
	8	Labor consumption rate (LCR)	[wh/m³]	6.00	6.41
	9	Maximum crane proportionality factor	[WK/crane]	20.00	20.00
	10	Normal number of workers	[WK]	63.88	68.23
	11	**Required number of workers**	[WK]	**85.17**	**90.97**
	12	**Worker proportionality factor – Trapezoidal model**	[-]	**0.80**	**0.80**
	13	**Maximum number of workers**	[WK]	**106.46**	**113.71**
	14	Mean wage cost	[€/wh]	35.00	35.00
	15	Available work area	[m²]	3,000.00	3,000.00
	16	Daily performance	[m³/d]	127.75	127.75
Crane capacity	17	$NO_{C,AV}$	[cranes]	5.00	5.00
	18	$NO_{C,OPT}$	[cranes]	5.32	5.69
	19	CC_{RED}	[%]	6.07	12.06
	20	$\mathbf{LCR_{CC,INC}}$	[%]	**2.98**	**6.81**
	21	$\mathbf{\Delta LCR_{CC,INC}}$	[wh/m³]	**0.18**	**0.41**
Minimum work area	22	WA_{AV}	[m²/wk]	35.23	32.98
	23	WA_{MIN}	[m²/wk]	30.00	30.00
	24	WA_{RED}	[%]	0.00	0.00
	25	$\mathbf{LCR_{WA,INC}}$	[%]	**0.00**	**0.00**
	26	$\mathbf{\Delta LCR_{WA,INC}}$	[wh/m³]	**0.00**	**0.00**
New values	27	$\mathbf{LCR_{INC}}$	[%]	**2.98**	**6.81**
	28	$\mathbf{LCR_{INC}}$	[wh/m³]	**0.18**	**0.41**
	29	**New LCR**	[wh/m³]	**6.18**	**6.41**
Wages	30	Total paid working hours	[wh]	236,808.79	245,618.99
	31	Wage costs	[€]	8,288,307.61	8,596,664.61

Table 10.50 Calculation table for determining the added (Column C) and aggregated (Column D) losses of productivity

Addition

For the addition of the losses of productivity, we first calculate the average required number of workers and obtain a value of 85.17 (= 127.75 m³/d * 6.00 wh/m³ / 9.00 h/d)

(Row 11). Dividing it by the proportional factor for workers (Row 12) of 0.80 (used as a deterministic value) results in the maximum required number of workers, i.e. 106.46 (Row 13).

Dividing the maximum required number of workers (Row 13) by the maximum crane proportionality factor (Row 9) gives the optimum number of cranes $NO_{C,OPT}$ as 5.32. However, the number of cranes that can be installed from a site management point of view was specified (deterministically) as 5 (Row 17). We get the crane capacity reduction CC_{RED} [%] by using Eq. (10-37) and the labor consumption rate increase resulting from the reduced crane capacity $\Delta LCR_{CC,INC}$ [%] by using Eq. (10-38) (see Rows 18 and 19). The percentage increase in the labor consumption rate is related to the basic consumption rate (Cell C8), which results in a consumption rate increase of 0.18 wh/m³ in the deterministic calculation.

The second loss of productivity considered in this example refers to the available work area WA_{AV} per worker [m²/worker] (Row 22). For the deterministic addition, we obtain an available work area of 35.23 m²/worker (= 3,000.00 m² / 85.17 workers).[78] The available work area is thus greater than the minimum work area (= 30.00 m²/worker, see Row 23). We find neither a reduction in the work area WA_{RED} [%] nor an increase in the labor consumption rate $\Delta LCR_{WA,INC}$ [%] when applying Eq. (10-39) and Eq. (10-40), respectively (Rows 24 to 26).

The uncertainties associated with the input parameters also allow for combinations with which the minimum work area would be underrun, thus causing a consumption rate increase. Particularly in the course of aggregation, the adjustment (i.e. the increase in the number of workers) increasingly leads to a loss of productivity due to the reduction in the available work area.

The total percentage increase in the consumption rate (Row 27) is the sum of the two individual consumption rate increases (Rows 20 and 25). We calculate the absolute consumption rate increase (Row 28) from the basic consumption rate (Cell C8) and the total percentage increase. Row 29 shows the new consumption rate (Cell C8 + Cell C28).

In Row 30, we obtain the total of paid working hours after additive consideration of the two potential losses of productivity by multiplying the new consumption rate (Cell C29) with the reinforced concrete volume (Row 4). Multiplying this value by the mean wage cost will result in the wage cost of the addition in Cell C31. The labor consumption rates, paid working hours, and wage costs entered in Table 10.50 as deterministic values are only one possible result of combining the uncertain input parameters. The Monte Carlo simulation gives a total of 50,000 individual results (iterations) that can be merged into a histogram (see Fig. 10.139).

Addition of the two potential losses of productivity results in an average increased labor consumption rate of about 6.21 wh/m³. The minimum amounts to approximately 5.50 wh/m³, and the maximum is about 7.67 wh/m³ (after 50,000 iterations). The minimum is thus equivalent to the minimum input value of the basic labor consumption rate. Accordingly, the iterative process also included cases in which there were no or only very minor losses of productivity. In 90% of the cases simulated for the addition, the increased labor consumption rate ranged from 5.69 to 6.82 wh/m³.

[78] The available work area is related to the average number of workers. For a more detailed analysis, we would be able to differentiate between the ramp-up, main, and final construction phases, including the corresponding available work area and number of workers.

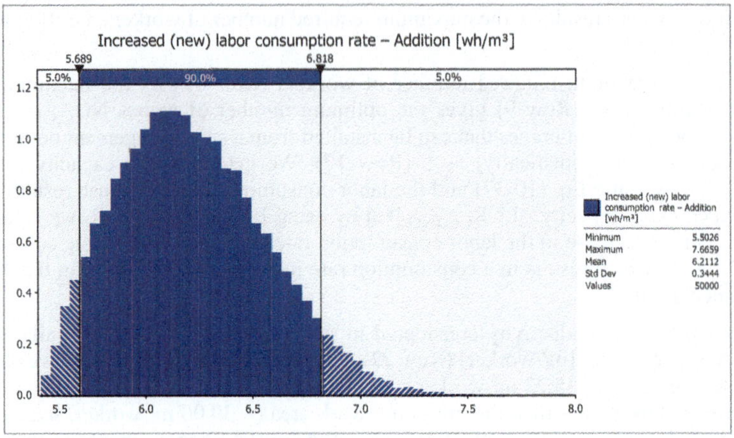

Fig. 10.139 Histogram – Increased (new) labor consumption rate – Addition

Aggregation

Aggregation essentially involves the same calculation steps as the addition procedure. The big difference, however, is that after adding up the productivity losses and calculating the new increased labor consumption rate, the latter will have an effect on the performance. It would not be possible to achieve the performance required for the short construction time on the basis of this increased labor consumption rate. This is why we need to adjust (increase) the number of required workers. However, this again will have an effect on the crane capacity reduction and the available work area, and thus also on the calculated losses of productivity or labor consumption rate increases.

The increased consumption rate shown in Cell D29 is re-entered into Cell D8 after each incremental iteration, and the calculation is repeated. In Excel, this results in a circular reference that is calculated until a termination criterion is reached.[79)]

In a Monte Carlo simulation, a value from within the range is randomly selected for each of the input parameters and used for performing the calculation. This process is repeated several thousand times, and these are repetitive (as opposed to incremental) iterations (see Section 6.2.4.1).

For each of the repetitive iterations in the Monte Carlo simulation, 100 incremental iterations (this number can be defined directly in Excel) are performed as part of the calculation procedure shown for the aggregation.

Fig. 10.140 shows the histograms representing the addition and aggregation of the losses of productivity.

The aggregation-related histogram shows a bimodal curve. This curve represents two different cases in the simulation: firstly, including only one productivity loss (due to reduced crane capacity) and, secondly, considering both productivity losses in the calculation.

The wider the ranges of the input parameters with an influence on the occurrence of losses of productivity are, the more likely it is that the simulation will also include states in

79) For the calculation, we defined a maximum of 100 incremental iterations as the termination criterion.

which no or only a minor increase in the labor consumption rate occurs.

If, for example, the maximum value of the available work area is set very high, the probability of labor consumption rate increases occurring due to the non-adherence to the minimum work area will be correspondingly lower.

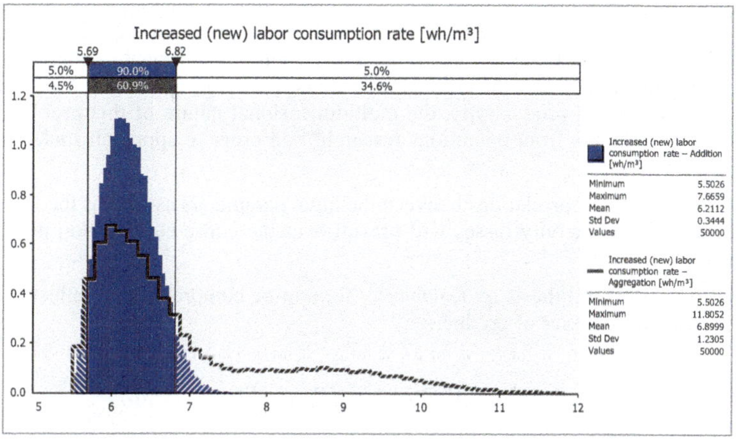

Fig. 10.140 Histogram – Increased (new) labor consumption rate – Comparison – Addition/aggregation

If the contractor had based their further considerations, pricing and process planning activities on a labor consumption rate of 6.82 wh/m³, for example, they would have assumed a chance/risk ratio of approximately 95% (chance) to 5% (risk) in the pure addition. However, the aggregation result indicates that the situation is much riskier than expected owing to the possible mutual reinforcement of effects. The histogram of the aggregation shows that the chance/risk ratio for the same deterministic value is approximately 65.4% (chance) to 34.6% (risk).

In this case, the key factors include not just the percentage ratio between chance and risk, but also the extremely right-skewed distribution and the wide value-based range of the increased consumption rate found when aggregating the losses of productivity.

10.4.7.5 Benefit for the Contractor

This example described the difference between the addition and aggregation of losses of productivity and demonstrated the calculation steps involved in each of the two approaches. Aggregation essentially relies on an incremental iterative calculation, where the results of an iteration always build upon the preceding calculation step on the basis of circular references. In contrast, the purely additive approach includes just a single calculation run that disregards feedback or reinforcement phenomena.

Since the input parameters for such a calculation are subject to uncertainties and have to be stated partly on the basis of subjective assessments, a probabilistic calculation (Monte Carlo simulation) was carried out in which the two considered losses of productivity (reduced crane capacity and work area reduction) were aggregated.

The histogram representing the increased labor consumption rate permits conclusions regarding chances and risks. In pricing, process planning, etc., decisions need to be made in favor of a specific consumption rate. Selecting this reference base is directly linked to a probabilistic conclusion regarding the associated chances and risks based on the spread of the input parameters. If we are able to narrow the range of individual input parameters, this will also alter the range of the aggregated labor consumption rate increase or loss of productivity.

The above-described methodology is also suitable for evaluating and subsequently reducing risks as well as for optimizing the utilization of resources. For instance, the increase in the daily working time could be examined as an additional or alternative work acceleration measure. If several production factors are to be optimized with a view to minimizing the loss of productivity, the multidimensional nature of this problem would require the use of tools from operations research[80)] in order to approach such a solution numerically.

In the above example, correlations between the input parameters as well as the fact that, in practice, not all productivity losses will prevail over the entire construction period were neglected.

We refer to the doctoral thesis by *Kummer*[81)] for a more comprehensive reflection on the topic of aggregating losses of productivity.

10.4.8 Determination of Stocking Quantities

The formwork stocking quantity is the quantity of formwork to be kept available on-site in order to ensure that workers are always supplied with a sufficient amount of formwork material and related equipment. This principle aims to ensure continuous work as a result of the selected construction sequence (i.e. either a continuous flow or a cycle-based approach). However, the formwork quantity kept in stock should not be overly large because this would incur additional costs, such as rental costs for the formwork equipment, and storage areas would be occupied by unnecessary formwork elements. The detailed calculation of stocking quantities belongs in PPH 4 (see Fig. 10.141).

The example outlined in this section differs from the example in section 10.3.7 in that considering the individual work sections and the (continuous flow) processes reaches a greater degree of detail.

80) Cf. Suhl/Mellouli (2013), p. 291ff.

81) Kummer (2015a)

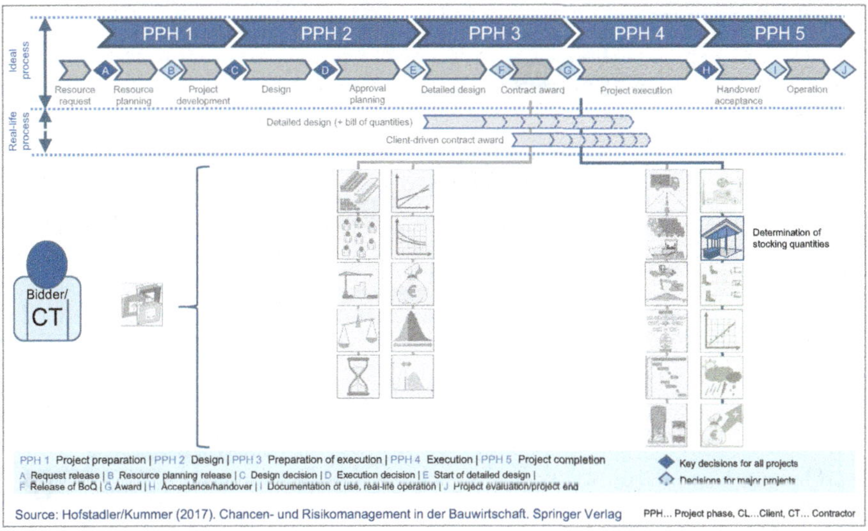

Fig. 10.141 Allocation of example in the project sequence

10.4.8.1 Basics

Calculating the stocking quantity is governed by a number of parameters, including the dimensions of structural components (such as those of floor slabs), the formwork performance, the duration of use of the formwork, the number of work sections, the selected production sequence, and the employed formwork system.

In this example, we determine the stocking quantity of the slab formwork required for a building construction project. The production sequence was assumed to follow the continuous flow principle.

In this setting, one of the work groups deals exclusively with formwork installation and stripping in the individual work sections. Reinforcing and concreting are taken care of by other teams.

When the formwork for a given work section i,j has been completed, the group of workers moves directly to the next work section i,j + 1 to commence the formwork operations there. The respective work section is represented by the symbol j while i designates the corresponding floor, where i = 1 to n_f and j = 1 to n_{ws}. Formwork performance thus has a significant influence on the stocking quantity.

Fig. 10.142 includes a diagram where the duration of the slab formwork operations is represented by bars marked D_{FI} (installation only). The bars labeled D_{RO} stand for the duration of reinforcing operations whereas those labeled D_{CO} represent the duration of concrete pouring. Fig. 10.142 does not include the bars representing the vertical structural members.

The reinforcing operations will commence with the offset $OF_{FO,RO}$ relative to the end of the formwork operations (see Fig. 10.143). Concreting of the work section will start after completion of the reinforcing operations, and the formwork stripping period D_{FS} will commence after the end of concrete pouring.

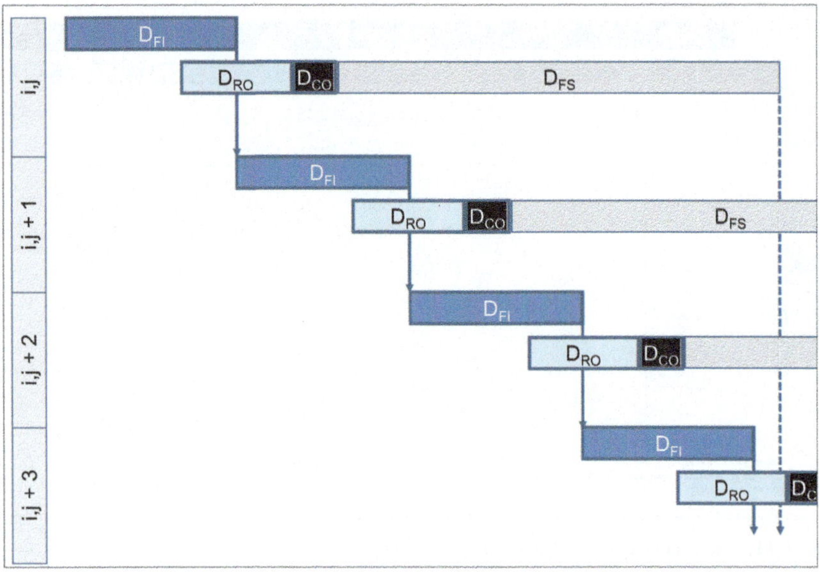

Fig. 10.142 Reinforced concrete operations for floor slabs – Continuous flow in formwork operations

The sequencing of formwork and reinforcing operations can either follow a normal pattern or an offset initial sequence. Fig. 10.143 illustrates an offset sequence of start times for a continuous flow process. Reinforcing operations commence before the end of formwork installation, offset by the dimension $OF_{FO,RO}$. This overlap during which the activities of formwork installation and reinforcing are carried out simultaneously is subject to a certain limit. This limit is referred to as critical approximation c_A [d]. The critical approximation measure ensures that reinforcing operations will not start before formwork has been installed on an appropriate minimum area. Furthermore, a "run-up" or clash of reinforcing with formwork operations should be prevented since this would cause disruptions to the construction process. The borderline case $OF_{FO,RO} = 0$ represents a normal sequence in which the reinforcing operations begin only upon the completion of formwork installation.[82)]

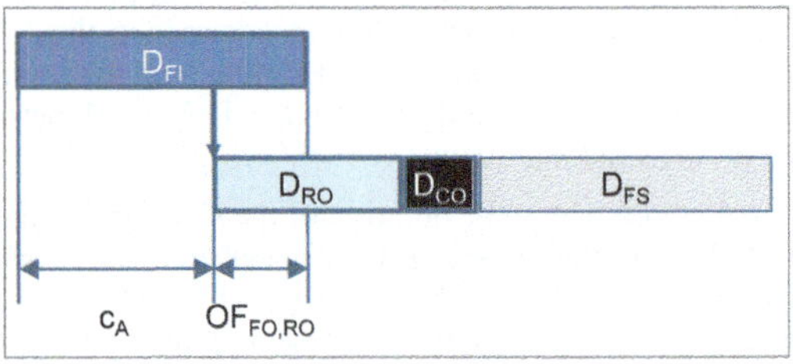

Fig. 10.143 Offset start-end relationships between formwork and reinforcing operations

[82)] Cf. Hofstadler (2007), p. 196f.

We use Eq. (10-41) for calculating the slab formwork stocking quantity for a continuous flow process. In the first term, we determine the work section area by dividing the standard floor area A_{SF} [m²] by the number of work sections per floor n_{ws} [-]. In the second term, we multiply the duration of formwork use $D_{FW,U}$ [d] by the daily formwork performance (installation only) PV_{FI} [m²/d].

$$SQ_{SF} = \frac{A_{SF}}{n_{ws}} + PV_{FI} \cdot D_{FW,U} \quad (10\text{-}41)$$

The duration of formwork use $D_{FW,U}$ [d] is the time during which the formwork is installed, and thus bound, in the work section. The formwork will not be available for any other use during this period. Additional formwork must be kept available for this period in a quantity that the formwork team is able to process during this time.[83)]

The duration of formwork use $D_{FW,U}$ comprises the sum of the times for reinforcement D_{RO} and concreting D_{CO}, the formwork stripping period[84)] D_{FS} multiplied by the reduction factor α_{av} less the offset between formwork operations and reinforcing $OF_{FO,RO}$ (Eq. (10-42)).

$$D_{FW,U} = \alpha_{av} \cdot D_{FS} + D_{RO} + D_{CO} - OF_{FO,RO} \quad (10\text{-}42)$$

We introduce the reduction factor α_{av} to take into account that concrete hardening also takes place on non-working days. This reduction factor[85)] reduces the stripping times in such a way that non-working days resulting from the selected working time model are taken into account. Only the number of working days will count towards calculating the stocking quantity.

However, these average values (which, however, do not reflect the exact conditions in the execution phase) are sufficiently accurate to draw conclusions regarding the duration of formwork use already in the process planning phase.

Fig. 10.144 shows the flow chart for this calculation, including the qualitative distribution functions of the input parameters and the outputs.

Additional workers must be employed for stripping if the scheduled formwork performance is to be maintained even when the first stripping operation starts. If no such plans have been made, and if the team size remains unchanged, the formwork performance will decrease due to the additional work required for stripping. This will also reduce the quantity to be kept in stock.

83) The possibility of stripping the slab formwork prior to that of the props can be taken into account by assuming different durations of use for the props and the slab formwork (e.g. panels and longitudinal beams including drophead shoring). See Hofstadler (2007), p. 302ff.

84) Formwork stripping times should be taken from the relevant codes, standards, and guidelines. In applicable standards, stripping times are stated in calendar days.

85) Cf. Hofstadler (1999), p. 120ff.

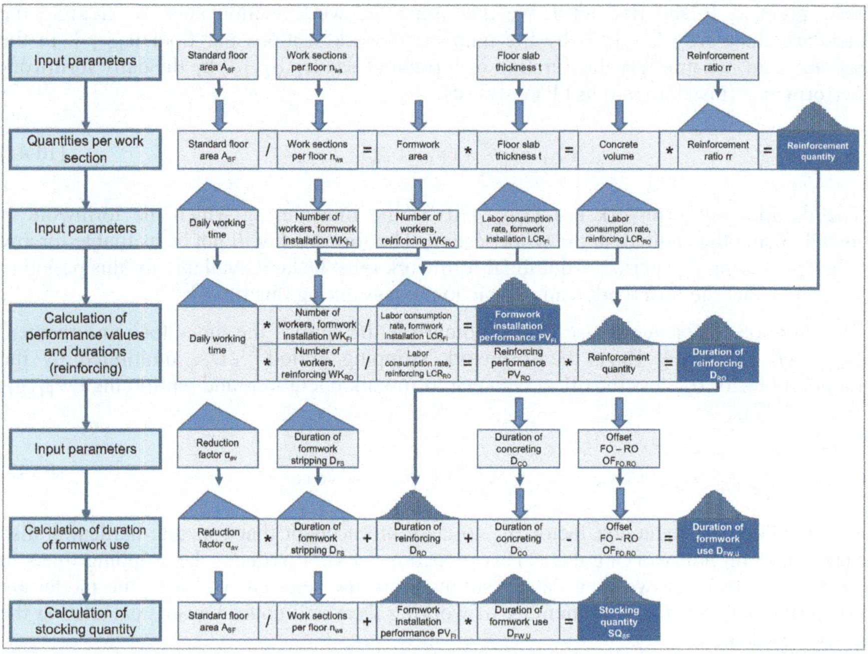

Fig. 10.144 Calculation flow chart – Determination of stocking quantity

10.4.8.2 Task

This example demonstrates how to calculate the stocking quantity required for slab formwork. For a three-story building whose floor slabs are each subdivided into four work sections, the stocking quantity is to be determined while considering the underlying conditions and existing uncertainties. The continuous flow principle is applied to the formwork operations for the reinforced concrete floor slabs in the individual work sections.

Table 10.51 includes the relevant input parameters. Column A of this table lists the parameter designations with their abbreviations. Column B contains the related units. Columns C, D, and E show the minimum, most likely, and maximum values for each parameter. If no values are assigned to the cells in Columns C and E, the input for this parameter should be taken to be a deterministic value. These values are transferred "as is" to Column F, which makes it easier to link the cells in Excel since only values from Column F are included in the subsequent calculation steps. Uncertain parameters are modeled as triangular distributions and entered in Columns F (based on the values in Columns C, D, and E).

No.	Parameter		Unit	MIN	MOST LIKELY	MAX	Distribution
0	A		B	C	D	E	F
1	Floor area	A_{SF}	[m²]		8,000.00		8,000.00
2	Floor slab thickness	t	[cm]		25.00		25.00
3	Number of floors	n_f	[-]		3.00		3.00
4	Number of work sections per floor	n_{ws}	[-]		4.00		4.00
5	Labor consumption rate, formwork installation	LCR_{FI}	[wh/m²]	0.20	0.25	0.32	0.26
6	Number of workers for formwork installation	WK_{FI}	[wh/h]		10.00		10.00
7	Labor consumption rate, reinforcing	LCR_{RO}	[wh/t]	12.00	14.00	16.00	14.00
8	Number of workers for reinforcing	WK_{RO}	[wh/h]		20.00		20.00
9	Reinforcement ratio in reinforced concrete floors	rr	[kg/m³]	90.00	100.00	120.00	103.33
10	Duration of concreting	D_{CO}	[d]		1.00		1.00
11	Daily working time	WT	[h/d]	8.00	8.00	9.00	8.33
12	Duration of formwork stripping	D_{FS}	[d]	5.00	5.00	6.00	5.33
13	Formwork installation to reinforcing offset	$OF_{FO,RO}$	[d]		2.00		2.00
14	Reduction factor	α_{av}	[-]	0.80	0.90	1.00	0.90

Table 10.51 Input parameters for calculating the stocking quantity of slab formwork

10.4.8.3 Calculation and Interpretation of Results

The quantities of formwork, reinforcement, and concrete per work section must first be calculated in order to determine the stocking quantity of slab formwork. Since the input parameters for the floor area, the number of work sections, and the slab thickness are defined as deterministic values, the formwork area and the concrete volume will also be deterministic.

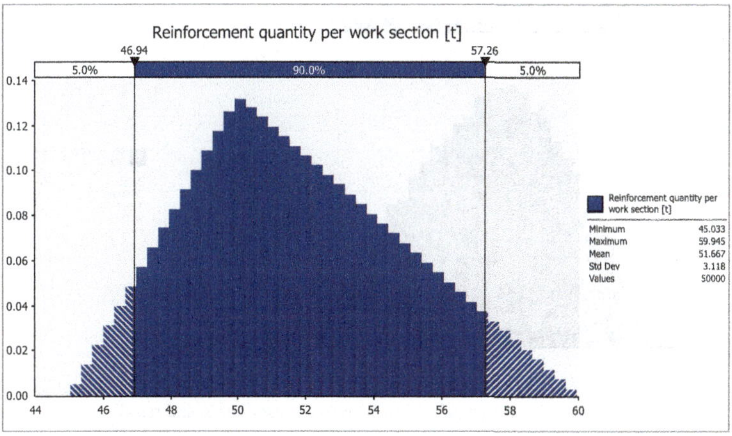

Fig. 10.145 Histogram – Reinforcement quantity per work section

The formwork area per work section is 2,000 m² (= 8,000 m² / 4)[86]. Multiplying this area by the floor slab thickness results in a concrete volume per work section of 500 m³ (= 2,000 m² * 0.25 m).

[86] Stop-end formwork is not considered.

The reinforcement ratio is an uncertain input parameter and was thus modeled as a triangular distribution. Multiplication by the concrete volume gives a triangular histogram representing the reinforcement quantity per work section (see Fig. 10.145).

In the next step, we determine the average performance values PV_{FI} [m²/d] for formwork installation and PV_{RO} [t/d] for reinforcing. For this purpose, we multiply the daily working time WT [h/d] by the number of workers and divide the result by the labor consumption rate. For the calculation of the formwork performance, we assume that additional workers are used for formwork stripping so that the average formwork performance can be retained.

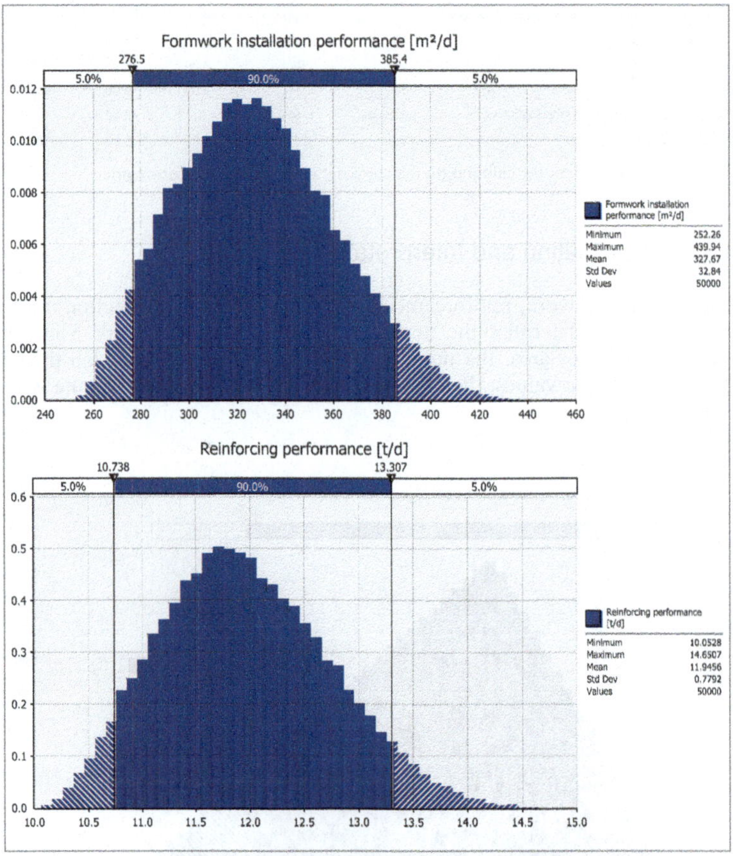

Fig. 10.146 Histograms – Performance values for formwork installation and reinforcing

The daily working time was set as a triangular distribution (8.00 | 8.00 | 9.00 h/d). The number of workers was defined as a deterministic value for both formwork and reinforcing operations (see Rows 6 and 8 in Table 10.51). The consumption rates of formwork and reinforcing operations are subject to uncertainties and were thus integrated into the calculation as triangular distributions (see Rows 5 and 7 in Table 10.51).

Fig. 10.146 shows a comparison of the histograms representing the formwork and reinforcing performance values. There is thus a 90% probability for the formwork performance to lie between approximately 277 and 385 m^2/d. The mean amounts to about 328 m^2/d.

We obtain a mean reinforcing performance of about 11.95 t/d, with a standard deviation of approximately 0.78 t/d. We find that a performance of 10.74 t/d is underrun and a performance of 13.31 t/d is exceeded in 5% of all cases.

If we divide the reinforcement quantity per work section (see Fig. 10.145) by the reinforcing performance, we get the histogram representing the duration of the reinforcing operations per work section as shown in Fig. 10.147.

The mean duration of reinforcing operations amounts to about 4.34 d while the 95% quantile is approximately 5.00 d and the 5% quantile is 3.74 d.

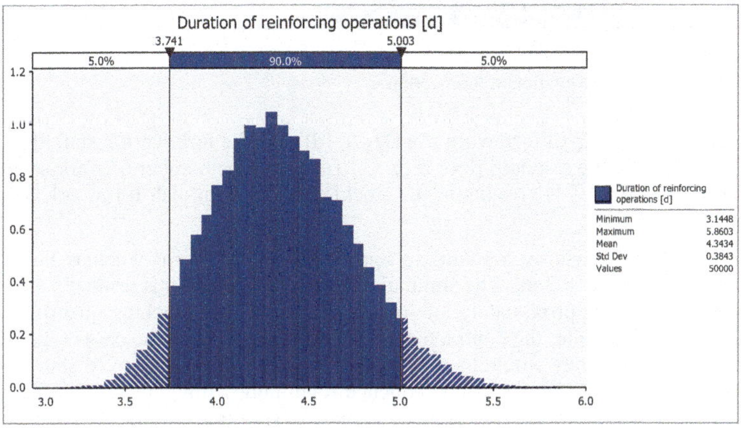

Fig. 10.147 Histogram – Duration of reinforcing operations

We insert the deterministic and probabilistic input parameters as well as the duration of reinforcing operations calculated as an intermediate result in Eq. (10-42) to determine the duration of formwork use. We obtain the duration of formwork use $D_{FW,U}$ [d] as a histogram resulting from the performed Monte Carlo simulation (see Fig. 10.148). The minimum duration (at 50,000 iterations) amounts to about 6.53 d while the maximum is approximately 10.23 d. We thus get a mean of 8.14 d with a standard deviation of 0.49 d.

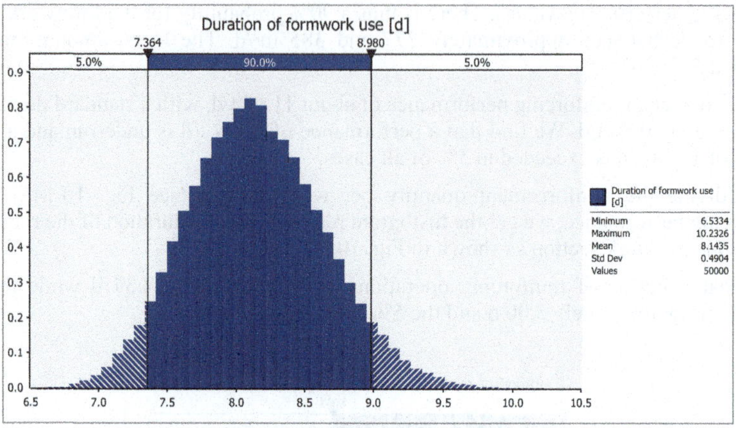

Fig. 10.148 Histogram – Duration of formwork use

Combining the duration of formwork use $D_{FW,U}$ [d] with the formwork installation performance PV_{FI} [m²/d], the standard floor area A_{SF} [m²], and the number of work sections per floor n_{ws} [-] in Eq. (10-41) results in the stocking quantity of slab formwork SQ_{SF} [m²] (see Fig. 10.149).

The mean stocking quantity amounts to approximately 4,670 m² whereas the standard deviation equals about 303 m². The simulation with 50,000 iterations results in a minimum stocking quantity of approximately 3,836 m² and a maximum stocking quantity of about 5,935 m². If, for example, the contractor wanted to cover 80% of the cases (chance) with its stocking quantity, they would have to stock approximately 4,923 m² of slab formwork on-site for the scheduled production sequence (continuous flow and four work sections per floor) and the planned utilization of resources (i.e. the number of workers, daily working hours, etc.).

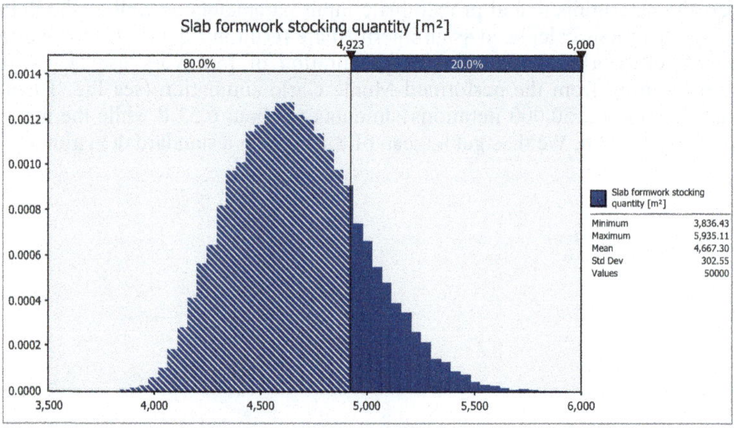

Fig. 10.149 Histogram – Slab formwork stocking quantity

10.4.8.4 Benefit for the Contractor

The detailed determination of the slab formwork stocking quantity is particularly important for the contractor to optimize their production system. On the one hand, the workers need to be supplied with sufficient formwork equipment and material; on the other hand, it is important to avoid exceedingly large stocking quantities for the sake of cost efficiency and in the interest of optimizing the construction process (in terms of rental costs and storage space, for instance).

The Monte Carlo simulation provides the option of combining deterministic and probabilistic input parameters in a single simulation. Within this framework, both certain and uncertain parameters can be systematically taken into account when determining the stocking quantity.

Choosing the most appropriate stocking quantity can be based on a selected chance/risk ratio. It is crucial for the contractor to take the existing uncertainties into consideration when deciding on a certain stocking quantity, particularly if the formwork operations are on the critical path.

10.4.9 Procurement Optimization

Optimizing the elementary production factors and processes is an essential tool for contractors to minimize costs and resource utilization while ensuring the same standard of workmanship.

In this project phase (PPH 4), the focus is predominantly on consistently supplying the site with the required materials (see Fig. 10.150). If there are several options for procuring construction materials, such as a number of ready-mixed concrete plants located near the site, and if several construction sites operated by the contractor need to be supplied at the same time, this leads to a multidimensional optimization problem. If we also consider uncertainties associated with the calculation parameters, this becomes a probabilistic multidimensional optimization problem that can no longer be solved analytically.

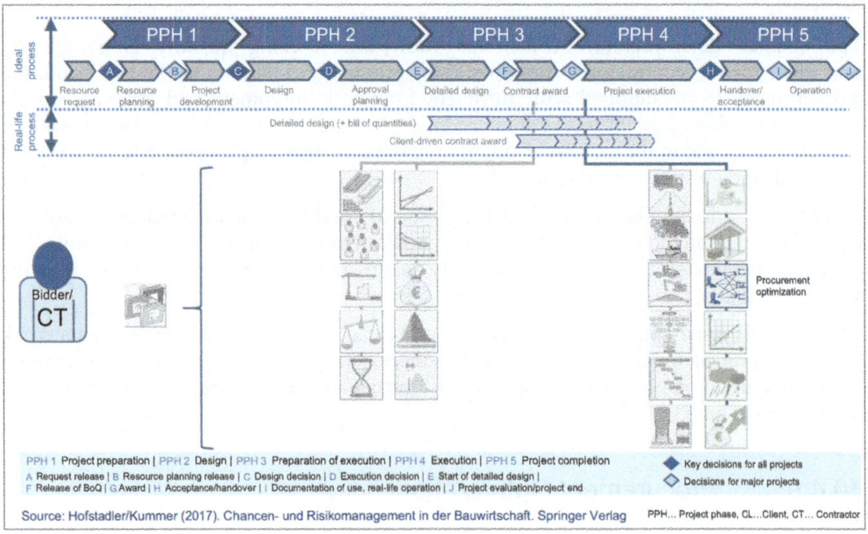

Fig. 10.150 Allocation of example in the project sequence

A similar problem arises when supplying construction sites with different materials from a variety of stocks. In this case, the optimization process involves minimizing the costs of purchase and/or transport (assuming identical material costs).

10.4.9.1 Basics

If the input parameters of a computation model are fraught with uncertainties, it may be of interest not only to present and analyze the different results based on the varying input variables in the form of a summarizing histogram, but also to obtain a recommendation for those controllable parameters that, if combined, would incur the lowest costs or losses of productivity, for example.

This is a (potentially multidimensional) optimization problem, the solution of which can be approximated numerically due to the stochastic influences with the aid of simulations and appropriate probabilistic optimization software[87)].

In this scenario, the variables to be optimized are subject to unavoidable scattering. Fig. 10.151 shows a schematic representation of such a scattering for two decision variables (X and Y).

[87)] Such as the RISKOptimizer optimization tool integrated in @Risk.

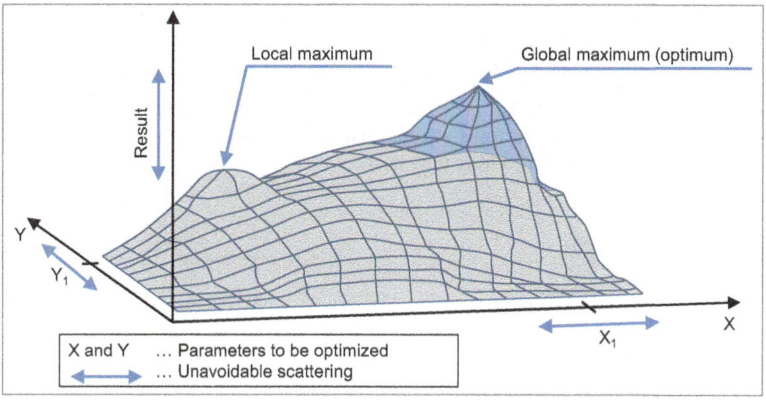

Fig. 10.151 Qualitative representation for determining a global optimum for two variables to be optimized, including unavoidable scattering[88)]

In order for an optimization algorithm to identify a global optimum[89)], it should also allow for a certain degree of deterioration of the result sought in the course of searching for possible solutions. Only a local optimum can usually be found if we exclusively consider improvements (see Fig. 10.151). For this purpose, we apply heuristic optimization methods (such as simulated annealing, threshold accepting, great deluge, and taboo search) or metaheuristics analogous to nature (such as genetic algorithms).[90)]

The problem with local searches is that a global optimum may not be found, and the algorithm may not identify better solutions anymore once a local optimum occurs. This can be likened to searching for the highest point in a three-dimensional terrain (see Fig. 10.151). In a local search, the algorithm will start at an arbitrary point (in the "terrain") and scan (almost) blindly to the next (neighboring) point. Only the "severity" or "gravity" is used as a search criterion, determining whether to move "up" or "down". This restricted search criterion can lead to a situation in which better solutions (i.e. higher points in the "terrain") can no longer be found. Improved optimization methods can revise decisions made in the search process (backtracking) or even allow for worse solutions to be found in order to move to another "solution area" where a better (global) optimum can potentially be identified.[91)]

10.4.9.2 Task

This example shows how probabilistic multidimensional optimization problems can be modeled and optimized numerically in the final step. We assume a situation in which a contractor was commissioned with reinforced concrete operations on three construction sites at the same time, and intends to carry out concreting operations on the same day. Each of the construction sites will require concrete of the same quality, but in different quantities.

88) Cf. Voigt (2008), Slide 62

89) Depending on the specific problem, an optimum can be a minimum (costs) or a maximum (productivity).

90) Cf. Suhl/Mellouli (2013), p. 293

91) Cf. ibid., p. 136ff.

The contractor can source concrete from three different ready-mixed concrete plants (see Fig. 10.152). They each charge a different price for concrete[92)] of the same quality, are located at different distances from the sites, and their production capacity is limited in each case. None of the plants can fully supply all three construction sites.

In this example, the optimization process aims to determine the combination of concrete deliveries that would incur the lowest total cost for all construction sites taken together, taking into account material and transport costs.

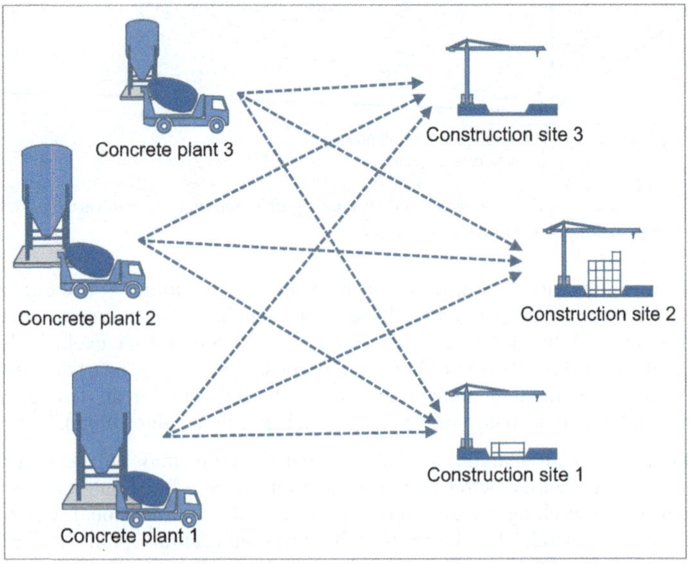

Fig. 10.152 Task – Procurement optimization

10.4.9.3 Modeling

To model the optimization example, we first need to define the boundary conditions, including the transport distances, the demand per construction site, the capacity per ready-mixed concrete plant, and the loading capacity per truck (see Fig. 10.153). Each of these values is deterministic in nature. The transport costs per kilometer and the concrete costs per mixing plant were identified as input parameters with uncertainties (modeled as triangular distributions). The quantities delivered per ready-mixed concrete plant and construction site are defined as variable values.

In the subsequent process steps, we need to alter the combination of these quantities such that the total cost (i.e. the target variable) reaches the possible minimum. The total costs include transport costs (depending on the number of transports, distances, and costs per kilometer) and concrete costs (depending on delivered quantities and concrete costs per mixing plant). In this example, the optimization relates to the mean total cost.

[92)] The prices charged by the respective ready-mixed concrete plant represent costs for the contractor.

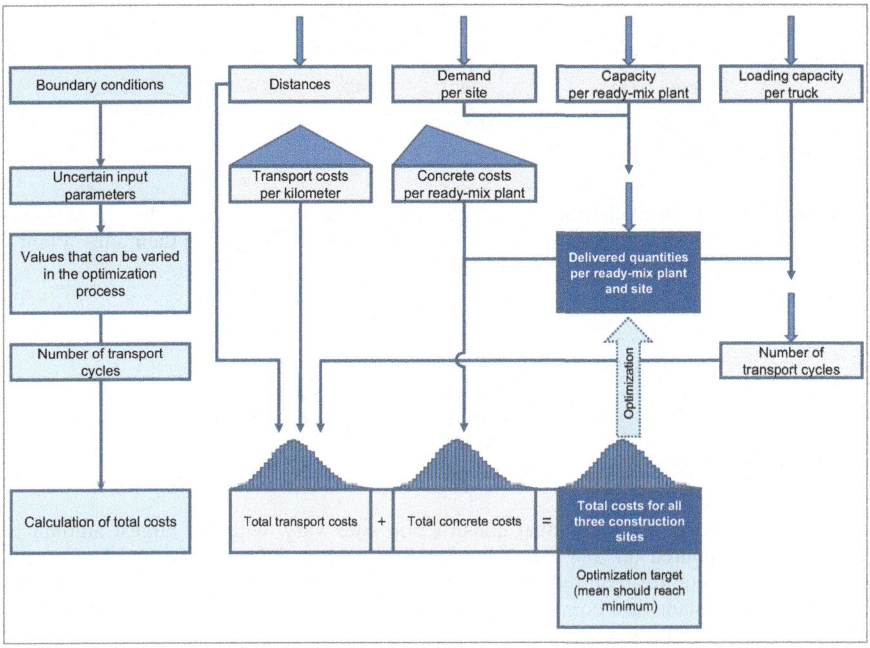

Fig. 10.153 Calculation flow chart

In this example, both the transport costs and the concrete costs per ready-mix plant are subject to uncertainties. The transport costs per kilometer are modeled as a triangular distribution (MIN: €1.00/km; MOST LIKELY: €1.50/km; MAX: €2.00/km). The distances between the ready-mixed concrete plants and the construction sites are integrated into the model as fixed parameters (deterministic values) (see Table 10.52). The transport costs per kilometer already include the outward and return journey and are thus only multiplied by the simple distance (one direction).

	Distance [km]	From		
		Concrete plant 1	Concrete plant 2	Concrete plant 3
To	Construction site 1	15.00	25.00	50.00
	Construction site 2	30.00	20.00	35.00
	Construction site 3	40.00	35.00	10.00

Table 10.52 Distances between ready-mixed concrete plants and construction sites

For each plant, a range was defined for the costs of concrete. These ranges already account for the capacity utilization of the respective plant as well as for additional costs of potentially required longer unloading times and/or overtime surcharges, etc. The concrete costs are modeled as triangular distributions for each ready-mix plant (see Table 10.53).

No.	Costs of concrete [€/m³] (same grade)				
	Ready-mix plant	MIN	MOST LIKELY	MAX	Distribution
0	A	B	C	D	E
1	Concrete plant 1	90.00	100.00	120.00	103.33
2	Concrete plant 2	80.00	85.00	95.00	86.67
3	Concrete plant 3	70.00	70.00	85.00	75.00

Table 10.53 Costs of concrete – Triangular distributions

The capacity of each of the three plants is limited (see Table 10.54), which is why the contractor will be forced to select more than one source for supplying their sites. Plant 3 has the lowest capacity while Plant 1 has the highest.[93)]

No.	Plant capacity	
	Ready-mix plant	[m³]
0	A	B
1	Concrete plant 1	250.00
2	Concrete plant 2	180.00
3	Concrete plant 3	100.00

Table 10.54 Capacities of ready-mixed concrete plants

The requirements of the individual construction sites vary, with the largest amount of concrete to be poured on Site 3 and the smallest on Site 2 (see Table 10.55). Overall, 440 m³ of concrete are required. This volume can be provided only if the three ready-mixed concrete plants are combined. The contractor's aim is to identify the most cost-effective order combination.

No.	Site demand	
	Construction site	[m³]
0	A	B
1	Construction site 1	165.00
2	Construction site 2	75.00
3	Construction site 3	200.00

Table 10.55 Concrete demand of construction sites

Irrespective of the plant, the capacity per truck mixer invariably amounts to 7.00 m³. If the required order quantity per plant exceeds a multiple of the truck mixer capacity, this will require a further transport cycle for the excess concrete volume (only an integer number of transport cycles is permitted).

Table 10.56 shows the ordering matrix for the individual concrete deliveries. Cells B1 to B3, C1 to C3, and D1 to D3 show the individual concrete volumes ordered per plant and construction site. However, this combination of orders is not the order that would incur the lowest total cost. Rather, it is a random combination that meets all the required conditions. The sum of the concrete volumes delivered to the sites (Column E) corresponds exactly to the respective demand (Column F). Nor do ordered deliveries (Row 4) exceed the capacities of the three ready-mixed concrete plants (Row 5).

93) The stated capacities are those that the plants have committed for the construction sites concerned since they also have to deliver concrete to other sites located within their supply area.

No.	Order	From			Total site delivery	Site demand
		Concrete plant 1	Concrete plant 2	Concrete plant 3		
0	A	B	C	D	E	F
1	Construction site 1	75.00	60.00	30.00	165.00	165.00
2	Construction site 2	15.00	40.00	20.00	75.00	75.00
3	Construction site 3	70.00	80.00	50.00	200.00	200.00
4	Total delivery	160.00	180.00	100.00		
5	Capacity	250.00	180.00	100.00		

Table 10.56 Ordering matrix

We then apply the probabilistic optimization method to reduce the values in Cells B1 to B3, C1 to C3, and D1 to D3 accordingly to achieve the minimum total cost (comprising material and transport costs). This optimization is geared towards the mean total cost, which is necessary because the uncertainties incorporated in the input parameters will result in a histogram rather than a distinct total cost figure.[94)]

For the order combination randomly determined in Table 10.56, we calculate the number of transports from the ready-mixed concrete plants to the construction sites according to Table 10.57, taking into account the available capacity of 7.00 m³ per truck mixer. In this context, it should be noted that not all transports are filled completely. Otherwise, the capacities of individual mixing plants may be exceeded or the delivery to one of the construction sites may be too large.

No.	Number of transport cycles	Concrete plant 1	Concrete plant 2	Concrete plant 3
0	A	B	C	D
1	Construction site 1	11.00	9.00	5.00
2	Construction site 2	3.00	6.00	3.00
3	Construction site 3	10.00	12.00	8.00

Table 10.57 Number of transport cycles

For the deterministic calculation based on the mean values, we obtain the number of transports per distance between the ready-mix plant and the construction site according to Table 10.58. The mean of the triangular distribution (MIN: €1.00/km; MOST LIKELY: €1.50/km; MAX: €2.00/km) representing the transport costs per kilometer is equal to the most likely value of €1.50/m³ owing to the symmetry of the distribution. In the deterministic calculation, we multiply this value by the distances listed in Table 10.52. We thus get, for example, transport costs between Plant 3 and Site 2 (Cell D2) of €52.50 (= €1.50/km * 35 km).

No.	Transport costs per truck	From		
		Concrete plant 1	Concrete plant 2	Concrete plant 3
0	A	B	C	D
1	Construction site 1	€22.50	€37.50	€75.00
2	Construction site 2	€45.00	€30.00	€52.50
3	Construction site 3	€60.00	€52.50	€15.00

Table 10.58 Transport costs per distance – Deterministic calculation

[94)] Optimization for a specific quantile or other statistical value would also be possible.

The total transport costs result from multiplying the transport costs per truck (see Table 10.58) with the number of transports (see Table 10.57). For this purpose, we first calculate the products of the cells with the same designation, and then add up these products. Overall, we obtain the following deterministic transport costs on the basis of the mean values:

Plant 1: 22.50 * 11 + 45.00 * 3 + 60.00 * 10 +
Plant 2: 37.50 * 9 + 30.00 * 6 + 52.50 * 12 +
Plant 3: 75.00 * 5 + 52.5 * 3 + 15.00 * 8

= €2,782.50

For the deterministic calculation of the costs of concrete, we also multiply the mean values of the triangular distributions of the concrete costs per plant (see Fig. 10.153, Column E) by the order quantities (see Table 10.56, Cells B1 to B3, C1 to C3, and D1 to D3), and sum up the results. The deterministic total costs of concrete are thus €39,633.33 [= 103.33 * (75 + 15 + 70) + 86.67 * (60 + 40 + 80) + 75.00 *(30 + 20 + 50)].

The total (means-based) deterministic costs of the random combination of the delivered quantities thus amount to €42,415.83 (= €2,782.50 + €39,633.33). Identical order situations may incur varying costs (within the defined ranges) owing to the uncertainties associated with the concrete and transport costs.

The optimization process aims to minimize costs by altering the order combination, taking into account the uncertainties of the input parameters.

10.4.9.4 Calculation and Interpretation of Results

To arrive at an ideal order scenario that minimizes costs and meets the defined conditions, different delivered quantities per construction site and ready-mixed concrete plant are combined with one another in the course of the probabilistic optimization process in order to identify the combination that would incur the lowest total cost.

This procedure starts from the randomly selected order quantities in Table 10.56 (Cells B1 to B3, D1 to D3, and E1 to E3).

Fig. 10.154 shows how the progression of the optimization process is documented for 1,000 trials (on the abscissa). The mean total cost values are plotted on the ordinate. Horizontal progressions of the optimization process indicate that no further cost reductions were achieved in the corresponding trials. A vertical leap in the curve is attributable to identifying a combination of orders that would lead to lower total costs. What is key in this analysis is whether the optimization process resulted in only a local or in a global minimum (see Fig. 10.151).

This diagram shows that, for this simple multidimensional optimization problem, a stable total cost minimum was identified after only a few trials. In the remaining number of trials (approximately 900), it was no longer possible to improve the result by varying the order quantities.[95)]

[95)] The optimization process may easily take several hours in the case of complex problems with many possible variations (number of adjustable cells).

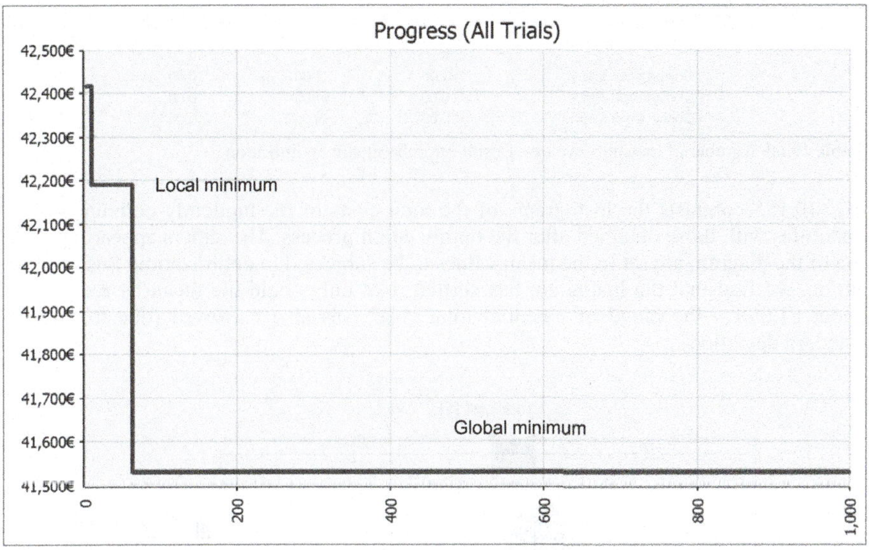

Fig. 10.154 Optimization progress

Table 10.59 shows the order quantities per ready-mixed concrete plant and construction site resulting from the probabilistic optimization process. This ideal combination of orders thus requires Plant 3 to supply its full output to Site 3. Also, Plant 1 should ideally supply almost its entire volume to Site 1. Only 5 m³ of concrete should be delivered from Plant 2, whereas, for practical reasons, the full order quantity for Site 1 should be sourced from Plant 1. Site 2 is also completely supplied by Plant 2, and the remaining 100 m³ required for Site 3 are also provided by Plant 2.

No.	Order	From			Total site delivery	Site demand
		Concrete plant 1	Concrete plant 2	Concrete plant 3		
0	A	B	C	D	E	F
1	Construction site 1	160.00	5.00	0.00	165.00	165.00
2	Construction site 2	0.00	75.00	0.00	75.00	75.00
3	Construction site 3	0.00	100.00	100.00	200.00	200.00
4	Total delivery	160.00	180.00	100.00		
5	Capacity	250.00	180.00	100.00		

Table 10.59 Ordering matrix – Result of probabilistic optimization

The number of transports is derived from Table 10.60 by dividing the optimized order quantities by the loading capacity per truck mixer (= 7.00 m³/truck). The number of transports will be rounded up if an integer multiple of this loading capacity is exceeded (in Cell C3: 100.00 m³ / 7.00 m³/truck = 14.29 trucks; rounded up to 15 transport cycles).

No.	Number of transport cycles	Concrete plant 1	Concrete plant 2	Concrete plant 3
0	A	B	C	D
1	Construction site 1	23.00	1.00	0.00
2	Construction site 2	0.00	11.00	0.00
3	Construction site 3	0.00	15.00	15.00

Table 10.60 Number of transport cycles – Result of probabilistic optimization

Fig. 10.155 contrasts the histograms of the total costs of the randomly combined order quantities with those obtained after the optimization process. The sliders appearing at the top of the diagram are set to the mean values to be subjected to optimization. Post-optimization, we find that the histogram has shifted: Not only could the mean be reduced by about €1,000 – the range of potential total costs was also narrowed (due to a lower standard deviation).

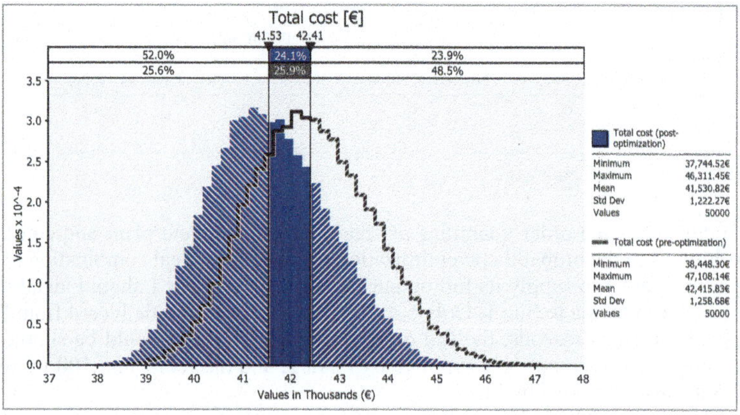

Fig. 10.155 Comparison of total cost pre- and post-optimization

This example refers to only a single concreting day. The greater the frequency of such cross-site concreting days occurring during the construction periods of the individual projects, the more significant the potential for savings will be.

Relative to the mean, the costs could be reduced by more than 2% merely by altering the order combination of the quantities. Assuming two concreting days per week, a construction time of 100 days would result in a savings potential of approximately €40,000 (= €1,000 * 100 d / 5 d / week * 2) in relation to the mean value of the cost distribution.

10.4.9.5 Benefit for the Contractor

For the contractor, procurement optimization is an essential source for taking advantage of existing opportunities and minimizing costs through an appropriately optimized order combination. The methodology demonstrated in this example covered several construction sites and ready-mixed concrete plants, but it can also be applied to other procurement optimization processes.

Using numerical, probabilistic optimization methods provides the option of considering input parameter uncertainties in addition to the multidimensional change of the scattering parameters (in this example, the combination of the order quantities). In the example, minimizing the total cost (including transport and material costs for all three construction sites) relative to the mean was defined as the target variable.

As a result, specific recommendations for action (i.e. order quantities per plant and construction site) were determined, which can be implemented directly after checking their viability against construction practice and ascertaining their plausibility.

Multidimensional probabilistic optimization processes are beneficial in that existing potentials can be utilized in the best possible way. This approach increases the chances for success and reduces existing risks.

The flexible computation model makes it possible to respond directly to changing boundary conditions, such as concrete costs, transport costs per kilometer, or concrete demand, and to run a new optimization sequence. The consequences of changed boundary conditions become directly visible since the optimization process will result in a new ordering strategy and new total cost histograms.

10.4.10 Trend Analyses

Trend analyses serve the purpose of drawing conclusions regarding the future development of the parameter under consideration on the basis of historical or progress data, such as in relation to an ongoing project, using mathematical/statistical methods. In construction projects, trend analyses are particularly suitable for target-actual comparisons in the course of the construction process (PPH 4, see Fig. 10.156).

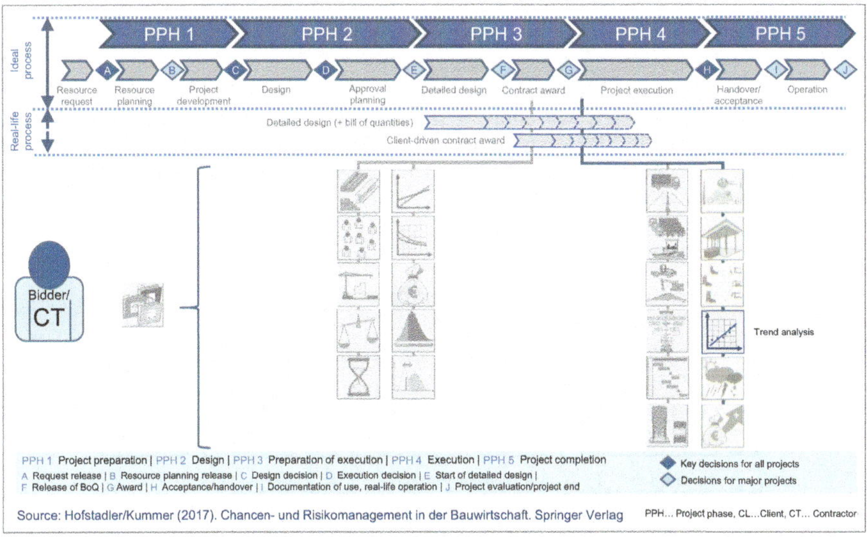

Fig. 10.156 Allocation of example in the project sequence

10.4.10.1 Basics

The basics of trend analysis and linear regression as a special type of regression analysis have already been explained in Section 9.4.1.1. The following sections describe a practical example to demonstrate how to perform probabilistic trend analyses.

10.4.10.2 Task

For a building construction project, a supporting target-actual comparison is carried out for reinforced concrete operations. In the first step, the scheduled project progress is calculated for the target, and displayed in a diagram.

The actual progress deviates from the target and is also incorporated in the graphical representation. In this example, we assume that the target and actual quantities are identical or that the actual quantities have already been delimited from the target quantities.

Trend analyses are used to determine whether the required work or service can be provided within the specified time or what the chances are of completing the work or service on time. The concrete volume is taken to be the indicative quantity for reinforced concrete operations. For the project considered in this example, a concrete volume of 38,275 m³ was specified.

The total labor consumption rate of reinforced concrete operations was calculated at 6.10 wh/m³, and the specified construction time is 400 days. The daily working time was set at 8.00 h/d.

This project-related information makes it possible to initially calculate the average number of workers over the entire construction period:

$$WK_{AVG} = \frac{38{,}275\ m^3 \cdot 6.10\ wh/m^3}{400\ d \cdot 8.00\ h/d} \approx 73\ WK$$

For the sake of simplicity, we apply a trapezoidal distribution to the utilization of resources (see Fig. 10.157 for the trapezoidal model). The proportional factor for workers f_{WK} [-] represents the relationship between the average (WK_{AVG}) and the maximum (WK_{MAX}) number of workers.

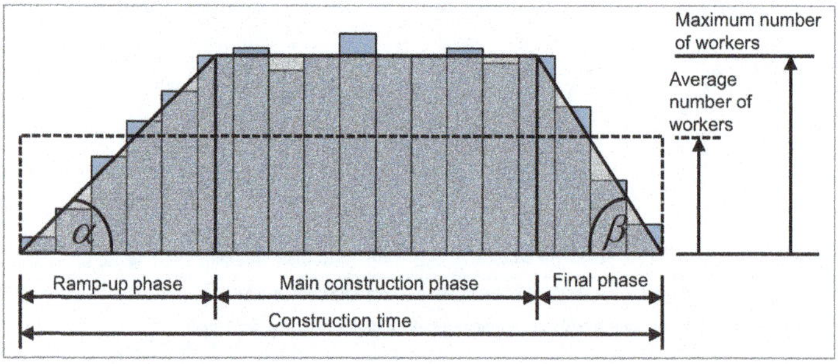

Fig. 10.157 Trapezoidal distribution of workers over the construction period – Qualitative representation

For the building construction project considered in this example, we assume a proportional factor for workers of 0.80, resulting in a maximum number of workers of approximately 91:

$$WK_{MAX} = \frac{WK_{AVG}}{f_{WK}} = \frac{73.0\ WK}{0.80} \approx 91\ WK$$

We obtain the share of the main construction phase D_{MP} [d] in the total time for reinforced concrete operations D_{RCO} [d] from the following equation for the trapezoidal distribution shown qualitatively in Fig. 10.157:

$$D_{MP} = \frac{2 \cdot D_{RCO} \cdot WK_{AVG}}{WK_{MAX}} - D_{RCO} \qquad (10\text{-}43)$$

Inserting the corresponding values in Eq. (10-43) results in a main phase of 240 days.

$$D_{MP} = \frac{2 \cdot 400\ d \cdot 73\ WK}{91\ WK} - 400\ d \approx 240\ d$$

The ramp-up phase D_{RP} [d] is determined according to Eq. (10-44) while the final phase D_{FP} [d] is calculated as one-third of the ramp-up phase.

$$D_{RP} = \frac{3}{4} \cdot (D_{RCO} - D_{MP}) \qquad (10\text{-}44)$$

This results in a ramp-up phase of approximately 120 days [= 3/4 * (400 d - 240 d)] and a final phase of about 40 days (= 1/3 * 120 d) for the reinforced concrete operations in the building construction project under consideration.

Each of these phases is also associated with a different daily performance owing to the different number of workers used in the individual phases of the reinforced concrete operations. The average daily performance in the ramp-up and final phases of reinforced concrete operations $PV_{RCO,RP\&FP}$ [m³/d] is calculated using the following equation:

$$PV_{RCO,RP\&FP} = \frac{WK_{MAX} \cdot WT}{2 \cdot LCR_{RCO}} \qquad (10\text{-}45)$$

For the ramp-up and final phases, we thus obtain an average daily performance of approximately 60 m³/d [= 91 workers * 8.00 h/d / (2 * 6.10 wh/m³)].

We use the performance equation and insert the maximum number of workers in order to determine the average daily performance during the main construction phase $PV_{RCO,MP}$ [m³/d]:

$$PV_{RCO,MP} = \frac{WK_{MAX} \cdot WT}{LCR_{RCO}} \qquad (10\text{-}46)$$

For the main phase, we get an average performance of approximately 119 m³/d [= 91 workers * 8.00 h/d / (2 * 6.10 wh/m³)].

Calculating the durations of the three construction phases as well as the corresponding performance values makes it possible to plot the target progress of the reinforced concrete operations over the construction period in a diagram (see Fig. 10.158).

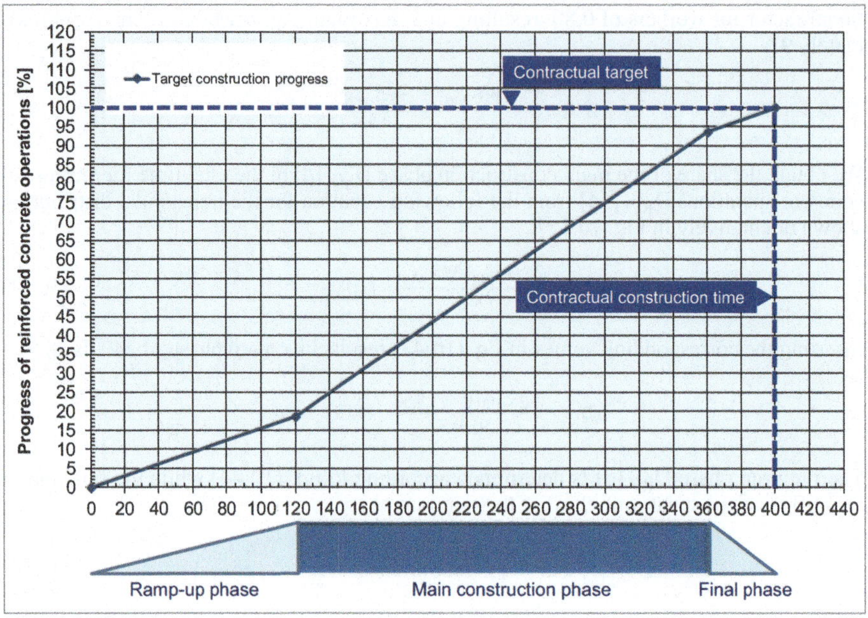

Fig. 10.158 Target construction progress

In the course of the construction work, records of the already completed installation of the indicative quantity of the reinforced concrete operations will be continuously updated. Table 10.61 shows the volume of concrete poured during the first 240 working days. This installed quantity was recorded continuously at intervals of 20 working days (4 weeks). The shorter the selected intervals, the greater the number of data points will be that the regression or trend analyses can be based on.

Actual construction progress				
Coordinate points	Construction time	Poured concrete volume	Performance value	Share of poured concrete volume in total volume
A	B	C	D	E
0	0.00 d	0.00 m³		0.00%
1	20.00 d	550.00 m³	27.50 m³/d	1.44%
2	40.00 d	1,640.00 m³	54.50 m³/d	4.28%
3	60.00 d	2,580.00 m³	47.00 m³/d	6.74%
4	80.00 d	4,920.00 m³	117.00 m³/d	12.85%
5	100.00 d	7,200.00 m³	114.00 m³/d	18.81%
6	120.00 d	10,000.00 m³	140.00 m³/d	26.13%
7	140.00 d	14,210.00 m³	210.50 m³/d	37.13%
8	160.00 d	17,500.00 m³	164.50 m³/d	45.72%
9	180.00 d	20,780.00 m³	164.00 m³/d	54.29%
10	200.00 d	24,060.00 m³	164.00 m³/d	62.86%
11	220.00 d	26,245.00 m³	109.25 m³/d	68.57%
12	240.00 d	28,430.00 m³	109.25 m³/d	74.28%

Table 10.61 Actual construction progress

The coordinate points of the documented actual construction progress are entered in Fig. 10.159. The abscissa shows the construction time, and the ordinate shows the cumulative progress of the reinforced concrete operations as a percentage. This diagram also includes the target construction progress.

The diagram reveals that the construction progress initially lagged behind the target in the ramp-up phase, but it subsequently accelerated beyond the target at the end of the ramp-up phase and until the end of the first half of the main phase.

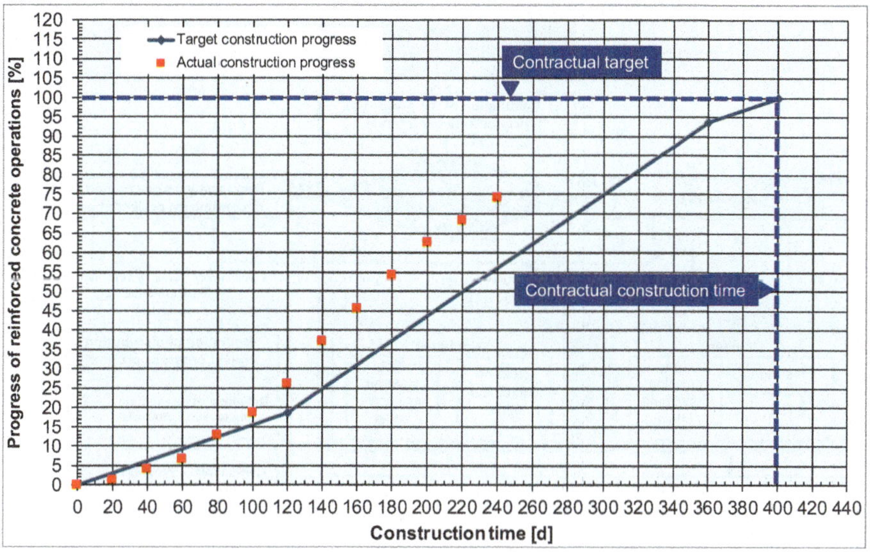

Fig. 10.159 Target-actual comparison – Construction progress

Our aim is to provide a forecast of the subsequent construction progress by means of trend analyses based on the work completed so far. In so doing, we rely on both deterministic and probabilistic methods. We also make a distinction between an interval-based and an overall analysis.

10.4.10.3 Modeling

Fig. 10.160 shows the workflow for the different types of trend analysis. Firstly, we need to distinguish if the interval at which the actual data is collected is fixed (for instance every 20 days) or variable. Irrespective of this consideration, we determine the proportional progress per interval. We then need to decide whether or not to differentiate between the ramp-up, main, and final construction phases.

Either of the approaches allows for deterministic trend analyses (linear or nonlinear trend analyses, crossing the coordinate origin or not) as well as probabilistic forecasts. For the probabilistic calculations, it is important to consider whether or not a distinction is to be made between the individual project phases for the purpose of defining the distributions of the proportional project progress. If we use a variable interval for documenting the actual progress, an additional distribution must be defined for the observation interval. In addition, a correlation between the observation interval and the project progress should be considered if required. Accordingly, any decision in favor of a variable interval will inevitably integrate additional uncertainties into the model, which can be excluded relatively easily by a constant observation interval.

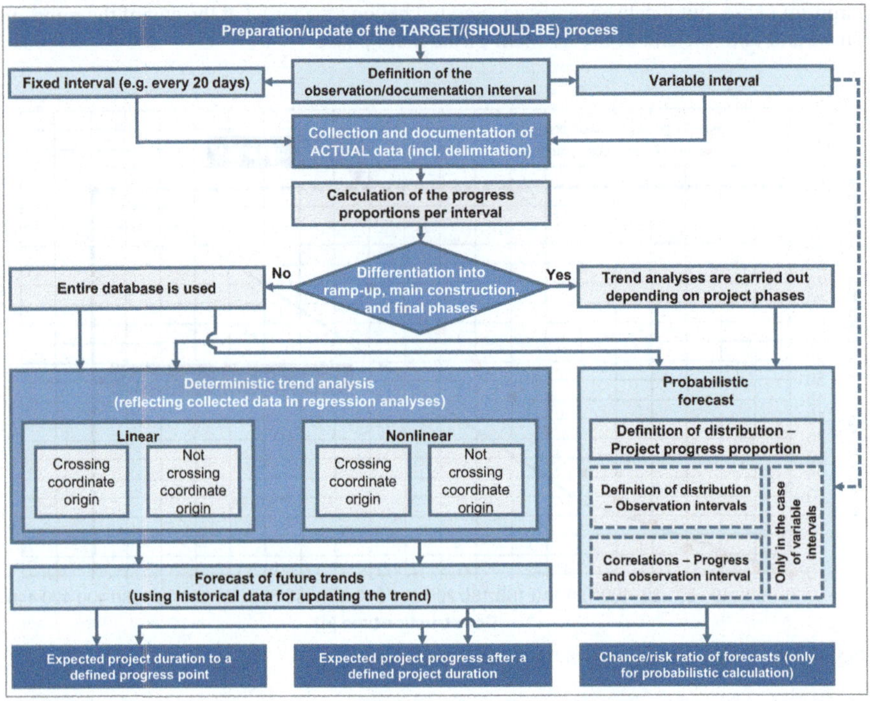

Fig. 10.160 Flow chart – Trend analyses and forecasts based on historical data

Deterministic trend analyses involve a forecast of future project progress based on historical data and the continuous update of the trend curve. Such analyses allow for specifying the project duration expected up to a defined milestone or, conversely, the project progress expected after a certain duration. Only probabilistic analyses will make it possible to also specify a chance/risk ratio for the derived forecast.

10.4.10.4 Calculation and Interpretation of Results

In the following sections, deterministic and probabilistic trend analyses are demonstrated, and a technically sound interpretation of the results is integrated into the process.

Linear trend analyses

Deterministic trend analyses (such as linear regression) can be used to perform a project completion forecast on the basis of historical data. This linear regression can be carried out such that the regression line passes through the coordinate origin or is adjusted to the existing data points in the best possible way independently of this definition.

Fig. 10.161 shows two regression lines based on the actual construction progress and extended up to a construction time of 400 days. For either of these lines, the equation is shown directly in the diagram.

In accordance with these two linear regressions, a forecast can be made for the completion of the reinforced concrete operations (i.e. 100% construction progress provided that there have been no quantity changes compared to the contractual target). Each of the two straight lines indicates a potential time to completion between approximately 320 and 350 days.

Compared to the actual construction progress, however, both regression lines can only be regarded as very rough approximations. A forecast for the next or next but one measuring point (i.e. 260 days after the start of construction work) could be sufficiently accurate using linear regression, but predictions based on linear regression should be interpreted with caution when it comes to determining the total time to completion.

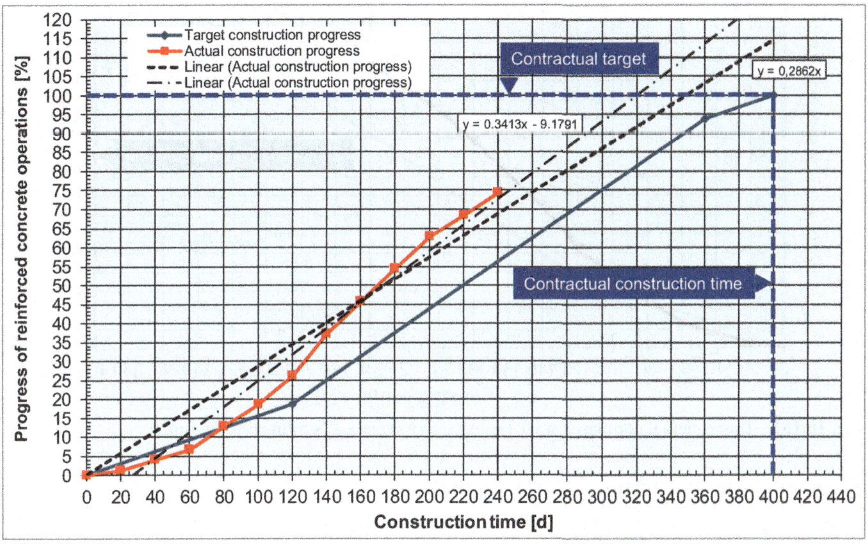

Fig. 10.161 Target-actual comparison – Construction progress – Linear regressions

Nonlinear trend analyses

Additional nonlinear regression analyses are performed in order to better approximate the sloped curve of the documented construction progress. Fig. 10.162 shows two results of these nonlinear regressions.

Both polygons are highly consistent with the existing database, and the two curves are also nearly congruent for the purpose of forecasting construction progress up to a construction time of approximately 270 days.

However, both curves appear to be unsuitable for longer-term predictions. The third-order polygon passes through the coordinate origin and shows a decrease in progress from a construction time of about 290 days. Such a progress trend is not plausible under the assumption of project completion when excluding dismantling and demolition activities during the progress of the project. This is why this trend curve does not allow for a long-term forecast.

The fifth-order polygon is also fairly consistent with the documented actual construction progress, but the trend curve shows a very steep increase in construction progress starting at about 290 days. Accordingly, we should expect completion of the reinforced concrete operations already after fewer than 320 days. This trend analysis also fails to provide the basis for a reliable long-term forecast. Contrary to the target construction progress, a very steep increase is entered in the diagram for the nonlinear regression in the final phase of the reinforced concrete operations. In practical terms, however, we should expect construction progress to level off towards the end of the construction period.

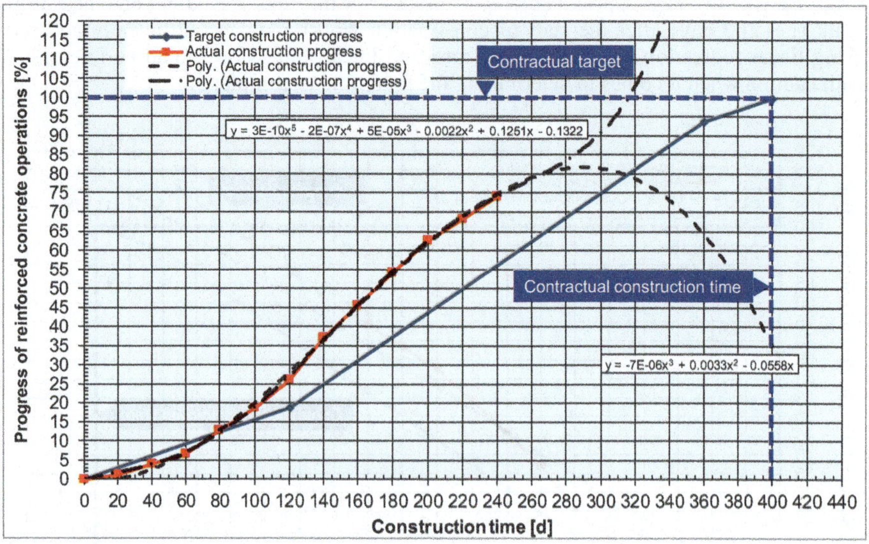

Fig. 10.162 Target-actual comparison – Construction progress – Nonlinear regressions

Trend analyses using Monte Carlo simulations

Another option for performing trend analyses or forecasts of future construction progress on the basis of the documented actual progress is to perform a Monte Carlo simulation in which the construction progress is calculated at certain intervals taking into account previous progress variations.[96)]

For this purpose, we first determine the differences of the percentage construction progress between the individual documented coordinate points (see Table 10.62, Column F).

96) In this example, the observation interval was defined as 20 (working) days.

Actual construction progress					
Coordinate points	Construction time	Poured concrete volume	Performance value	Share of poured concrete volume in total volume	Δ concrete proportion
A	B	C	D	E	F
0	0.00 d	0.00 m³		0.00%	
1	20.00 d	550.00 m³	27.50 m³/d	1.44%	1.44%
2	40.00 d	1,640.00 m³	54.50 m³/d	4.28%	2.85%
3	60.00 d	2,580.00 m³	47.00 m³/d	6.74%	2.46%
4	80.00 d	4,920.00 m³	117.00 m³/d	12.85%	6.11%
5	100.00 d	7,200.00 m³	114.00 m³/d	18.81%	5.96%
6	120.00 d	10,000.00 m³	140.00 m³/d	26.13%	7.32%
7	140.00 d	14,210.00 m³	210.50 m³/d	37.13%	11.00%
8	160.00 d	17,500.00 m³	164.50 m³/d	45.72%	8.60%
9	180.00 d	20,780.00 m³	164.00 m³/d	54.29%	8.57%
10	200.00 d	24,060.00 m³	164.00 m³/d	62.86%	8.57%
11	220.00 d	26,245.00 m³	109.25 m³/d	68.57%	5.71%
12	240.00 d	28,430.00 m³	109.25 m³/d	74.28%	5.71%

Table 10.62 Actual construction progress including differences

These differences are then used to perform a data fitting (KS test). In so doing, we define the lower limit of the distribution as being equal to 0% construction progress. We also determine that an upper limit of the distribution exists, but that this limit will have to be identified during the data fitting.

The differences taken from Column F of Table 10.62 are displayed as a histogram in Fig. 10.163. This diagram also shows the fitted distribution, which for this database is a triangular distribution (KS test). The distribution of the percentage increase in the amount of concrete placed was fitted on the basis of historical data. It is used to calculate potential future construction progress trends.

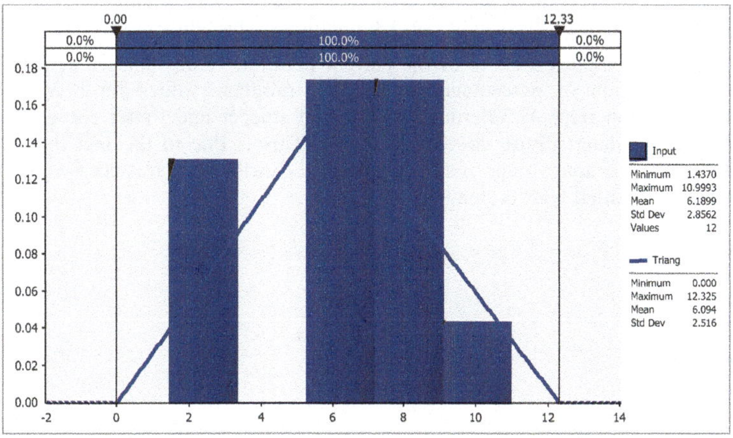

Fig. 10.163 Data fitting of the percentage progress increase of reinforced concrete operations based on historical data

For the forecast of future trends, the y values (i.e. the percentage progress of the reinforced concrete operations) of future coordinate points are calculated. The x values (construction times) correspond to the pattern of the documented construction progress and follow a 20-day interval.

Each of the coordinate points of the total poured concrete volume builds upon the construction progress at the preceding coordinate point. In the course of the Monte Carlo simulation, a random value is selected from the fitted distribution, which represents the percentage increase in construction progress. If several future coordinate points are calculated, it is crucial that a separate fitted distribution is stored for each coordinate point to be calculated. Even though these distributions all rely on the same database and provide the same fitting result, different numerical values are selected from the range of the distributions in any iteration step of the simulation. If this is not the case and if the same fitted distribution is accessed for each progress calculation, this will result in a straight line with a constant gradient in each iteration step because any of these steps integrates the same percentage increases for all future coordinate points.

Any increase in construction progress that varies within certain limits appears to be more realistic and is also more in line with the recorded historical data than a constant increase in the poured concrete volume.

The simulation results in distributions for the expected construction progress for each of the selected 20-day intervals. We can use these distributions to derive statistical parameters such as the 5% and 95% quantiles, the mean, or the standard deviation.

If these values are added to the diagram showing the actual construction progress documented so far, this will result in a scattering of the potential future progress trends on the basis of the entirety of the historical project data pertaining to the increase in the installed reinforced concrete volume.

Fig. 10.164 shows the documented actual construction progress for construction times from 0 to 240 days. The other curves starting from a construction time of 240 days correspond to the mean, the mean plus/minus a standard deviation, and the 5% and 95% quantiles as the results of the Monte Carlo simulation of the future construction progress.

This figure also includes a curve of the current iteration value. This value represents a random combination of percentage construction progress values for a Monte Carlo simulation iteration step. An alternating pattern of steeper and flatter curves can occur between the calculated future coordinate points. This is due to the fact that, for each coordinate point, random values of several identical distributions are accessed, rather than one and the same fitted distribution.

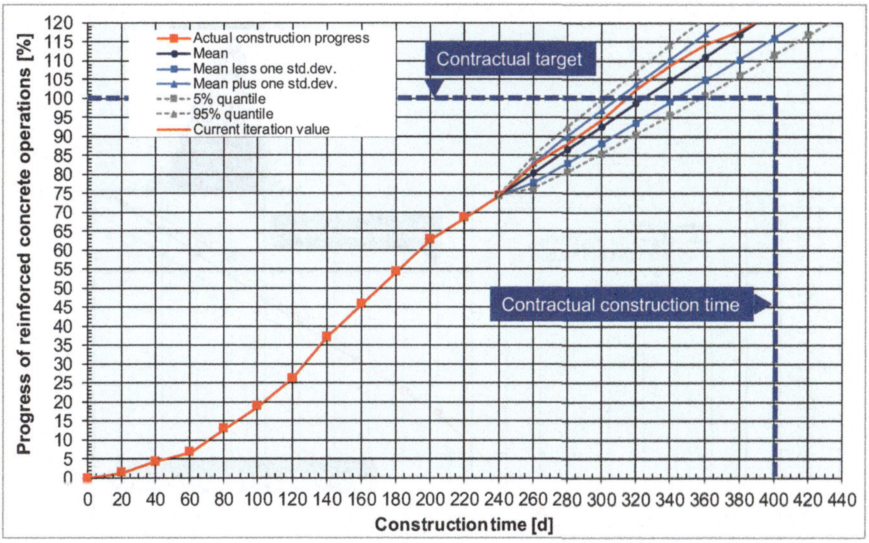

Fig. 10.164 Actual construction progress with forecast of subsequent construction progress

Of particular interest when interpreting the diagram shown is the question of the expected construction time until the contractually agreed target will be reached (i.e. 100% construction progress). For this purpose, the results of the Monte Carlo simulation are evaluated by linear interpolation between the construction time intervals (i.e. 20 days) in such a way that the individual iteration results for achieving the contractual target can be presented in the form of a histogram (see Fig. 10.165). We find that the 5% and 95% quantile curves from the diagram correspond exactly to the sliders displayed in the histogram.

Fig. 10.166 also separately shows the histogram for the predicted construction time for achieving the contractual target (i.e. 100% construction progress), revealing that the time to completion of the reinforced concrete operations will lie between approximately 300 and 357 days with a probability of 90%.

When interpreting the results of the Monte Carlo simulation, it may also be of interest to know how the reinforced concrete operations are expected to progress up to a certain point during the construction period.

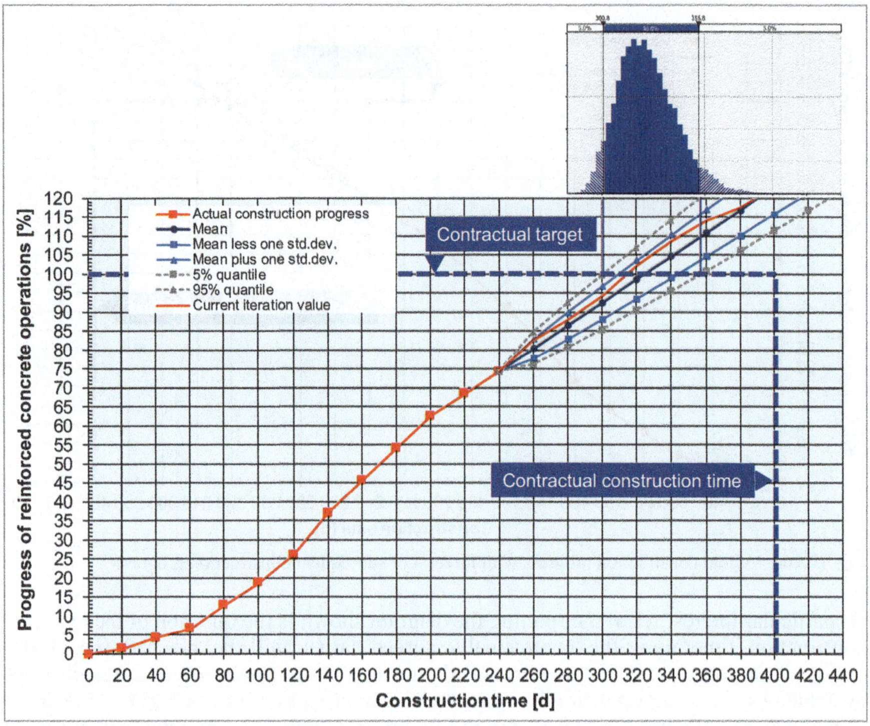

Fig. 10.165 Actual construction progress including forecast of subsequent construction progress – Histogram for the construction time upon achieving the contractual target

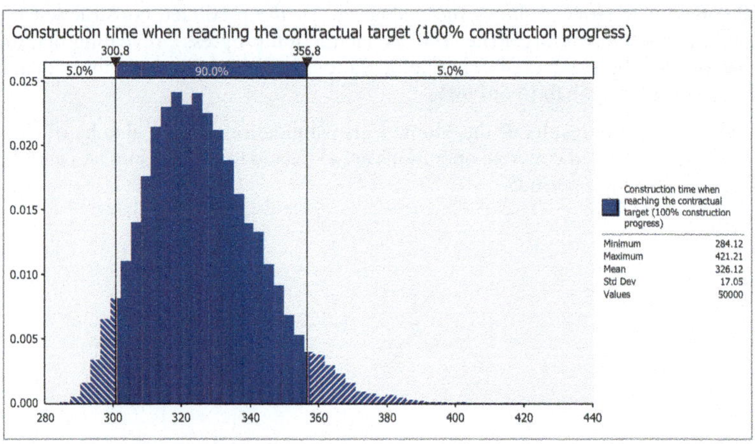

Fig. 10.166 Histogram representing the construction time when reaching the contractual target (100% construction progress)

By way of example, we determine the construction progress predicted after 290 days (see Fig. 10.167). This query also uses linear interpolation between the calculated future coordinate points to obtain a histogram based on the potential progress curves taken from the Monte Carlo simulation.

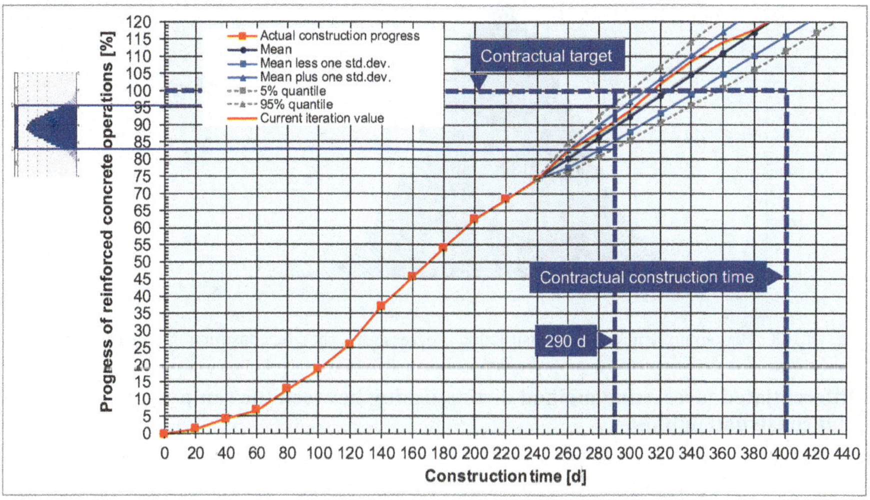

Fig. 10.167 Actual construction progress including forecast of subsequent construction progress – Histogram of construction progress for a specific construction time (290 d)

Fig. 10.168 includes the histogram representing the forecast 290-day construction progress. The histogram shows that, under the assumptions made, a degree of completion of over 96% will only be exceeded with a 5% probability. Based on historical data, the probability that less than 83% of the reinforced concrete operations will be completed after 290 days (corresponding to 72.5% of the specified construction time) also amounts to approximately 5%.

The greater the number of available data points and the nearer into the future the forecast horizon is, the higher the accuracy of the forecasts will be.

The above calculations rely on the fact that the installed concrete volume is expected to show the same behavior for the future coordinate points as documented for the entirety of historical data.

Further detailing makes it possible to consider, for the final phase, only those percentage increases in progress for data fitting that were documented in the ramp-up phase. This approach takes into account the fact that the final phase will be characterized by a continuous reduction in the utilization of resources, and thus also in the average daily performance.

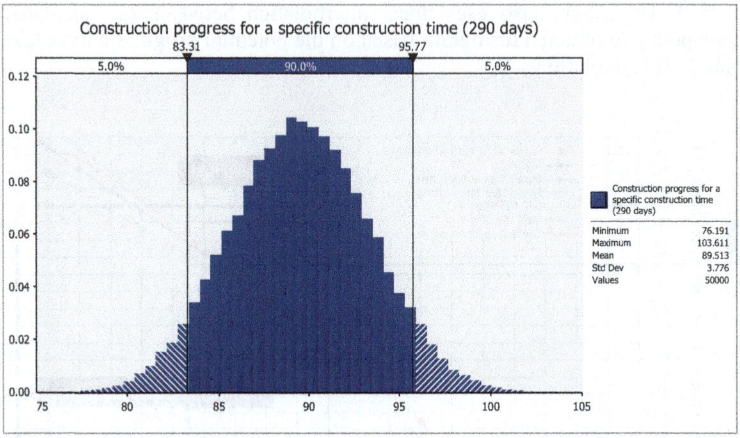

Fig. 10.168 Histogram representing construction progress for a specific construction time (290 days)

Differentiated forecasting method distinguishing between the ramp-up, main, and final construction phases

For the forecasts shown above, we assumed that the increase in the amount of concrete poured would lie within certain ranges based on the previously documented construction progress. This approach failed to take account of the fact that different progress gradients should be expected in the respective construction phases (ramp-up, main, and final) both in the target and in the actual consideration. It is particularly the final phase that has so far been neglected in the previous considerations.

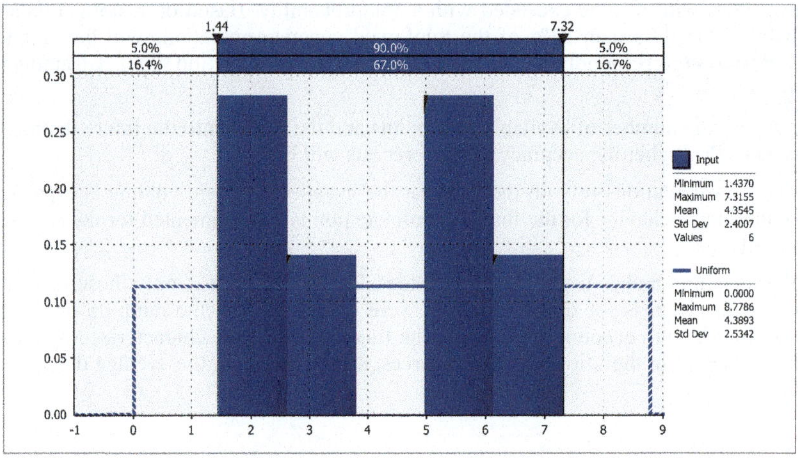

Fig. 10.169 Data fitting of the percentage progress increase of reinforced concrete operations based on historical ramp-up phase data (0 to 120 d)

For a further probabilistic forecast, two data fittings of the documented percentage construction progress are performed in the first step. The first data fitting is based only on the percentage progress of the reinforced concrete operations in the ramp-up phase (reaching from the start of construction to a construction time of 120 days). The best possible approximation to this limited database is a rectangular distribution (KS test), see Fig. 10.169.

The second data fitting is based on the remaining documented percentage progress (120 to 240 days).[97] In this case, the best approximation is a triangular distribution (KS test), see Fig. 10.170.

For both data fittings, we determined that the minimum distribution is 0% (corresponding to a complete standstill of construction work). The maximum results from the data fitting, where an existing upper limit to the distribution was assumed as a restriction (no open-ended distribution allowed).

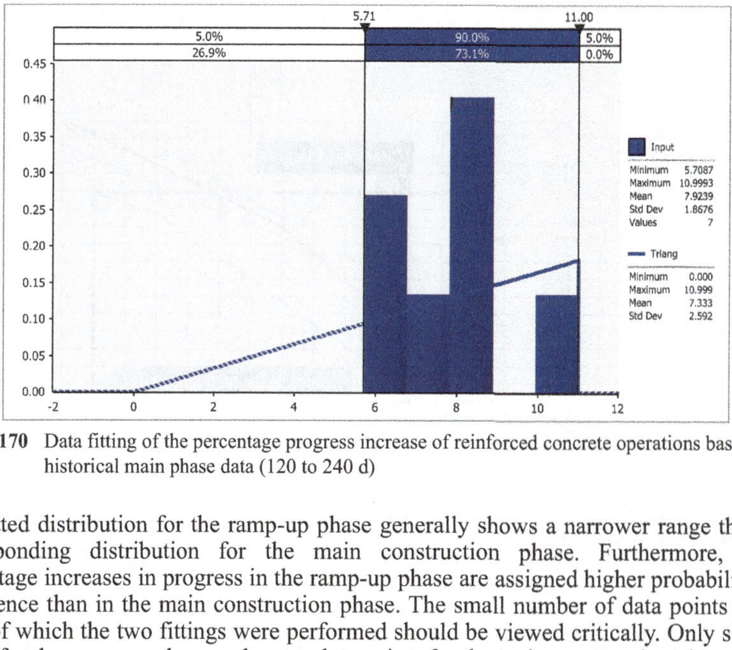

Fig. 10.170 Data fitting of the percentage progress increase of reinforced concrete operations based on historical main phase data (120 to 240 d)

The fitted distribution for the ramp-up phase generally shows a narrower range than the corresponding distribution for the main construction phase. Furthermore, lower percentage increases in progress in the ramp-up phase are assigned higher probabilities of occurrence than in the main construction phase. The small number of data points on the basis of which the two fittings were performed should be viewed critically. Only six data points for the ramp-up phase and seven data points for the main construction phase would permit only a rough approximation of the actual distribution of the percentage increase in progress of the two phases. The greater the number of documented progress increases (i.e. shorter intervals between documentation times) the data fitting can be based on, the more accurately the theoretical distributions can be adjusted to the database.

In the following steps, we use the two distributions as the basis for forecasting the subsequent progress of the project. Again, in this simulation, each of the data points lying in the future is assigned its own distribution function for the percentage project progress in order to be able to simulate random project progress within the defined ranges.

97) The percentage increase in progress at a construction time of 120 days is thus incorporated into the data fitting of both the ramp-up and the main construction phase.

Since the project progress remains above the scheduled target up to a construction time of 240 days, we assume that the final phase will already be initiated after 280 days, such as by reducing the number of key pieces of equipment used and/or the number of workers employed on-site. For the simulation, this means that up to a construction time of 280 days, we expect a project progress based on the data fitting for the main construction phase. After this point in time, the fitted distribution of the ramp-up phase is used to determine the further progress of the project.

Fig. 10.171 illustrates the shift from the main to the final phase by a dip in the forecast curves visible at a construction time of 280 days.

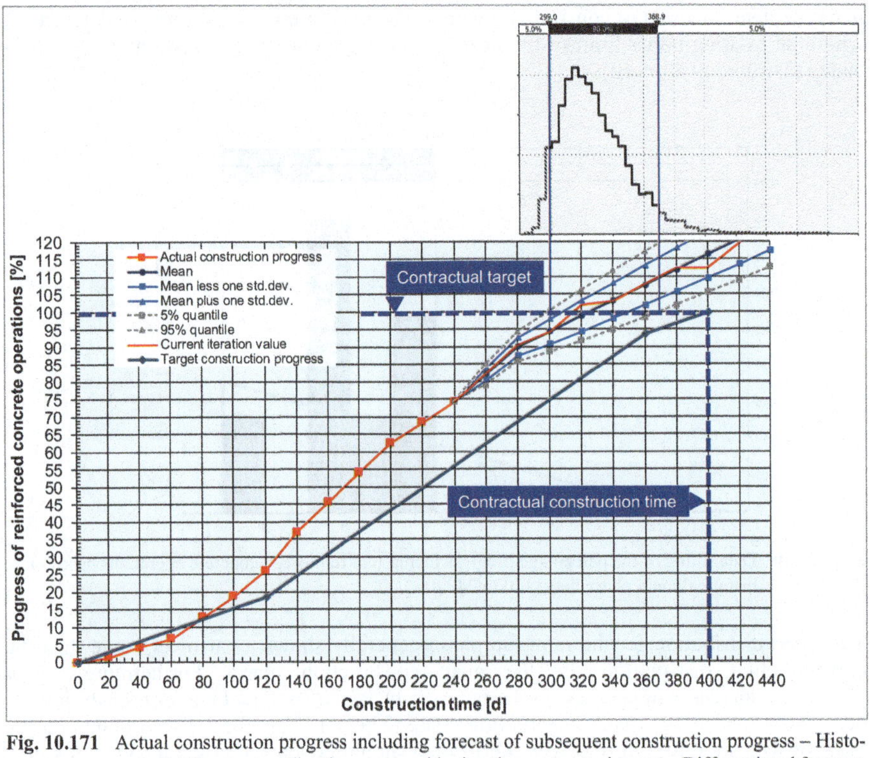

Fig. 10.171 Actual construction progress including forecast of subsequent construction progress – Histogram for the construction time upon achieving the contractual target – Differentiated forecast method

Overall, we find that considering the different phases in the simulation enables us to predict a project progress trend that better approximates the scheduled target progress as well as day-to-day construction reality in qualitative terms.

Fig. 10.172 compares the two histograms representing the predicted construction time passed when achieving the contractual target (100% construction progress). Analyzing the histogram derived from the differentiated forecast method delivers a probability of approximately 99.3% for the contractually agreed construction time of 400 days being adhered to.

The maximum construction time obtained in the first simulation (without differentiating into main and final phases) after 50,000 iterations (Latin Hypercube sampling method) amounts to approximately 421 days (see the key in the top right corner of Fig. 10.172). Applying differentiated forecast methods results in a maximum of about 482 days (see the key in the bottom right corner of Fig. 10.172).

The differentiated forecast method thus yields a wider range of possible progress increases and thus also a potentially wider scope of works or services.

The differentiated method depends on the future point in time at which the shift from the main construction phase to the final phase is defined. Depending on when the final phase is actually initiated, for instance through the reduction in equipment and/or workforce deployment, steeper or flatter trend curves may result for the predicted construction progress. The manageable transition point from the main construction phase to the final phase makes it possible to respond to an exceedingly fast or slow construction progress in the execution phase, and thus to intervene in a controlled manner.

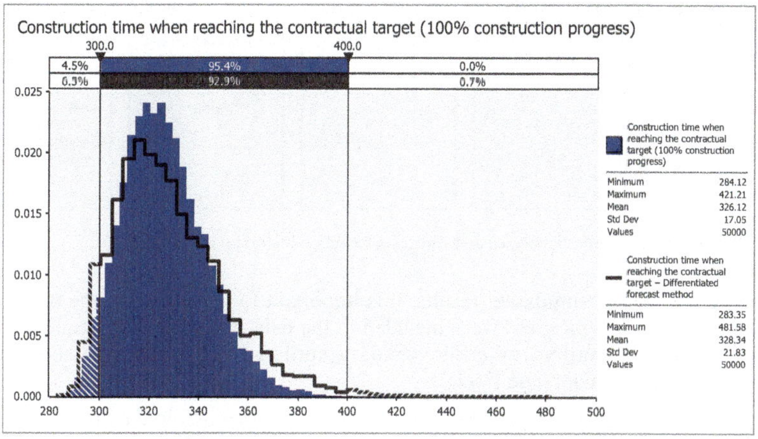

Fig. 10.172 Comparison of histograms representing the construction time when reaching the contractual target (100% construction progress)

In this example, no quantity changes are taken into account that would trigger an upward or downward adjustment of the contractually agreed target. If, for example, the volume of reinforced concrete increases by 10% compared to the specified quantity, the diagrams would show a progress of 110% instead of 100% on the ordinate. If the construction time is not adjusted despite the change in quantity, the probabilistic trend analysis can be used to draw direct conclusions regarding the chance/risk ratio associated with the quantity increase with a view to ensuring compliance with the specified construction time.

10.4.10.5 Benefit for the Contractor

Fig. 10.173 illustrates the benefits of trend analyses and probabilistic forecasting methods for the contractor. These methods constitute essential controlling instruments in the execution of construction projects. They focus not only on activities and processes completed in past periods. Instead, they also use project-specific historical data derived from the past to draw conclusions regarding possible future trends. Applying probabilistic forecasting methods not only allows for individual deterministic numerical values to be used as point estimators. Such methods also permit assessments relative to the expected chance/risk ratio. If, for example, the risk of a specified construction time being exceeded is too high, countermeasures must be initiated in order to achieve a turnaround in the progress of the project.

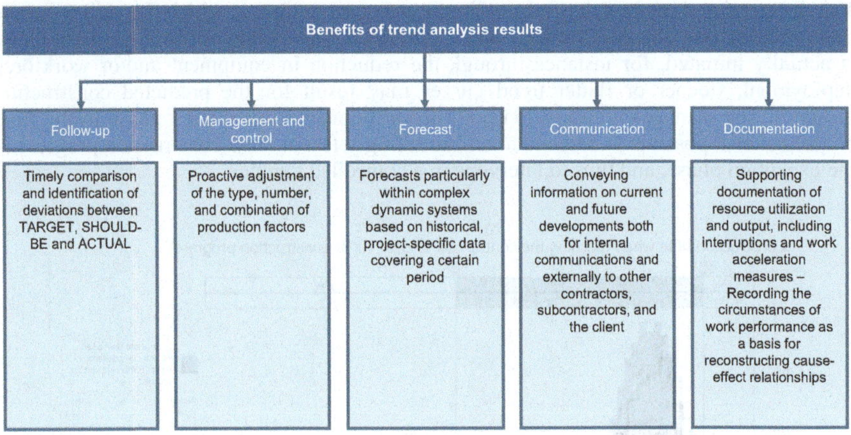

Fig. 10.173 Benefits of trend analyses and forecasts based on historical data

When interpreting the simulation results, it is important to note whether the documented percentage changes in progress were included in the data fitting, and subsequently in the simulation, in their entirety or differentiated according to the individual construction phases (i.e. ramp-up, main, and final).

In the above example, all available data points were incorporated in the data fitting of the percentage progress change. If, in the course of the project, individual, particularly small or large percentage changes can be traced back to a specific cause that is no longer expected to occur during further execution, an elimination of this data record from the forecast calculation or from the fitting may be considered in order to limit the range of possible changes in progress and to obtain more realistic forecasts for the period considered.

The procedure explained above and the methodology used for reviewing the progress of reinforced concrete operations can also be transferred to other works or problems, such as with regard to cost control or consumption of paid working hours[98)].

Forecast accuracy will essentially depend on the quality and delimitation of the underlying documentation. If the latter is not done at regular intervals, as in the given example, a second data fitting for the differences of the observation times must be integrated into the calculation. In so doing, considerations regarding possible correlations between the interval of observation times and the percentage of project progress should also be taken into account. A strong positive correlation coefficient is assumed in this case because

[98)] In this case, the actual number of paid working hours is usually easy to glean from the daily work reports on a continuous basis.

longer intervals between two measuring or observation points also tend to be associated with a more considerable construction progress.

If the trend analysis shows that the project can probably be completed much earlier, the speed of work can be reduced unless a shorter construction period is rewarded accordingly by the client in line with the principle of value engineering. When reducing this progress rate, the combination of production factors will be optimized, including with regard to team sizes, stocking quantities, or pieces of equipment used.

Appropriate countermeasures must be implemented if the trend analysis shows that the completion of the project will take longer than the contractually specified construction period. Such measures can include, for example, an extension of the main construction phase and/or the use of additional equipment or staff. In this respect, it should be noted that losses of productivity may occur if specific limits for the optimal utilization of the production factors are violated, including the minimum work area or available crane capacity.

Overall, the results of probabilistic trend analyses support internal as well as external communications regarding current and future developments while opening up the potential (thanks to a transparent discussion of uncertainties) to contribute to a partnership-based approach to project management.

10.4.11 Weather Impact

A construction site constitutes a specific type of production site that is temporary in nature. The daily achievable performance during the execution of a construction project (PPH 4, see Fig. 10.174) will be determined by internal and external influences. One of these factors is the outdoor temperature, which has a significant effect on the ability of workers to perform their activities in an outdoor environment. If temperatures deviate from an optimal value in either upward or downward direction, losses of productivity will occur that trigger an increase in the labor consumption rate.

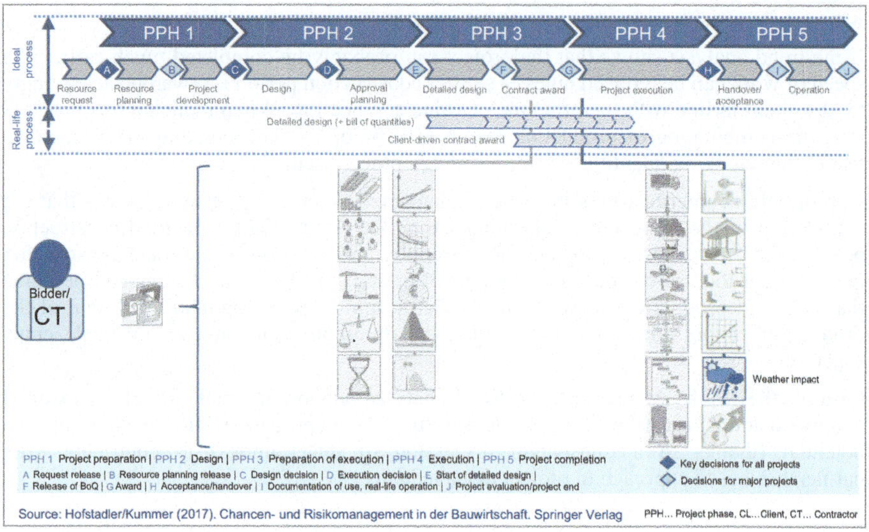

Fig. 10.174 Allocation of example in the project sequence

We calculate an example in order to demonstrate how the choice of the construction start date will affect the total number of paid working hours. Probabilistic calculations are performed to address the uncertainties included in the equations and to illustrate the results in histograms and curves. These calculations rely on Monte Carlo simulations, which make it possible to assign distribution functions to individual input parameters used in the calculations.

We show how a beneficial choice of the construction start date with respect to the prevailing seasonal conditions can unleash saving potentials. In the event of any disruptions or delays, potential effects on the labor consumption rate can be assessed and forecast early in the process. In this context, the timely detection of such disruptions or delays and their systematic assessment in terms of chances and risks paves the way to identifying potentially required countermeasures.

10.4.11.1 Basics

Weather conditions influence the performance capacity of the production factors. In the case of outdoor activities, prevailing external temperatures are one of the factors that determine the ability of workers to perform. At relatively low temperatures, people who carry out work on the construction site are no longer able to do so at their "normal" productivity level because, for instance, their movement patterns are obstructed by warmer work wear (such as thicker gloves). Productivity is also reduced by breaks for warming up or repositioning clothes. In addition, on-site heating equipment may create additional obstacles to the movement patterns of workers. When working at low temperatures, the accuracy and swiftness with which individual work steps are completed is reduced, which leads to losses of productivity.

In the hot season, productivity is lowered, for instance, by additional breaks to relax and drink, which lead to a general decrease in the performance capacity of workers. Higher temperatures may also necessitate wearing of protective gloves when working with steel, for example, which, again, can potentially slow down movement patterns. Personal protective equipment has a disruptive effect and increases the heat load.

In the literature, information on the optimal daytime temperature for different types of work can be found in the works of *Koehn/Brown*, *Thomas/Yiakoumis*, *Oglesby et al.*, and *Fetzner*, to name but a few sources.

The analyses conducted by *Koehn/Brown* suggest an optimal temperature range between 10°C and 21.1°C, independently of relative humidity.[99)]

Thomas/Yiakoumis state an optimal daytime temperature range for labor productivity between 10°C and 15.6°C at a relative humidity between 5% and 75%.[100)]

Oglesby et al. specify an optimal daytime temperature ranging from 10°C to 21°C.[101)]

Fetzner differentiates between specific activities for the purpose of defining optimal daytime temperatures. He states 14°C for masonry operations, 10.5°C for concreting, and 6.5°C for formwork and reinforcing operations.[102)]

Suboptimal daytime temperatures influence the labor productivity of workers and have an effect on the required paid working hours [wh], for instance on the number of hours needed to produce one cubic meter of reinforced concrete. This ratio of paid working hours to the unit of quantity produced is referred to as labor consumption rate [wh/m³].

Labor consumption rates and productivity are directly correlated to one another. Productivity is reduced if the labor consumption rate increases, and vice versa. The relative change in productivity ΔPL [%] can be determined using Eq. (10-47). For this purpose, the difference between the baseline and the actual productivity is calculated and divided by the baseline productivity. The baseline productivity follows from the reciprocal of the target labor consumption rate LCR_{TARG} [wh/m³], while the actual productivity derives from the reciprocal of the actual labor consumption rate LCR_{ACT} [wh/m³].[103)]

$$\Delta PL = \left(\frac{\frac{1}{LCR_{TARG}} - \frac{1}{LCR_{ACT}}}{\frac{1}{LCR_{ACT}}} \right) \cdot 100\% \qquad (10\text{-}47)$$

In 2012/2013, an expert survey was conducted, at Graz University of Technology, among 35 Austrian and German respondents from construction trades and industry who had an average professional experience of 17 years (minimum: 5 years; maximum: 43 years). One of the outcomes of this survey was the determination of the optimal daytime temperature at 18°C (robust mean estimator), where workers enjoy ideal conditions to achieve their normal productivity level.[104)]

99) Cf. Koehn/Brown (1985), p. 129ff.
100) Cf. Thomas/Yiakoumis (1987), p. 623ff.
101) Cf. Oglesby et al. (1989), p. 263f.
102) Cf. Fetzner (2007), p. 170ff.
103) Cf. Hofstadler (2014), p. 35
104) Cf. ibid., p. 231ff.

In addition, a curve was determined for the increase in the labor consumption rate during reinforced concrete operations that occurs if the daytime temperature deviates from this optimal value (see Fig. 10.175).

Robust mean estimators are determined (using *Huber's* M-estimator method) to prevent biased outcomes and to avoid the deletion of outliers indicated by respondents in the survey. This approach ensures that outliers and extreme values are included in the averaging only with a low weight. It is particularly well-suited for small databases.

We determine the labor consumption rate increase $LCR_{INC,TMP}$ [%] derived from the M-estimators stated in the expert responses by inserting the temperature TMP [°C] in Eq. (10-48). This approach is valid for -14°C ≤ TMP ≤ 17.5°C and 18.5°C ≤ TMP ≤ 38°C.[105)]

$$\Delta LCR_{INC,TMP} = 0.000466 \cdot TMP^3 + 0.048878 \cdot TMP^2 - 2.011685 \cdot TMP + 17.597270 \qquad (10\text{-}48)$$

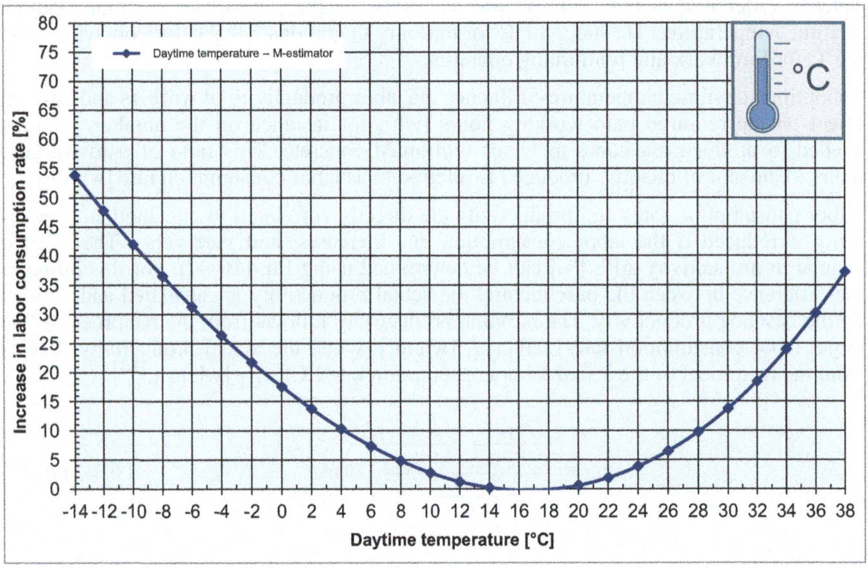

Fig. 10.175 Labor consumption rate increases depending on the mean daytime temperature – Reinforced concrete operations[106)]

10.4.11.2 Task

The following sections discuss a worked example that uses probabilistic input parameters to consider losses of productivity or labor consumption rate increases due to suboptimal mean daytime temperatures.

[105)] Cf. Hofstadler (2014), p. 434f.

[106)] Cf. Hofstadler (2014), p. 434

This example relates to the Opera House project in Linz, Austria, and exclusively analyzes reinforced concrete operations. The total labor consumption rate and the specified construction time were assumed by the authors (see Table 10.63).[107)]

No.	Item	Value
0	A	B
1	Concrete volume	38,275.00 m³
2	Total labor consumption rate	6.00 wh/m³
3	Construction time – reinforced concrete operations	400.00 d

Table 10.63 Deterministic input parameters

This analysis deals with the effects that a varying start of construction in different months of the year would have on the total labor consumption rate and, subsequently, on the total number of paid working hours and the chance/risk ratio. For this purpose, we base our calculation on historical temperature data. This data for the city of Linz (1971-2000 database) were provided by the Austrian *Zentralanstalt für Meteorologie und Geodynamik* (ZAMG; Central Agency for Meteorology and Geodynamics), and are publicly accessible online.[108)] Fig. 10.176 shows the mean daytime temperature curve on a monthly basis, means of daily minima and maxima, and the absolute minima and maxima.

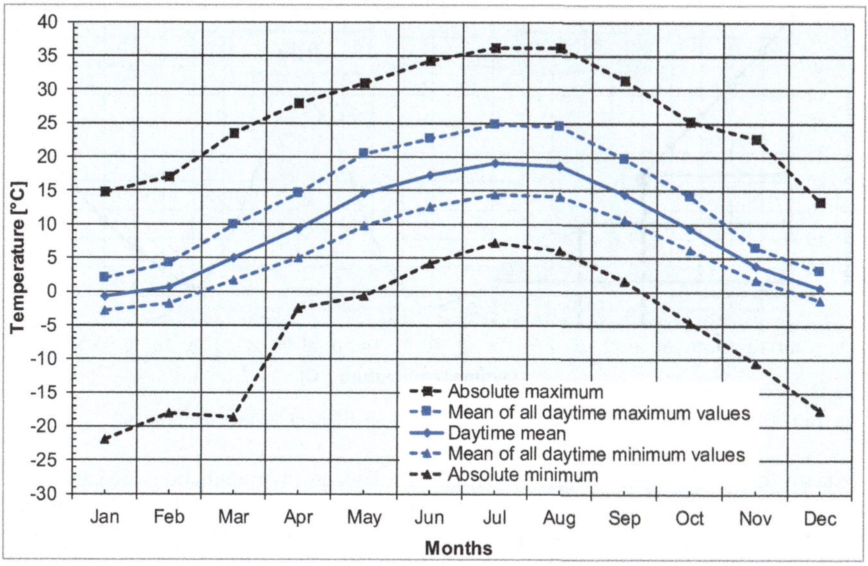

Fig. 10.176 Temperature curve for the Linz-Stadt measuring station

107) Cf. Kummer/Hofstadler (2015), p. 851ff.

108) http://www.zamg.ac.at/fix/klima/oe71-00/klima2000/klimadaten_oesterreich_1971_frame1.htm. Accessed December 30, 2020

10.4.11.3 Modeling

In the probabilistic calculation, normal distributions are used to represent the temperature curve, where the daily average per month (see Fig. 10.177) forms the mean value. For the sake of simplicity, we define the mean difference between the daily average and the mean of the daily minima and maxima as the standard deviation. If all temperature measurement data records are available within one period considered, these can be used directly for the calculations or be subjected to data fitting.

Fig. 10.177 illustrates the different effects of the daytime temperature on the labor consumption rate increase. Whereas, in January, when temperatures are low, larger increases in the labor consumption rate are expected due to falling below the optimal temperature range, the range of temperatures expected in July has a more favorable effect on labor productivity. This is why labor consumption rate increases will be smaller in July.

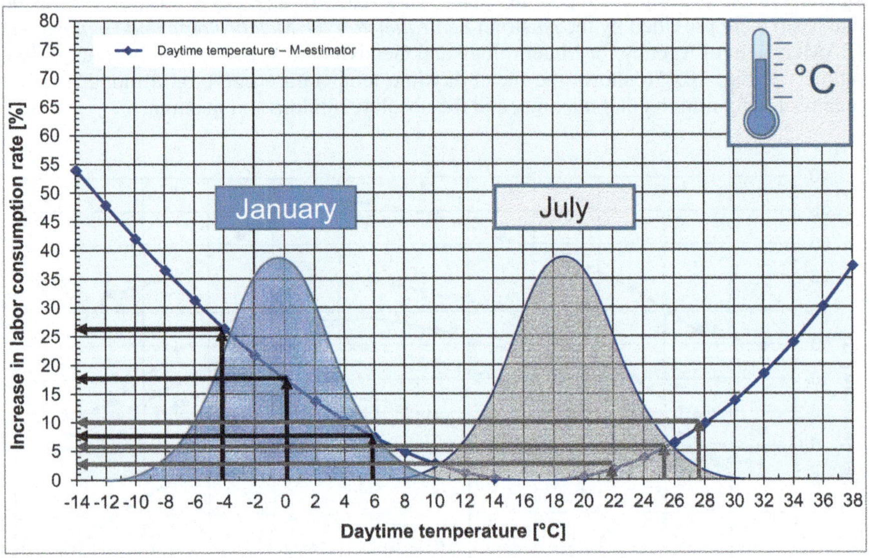

Fig. 10.177 Labor consumption rate increases depending on different daytime temperatures

To show the effects of different construction start dates, twelve simulations are carried out for all months of the year (start of construction in January, February, March, etc.). In Fig. 10.178, the constant construction times (400 days) are represented by bars in relation to different construction start dates. In this diagram, each of the bars represents one of the twelve simulations performed. We simplify the model and assume that each month has an average of 4.33 weeks with five working days each (= 21.65 days/month).[109)]

Fig. 10.179 shows the flow chart for this calculation. Starting from the historical temperature data, normal mean temperature distributions are modeled for each month.

Each of these different temperature curves has a distinct influence on the labor consumption rate increase per month. In months in which particularly low or particularly

[109)] It would be possible to develop a refined model incorporating public holidays, seasonal closure periods, etc., but this would require a much greater modeling effort.

high temperatures are expected, the effect on the consumption rate will obviously be higher than in months in which the expected temperatures come close to the optimal range. If we vary the construction start date in each of the twelve simulations, several periods with unfavorable temperature levels may occur during construction work, depending on the specified construction time. This leads to a greater labor consumption rate increase in overall terms.

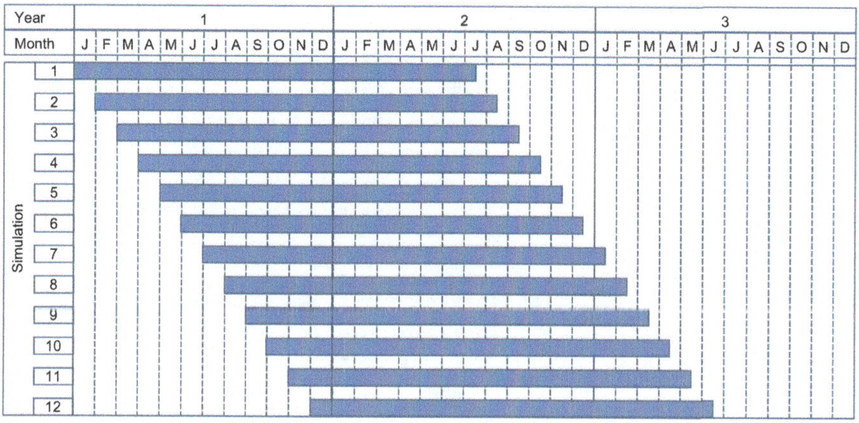

Fig. 10.178 Extension of construction time over the months depending on varied construction start dates

We use Eq. (10-48) for calculating the percentage consumption rate increases for each month. The monthly consumption rate increases can be used as a starting point for determining the mean labor consumption rate increases per simulation depending on the given construction time (which is a deterministic value).

The total construction time amounts to approximately 18.48 months [= 400 d / (4.33 weeks/month * 5 d/week)]. If the construction time extends over the entire month under consideration, the total percentage increase in the consumption rate of the month is included in the subsequent calculation. The last month is not considered in its entirety but only at approximately 47.5% {= [400 d / (4.33 weeks/month * 5 d/week) - 18 months] * 100%} when determining the mean labor consumption rate. This is why, for any last month, we consider the calculated labor consumption rate increase only with the corresponding percentage in the further calculation steps.

In each of the simulations, this mean increase is added to the basic labor consumption rate (excluding losses of productivity). We then multiply the resulting increased labor consumption rate by the concrete volume to be produced (deterministic value) for each of the simulations, which gives the total number of paid working hours depending on the relevant construction start date.

For easier interpretation of the results, we present the twelve histograms for the individual paid working hour totals in a diagram including selected statistical parameters (mean value, quantiles, etc.) in order to determine the month to be selected for the construction start date so as to be able to perform the work consuming the lowest paid working hour total.

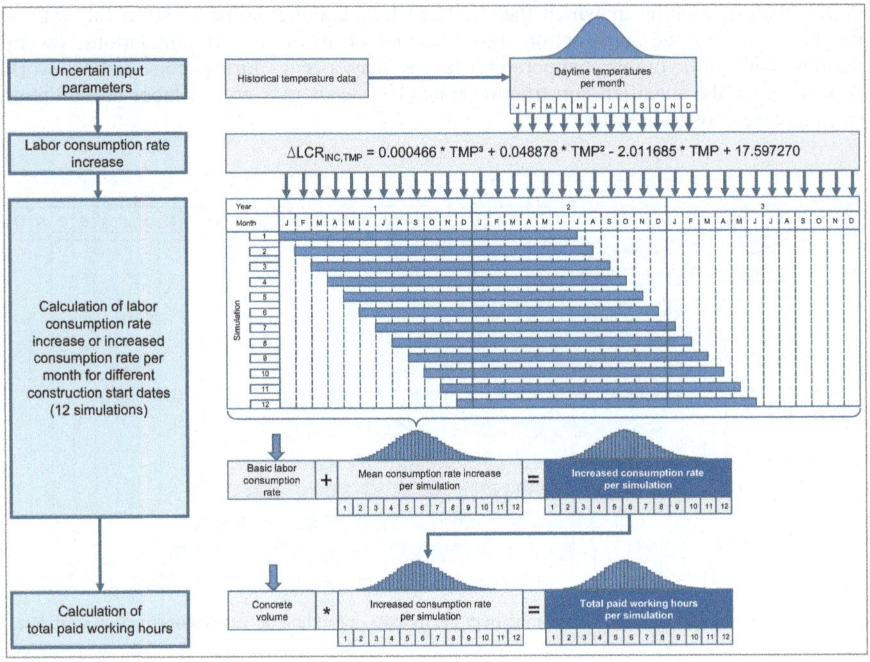

Fig. 10.179 Calculation flow chart

10.4.11.4 Calculation and Interpretation of Results

For the probabilistic calculation, normal temperature distributions were applied for each month, and the mean increased labor consumption rates were calculated for the reinforced concrete operations. Multiplication by the concrete volume results in the paid working hour totals expressed as histograms depending on the month in which construction commences. Fig. 10.180 shows the curves of the simple standard deviation from the mean as well as the 5% and 95% quantiles. This diagram also includes the histogram of the paid working hour total for a construction start in April. The 5% quantile is 239,384 wh while the 95% quantile amounts to 247,832 wh.

Accordingly, we expect the smallest paid working hour total if construction begins in April, and the greatest if work starts in October. The difference of the paid working hour totals relative to the mean values would be approximately 7,841 wh (= 251,058 wh-243,217 wh). At a mean wage price of €40.00/wh, this is equivalent to a wage difference of €313,640.00 (= 7,841 wh* €40.00/wh).

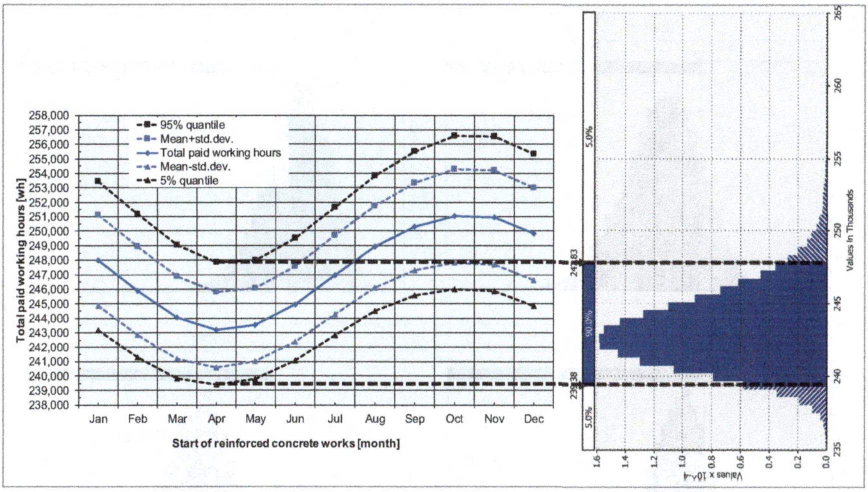

Fig. 10.180 Total number of paid working hours depending on the construction start date

Assuming a paid working hour total of 245,000 wh, the chance/risk ratio will also change depending on the start date of construction. Fig. 10.181 shows the chance/risk ratios for selected start months. We notice clear differences in the chances and risks that a company would take depending on the construction start date if it were to base its costing of the reinforced concrete operations on a paid working hour total of 245,000 wh. Whereas the chance of underrunning this figure amounts to about 77% if construction commences in April, the corresponding chance shrinks to just 2% if construction begins later, for instance in October.

If the contractor assumes that construction will begin in April and if work cannot start before October due to delays, this will significantly change the chance/risk ratio for the contractor, unless the relevant costing is adjusted to the changed circumstances of work performance.

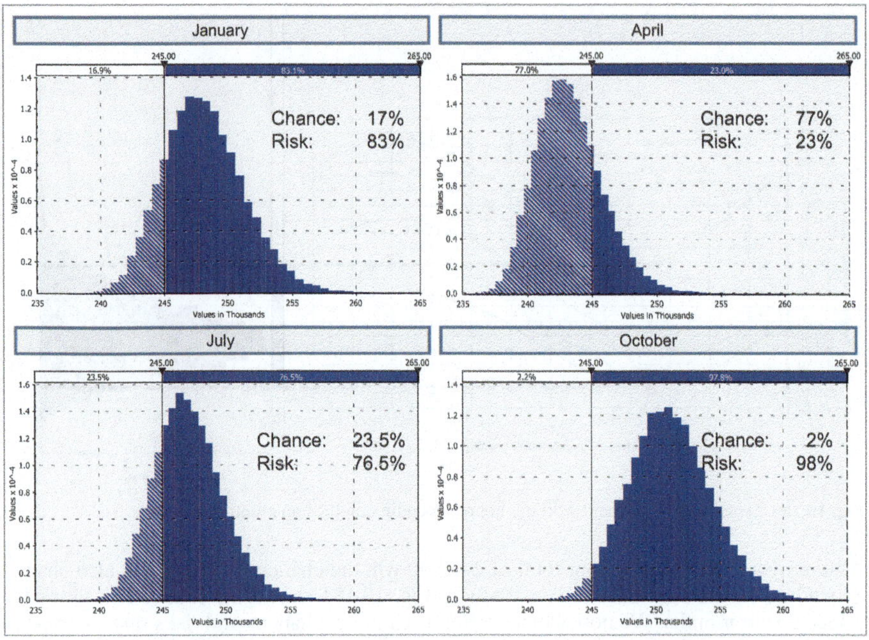

Fig. 10.181 Chance/risk ratios for the total number of paid working hours depending on the construction start date

10.4.11.5 Benefit for the Contractor

Any deviation from the optimal mean daytime temperature will lead to losses of productivity and/or labor consumption rate increases. Taking into account the functional relationship noted in the expert survey, the method described above is suitable for adequately integrating the effects of temperature on the ability of workers to perform into practical calculations.

Incorporating historical and locally adjusted temperature data makes it possible to analyze the influence of the outdoor temperature on the total number of paid working hours. Monte Carlo simulations can be carried out to systematically take into account the associated uncertainties in the equations. Histograms help to assess the chances and risks for estimated paid working hour totals and to transparently present and communicate the effects of postponing the start of construction. Decisions can be made on the basis of a selected chance/risk ratio while making it possible to identify potentially required work acceleration measures at an early stage in the process.

10.4.12 Determination of Additional Costs

For the bidder (and subsequent contractor), the bill of quantities forming part of the structural building specification is the most important basis upon which to proceed with costing and pricing as well as process planning and work scheduling. In this document, the client must describe all circumstances and conditions for works to be performed so as to enable the bidder to prepare their proposal in accordance with the objective explanatory content of the bid documents. This means that it is in the client's sphere of responsibility to specify the works to be performed in sufficient detail. The Austrian OENORM B 2110 and B 2118 standards define the concept of sphere as a "contractually or legally defined risk area of the relevant party to the contract". If, for instance, the client merges several types of work or service for different structural components, trades, soil types, rebar diameters, etc. into a single item, and if the costing assumptions for such combined items are not disclosed, the bidder will be forced to make their own free costing assumptions in order to be able to perform their costing exercise at all.

Any change in the proportions of the works included in the merged item compared to the bidder's original pricing assumptions will lead to a changed pricing structure associated with potentially higher or lower costs compared to the bid price.

The example given for determining the additional costs can be assigned to PPH 4 (see Fig. 10.182) and is based on the example discussed in Section 10.3.1.

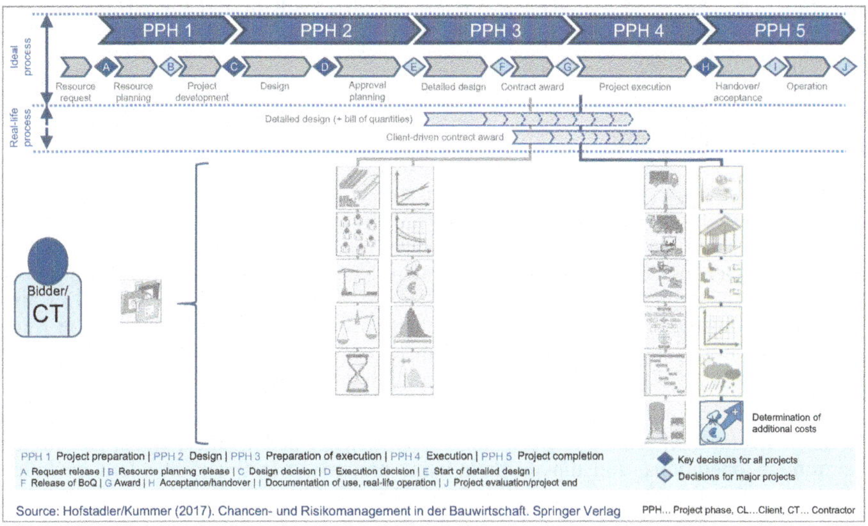

Fig. 10.182 Allocation of example in the project sequence

10.4.12.1 Task

The structural building specification constitutes the main basis for the bidder/contractor to carry out their costing and pricing exercise. With regard to the example concerning the changed quantity structure of the combined steel rebar price item, this means that the information must be assessed or interpreted according to how these explanations were to be understood by the bidder if the situation were assessed objectively. Accordingly, the bidder or cost surveyor can be confident with regard to the documents provided. They will consider their content combined with the communicated or expected conditions and circumstances under which the work is to be performed for the purpose of their costing and pricing exercise. In this respect, these documents are of paramount importance.

Thus, the objective explanatory content of the specification documents is crucial. What the explaining party wanted to express or what the recipient of the explanation understood by it is irrelevant. What really matters is how the explanation should be understood on the basis of an objective evaluation of facts.

The client needs to specify, for instance, a mean rebar diameter in their building specification if they wish to prevent the bidder from determining their own mean rebar diameter for the concrete steel reinforcement and using this diameter as a basis for their costing exercise. The same applies to the distribution of bend shapes and reinforcement ratios of individual structural components.

In the example (following on from Section 10.3.1), the bidder and subsequent contractor had to make their own assumptions regarding the diameter and bend shape distributions. The bidder based their assumptions on the recommendations published by the VÖBV. In the construction phase, however, the diameter and bend shape distributions changed compared to the contractor's original assumptions. This change originates from the client's sphere because the structural concept and reinforcement plans were prepared by representatives of the client, and results in a subsequent change in the contractor's pricing structure. The change in the quantity proportions also leads to a change in the circumstances of work performance and thus to additional costs that the contractor will be claiming from the client.

10.4.12.2 Modeling

The contractor's bid states a wage proportion of €416.54/t for the reinforcing operations (steel rebar) item (see the arrow in Fig. 10.183). Initially, this deterministic value is not linked to a specific chance/risk ratio. To assess the additional cost claim, we need to determine the chance/risk ratio that the contractor has taken in their bid. For this purpose, we refer to the probabilistic calculation discussed in Section 10.3.1. Fig. 10.183 also includes the histogram representing the calculation of the combined price for the wage proportion. This calculation is based on a diameter and bend shape distribution selected by the contractor as well as on the information on labor consumption rates provided by the VÖBV.

In the course of their pricing exercise, the bidder thus decided to take a higher risk for the reinforcing operations, and moved their wage proportion [€/t] one standard deviation away from the mean (corresponding to €416.54/t, see Fig. 10.183). The contractor thus took a chance/risk ratio of approximately 16% to 84%.

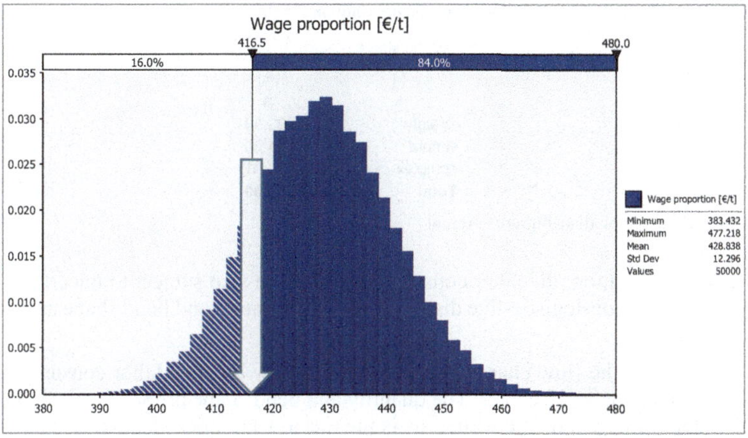

Fig. 10.183 Histogram of the wage proportion – Selection of the chance/risk ratio for the bid

In the execution phase, the quantity proportions changed fundamentally compared to the plausible assumptions made in the bid. A larger share of small rebar diameters was installed, and the proportion of simple and complex bends also increased in the bend shape distribution.

For the modeling of the changed price base, the diameter and bend shape distributions are used as deterministic values because these can be taken from the delivery note during construction and included in the calculations without uncertainties.

Table 10.64 lists the actual quantity proportions for the diameter distribution while Table 10.65 includes the corresponding figures for the bend shape distribution.

Rebar diameter distribution – Actual		
No.	Diameter [mm]	Quantity proportions [%]
0	A	B
1	8	0.50
2	10	3.00
3	12	15.00
4	14	20.00
5	16	30.00
6	20	20.00
7	26	10.00
8	30	1.00
9	36	0.50
10	40	0.00
11	**Total**	**100.000**

Table 10.64 Rebar diameter distribution – Actual

Bend shape distribution – Actual		
No.	Bend shape	Quantity proportions
	[-]	[%]
0	A	B
1	straight	33.89
2	simple	44.20
3	complex	21.91
4	**Total**	**100.00**

Table 10.65 Bend shape distribution – Actual

In the actual calculation, the labor consumption rates are also subject to uncertainties and are thus applied as log-logistic-like distributions per diameter and bend shape as described in Section 10.3.1.

Fig. 10.184 shows the flow chart for calculating the new blended labor consumption rate and the associated new price base. The quantity and bend shape proportions are integrated into the model as deterministic values, as is the mean wage price. The labor consumption rates per rebar diameter and bend shape are modeled as log-logistic-like distributions (as in Section 10.3.1). Otherwise, the calculation procedure remains unchanged compared to the sequence shown in Fig. 10.4 (see page 435).

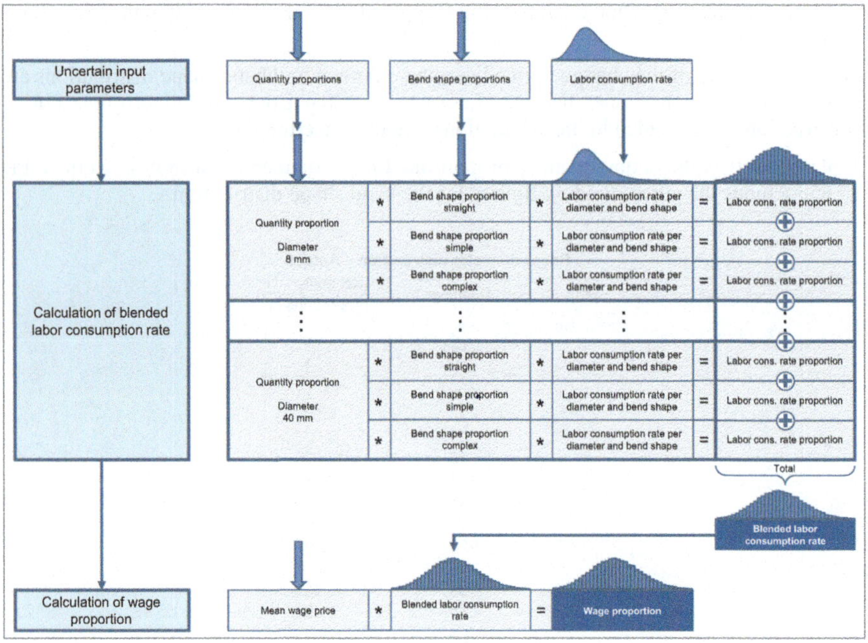

Fig. 10.184 Calculation flow chart

10.4.12.3 Calculation and Interpretation of Results

To determine the new price base of the wage proportion, we first need to calculate the new blended labor consumption rate of the changed diameter and bend shape distribution. For this purpose, we multiply the (deterministic) actual quantity proportions per diameter and bend shape by the labor consumption rate distributions. In the actual analysis, this means that only the consumption rates (log-logistic-like distributions) are fraught with uncertainties. Multiplication of the distributions then results in 30 histograms for the actual consumption rate proportions (three bend shape distributions multiplied by ten diameters).

This approach is applied to calculate the proportions of labor consumption rates in the form of a histogram for each of the three bend shapes (straight, simple, and complex) of the individual diameters (8 to 40 mm). These values are shown in Columns B to D (Rows 1 to 10) of Table 10.66. Column E then calculates the proportions of labor consumption rates per diameter, and Cell E11 determines the histogram representing the actual blended labor consumption rate.

No.	Diameter	Blended labor consumption rate Bend shape straight	simple bend	complex bend	Proportional consumption rate	Mean wage price	Wage proportion
	[mm]	[wh/t]	[wh/t]	[wh/t]	[wh/t]	[€/wh]	[€/t]
0	A	B	C	D	E	F	G
1	8	0.03	0.08	0.06	0.17		
2	10	0.13	0.35	0.30	0.77		
3	12	0.48	1.25	1.08	2.81		
4	14	0.51	1.30	1.03	2.83		
5	16	0.63	1.46	1.21	3.30		
6	20	0.29	0.68	0.57	1.54		
7	26	0.12	0.28	0.25	0.66		
8	30	0.01	0.03	0.02	0.06		
9	36	0.01	0.01	0.01	0.03		
10	40	0.00	0.00	0.00	0.00		
11	**Total**				**12.18**	**40.00**	**487.03**

Table 10.66 Calculation of the actual blended labor consumption rate

The actual labor consumption rate is higher than the blended labor consumption rate included in the bid owing to the changed quantity proportions of the diameters and bend shapes. Comparing the two histograms reveals the different ranges in general but also the different value ranges (see Fig. 10.185).

The fact that the uncertainties associated with the diameter and bend shape distributions have been eliminated in the actual condition so that deterministic values can be applied results in a narrower range of results for the actual blended labor consumption rate. The histogram representing the actual situation also lies in a different value range owing to the changed quantity distributions. The blended labor consumption rate has increased in overall terms because a larger quantity of smaller rebar diameters and a higher proportion of bent rebar were installed.

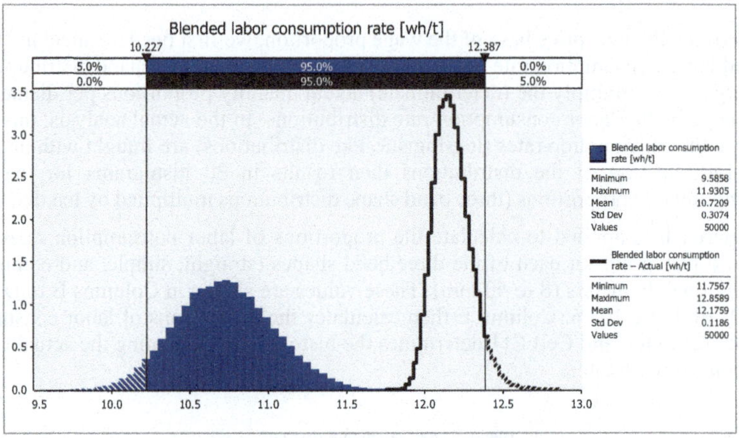

Fig. 10.185 Comparison – Histograms of blended labor consumption rates

To calculate the actual wage proportion, we multiply the blended labor consumption rate by the average wage price taken from the original calculation (i.e. €40/wh). The two histograms can also be compared with respect to the wage proportion (see Fig. 10.186). The result is similar to that obtained for the blended labor consumption rates. This is due to the fact that each of the two histograms representing the blended labor consumption rates was only multiplied by a deterministic value (€40/wh). No further uncertainties are included in the simulation.

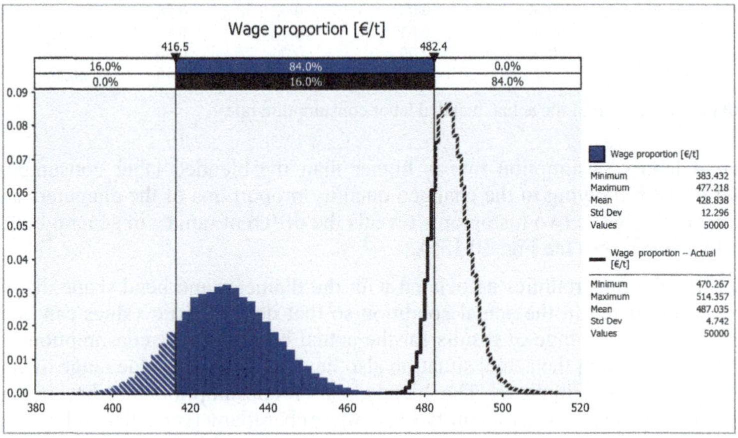

Fig. 10.186 Comparison – Histograms representing the wage proportion

In Fig. 10.186, the sliders at the top of the diagram are each set such that the same chance/risk ratio is obtained for both histograms. The contractor has taken a relatively high risk of approximately 84% with their bid. In the interest of commercial viability, this chance/risk ratio should also be taken into account when determining additional costs.

Accordingly, we would consider an actual wage proportion of €482.40/t to be viable. Compared to the bid, this is equivalent to an increase of €65.86/t (= €482.40/t - €416.54/t). Relative to the specified reinforcement quantity of 3,500 t, this would correspond to additional costs of €230,510 (= €65.86/t * 3,500 t).

This calculation did not take into account the effects of the changed diameter and bend shape distribution on the equipment and material costs and on the costs of cutting and bending. Nor did we consider any other losses of productivity that may have occurred as a result of the changed reinforcement layout.

10.4.12.4 Benefit for the Contractor

When determining additional costs, the contractor will always be faced with the challenge of presenting and processing the changed circumstances of or disruptions to work performance from the client's sphere in such a way that enables the client to review this documentation quickly and with as little effort as possible. For this purpose, the key points need to be presented in a consistent and plausible manner in order to increase the chances for the client to accept and cover the additional costs.

Preparing and justifying additional cost claims also requires a closer look at the nature and scope of work owed under the contract (or the bid pricing). The costing and pricing assumptions need to be checked for appropriateness and reasonableness. Sections 10.3.1 and 9.3.3 describe the way in which the uncertainties associated with the costing and pricing phase can be integrated into the calculations and how the bid prices can be assessed for their appropriateness and plausibility.

Comparing the two histograms for the blended labor consumption rate and the wage proportion as well as disclosing the costing and pricing bases provide a sound basis for decision making in relation to the bid price as well as for the continuous update of the contract while retaining the chance/risk ratio chosen for the additional cost claim.

In any case, the presentation and communication of the existing uncertainties on the target and actual sides will increase the transparency of additional cost claims while helping to improve the communication process between the parties to the contract. In addition, the results of the probabilistic calculations highlight the possible ranges and negotiation margins resulting from the uncertainties in the considerations.

10.5 Project Phase 5 (PPH 5)

No examples are currently included in this book that would be grouped under Project Phase 5. However, Monte Carlo simulations appear to be suitable for operational and strategic facility management applications. In real estate rental and operations, for instance, lost rental income and its repercussions on the profitability of the property can be assessed together with the associated uncertainties.

10.6 Project Portfolio Management across all Phases

Rather than assigning portfolio management directly to a specific project phase, we should consider it to be a management tool to be applied across all phases (see Fig. 10.187). This method jointly considers several different projects in a collection (project portfolio).

Our aim is to obtain an overview of the chance/risk ratio for the entire portfolio and, conversely, to identify and evaluate the influence of specific projects on the portfolio.

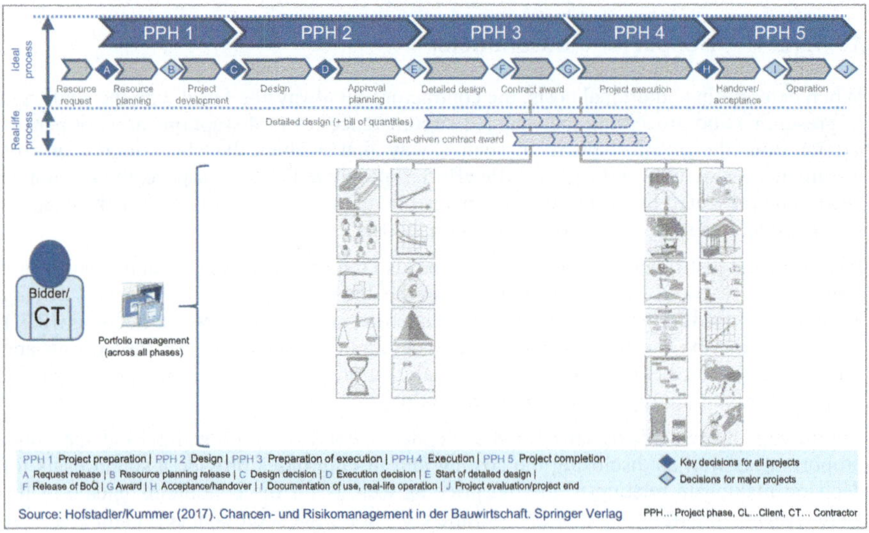

Fig. 10.187 Allocation of example in the project sequence

10.6.1 Basics

In the engineering sciences, the concept of resilience refers to the capability of technical systems to not fail completely in the event of a partial breakdown, but to retain the functionality of essential system components instead. In this sense, any major project that a single company or consortium is unable to financially cushion by other projects poses a risk in terms of resilience.

Any substantial loss incurred in the course of a major contract may put the continued existence of the company at risk.

Elements of project portfolio management are utilized to investigate the resilience of a business or of one of its departments with respect to current and potential future projects. In this context, a (project) portfolio comprises a set of projects that are being jointly managed in order to yield a greater benefit for the business or unit compared to independently managing each individual project. This exercise aggregates similar projects that can be allocated to a defined part of the organization.

Project portfolio management[110)] focuses on projects of a single management unit and deals with the permanent planning, prioritization, overall management, and monitoring of these projects. The related key tasks include:

- Assessment of future projects
- Continuous monitoring of projects under construction
- Information and knowledge management across projects

Unlike managing individual projects, project portfolio management does not end with the completion of the considered project. Rather, it is a permanent management task that must be repeated in each subsequent cycle.

10.6.2 Task

Fig. 10.188 shows a schematic representation of the structure of a company (or a group of [construction] companies). This company is subdivided into various units that each manage projects of various sizes (project volumes) and at various stages of progress. In the first step, the chance/risk ratio of the existing project portfolio can be determined if a probabilistic cost estimate was previously performed for each of the projects, or if the chance/risk ratio can at least be stated that was assumed for the individual projects on the basis of the bid price.

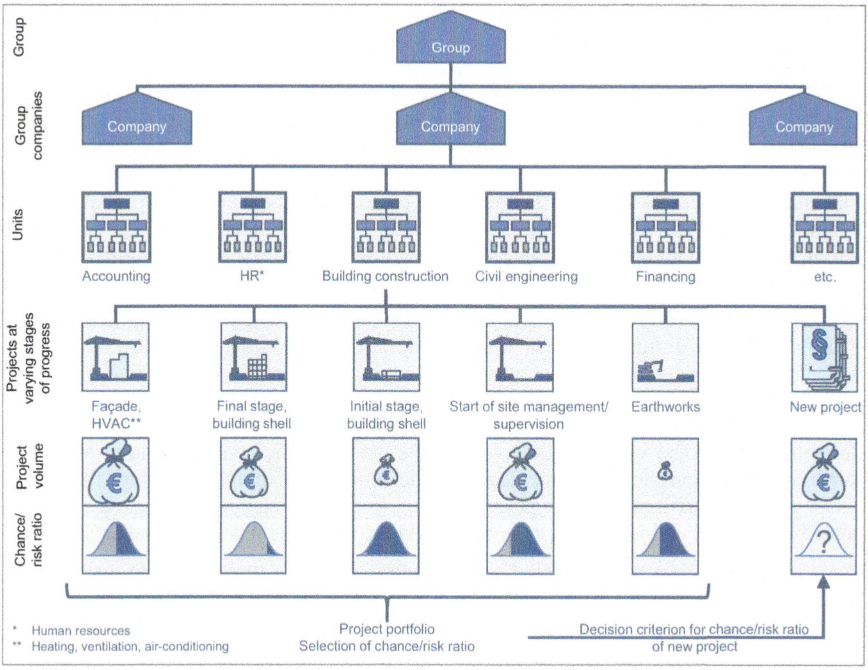

Fig. 10.188 Selecting the chance/risk ratio for new projects[111)]

110) Also referred to as strategic multi-project management.

It is then possible to incorporate the probabilistic cost estimate in the portfolio for each new project and to determine the total break-even bid price[112] from within the given range by performing a reverse calculation or selecting a chance/risk ratio.

At this stage, project portfolio management considers projects managed by identical units, but it can also be extended in the vertical direction, for instance by forming a portfolio comprised of several units or companies.

Cost calculations for projects in the execution phase must be continuously updated to allow for project-related conclusions regarding the current chance/risk ratio.

10.6.3 Modeling

Fig. 10.189 shows a flow chart in which the chance/risk ratio of a project portfolio is determined in the first step (controlling instrument). This ratio can then be applied to calculate the bid price of a new project on the basis of a target chance/risk ratio (management instrument).

Initially, all current projects managed by the unit or company must be listed with their contract volume or total amount. For projects in later completion phases, the contract amount needs to be adjusted to consider any quantity over- or underruns and additions to the contract. The total is derived from the deterministic contract volume data. This amount is subsequently used to calculate the chance/risk ratio of the current project portfolio.

Each project needs to be checked with regard to the existence of a probabilistic cost estimate (adjusted to project progress). If such a probabilistic cost estimate is available, it should be directly applied to modeling the cost distribution for the respective project.[113] If no such estimate exists, further information or qualified estimates of the current project situation will be necessary. This information comprises potential cost deviations (minimum and maximum values) as well as the probability (expressed as a percentage) of over- or underrunning the (adjusted) bid price. The sum of these two percentages, which indicate the chance/risk ratio for the individual projects, must equal 100% (probability and counterprobability). Depending on the distribution function selected for the cost distribution per project, additional or complementary information may also be required, such as form or shape parameters.

If no probabilistic calculation of the costs is available, we suggest the application of a PERT distribution at this point, whose default definition includes a minimal, a most likely, and a maximum value. In this example, however, there is an additional option for defining a PERT distribution. This approach also necessitates the definition of a minimum and a maximum, but the most likely value is substituted by a quantile value. This value is defined such that the value is equivalent to the deterministic contract volume while the quantile corresponds to the percentage for the probability of falling below this value. Thus, the chance/risk ratio of the project as well as the expected range of costs are represented by a distribution function to be applied in the subsequent calculation steps.

Provided all cost curves are modeled as distribution functions, the sum of the cost distributions can be determined using a Monte Carlo simulation. This addition results in a histogram whose range also contains the deterministic value of the total derived from the

[111] Cf. Kummer/Hofstadler (2018), 22ff.

[112] This is the bid price of a new project that must not be underrun in order not to miss the target chance/risk ratio owing to a higher risk.

[113] Such as from probabilistic trend analyses on the basis of current project progress data and the derived forecast.

contract volumes. The probability of over- or underrunning this value within the range of the derived histogram corresponds to the chance/risk ratio of the current project portfolio.

For any new project that is to be integrated in the product portfolio by submitting a bid, a probabilistic cost distribution must be determined and added up with the other cost distributions of current projects. This step results in a histogram of the cost distribution for the new portfolio.

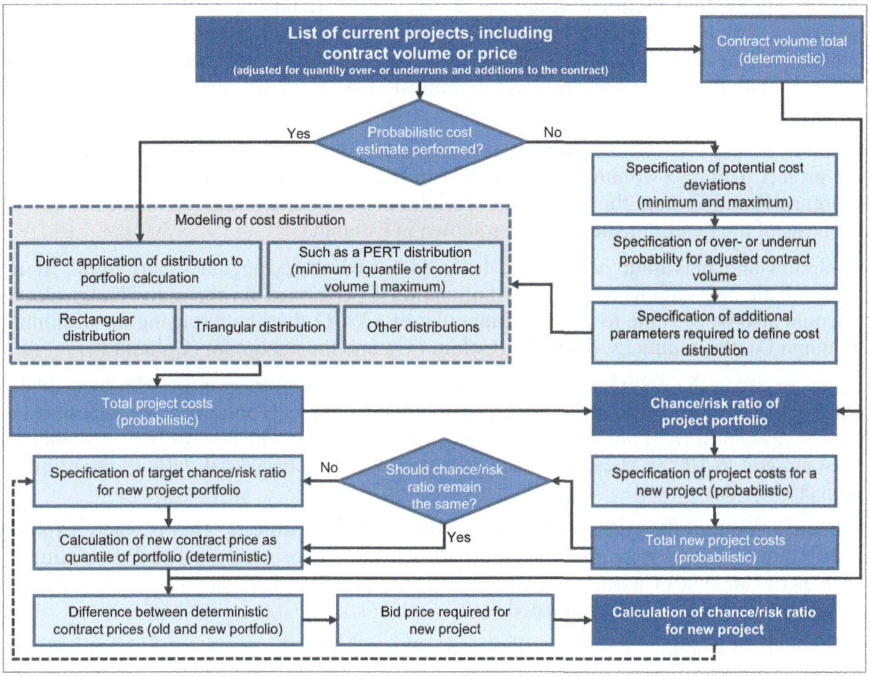

Fig. 10.189 Flow chart – Chance/risk ratio – Project portfolio

In the next step, a decision must be made as to whether the previous chance/risk ratio should still be used after adding the new project to the portfolio, or if this ratio needs to be adjusted. If such an adjustment is intended, the new ratio should be stated, and the histogram representing the new total should be used to determine the corresponding quantile of bid prices for the new portfolio.

The difference between the (deterministic) bid prices of the original and the new portfolio results in the (minimum) bid price for the new project required to achieve the risk-diversified target chance/risk ratio for the entire new portfolio.

Finally, entering the derived deterministic bid price into the cost distribution of the new project allows for determining the corresponding project-specific chance/risk ratio.

If the derived chance/risk ratio or bid price for the new project is unacceptable, for instance because of an exceedingly high risk for the single project, the target chance/risk ratio must be adjusted for the portfolio, or the project should not be added to the portfolio at all.

10.6.4 Calculation and Interpretation of Results

A worked example should demonstrate for a business unit how the chance/risk ratio can be determined for the portfolio of current projects and how a new project can be integrated into the portfolio.

Cells 1 to 5 of Table 10.67 list the projects currently under construction with their designations (Column A) and current progress levels (Column B). Column C specifies the deterministic individual contract volumes or bid prices (adjusted for quantity over- or underruns and/or additions to the contract). The total of these contract volumes is calculated in Cell C6. Thus, the project portfolio is comprised of projects with various levels of progress and different sizes (project volumes). Likewise, the individual projects are associated with different chance/risk ratios.

Columns D and F contain the possible absolute cost deviations from the contract volume per project. Contract volumes must lie within the range specified in Columns D and F. Columns E and G state the percentage deviations of minimum and maximum values on the basis of the deterministic bid prices shown in Column C.[114)]

Minimum and maximum values can be used to define rectangular distributions if no probabilistic calculations of cost distributions were performed for the individual projects. It would also be possible to define a triangular or a PERT distribution using the minimum (Column D), the contract volume[115)] (Column C), and the maximum (Column F).

However, since the contract volume need not necessarily correspond to the most likely value (mode) of, for instance, a PERT distribution, Columns H and I provide the option to specify probabilities of contract volume over- or underruns. The total of both percentages must amount to 100%. These percentages also reflect the chance/risk ratio of the relevant project.

Column J contains the cost distributions of the individual projects. This is where existing distributions derived from probabilistic cost calculations or newly defined distributions, such as rectangular or triangular distributions, can be entered as far as they are available and/or have been updated or adjusted.

In the worked example, PERT distributions determined by the contents of Columns C, D, F, and H are consistently used to demonstrate the applied methodology. In the @Risk program, we use the following function for generating the input distributions: =RiskPertAlt(„min";D;H;C; „max";F). The letters D, H, C, and F correspond to the values in Rows 1 to 5 of the respective columns in Table 10.67. This definition of the distribution by means of a quantile instead of the most likely value makes it possible to directly integrate the estimated probabilities for contract volume over- or underruns into the cost distribution.

114) In Cell E2, for example, this would result in a negative cost deviation of 20%.
[= (1 - €12,000,000 / €15,000,000) * 100%]

115) In this case, the contract volume or amount should be considered to be a net figure (excluding VAT and the total markup).

No.	Project designation	Project progress	Contract volume/ bid price*	Potential cost deviations*				Underrun probability	Overrun probability	Project cost distribution
				Minimum		Maximum		Chance	Risk	
		[%]	[€]	[€]	[%]	[€]	[%]	[%]	[%]	[€]
0	A	B	C	D	E	F	G	H	I	J
1	Project A	80.00%	30,000,000	28,000,000	-6.67%	32,000,000	6.67%	0.40%	0.60%	30,192,650
2	Project B	60.00%	15,000,000	12,000,000	-20.00%	15,500,000	3.33%	0.55%	0.45%	14,814,020
3	Project C	35.00%	1,200,000	1,100,000	-8.33%	2,000,000	66.67%	0.10%	0.90%	1,398,542
4	Project D	30.00%	17,500,000	15,000,000	-14.29%	21,000,000	20.00%	0.35%	0.65%	17,989,584
5	Project E	5.00%	800,000	700,000	-12.50%	1,000,000	25.00%	0.50%	0.50%	805,251
6	**Total**		**64,500,000**					**0.00%**	**100.00%**	**65,200,047**

* Adjusted for quantity over- or underruns and additions to the contract

Table 10.67 Calculation table – Project portfolio

For Project C, the PERT distribution would thus be defined using the following command: =RiskPertAlt(„min";1,100,000;0.10;1,200,000;„max";2,000,000). Fig. 10.190 shows the corresponding cost distribution for Project C. The sliders visible at the top of the diagram show that the probability for the total contract amount of €1.2 million to be underrun is 10% (chance, see Cell H3) while its overrun probability is 90% (risk, see Cell I3). The limits (i.e. the minimum and the maximum) of the distribution are shown in the key on the right-hand side of the diagram. They correspond to the values in Table 10.67 (Cells D3 and F3).

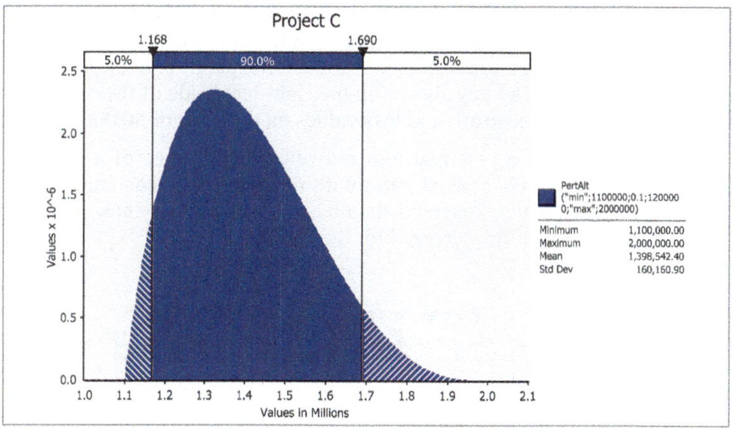

Fig. 10.190 Cost distribution – Project C

Cell J6 of Table 10.67 calculates the sum of the cost distributions of the projects under construction. A subsequently performed Monte Carlo simulation (50,000 iterations; Latin Hypercube sampling method) serves as a basis to derive the histogram of the total and to enter the deterministic contract volume total (Column C6) (see Fig. 10.191). For the entire portfolio of currently executed projects (i.e. A to E), a chance of about 32% results for the actual costs being lower than the invoiced contract volume (Cell C6). On the other hand, there is a risk of approximately 68% (counterprobability) for actual costs to exceed the invoiced contract volume.

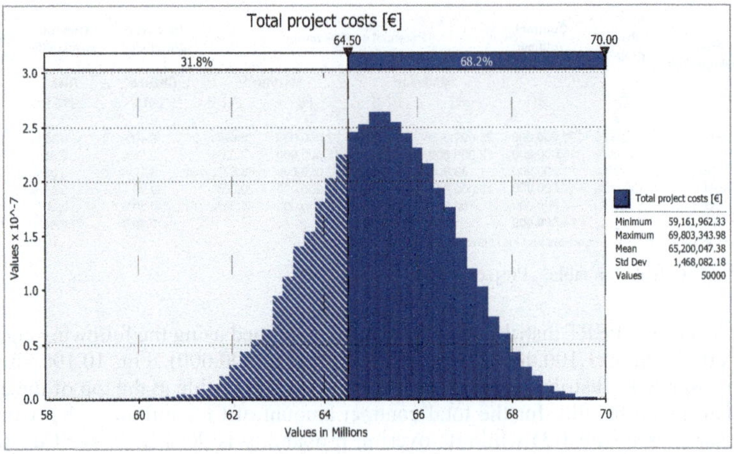

Fig. 10.191 Histogram – Cost distribution total (Projects A to E)

The difference between the contract volume total (the deterministic value in Cell C6) and the cost distribution total (Cell J6) results in a histogram that shows the potential profit or loss relative to the entire portfolio (see Fig. 10.192). Again, this histogram shows the chance/risk ratio for the portfolio (32% to 68%). On the abscissa, the amounts of possible profits are represented by positive values while the amounts of possible losses are represented by negative values. The key shown on the right-hand side of the diagram makes it possible to gather the extreme profit and loss values on the basis of 50,000 iterations.

In the example, the risk of a potential loss outweighs the chance of a potential profit, resulting in an unfavorable chance/risk ratio with respect to the economic success of the unit or company. Consequently, this trend should be reversed for any new projects, and the proportion of chances should be increased for the entire portfolio.

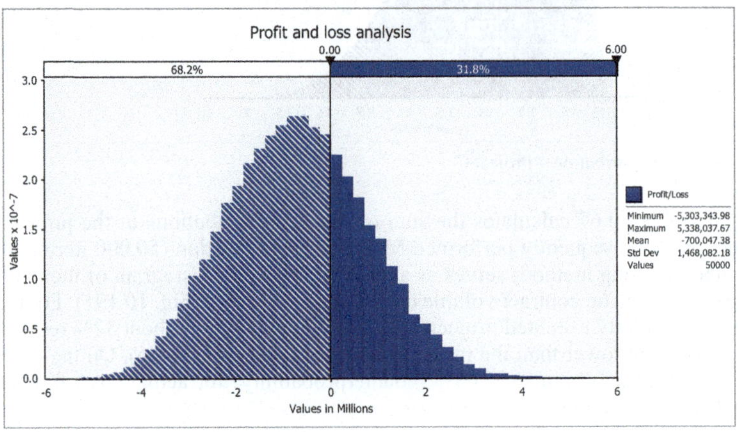

Fig. 10.192 Histogram – Profit and loss analysis

This example also examines the effects that a new project would have on the chance/risk ratio of the portfolio as well as the bid price that would be required in order to achieve a defined ratio.

Table 10.68 is identical to Table 10.67 up to and including Row 6. In Row 7, however, an additional project ("Project NEW") is added. For this new project, a probabilistic cost estimate or specification of a cost distribution is assumed. This distribution is stored in Cell J7 of Table 10.68.

In this example, we assumed a simplified PERT distribution with the following cost values:

- Minimum: €9,000,000
- Most likely: €10,000,000
- Maximum: €12,500,000

This distribution is subsequently added to the cost distribution total (Cell J6) and gives a new total in Cell J8. In the next step, the new (target) chance/risk ratio of the new project portfolio (i.e. Projects A to E plus Project NEW) is stated in Cells H8 and I8. The total of the two percentages must amount to 100%. In this example, a new target chance/risk ratio of 48% to 52% was specified to increase the proportion of chances for the entire portfolio.

No.	Project designation	Project progress	Contract volume/ bid price*	Potential cost deviations*				Underrun probability	Overrun probability	Project cost distribution
				Minimum		Maximum		Chance	Risk	
		[%]	[€]	[€]	[%]	[€]	[%]	[%]	[%]	[€]
0	A	B	C	D	E	F	G	H	I	J
1	Project A	80.00%	30,000,000	28,000,000	-6.67%	32,000,000	6.67%	0.40%	0.60%	30,192,650
2	Project B	60.00%	15,000,000	12,000,000	-20.00%	15,500,000	3.33%	0.55%	0.45%	14,814,020
3	Project C	35.00%	1,200,000	1,100,000	-8.33%	2,000,000	66.67%	0.10%	0.90%	1,398,542
4	Project D	30.00%	17,500,000	15,000,000	-14.29%	21,000,000	20.00%	0.35%	0.65%	17,989,584
5	Project E	5.00%	800,000	700,000	-12.50%	1,000,000	25.00%	0.50%	0.50%	805,251
6	**Total**		**64,500,000**					**31.83%**	**68.17%**	**65,200,047**
7	Project NEW	0.00%	10,875,985	9,000,000	-17.25%	12,500,000	14.93%	**82.28%**	**17.72%**	10,250,000
8	**New total**		**75,375,985**					**48.00%**	**52.00%**	**75,450,047**

* Adjusted for quantity over- or underruns and additions to the contract

Table 10.68 Calculation table – Project portfolio with new project

Calculating the 48% quantile of the new cost distribution total (Cell J8) makes it possible to derive the deterministic value for the new contract volume total. Fig. 10.194 shows the new (target) chance/risk ratio for the project portfolio that was set using the sliders. The deterministic value indicating the boundary between the two areas at the top of the diagram corresponds to the value in Cell C8 of Table 10.68 (€75,375,985).

Calculating the difference between the contract volume totals with and without Project NEW (Cells C8 and C6) results in the minimum contract volume for Project NEW in Cell C7. This total is equivalent to the bid price that must not be underrun on the basis of the probabilistic cost distribution of the new project (Cell J7) in order to achieve the target chance/risk ratio for the new portfolio. If a higher bid were submitted for the project, this would increase the proportion of chances for the entire portfolio.

Cells D7 and F7 show the minimum and the maximum of the cost distribution stored in Cell J7 for Project NEW. The percentage deviations refer to the deterministic bid price in Cell C7.

Finally, Cells H7 and I7 are used to state the chance/risk ratio for Project NEW. This step proves that the proportion of chances for this project (with a corresponding project volume) must be defined at a relatively high level in order to implement the intended change in the chance/risk ratio of the entire portfolio.

In general, major projects (involving a large contract amount) will have a greater impact on the overall portfolio, and the chance/risk ratio of such projects will have a more significant effect on the portfolio than that of smaller projects.

The target chance/risk ratio of the portfolio must be adjusted if the bid price determined for the new project using the above approach appears to be exceedingly high (in relation to the market price level). However, the lower the bid price for a new project is, the smaller the increase in the proportion of chances for the portfolio will be.

This principle applies until a specific bid amount (break-even amount) is underrun, and the addition of the new project to the portfolio would result in an increase in the proportion of risks for the portfolio.

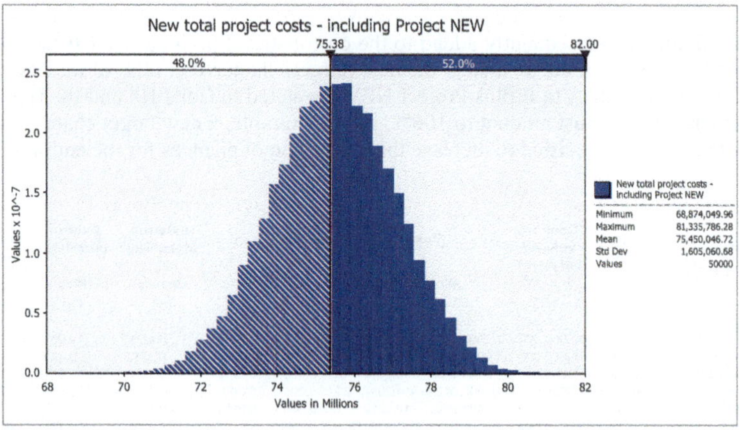

Fig. 10.193 Histogram – New cost distribution total, including Project NEW

10.6.5 Benefit for the Contractor

The method of managing and controlling the chance/risk ratio of a full set of current and new projects (portfolio) offers the advantage that bid prices of new projects can be proactively modified in iterative steps if the underlying database[116)] is adjusted accordingly. In this process, the chance/risk ratio determined for the portfolio is used as a key management instrument.

The system outlined above makes it possible to effectively aggregate projects at various stages of progress, with different contract volumes and chance/risk ratios in a single portfolio, which then permits conclusions regarding the parameters to be set for individual new projects.

Project portfolio management can thus be used both as a controlling instrument (monitoring the effects of a selected bid price on the chance/risk ratio of the portfolio) and as a management instrument (determining a minimum bid price to achieve a defined chance/risk ratio for the portfolio).

116) Bid prices account for quantity over- and underruns as well as additions to the contract. The cost minima and maxima and/or chance/risk ratios of the individual projects are stated relative to the current project progress.

10.7 Summary

This chapter outlined practical examples of performing Monte Carlo simulations from the bidder's/contractor's vantage point. The examples discussed in this chapter primarily relate to PPH 3 and 4. However, the described methods and tools can also be applied to other project phases, and they can also be used by clients if required.

We outlined the examples in a transparent and plausible manner while basing them on a uniform system and structure. The calculation tables were explained in the relevant sections and complemented by graphical representations of the calculation processes. We put particular emphasis on the interpretation of the simulation results and on demonstrating their benefits for the bidder or subsequent contractor.

10.8 References

Brecheler, Winfried et al. (1998). Baubetriebslehre – Kosten- und Leistungsrechnung – Bauverfahren. Braunschweig/Wiesbaden. Vieweg & Sohn Verlagsgesellschaft mbH. (ISBN 3-528-07708-5)

Busch, Thorsten (2004). Systematisches Projektrisikomanagement in der Angebotsphase. In: Tagungsband – 2. Grazer Baubetriebs- und Bauwirtschaftssymposium – Risikomanagement in der Bauwirtschaft. Ed.: Institute of Construction Management and Economics, Graz University of Technology. 49-62. Graz. Verlag der Technischen Universität Graz. (ISBN 3-901351-91-4)

Fetzner, Torsten (2007). Ein Verfahren zur Erfassung von Minderleistungen aufgrund witterungsbedingter Bauablaufstörungen. Doctoral thesis. Darmstadt. Institut für Baubetrieb der Technischen Universität Darmstadt.

Hofstadler, Christian (1999). Zur exakten Ermittlung der Vorhaltemenge von Schalung und Rüstung für die Herstellung von Stahlbetondecken und die Auswirkungen des Frühausschalens auf Bauzeit und Baukosten. Doctoral thesis. Graz. Institute of Construction Management and Economics, Graz University of Technology.

Hofstadler, Christian (2007). Bauablaufplanung und Logistik im Baubetrieb. Berlin, Heidelberg. Springer Verlag (ISBN-13 978-3-540-34320-2)

Hofstadler, Christian (2008). Schalarbeiten – Technologische Grundlagen, Sichtbeton, Systemauswahl, Ablaufplanung, Logistik und Kalkulation. Berlin, Heidelberg. Springer Verlag (ISBN 978-3-540-85178-3)

Hofstadler, Christian (2009). Beschaffungslogistik für die Phase Bauwerk-Rohbau – Berechnung der Anzahl der Transporte für Stahlbetonarbeiten unter Anwendung der Monte-Carlo-Simulation. In: Festschrift anlässlich des 30-jährigen Bestehens des Instituts für Baubetrieb der Technischen Universität Darmstadt – Baubetriebliche Aufgaben. Fortschritt-Berichte VDI, Reihe 4 Bauingenieurwesen, Nr. 211. Ed.: Motzko, Christoph. 144-158. Düsseldorf. VDI-Verlag. (ISBN 978-3-18-321104-3)

Hofstadler, Christian; Franzl, Gerald (2011). Bewehrungsarbeiten im Baubetrieb. Ed.: Verband Österreichischer Biege- und Verlegetechnik (Austrian Association for Bending and Installation of Concrete Reinforcement) (ISBN 978-3-200-02149-5)

Hofstadler, Christian (2014). Produktivität im Baubetrieb – Bauablaufstörungen und Produktivitätsverluste. Berlin, Heidelberg. Springer Verlag (ISBN 978-3-642-41632-3)

Hofstadler, Christian; Kummer, Markus (2014). Systematischer Umgang mit Produktivitätsrisiken in der Auftragskalkulation. In: Tagungsband – 12. Grazer Baubetriebs- und Bauwirtschaftssymposium – Risiken im Bauvertrag – Baubetriebliche, bauwirtschaftliche und rechtliche Aspekte. Eds.: Heck, Detlef; Mauerhofer, Gottfried; Hofstadler, Christian. 53-90. Graz. Verlag der Technischen Universität Graz. (ISBN 978-3-85125-312-2)

Hofstadler, Christian; Kummer, Markus (2017a). Chancen- und Risikomanagement in der Bauwirtschaft – Für Auftraggeber und Auftragnehmer in Projektmanagement, Baubetrieb und Bauwirtschaft. Berlin, Heidelberg. Springer-Verlag. (ISBN 978-3-662-54318-4)

Hofstadler, Christian; Kummer, Markus (2019). Systematische Berücksichtigung von Unsicherheiten in der Preisermittlung – Einsatz der Monte-Carlo-Simulation. In: Aktuelle Entwicklungen in Baubetrieb, Bauwirtschaft und Bauvertragsrecht – 50 Jahre Institut für Baubetrieb und Bauwirtschaft der TU Graz. Ed.: Hofstadler, Christian. 739-768. Wiesbaden. Springer Vieweg (ISBN 978-3-658-27430-6)

Joseph Vögele AG – Product Data Sheet – Vögele tracked paver SUPER 1900-3i – Highway Class. Available online at http://media.voegele.info/media/03_voegele/produkte/broschueren/02_raupenfertiger/layout_2016/super_1900_3i_1/SUPER_19003i_-DE_2556102_mPW_0916_Lay2016.pdf. Accessed August 6, 2016.

Jurecka, Walter (1975). Kosten und Leistungen von Baumaschinen. Vienna, New York. Springer Verlag. (ISBN 3-211-81324-1)

Koehn, Enno; Brown, Gerald (1985). Climatic Effects on Construction. In: Journal of Construction Engineering and Management, 111(2), June, 1985. 129-137. American Society of Civil Engineers (ASCE). (ISSN 1943-7862)

Kummer, Markus (2014). Additive Berücksichtigung von Produktivitätsverlusten. In: Tagungsband – 25. BBB-Assistententreffen in Graz – Fachkongress der wissenschaftlichen Mitarbeiter der Bereiche Bauwirtschaft, Baubetrieb und Bauverfahrenstechnik. Eds.: Heck, Detlef; Mauerhofer, Gottfried; Hofstadler, Christian. 171-186. Graz. Verlag der Technischen Universität Graz. (ISBN 978-3-85125-356-6)

Kummer, Markus (2015a). Aggregierte Berücksichtigung von Produktivitätsverlusten bei der Ermittlung von Baukosten und Bauzeiten – Deterministische und probabilistische Betrachtungen. Doctoral thesis. Graz. Institute of Construction Management and Economics, Graz University of Technology.

Kummer, Markus (2015b). Aggregierte Berücksichtigung von Produktivitätsverlusten. In: Tagungsband – 26. BBB-Assistententreffen – Fachkongress der wissenschaftlichen Mitarbeiter der Bereiche Baubetrieb, Bauwirtschaft und Bauverfahrenstechnik. Ed.: Berner, Fritz. 167-183. Stuttgart. Institut für Baubetriebslehre der Universität Stuttgart. (ISBN 978-3-9814355-7-3)

Kummer, Markus; Hofstadler, Christian (2015). Influence of mean daily temperature on the selection of the construction start date. In: Implementing Innovative Ideas in Structural Engineering and Project Management – ISEC-8. Eds.: Saha, Swapan; Zhang, Y. X.; Yazdani, Siamak; Singh, Amarjit. 851-856. Fargo. ISEC Press. (ISBN 978-0-9960437-1-7)

Kummer, Markus; Hofstadler, Christian (2018). Portfolio Management in the Construction Industry – Pricing Strategies Considering the Chance/Risk Ratio for Several Projects. In: The Tenth International Conference on Construction in the 21st Century (CITC-10). Eds.: Ahmed, Syed M.; Shah, Attaullah; Azhar, Salman; Smith, Norma A.; Campbell, Shaunna C.; Mahaffy, Kelly; Saul, Amelia. 22-30. Greenville, North Carolina, USA. (ISBN 978-1-7324416-1-3)

Kummer, Markus; Hofstadler, Christian (2019). Chances and risks of combined pricing. In: Interdependence between Structural Engineering and Construction Management – ISEC-10. Eds.: Ozevin, Didem; Ataei, Hossein; Modares, Mehdi; Gürgün, Asli Pelin; Yazdani, Siamak; Singh, Amarjit. CPM-15. Fargo. ISEC Press. (ISSN 2644-108X; ISBN 978-0-9960437-6-2)

Oberndorfer, Wolfgang; Jodl, Hans Georg et al. (2001). Handwörterbuch der Bauwirtschaft – ON V 208 – 2nd ed. Vienna. ON Österreichisches Normungsinstitut. (ISBN 3-85402-072-4)

Oglesby, Clarkson H.; Parker, Henry; Howell, Gregory (1989). Productivity Improvement in Construction. New York. McGraw-Hill Inc. (ISBN 0-07-047802-3)

OENORM B 2110 (2013-03-15 edition) Allgemeine Vertragsbestimmungen für Bauleistungen – Werkvertragsnorm.

Raaber, Norbert (1979). Eine Analyse des Baupreisrisikos. Lecture held on March 15, 1979 at Graz University of Technology.

Scheuer, Christian (1997). Zur Leistungsermittlung von Maschinen und Geräten im Baubetrieb – Hauptgruppen 4 bis 7 nach ÖBGL 1996. Master's thesis. Graz. Institute of Construction Management and Economics, Graz University of Technology.

Spranz, Dieter (2003). Arbeitsvorbereitung im Ingenieurhochbau. Berlin. Bauwerk Verlag. (ISBN 978-3-89932-023-7)

Suhl, Leena; Mellouli, Taïeb (2013). Optimierungssysteme – Modelle, Verfahren, Software, Anwendungen – 3rd ed. Berlin, Heidelberg. Springer Verlag (ISBN 978-3-642-38936-8)

Thomas, Randolph; Yiakoumis, Iacovos (1987). Factor Model of Construction Productivity. In: Journal of Construction Engineering and Management, 113(4), December 1987. 623-639. American Society of Civil Engineers (ASCE). (ISSN 1943-7862)

VÖBV – Verband Österreichische Biege- und Verlegetechnik (2010). Wirtschaftliches Bewehren. Available online at: http://www.voebv.at/pdf/VOEBV%20Wirtschaftliches%20Bewehren.pdf. Accessed October 1, 2016.

Voigt, Matthias (2008). Grundlagen der Probabilistik. Vortrag im Zuge des 1. Dresdner Probabilistik-Workshops der Professur für Turbomaschinen und Strahlantriebe am Institut für Strömungsmechanik. Technische Universität Dresden. October 9-10, 2008. Available online at http://www.probabilistik.de/vortrag/Vortrag0102_Voigt_TUD.pdf. Accessed November 9, 2016

ZAMG: http://www.zamg.ac.at/fix/klima/oe71-00/klima2000/klimadaten_oesterreich_1971_frame1.htm. Accessed August 20, 2016.

Figures

C. Hofstadler and M. Kummer, *Chances and Risks in Construction Management and Economics*, Springer Tracts in Civil Engineering,
https://doi.org/10.1007/978-3-030-64014-9

Tables

C. Hofstadler and M. Kummer, *Chances and Risks in Construction Management and Economics*, Springer Tracts in Civil Engineering,
https://doi.org/10.1007/978-3-030-64014-9

Index

C. Hofstadler and M. Kummer, *Chances and Risks in Construction Management and Economics*, Springer Tracts in Civil Engineering,
https://doi.org/10.1007/978-3-030-64014-9

D

E

F

G

H

I

K

L

M

N

O

P

Q

The manufacturer's authorised representative in the EU is Springer Nature Customer Service Centre GmbH, Europaplatz 3, 69115 Heidelberg, Germany. If you have any concerns regarding our products, please contact ProductSafety@springernature.com

Printed and bound by CPI Group (UK) Ltd, Croydon, CR0 4YY
22/09/2026
02221048-0016